LOW TEMPERATURE PHYSICS - LT 13

Volume 2: Quantum Crystals and Magnetism

LOW TEMPERATURE PHYSICS-LT 13

Volume 1: Quantum Fluids

Volume 2: Quantum Crystals and Magnetism

Volume 3: Superconductivity

Volume 4: Electronic Properties, Instrumentation, and Measurement

LOW TEMPERATURE PHYSICS - LT 13

Edited by

K. D. Timmerhaus
University of Colorado
Boulder, Colorado and
National Science Foundation
Washington, D.C.

W. J. O'Sullivan
University of Colorado
Boulder, Colorado

and

E. F. Hammel
Los Alamos Scientific Laboratory
University of California
Los Alamos, New Mexico

Volume 2: Quantum Crystals and Magnetism

PLENUM PRESS • NEW YORK-LONDON

Library of Congress Cataloging in Publication Data

International Conference on Low Temperature Physics, 13th, University of Colorado, 1972.
    Low Temperature physics—LT 13; [proceedings]

  Includes bibliographical references.
  CONTENTS: v. 1. Quantum fluids.—v. 2. Quantum crystals and magnetism.—v. 3. Superconductivity.—v. 4. Electronic properties, instrumentation, and measurement.
  1. Low temperatures—Congresses. 2. Free electron theory of metals—Congresses. 3. Energy-band theory of solids—Congresses. I. Timmerhaus, Klaus D., ed. II. O'Sullivan, William John, ed. III. Hammel, E. F., 1918- ed. IV Title.
QC278.I512 1972   536'.56   73-81092
ISBN 0-306-35122-6 (v. 2)

The proceedings of the XIIIth International Conference on Low Temperature Physics, University of Colorado, Boulder, Colorado, August 21-25, 1972, will be published in four volumes, of which this is volume two.

© 1974 Plenum Press, New York
A Division of Plenum Publishing Corporation
227 West 17th Street, New York, N.Y. 10011

United Kingdom edition published by Plenum Press, London
A Division of Plenum Publishing Company, Ltd.
4a Lower John Street, London W1R 3PD, England

All rights reserved

No part of this book may be reproduced, stored in a retrieval system, or transmitted, in any form or by any means, electronic, mechanical, photocopying, microfilming, recording, or otherwise, without written permission from the Publisher

Printed in the United States of America

Contents*

Quantum Crystals

1. Plenary Topics

Quantum Crystals: Theory of the Phonon Spectrum. *Heinz Horner* 3
Quantum Solids and Inelastic Neutron Scattering. *V. J. Minkiewicz, T. A. Kitchens, G. Shirane, and E. B. Osgood* 14
Magnetic and Thermal Properties of Solid and Liquid ^{3}He Near the Melting Curve. *D. M. Lee* 25

2. Helium Lattice Dynamics

2.1 Specific Heat and Sound

Specific Heat of Solid ^{3}He. *S. H. Castles, W. P. Kirk, and E. D. Adams* 43
The Temperature Dependence of the Longitudinal Sound Velocity of Single Crystals of HCP ^{4}He. *J. P. Franck and R. A. D. Hewko* 48
Lifetimes of Hypersonic Phonons in Solid ^{4}He. *P. Leiderer, P. Berberich, S. Hunklinger, and K. Dransfeld* 53
Sound Wave Propagation and Anharmonic Effect in Solid ^{3}He and ^{4}He. *D. Y. Chung, C.-C. Ni, and Y. Li* 62

2.2 Heat Transport in Isotope Mixtures

NMR Measurements on ^{3}He Impurity in Solid ^{4}He. *M. G. Richards, J. Pope, and A. Widom* 67
Calculation of the Diffusion Rate of ^{3}He Impurities in Solid ^{4}He. *A. Landesman and J. M. Winter* 73
NMR Study of Solid ^{3}He–^{4}He Mixtures Rich in ^{3}He. *M. E. R. Bernier* 79
Phonon Scattering by Isotopic Impurities in Helium Single Crystals. *D. T. Lawson and H. A. Fairbank* 85
The Orientation and Molar Volume Dependence of Second Sound in HCP ^{4}He Crystals. *K. H. Mueller, Jr. and H. A. Fairbank* 90
Structure of Phase-Separated Solid Helium Mixtures. *A. E. Burgess and M. J. Crooks* 95

* Tables of contents for Volumes 1, 3, and 4 and an index to contributors appear at the back of this volume.

2.3 Scattering and Vacancies

Raman Scattering from Condensed Phases of ^{3}He and ^{4}He. *C. M. Surko and R. E. Slusher* 100

Single-Particle Excitations in Solid Helium. *T. A. Kitchens, G. Shirane, V. J. Minkiewicz, and E. B. Osgood* 105

Single-Particle Density and Debye–Waller Factor for BCC ^{4}He. *A. K. McMahan and R. A. Guyer* 110

Thermal Defects in BCC ^{3}He Crystals Determined by X-Ray Diffraction. *R. Balzer and R. O. Simmons* 115

Possible Bound State of Two Vacancy Waves in Crystalline ^{4}He. *William J. Mullin* 120

Multiphonon and Single-Particle Excitations in Quantum Crystals. *Heinz Horner* 125

3. Helium-3 Nuclear Magnetism

What is the Spin Hamiltonian of Solid BCC ^{3}He? *L. I. Zane* 131

Magnetic Properties of Liquid ^{3}He below 3 m°K. *D. D. Osheroff, W. J. Gully, R. C. Richardson, and D. M. Lee* 134

Properties of ^{3}He on the Melting Curve. *W. P. Halperin, R. A. Buhrman, W. W. Webb, and R. C. Richardson* 139

Low-Temperature Solid ^{3}He in Large Magnetic Fields. *R. T. Johnson, D. N. Paulson, R. P. Giffard, and J. C. Wheatley* 144

High-Magnetic-Field Behavior of the Nuclear Spin Pressure of Solid ^{3}He. *W. P. Kirk and E. D. Adams* 149

The Interaction of Acoustic Phonons with Nuclear Spins in Solid ^{3}He. *Kenneth L. Verosub* 155

4. Helium Monolayers

Motional Narrowing in Monolayer ^{3}He Film NMR. *R. J. Rollefson* 161

Neon Adsorbed on Graphite: Heat Capacity Determination of the Phase Diagram in the First Monolayer from 1 to 20°K. *G. B. Huff and J. G. Dash* ... 165

NMR in ^{3}He Monolayers Adsorbed on Graphite below 4.2°K. *D. P. Grimmer and K. Luszczynski* 170

Thermodynamic Functions for ^{4}He Submonolayers. *R. L. Elgin and D. L. Goodstein* 175

Elastic Properties of Solid ^{4}He Monolayers from 4.2°K Vapor Pressure Studies. *G. A. Stewart, S. Siegel, and D. L. Goodstein* 180

5. Molecular Solids

Sound Velocities in Solid HCP H_2 and D_2. *R. Wanner and H. Meyer* 189

Polarizability of Solid H_2 and D_2. *Barnie Wallace, Jr. and Horst Meyer* ... 194

Scattering of Neutrons by Phonons and Librons in Solid o-Hydrogen. *A. Bickermann, F. G. Mertens, and W. Biem* 198

Determination of the Crystal Structures of D_2 by Neutron Diffraction. *R. L. Mills, J. L. Yarnell, and A. F. Schuch* 203

Velocity and Absorption of High-Frequency Sound Near the Lambda Transitions in Solid CD_4. *R. P. Wolf and F. A. Stahl* 210

Growth and Neutron Diffraction Experiments on Single Crystals of Deuterium. *Joseph Votano, R. A. Erickson, and P. M. Harris* 215

Proton Resonance in Highly Polarized HD. *H. M. Bozler and E. H. Graf* 218

Thermal Conductivity of Solid HD Containing Isotopic Impurities. *J. H. Constable and J. R. Gaines* 223

Self-Consistent Calculations of the Lattice Modes of Solid Nitrogen. *J. C. Raich and N. S. Gillis* 227

6. Other Topics in Quantum Crystals

Ionic Mobilities in Solid Helium. *G. A. Sai-Halasz and A. J. Dahm* 233

Vapor Pressure Ratios of the Neon Isotopes. *G. T. McConville* 238

Investigation of Phonon Radiation Temperature of Metal Films on Dielectric Substrates. *J. D. N. Cheeke, B. Hebral, and Catherine Martinon* 242

Surface Thermal Expansion in Noble Gas Solids. *V. E. Kenner and R. E. Allen* 245

Pair Potentials for van der Waals Solids from a One-Electron Model. *S. B. Trickey, F. W. Averill, and F. R. Green, Jr.* 251

Magnetism

7. Plenary Topics

Magnetic Interaction Effects in Dilute Alloys, *R. F. Tournier* 257

Magnetic Phase Transitions at Low Temperatures. *W. J. Huiskamp* 272

8. Phase Transitions

Magnetic Equations of State in the Critical Region. *Sava Milošević, Douglas Karo, Richard Krasnow and H. Eugene Stanley* 295

Properties of $PrPb_3$ in Relation to Other $L1_2$ Phases of Pr. *E. Bucher, K. Andres, A. C. Gossard, and J. P. Maita* 322

On Nuclear Magnetic Ordering Phenomena in Van Vleck Paramagnetic Materials. *K. Andres* 327

Electronic Magnetic Ordering Induced by Hyperfine Interactions in Terbium and Holmium Gallium Garnets. *J. Hammann and P. Manneville* 328

Low-Field Magnetic Properties of $DyVO_4$ and $TbPO_4$. *H. Suzuki, T. Ohtsuka, and T. Yamadaya* 334

Magnetic Ordering of the Delocalized Electron Spin in DPPH. *S. Saito and T. Sato* 338

Tricritical Susceptibility in Ising Model Metamagnets and Some Remarks Concerning Tricritical Point Scaling. *Fredric Harbus, H. Eugene Stanley, and T. S. Chang* 343

Double-Power Scaling Functions Near Tricritical Points. *T. S. Chang, A. Hankey, and H. E. Stanley* 349

Low-Temperature Spontaneous Magnetization of Two Ferromagnetic Insulators: $CuRb_2Br_4 \cdot 2H_2O$ and $CuK_2Cl_4 \cdot 2H_2O$. *C. Dupas, J.-P. Renard, and E. Velu* 355

Heat Capacity Measurements on $KMnF_3$ at the Soft Mode and Magnetic Phase Transitions. *W. D. McCormick and K. I. Trappe* 360

Magnetic and Magnetoopical Effects at the Phase Transition in Antiferromagnetic Ferrous Carbonate. *V. V. Eremenko, K. L. Dudko, Yu. G. Litvinenko, V. M. Naumenko, and N. F. Kharchenko* 365

9. Low-Dimensional Systems

Fluctuations in One-Dimensional Magnets: Low Temperatures and Long Wavelengths. *R. J. Birgeneau, G. Shirane, and T. A. Kitchens* 371

Nuclear Relaxation in a Spin $\frac{1}{2}$ One-Dimensional Antiferromagnet. *E. F. Rybaczewski, E. Ehrenfreund, A. F. Garito, A. J. Heeger, and P. Pincus* 373

Low-Dimensional Magnetic Behavior of $Cu(NH_3)_2 \cdot Ni(CN)_4 \cdot 2C_6H_6$. *Hisao Kitaguchi, Shoichi Nagata, Yoshihito Miyako, and Takashi Watanabe* 377

Electron Paramagnetic Resonance in $Cu(NO_3)_2 \cdot 2.5H_2O$. *Y. Ajiro, N. S. VanderVen, and S. A. Friedberg* 380

One-Magnon Raman Scattering in the Two-Dimensional Antiferromagnet K_2NiF_4. *D. J. Toms, W. J. O'Sullivan, and H. J. Guggenheim* 384

NMR Study of a Two-Dimensional Weak Ferromagnet, $Cu(HCOO)_2 \cdot 4D_2O$, in Magnetic Fields up to 10 kOe. *A. Dupas and J.-P. Renard* 389

10. Ferromagnetism

Heisenberg Model for Dilute Alloys in the Molecular Field Approximation. *Michael W. Klein* 397

Spin Polarization of Electrons Tunneling from Thin Ferromagnetic Films. *R. Meservey and P. M. Tedrow* 405

Proximity Effect for Weak Itinerant Ferromagnets. *M. J. Zuckermann* 410

Onset of Ferromagnetism in Alloys at Low Temperatures. *B. R. Coles, A. Tari, and H. C. Jamieson* 414

Distribution of Atomic Magnetic Moment in Ferromagnetic Ni–Cu Alloys. *A. T. Aldred, B. D. Rainford, T. J. Hicks, and J. S. Kouvel* 417

Low-Temperature Resistance Anomalies in Iron-Doped V–Cr Alloys. *R. Rusby and B. R. Coles* 423

Magnetic Properties of $(Ge_{1-x}Mn_x)$ Te Alloys. *R. W. Cochrane and J. O. Ström-Olsen* 427

11. Dilute Alloys

Electron Spin Relaxation through Matrix NMR in Dilute Magnetic Alloys. *H. Alloul and P. Bernier* 435

NMR Experimental Test for the Existence of the Kondo Resonance. *F. Mezei and G. Grüner* 444

Nuclear Orientation Experiments on Dilute $Au_{1-x}Ag_x$ Yb Alloys. *J. Flouquet and J. Sanchez* 448

Temperature Dependence of the High-Frequency Resistivity in Dilute Magnetic Alloys. *H. Nagasawa and T. Sugawara* 451

Spin-Dependent Resistivity in Cu:Mn. *G. Toth* 455

Properties of Dilute Transition-Based Alloys with Actinide Impurities: Evidence of Localized Magnetism for Neptunium and Plutonium. *E. Galleani d'Angliano, A. A. Gomès, R. Jullien, and B. Coqblin* 459

New Treatment of the Anderson Model for Single Magnetic Impurity. *J. W. Schweitzer* 465

Concentration Dependence of the Kondo Effect in a Random Impurity Potential. *M. W. Klein, Y. C. Tsay, and L. Shen* 470

Concentration Effects in the Thermopower of Kondo Dilute Alloys. *K. Matho and M. T. Béal-Monod* 475

Resistance and Magnetoresistance in Dilute Magnetic Systems. *J. Souletie* 479

The Linear Variation of the Impurity Resistivity in *Al*Mn, *Al*Cr, *Zn*Fe, and Other Dilute Alloys. *E. Babić and C. Rizzuto* 484

Hall Effect Induced by Skew Scattering in *La*Ce. *A. Fert and O. Jaoul* 488

Kondo Effect in *Y*Ce under Pressure. *W. Gey, M. Dietrich, and E. Umlauf* 491

The Low-Temperature Magnetic Properties of Zn–Mn Single Crystals. *P. L. Li and W. B. Muir* 495

Influence of Lattice Defects on the Kondo Resistance Anomaly in Dilute *Zn*Mn Thin-Film Alloys. *H. P. Falke, H. P. Jablonski, and E. F. Wassermann* 500

Evidence for Impurity Interactions from Low-Temperature Susceptibility Measurements on Dilute *Zn*Mn Alloys. *W. Schlabitz, E. F. Wassermann, and H. P. Falke* 503

Low-Temperature Electrical Resistivity of Palladium–Cerium Alloys. *J. A. Mydosh* 506

Spin and g Factor of Impurities with Giant Moments in Pd and Pt. *G. J. Nieuwenhuys, B. M. Boerstoel, and W. M. Star* 510

Specific Heat Anomalies of *Pt*Co Alloys at Very Low Temperatures. *P. Costa-Ribeiro, M. Saint-Paul, D. Thoulouze, and R. Tournier* 520

Resistivity of Paramagnetic Pd–Ni Alloys. *R. Harris and M. J. Zuckermann* 525

Anomalous Low-Temperature Specific Heat of Dilute Ferromagnetic Alloys. *M. Héritier and P. Lederer* 529

12. Theory

Charge Transfer and Spin Magnetism of Binary Alloys. *H. Fukuyama* 537
Coherent Potential Approximation for the Impure Heisenberg Ferromagnet. *R. Harris, M. Plischke, and M. J. Zuckermann* 540
Resistivity of Nearly Antiferromagnetic Metals. *P. Lederer, M. Héritier, and A. A. Gomès* 543
Magnetic Properties of Electrons in a Narrow Correlated Energy Band. *C. Mehrotra and K. S. Viswanathan* 547
Magnetism and Metal–Nonmetal Transition in Narrow Energy Bands: Applications to Doped V_2O_3. *M. Cyrot and P. Lacour-Gayet* 551
Determination of Crossover Temperature and Evidence Supporting Scaling of the Anisotropy Parameter of Weakly Coupled Magnetic Layers. *Luke L. Liu, Fredric Harbus, Richard Krasnow, and H. Eugene Stanley* 555
A Corrected Version of the t-Approximation for Strong Repulsion. *G. Horwitz and D. Jacobi* 561

13. Magnetism and Superconductivity

Anomalous Behavior of the Kondo Superconductor $(La_{1-x}Ce_x)Al_2$. *G. v. Minnigerode, H. Armbrüster, G. Riblet, F. Steglich, and K. Winzer* 567
Magnetic Impurities in Superconducting La_3Al Alloys. *Toshio Aoi and Yoshika Masuda* 574
Ce Impurities in Th-Based Superconducting Hosts. *J. G. Huber and M. B. Maple* 579
Heat Capacity of *Th*U Alloys at Low Temperatures. *C. A. Luengo, J. M. Cotignola, J. Sereni, A. R. Sweedler, and M. B. Maple* 585
Superconducting Critical Field Curves for Th–U. *H. L. Watson, D. T. Peterson, and D. K. Finnemore* 590
c^2 Contributions to the Abrikosov–Gor'kov Theory of Superconductors Containing Magnetic Impurities. *D. Rainer* 593
The Effects of Some $3d$ and $4d$ Solutes on the Superconductivity of Technetium. *C. C. Koch, W. E. Gardner, and M. J. Mortimer* 595
Specific Heat of SnTe between 0.06 and 30°K under Strong Magnetic Field. *M.P. Mathur, M. Ashkin, J. K. Hulm, C. K. Jones, M. M. Conway, N. E. Phillips, H. E. Simon, and B. B. Triplett* 601
Direct Evidence for the Coexistence of Superconductivity and Ferromagnetism. *R. D. Taylor, W. R. Decker, D. J. Erickson, A. L. Giorgi, B. T. Matthias, C. E. Olsen, and E. G. Szklarz* 605

14. Small Particles, Heat Capacity, and Paramagnetism

Electric and Magnetic Moments of Small Metallic Particles in the Quantum Size Effect Regime. *F. Meier and P. Wyder* 613

High-Frequency Relaxation Measurements of Magnetic Specific Heats. *A. T. Skjeltorp and W. P. Wolf* 618

Some Recent Results on Paramagnetic Relaxations. *C. J. Gorter and A. J. van Duyneveldt* 621

Low-Temperature Heat Capacity of α- and β-Cerium. *M. M. Conway and Norman E. Phillips* 629

A Measurement of the Magnetic Moment of Oxygen Isolated in a Methane Lattice. *J. E. Piott and W. D. McCormick* 633

Contents of Other Volumes 637
Index to Contributors 660
Subject Index 667

QUANTUM CRYSTALS

1

Plenary Topics

Quantum Crystals: Theory of the Phonon Spectrum

Heinz Horner*

Institut für Festkörperforschung
Kernforschungsanlage Jülich, Jülich, West Germany

Introduction

Quantum crystals are crystals with large zero-point motions, caused by a light mass and a weak interaction of the lattice particles. Among these are the solid phases of the quantum liquids ^{4}He and ^{3}He, molecular hydrogen, and solid neon. The existence of large zero-point motions can cause striking effects, e.g., the existence of a sizable nuclear exchange interaction in ^{3}He or a wavelike propagation of vacancies or isotopic impurities.

In this paper I explore the lattice dynamic aspects of quantum crystals. We are actually investigating strong anharmonicities which could be found in other crystals as well as near melting or near structural phase transitions. It turns out, however, that the anharmonicities in a quantum crystal can be much stronger than, for instance, those in one of the heavier rare gas crystals near melting.

A rather interesting aspect has come up quite recently. Inelastic neutron scattering experiments in both the solid[1] and the liquid[2] have revealed striking similarities, and we might ask the question: Does the solid show liquidlike behavior[3] or is it the other way around?[4]

Let me list the problems which have to be faced if a microscopic theory is intended. First, we note an expansion of the lattice due to the zero-point motions in much the same way as ordinary thermal expansion due to thermal vibrations. This expansion can actually be so large that even the next-neighbor distance would be beyond the inflection point of the interaction potential. If we try to start our theory with the harmonic approximation, we end up with imaginary frequencies—in other words, with an unstable crystal. This difficulty has, however, been overcome by the renormalized harmonic approximation[5] in which the harmonic coupling constants are averaged over the zero-point motions.

This brings another difficulty. The zero-point motions are actually large enough that there is a fair chance that two lattice particles can approach each other within the hard-core radius. This means short-range correlations have to be an essential part of our theory. There are actually several ways to accomplish this: for instance, Jastrow factors[6] or one or the other forms of a scattering matrix.[7] For the moment, however, we adopt a slightly more general point of view.[8]

---

* Partially supported by U.S. Atomic Energy Commission at Brookhaven National Laboratory, Summer 1972.

Renormalized Phonon Theory

Let me go through the essential steps of the theory without going into too much detail. Assume the crystal is under the influence of some external forces $\mathbf{f}_i$ representing the external pressure or some small disturbance, eventually time dependent. The equation of motion of the position operator $\mathbf{x}_i$ of any particular particle is then

$$- m \frac{\partial^2}{\partial t^2} \mathbf{x}_i = \sum_j{}' \nabla V(\mathbf{x}_i - \mathbf{x}_j) + \mathbf{f}_i \tag{1}$$

where m is the mass and $V(\mathbf{r})$ is the interaction of the lattice particles. Let

$$\mathbf{d}_i(t) = \langle \mathbf{x}_i(t) \rangle \tag{2}$$

be the expectation value of $\mathbf{x}_i$, i.e., the average position of particle i, in the presence of the external force, then

$$- m \frac{\partial^2}{\partial t^2} \mathbf{d}_i(t) = \mathbf{K}_i(t) + \mathbf{f}_i(t) \tag{3}$$

where

$$\mathbf{K}_i(t) = \sum_j{}' \langle \nabla V(\mathbf{x}_i - \mathbf{x}_j) \rangle = \sum_j{}' \int d^3r\, g_{ij}(\mathbf{r}) \nabla V(\mathbf{r}) \tag{4}$$

is the average internal force on particle i due to the presence of the particles labeled by $j \neq i$. It has been expressed by the pair correlation function for a distinct pair of particles

$$g_{ij}(\mathbf{r}) = \langle \delta(\mathbf{r} - \mathbf{x}_i + \mathbf{x}_j) \rangle \tag{5}$$

where again the expectation value is in the presence of the external forces. Therefore, g_{ij} might be time dependent. In the absence of time-dependent external forces, the left-hand side of Eq. (3) actually vanishes and we recover an expression for the equation of state.

One quantity of primary interest is the displacement correlation function

$$\mathbf{D}_{ij}(t, t') = \delta \mathbf{d}_i(t)/\delta \mathbf{f}_j(t')$$
$$= \langle \mathbf{x}_i(t) \mathbf{x}_j(t') \rangle - \langle \mathbf{x}_i(t) \rangle \langle \mathbf{x}_j(t') \rangle \tag{6}$$

It describes how a disturbance $\delta \mathbf{f}_j(t')$ propagates through the crystal causing a change $\delta \mathbf{d}_i(t)$ of the expectation value of the position of particle i at time t. Since such a disturbance propagates as a phonon, at least in a harmonic crystal; this quantity is called the phonon propagator. It also contains information about equilibrium properties; for instance, $\mathbf{D}_{ii}(0,0)$ gives the mean square fluctuations of particle i around its equilibrium position described by $\mathbf{d}_i$.

Let me return for a moment to the pair correlation function $g_{ij}(\mathbf{r})$, which is one of the crucial quantities to calculate. We already have several pieces of information, for instance: (1) It has to be normalized to unity; (2) the first moment gives the average distance between particle i and j, which is also given by $\mathbf{d}_i - \mathbf{d}_j$; (3) its second moment gives the mean square fluctuations of this distance, which can also be expressed by $\mathbf{D}_{ij}(00)$; (4) its asymptotic form at small distances is that of the scattering problem of a pair of particles interacting with the true two-particle interaction. These pieces of information actually turn out to be sufficient to determine $g_{ij}(\mathbf{r})$ for given $\mathbf{d}_i$ and $\mathbf{D}_{ij}$.

Let me now come back to the displacement correlation function. Using Eqs. (6) and (3), we can find an equation of motion having in mind that $g_{ij}(\mathbf{r})$, and with it $\mathbf{K}_i(t)$, is a function of $\mathbf{d}_i(t)$ and $\mathbf{D}_{ij}(t,t')$

$$-m\frac{\partial^2}{\partial t^2}\mathbf{D}_{ij}(t,t') = \mathbf{1}\delta_{ij}\,\delta(t-t') + \sum_l \int d\tau\,\mathbf{M}_{il}(t\tau)\,\mathbf{D}_{lj}(\tau t') \quad (7)$$

where we have introduced the self-energy

$$\mathbf{M}_{ij}(tt') = \delta\mathbf{K}_i(t)/\delta\mathbf{d}_j(t')\big|_{\text{tot}} \quad (8)$$

We have for the moment considered $\mathbf{D}_{ij}(t,t')$ as a function of $\mathbf{d}_i(t)$ and the derivative has to be taken with respect to the explicit dependence of $g_{ij}(\mathbf{r})$ as well as with respect to the implicit dependence through the width given by $\mathbf{D}_{ij}$.

In physical terms the self-energy, a generalization of the dynamic matrix, is given by the change in the internal force on a particular particle, provided the equilibrium positions of some other particles are changed. The simplest assumption we can make about $g_{ij}(\mathbf{r})$ is that it is some function $g(\mathbf{r} - \mathbf{d}_i(t) + \mathbf{d}_j(t))$. Inserting this in Eqs. (4) and (8) and integrating by parts, the self-energy would simply be the second derivative of the interaction averaged over the pair distribution function. In the limit where $g(\mathbf{r}) \to \delta(\mathbf{r})$ we recover, obviously, the harmonic approximation. In the case of quantum crystals, however, this averaging yields real phonon frequencies.

In general, the functional dependence of $g_{ij}(\mathbf{r})$ on $\mathbf{d}_i(t)$ is more complicated, and even for fixed widths this means neglecting the implicit dependence through the $\mathbf{D}_{ij}$, it changes its shape for varying $\mathbf{d}_i(t)$ as shown in Fig. 1.

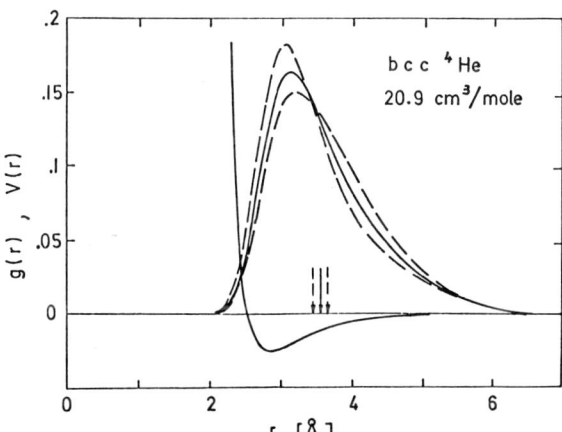

Fig. 1. Pair distribution function for three mean interparticle distances, indicated by arrows. Also shown is the interatomic potential for helium.

The way to proceed in a calculation would be to find a form for g_{ij} such that the conditions mentioned above are met and to calculate $g_{ij}(\mathbf{r})$ and $\mathbf{D}_{ij}(t)$ self-consistently.

Residual Anharmonicities

From this scheme, neglecting the dependence of $g_{ij}(\mathbf{r})$ on $\mathbf{D}_{ij}$ in calculating the self-energy, we obtain phonons without damping. Furthermore, the phonon, in this approximation, is a pure displacement motion.

If we include the dependence of $g_{ij}(\mathbf{r})$ on $\mathbf{D}_{ij}$ in lowest order, we obtain the bubble diagram (Fig. 2a) well known from ordinary anharmonic theory. The difference is, however, that the harmonic phonon frequencies in the intermediate lines are replaced by anharmonic ones, and the third-order coupling constants are replaced by renormalized vertices in very much the same way as the dynamic matrix was replaced by the renormalized harmonic vertex discussed above. As is well known, this diagram is responsible for phonon damping and for an additional anharmonic shift in the frequency.

We might take a slightly different point of view and say the third-order coupling constant represents a coupling between the one-phonon process (Fig. 2b) and the two-phonon process (Fig. 2c). This latter has a broad frequency distribution extending out to twice the maximum phonon frequency. This means that in the presence of this coupling the frequency distribution of the displacement response function now has not only a more or less sharp peak at the shifted phonon frequency, but in addition a tail ranging up to twice the maximum frequency and resembling the two-phonon frequency distribution. This is shown in Fig. 3 by the dashed lines.

The existence of this tail tells us that the true elementary excitation, represented by the sharp structure only, is no longer a pure displacement motion in the presence of the coupling. If we make a simple picture of a quantum solid where each particle has a Gaussian wave function near its lattice site, then a phonon in the absence of the coupling would be a collective oscillation of the rigid wave functions. In the presence of the coupling the wave functions also change their width in such a way that they narrow if neighboring particles move toward each other and widen if they move apart. This means that the actual elementary excitation is in general a coupled displacement and width fluctuation motion.

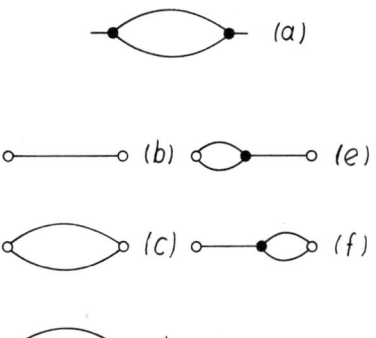

Fig. 2. (a) Bubble diagram contributing to phonon damping. (b–g) Diagrams for neutron scattering. (b) Bare single-phonon scattering $S_1^{(0)}(Q,\omega)$; (c–d) multiphonon processes $S_2^{(0)}(Q,\omega)$; (e–g) Interference terms $S_{\text{int}}(Q,\omega)$.

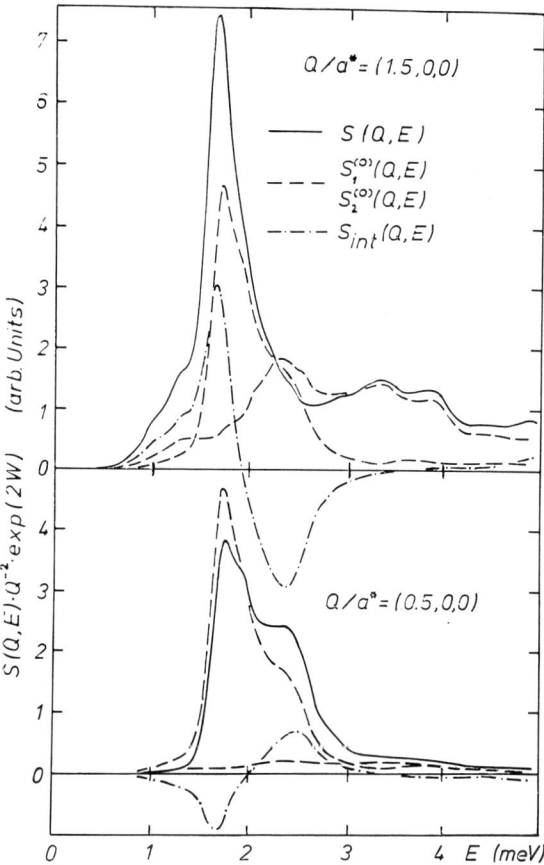

Fig. 3. Contributions to $S(Q, \omega)$ for two equivalent Q (see caption to Fig. 2.)

Neutron Scattering

As is well known,[10] the neutron scattering cross section in a crystal, in second Born approximation, is proportional to the dynamic scattering function

$$S(Q,\omega) = \frac{1}{2\pi N} \exp[-2W(Q)]$$

$$\times \int_{-\infty}^{\infty} dt \exp(-i\omega t) \sum_{ij} \exp[i\mathbf{Q}\cdot(\mathbf{R}_i - \mathbf{R}_j)]$$

$$\times \langle\!\langle \exp[i\mathbf{Q}\cdot\mathbf{u}_i(t)] \exp[-i\mathbf{Q}\cdot\mathbf{u}_j(0)] \rangle\!\rangle \qquad (9)$$

where $\exp[-2W(Q)] = |\langle \exp(i\mathbf{Q}\cdot\mathbf{u}_i)\rangle|^2$ is the Debye–Waller factor. The double bracket stands for the cumulant of the corresponding expectation value plus one responsible for Bragg scattering. The usual way to evaluate the cumulant is to expand the exponentials. The resulting series is conveniently represented by diagrams. In a harmonic crystal there would be, besides Bragg scattering, the one-phonon (Fig. 2b), two-phonon (Fig. 2c), and three- and higher-order processes (Fig. 2d). In an anharmonic crystal there are, in addition, the so-called interference diagrams (Fig. 2e–g) representing an interference between a one- and a two-phonon process. In addition, there are of course more complicated anharmonic diagrams. The structure of the diagrams in the theory for quantum crystals is exactly the same, but the third-order coupling constants are, as previously, replaced by the renormalized quantities.

There are two sum rules for the dynamic scattering function. The Platzcek sum rule[9]

$$(2\pi)^{-1}\int d\omega\, \omega S(Q,\omega) = Q^2/2m \qquad (10)$$

states that the average energy transfer of the neutron is equal to the free recoil energy of the scattering lattice particle. The ACB sum rule[10] states that a fraction given by the Debye–Waller factor is contributed by the single-phonon scattering

$$(2\pi)^{-1}\int d\omega\, \omega S_1(Q,\omega) = e^{-2W(Q)} \cdot Q^2/2m \qquad (11)$$

This sum rule holds for S_1 representing the bare one-phonon line (Fig. 2b) as well as for an S_1 including any number of one-multiphonon interference terms (Fig. 2e–g). This means the interference terms give a vanishing contribution to the sum rule, or in other words, they redistribute intensity within the phonon line. Since quantum crystals are rather anharmonic, these interference terms are certainly important and can lead to sizable distortions of the lines, as seen in Fig. 3. This picture also shows the multiphonon background and the interference contribution, where multiphonon processes of all orders were included in the calculation and the lowest-order interference terms (Fig. 2e–g).

Figure 4 shows that the multiphonon background and the interference terms can produce rather sharp structures besides the single-phonon line, as seen, for instance, at $Q/a^* = 0.75$ (a^* is the lattice constant in reciprocal space). Such structures are, however, not resolved in the experiments, due to the finite resolution. This has to be kept in mind if experimental phonon dispersion curves are compared to the theory, as done in Fig. 5, and the discrepancy at high frequencies is certainly mainly due to this fact. For lower-frequency phonons this problem does not usually arise and the remaining discrepancies are likely to be due to uncertainties in the calculation.

The problem of the integrated intensity in the one-phonon line has created some excitement recently. Calculations of the Debye–Waller factor[4,11] show no significant deviations from a Gaussian in the region of interest. This is in striking disagreement with the "effective Debye–Waller factor" which had been obtained from the experiments using the ACB sum rule after subtracting the background from the scattering. This discrepancy is still present for low-energy phonons which are experimentally well defined and where no problem appears to exist concerning how to subtract the background.

Obviously, the sharp phonon line without background cannot be identified with the theoretical "one-phonon scattering" $S_1(Q,\omega)$, which is defined via the displace-

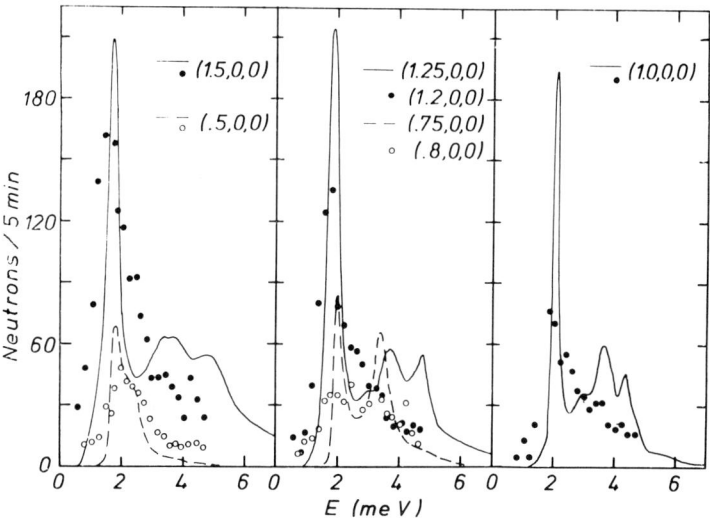

Fig. 4. Scattering function at equivalent points in different Brillouin zones for bcc ^{4}He. Experiments from Ref. 1.

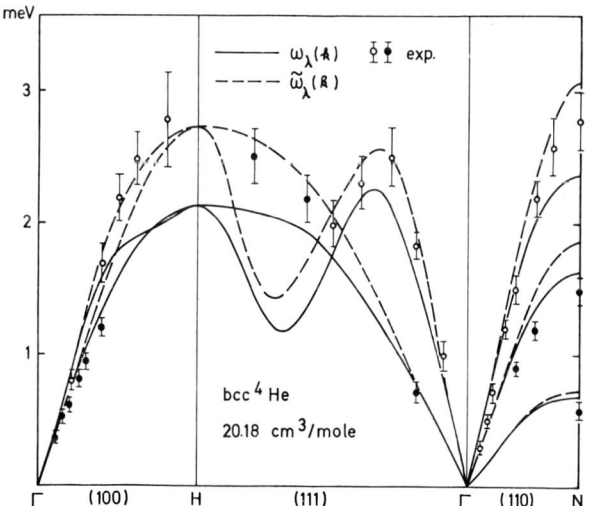

Fig. 5. Phonon dispersion curves in bcc ^{4}He. $\omega_\lambda(k)$ is the frequency of the maximum, $\tilde{\omega}_\lambda(k)$ the average frequency of the one-phonon line. The experimental points at high frequencies are too high due to unresolved two-phonon processes.

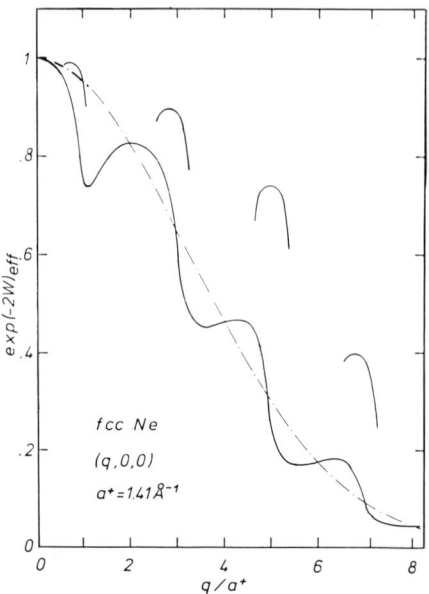

Fig. 6. Effective Debye–Waller factor for neon as determined from Eq. (11). Only the contributions from the sharp one-phonon peak have been included in S, as has been done in the experiment. The light lines include some two-phonon processes which are possibly not resolved in the experiment.

ment correlation function. We had seen that the spectral distribution of this correlation function has a tail besides the sharp phonon peak. This tail obviously has to be included in the ACB sum rule. The situation is even worse, however. As mentioned above, the interference terms redistribute intensity within the phonon line, including the tail. This means the intensity in the sharp line can actually grow at the cost of the tail, even giving the tail negative weight, or the other way around. Since the experiment obviously can identify the sharp line only, the discrepancy in the integrated intensity is no longer surprising.

This fact could be viewed from a different point of view. The elementary excitation is not a pure displacement motion, but rather a coupled mode containing fluctuations of the width of the single-particle wave functions. The neutron now couples in two ways to the elementary excitations: via displacement coupling, described by the term linear in **u** in an expansion of the exponential in Eq. (9); and second, it couples to the width via the quadratic term in **u** in Eq. (9). Depending on the value of **Q**, there can be constructive or destructive interference resulting in unexpectedly high or low intensities. The interference pattern follows the periodicity in reciprocal space, as seen for neon in Fig. 6 and for helium in Fig. 7. We see reasonable agreement with experiment, especially if we keep in mind that some of the lines might contain contributions from two-phonon processes. These oscillations have also been observed in molecular solid hydrogen.[12]

It is worthwhile to note that essentially the same results, that is, the dip near $Q = 1.5 \text{ Å}^{-1}$ and the hump near $Q = 2.5 \text{ Å}^{-1}$, have been observed in superfluid helium.[2] This similarity is indeed rather puzzling. To push this similarity even further, we show the neutron cross section for several **Q** along (111) in solid helium in Fig. 8. We find essentially a phonon–roton spectrum and a behavior of the two-phonon

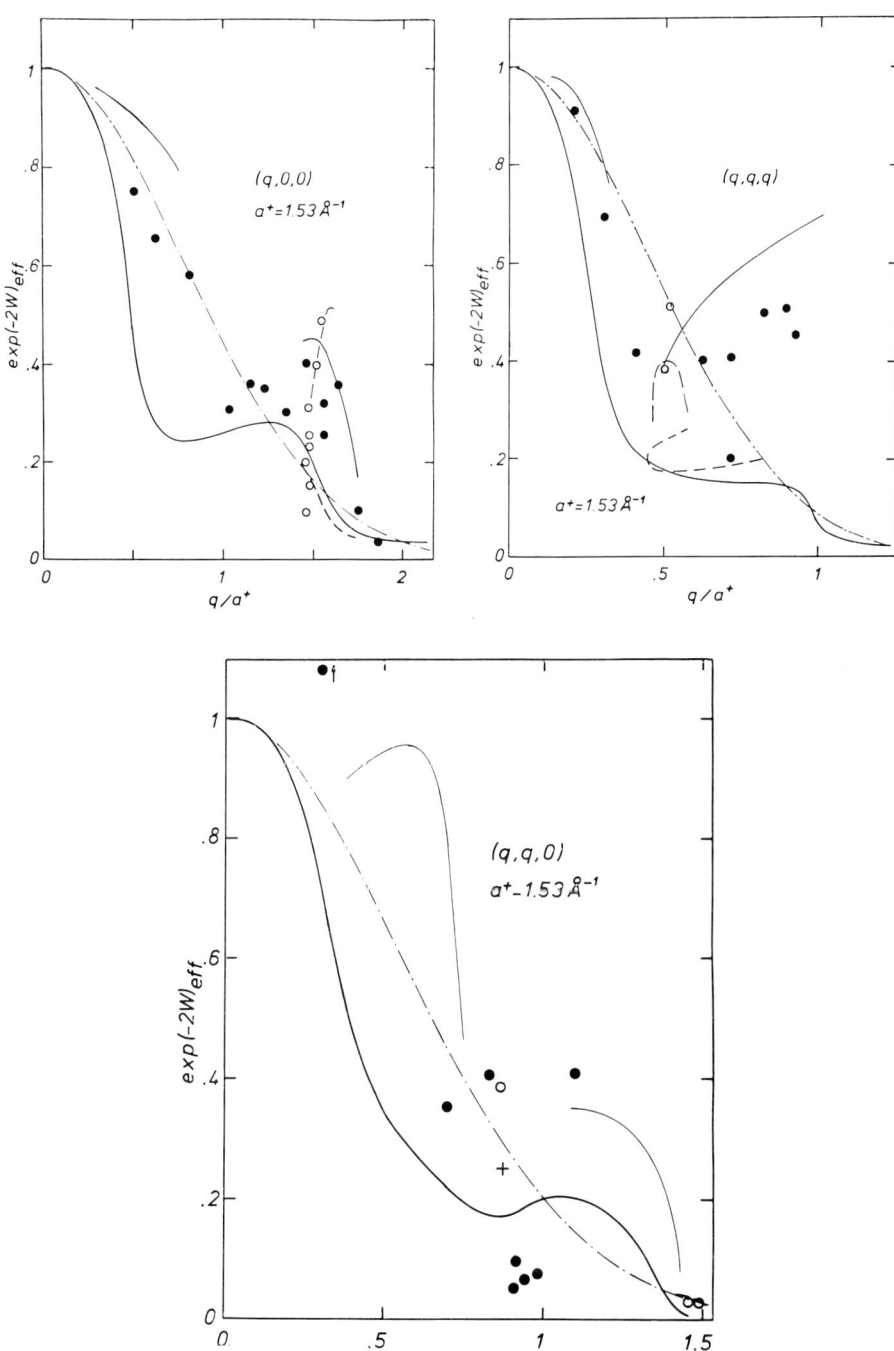

Fig. 7. Effective Debye–Waller factor (see caption to Fig. 6) for bcc ^{4}He in several directions. Solid lines and closed dots: longitudinal phonons. Dashed lines and open dots: transverse phonons. Experiments from Ref. 1.

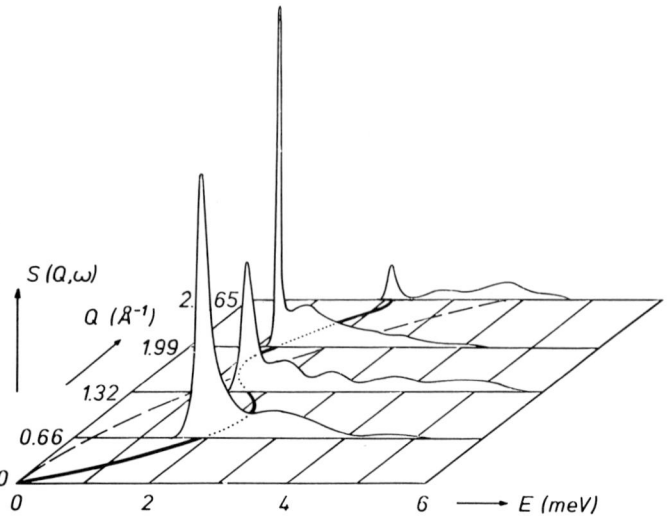

Fig. 8. Scattering function along (111) direction in bcc ^{4}He showing a phonon–roton type of scattering and the onset of free-particle-type scattering.

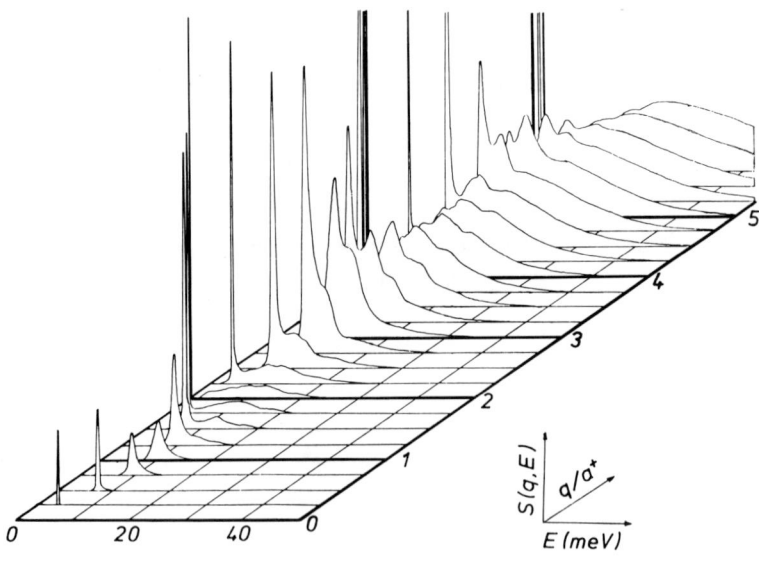

Fig. 9. Scattering function along (100) direction in fcc ^{4}He showing the growing importance of multiphonon scattering and the transition to the free-particle scattering with increasing Q.

scattering almost indistinguishable from the corresponding experiments in the liquid. At higher **Q** vectors we see the onset of free-particle-like scattering, again resembling very much the experiments in the liquid.

Conclusion

Within the last few years experiments and theory have provided us with a rather consistent picture of the lattice dynamics of quantum solids. We can describe neutron experiments with reasonable accuracy, including the anomalous line shapes and intensities as well as the transition to free-particle scattering (Fig. 9).

There is a rather striking similarity between solid and liquid helium. This might help us to find a better understanding of the phonon–roton spectrum; perhaps the theory outlined in this paper could be modified to be applicable to the liquid as well.

Acknowledgments

It is a pleasure to thank Drs. J. D. Axe, M. Blume, K. Kehr, T. A. Kitchens, V. J. Minkiewicz, L. H. Nosanow, L. Passell, J. Skalyo and, last but not least, N. R. Werthamer for valuable discussions.

References

1. E.B. Osgood, V.J. Minkiewicz, T.A. Kitchens, and G. Shirane, *Phys. Rev. A* **5**, 1537 (1972); and to be published.
2. R.A. Cowley and A.D.B. Woods, *Can. J. Phys.* **49**, 177 (1971).
3. N.R. Werthamer, *Phys. Rev. Lett.* **28**, 1102 (1972).
4. H. Horner, *Phys. Rev. Lett.* **29**, 556 (1972).
5. M. Born, *Festschrift der Akademie der Wissenschaften, Göttingen*, Springer, Berlin (1951); D.J. Hooton, *Ann. Physik* **142**, 42 (1955); T.R. Koehler, *Phys. Rev. Lett.* **17**, 89 (1966); P. Choquard, *The Anharmonic Crystal*, Benjamin, New York (1967); H. Horner, *Z. Physik* **205**, 72 (1967); N.R. Werthamer, *Am. J. Phys.* **37**, 763 (1969).
6. L.H. Nosanow, *Phys. Rev.* **146**, 120 (1966); T.R. Koehler and N.R. Werthamer, *Phys. Rev. A* **3**, 2074 (1971).
7. H.R. Glyde and F.C. Khanna, *Can. J. Phys.* **49**, 2997 (1971); B.H. Brandow, *Phys. Rev. A***4**, 422 (1971), and references cited therein.
8. H. Horner, *Z. Physik* **242**, 432 (1971); *J. Low Temp. Phys.* **9**, 547 (1972).
9. G. Placzek, *Phys. Rev.* **86**, 377 (1952).
10. V. Ambegaokar, J.M. Conway, and G. Baym, *Lattice Dynamics*, R.F. Wallis, ed. Pergamon, London (1965).
11. V.F. Sears and F.C. Khanna (to be published); A.K. McMahan and R.A. Guyer, this volume.
12. M. Nielsen, private communication.

Quantum Solids and Inelastic Neutron Scattering*

V. J. Minkiewicz†

University of Maryland, College Park, Maryland

T. A. Kitchens and G. Shirane

Brookhaven National Laboratory, Upton, New York

and

E. B. Osgood

Stevens Institute of Technology, Hoboken, New Jersey

Introduction

For large molar volume the solids of ^{4}He have the unique feature that, in contrast to normal systems, the average random excursion of an atom from its equilibrium position is a significant fraction ($\sim 1/3$) of the interatomic spacing. The weak interatomic interaction, which is known from measurements on the gas phase, and the light mass combine to produce crystals whose lattice dynamics cannot be described within the context of classical theory. For these reasons helium crystals are expected to be ideal systems with which one can test the subtle predictions of the quantum mechanical many-body theory of crystal dynamics.

In recent years the overall experimental effort in inelastic neutron scattering from these systems has made considerable progress, and a few brief introductory remarks are perhaps necessary. The early work in the hcp phase revealed that much of the phonon spectrum was not unusual.[1,2] The results of these experiments were reasonably consistent with theory, in that well defined single-phonon scattering was often observed. There were exceptions, of course, one of the most notable being the unusual "anomalous neutron groups" observed by Reese *et al.*[2] in the higher-energy region of the phonon spectrum.

The analysis of the scattering experiments in the hcp phase was complicated by the fact that there are two atoms in the unit cell; the bcc phase, on the other hand, has one. It was therefore decided that experiments should be done, if possible, in this phase. The initial experiment in the bcc phase[3] collected data that appeared to disagree with the theoretical results. More recent improved measurements[4,5] largely substantiated these results, and in addition presented data that clarified certain puzzling features of the earlier work and established new results that provide further tests for the dynamic theory of quantum solids. The purpose of this paper is

---

* Work supported by the U.S. Atomic Energy Commission.
† Partially supported by the Advanced Research Projects Agency.

to review these recent experimental results and to suggest those areas where further experimentation should be done to enhance our current concept of the dynamics of these solids.

Phonon Dispersion in the BCC Phase

The results of the experiments showed that the phonon spectrum over much of the Brillouin zone behaved as was expected. Phonon groups were often observed that appeared to be the type that is customarily compared with the theoretical calculations for the dispersion relation. Figure 1 contains two typical phonon groups of this type. The solid lines in the figure are reproductions of a computer calculation that convolutes the resolution function of the spectrometer with the dispersion curves of bcc helium as determined by a force-constant fit. The cross-section used in the convolution assumed that the excitations were infinitely long lived. The fact that the solid lines in the figure faithfully reproduce the data indicates that the observed linewidth of these phonon groups is primarily caused by the resolution of the instrument. An upper limit for the intrinsic linewidth of these groups can be fixed at approximately one-third of the observed FWHM, or ~ 0.03 meV. The data of the type given in Fig. 1 are typical of the observed scattered neutron groups in most crystals. However, for solid helium, as we shall see, the results at higher energy transfers are not of this character and their interpretation is more difficult.

The data for the phonon spectrum of bcc helium are given in Fig. 2.[3,4] The solid lines in the figure are theoretical calculations for the spectrum by Glyde.[6] The lowest-order self-consistent phonon approximation was used in the calculation. Attempts to improve the theory by including cubic anharmonic interactions tend to reduce the agreement between theory and experiment shown in the figure. One

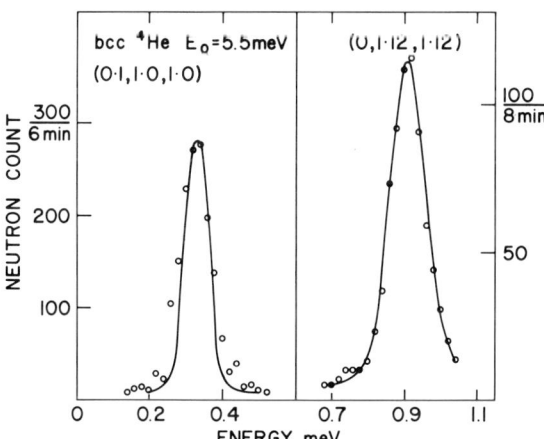

Fig. 1. Representative examples of two high-quality scans taken with high resolution. Data of the type shown were used to quantitatively describe the Debye–Waller factor. The solid lines are a reproduction of a computer calculation of the convolution of the resolution function with an infinitely long-lived excitation. The data were taken from Ref. 4.

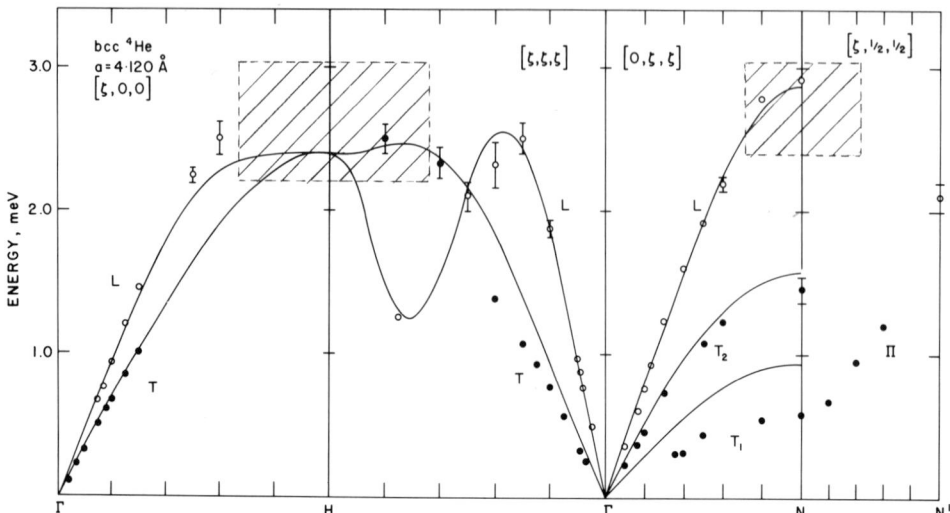

Fig. 2. The data for the dispersion curves of bcc helium. The cross-hatched regions illustrate those parts of the dispersion curves where there existed an uncertainty in defining either a peak position in the phonon group or an ambiguity in identifying the group as single-phonon scattering.

of the striking features of the data is that they do not extend over the entire zone. The shaded areas in the figure and the incomplete branches in the spectrum reflect both an uncertainty in being able to define a peak in certain profiles and in deciding whether a given group was in fact "single-phonon scattering."

The results of experimentation in the higher-energy regions of the phonon spectrum[3] were difficult to interpret. The profiles of these groups were generally broad, and had contours that were not always identical at equivalent positions in reciprocal space. In the second zone the groups were often found to be severely distorted from the normal Gaussian-like contour. The data given in Fig. 3 illustrate the gradual distortion[2] in the profiles observed for the longitudinal branch in the [100] direction as the scattering vector was increased from the first to the second zone.

These results strongly suggest that the conventional concept of the dispersion relation should be changed, and that the proper method of analyzing them is generally not by assigning "peak positions" and "widths" but by detailed computer convolutions of the resolution function of the spectrometer with theoretical results for the scattering function. The theoretical results for it are unfortunately difficult to calculate. It has been suggested that the origin of the distortion in the line profiles in the figure is to be found in the interference between the single-phonon and multiphonon scattering amplitudes.[7] A comparison between theory and experiment for solid helium will be especially interesting.

The Intensity

In addition to the strong distortion observed in the profiles for particular scattering vectors, the experiments revealed an unexpected dependence of the

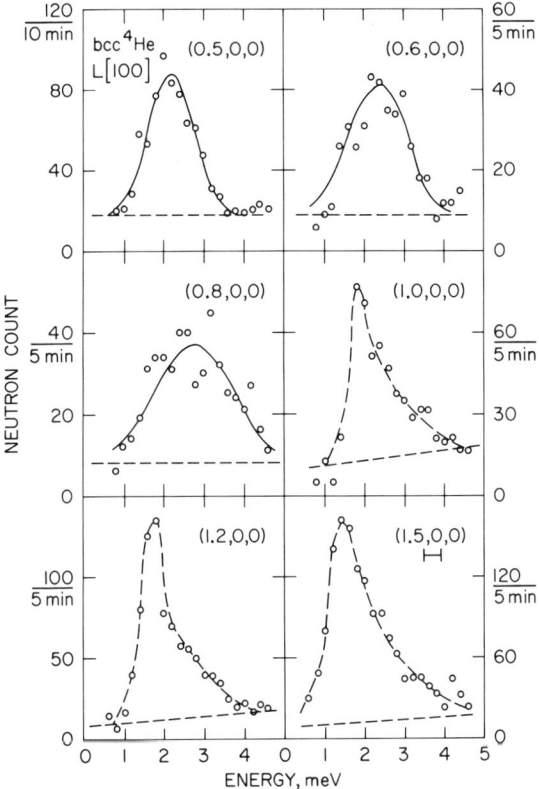

Fig. 3. Representative phonon groups for scattering vectors along the [100] direction, illustrating the gradual unexpected severe distortion that occurs in the group profile. The data were taken from Ref. 3.

scattering cross section on the magnitude of the scattering vector. The essential feature of the dependence of the cross section on $|\mathbf{Q}|$ is that it shows variations that are not expected from the classical form of the Debye–Waller factor. The data shown in Fig. 4 illustrate what is actually observed. The profiles on the bottom half are results taken at symmetry-related positions. The two groups are profiles of a LA [001] mode taken close to and on opposite sides of the (001) Bragg position. The top half illustrates what is expected from conventional theory at both positions. The cross section at the position A should be roughly 20% *smaller* than at B. In fact, one observes that the cross section at A is *larger* than at B by approximately 200%. Clearly the profile of these groups, apart from their intensity, are completely normal. The solid lines in the figure are computer calculations for their line shape and show that their linewidths are primarily caused by instrumental resolution. One of the natural ways to discuss the unexpected behavior of the intensity of the cross section for these phonons is via the sum rule developed by Ambegaokar,

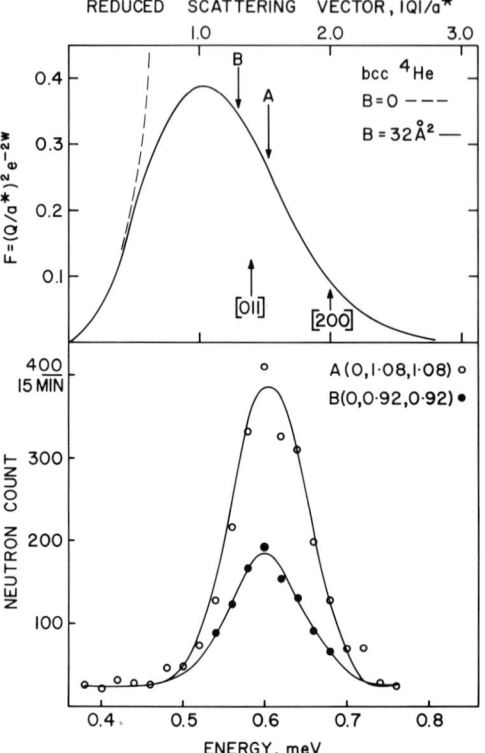

Fig. 4. Top: Normal Q dependence of the single-phonon cross section for solid helium from conventional theory. Bottom: two phonon groups taken at equivalent positions in reciprocal space in the bcc phase.

Conway, and Baym[8] (ACB sum rule):

$$\int_{-\infty}^{+\infty} d\omega\, \omega S_p^i(\mathbf{Q}, \omega) = (\mathbf{Q} \cdot \boldsymbol{\xi}_i)^2 |d(\mathbf{Q})|^2 / M \qquad (1)$$

where $S_p^i(\mathbf{Q}, \omega)$ is the single-phonon scattering function for branch i, $\mathbf{Q}$ is the scattering vector, ω is the energy transfer, and $|d(\mathbf{Q})|^2$ is the Debye–Waller factor. We note that $S_p^i(\mathbf{Q})$ contains the cross terms, or the interference terms, introduced into cross section by the one-phonon and multiphonon scattering amplitudes. Detailed balance can be used to restrict the integral in Eq. (1) to phonon creation to give

$$\int_0^\infty d\omega\, \omega S_p^i(\mathbf{Q}, \omega)(1 - e^{-\beta\omega}) = (\mathbf{Q} \cdot \boldsymbol{\xi}_i)^2 |d(\mathbf{Q})|^2 / 2M \qquad (2)$$

If we define the integral in Eq. (2) to be the quantity $M_i^1(\mathbf{Q})$, we note that it can be constructed from the data, and then either $M_i^1(\mathbf{Q})$, $M_i^1(\mathbf{Q})/(\mathbf{Q} \cdot \boldsymbol{\xi}_i)^2$, or $M_i^1(\mathbf{Q})/(\mathbf{Q} \cdot \boldsymbol{\xi}_i)^2 e^{-2W_h}$ can be used to present the experimental results. These three equivalent

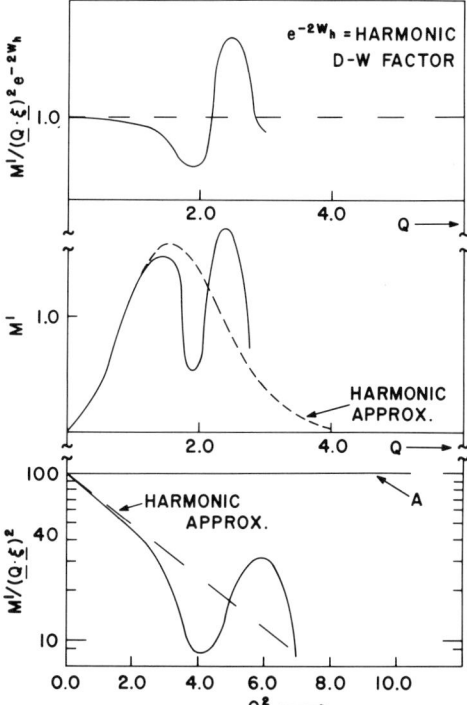

Fig. 5. The three equivalent ways in which the data for single-phonon scattering can be presented. The data are integrated by computer via the ACB sum rule. The line A in the bottom figure represents Placzek's sum rule.

ways of presenting the results are schemetically illustrated in Figure 5. The dashed lines in the figure illustrate what is expected for the coherent single-phonon cross section using the normal Gaussian Debye–Waller approximation, $|d(\mathbf{Q})|^2 = \exp[-2BQ/4\pi)^2]$. The solid lines qualitatively illustrate, as will be seen, what is actually observed for the scattering from solid helium. The data in the experiments were in effect reduced via the ACB sum rule to give an effective Debye–Waller factor or, alternatively, a "structure factor."

The illustration given in the bottom part of the figure is particularly instructive. The solid line A in the figure represents what one would get if all the scattering, that is, the sum of the single-phonon and multiphonon parts, is contained in the observations. It represents the Placzek[9] sum rule on the total scattering function

$$\int_{-\infty}^{+\infty} d\omega\, \omega S(\mathbf{Q}, \omega) = Q^2/2M \qquad (3)$$

where $S(\mathbf{Q}, \omega) = S_p(\mathbf{Q}, \omega) + S_m(\mathbf{Q}, \omega)$, and $S_m(\mathbf{Q}, \omega)$ is multiphonon scattering function. The figure clearly suggests that one might immediately anticipate difficulties in uniquely identifying the single-phonon part of the scattering. The multiphonon

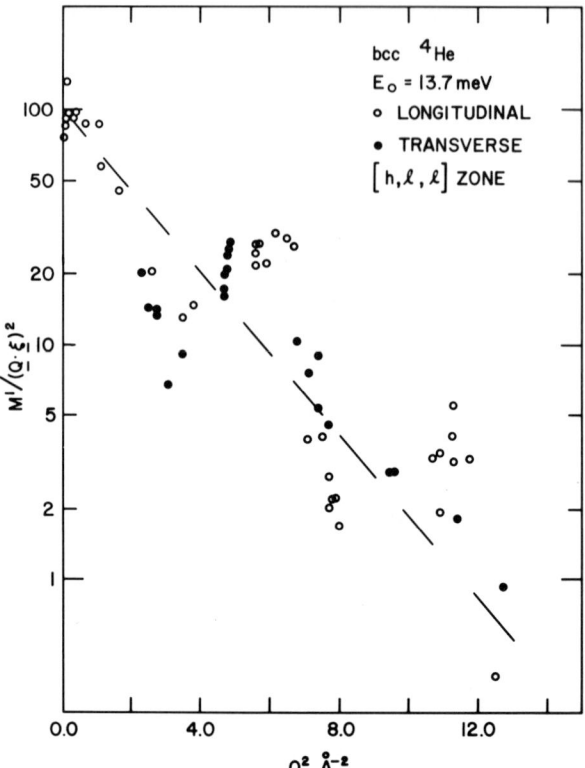

Fig. 6. The results of the Debye–Waller factor for bcc helium. The results should be taken as a qualitative description. Two oscillations are present in the data. The dashed line is the result expected from conventional theory. The data were taken from Ref. 4.

contribution to the total scattering for reasonable scattering vectors for solid helium is as intense as the single-phonon part. The experiments provided abundant evidence to support this observation. It was often observed that this multiple-phonon scattering bollixed the determination of the single-phonon part.

The presence of the multiple-phonon scattering had an immediate effect on the reliability with which one could construct the effective Debye–Waller factor.[3] In this experiment those groups were identified as single-phonon scattering if they were observed at energy transfers less than 1.1 meV, and they were collected under conditions in which the spectrometer was operated with high energy resolution. After careful consideration it was decided that selected scans could be used to *qualitatively* describe the single-phonon cross section. A quantitive description, on the other hand, could only be obtained with confidence with the high-resolution results. The data for the quantitive description of the single-phonon scattering used results similar to those given in Fig. 4.

The intensity data that were used to construct a qualitative description of the effective Debye–Waller factor are given in Fig. 6. The quantity $M^1/(\mathbf{Q}\cdot\boldsymbol{\xi})^2$ is plotted on a log scale vs. Q^2. If the Gaussian approximation for the Debye–Waller factor for solid helium were correct, i.e., that $|d(\mathbf{Q})|^2 = \exp[-2B(Q/4\pi)^2]$, the data would lie on a straight line, the slope of which would determine the quantity $B = 3h^2/2Mk\theta$, where θ is the Debye temperature. The dashed line in the figure is the calculated slope for bcc helium using $\theta = 22.5°K$.[10] It is clear that the effective Debye–Waller factor displays oscillations about the expected result. The errors in the measurement are illustrated by the scatter in the data. The data were collected for scattering vectors in the $[0\bar{1}1]$ zone and in this zone to within experimental accuracy they lie on a universal curve that depends only on $|\mathbf{Q}|$. The figure contains representative results from all branches in the phonon spectrum along the principal symmetry directions that can be observed in this zone. We also point out that the data in the figure are a composite of results taken on five different crystals. The data were consistent among all crystals.

The entire lowest transverse mode T_1 along the [011] direction cannot be observed in the $[0\bar{1}1]$ zone. This mode has its polarization vector along the $[0\bar{1}1]$ direction. The results of experiments on this mode on a crystal oriented in a [100] zone are shown in Fig. 7. The dashed line is again the expected slope with $\theta = 22.5°K$. The two data points with $Q^2 > 12.0\,\text{Å}^{-2}$ were taken on crystals in the $[0\bar{1}1]$ zone,

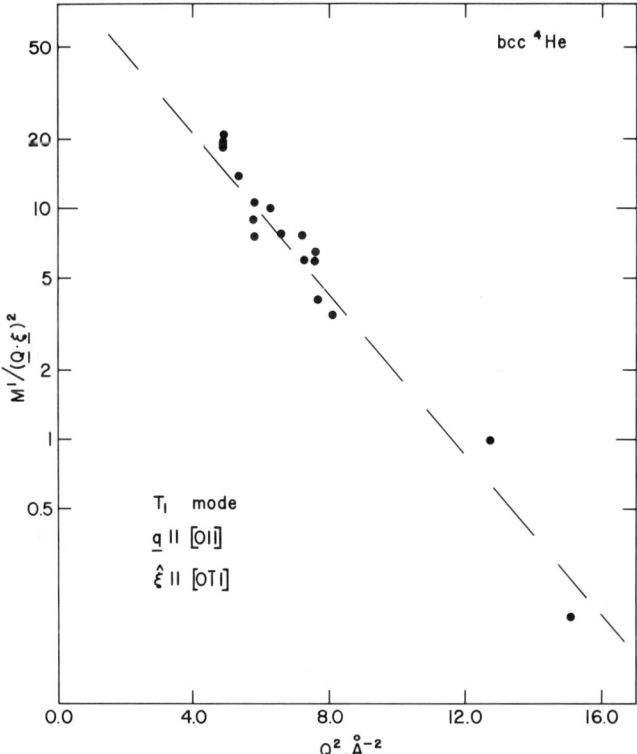

Fig. 7. The results for the T_1 mode. As is apparent, the data taken for this mode behave in the normal manner. The data were collected in the [100] zone.

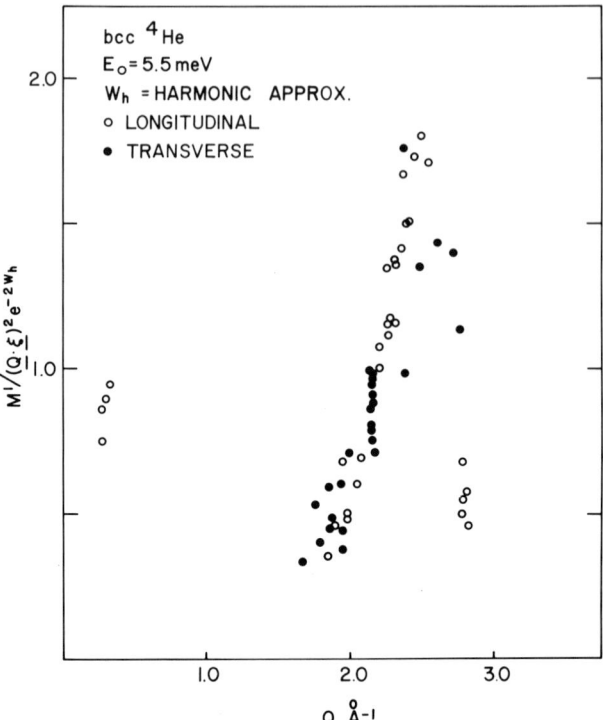

Fig. 8. A quantitative description of the first oscillation in the effective Debye–Waller factor for bcc helium.

and were included in the figure to compare theory and experiment over a wider range in Q. The important feature in the figure is that, surprisingly, the data in this mode do *not* show the oscillatory behavior observed in the other. It appears to behave normally. It was concluded that the effective Debye–Waller factor for bcc helium depends not only on the magnitude of the scattering vector but also on its direction.

The data which quantitatively reflect the effective Debye–Waller factor for bcc helium are given in Fig. 8. The quantity $M_i^1(\mathbf{Q})/(\mathbf{Q}\cdot\boldsymbol{\xi}_i)^2 e^{-2W_h}$ is plotted vs. Q. If the Gaussian approximation were correct, the data would be independent of Q, and lie on the straight line $M_i^1(\mathbf{Q})/(\mathbf{Q}\cdot\boldsymbol{\xi}_i)^2 e^{-2W_h} = 1.0$. The first oscillation shown in Fig. 6 is clearly visible. The four data for $Q \sim 0.2$ Å^{-1} are carefully selected data taken in forward scattering with the incident energy of 13.7 meV. They were taken in the first Brillouin zone at scattering vectors for which the single-phonon cross section continues to dominate Placzek's[9] sum rule. The purpose for including these data was a very important one; they served to establish the $Q = 0$ limit for the abscissa.

There is one further aspect of our measurements that is important; they show that the effective Debye–Waller factor behaves very much the same in both the bcc and hcp phases. Details of the hcp data will be given elsewhere.[4]

Concluding Remarks

The lattice excitations of bcc solid helium have been studied by inelastic neutron scattering. The data in the higher-energy region of the phonon spectrum show that the groups are often severely distorted from the normal Gaussian-like contour. To properly analyze these data, it is essential to convolute the resolution function of the instrument with theoretical results for the total scattering function. We emphasize here that further investigation of the full scattering law $S(\mathbf{Q}, \omega)$ of these solids will be extremely rewarding.

The phonon cross section was observed to be unusual. The ACB sum rule was used as a means to present the intensity data, and to construct an effective Debye–Waller factor. This effective Debye–Waller factor has "oscillations"; two such oscillations are clearly present in the data. The results show that this unexpected behavior does not depend on the energy transfer as was previously suggested,[3] but depends solely on the scattering vector.[4] This Debye–Waller factor was constructed from data taken on branches along the symmetry direction in the $[0\bar{1}1]$ zone. The cross section for the T_1 branch in the [011] direction does not behave abnormally; the Debye–Waller factor derived from it behaves as is expected from conventional theory. The unusual behavior of the cross section was observed for phonon groups that, to within the resolution of the instrument, have the normal Gaussian profile that is commonly observed in most systems. Specifically, it is suggested that a theory that purports to explain this phenomenon must agree with data of the type given in Fig. 4.

It is also established[3,4] that the multiple-phonon component can severely complicate neutron scattering observations in the higher energy region of the phonon spectrum. Experiments have been performed in the hcp phase of helium on crystals with a molar volume of 16.0 cm^3/mole by Reese et al.[1] Their measurements contain evidence for the existence of "anomalous neutron groups" at this molar volume. It is suggested[4] that those "anomalous groups" are the counterpart of the multiphonon scattering at that density.

There remains a great deal more work to be done in the quantum region of the phase diagram of solid helium. The solids in this region are unique; it is generally accepted that they will yield new, exciting results that will greatly modify our current concepts of the dynamics of a quantum solid.

Acknowledgments

The many conversations with our colleagues N. R. Werthamer, T. Koehler, H. R. Glyde, H. Horner, M. Blume, P. P. Craig, V. J. Emery, E. P. Lipschultz, L. Passell, V. Korenman, and R. Prange are greatly appreciated.

References

1. V.J. Minkiewicz, T.A. Kitchens, F.P. Lipschultz, and G. Shirane, *Phys. Rev.* **174**, 267 (1968); F.P. Lipschultz, V.J. Minkiewicz, T.A. Kitchens, G. Shirane, and R. Nathans, *Phys. Rev. Lett.* **19**, 1307 (1967).
2. R.A. Reese, S.K. Sinha, T.O. Brun, S.K. Sinha, and C.R. Telford, *Phys. Rev. A* **3**, 1688 (1971); T.O.

Brun, S.K. Sinha, C.A. Swenson, and C.R. Tilford, in *Neutron and Inelastic Scattering*, International Atomic Energy Agency, Vienna (1968), Vol. VI.
3. E.B. Osgood, V.J. Minkiewicz, T.A. Kitchens, and G. Shirane, *Phys. Rev. A* **5**, 1537 (1972).
4. V.J. Minkiewicz, T.A. Kitchens, G. Shirane, E.B. Osgood, *Phys. Rev.* (to be published).
5. T.A. Kitchens, G. Shirane, V.J. Minkiewicz, and E.B. Osgood, *Phys. Rev. Lett.* (to be published).
6. H.R. Glyde, *J. Low Temp. Phys.* **3**, 559 (1970), and private communication based on H.R. Glyde, *Can. J. Phys.* **49**, 761 (1971).
7. N.R. Werthamer, *Phys. Rev. Lett.* **28**, 1102 (1972).
8. V. Ambegaokar, J.M. Conway, and G. Baym, *J. Phys. Chem. Solids Suppl.* **1**, 261 (1965); in Proc. *Intern. Conf. Lattice Dynamics*, Pergamon, New York (1965).
9. G. Placzek, *Phys. Rev.* **86**, 377 (1952).
10. B.J. Alder, W.R. Gardner, J.K. Hoffer, N.E. Phillips, and D.A. Young, *Phys. Rev. Lett.* **21**, 732 (1966); J.K. Hoffer, Ph.D. Thesis, University of California, Berkeley, 1968, unpublished.

Magnetic and Thermal Properties of Solid and Liquid ^{3}He Near the Melting Curve*

D. M. Lee

Laboratory of Atomic and Solid State Physics
Cornell University, Ithaca, New York

Progress in the study of bcc solid ^{3}He and of properties of ^{3}He along the melting curve has been extremely rapid in the past few years. Pomeranchuk cooling and high-field superconducting magnets have made the temperature region below 15 m°K and the magnetic field region above 20 kG accessible to the low-temperature experimentalist. More sophisticated and sensitive measuring techniques such as the sensitive capacitative strain gauges developed by the Florida group and elsewhere, X-ray determination of crystal orientation, and superconducting quantum interference detectors have also contributed in various ways to the rapid progress in this field.

Since it is impossible in the space given to provide an exhaustive coverage of all the recent developments in solid ^{3}He, we wish to point out that there are two excellent review articles, one by Guyer et al.[1] and the other by Trickey et al.[2] The former article treats in detail nuclear magnetic resonance studies and relevant theories of solid ^{3}He, whereas the latter article is a general survey of solid helium with emphasis on results other than those obtained using NMR.

The present discussion will be confined to only a few aspects of the ^{3}He problem. The subjects treated here will include: (1) phonons and thermal properties of bcc solid ^{3}He mainly below 1°K; (2) influence of exchange on the bulk properties of bcc solid ^{3}He; (3) properties of bcc solid ^{3}He and its melting curve in high magnetic fields; (4) exotic new properties of liquid and solid ^{3}He along the melting curve obtained in Pomeranchuk cooling experiments.

Various investigations determining the Debye temperature of bcc solid ^{3}He have shown anomalous behavior of the Debye θ at the lowest temperatures. Among these studies are the specific heat measurements of Pandorff and Edwards[3] and Sample and Swenson,[4] the thermal conductivity determinations of Thomlinson[5] ($k = C_v v \lambda / 3$), and the $(\partial P / \partial T)_V = \gamma C_v / V$ measurements of Adams and co-workers.[6] All of these studies indicate that at low temperatures the Debye θ starts to decrease with decreasing temperature. More recent work on the specific heat of bcc solid ^{3}He by Castles et al.[7] is discussed elsewhere in these proceedings. The interpretation of this drop in θ has been under active consideration by a number of investigators. Henriksen et al.[6] suggested that the anomalous θ behavior was a lattice effect associated with the high degree of elastic anisotropy in bcc ^{3}He and accompanying anharmonicity.

---

* Supported by the National Science Foundation (Grant GH 33637) through the Cornell Materials Science Center.

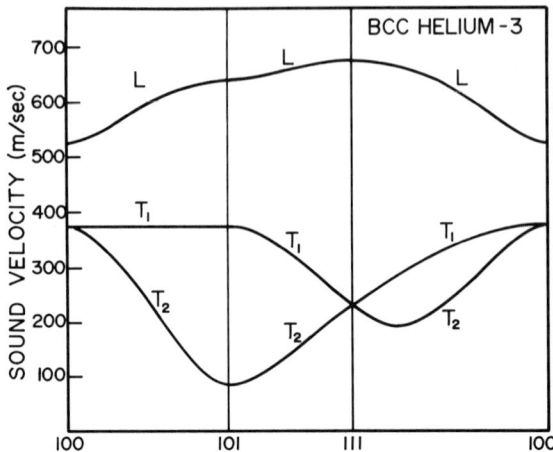

Fig. 1. Sound velocities as a function of crystal direction in bcc solid ^{3}He for the two transverse and the longitudinal modes as obtained by Greywall.[13b] Notice the low transverse velocity in the [110] direction.

Horner[8] pursued this approach theoretically and his results compared qualitatively with the experiments. Glyde[9] and later Glyde and Khanna[10] showed that the θ anomaly was dependent on whether the interatomic potential was replaced by a t-matrix or whether the Jastrow effective potential was used. Finally, Guyer[11] attempted to link the low-temperature anomaly in θ to the high-temperature specific heat anomaly[4] which had previously been attributed to vacancies. As yet none of these approaches has yielded a full understanding of the experimental results and further work in this area should certainly be anticipated.

The hypothesis that bcc solid ^{3}He has a high degree of elastic anisotropy has been strikingly confirmed by Wanner[12] and Greywall and Munarin.[13a] Wanner realized that for crystals with large anisotropy the wave vector of a sound wave can deviate by a large angle from the direction of energy flow. In a long, thin chamber sound will propagate for only a small range of directions for certain modes. Wanner applied this idea quantitatively to the early longitudinal and transverse sound data on bcc solid ^{3}He of Lipschultz and Lee[14] and to some thermodynamic results and data to derive the elastic constants of bcc solid ^{3}He. Figure 1 shows the directly measured sound velocities found by Greywall[13b] for the various modes as a function of direction in the crystal. It is evident from the figure that bcc solid ^{3}He is elastically highly anisotropic. The lowest transverse velocity is in the [110] direction. Wanner's analysis gives for this velocity a magnitude of 123 m/sec at 24 cm^3/mole and he points out that this is remarkably close to the observed velocity of second sound in bcc solid ^{3}He of 127 m/sec. This indicates the possibility that a resonance effect between first and second sound proposed by Guyer[15] might lead to a large attenuation of [110] slow transverse sound. In the direct observation of the sound velocities in the bcc phase as a function of crystal direction by Greywall and Munarin[13a] the method used to establish the crystal orientation was X-ray diffraction, which had the added advantage of making possible a determination of crystal quality. The results of this direct experiment are in quite good agreement with the indirect estimates made by Wanner and again confirm that bcc solid ^{3}He does indeed show a high degree of anisotropy. Further theoretical and experimental studies will be required to establish the relationship, if any, between the observed elastic anisotropy and the low-temperature specific heat anomaly of bcc solid ^{3}He. Greywall and Munarin[13a] and Greywall[13b] give bcc ^{3}He θ_0 (elastic) = 19.5°K($^{+2}_{-3}$) and therefore confirm the low-temperature θ anomaly.

So far we have not discussed the magnetic degrees of freedom of solid ^{3}He. The fact that there is an exchange interaction leads to a whole array of interesting phenomena in solid ^{3}He. This exchange interaction is important because of the quantum solid nature of solid ^{3}He, that is, the high degree of overlap of the wave function of the ^{3}He atoms occupying adjacent lattice sites. For bcc ^{3}He near the melting curve the strength of the exchange interactions is on the order of 1 m°K, which is several orders of magnitude larger than the strength of the interaction between nuclear magnetic dipoles. Thus the effects of exchange are to be found in temperature regions easily accessible by modern techniques. A detailed discussion of exchange in solid ^{3}He is given in an excellent review by Guyer.[16] It is customary to discuss solid ^{3}He in terms of the Heisenberg Hamiltonian

$$\mathcal{H} = -2 \sum_{i<j} J_{ij} \mathbf{I}_i \cdot \mathbf{I}_j$$

which takes into account nearest-neighbor, next-nearest-neighbor, and higher-order exchange. $J < 0$ favors antiparallel spin alignment, whereas $J > 0$ favors parallel spin alignment.

In the nearest-neighbor approximation we can write

$$\mathcal{H} = -2J \sum_{i<j} \mathbf{I}_i \cdot \mathbf{I}_j$$

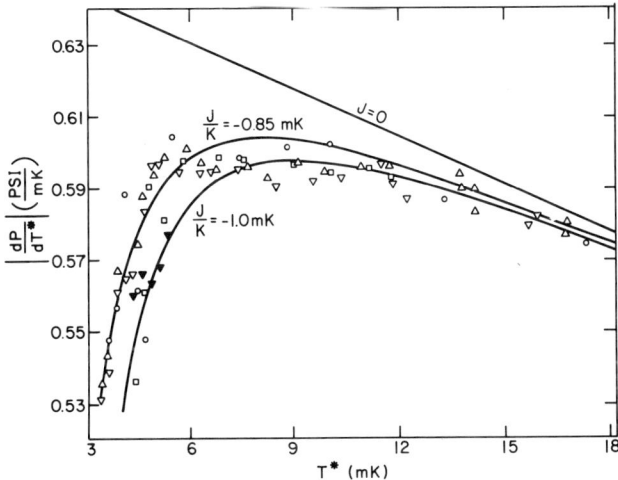

Fig. 2. Melting curve slope measured in a Pomeranchuk cell by Johnson et. al.[24] Solid curves indicate theoretical curves for various values of J/k.

In this case we have a simplified situation where we deal with a single parameter J whose magnitude is related directly to the temperature at which a magnetic ordering transition in the solid should occur. Studies of exchange and attempts to determine the sign and magnitude of J have occupied experimentalists and theorists for a number of years. For a more complete discussion of these results we refer to the review papers mentioned previously.[1,2] A brief discussion of some of the more recent experimental measurements of J follows. The earliest reliable studies of J involved measurements of transient effects in nuclear magnetic resonance experiments such as those of Richardson et al.[17] These results showed a rapidly decreasing J with increasing density with a maximum $|J/k|$ of about 0.7 at 24 cm^3/mole. As a result of improved cooling techniques it has recently become possible to determine J directly from low-temperature nuclear magnetic susceptibility measurements. The nuclear susceptibility would be expected to obey a Curie–Weiss law and by evaluation of the Weiss θ it is possible to obtain the magnitude and the sign of J. Baker et al.[18] have developed a high-temperature series expansion for the partition function corresponding to the nearest-neighbor Heisenberg Hamiltonian with an added Zeeman term to take account of the applied magnetic field. From this expansion they obtain a series expansion for the magnetic susceptibility χ,

$$\chi = \frac{N\mu^2}{VkT}\left[1 + 4\frac{J}{kT} + 12\left(\frac{J}{kT}\right)^2 + \cdots\right]$$

In the Curie–Weiss approximation $\theta = 4J/k$. Nuclear magnetic resonance measurements of χ were performed recently by Sites et al.,[19] Kirk et al.,[20] Pipes and Fairbank,[21] and Anderson et al.[22] as analyzed by Johnson and Wheatley.[23] The first two references gave values of J/k of -0.75 ± 0.08 and -0.73 ± 0.10, respectively. The latter two references gave somewhat larger values of J/k for the molar volume 24 cm^3/mole. Kirk et al.[20] also extended their measurements down to volumes as small as 21 cm^3/mole and again verified the tendency for $|J|$ to decrease rapidly with increasing density. The sign of J indicates an antiferromagnetic type of ordering with a predicted Néel temperature of about 2.0 m°K as derived from the expression of Baker et al.,[18] $T_N = -2.748\, J/k$.

Another method for obtaining $|J|$ is by measurement of the ^{3}He melting curve to very low temperatures and then applying the Clausius–Clapeyron equation

$$\frac{dP}{dT} = \frac{S_l - S_s}{V_l - V_s}$$

and the expansion of Baker et al.[18] for the entropy. Johnson et al.[24] measured the melting curve of ^{3}He in a Pomeranchuk cell down to 3 m°K and calculated a value of J/k of -0.85. Their data, shown in Fig. 2, show considerable nuclear ordering has taken place in the solid at the lowest temperatures reached.

Finally, Adams and co-workers[25,26] have developed a powerful method for obtaining accurate values of $|J|$, using a sensitive strain gauge to determine the variation of pressure at constant volume due to exchange. At sufficiently low temperatures the phonon contributions to this pressure variation, which was discussed earlier, will be small. The expression for the pressure derived from the partition

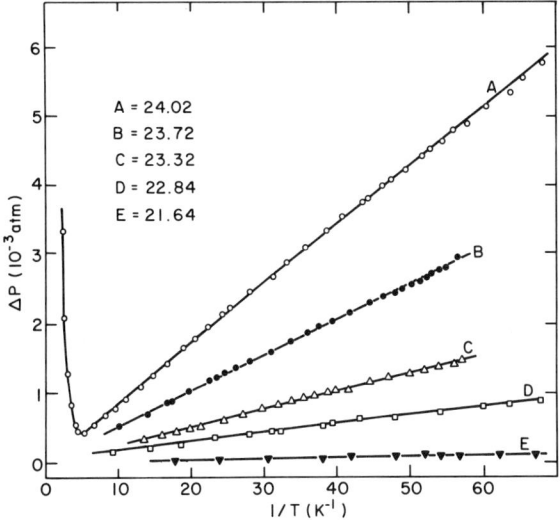

Fig. 3. Magnetic pressure plotted vs. $1/T$ for various molar volumes as obtained by Panczyk and Adams.[26] The 24.02 cm^3/mole curve is extended to high temperatures to illustrate phonon contributions.

function expansion of Baker et al.[18] is

$$P_{\text{exchange}} = \frac{3R}{V}\left(\frac{J}{k}\right)^2 \frac{\partial \ln|J|}{\partial \ln V} \frac{1}{T}\left(1 - \frac{J}{2KT} + \cdots\right)$$

Only the first two terms in the parentheses are considered. Some of the data of Panczyk and Adams[26] are shown in Fig. 3. The pressure variation under constant volume is plotted as a function of $1/T$ for various densities. At the lowest density ($V = 24.02$ cm^3/mole) the high-temperature phonon contribution as well as the low-temperature exchange contribution is shown. These measurements yield values of $|J|/k$ ranging from 0.718 for $V = 24$ cm^3/mole to 0.057 at $V = 21$ cm^3/mole. These determinations of $|J|/k$ are of considerably higher precision than those obtained by any other method. For the lowest density (24 cm^3 molar volume), the predicted magnetic ordering temperature is again 2.0 m°K, assuming the analysis based on the nearest-neighbor Heisenberg Hamiltonian is correct.

Elsewhere in these proceedings Castles et al[7] report specific heat measurements on solid ^{3}He which give another method of evaluating J/k.

So far we have considered only the nearest-neighbor Heisenberg Hamiltonian. Higher-order exchange processes may also be important in discussing the spin-dependent properties of bcc solid ^{3}He. In addition to the pair interactions included in the Heisenberg Hamiltonian, triple exchange may also play a significant role. Thouless[27] first showed that triple exchange is inherently ferromagnetic. Later Guyer and Zane[28] estimated the relative magnitude of pair exchange versus triple exchange. Zane[29] has recently used a triple exchange term

$$\mathscr{H}_{\text{triple}} = 2J_3 \sum_{i<j<k} (\mathbf{I}_i \cdot \mathbf{I}_j + \mathbf{I}_j \cdot \mathbf{I}_k + \mathbf{I}_k \cdot \mathbf{I}_i)$$

corresponding to a cyclic permutation of three particles labeled i, j, and k. Zane has calculated the influence of this term on the exchange pressure. McMahan[30] has considered the role of triple exchange and has calculated the higher-order exchange frequencies using a Monte Carlo method. Following McMahan's discussion, we write the total Hamiltonian as follows:

$$\mathcal{H} = -2J_1 \sum_{i<j} \mathbf{I}_i \cdot \mathbf{I}_j - 2J_2 \sum_{i<j} \mathbf{I}_i \cdot \mathbf{I}_j$$
$$+ 2J_3 \sum_{i<j<k} (\mathbf{I}_i \cdot \mathbf{I}_j + \mathbf{I}_j \cdot \mathbf{I}_k + \mathbf{I}_k \cdot \mathbf{I}_i)$$

where the three terms represent first- and second-neighbor and triple exchange terms, respectively. The terms in the above Hamiltonian can be rearranged, leading to the expression

$$\mathcal{H} = -2\underbrace{(J_1 - 6J_3)}_{\mathbf{J}_1} \sum_{i<j} \mathbf{I}_i \cdot \mathbf{I}_j - 2\underbrace{(J_2 - 4J_3)}_{\mathbf{J}_2} \sum_{i<j} \mathbf{I}_i \cdot \mathbf{I}_j$$

where the terms represent first- and second-neighbors, respectively, and which resembles in form a first- and second-neighbor pairwise exchange Heisenberg Hamiltonian with renormalized coefficients $\mathbf{J}_1$ and $\mathbf{J}_2$. $\mathbf{J}_1$ will favor antiferromagnetism, and $\mathbf{J}_2$ will favor ferromagnetism if $|J_2| < |4J_3|$.

Recent measurements on magnetic pressure by Kirk and Adams[31] using the previously described capacitance strain gauge technique in high magnetic fields have made possible an experimental test of the above analysis. If the expansion of Baker et al.[18] for the simple Heisenberg nearest-neighbor Hamiltonian is applied to a calculation of pressure with an applied magnetic field present, one obtains the following series expansion for P:

$$P = \frac{RT}{V} \frac{\partial \ln|J|}{\partial \ln|V|} \left\{ 3\left(\frac{J}{kT}\right)^2 - \frac{3}{2}\left(\frac{J}{kT}\right)^3 + \cdots \right.$$
$$+ \left(\frac{\mu H}{kT}\right)^2 \left[2\left(\frac{J}{kT}\right) + 12\left(\frac{J}{kT}\right)^2 + 52\left(\frac{J}{kT}\right)^3 + \cdots \right]$$
$$+ \left(\frac{\mu H}{kT}\right)^4 \left[-1.33\left(\frac{J}{kT}\right) - 23\left(\frac{J}{kT}\right)^2 + \cdots \right] \right\}$$

Note that the term involving $(\mu H/kT)^2 (2J/kT)$ allows the sign of J to be determined. In their measurements Kirk and Adams found that their experimental data did not agree with the above expression for P based on the nearest-neighbor Heisenberg Hamiltonian, although their measurements did give evidence that the effective J was negative, leading to antiferromagnetic ordering in agreement with the nuclear susceptibility results. The data of Kirk and Adams are shown in Fig. 4, where they are compared with the expansion for P based on the nearest-neighbor Heisenberg Hamiltonian and with the theoretical expression found by Zane which included the contribution of triple exchange to the magnetic pressure P. Zane's expression is complicated and will not be given here. However, it contains a parameter which is

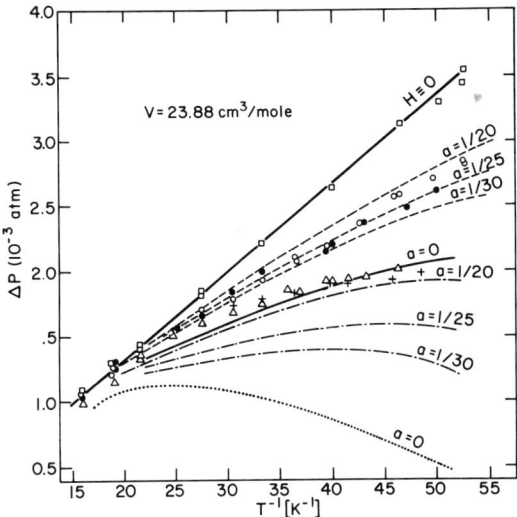

Fig. 4. The magnetic pressure data of Kirk and Adams[31] in high magnetic fields compared with the theory of Zane[29] for various values of the parameter a.

the ratio of triple exchange to pair exchange. By adjusting this parameter a, it is possible to fit the data of Kirk and Adams as shown in Fig. 4. The value of a of 1/20, which is rather a small fraction, provides a reasonably good fit. An encouraging result emerged from the experiments of Kirk and Adams. In fields of 60 kG and for volumes of 24 cm^3/mole thermal equilibrium times in the solid were found to be less than 1 min, so that it is almost certain that their sample was in thermal equilibrium with their thermometer.

The important qualitative conclusion to be drawn here is that the data of Kirk and Adams show deviations from the nearest-neighbor Heisenberg model in a direction which favors a ferromagnetic contribution from higher-order exchange. For further details of these fascinating results we refer to the work by Zane[32] and the work by Kirk and Adams[33] given elsewhere in these proceedings. Some early measurements were also made of $(\partial P/\partial T)_V$ in solid ^{3}He by Osgood and Garber[34]. Their results showed negligible effect of magnetic field on the magnetic pressure, in disagreement with the Kirk and Adams[31] data. Kirk and Adams have discussed this discrepancy and have put forward the hypothesis that the negative result obtained by Osgood and Garber might have been caused by a magnetoresistive effect in their resistance thermometer. Other hypotheses have been suggested to explain the discrepancy, including a hypothesis of Osgood[35] indicating possible thermal relaxation problems.

The melting curve of ^{3}He is expected to be strongly influenced by the magnetic field at low temperatures and strong fields ($\mu H \cong kT$). The slope of the melting curve is given by the Clausius–Clapeyron equation which, at low enough temperatures, can be written $dP/dT \cong S_{\text{solid}}/\Delta V$, since the entropy of the liquid is negligible compared with the entropy of the solid, which is expected to be of order $R \ln 2$. Following

Trickey et al.,[2] we write the high-temperature expansion

$$\frac{dP}{dT} \propto \frac{S_{\text{solid}}}{R} = \ln 2 - \frac{3}{2}\left(\frac{J}{kT}\right)^2 + \left(\frac{J}{kT}\right)^3 + \cdots$$

$$- \frac{1}{2}\left(\frac{\mu H}{kT}\right)^2 \left[1 + 8\frac{J}{kT} + 36\left(\frac{J}{kT}\right)^2 + \cdots \right] + \cdots$$

based on the next-nearest-neighbor Heisenberg Hamiltonian. Notice carefully that the leading two terms in the square bracket multiplying $(\mu H/kT)^2$ are $[1 + 8(J/kT)]$, whereas in the expansion for P measured at constant volume mentioned in conjunction with the Kirk and Adams experiment, the leading two terms multiplying $(\mu H/kT)^2$ are $[2(J/kT) + 12(J/kT)^2]$. Trickey et al.[2] point out that at high temperatures, where the expansion can be valid, J/kT will be small with respect to one so that the melting curve measurement will be less sensitive than magnetic pressure measurements to exchange effects in applied fields. Nevertheless, if sufficiently high fields and low temperatures are attained, deviations from a simple Heisenberg nearest-neighbor exchange model can be studied by means of melting curve measurements. Two groups, both using Pomeranchuk cooling, have recently studied the melting curve in magnetic fields. Osheroff et al.[36] made measurements at temperatures above 5 m°K and fields up to 13.4 kG, using NMR thermometry. Their results agree with the Heisenberg nearest-neighbor expansion, although the deviation from the zero-field melting curve was rather small and hence the test was not very sensitive. Only at higher fields is it convenient to observe possible deviations from the simple Heisenberg result, as mentioned previously. Johnson et al.[37] made measurements in fields as high as 63.6 kG and temperatures down to 5 m°K. They used a ^{54}Mn γ-ray thermometer to determine their temperatures. This gave two kinds of difficulties, namely γ-ray heating causing a possible error in temperature due to thermal boundary resistance between thermometer and sample, and incomplete saturation and domain alignment of the host iron at the lower fields. The saturation effect was overcome to a degree by assuming that the highest-field results agreed with the Heisenberg nearest-neighbor expansion. Between 31.8 and 63.6 kG[38] there was reasonable agreement between the observed melting curves and the Heisenberg nearest-neighbor theory, although discrepancies were observed at lower fields.

It was pointed out above that melting curve measurements are not as sensitive to deviations in the behavior of exchange effects as measurements of magnetic pressure at constant volume. Johnson et al.[38] feel that their results are not precise enough to see the kind of deviations from the Heisenberg nearest-neighbor high-temperature expansion observed by Kirk and Adams. They also bring up the point that the theory of Zane[29] based on triple exchange has not been worked out at low enough temperatures to make adequate comparisons with their results at the lowest end of their temperature range.

Interesting results have recently been obtained by Johnson et al.[39] on bulk nuclear polarization of solid ^{3}He in a Pomeranchuk cell. These authors developed the capability of performing NMR measurements at 54.5 kG at temperatures below 5 m°K to measure the nuclear polarization and observed a maximum average polarization in their cell of 47%. Since the cell always contained a mixture of liquid and solid

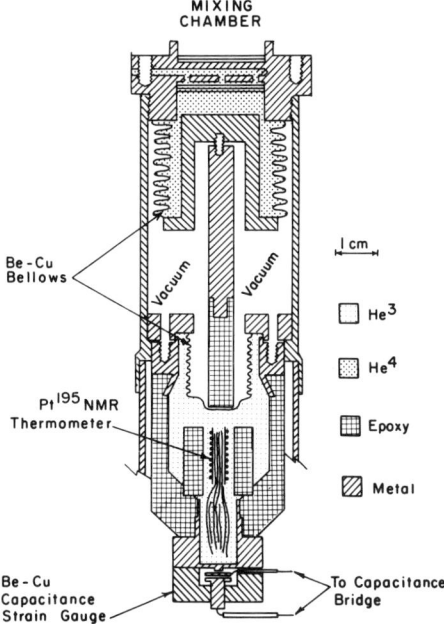

Fig. 5. Bellows apparatus used for Pomeranchuk cooling and melting pressure measurements of ^{3}He by Osheroff et al.[36]

^{3}He, the actual nuclear polarization in the solid was greater than this value. Perhaps some day this novel method of achieving polarized ^{3}He will be utilized by high-energy physicists. The authors suggest possible improvements in cell design which will maximize the amount of solid ^{3}He in the Pomeranchuk cell and thereby improve the average nuclear polarization. In conjunction with the above measurements, Johnson et al.[39] also made some interesting transient measurements on their solid ^{3}He sample. They conclude that the thermal equilibrium time for solid ^{3}He near the melting curve in these high fields was shorter than 5 min. The mechanism for this rapid thermal relaxation cannot be understood in terms of diffusion since relaxation by spin diffusion will require an extremely long time ($D \simeq 10^{-7}$ cm^2/sec).[1] Giffard et al.[40] have performed nuclear magnetic relaxation measurements at higher densities and temperatures and on the basis of their results suggest that crystal defects distributed throughout the ^{3}He sample may be responsible for the spin relaxation.

These rather short relaxation times were also observed by Osgood and Goodkind[41] who found relaxation times of less than 10 min at 7.5 m°K. The short spin relaxation times are no doubt related to the short thermal relaxation times noted by Kirk and Adams[33] for 24 cm^3 molar volumes, since most of the energy content of the solid is in the spin system in this temperature range. Further details of the work of Johnson et al.[42] are reported elsewhere in the proceedings of this conference.

Melting pressure measurements have recently been performed in a Pomeranchuk cell by Osheroff et al.[36] The compression was accomplished by a bellows arrangement

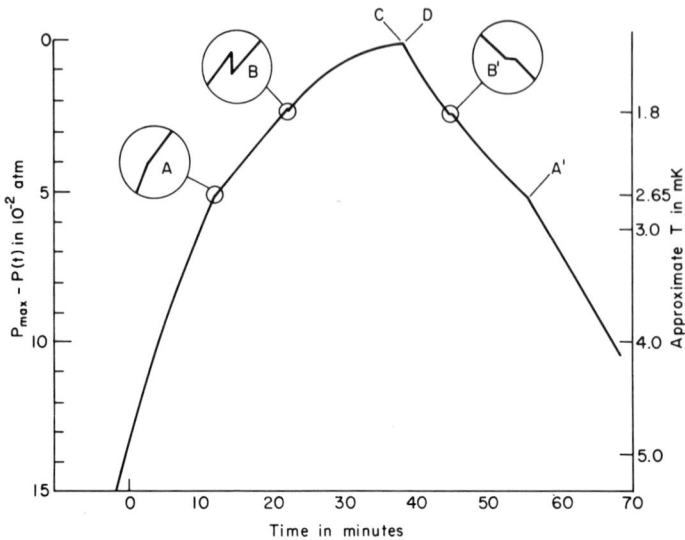

Fig. 6. Pressure vs. time curve obtained by Osheroff et al.[36] for a constant compression rate to D followed by a constant decompression rate.

as shown in Fig. 5. Inside the cell was a Pt NMR thermometer, and attached to the bottom of the cell was a sensitive capacitance strain gauge. For reasons which are not understood, the thermometer began to lose thermal contact with the ^{3}He sample below 3.5 m°K. In Fig. 6 is a plot of pressure vs. time for a typical run in which the volume of the cell was decreased at a constant rate to D and then increased at a constant rate. Interesting reproducible features labeled A, A', B, and B' were observed below 3 m°K. On the basis of an extrapolated melting curve, A and A' are thought to occur at 2.65 m°K. These features are thought to be associated with possible phase changes along the melting curve. Feature B always occurred at pressures greater than B', indicating a supercooling effect, whereas the A feature did not exhibit supercooling to any degree. Features B and B' were suppressed strongly to lower temperatures by magnetic fields, whereas A and A' were only slightly affected. It was difficult to determine the nature of these transitions from the melting pressure measurements because it was not possible to perform simultaneous temperature measurements. This difficulty can be appreciated by noting that the change of pressure in a Pomeranchuk cell with time is given by $dP/dt = (dP/dT)(dT/dt)$. Now if the heat leak into the cell from the dilution refrigerator is presumed to be a constant $\dot{Q}$, then the appropriate heat balance equation is $\dot{Q} \cong L\,dn/dt + (n_l C_l + n_s C_s)(dT/dt)$, where L is the latent heat and dn/dt corresponds to the number of moles changing from liquid to solid per unit time, which is roughly constant for a constant compression rate neglecting compressibility effects. Solving for dP/dt then gives

$$\frac{dP}{dt} \simeq \frac{dP}{dT} \frac{\dot{Q} - L(dn/dt)}{n_l C_l + n_s C_s}$$

The changes in slope seen at A in Fig. 6 can then be interpreted in the following three ways: (1) A first-order transition occurs in solid ^{3}He giving rise to a discontinuous change of slope dP/dT in the true melting curve as well as a discontinuous change in L. (2) A second-order phase transition occurs in solid ^{3}He, possibly of the predicted antiferromagnetic type with the attendant specific heat anomaly. (3) A second-order phase transition occurs in liquid ^{3}He, again with a resulting specific heat anomaly. Osheroff et al.[36] used the first choice in interpreting their data because the melting pressure anomaly at A had about the same slope ratio for a variety of compression rates. Horner and Nosanow[43] suggested that the second choice was the appropriate one and Vvedensky[44] suggested the third choice. Adams and Nosanow[45] point out that if A is in the solid, one might expect it to be at a higher temperature than 2 m°K on the basis of higher-order exchange effects. These discussions depend to some extent on the model used for heat transfer in solid helium. It is therefore of interest to point out that Johnson and co-workers[46] have made measurements of the spin diffusion coefficient of solid ^{3}He in a Pomeranchuk cell and have found that the spin diffusion coefficient decreases markedly at the lowest temperatures. The melting pressure results of Osheroff et al.[36] have recently been confirmed by Halperin et al.[47] using an elegant stressed-diaphragm Pomeranchuk cell. This cell is particularly suited to investigating solid ^{3}He, since its design makes it possible to fill the cell with solid without catastrophic heating. The work is described elsewhere in these proceedings. The La Jolla group of Johnson, Wheatley, and co-workers[48] has also recently observed the melting pressure changes at A and B using the very effective Pomeranchuk cell design described in their earlier work.[24]

Recently nuclear magnetic resonance measurements on the ^{3}He sample in a Pomeranchuk cell have been performed by Osheroff et al.[49] and are reported elsewhere in these proceedings. Since both liquid and solid ^{3}He are present in the cell, it is desirable to devise a technique which enables one to separate the contributions of the liquid and solid ^{3}He to the NMR signal. To accomplish this purpose a gradient in the applied field was established parallel to the axis of the rf coil, leading to an inhomogeneously broadened CW NMR line. As mentioned previously, solid ^{3}He obeys a Curie–Weiss law (at least at higher temperatures), whereas liquid ^{3}He is a degenerate Fermi liquid and therefore has a nearly constant nuclear susceptibility. Thus the signal will grow rapidly in regions where solid is being formed during compression but the signal will remain constant in regions that remain in the liquid phase. It was observed during compression that the solid formed in limited regions of the coil, and that this solid formation was manifested by peaks forming on the NMR profile. This type of behavior was also seen by Johnson et al.[39] in their bulk nuclear polarization experiments. In regions of the coil where no solid formed the signal remained constant. Figure 7 shows a sequence of profiles during a compression where only a single solid peak formed and also a sequence of profiles where two solid peaks formed in or near the coil. Interesting features are illustrated in Fig. 7 that are associated with the A and B melting pressure phenomena. As the temperature is decreased below the A transition the liquid signal seems to shift progressively farther away from the solid peak until the B transition is reached. As the sample is cooled through B the liquid signal suddenly moves back to its original position and drops by about a factor of two in amplitude. The liquid signal was closely monitored during a slow passage through the B transition and it was found that the "B" phase tended to start forming near the bottom of the cell and the interface rose through the coil until the transition was complete. The process was reversed at the B' transition. On the other

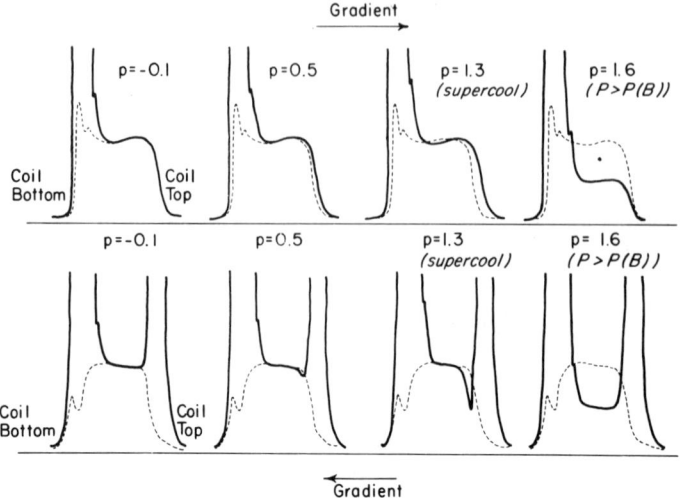

Fig. 7. Sequences of nuclear magnetic resonance profiles of ^{3}He in a Pomeranchuk cell obtained for various temperatures in a field gradient by Osheroff et al.[49]

$$p(t) = [P(t) - P(A)]/[P(B') - P(A)]_{H=850\,G}$$

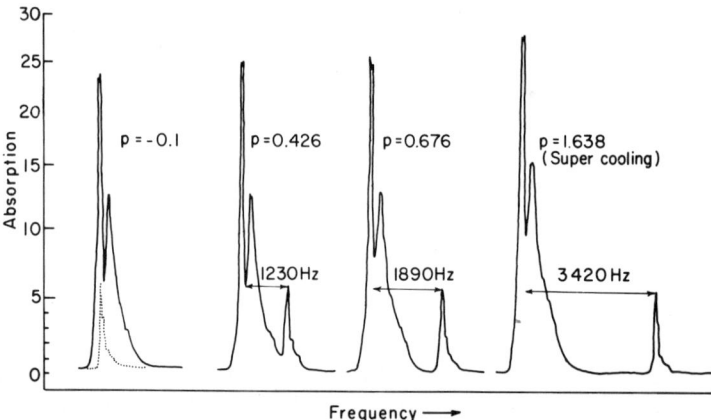

Fig. 8. Sequence of ^{3}He nuclear magnetic resonance peaks in a Pomeranchuk cell in zero applied field gradient obtained by Osheroff et al.[49] showing how the liquid line splits away from the solid line between the A and the B transitions.

$$p(t) = [P(t) - P(A)]/[P(B') - P(A)]_{H=850\,G}$$

hand, Halperin et al.[47] have measured the magnetization of the ^{3}He sample using a Squid technique in their Pomeranchuk cell, but saw no abrupt change when going through B. The magnetization data of Halperin et al. can be reconciled with the NMR results at B only if we assume that part of the nuclear susceptibility has moved to a different characteristic frequency in such a way as to keep the magnetization constant.

Interesting transient phenomena have been seen during heating pulses in the liquid signal at temperatures well below the B transition. Upon application of heat

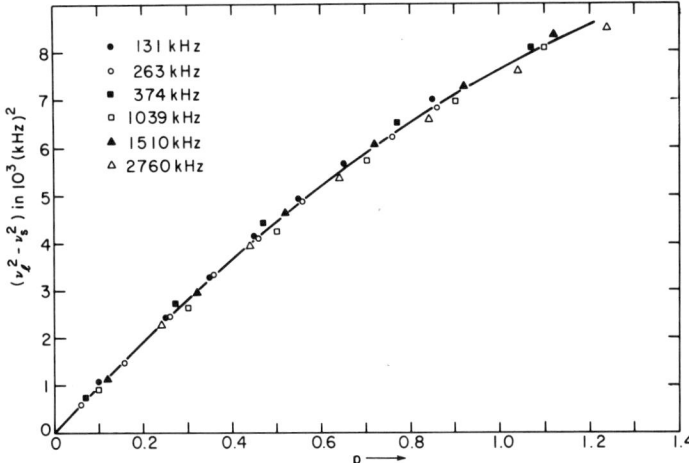

Fig. 9. Plot of the shifted liquid line obtained by Osheroff et al.[49] The difference between the squares of the liquid and solid frequencies is plotted against a normalized pressure difference between the observed pressure and the pressure at A.

$$p(t) = [P(t) - P(A)]/[P(B') - P(A)]_{H=850\,G}$$

the liquid susceptibility drops to zero, but when the heating is stopped the signal recovers to its original size in a wavelike fashion. The transient experiments are discussed in more detail by Osheroff et al.[49] elsewhere in these proceedings.

The frequency shift of the liquid ^{3}He NMR signal was studied more carefully by Osheroff et al.[49] by removing the field gradient. At temperatures above the A transition a single, fairly narrow, irregularly shaped peak is observed. As the temperature is decreased through A a line roughly equal in amplitude to the original all-liquid signal splits off and shifts progressively to higher frequencies as the temperature is reduced, as shown in Fig. 8. At B the shifted line abruptly disappears. The shift was studied as a function of a parameter p proportional to $P - P(A)$ for various Larmor frequencies from 131 to 2760 kHz. The results are shown in Fig. 9, where $v^2_{\text{liquid}} - v^2_{\text{solid}}$ is plotted against p. The data appear to fit a universal curve. Such a result would be obtained if, in addition to the applied field, there were an orthogonal internal field which depended on the temperature but not on the applied field. Extrapolating the graph of Fig. 9 gives a maximum value of this orthogonal field of 31 G. In solids such an orthogonal field is a feature of the spin-flopped antiferromagnetic phase. The value of T_1 was obtained for the solid peak and for the shifted line. The shifted peak was found to have a value of T_1 of about 2 sec, which is a factor of ten longer than T_1 for the solid line, providing further evidence that the shifted line is associated with the liquid.

On the basis of the above results the A and B transitions must certainly involve liquid ^{3}He. Since the experiments of Osheroff et al.[49] were performed in a Pomeranchuk cell, it cannot be completely ruled out that the solid plays some role in the transitions. Thus it would be desirable to perform measurements on liquid ^{3}He off the melting curve using a paramagnetic salt such as cerium magnesium nitrate or perhaps diluted cerium magnesium nitrate. Such an experiment is presently underway at La Jolla.[50] It is hoped that it will be possible to determine the complete phase diagram associated with the new phases. Another important test to be made is a test for superfluidity. There has been a considerable amount of literature addressed to the possibility of a Bardeen–Cooper–Schrieffer[51] type of pairing transition in liquid ^{3}He. A good summary of this literature is given by Sessler.[52] Anderson and Morel[53] and later Balian and Werthamer[54] have suggested a p-wave pairing transition to explain the Knight shift in superconductors. Leggett[55] has applied this to liquid ^{3}He and has shown that the susceptibility below such a transition starts to fall with temperature, reaching a value of one-third the Fermi degenerate value. It is unknown whether this is in any way related to the observed sudden drop in nuclear susceptibility at B observed by Osheroff et al.[49] Any relationship between the results at A and the phenomena observed by Peshkov[56] at low pressures can only be established by additional phase diagram data.

A further question must now be considered. Much of the earlier part of this discussion has been devoted to exchange in the solid. If the A and B features involve transitions in the liquid, what has become of the predicted nuclear magnetic transition in solid ^{3}He? Osheroff et al.[49] observed that the maximum height of the susceptibility peaks was a factor of 60 greater than the Fermi degenerate susceptibility of liquid ^{3}He when the compression was carried out in such a fashion as to form a maximum amount of solid at the lowest temperatures. This large susceptibility ratio indicates a substantial deviation from the Curie–Weiss law observed at a higher temperature. A susceptibility ratio of about 35 would be observed for a simple antiferromagnetic transition at 2 m°K in the solid. At a temperature sufficiently far below the solid magnetic transition the entropies of liquid and solid ^{3}He will become equal and the slope of the melting curve will be zero. Pomeranchuk cooling will no longer be possible at this temperature. If the maximum pressure of the melting curve corresponds to the maximum pressure observed by Osheroff et al[36] the expression

$$P_{max} - P(A) \approx \int (S_{solid}/\Delta V)\, dT$$

obtained by integrating the Clausius–Clapeyron equation, gives a value which is too large for a phase transition at A (2.65 m°K) in solid ^{3}He unless the solid entropy is constant over a broad temperature range below A. Another possible alternative is that the solid transition temperature is below the temperature of the B transition. Any model for the solid-phase transition will be required to explain the large value of P_{max}, the large susceptibility ratio, and the drop in the spin diffusion coefficient.

In conclusion, we quote from Sessler's address[52] at the London Conference on Low Temperature Physics in 1962. In that address he referred to the superfluid transition in ^{3}He and stated, "The experimentalists must find it discouraging to have the theorists continually lowering T_c. An effort has been made here to make it clear that the theory is not now capable of quantitatively evaluating T_c but—and this has been the spirit in which the theoretical estimates have been made—the best calculations give some basis for hoping that T_c is within an experimentally attainable range and thus should serve to encourage the experimentalists to even more heroic efforts." As an experimentalist I should like to encourage the low-temperature theorists to keep their spirits high and to urge them to make the required heroic efforts to understand and explain the fascinating and complex picture which is unfolding in the ^{3}He problem.

Acknowledgments

I wish to thank W. J. Mullin, R. C. Richardson, D. D. Osheroff, D. T. Lawson, J. D. Reppy, and R. H. Silsbee for many interesting and informative discussions bearing on this manuscript.

References

1. R.A. Guyer, R.C. Richardson, and L.I. Zane, *Rev. Mod. Phys.* **43**, 532 (1971).
2. S.B. Trickey, W.P. Kirk, and E.D. Adams, *Rev. Mod. Phys.* (to be published).
3. R.C. Pandorff and D.O. Edwards, *Phys. Rev.* **169**, 222 (1968).
4. H.H. Sample and C.A. Swenson, *Phys. Rev.* **158**, 188 (1967).
5. W.C. Thomlinson, *Phys. Rev. Lett.* **23**, 1330 (1969).
6. P.N. Henriksen, M.F. Panczyk, S.B. Trickey, and E.D. Adams, *Phys. Rev. Lett.* **23**, 518 (1969).
7. S.H. Castles, W.P. Kirk, and E.D. Adams, this volume.
8. H. Horner, *Phys. Rev. Lett.* **25**, 147 (1970).
9. H.R. Glyde, *Can. J. Phys.* **49**, 761 (1971).
10. H.R. Glyde and A. Khanna, *Can. J. Phys.* **49**, 2997 (1971).
11. R.A. Guyer, *Low Temp. Phys.* **6**, 251 (1972).
12. R. Wanner, *Phys. Rev.* **A3**, 448 (1971).
13a. D.S. Greywall and J.A. Munarin, *Phys. Rev. Lett.* **24**, 1282 (1970); *Phys. Rev. Lett.* **25**, 261 (E) (1970).
13b. D.S. Greywall, *Phys. Rev.* **A3**, 2106 (1971).
14. F.P. Lipschultz and D.M. Lee, in *Proc. 10th Intern. Conf. Low Temp. Phys., 1966*, VINITI, Moscow (1967), Vol. I, p. 309.
15. R.A. Guyer, *Phys. Rev.* **148**, 789 (1966).
16. R.A. Guyer, in *Solid State Physics*, Academic Press, New York (1969), Vol. 23, p. 412.
17. R.C. Richardson, E. Hunt, and H. Meyer, *Phys. Rev.* **138**, A1326 (1965).
18. G. Baker, H.E. Gilbert, J. Eve, and G.S. Rushbrooke, *Phys. Rev.* **164**, 800 (1967).
19. J.R. Sites, D.D. Osheroff, R.C. Richardson, and D.M. Lee, *Phys. Rev. Lett.* **23**, 835 (1969).
20. W.P. Kirk, E.B. Osgood, and M.C. Garber, *Phys. Rev. Lett.* **23**, 833 (1969).
21. B. Pipes and W.M. Fairbank, *Phys. Rev. Lett.* **23**, 520 (1969).
22. A.C. Anderson, W. Reese, and J.C. Wheatley, *Phys. Rev. Lett.* **7**, 366 (1961).
23. R.T. Johnson, and J.C. Wheatley, *Phys. Rev.* **1**, 1836 (1970).
24. R.T. Johnson, O.G. Symko, and J.C. Wheatley, *Phys. Rev. Lett.* **23**, 1017 (1969).
25. R.A. Scribner and E.D. Adams, *Rev. Sci. Instr.* **41**, 287 (1970).
26. M.F. Panczyk and E.D. Adams, *Phys. Rev.* **187**, 321 (1969).
27. D.J. Thouless, *Proc. Phys. Soc. (London)* **86**, 893 (1965).
28. R.A. Guyer and L.I. Zane, *Phys. Rev.* **188**, 445 (1969); and *Phys. Rev. Lett.* **24**, 660 (1970).
29. L.I. Zane, *Phys. Rev. Lett.* **28**, 420 (1972).
30. A.K. McMahan, *J. Low Temp. Phys.* **8**, 115 (1972), and private communication.
31. W.P. Kirk and E.D. Adams, *Phys. Rev. Lett.* **27**, 392 (1971).
32. L.I. Zane, this volume.
33. W.P. Kirk and E.D. Adams, this volume.
34. E.B. Osgood and M. Garber, *Phys. Rev. Lett.* **26**, 353 (1971).
35. E.B. Osgood, private communication.
36. D.D. Osheroff, R.C. Richardson, and D.M. Lee, *Phys. Rev. Lett.* **28**, 885 (1972).
37. R.T. Johnson, R.E. Rapp, and J.C. Wheatley, *J. Low Temp. Phys.* **6**, 445 (1971).
38. R.T. Johnson, R.E. Rapp, J.C. Wheatley (to be published).
39. R.T. Johnson, D.N. Paulson, R.P. Giffard, and J.C. Wheatley (to be published).
40. R.P. Giffard, W.S. Truscott, and J. Hatton, *J. Low Temp. Phys.* **4**, 153 (1971).
41. E.B. Osgood and J.M. Goodkind, *Phys. Rev. Lett.* **18**, 894 (1967).
42. R.T. Johnson, D.N. Paulson, R.P. Giffard, and J.C. Wheatley, this volume.
43. H. Horner and L.H. Nosanow, *Phys. Rev. Lett.* **29**, 88 (1972).
44. V.L. Vvedensky, private communication.
45. E.D. Adams and L.H. Nosanow (to be published).

46. R.T. Johnson, O.G. Symko, and J.C. Wheatley, *Phys. Lett.* **394**, 173 (1972).
47. W.P. Halperin, W.W. Webb, R.A. Buhrman, and R.C. Richardson, this volume,
48. R.T. Johnson and J.C. Wheatley, private communication.
49. D.D. Osheroff, W. Gully, R.C. Richardson, and D.M. Lee, this volume.
50. J.C. Wheatley, private communication.
51. J. Bardeen, L.N. Cooper, and J.R. Schrieffer, *Phys. Rev.* **108**, 1175 (1957).
52. A.M. Sessler, in *Proc. 8th Intern. Conf. Low Temp. Phys.* Butterworth, Washington, D.C. (1963), p. 11.
53. P.W. Anderson and P. Morel, *Phys. Rev.* **123**, 1911 (1961).
54. R. Balian and N.R. Werthamer, *Phys. Rev.* **131**, 1553 (1963).
55. A.J. Leggett, *Phys. Rev. Lett.* **14**, 537 (1965).
56. V.P. Peshkov, in *Proc. 9th Intern. Conf. Low Temp. Phys. 1964*, Plenum Press, New York (1965), p. 79.

2
Helium Lattice Dynamics

Specific Heat of Solid ^{3}He*

S. H. Castles, W. P. Kirk, and E. D. Adams

Physics Department, University of Florida
Gainesville, Florida

The specific heat of bcc ^{3}He has been investigated from 25 m°K to near the melting curve. There are two regions of interest. At the lowest temperatures the major contribution to the specific heat is due to the onset of nuclear-spin ordering and the exchange energy J can be obtained. These are the first specific heat measurements of solid ^{3}He that have gone to low enough temperatures with sufficient resolution to investigate nuclear-spin ordering.

At higher temperatures the interest is in the behavior of the lattice specific heat, which can be characterized by the Debye temperature θ_D. It has been observed that for the heavy rare gas solids as well as for ^{4}He, θ_D approaches a constant θ_0 for temperatures less than $T/\theta_0 \approx 0.02$. However, specific heat measurements[1,2] on bcc ^{3}He showed that θ_D reached a maximum and then at temperatures below $T/\theta_{max} \approx 0.02$ decreased in value. This decrease is referred to as "the low-temperature anomaly." Since a similar decrease in θ_D which now appears to be spurious had sometimes been observed in ^{4}He, some doubt has persisted as to the validity of the effect in ^{3}He. However, subsequent $(\partial P/\partial T)_V$ and thermal conductivity measurements[3,4] have also indicated a decrease in θ_D in bcc ^{3}He. Here we present heat capacity data which confirm the decrease in θ_D in bcc ^{3}He below about $T/\theta_{max} \approx 0.02$.

The beryllium copper calorimeter held approximately 0.9 cm^3 of ^{3}He and contained a copper wire brush with about 250 cm^2 surface area to provide thermal contact with the sample. The calorimeter was equipped with a capacitance strain gauge used to determine the molar volume and to calibrate the resistance thermometers against the melting pressure of ^{3}He. In order to cool the calorimeter to 25 m°K, good thermal contact between it and the cooling source was provided by a relatively large ($A/L = 5 \times 10^{-3}$ cm), 99.999% pure Sn superconducting heat switch. When in the superconducting state this heat switch provided excellent thermal isolation for temperatures less than about 500 m°K. However, above these temperatures the large (T^3) temperature dependence of the heat leak through the switch rapidly caused the heat leak to become large enough to destroy the quality of the heat capacity measurements. Therefore it was necessary to use a smaller heat switch when higher temperature data were taken.

The specific heat was measured by the heat pulse method. A correction was made for the temperature drifts before and after each heat pulse by measuring the drift rate and using this rate to extrapolate the initial and final temperatures to the center of the heat pulse. Each heat pulse typically lasted for 20 sec, after which suffi-

---

* Work supported in part by the National Science Foundation.

cient time (about 5 min at 30 m°K, decreasing to about 30 sec at 100 m°K) was allowed to establish equilibrium in the calorimeter before measurement of the final temperature drift. The temperature drift correction was relatively small.

The data analysis was performed in a manner similar to that of Collan et al.[5] A functional form was assumed for the specific heat, such as

$$C(T) = aT^{-2} + bT + cT^3 \qquad (1)$$

The fitting parameters a, b, and c were obtained from a least-squares fit to the data, using the integral of this functional form

$$Q_{\text{meas}} = a(T_i^{-1} - T_f^{-1}) + \tfrac{1}{2}b(T_f^2 - T_i^2) + \tfrac{1}{4}c(T_f^4 - T_i^4) \qquad (2)$$

where T_i is the initial temperature, T_f the final temperature, and Q_{meas} the heat added during each heat pulse. With the fitted parameters calculated values of C and Q, C_{calc} and Q_{calc}, are given by Eqs. (1) and (2). Any functional form can be tried for $C(T)$, and several with up to nine fitting parameters have been used.

The advantage of this procedure is that large temperature intervals can be used with no "curvature correction." The temperature intervals used here varied from 20% to 50% of the temperature at the midpoint of the interval.

All the data (for a given sample) are fitted simultaneously by Eq. (2), so there are no individual specific heat data points. Therefore, following Collan et al., the measured specific heat at the midpoint of the heating interval is taken to be

$$C_{\text{meas}} = (Q_{\text{meas}}/Q_{\text{calc}}) C_{\text{calc}} \qquad (3)$$

The experimental points presented here were calculated in this manner.

Specific heat measurements have been made on five bcc ^{3}He samples, with molar volumes determined from the data of Grilly.[6] All of the samples were annealed just below the melting curve. However, the 24.40 cm^3/mole sample, which melts at 35 m°K, was used to calibrate the resistance thermometers along the melting curve below 35 m°K and was not reannealed. The ^{4}He impurity concentration of all the samples was 2 ppm, except for the 21.67 cm^3/mole sample, which contained less than 10 ppm ^{4}He.

The heat capacity data for two of the samples and for the empty sample chamber are presented in Fig. 1. The circles are for the empty sample chamber, the triangles are for the 24.40 cm^3/mole sample, and the squares are for the 21.67 cm^3/mole sample. The 24.40 cm^3/mole sample is typical of the high-molar-volume bcc ^{3}He samples with their large nuclear-spin-ordering specific heat. Because of the strong molar volume dependence of J, the nuclear contribution is not seen for the smaller molar volume samples.

At this time specific heat of the empty sample chamber has been measured only between about 70 and 300 m°K. For the results presented here the specific heat of the ^{3}He samples above and below this temperature interval were obtained by assuming a straight-line extrapolation of the specific heat of the empty sample chamber, as indicated in Fig. 1. Further measurements on the empty sample chamber will be made in the future; however, no substantial changes of the ^{3}He specific heat are expected as a result of these measurements. [All of the $\theta_D(T)$ results presented here are for $T > 70$ m°K.]

Specific Heat of Solid ^{3}He

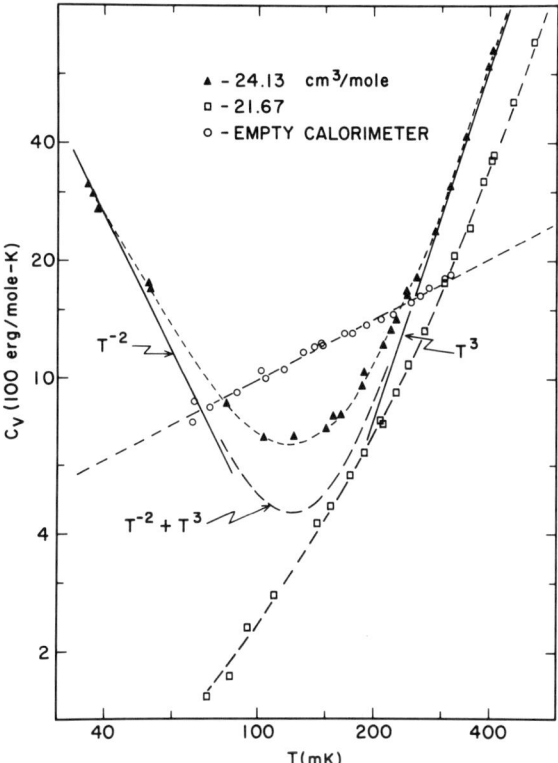

Fig. 1. C_V vs. T for two bcc ^{3}He samples and for the empty sample chamber. The T^{-2} line is the specific heat due to nuclear-spin ordering for $V = 24.13$ cm^3/mole. The $T^{-2} + T^3$ line is the specific heat expected for a Debye solid with $\theta_D = \theta_{max}$.

The T^{-2} line in Fig. 1 represents the specific heat due to nuclear-spin ordering with $|J|/k = 0.67$ m°K, which is the value given by Panczyk and Adams[7] for $V = 24.13$ cm^3/mole. At this time no attempt has been made to calculate $|J|/k$ values from specific heat measurements. However, preliminary indications are that they will be in substantial agreement with the values determined by Panczyk and Adams.

The T^3 line in Fig. 1 was drawn tangent to the $V = 24.13$ cm^3/mole data. It represents the specific heat one would expect from a Debye solid with $\theta_D = 19.75$, which is the maximum value of θ_D obtained by the 24.13 cm^3/mole sample. The dashed line is the sum of the T^3 and T^{-2} lines. The difference between this line and the experimental results represented by the triangles can be thought of as the "excess" specific heat corresponding to the decrease of θ_D below its maximum value.

The nuclear-spin ordering contribution, calculated using the $|J|/k$ values of Panczyk and Adams, was subtracted from the data. From the remaining specific heat, values of θ_D were calculated, with the results shown in Fig. 2. Smoothed results of Sample and Swenson[1] and of Pandorf and Edwards[2] are shown for comparison. In the region of overlap the comparison of the various results is reasonably satis-

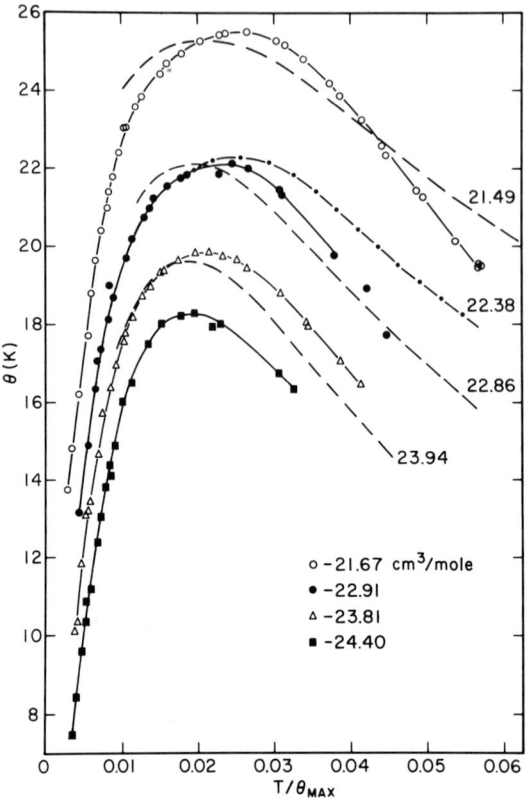

Fig. 2. Debye θ versus T/θ_{max} for bcc ^{3}He. The dashed curves are from the data of Sample and Swenson. The dot-dash curve is from the data of Pandorf and Edwards.

factory. The previously observed decrease in θ_D below $T/\theta_{max} \approx 0.02$ is seen in our results. Furthermore, no leveling off of θ_D has occurred in our results, which extend much lower than previous ones. Greywall[8] has determined the elastic θ_0 from sound velocity measurements on a sample with $V = 21.64$ cm^3/mole and found $\theta_0 = 19.5^{+2}_{-3}$°K. This value is somewhat above that obtained in our measurements for $V = 21.67$ cm^3/mole at the lowest temperature.

As mentioned, a similar, though spurious, decrease in θ_D has been observed in ^{4}He. We have taken various precautions to eliminate spurious effects. As a further check of our apparatus and techniques we plan to study a ^{4}He sample.

Speculations on possible theoretical explanations of the low-temperature anomaly have centered in two areas. The first of these has been based on the assumption that the ^{3}He nuclear-spin system is somehow responsible for the observation of the anomaly in ^{3}He but not in ^{4}He. However, there have been no calculations that substantiate this idea. (Whether or not the anomaly exists in hcp ^{3}He is still open to question. We intend to investigate this matter in the near future.) The second possibility is that the phonons themselves are responsible for the low-temperature anomaly.[3] Horner[9] has calculated the phonon spectrum for bcc ^{3}He at four molar volumes, employing the self-consistent harmonic approximation corrected for cubic anharmonicities. His calculations, which showed a small decrease in θ_D as a result of anomalous dispersion, have the qualitative behavior seen but do not agree quantitatively. Whether anomalous dispersion is found in other calculations[10,11] depends on the details of these calculations. Thus the present theories of lattice dynamics are not sufficiently refined to resolve the question of the low-temperature θ_D anomaly in bcc ^{3}He.

Acknowledgment

We acknowledge useful discussions with Professor S. B. Trickey.

References

1. H.H. Sample and C.A. Swenson, *Phys. Rev.* **158**, 188 (1967).
2. R.C. Pandorf and D.O. Edwards, *Phys. Rev.* **169**, 222 (1968).
3. P.N. Henriksen, M.F. Panczyk, S.B. Trickey, and E.D. Adams, *Phys. Rev. Lett.* **23**, 518 (1969).
4. W.C. Thomlinson. *Phys. Rev. Lett.* **23**, 1330 (1969).
5. H.K. Collan, T. Heikkila, M. Krusius, and G.R. Pickett, *Cryogenics* **10**, 389 (1970).
6. E.R. Grilly, *J. Low Temp. Phys.* **4**, 615 (1971).
7. M.F. Panczyk and E.D. Adams, *Phys. Rev.* **187**, 321 (1969).
8. D.S. Greywall, *Phys. Rev. A***3**, 2106 (1971).
9. H. Horner, *Phys. Rev. Lett.* **25**, 147 (1970).
10. H.R. Glyde and F.C. Khanna, *Can. J. Phys.* **49**, 2997 (1971).
11. T.R. Koehler and N.R. Werthamer, *Phys. Rev. A***5**, 2230 (1972).

The Temperature Dependence of the Longitudinal Sound Velocity of Single Crystals of HCP ^{4}He

J. P. Franck and R. A. D. Hewko

University of Alberta
Edmonton, Canada

We report on measurements of the longitudinal sound velocity of hcp ^{4}He as a function of temperature from 0.75°K to the melting point. Measurements were made on single crystals at densities corresponding to melting pressures of 87 and 120 bar, as well as some preliminary measurements at 57 bar.

Experimental

The crystals were grown at constant pressure over a period of 24 hr. The crystals were subsequently annealed for a further 24 hr in a temperature gradient of about 1 deg. This annealing procedure quite frequently produced crystals of very high quality, showing a large number of ultrasonic echoes (we observed up to 130 echoes near the melting point). We believe that many of these crystals were single.

The crystals were grown inside a beryllium–copper pressure cell. Built into this cell was an ultrasonic etalon using a 5-MHz, longitudinally cut quartz transducer. A pulse reflection method was used, with a path length of about 1 cm using the transducer both as emitter and receiver. The reflector was made from copper, polished flat to two wavelengths of sodium light; it formed the bottom of the cell.

An absolute value for the sound velocity was obtained from a least-squares fit to the arrival times of successive echoes. The estimated error in this measurement is 2%. Small variations in the sound velocity as a function of temperature were observed by using a phase comparison method with a sensitivity of about one part in 10^4.

Results

In all our measurements we found that the acoustical attenuation, as indicated by the number and amplitude of echoes, increases sharply with falling temperature, down to our lowest temperature of 0.75°K. We further find that the velocity always increases with falling temperature according to

$$c = c_0 - AT^n \tag{1}$$

The exponent n usually lies very close to 4, although sometimes smaller and, more particularly, larger values have been observed. A typical result for a crystal grown at a pressure of 120 bar is shown in Fig. 1.

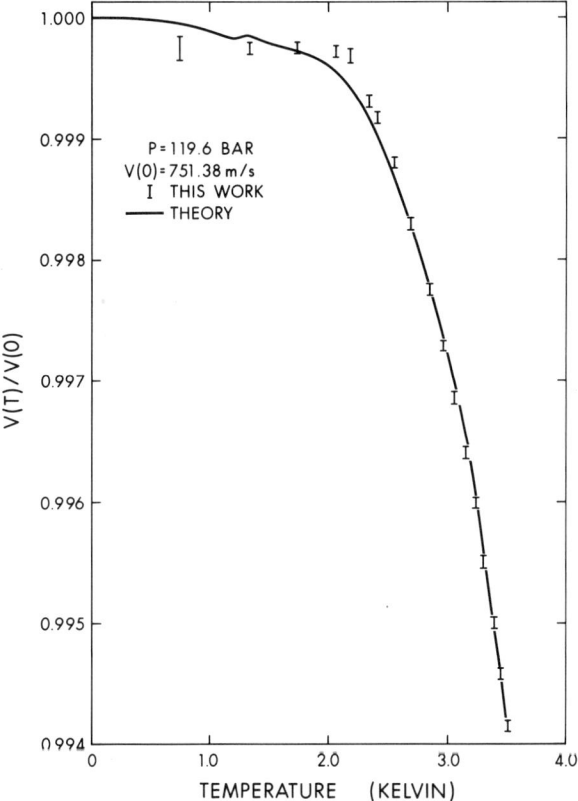

Fig. 1. Longitudinal velocity of sound (this work) compared to theory of Niklasson[4] at 120 bar.

In many, although not all, of the crystals measured we observed a rather sharp flattening out of the velocity at low temperatures. The position where this effect starts we refer to as the "knee." The knee is quite clearly seen in the crystal of Fig. 1. In Fig. 2 we have plotted the position of the knee in a P–T diagram. As can be seen, a fairly wide spread is observed for the position of the anomaly for crystals of the same density.

Discussion

For a full discussion it would be necessary to know the orientation of the crystals. This information could not be obtained in this experiment. We will therefore use only expressions valid for an isotropic solid. We will begin by discussing the major aspect of the temperature variation, connecting it with other data on hcp ^{4}He. After that we will go on to a possible explanation for the low-temperature anomaly.

The sound velocity for an isotropic substance is given by

$$C = (\rho K_s)^{-\frac{1}{2}} \qquad (2)$$

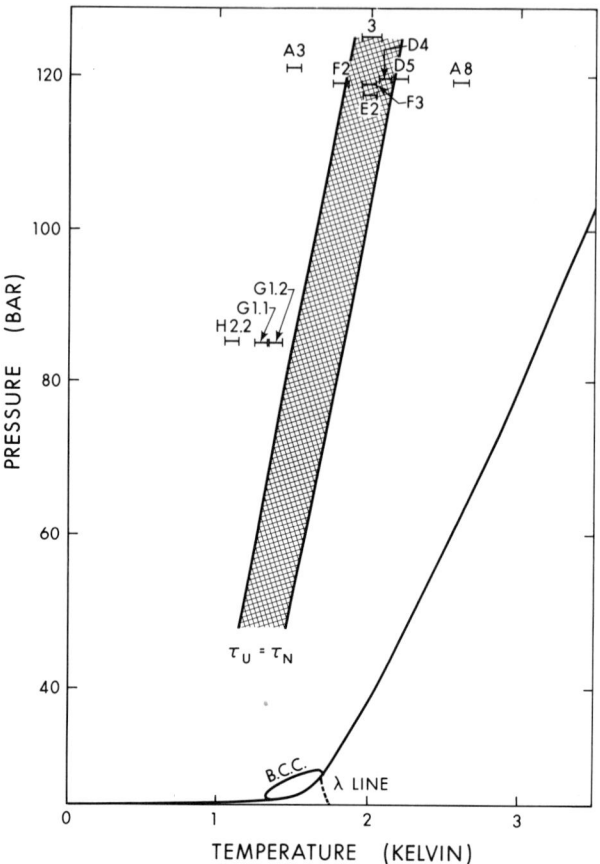

Fig. 2. Position of "knee" as a function of temperature and pressure for hcp ^{4}He showing where Umklapp and normal relaxation times are equal.

where ρ is the density and K_s is the adiabatic compressibility. Expression (2) does not give a good value for the absolute value of c; it should, however, give a good representation for the temperature variation. Since ρ is held constant in our experiment, we therefore interpret the temperature variation of c through that of K_s. By straightforward thermodynamic arguments one obtains from Eq. (2) the relation

$$C - C_0 = \frac{1}{2\rho C_0} \frac{1}{K_T} - \frac{1}{K_{T0}} + \frac{TV}{C_V}\left(\frac{\partial P}{\partial T}\right)_V^2 \qquad (3)$$

where K_T is the isothermal compressibility; the subscript 0 in K_{T0} refers to 0°K. The right-hand side of this expression contains experimentally known data (see Jarvis et al.[1]). In Fig. 3 we have compared in this way data on a crystal grown at 57 bar with the data of Jarvis et al.; we observe good agreement.

Instead of using measured data on K_T and $(\partial P/\partial T)_V$ one can also attempt to derive these data from the equation of state. Generally in this approach a reduced

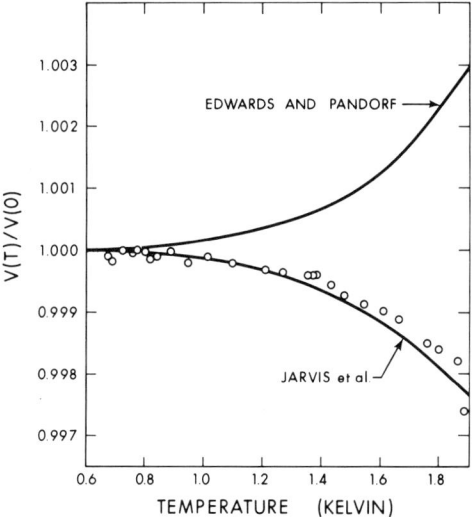

Fig. 3. Longitudinal velocity of sound (this work) compared to compressibility data of Jarvis et al.[1] and reduced equation of state (data from Edwards et al.[6]).

equation of state is assumed of the form

$$S = S(T/\theta(V)) \qquad (4)$$

where one normally chooses $\theta(V)$ as the Debye temperature. With this assumption one obtains, after some thermodynamic arguments, the relation

$$C - C_0 = \frac{1}{2MC_0}(U - U_0)\gamma\left(1 + \gamma - \frac{d\ln\gamma}{d\ln V}\right) \qquad (5)$$

where M is the molar weight, U is the internal energy, and γ is the Gruneisen parameter:

$$\gamma = -d(\ln\theta)/d(\ln V) \qquad (6)$$

From specific heat measurements[2] γ is known to be close to 2.6 at our densities, and $|d(\ln\gamma)/d(\ln V)| \approx 1$. We therefore see that the relation (5) predicts a rise of the velocity with rising temperature, contrary to our results and the prediction of Eq. (3). From this we must conclude that the assumption of a reduced equation of state, Eq. (4), is incorrect. A similar conclusion was reached previously by Ahlers[2] from specific heat measurements.

We now turn to the low-temperature anomaly. In Fig. 2 we compare the position of the "knee" with the normal and Umklapp relaxation times τ_N and τ_U taken from the work of Hogan et al.[3] It is evident that the anomaly occurs close to the point where $\tau_N = \tau_U$. At this point heat losses of the sound wave will change from diffusive to wavelike (second sound). This change in the loss mechanism will in turn influence the sound velocity.

Several authors have derived expressions for the sound velocity assuming a coupling to the temperature field.[4,5] Niklasson[4] gives the following expression:

$$\left(\frac{C}{C_T}\right)^2 = 1 + \left(\frac{C_p}{C_V} - 1\right) \frac{S^2(S^2 - 1)C_T^2 Q^2 + 4S^4 \Gamma_0 \Gamma_2^0}{(S^2 - 1)^2 C_T^2 Q^2 + 4S^4 (\Gamma_2^0)^2} \quad (7)$$

where C_T is the isothermal velocity under the assumption of no coupling to the temperature field; S is the ratio of first- to second-sound velocity; Q is the wave vector; and Γ_0 and Γ_2^0 are constants that can be expressed in terms of the normal and Umklapp relaxation times. We have evaluated this expression by assuming for C_T the form

$$C_T = a + bT^4$$

and obtaining the constants a and b by fitting C from Eq. (7) to the experimental data at two points. For the ratio S we choose the value for the isotropic case, $S = \sqrt{3}$. The results of such a calculation are shown in Fig. 1. As can be seen from Fig. 1, the expression (7) gives a rather sharp break in the sound velocity of the type observed at nearly the correct temperature. Small deviations between the experimental and theoretical curve are not unexpected in view of the fact that Eq. (7) holds only for the isotropic case and of experimental uncertainties in the relaxation times used to obtain Γ_0 and Γ_2^0.

Conclusion

The longitudinal sound velocity of hcp ^{4}He as a function of temperature is found to be in agreement with measured thermodynamic data. The data provide arguments against the existence of a simple reduced equation of state for hcp ^{4}He. A low-temperature anomaly is formed and believed to be due to a coupling between first and second sound.

References

1. J.F. Jarvis, D. Ramm, and H. Meyer, *Phys. Rev.* **170**, 320 (1968).
2. G. Ahlers, *Phys. Rev. A* **2**(4), (1970).
3. E.M. Hogan, R.A. Guyer, and H.A. Fairbank, *Phys. Rev.* **185**, 356 (1969).
4. G. Niklasson, *Ann. Phys.* **59**, 263 (1970).
5. V.L. Gurevich and A.L. Efros, *Soviet Phys.—JETP* **24**, 1146 (1967).
6. D.O. Edwards and R.C. Pandorf, *Phys. Rev.* **140**, 3A, 816 (1965).

Lifetimes of Hypersonic Phonons in Solid ^{4}He*

P. Leiderer, P. Berberich, S. Hunklinger, and K. Dransfeld

Physik Department der Technischen Universität München
Munich, Germany

Introduction

The quantum liquid "superfluid helium" shows several outstanding properties as compared to normal liquids. In contrast to this the quantum crystal "solid helium" (in the following we shall only consider ^{4}He) seems to exhibit no remarkable differences with respect to ordinary solids. For example, solid helium was found to have well-defined crystal structures,[1] specific heat and sound velocity measurements remained without surprise,[2,3] and neutron scattering experiments led to the usual dispersion spectrum for phonons.[4-6] After initial difficulties due to the large zero-point motion the theoretical treatment of these phonons has been quite successful. Using techniques like self-consistent field methods, the calculations for the phonon frequencies are in fair agreement with all these experiments.[7]

If one considers the attenuation of the phonons, the problem is not yet settled. Because of the large zero-point energy the atoms move far beyond the harmonic range of the interatomic potential. Therefore, at first glance, one might expect such a crystal to be highly anharmonic and, as a consequence, to show considerably higher damping of lattice waves than classical crystals. Jäckle and Kehr,[8] however, showed that the classical theory of ultrasonic absorption should also be applicable to anharmonic crystals if the actual interatomic potential is replaced by a renormalized one. The important parameters entering into such a calculation are the higher-order elastic constants of the crystal, which roughly are represented by the Grüneisen constant. The value of about 3 of the Grüneisen constant in solid helium[5,9,10] is quite close to that of heavier rare gas solids and of many other dielectrics.

What is to be expected for the attenuation of sound at low temperatures ($T \ll \theta$) from the classical theory? Under the basic assumption that the attenuation of ultrasonic phonons in an ideal crystal is caused by their interaction with thermal phonons, the following expression for the inverse lifetime (in sec^{-1}) of longitudinal ultrasonic phonons has been derived[11]:

$$\tau^{-1} = \frac{\hbar \gamma^2}{32\pi^2 \rho v_l^5} \left(\frac{4\pi^4}{15} \right) \left(\frac{k_B T}{\hbar} \right)^4 \omega \tan^{-1}(\omega \tau_{th}) \qquad (1)$$

where γ is an effective Grüneisen constant, ω and v_l are the ultrasonic angular frequency and longitudinal velocity, respectively, and τ_{th} is the lifetime of longitudinal thermal phonons.

* Supported by the Deutsche Forschungsgemeinschaft.

For $\omega\tau_{th} \ll 1$ Eq. (1) leads to

$$\tau^{-1} \propto \omega^2 \qquad (2)$$

The temperature dependences of τ_{th} and of the prefactor nearly compensate in this regime.

At low temperatures the mean free path of the thermal phonons becomes much longer than the wavelength of the ultrasonic phonons, i.e., $\omega\tau_{th} \gg 1$. Then the arctan is near $\pi/2$, resulting in

$$\tau^{-1} \propto \omega T^4 \qquad (3)$$

This is known as the Landau–Rumer relation.

Our experiment was conducted in the temperature range from 1 to 2°K. It is known from measurements of the thermal conductivity that the mean lifetime of thermal phonons in this region is limited both by normal and Umklapp processes and is

$$\langle \tau_{th} \rangle > 10^{-9} \quad \text{sec} \qquad (4)$$

The frequency of sound in our experiment is

$$\nu \geq 1.5 \quad \text{GHz}$$

Assuming Eq. (4) also to be valid for the longitudinal thermal phonons, we get

$$\omega\tau_{th} \gg 1$$

and we should therefore expect relation Eq. (3) to apply.*

This relation is well satisfied for many classical dielectric crystals. In order to compare different substances, reduced coordinates T/θ and ω/ω_D have been used, where θ and ω_D are Debye temperature and Debye frequency, respectively. Then from Eq. (1) we obtain

$$\tau^{-1} \propto \gamma^2 \left(\frac{\rho^2}{m^5}\right)^{1/3} \left(\frac{\omega}{\omega_D}\right) \left(\frac{T}{\theta}\right)^4 \tan^{-1}(\omega\tau_{th}) \qquad (5)$$

Some experimental results are shown in Fig. 1. We have plotted the inverse lifetimes of the hypersonic phonons multiplied by a normalizing factor due to Eq. (5), versus the reduced temperature. In our temperature regime, for $T/\theta < 0.075$, the inverse lifetime has the same temperature dependence in KCl,[14] NaCl,[15] SiO_2,[16] and Al_2O_3,[17] namely

$$\tau^{-1} \propto T^4$$

The difference in the absolute value is only due to the different Grüneisen parameters, which for the quoted substances vary from 1 to 3. The dashed line in Fig. 1 shows the behavior expected for solid helium from this classical theory.

At this point some reservation has to be made: Assuming a pressure of the solid

---

* Equation (1) does not include effects arising from the curvature of the phonon dispersion curve. Therefore it is only valid if $\omega\tau_{th} < 2(\theta/T)^2$, which in solid helium is fulfilled for our experimental frequencies if $T > 1°K$.[13]

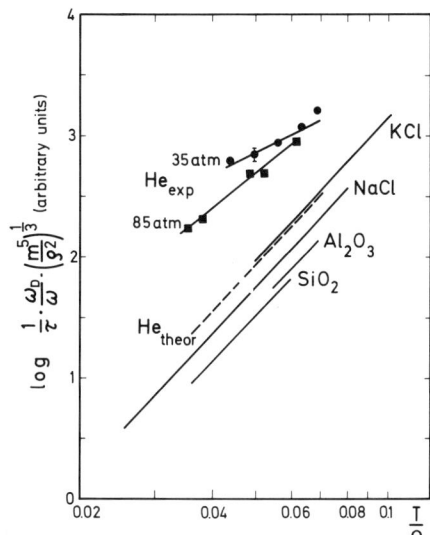

Fig. 1. Reduced inverse lifetime of hypersonic phonons in dielectric crystals. The data for KCl, NaCl, SiO$_2$, and Al$_2$O$_3$ are taken from experiment. The dashed line indicates the expected behavior of solid helium. Circles and squares represent the experimental data of solid helium.

helium sample of 35 atm, as in most of our experiments, the crystal already melts at 2°K, whereas the Debye temperature is 28°K. So the reduced melting temperature is $T_M/\theta = 0.075$, more than a factor of ten lower than for other substances. Very near the melting temperature a deviation from the classical behavior would not be surprising. Slightly below T_M/θ, however, all the thermodynamic properties of solid helium can be reasonably well approximated in terms of the low-temperature limits. From this consideration one should also expect the Landau–Rumer theory to be applicable in this regime.

Until now experiments on the attenuation of phonons in solid helium which can be compared with this theoretical prediction are quite rare. From neutron scattering experiments it is known that well-defined phonons exist up to the highest frequencies in the dispersion spectrum, although some branches show an anomalous broadening. For ultrasonic frequencies near 10 MHz, measurements by Vignos and Fairbank,[3] Lipschultz and Lee,[18] and Crepeau et al.[19] resulted in attenuation values of 0.2–0.7 cm^{-1}, which is slightly higher than in liquid helium near 1°K, both for longitudinal and transverse sound waves in hcp as well as in bcc ^{4}He.

As the authors themselves indicate, however, until now such conventional ultrasonic measurements using transducers only give an upper limit for the intrinsic absorption in solid helium, since they are subject to large parasitic attenuation effects. Because of the pronounced acoustical anisotropy of solid helium, sound waves generally do not propagate perpendicular to the transducer surface and are therefore partially absorbed at the chamber walls.[20] Furthermore, ultrasonic measurements require a high degree of crystal perfection over the whole sample volume. Finally, the mechanical contact between the transducer and the sample must not vary with temperature.

All these difficulties can be avoided or become unimportant if instead of using transducers the ultrasonic phonons are generated and detected by means of light scattering. With such a technique we could measure for the first time the intrinsic attenuation of ultrasonic phonons in solid ^{4}He and also determine the velocity of these phonons in the GHz range.

Experimental

Generation of the phonon field was achieved by means of stimulated Brillouin scattering using a giant pulse ruby laser.[21] The wavelength of these phonons is nearly constant and given by

$$\lambda_S = \lambda_L/2n$$

where λ_L is the wavelength of the laser light and n is the refractive index of the sample.

Due to the strongly pressure-dependent sound velocity, this corresponds to phonon frequencies of 1.5 and 2 GHz for helium crystals at 35 and 85 atm, respectively. The velocity can be determined from the frequency shift of the backscattered Brillouin light with respect to the laser light, which is equal to the phonon frequency v_S:

$$v_L - v_B = v_S = 2nv_S/\lambda_L$$

In order to determine the absorption we employed a method which is a further improvement of a technique first used by Winterling and Heinicke.[22]

The setup is schematically shown in Fig. 2. Light from a second ruby laser which is weak enough not to disturb the phonon population is directed onto the hypersonic wave generated by the giant pulse. This probing light is partly backscattered by the hypersonic phonons. The backscattered intensity dies away as the phonons decay. Thus from the time-dependent ratio of backscattered to incident light the phonon lifetime can be obtained.

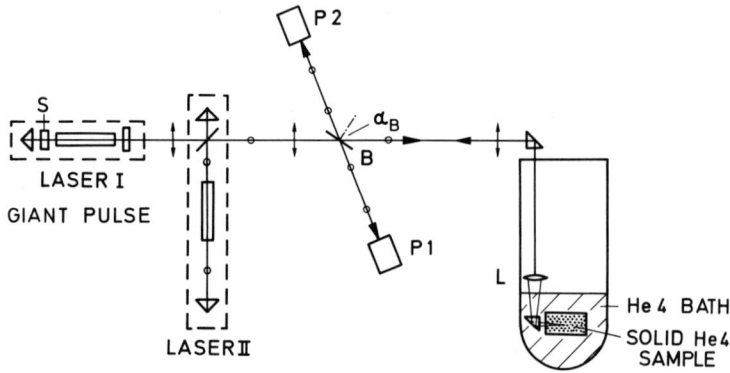

Fig. 2. Experimental setup for measuring phonon lifetimes in solid helium. Polarization of laser I (indicated by arrows) is in the plane of the page; polarization of laser II is perpendicular to it (indicated by circles). The laser light is focused into the sample chamber with lens $L(f = 20$ cm). Photocells $P1$ and $P2$ record incident and backscattered light, respectively. Registration of the giant pulse light is strongly suppressed because the beam splitter B is mounted at the Brewster angle α. The electrooptical shutter S is used for synchronization of the two lasers.

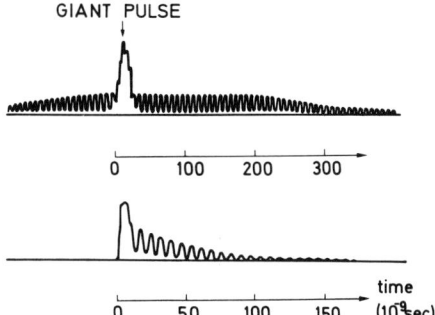

Fig. 3. Upper: Typical incident light intensity registered by photocell $P1$ of Fig. 2. The giant pulse is suppressed by a factor of 500. The long, regularly spiked train of pulses is due to the second laser. Lower: Backscattered light intensity registered by photocell $P2$.

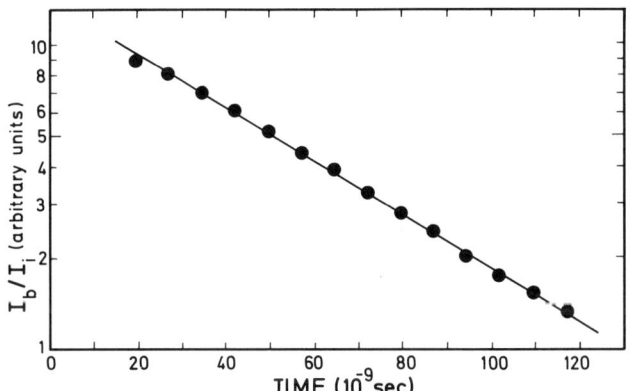

Fig. 4. Ratio of backscattered to incident light I_b/I_i plotted for intensities of Fig. 3.

The upper half of Fig. 3 shows a typical oscilloscope trace of the incident light; the lower curve represents the backscattered light intensity. The ratio of backscattered to incident light is plotted in Fig. 4 as a function of time. The straight line indicates an exponential decrease of the number of phonons and yields a well-defined lifetime, which is 50 nsec in this case. In this way the absolute value of the lifetime may be obtained from the record of a single laser pulse.

The advantages of this method are evident: (a) No transducer is needed, so coupling problems are eliminated. (b) The crystal has to be of high quality only in the small volume where the laser light is focused. (c) Because of the short distance through which the phonons travel during measurement ($< 200 \,\mu\text{m}$) "walk off" effects are negligible.

Nevertheless, as the method described above is new, we tested it with liquid helium, where the attenuation near 1 GHz is known from a number of experiments. We obtained exactly the same results as, for instance, St. Peters et al.,[23] who used a spontaneous Brillouin scattering technique.

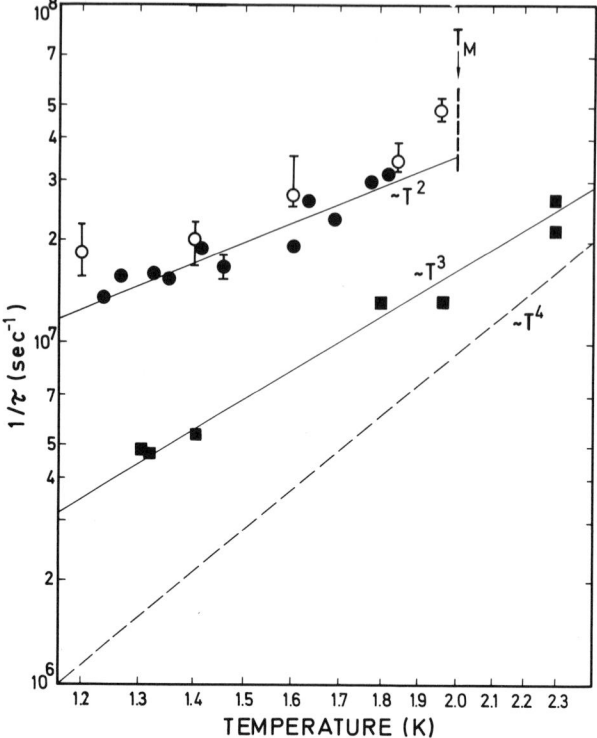

Fig. 5. Inverse lifetime of longitudinal phonons in solid ^{4}He. Open and full circles represent two different hcp crystals at 35 atm; T_M denotes the melting temperature of these crystals. The squares are data from an 85-atm hcp crystal. The wavelength of the phonons is 3400 Å, which corresponds to phonon frequencies of 1.55 and 2.02 GHz for 35 and 85 atm, respectively. The dashed line indicates the expected T^4 temperature dependence.

Results

As to the *velocity* of 1.5-GHz longitudinal phonons in solid helium, our results were found to be in agreement with the previous data of Vignos and Fairbank,[3] Crepeau *et al.*,[19] and several other groups, obtained by ultrasonic techniques at lower frequencies.

Regarding the *attenuation* of these phonons, however, the results were unexpected. Some data are shown in Fig. 5. We have plotted the inverse phonon lifetimes for three different hcp crystals at pressures of 35 and 85 atm in the temperature range from 1.2 to 2.3°K.

The data for the two 35-atm crystals are representative also of several other experimental runs under equal conditions. Apart from the point at 1.95°K, very near the melting temperature of ^{4}He at 35 atm, the inverse phonon lifetime (in sec^{-1}) of these crystals approximately follows the relation

$$\tau^{-1} = 10^7 T^P, \quad 1.5 \leqq P \leqq 2; \quad 1.2 \leqq T \leqq 1.95°K$$

at a phonon frequency of 1.5 GHz.

The data for the 85-atm crystal are quite recent and somewhat preliminary. They can be represented by

$$\tau^{-1} = 2 \times 10^6 T^3 \qquad 1.3 \leqq T \leqq 2.3°K$$

Discussion

Two results may be derived from Fig. 5: In the first place, the observed temperature dependence is weaker than the expected T^4 law, the disagreement being higher for the low-pressure crystals. In this context it is interesting to note that also for thermal phonons the relaxation time for normal processes shows a weaker temperature dependence than one would expect classically: Poiseuille flow and second-sound data[12,24] lead to a $\tau_N^{-1} \propto T^3$ dependence between 0.3 and 0.8°K compared to the relationship $\tau_N^{-1} \propto T^5$ expected classically.[25]

Second, the lifetimes of the phonons increase as the pressure is raised. According to the classical theory [see Eq. (1)], this increase should vary as v_l^4 if the wave vector is held constant as in our experiment, with a small correction due to the change in density. The mean longitudinal velocity changes from 520 m/sec at 35 atm to 680 m/sec at 85 atm, leading to an expected increase of τ by a factor of three, which is in fair accordance with our observation. Since the orientation of the crystal has not been determined so far and the attenuation is supposed to be anisotropic, this agreement may be somewhat fortuitous.

For a comparison with classical crystals we have plotted our data also in Fig. 1. Here a further discrepancy becomes obvious: The experimental reduced lifetimes in solid helium are much shorter than expected, more than one order of magnitude at the lowest temperatures. (This also holds for bcc ^{4}He, where the lifetimes turned out to be comparable to that in the hcp phase in the immediate neighborhood.) Again the discrepancy is somewhat smaller for the higher-pressure crystal.

In order to explain these discrepancies, one could either assume that they are due to an improper choice of the parameters entering into the calculations, or that an additional damping mechanism comes into play in solid helium.

Let us first consider how the parameters had to be changed in order to obtain agreement with classical theory.

(a) The high absolute value of the inverse phonon lifetime would require the anharmonic parameter γ of Eq. (1) to have a value exceeding ten at 1.2°K. This is more than a factor of three larger than the Grüneisen constant as determined from specific heat,[9] sound velocity,[10] and neutron scattering experiments.[5]

(b) The weak temperature dependence of τ could be explained if the condition $\omega\tau_{th} \gg 1$ which is necessary for a Landau–Rumer behavior were not fulfilled, but only $\omega\tau_{th} \gtrsim 1$. This would lead to relaxation times of the order $\tau_{th} \sim 10^{-10}$ sec, which is more than two orders of magnitude smaller than the observed mean thermal value $\langle\tau_{th}\rangle$ at 1.2°K.[12]

Therefore it seems to be more probable that some other damping mechanism contributes to the phonon attenuation in solid helium, although at the moment there exist only speculations about its origin:

(a) One possibility could be nonlinear effects, like the generation of harmonics, because of the relatively high amplitude of the sound wave in our experiment.

Applying our technique to liquid helium and quartz, however, where the buildup time for shock wave formation should be comparable to that in solid helium,[26] we did not observe such effects.

(b) Furthermore, some contribution from zero-point phonons[27] could decrease the phonon lifetime, although we are lacking a model for this mechanism so far.

(c) One might also think of an interaction between acoustic Brillouin phonons and some other excitations present in solid helium. For instance, regarding the similarity of the excitation spectra of liquid and solid helium,[28] one could assume the damping to be due to a relaxation process between acoustic and optical phonons similar to the phonon–roton relaxation in liquid helium. This idea is supported by the pronounced similarity we found for the attenuation curves in liquid helium near 24 atm and in the solid near 30 atm. In order to clear up these questions, experiments at lower temperatures and various frequencies are in progress.

Summary

Using a light scattering technique, we have determined the velocity and the lifetime of longitudinal GHz phonons in solid ^{4}He. Comparing our results for the phonon lifetime with the classical theory, we have found a much smaller absolute value and a weaker temperature dependence than expected. This disagreement seems to decrease for crystals at higher pressure.

Acknowledgment

We are grateful to Professor R. Nava for helpful discussions.

References

1. J. Wilks, *Properties of Liquid and Solid Helium,* Clarendon Press, Oxford (1967), p. 589.
2. D.O. Edwards and R.C. Pandorf, *Phys. Rev.* **140**, A816 (1965); **144**, 143 (1966).
3. J.H. Vignos and H.A. Fairbank, *Phys. Rev.* **147**, 185 (1966).
4. V.J. Minkiewicz, T.A. Kitchens, F.P. Lipschultz, R. Nathans, and G. Shirane, *Phys. Rev.* **174**, 267 (1968).
5. R.A. Reese, S.K. Sinha, T.O. Brun, and C.R. Tilford, *Phys. Rev. A* **3**, 1688 (1971).
6. E.B. Osgood, V.J. Minkiewicz, T.A. Kitchens, and G. Shirane, *Phys. Rev. A* **5**, 1537 (1972).
7. N.R. Werthamer, *Am. J. Phys.* **37**, 763 (1969).
8. J. Jäckle and K.W. Kehr, *Phys. Rev. Lett.* **24**, 1101 (1970).
9. G. Ahlers, *Phys. Lett.* **24A**, 152 (1967).
10. R. Wanner and J.P. Franck, *Phys. Rev. Lett.* **24**, 365 (1970).
11. P.C.K. Kwok, in *Solid State Physics,* F. Seitz, D. Turnbull, and H. Ehrenreich, eds., New York, London (1967), Vol. 20, p. 213.
12. E.M. Hogan, R.A. Guyer, and H.A. Fairbank, *Phys. Rev.* **185**, 356 (1969).
13. H.J. Maris, *Phil. Mag.* **9**, 901 (1964); J.B. Thaxter and P.E. Tannenwald, *IEEE Trans. on Sonics and Ultrasonics* **SU13**, 61 (1966).
14. R.P. Auyang, Thesis, Cornell University, 1968.
15. L.G. Merkulov, R.V. Kovalenok, and E.V. Konovodchenko, *Soviet Phys.—Solid State* **13**, 968 (1971).
16. M.F. Lewis and E. Patterson, *Phys. Rev.* **159**, 703 (1967).
17. J. de Klerk, *Phys. Rev.* **139**, A1635 (1965).
18. F.P. Lipschultz and D.M. Lee, *Phys. Rev. Lett.* **14**, 1017 (1965).
19. R.H. Crepeau, O. Heybey, D.M. Lee, and S.A. Strauss, *Phys. Rev. A* **3**, 1162 (1971).
20. R. Wanner, *Phys. Rev. A* **3**, 448 (1971).

21. R.J. Chiao, C.A. Townes, and B.P. Stoicheff, *Phys. Rev. Lett.* **12**, 592 (1964).
22. G. Winterling and W. Heinicke, *Phys. Lett.* **27A**, 329 (1968).
23. R.L.St. Peters, T.J. Greytak, and G.B. Benedek, *Optics Commun.* **1**, 412 (1970).
24. C.C. Ackerman and R.A. Guyer, *Ann. Phys. (N.Y.)* **50**, 128 (1968).
25. H.E. Jackson and C.T. Walker, *Phys. Rev. B* **3**, 1428 (1971).
26. R.T. Beyer and S.V. Letcher, *Physical Ultrasonics,* Academic Press, New York (1969), p. 202.
27. G. Meissner, private communication.
28. O.W. Dietrich, E.H. Graf, C.H. Huang, and L. Passel, *Phys. Rev. A* **5**, 1377 (1972).

Sound Wave Propagation and Anharmonic Effect in Solid ^{3}He and ^{4}He

D. Y. Chung*

*Physics Department, Howard University
Washington, D.C.*

C.-C. Ni

*Applied Mechanics Branch, Ocean Technology Division
Naval Research Laboratory, Washington, D.C.*

and

Y. Li

*Department of Physics, Rutgers
The State University, Newark, New Jersey*

Introduction

In the treatment of lattice dynamics of quantum solids at low temperatures the harmonic approximation is not applicable because the zero-point motion of such an atom is a large fraction of the near-neighbor distance. Various theoretical models which take anharmonicity into consideration have been proposed to study thermal and elastic properties of rare gas solids. Theoretical calculations of the second-order elastic constants (SOEC)[1] and velocity contours of sound wave propagation in bcc ^{3}He[2] and hcp ^{4}He[3] crystals have been reported. Experimental determination of SOEC of these crystals[4] through sound velocity measurements yielded results which in varying degrees are in agreement with the theoretical ones. However, the verification of the anisotropic behavior of sound waves in these solids became possible when the recently improved techniques for growing large single crystals were achieved. In view of this advancement, this paper is primarily concerned with the information which may be obtained about the effect of anharmonicities on the physical properties of these crystals through the study of the third-order elastic constants (TOEC).

To investigate a method for the determination of the TOEC, we take advantage of the similarity in elastic properties among the crystals with the same crystallographic structure, such as cubic Ar, Al, or Cu for bcc ^{3}He and hexagonal Mg and Zn for hcp ^{4}He. Using the phenomenological theory of nonlinear acoustics developed by Ljamov[5] and Musgrave's[6] method for calculating the velocity and wave surfaces for anisotropic crystals, we have studied the pressure dependence of the velocity

---

* Partially supported by the National Science Foundation.

contour in cubic and hexagonal systems. Using the nonlinear properties of these crystals, such calculations can be employed to determine the favorable directions for the TOEC determination of bcc ^{3}He and hcp ^{4}He.

Method of Calculation

According to the phenomenological theory of elasticity,[5] the wave equation for elastic waves including the third-order anharmonic term can be written as

$$\rho \frac{\partial^2 u_j}{\partial t^2} = (C_{ijkl} + C_{ijklqr} u_{qr}) \frac{\partial^2 u_k}{\partial x_i \partial x_j} \quad (1)$$

The summation convention is implied here. Here ρ is the density of the medium, and C_{ijkl} and C_{ijklqr} are the SOEC and TOEC, respectively. Other notations are conventional. Ljamov's theory of nonlinear acoustics introduces the acoustic Mach number M_r into the nonlinear wave equation by the following definition:

$$u_{qr} = n_q M_r = \frac{\partial u_r}{\partial x_q} = \frac{\dot{u}_r}{V_q} \quad (2)$$

where $\dot{u}_r$ is rth particle velocity and V_q is the sound phase velocity in qth direction. The plane wave solution of Eq. (1) with the substitution in Eq. (2) leads to the solution of the secular equation by Musgrave's method; i.e.,

$$\begin{vmatrix} \Gamma^*_{11} - \rho V^2 & \Gamma^*_{12} & \Gamma^*_{13} \\ \Gamma^*_{12} & \Gamma^*_{22} - \rho V^2 & \Gamma^*_{23} \\ \Gamma^*_{13} & \Gamma^*_{23} & \Gamma^*_{33} - \rho V^2 \end{vmatrix} = 0 \quad (3)$$

where $\Gamma^*_{jk} = \Gamma_{jk} + N_{jkr} M_r$, $\Gamma_{jk} = C_{ijkl} n_i n_l$, and $N_{jkr} = C_{ijklqr} n_i n_l n_q$. The n's are the direction cosines of the wave vector. From the definition given in Eq. (2), it is realized that different amplitude of external loading, static or dynamic, can be expressed in terms of this dimensionless Mach number M_r. By knowing the SOEC and TOEC of a given crystal, the wave velocity which would be measured under loading condition in a given direction can be calculated by solving the cubic equation in ρV^2 resulting from Eq. (3). The three roots correspond to the three modes of propagation in this direction. We have developed a computer program to calculate the velocity contours using the Mach number M_r as a parameter.

Results

Many calculations made by one of the authors* have demonstrated that the nonlinear behavior which occurs due to external force loading of a crystal depends very strongly on the crystal symmetry. It is reasonable to extend these qualitative nonlinear properties from crystals with bcc and hcp structures to solids ^{3}He and ^{4}He having the same structures. Typical results are given in Figs. 1 and 2. In these figures dimensionless quantities $(V - V_0)/V_0 = \Delta V/V_0$ are used to show the degree

---

* David Y. Chung.

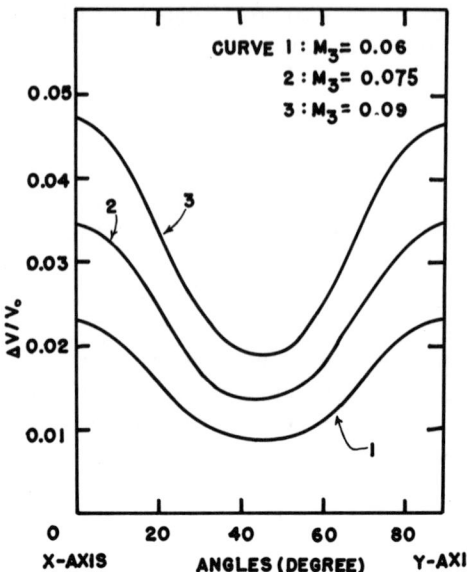

Fig. 1. The fractional change in longitudinal sound velocity ($\Delta V/V_0$) as a function of crystal orientations for several Mach numbers in a cubic crystal. The external loading is applied perpendicular to the sound wave propagation.

of such nonlinearity along different directions in a given plane. The subscript zero designates the sound velocity V corresponding to $M_r = 0$. From these curves one notes that the variation $\Delta V/V_0$ for a given Mach number is anisotropic. It shows that the most pronounced nonlinear effect is along the (100) direction or equivalent of a cubic crystal, and the c axis of a hexagonal crystal. The largest variation directions in different planes of symmetry for the above-mentioned crystal structures are summarized in Table I.

Discussion

The qualitative description of the anisotropic characteristics of velocity change with respect to the assigned Mach numbers is of interest. To visualize its importance,

Table I. The "Favorable" Directions for Observing the Nonlinear Effect in BCC and HCP Crystals*

		Longitudinal	Transverse (1)	Transverse (2)
BCC ^{3}He	{110} plane	(100)	(100)	(110)
	{111} plane	($1\bar{1}2$)	($1\bar{1}2$)	($1\bar{1}0$)
HCP ^{4}He	Any plane including c axis	c axis	c axis	60° from c axis

* All planes and directions listed include their equivalents by symmetry.

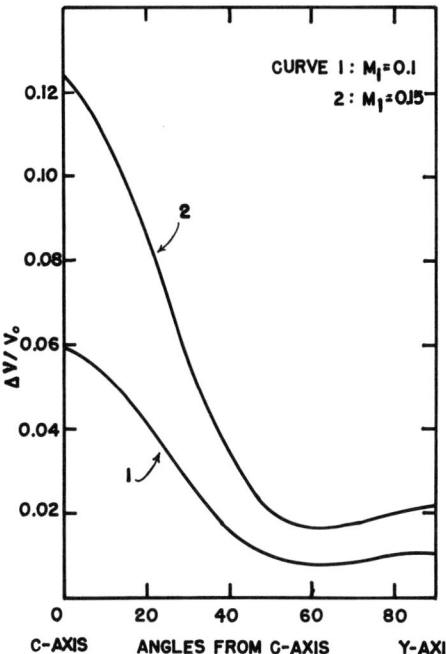

Fig. 2. The fractional change in longitudinal sound velocity ($\Delta V/V_0$) as a function of crystal orientation (any plane including c axis) for several Mach numbers in a hexagonal crystal. The external loading is applied perpendicular to the sound wave propagation.

we recall the definition of the Mach number M_r in Eq. (2), which is expressed as being directly proportional to the strains u_{qr} for a given direction n_q. The cause of the strains is immaterial—it can be static or dynamic. Therefore the third-order term in the wave equation of Ljamov's nonlinear acoustic theory can be regarded as the finite mechanical effect due to internal and/or external disturbances. Then the physical meaning of the calculated results shown in Figs. 1 and 2 can be interpreted as the velocity shift due to externally applied forces or acoustic pressure. Thus it suggests an experimental method which can be used to determine TOEC of a solid by measuring the velocity change as a function of sound amplitude. Second harmonic generations or interactions of two sound waves in the directions listed in Table I can then be utilized for measuring the nonlinear effect as well as the TOEC in solids ^{3}He and ^{4}He. It should be interesting to see the interactions between two waves which are propagated perpendicular to each other. With two pairs of transducers (it is not necessary to have the same frequency for different pairs), if the interaction is large enough, the variation of the amplitude of one wave should change the velocity of the other. For the highly anharmonic crystals, such as solid ^{3}He and ^{4}He, this effect might be observable in the "favorable" directions. On the other hand, the additional experimental arrangement is relatively simple as compared with the usual method of applying the uniaxial pressure. The difficulty might be in growing a single crystal with the desired orientation. However, the exact orientation is not strictly required as long as it is precisely known (by X-ray or any other means) and is close to one of those listed in Table I. Although the calculations of TOEC may be more involved, this can be done.

Conclusion

In order to study the nonlinear effect in the highly anharmonic crystals like solid ^{3}He and ^{4}He, the method of harmonic generations and interaction of waves can be used. The nonlinear theory of Ljamov has been used to predict the "favorable" directions for the study of these effects. Further studies of TOEC could provide more information for the interactions and interatomic potential of these quantum solids at low temperatures.

Acknowledgment

The authors wish to thank Dr. C.M. Davis, Jr. for many valuable discussions.

References

1. H.R. Glyde, *Can J. Phys.* **49**, 769 (1971); R.A. Guyer, in *Solid State Physics,* F. Seitz, D. Turnbull and H. Ehrenreich, eds., Academic Press, New York (1969), Vol. 23, p. 413.
2. R. Wanner, *Phys. Rev. A* **3**, 448 (1971).
3. R.H. Crepeau and D.M. Lee, *Phys. Rev. A* **6**, 516 (1972).
4. J.P. Franck and R. Wanner, *Phys. Rev. Lett.* **25**, 345 (1970); R.H. Crepeau, O. Heybey, D.M. Lee, and S.A. Strauss, *Phys. Rev. A* **3**, 1162 (1971); D.S. Greywall, *Phys. Rev. A* **3**, 2106 (1971).
5. V.E. Ljamov, T.H. Hsu, and R.M. White, *J. Appl. Phys.* **43**, 800 (1972); V.E. Ljamov, *Soviet Phys.—Solid State* (to be published).
6. M.J.P. Musgrave, *Crystal Acoustics,* Holden-Day, San Francisco (1970).

NMR Measurements on ^{3}He Impurity in Solid ^{4}He*

M. G. Richards, J. Pope, and A. Widom †

School of Mathematical and Physical Sciences
University of Sussex, Brighton, England

Three questions of particular interest in relation to isotopic impurities in solid helium are as follows.

(a) Does the impurity move about at low temperatures by exchanging places with host atoms, and if so, what is the tunneling frequency compared with that between a pair of neighboring host atoms?

(b) Assuming the impurities are mobile, what is the form of their motion? In particular, is there some concentration below which the impurities move coherently,[1] i.e., like a gas, through the host background?

(c) How do the impurities interact with each other?

By carrying out NMR experiments on ^{3}He nuclei in the case of ^{3}He impurity in solid ^{4}He, we can get strong indications of the answers to all these questions. The relaxation times T_1 and T_2 and the spin diffusion coefficient D all provide information of a dynamic kind which is relevant to the problems discussed.

The data presently available for the case of ^{3}He impurity consist of measurements[2,3]‡ from Cornell of T_1 and T_2 at various frequencies, molar volumes, and ^{3}He concentrations down to 10^{-2}, and measurements of T_1, T_2, and D at 5 MHz and molar volume 21 cm^3 at concentrations down to 5×10^{-4}, which are reported here. In all cases the data are obtained at about 0.5°K, where the processes concerned are independent of temperature.

We now discuss a model which we believe must be correct at sufficiently low impurity concentration and then, by examining the data available at present, conclude that the model is correct for impurity concentrations below about 10^{-3}.

Ignoring for the present any long-range interaction between impurities, then, providing the impurity can tunnel into neighboring lattice sites, we would expect the excitations of the crystal to include an energy band representing wavelike states in which impurity atoms are not localized but are replaced by "impuritons" which behave like a gas of particles weakly interacting with each other and with the ^{4}He background (phonons). The situation is in some respects analogous to the motion of conduction electrons in a metal and also to the behavior of ^{3}He impurity in liquid ^{4}He, where however, there is no band of states because of the absence of a periodic

---

* Work supported by the Science Research Council.
† Permanent address: Physics Department, Northeastern University, Boston, Massachusetts.
‡ Measurements of T_1 at x_3 below 10^{-2} are reported in Ref. 3 but there is a 50% uncertainty in the average sample concentrations, a strong possibility of concentration gradients in the samples, and large scatter in the data.

potential in which the impurities move. The gaslike behavior of ^{3}He impurity in liquid ^{4}He is well established and we suggest that many of the manifestations of the impuriton model will be common to the liquid and the solid.

At a temperature well below the Debye temperature for the solid the phonons cease to scatter impuritons[1,4,5] and impuriton–impuriton scattering dominates. In this regime we can apply kinetic theory to calculate the diffusion coefficient. An approach[6] that is based on response functions and considers the band of states yields the same result

$$D \sim v\lambda/3$$

where $v \sim \sqrt{zJ_{34}}a$ is the impuriton group velocity and $\lambda \sim (n_3 R^2)^{-1} \sim a^3/x_3 R^2$ is the mean free path for the impuritons. The number of nearest neighbors in the crystal is z (12 in the case considered since all the measurements discussed refer to hcp ^{4}He); $2z\hbar J_{34}$ is the impuriton band thickness, n_3 is the ^{3}He number density, x_3 is the impurity molar fraction, and R^2 is the impuriton–impuriton cross section. Hence

$$D \sim \sqrt{zJ_{34}}a^4/3R^2 x_3 \tag{1}$$

For T_2 we have to calculate the effect of two-particle collisions that take place incoherently at an average rate $(\tau_{\text{coll}})^{-1}$. In each collision the transverse spin magnetization dephases by $\Delta\phi$:

$$1/T_2 = \overline{\Delta\phi^2}(\tau_{\text{coll}})^{-1} \tag{2}$$

provided $\Delta\phi \ll 1$.

We have

$$\overline{\Delta\phi^2} \sim (\hbar^2\gamma^4/R^6)(R^2/v^2) \quad \text{and} \quad \tau_{\text{coll}} \sim 3D/v^2$$

Therefore

$$1/T_2 \sim \hbar^2\gamma^4/3R^4 D = \hbar^2\gamma^4 x_3/\sqrt{z}R^2 a^4 J_{34} \tag{3}$$

T_1 presents a much more difficult problem because more detailed knowledge about the motion of the impuritons (in the form of correlation functions) must be calculated or assumed.

However, in the dilute limit where only two-particle collisions are considered, $T_1^{-1}(\omega)$ is zero at $\omega > 4zJ_{34}$ since this is the maximum energy change for a pair of impuritons. At higher frequencies there will be a contribution from ^{3}He–^{3}He tunneling. However, at least three impurities are required for this since a pair of impurities have well-defined singlet and triplet states. This leads to

$$1/T_1 = zx_3^2/T_1(\text{triple}) \tag{4}$$

where T_1(triple) is the relaxation time for a ^{3}He atom in a group of three impurities. Figure 1 gives an approximate form for the combined spectral density curve. We now examine the data to find evidence that the model is correct. Figure 2(a) shows D as a function of x_3 at molar volume 21 cm^3. We estimate

$$Dx_3^n = (1.2 \pm 0.3) \times 10^{-11} \quad \text{cm}^2/\text{sec}, \quad n = 1.0 \pm 0.2$$

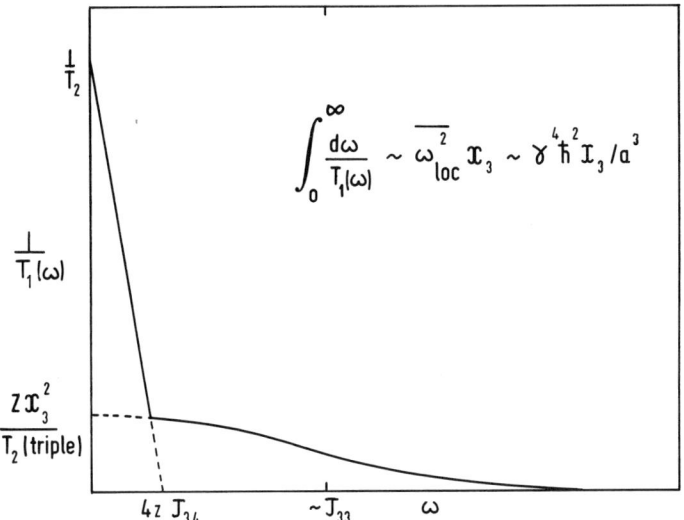

Fig. 1. An illustrative representation of the spectral density $J(\omega) = T_1^{-1}(\omega)$ for ^{3}He in solid ^{4}He in the impuriton regime.

Figure 2(b) shows T_2 as a function of x_3 at the same molar volume and frequency (5 MHz). The dashed lines give the power laws predicted by a theory[7] in which the impurities interact by a long-range r^{-3} coupling.

Figure 2(c) shows D/T_2 and, for comparison, the value[8] for pure ^{3}He gas at 4°K, where it is assumed, $T_1 = T_2$.

For x_3 less than about 2×10^{-3}, the x_3 dependences are as expected in the impuriton model. From D/T_2 we obtain a value of R very close to a, the lattice spacing. This suggests that the impuriton–impuriton interaction is short-ranged as would be expected from detailed calculations[9] of the distortion in the wave functions near an isotopic impurity in solid helium.

Putting $R = a$, we obtain a value of $\sim 10^4$ sec^{-1} for J_{34}. This is of order $10^{-3} J_{33}$, where J_{33} is the exchange frequency for pure solid ^{3}He at the same molar volume[4]. This is so low that one might wonder if one is in the motionally narrowed region at all. The condition required is $\Delta\phi \ll 1$ in Eq. (2), i.e., $\omega_{loc}\tau_{34} \ll 1$ (assuming $R \approx a$) where

$$\omega_{loc} \sim \gamma^2\hbar^2/a^3 \sim 10^4 \quad \text{and} \quad \bar{\tau}_{34} \sim zJ_{34} \sim 10^5$$

Hence the NMR line should be motionally narrowed at molar volume 21 cm^3, but on compressing the solid and thereby further reducing J_{34}, one might expect T_2 to level off at its "rigid lattice" value. There are no data to test this assertion in the region where Fig. 2 indicates our model is correct. However, as Fig. 3 (taken from Ref. 3) shows, we do see precisely this effect at $x_3 = 10^{-2}$. Moreover the plateau value of 3 ± 1 msec agrees well with the calculated[10] value of 2 msec obtained for a random distribution of static impurity spins. Since $\omega_{loc}\tau_{34} \sim 1$ at the knee, $\tau_{34} \sim 10^{-4}$ sec at that molar volume. The subsequent increase of T_2 by a factor of about ten on increasing the molar volume to 21 cm^3 suggests $\tau_{34}(21 \text{ cm}^3) \sim 10^{-5}$ sec and $J_{34}(21 \text{ cm}^3) \sim 10^4$ sec^{-1}, in agreement with the value estimated before.

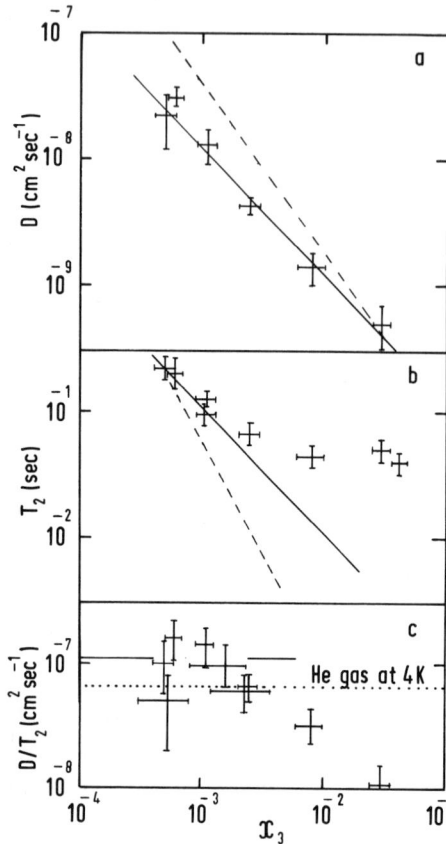

Fig. 2. The spin diffusion coefficient D and the spin–spin relaxation time T_2 for ^{3}He in solid ^{4}He vs. ^{3}He concentration x_3. The operating frequency is 5 MHz and the molar volume is 21 cm^3. The dashed lines refer to the theory of Ref. 7 and the full lines to the impuriton theory.

The T_1 data are given in Fig. 4. Since the frequency (again 5 MHz) is well above $4zJ_{34}/2\pi$, the data will probably not relate to ^{3}He–^{4}He tunneling but to rare encounters of an impurity with an impurity pair. The line drawn corresponds to $T_1 x_3^2 = 10^{-3}$ sec, consistent with T_1(triple) $= 10^{-2}$ sec. This is shorter than might be expected since T_1 in pure solid ^{3}He at this molar volume and frequency is ~ 60 msec.[11] All the data in Figs. 2 and 4 were taken at 0.5°K.

What is needed are T_1 data at lower frequencies and $x_3 < 2 \times 10^{-3}$. These would show up the step in Fig. 1 and allow a clearcut evaluation of J_{34}. Experiments along these lines are being currently undertaken at Sussex.

We conclude by returning to the original three questions.

(a) D and T_2 data suggest a tunneling frequency of about 10^4 sec^{-1} at a molar volume 21 cm^3. Thus $J_{34} \sim 10^{-3} J_{33}$ at the same molar volume. This result is difficult to reconcile with the current understanding[12] of solid ^{3}He, where it has been supposed that $J_{34} \sim J_{33}$.

(b) For impurity concentrations below about 2×10^{-3}, ^{3}He seems to move like a gas through ^{4}He solid. At molar volume 21 cm^3 the velocity is $\sim 10^{-3}$ cm/sec.

(c) The model discussed assumes that the impuritons only interact at short range. However this result is justified *a posteriori*, first by the observed x_3 dependence of D and T_2, and second by the deduction of a scattering cross section of $\sim a^2$.

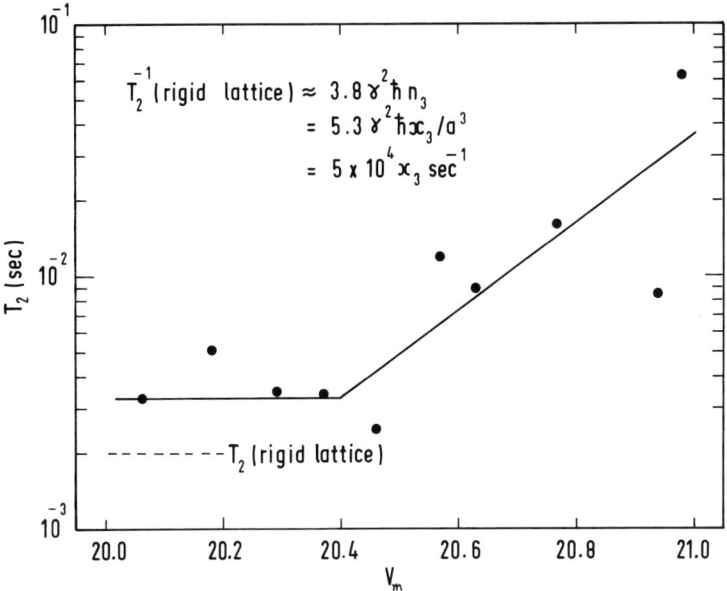

Fig. 3. T_2 vs. molar volume V_m. The data are taken from Ref. 3 and the theoretical rigid lattice value from Ref. 10.

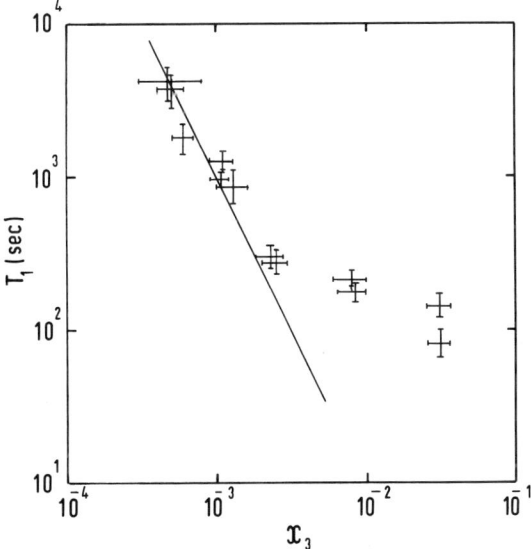

Fig. 4. The spin lattice relaxation time T_1 for ^{3}He in solid ^{4}He vs. x_3, the ^{3}He concentration. The operating frequency is 5 MHz and the molar volume is 21 cm^3. The line drawn represents $T_1 x_3^2 = 10^{-3}$ sec.

Experiments[5] on the isotopic phase separation curve and on samples with small ^{4}He impurity are appropriate for measuring the strength, as opposed to the range, of the impuriton–impuriton coupling.

References

1. A.F. Andreev and I.M. Lifshitz, *Sov. Phys.—JETP* **29**, 1107 (1969); R.A. Guyer and L.I. Zane, *Phys. Rev. Lett.* **24**, 660 (1970).
2. D.S. Myoshi, R.M. Cotts, A.S. Greenberg, and R.C. Richardson, *Phys. Rev. A* **2**, 870 (1970).
3. A.S. Greenberg, W.C. Thomlinson, and R.C. Richardson, *Phys. Rev. Lett.* **27**, 179 (1971); *J. Low Temp. Phys.* **8**, 3 (1972).
4. R.A. Guyer, R.C. Richardson, and L.I. Zane, *Rev. Mod. Phys.* **43**, 532 (1971).
5. M. Bernier and A. Landesman, *J. de Physique* (Suppl.) **32**, C5a, 213 (1971).
6. A. Widom and M.G. Richards, *Phys. Rev. A* **6**, 1196 (1972).
7. A. Landesman and J.M. Winter, this volume.
8. K. Luszczynski, R.E. Norberg, and J.E. Opfer, *Phys. Rev.* **128**, 186 (1962); R. Chapman and M.G. Richards, unpublished.
9. H.R. Glyde, *Phys. Rev.* **177**, 202 (1969).
10. P.W. Anderson, *Phys. Rev.* **82**, 342 (1951).
11. M.G. Richards, J. Hatton, and R.P. Gifford, *Phys. Rev.* **139**, A91 (1965).
12. L.H. Nosanow, private communication.

Calculation of the Diffusion Rate of ^{3}He Impurities in Solid ^{4}He

A. Landesman and J. M. Winter

Service de Physique du Solide et de Résonance Magnétique
Centre d'Etudes Nucléaires de Saclay, Gif-sur-Yvette, France

Introduction

Let us consider a system of a small concentration x of ^{3}He in solid ^{4}He (x is supposed to be small enough so that the Pauli exchange between ^{3}He particles may be neglected).

The system is described in terms of pseudospins S_i.[1] S_i^z has the value $+1/2$ if the site i is occupied by a ^{3}He, whereas it has the value $-1/2$ if there is a ^{4}He. The Hamiltonian of the system has the form

$$\hbar \mathcal{H} = -2\hbar \sum_{i<j} K_{ij} S_i^z S_j^z + \hbar J' \sum_{i<j}{}' (S_j^+ S_i^- + S_j^- S_i^+) \quad (1)$$

The term J' describes the tunneling of ^{3}He and is assumed to exist only between nearest neighbors. The Ising part is due to the interaction between ^{3}He impurities. Sometimes, for simplicity, it will be assumed that K_{ij} exists only for nearest neighbors with the value K. We always have the condition $K \gg J'$.

Calculation of the Diffusion Rate Using a Moment Expansion

The diffusion of ^{3}He is related to the behavior of the correlation function $\Gamma_{jk}^0(t)$, which is the probability of having both a ^{3}He at site j at time zero and a ^{3}He atom at site k at time t:

$$\Gamma_{jk}^0 = \text{Tr}\left[1/2 + S_k^z(t)\right]\left[(1/2 + S_j^z)\rho_0\right] \quad (2)$$

with

$$\rho_0 = \exp\left[-\hbar\beta\mathcal{H} - \beta\mu \sum_i (1/2 + S_i^z)\right]/Z$$

$$Z = \text{Tr}\left\{\exp\left[-\beta\mathcal{H} - \beta\mu \sum_i (1/2 + S_i^z)\right]\right\}$$

μ is a chemical potential defined by

$$x = \text{Tr}\left[(1/2 + S_i^z)\rho_0\right] \quad \text{and} \quad S_k^z(t) = e^{i\mathcal{H}t} S_k^z e^{-i\mathcal{H}t}$$

We define the space Fourier transform $\Gamma_q^0(t)$ as

$$\Gamma_q^0(t) = \sum_k (\exp i\mathbf{q}\cdot\mathbf{r}_{jk})\left[\Gamma_{jk}^0(t) - x^2\right] \quad (3)$$

The term x^2 is added because with this definition Γ_q^0 tends to zero when t is very large.

The behavior of $\Gamma_q^0(t)$ will be considered only in the high-temperature limit, i.e., well above the phase separation, where $\mathscr{H}$ may be neglected in ρ_0. This function may be estimated using a moment expansion[2]; the correlation function is written

$$\Gamma_q^0(t) = x(1-x)[1 - 1/2 m_2(\mathbf{q}) t^2 + 1/4! m_4(\mathbf{q}) t^4 + \cdots] \qquad (4)$$

For small vectors of amplitude q, in the hcp phase, and assuming K_{ij} to be limited to nearest neighbors, one finds

$$m_2(q) = 4 J'^2 a^2 q^2$$
$$m_4(q) = x(1-x)(224 J'^2 K^2 + 304 J'^4) a^2 q^2 + 4 J'^4 a^2 q^2 \qquad (5)$$

A similar result is found for a bcc lattice, except for the fact that the fourth moment has no term without the factor $x(1-x)$. (Indeed, it can be shown that the term $4 J'^4 a^2 q^2$ is present because the hcp is not a Bravais lattice.)

If for K we take a value deduced from the relaxation measurements in solid ^{3}He with a small concentration of ^{4}He, and if for J' we take a value of the order of the direct exchange for solid ^{3}He at the same molar volume, the ratio K^2/J'^2 is found to be very large and it is impossible to reach a situation where $xK^2 \ll J'^2$. In the fourth moment the only important term is the term $224 xJ'^2 K^2$. Thus we have the relation $m_4(q) \gg [m_2(q)]^2$. If we assume an exponential variation for $\Gamma_q^0(t)$ of the form $\Gamma_q^0(t) = x(1-x)\exp(-Dq^2 t)$, the diffusion coefficient D may be calculated; one finds

$$D = \eta J'^2 a^2 / K [x(1-x)]^{1/2} \qquad (6)$$

η is a numerical coefficient.

We notice that the interaction between ^{3}He particles has the effect of strongly reducing the diffusion coefficient and introduces a dependence on the concentration.

Since a moment calculation using only two moments is far from being convincing, it will be quite useful to find another approach to calculate the correlation function.

Let us finally remark that the calculations may be extended to situations where q is not small. Providing $xK^2 \gg J'^2$, one finds that $m_4(q)$ is larger than $[m_2(q)]^2$; an exponential decay is expected, with a time constant τ_q of the order $1/\tau_q \simeq J'^2/K\sqrt{x}$.

Calculation of the Diffusion Rate Using the Kubo Equation

The calculation of time-dependent correlation function Δ has made great progress because this problem was involved in the understanding of the dynamics of critical phenomena. For instance, Kubo established a very simple equation for deriving the time constant of a conservative or critical variable.[3] Such a variable X_q is characterized by the fact that the time constant τ_q for its evolution is long compared to the time constants for the evolution of the variables involved in its time derivative $\partial X_q / \partial t$. In our problem the variable S_q^z

$$S_q^z = \sum_j (\exp i\mathbf{q} \cdot \mathbf{r}_j) S_j^z$$

belongs to this category. Its time derivative has the value

$$dS_q^z/dt = iJ' \sum_{k,j}{}' (\exp i\mathbf{q}\cdot\mathbf{r}_j)(S_k^+ S_j^- - S_k^- S_j^+)$$

and the evolution of transverse operators is very fast due to the presence of the large Ising interaction.

The Kubo equation for the time constant τ_q for the evolution of $\Gamma_q^0(t)$ is

$$\frac{1}{\tau_q} = \frac{1}{\langle S_q^z S_{-q}^z \rangle} \int_0^{\theta_q} \left\langle \frac{dS_q^z(t)}{dt} \frac{dS_{-q}^z}{dt} \right\rangle dt \tag{7}$$

$\theta_q \ll \tau_q$, but since S_k^+ and S_j^- are varying very rapidly, θ_q may be replaced by ∞; $1/\tau_q$ is a function of a four-spin correlation function because

$$\left\langle \frac{dS_q^z(t)}{dt} \frac{dS_{-q}^z}{dt} \right\rangle = J'^2 \sum_{k,j,i,l}{}'' (\exp i\mathbf{q}\cdot\mathbf{r}_{ij}) \langle (S_k^+ S_j^- - S_k^- S_j^+)(t)(S_i^+ S_l^- - S_i^- S_l^+) \rangle$$

The second moment of this function is related to the fourth moment of Γ_q^0 and again, except for very low x, we may consider only the effect of the Ising interaction.

We have the simple relation

$$S_j^+(t) = S_j^+ e^{i(\omega_i - \omega_j)t} \quad \text{with} \quad \omega_j = \sum_k K_{jk} S_k^z$$

and the correlation function for the derivative becomes

$$\left\langle \frac{dS_q^z(t)}{dt} \frac{dS_{-q}^z}{dt} \right\rangle = 2J'^2 \sum_{i,j}{}' \langle S_i^+ S_j^- S_i^- S_j^+ (1 - \cos\mathbf{q}\cdot\mathbf{r}_{ij}) e^{i(\omega_i - \omega_j)t} \rangle$$

giving for $1/\tau_q$

$$\frac{1}{\tau_q} = 2J'^2 \sum_{ij}{}' \int_0^\infty \langle (1 - \cos\mathbf{q}\cdot\mathbf{r}_{ij}) e^{i(\omega_i - \omega_j)t} \rangle \, dt \tag{8}$$

The problem now is to evaluate the following function:

$$F_{ij}(t) = \left\langle \exp \sum_k i(K_{ik} - K_{jk}) S_z^k t \right\rangle$$

For large $x(1/z < x < \frac{1}{2}$, z being the number of nearest neighbors) this function may be calculated exactly, because the problem of evaluating a transverse correlation function in a dense system of spins with an Ising interaction in an external field has been solved.[4] (The magnetic field appears by the presence of μ in ρ_0.) One finds

$$F_{ij}(t) = \exp\left[-2t^2 x(1-x) \sum_k (K_{ik} - K_{jk})^2\right]$$

A value for D is deduced, $D \simeq a^2 J'^2/K \sqrt{x}$, in agreement with the moment calculation. However, for such values of x the direct exchange may no longer be neglected and this result has no physical meaning.

If x is not too large, $J'^2/K^2 \ll x \ll 1/z$, the Gaussian variation for $F_{ij}(t)$ is no longer correct. This fact may be shown by looking at the moments of F_{ij}; the second moment M_2 varies as x, but in the fourth moment M_4 there are terms in x and

x^2; for very small x, $M_2^2 \ll M_4$. The problem bears some resemblance to the calculation of the nuclear free precession in dilute system.[5] The calculation is possible only if the variation of K_{jk} with the distance r_{jk} is known. Following the calculation of Ref. 6, it is found that

$$F_{ij}(t) = e^{-x\lambda(t)}$$

with

$$\lambda(t) = \int d^3r_k [1 - \exp(it[K(r_{ik}) - K(r_{jk})])]$$

In the dilute limit, providing the range of $K(r)$ is large enough, r_{ik} and r_{jk} are always large compared to r_{ij} for nearest neighbors and

$$\lambda = \int_{-1}^{+1} \sin\theta\, d\theta \int_0^\infty \frac{r^2\, dr}{a^3}\left[1 - \cos\left(at\cos\theta\,\frac{\partial K}{\partial r}\right)\right]$$

If for $K(r)$ the form $K(r) = K(a/r)^n$ is assumed, one obtains

$$F_{ij} = \exp[-\alpha x(Kt)^{3/(n+1)}]$$

where α is a numerical coefficient. The diffusion constant becomes

$$D \sim \frac{a^2 J'^2}{x^{(n+1)/3} K}$$

If, as suggested in Ref. 7, the value $n = 3$ is taken, D varies as $x^{-4/3}$.

The Correlation Function in the Limit of Very Small Concentration

If x is very small, so that the damping due to the interaction may be neglected, the correlation function $\Gamma_q^0(t)$ may be calculated using the model of the free mass wave. We give an indication of the calculation, although we believe that this model will be valid for regions of x where it is impossible to have experimental results.

We start from the equation $S_j^z = \frac{1}{2} - S_j^- S_j^+$. We obtain $S_q^z = \sum_{q'} S_{q-q'}^+ S_{q'}^-$. In the limit of free mass wave and for a Bravais lattice $S_q^+(t) = S_q^+ e^{i\omega_q(t)}$,

$$\omega_q = zK - J' \sum_{\langle ij \rangle} \cos \mathbf{q}\cdot\mathbf{r}_{ij}$$

and the correlation function may be written

$$\Gamma_q^0(t) = \sum_{q'} \exp[i(\omega_{q'-q} - \omega_{q'})t]$$

For small q we obtain

$$\Gamma_q^0(t) \simeq \sum_{q'} \cos[iq(\partial/\partial q')(\omega_{q'})t]$$

The time variation becomes very slow when q is small, but there is no reason to find an exponential decay. The situation is slightly more subtle for the hcp lattice because there are two branches for the mass waves for a given q. It is possible to find good creation operators which are now combinations of creation operators for the two branches. The calculation is straightforward but tedious.

In all our considerations the damping due to the coupling with the phonons is neglected.

Comparison With Experimental Results

The diffusion coefficient for dilute solution has been measured by Richards et al.[8] They found a value of D varying with x, and interpreted the results by the equation $D \sim J'a^4/R^2 x$, but a variation as $x^{-4/3}$ will fit the results as well. Since $R \simeq a$, the value of J' they deduced was two orders of magnitude smaller than expected (expected means the value of the direct exchange in ^{3}He with the same molar volume). Taking[1] $J' \simeq 1.0$ MHz and $K \simeq 300$ MHz, our equation is in much better agreement with the results.

The relaxation times T_1 and T_2 were also measured by Richards et al.[8] and Greenberg et al.[9] If a crude model of a modulation of the dipolar field by the motion of ^{3}He is taken, for T_2 we obtain the relaxation

$$1/T_2 \simeq x M_2 \tau_{34}$$

where M_2 is the second moment for dipolar interactions in a dense lattice; and $1/\tau_{34}$ is the tunneling frequency, which will be assumed to be equal to $1/\tau_q$ for large q. We predict $1/\tau_{34} \simeq J'^2/K x^{4/3}$; thus

$$1/T_2 \simeq x^{7/3} M_2 K/J'^2$$

Again the authors[9] noticed that for a fixed x the tunneling frequency was much too small. Taking the molar volume 21 cm^3, they found $1/\tau_{34} \simeq J/10$; assuming $K \simeq 300$ MHz, using our equation, we obtain $J' \simeq 1.4$ MHz, a value of the order of J.

Another observation is in good agreement with the model: For a fixed x the variation of $1/\tau_{34}$ with the molar volume fits the variation of J^2 better than the variation of J. (This result suggests that K does not vary as fast as J or J' with the molar volume, an expected result.)

Let us remark, however, that the calculations that we present here for T_2 are crude; more sophisticated approaches based on the calculation of the correlation functions of the dipolar field can be made, leading essentially to the same results.

Conclusion

Using our model of a fictitious spin, we were able to calculate the diffusion rate and the relaxation time in dilute solution of ^{3}He in ^{4}He.

The important result is that, except for very low concentration, the presence of the interaction between the ^{3}He particles strongly reduces the diffusion rate. This result may be described as follows. The diffusion of ^{3}He is due to a tunneling between two equivalent potential wells; the other impurities give a large broadening of the levels in the two wells, thus decreasing the probability of transition.

The diffusion constant depends upon the concentration in a very peculiar way, and this variation is a function of the range of the interaction. By making accurate measurements of D as a function of x, it would be possible in principle to get information about the spatial variation of this interaction.

Finally, let us add that a similar model may be used to describe the motion of vacancies in solid helium. The only difference is that we do not need a chemical potential μ because the number of vacancies is not fixed, but we must add a term $h \sum_i (S_i^z + \frac{1}{2})$ in the Hamiltonian for describing the energy of formation of a vacancy. Such a term plays for our spin system the role of an external magnetic field. Also, the condition $K \gg J'$ may not hold any more for vacancies.

References

1. M. Bernier and A. Landesman, *J. Phys.* **32**(C5a), 213 (1971); M. Bernier, Thesis, Orsay, France 1971, unpublished.
2. A.G. Redfield and W.N. Yu, *Phys. Rev.* **169**, 443 (1968); **177**, 1018 (1969).
3. J. Villain, *J. Phys.* **32**(C5a), 169 (1971); **32**(C1), 310 (1971); H. Mori, *Progr. Theor. Phys.* **34**, 399 (1965).
4. J.M. Winter, *Ann. de Phys.* **6**, 167 (1971).
5. P.W. Anderson, *Phys. Rev.* **82**, 342 (1951).
6. A. Abragam, *The Principles of Nuclear Magnetism*, Oxford Press (1961), pp. 126–128.
7. R.A. Guyer, R.C. Richardson, and L.I. Zane, *Rev. Mod. Phys.* **43**, 532 (1971).
8. M.G. Richards and J. Pope, private communication and this volume.
9. A.S. Greenberg, W.C. Thomlinson, and R.C. Richardson, *Phys. Rev. Lett.* **27**, 700 (1971); *J. Low Temp. Phys.* **8**, 3 (1972).

NMR Study of Solid ^{3}He–^{4}He Mixtures Rich in ^{3}He

M. E. R. Bernier

Service de Physique du Solide et de Résonance Magnétique
CEN—Saclay, Gif-sur-Yvette, France

We study by pulsed NMR techniques the effect of ^{4}He impurities on spin–lattice and transverse relaxation times of solid ^{3}He.[1,2] Experiments are made in the temperature range $1.2 < T < 0.27°K$, where the spin–lattice relaxation time T_1 is sensitive to the concentration x of ^{4}He impurity in the sample. T_1 has been measured as a function of the temperature, the concentration x of ^{4}He impurity, and the NMR frequency for several molar volumes in both bcc and hcp phases.

In the bcc phase, according to the variation of T_1 both with T and x, the temperature range can be divided into three parts: (I) At high temperatures ($T \sim 1°K$) T_1 is independent of T and x for $x < 6.8 \times 10^{-3}$. (II) For temperatures lower than $0.6°K$, T_1 is strongly temperature dependent and decreases when x increases. (III) For still lower temperatures[2] T_1 is temperature independent and decreases when x increases. This plateau has been observed in Refs. 3 and 4 for samples of low ^{4}He content and the onset temperature for this regime increases when x increases.

The transverse relaxation time T_2 is independent of both temperature and impurity concentration for the regimes II and III. From these T_2 values we deduce the exchange interaction in the same manner as for pure ^{3}He.

In the hcp phase T_1 at first decreases when x increases and then increases when x goes on increasing, as shown in Fig. 6 of Ref. 5 for the molar volume $V = 19.25 \text{ cm}^3$. This behavior is masked in the bcc phase by the onset of the temperature-independent plateau at high impurity concentration.

The effect of ^{4}He impurities on the spin–lattice relaxation may be understood by taking into account both the tunneling of ^{4}He atoms through the ^{3}He crystal with characteristic frequency J' of the order of the exchange frequency J and the interaction K between ^{4}He impurities responsible for the phase separation ($K \gg J, J'$). The ^{4}He impurity system is then described by a ^{3}He–^{4}He tunneling Hamiltonian $\hbar\mathcal{H}_t$ and a Hamiltonian $\hbar\mathcal{H}_a$ which accounts for the phase separation.[1,6,7]

As a first step we postpone the problem of the real spin of ^{3}He and describe the occupation of the lattice sites by a fictitious spin $1/2$. Then, treating the system of ^{4}He impurities and phonons as isolated, we study the time evolution of the density matrix of the system and calculate the energy transfer from ^{4}He impurities to the lattice by Raman scattering of phonons. The constant appearing in the Hamiltonian coupling ^{4}He to the phonons is related to the Rayleigh lifetime of phonons and is deduced from thermal conductivity experiments.[13,14]

This is then generalized by introducing the real spin of ^{3}He and taking into account the coupling between the Hamiltonians $\hbar\mathcal{H}_{ex}$ and $\hbar\mathcal{H}_a$ due to $\hbar\mathcal{H}_t$. We

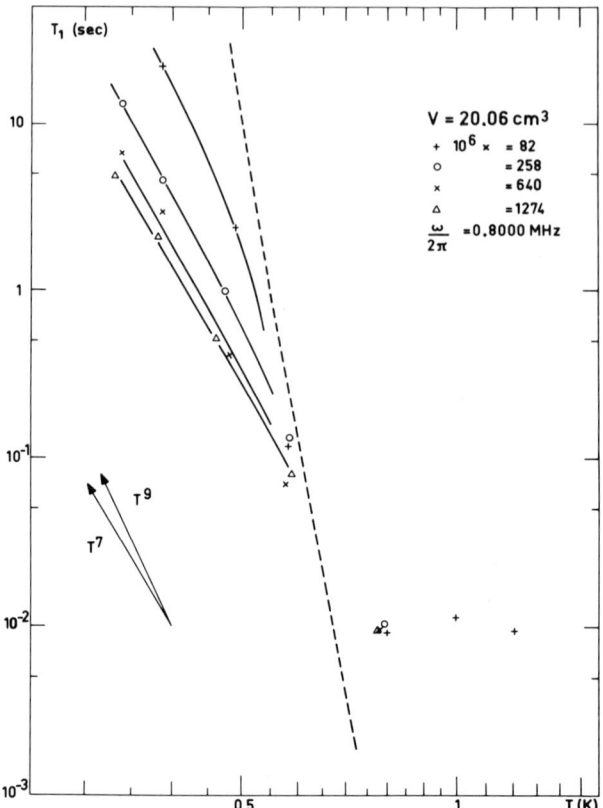

Fig. 1. Spin–lattice relaxation time T_1, measured at a Larmor frequency of 0.8000 MHz, as a function of the lattice temperature for several samples of ^{3}He with a molar volume of $V = 20.06$ cm^3 and different concentrations x of ^{4}He.

find that T_1^{-1}, when the relaxation is due to ^{4}He impurities, is (regime II):

$$\frac{1}{T_1} = \frac{1}{T_1^\circ} \frac{x}{(1/4z)(\omega/J')^2 + \tfrac{3}{8}(J/J')^2 + x[1 - \tfrac{3}{8}(J/J')^2] + 2x^2(1-x)(K/J')^2}$$

where ω is the Larmor frequency, z is the number of nearest neighbors in the lattice, and T_1° an intrinsic relaxation time which depends on lattice properties; the calculation gives a T^{-9} variation for T_1° both for the bcc and hcp phases.

The coupling of $h\mathcal{H}_t$ and $h\mathcal{H}_a$ with the Zeeman being weak, the effect of impurities on T_1 becomes large only when the temperature is sufficiently low that the bottleneck in the relaxation of Zeeman energy to the lattice is between the exchange and lattice energy reservoirs. By comparing the temperature dependence of the relaxation rate due to the modulation of exchange by atomic jumps, which varies as $e^{-W/T}$ where W is an activation energy, with that of the relaxation by impurities, we obtain the temperature range of each mechanism. This describes the behavior of T_1 in the temperature range II which can then be divided into two subregions.

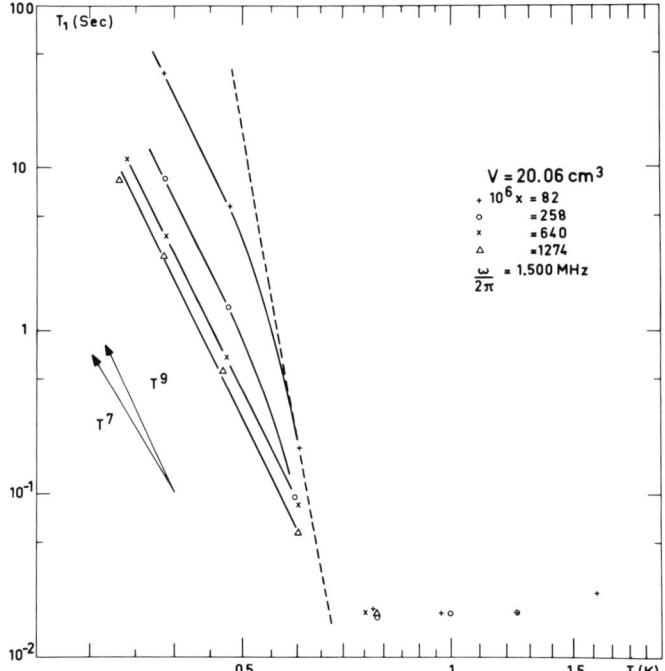

Fig. 2. Spin–lattice relaxation time T_1, measured at a Larmor frequency of 1.500 MHz, as a function of the lattice temperature for several samples of ^{3}He with a molar volume of 20.06 cm^3 and different concentrations x of ^{4}He.

In the high-temperature region T_1 varies as $e^{-W/T}$,* whereas in the low-temperature region T_1 varies as T^{-n}. This behavior is in agreement with the experimental results given in Figs. 1–3, and in Refs. 1 and 5. The experimental value of n is consistent with 9 for the bcc phase, but closer to 7 for the hcp phase. In Figs. 1–3 for $V = 20.06$ cm^3 the dashed curves show the spin–lattice relaxation time due to the modulation of exchange by atomic jumps for pure solid ^{3}He. The value of W used in calculating these curves is taken from Refs. 8 and 9; D_0 is deduced from Fig. 2 of Ref. 5 by assuming that the whole spin–lattice relaxation is due to this mechanism for the temperature at which the spin–lattice relaxation time begins to increase in region II, i.e., $1/T = 1.5°\text{K}^{-1}$. This gives $D_0 \simeq 7 \times 10^{-6}$ cm^2/sec for $V = 20.06$ cm^3. For $V = 21.00$ cm^3 (Fig. 3 of Ref. 5) the same procedure gives $D_0 \simeq 2 \times 10^{-5}$ cm^2/sec. These figures are in reasonable agreement with the values deduced in Ref. 9 from direct Zeeman lattice relaxation times measured at higher temperatures.

* The relaxation rate $(\tau_{\text{ex},L})_d^{-1}$ for the relaxation by atomic diffusion is given by

$$\frac{1}{(\tau_{\text{ex},L})_d} = \frac{12(z-1)}{z}\frac{D_0}{a^2}e^{-W/T}$$

where z is the number of nearest neighbors, a is the distance to the neighboring site, and D_0 is the diffusion coefficient.

Table I

	BCC		HCP
V, cm^3	21.00	20.06	19.25
$J/2\pi$, MHz	1.2	0.45	0.11
$J'/2\pi$, MHz	0.68	0.25	—
$K/2\pi$, MHz	~ 280	~ 200	~ 600

Figures 4 and 5 of Ref. 5 give the x dependence of T_1 at constant temperature in a temperature region where the relaxation is due to the impurities. For low ^{4}He concentrations $1/T_1$ is proportional to x. By using Eq. (1), deducing J from T_2, and calculating T_1° with the values of the physical constants deduced from thermal conductivity and heat capacity measurements,[7] we obtain the values of J' given in Table I.[7] The results for the two molar volumes measured suggest that J' is strongly volume dependent and the ratio J'/J is roughly constant and a bit smaller than one, as expected.

When the impurity concentration becomes large the leading term in the denominator of Eq. (1) is $2(K/J')^2 x^2$ and we expect T_1 to go through a minimum when x increases. This minimum exists in the hcp phase where data of Fig. 6 of Ref. 5 exhibit a decrease of T_1 when x increases at low x and then an increase of T_1 when x goes on increasing. From these data we get a rough estimate of K:

$$K/2\pi \simeq 600 \quad \text{MHz}, \qquad V = 19.25 \quad \text{cm}^3$$

The frequency dependence of T_1 at constant temperature is consistent with Eq. (1); the curves T_1 vs. ω^2 are straight lines[1] and for low impurity concentration the abscissas for $T_1 = 0$ are constant and equal to $(\omega^2)_0 \simeq -2.5$ MHz^2 for $V = 20.06$ cm^3. This value is in good agreement with the value of $J/2\pi = 0.45$ MHz for this molar volume. When x becomes large the term $x^2 K^2$ is no longer negligible and $(\omega^2)_0$ decreases; for $x = 1274 \times 10^{-6}$ we get $(\omega^2)_0 \simeq -6.2$ MHz^2. From Eq. (1) we then evaluate $K/2\pi \simeq 200$ MHz.

There is a different way of obtaining the interaction K. We study the ordinate $\ln(L_0)$ for $t = 0$ of the straight line representing $\ln(1 - \beta_Z/\beta_L)$ as a function of t for large t at low temperatures when measuring T_1 with two 90° pulses separated by a time t. Here β_Z and β_L stand for the inverse temperatures of the Zeeman and lattice energy reservoirs. L_0 is easy to evaluate when the spin–lattice relaxation is due only to the impurities and the coupling between the exchange and interaction energy reservoirs due to the tunneling is strong, i.e., for a solid rich in ^{4}He impurities. Then L_0 corresponds to the magnetization of the Zeeman system at $t = 0$ when the ensemble (Zeeman + exchange + impurities) is in internal equilibrium, but does not relax to the lattice with the characteristic time T_1 given by Eq. (1). For 90° pulses we get

$$L_0 = \frac{1}{1 + \tfrac{3}{2}z(J/\omega)^2(1-x) + 4z(J'/\omega)^2 x + 8z(K/\omega)^2 x^2(1-x)}$$

We emphasize the fact that this equation is only valid for strong coupling between the exchange and the impurities and we use it only for $x = 1274 \times 10^{-6}$. Taking

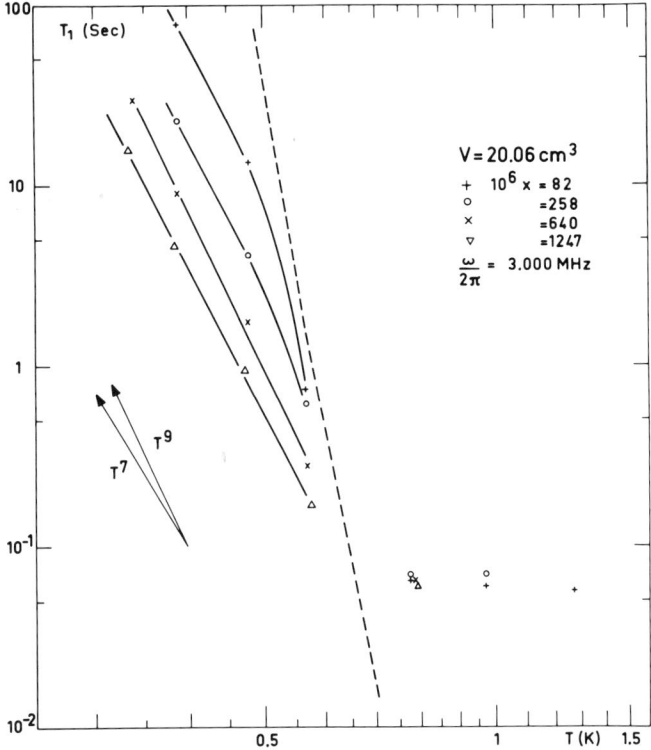

Fig. 3. Spin–lattice relaxation time T_1, measured at a Larmor frequency of 3.000 MHz, as a function of the lattice temperature for several samples of ^{3}He with a molar volume of 20.06 cm^3 and different concentrations x of ^{4}He.

into account the possibility that the pulses might be slightly different from 90° and using the values of the exchange and tunneling frequencies obtained previously, we find for $V = 20.06$ cm^3 and $V = 21.00$ cm^3 the values of $K/2\pi$ given in Table I. These values are the means of the determinations of K at different frequencies[1] which are in fair agreement considering the assumptions made and the precision of the measurements. It is worth noting the agreement of the value obtained at $V = 20.06$ cm^3 by this method and the value deduced from the frequency dependence of the spin–lattice relaxation time. From these determinations of K for two molar volumes it appears that K varies slowly with V, unlike J and J'. Preliminary results of new experiments at higher impurity concentration for three molar volumes in the bcc phase confirm this point.[2] The interaction K between ^{4}He impurities can be interpreted in a continuum model of the lattice[10] as the energy of a sphere distorting the elastic continuum, with the pressure field of a similar sphere located at a distance R of the first one. This kind of interaction is not expected to vary rapidly with the molar volume, as J and J' do, since K has nothing to do with the overlap of the atomic wave functions. Moreover, following the regular solution model, K is directly related to the phase separation temperature for $x = 1/2$ and this temperature is not very sensitive to the molar volume, as seen in Refs. 11 and 12. However, the value of $K/2\pi$ deduced from these experiments is of the order of 2000 MHz, which is an order of magnitude higher than our determination. The question arises whether the interaction we measure for small x is really identical to the parameter K which fits the phase separation curve for $x = 1/2$. The assumption that K is limited to nearest neighbors is not essential for analyzing our data; if we admit an R^{-3} variation, this would decrease the interaction between nearest neighbors by less than 20%.

Finally, we would like to mention that preliminary T_1 results in the temperature range III show a strong dependence on the impurity concentration, and that $1/T_1 \propto x^3$ for x ranging from 0.3×10^{-3} to 6.8×10^{-3} for several molar volumes. This point will be analyzed elsewhere.[2]

Acknowledgments

We wish to thank Dr. A. Landesman, Dr. J. Winter, and Dr. F.I.B. Williams for many helpful discussions.

References

1. M.E.R. Bernier, Thesis, Orsay, France, 1971 (CRNS-A.O. 3704).
2. M.E.R. Bernier and G. Deville (to be published).
3. R.P. Giffard, Thesis, Oxford, England, 1968.
4. E.R. Hunt, R.C. Richardson, T.R. Thompson, R.A. Guyer, and H. Meyer, *Phys. Rev.* **163**, 181 (1967).
5. M.E.R. Bernier, *J. Low Temp. Phys.* **3**, 29 (1970).
6. A. Landesman and M. Bernier, *Solid State Comm.* **8**, 2151 (1970).
7. M. Bernier and A. Landesman, *J. de Phys.* **32**(C5a), 213 (1971).
8. R.P. Giffard and J. Hatton, *Phys. Rev. Lett.* **18**, 1106 (1967).
9. H.A. Reich, *Phys. Rev.* **129**, 630 (1963).
10. R.A. Guyer, R.C. Richardson, and L.I. Zane, *Rev. Mod. Phys.* **43**, 532 (1971).
11. D.O. Edwards, A.S. McWilliams, and J.G. Daunt, *Phys. Rev. Lett.* **9**, 195 (1962).
12. M.F. Panczyk, R.A. Scribner, J.R. Gonano, and E.D. Adams, *Phys. Rev. Lett.* **21**, 594 (1968).
13. B. Bertman, H.A. Fairbank, R.A. Guyer, and C.W. White, *Phys. Rev.* **142**, 79 (1966).
14. B.K. Agrawal, *Phys. Rev.* **162**, 731 (1967).

Phonon Scattering by Isotopic Impurities in Helium Single Crystals*

D. T. Lawson† and H. A. Fairbank

Department of Physics, Duke University
Durham, North Carolina

We have utilized the enhancement of thermal conductivity by Poiseuille flow of the phonon gas to study phonon scattering in ^{4}He crystals containing small amounts of ^{3}He. Our samples differed from those of previous investigations in two respects: (1) they were single crystals of high quality, and (2) the ^{3}He concentrations studied were quite small, the largest being 2.0×10^{-5}.

Our sample chamber had an internal diameter of 2.72 mm and was essentially like that of Hogan et al.[1] except that the tip of the nucleation post was a flat surface covered with a piece of gold foil. The samples were grown from ensembles of crystallites nucleated on this surface, the foil having been prepared in such a way that it closely approximated the 100 face of a single crystal of gold.‡ The use of this apparatus resulted in a few discrete c-axis orientations occurring with high relative probabilities among our single-crystal samples. In addition, we found that the incidence of samples among these preferred angles could be influenced through growth techniques designed to exploit the large thermal conductivity and anisotropy[1] in our hexagonal close-packed crystals. It was this anisotropy—a factor of 20 near the thermal conductivity peaks for our growth pressure of 85.1 atm—which made it necessary to compare crystals of a common c-axis orientation. Previous experiments,[1] the Umklapp region temperature dependence of our data, and a computer simulation of our nucleation process all support the identification of our observed conductivity extrema as the true 0 and 90° c-axis orientations.

Crystal size and quality also may be inferred from the thermal conductivity data. Size was determined by comparing conductivity curves for various portions of the same sample. If the different parts of the sample shared a common Umklapp region curve *and* a common boundary-scattering-dominated region (provided the latter was also consistent with the known sample diameter), it was concluded that a single crystal filled the observed volume, 5 cm in length.

Figure 1 illustrates the reproducibility of the data. We have plotted the logarithm

---

* This work was supported in part by the National Science Foundation and the Office of Naval Research.
† Present address: Laboratory of Atomic and Solid State Physics, Cornell University, Ithaca, New York.
‡ Our foil was prepared by L. Meyer and co-workers. (See C. Barrett et al.[2]) We are grateful to Prof. H. Meyer for providing us with a piece of this foil and suggesting its usefulness in growing oriented helium crystals.

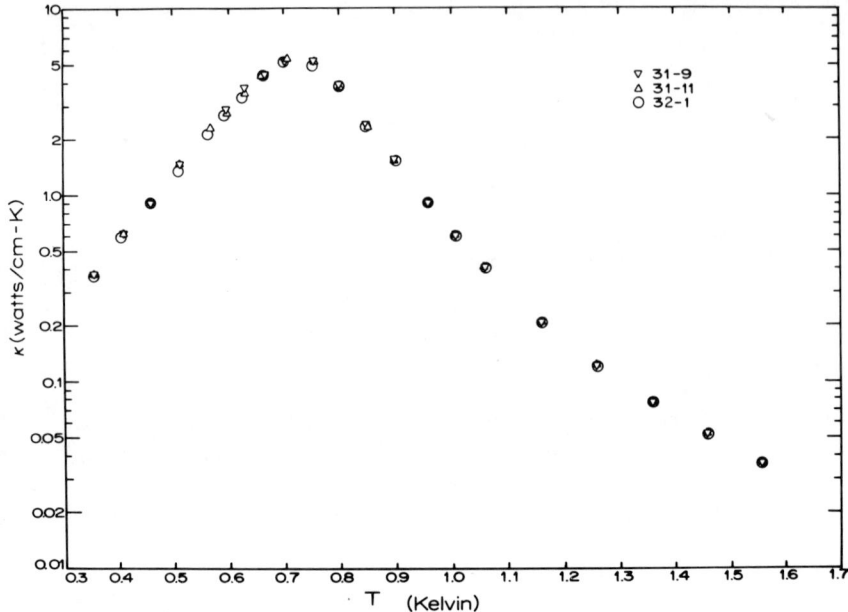

Fig. 1. Reproducibility of data. Thermal conductivity parallel to the c axis vs. temperature for three separately grown crystals of pure ^{4}He.

of thermal conductivity parallel to the c-axis as a function of temperature for three separately grown samples of pure ^{4}He. Each of our curves is reproducible within a few percent over the entire temperature range shown.

These measurements were extremely sensitive to phonon scattering by crystal defects—an isotope impurity concentration of 10^{-5} decreased the peak conductivity perpendicular to the c-axis by a factor of 1.9. The scattering rates for point defects and, coincidentally, spherically symmetric strain about a point defect,[3] have an ω^4 dependence on phonon frequency, which appears as a T^4 dependence in our data. Thus for small enough concentrations only the very peak in conductivity is affected.

In Fig. 2 we show thermal conductivity data points as a function of temperature for pure ^{4}He and for ^{3}He concentrations of 1.0×10^{-5} and 1.5×10^{-5}, all perpendicular to the c axis. Note that these points coalesce in both the Umklapp-process-dominated and boundary-scattering-dominated regions. The uppermost curve is drawn through the experimental points for the pure sample, while the other two curves are obtained as follows: We express the conductivity κ in terms of Debye theory specific heat C, sound velocity v, and an effective phonon mean free path λ_e:

$$\kappa = \tfrac{1}{3} C v \lambda_e$$

We then assume a very simple model in which the effective mean free path $\lambda_e(x, T)$ for nonzero concentration x is obtained by inverse addition of the experimentally determined $\lambda_e(0, T)$ and a mean free path characteristic of the isotope scattering, $\lambda_I(x, T)$, which we require to vary inversely with concentration and with the fourth power of the temperature, assuming that the scattering follows a Rayleigh frequency

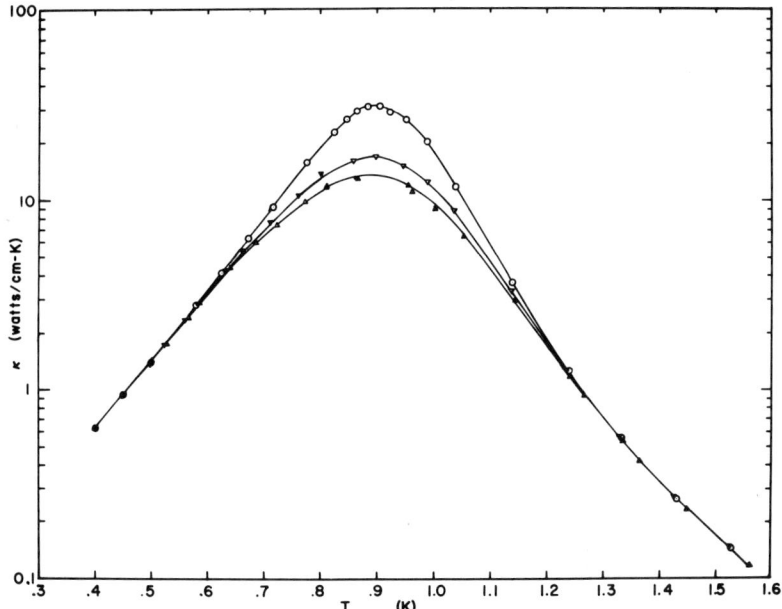

Fig. 2. Isotopic scattering data. Thermal conductivity perpendicular to the c axis vs. temperature for pure ^{4}He and for ^{3}He concentrations of 1.0×10^{-5} and 1.5×10^{-5}.

dependence:

$$\lambda_e^{-1}(x, T) = \lambda_e^{-1}(0, T) + \lambda_I^{-1}(x, T), \qquad \lambda_I(x, T) = \beta/xT^4$$

The constant $\beta = 1.17 \times 10^{-5}$ cm°K^4 was chosen to satisfy the $x = 1.0 \times 10^{-5}$ experimental data at 0.9°K and both lower curves were generated with these expressions. If β is plotted as a function of x and T rather than treated as a constant, we find the two curves for the data shown here differ by less than 10%. We thus obtain an isotope scattering strength directly from the experimental data without recourse to any intermediate parameterization of the pure ^{4}He curve. The conductivity data taken parallel to the c axis are less sensitive, but are consistent with the same value of β.

We compare our isotopic scattering result with a cross section derived from a calculation by Ziman[4] for the case of predominant normal processes with a small amount of mass defect scattering. Ziman's calculation yields an upper bound on scattering attributable to mass substitution alone. Our observed scattering strength is a factor of 2.7 larger than this upper bound.

The scaling of our isotopic scattering results with absolute isotope concentration and the reproducibility of our peak conductivities for growth rates differing by a factor of two indicate that point defects other than the ones we wished to study had an insignificant effect on our data. Other defects exhibit T^n, $n < 4$, dependences and could be detected quite sensitively since the observable changes in conductivity extended to lower temperatures. It was found that extraordinary care had to be taken in the growth and study of pure ^{4}He crystals perpendicular to the c axis because of the extreme sensitivity of the peak height to macroscopic strain in these samples.

In the phonon Poiseuille flow region—at temperatures just below that of the

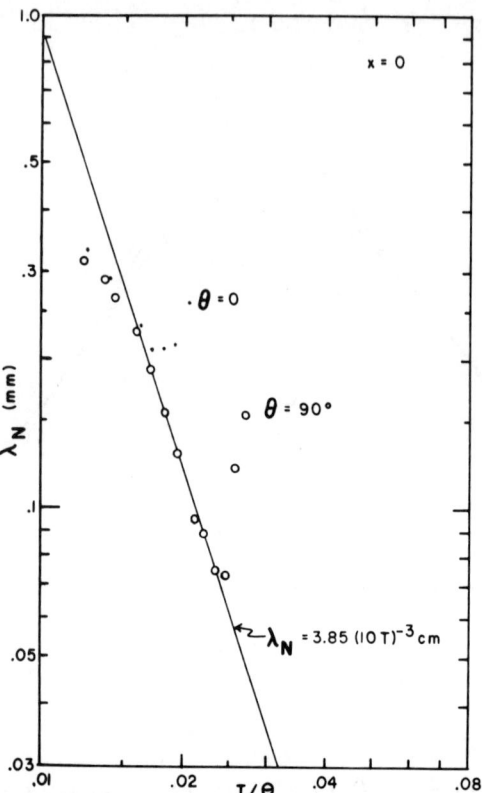

Fig. 3. Normal process mean free path. Experimental values of λ_N as a function of T/θ_D derived from thermal conductivity measurements made parallel and perpendicular to the c axis. The solid line indicates λ_N as determined from these data.

conductivity peak — the normal process mean free path λ_N can be determined directly from the thermal conductivity (see, e.g., Ref. 5):

$$\lambda_N(T) = \tfrac{5}{8} r^2 / \lambda_e(0, T)$$

where r is the sample radius. Our pure ^{4}He results include the observation that in this region increasing crystal quality simply increases that temperature interval over which $\kappa \propto T^6$, corresponding to $\lambda_N \propto T^{-3}$. There was no evidence of any change in the T dependence of the Poiseuille region as a function of crystal quality. For the pure ^{4}He curve shown here the T^6 dependence is observed over a temperature interval of almost 0.3°K.

In Fig. 3 we have plotted our value for λ_N, determined by using $\lambda_e(0, T)$ from the T^6 regions of our best-quality crystals, along with our data for the extreme c-axis orientations. Note that we observe no anisotropy in the normal process mean free path. It may be of interest to note that this temperature dependence implies a temperature-independent cross section for normal process collisions among the phonons.

Finally, we would like to note that it was possible to cause variations of 30% in the boundary-limited mean free path by varying the portion of the chemical impurities that were deposited on the sample chamber wall rather than elsewhere. We tentatively attribute this effect to interference with the onset of phonon specular reflection at the wall.

References

1. E.M. Hogan, R.A. Guyer, and H.A. Fairbank, *Phys. Rev.* **185**, 356 (1969).
2. C.S. Barrett, L. Meyer, and J. Wasserman, *J. Chem. Phys.* **45**, 834 (1966).
3. P. Carruthers, *Rev. Mod. Phys.* **33**, 92 (1961).
4. J.M. Ziman, *Can. J. Phys.* **34**, 1256 (1956); R. Berman, E.L. Foster, and J.M. Ziman *Proc. Roy. Soc.* **A237**, 344 (1956).
5. R.A. Guyer and J.A. Krumhansl, *Phys. Rev.* **148**, 778 (1966).

The Orientation and Molar Volume Dependence of Second Sound in HCP ^{4}He Crystals*

K. H. Mueller, Jr. and H. A. Fairbank

Physics Department, Duke University
Durham, North Carolina

While second sound has been shown to exist in NaF[1] and Bi,[2] solid helium[3-7] remains the only case where the phenomenon is well developed and free from resistive scattering. Furthermore, the elastic properties of helium are readily changed by large amounts with the application of modest pressures. Thus solid helium is ideally suited to test theoretical predictions for second sound in solids. The one major drawback is that sound velocities in solid helium are anisotropic and crystal orientation is not easily controlled.

In this work the propagation of heat pulses has been studied in hcp ^{4}He crystals grown at pressures of 25, 50, 80, and 130 atm. Pulse shapes were studied from the onset of second sound down to 80 m°K, where phonons cross the sample chamber ballistically. A boxcar integrator was used to improve signal-to-noise ratios and thereby make possible accurate measurements with small input pulses. Both carbon film and metal film transducers were used. The metal film detector was a superconducting cadmium film biased to the midpoint of its superconducting transition. At temperatures below about 200 m°K the carbon film has very poor response to the rapid temperature fluctuations in the crystal. The metal film transducers are far superior in the lower temperature range.

Since the mean free path for phonon–phonon interaction is strongly temperature dependent ($\propto T^{-3}$), at sufficiently low temperatures the phonons in a thermal pulse will cross the sample chamber ballistically. In this situation a thermal input pulse should be received as at least three separate pulses corresponding to the three acoustic phonon branches. Measurement of the velocities of these pulses permits the determination of crystal orientation since direct measurements of the first-sound velocity as a function of orientation have previously been made.[8,9]

Conventional first-sound velocity measurements and ballistic phonon (from a thermal pulse) velocity measurements do not measure the same quantity, but the relation connecting them is well known.[10] First-sound experiments measure a phase velocity (the component of the velocity with which the wavefront traverses the crystal in the direction normal to the transducer). Thermal pulse ballistic phonon experiments measure the wave velocity surface. This surface represents the expanding wavefront of a disturbance which excites all possible plane waves at the origin. In general, for an anistropic medium the path of the sound ray does not coincide with

---

* This work was supported by the Office of Naval Research and the National Science Foundation.

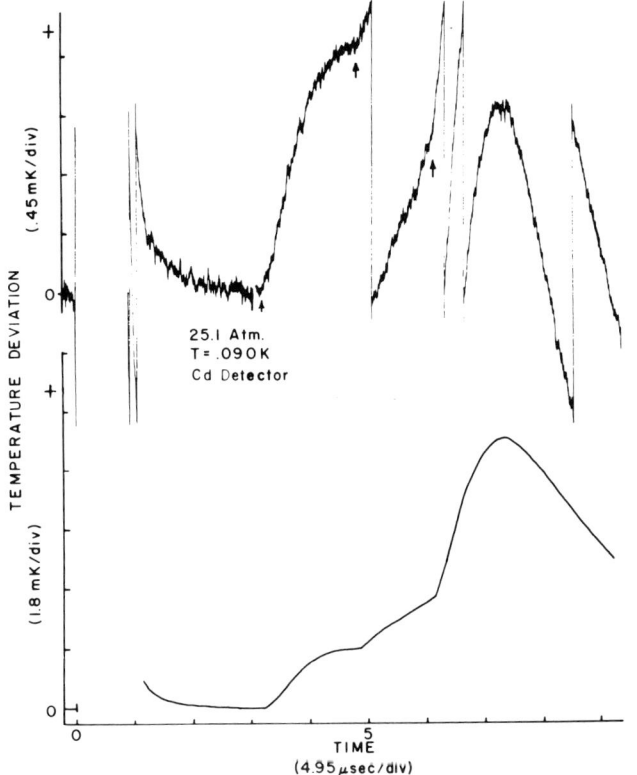

Fig. 1. Thermal pulse ballistic phonon data. The upper trace is an actual data recording (see text) while the lower trace is a reconstruction of the upper trace.

the normal to the wavefront and the phase velocity surface does not coincide with the wave velocity surface. It is, however, an easy matter to determine the wave surface from the experimentally measured phase velocities.

Figure 1 shows a typical thermal pulse at low temperatures where there is ballistic phonon propagation. Due to mechanisms which are not completely understood, the three pulses are not resolved, but one can clearly distinguish their arrival times. The upper trace is an actual recording of the data on an $x - y$ plotter using an automatic zero suppression device.[11] Whenever the recording starts to go off scale a dc voltage step is discontinuously applied to move the pen in the opposite direction by a fixed amount. The lower graph is a reconstructed plot of the upper trace with the amplitude reduced by a factor of four. In either case the three arrival times are quite distinct.

Using data such as this to determine a set of wave velocities for each crystal, one can determine the crystal orientation with modest accuracy ($\pm 5°$). Coupling this information with accurate measurements of the second-sound velocity (a time measurement made at higher temperatures between the peak of the first received pulse and the peak of the first echo), one can obtain the results shown in Fig. 2. Two curves are also shown which were calculated from first-sound velocity measurements.

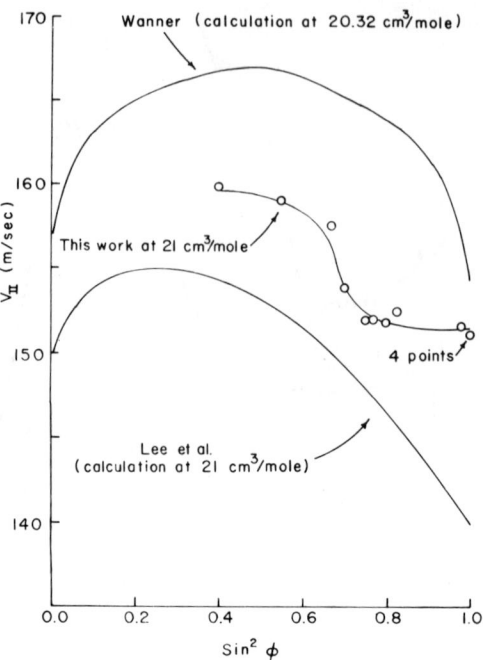

Fig. 2. Second-sound velocity in hcp ^{4}He as a function of orientation relative to the c axis. Here ϕ is the angle between the sound path and the c axis. The uncertainty in the data points is about twice their diameter.

The calculations used the expression for the second-sound velocity in an isotropic medium

$$V_{II}^2 = \frac{1}{3}\frac{V_{t1}^{-3} + V_{t2}^{-3} + V_{l}^{-3}}{V_{t1}^{-5} + V_{t2}^{-5} + V_{l}^{-5}} \tag{1}$$

and the first-sound velocities as a function of orientation relative to the c axis. In other words, a second-sound velocity was calculated in a particular direction, using the first-sound velocities in that same direction. This is clearly an oversimplification but it provides a reasonable estimate in trying to establish the range of the second-sound velocity variation with orientation.

In examining the molar volume dependence of the second-sound velocity, we have assumed a model in which the velocity of sound is independent of orientation. We calculate the sound velocities in an isotropic model which is "equivalent" to hcp helium (the isotropic model can be made to match either thermodynamic or elastic properties of hcp helium). By doing this we may use the simple formula (1) for the isotropic second-sound velocity. These computed results represent an "averaging" of the orientation-dependent second-sound velocity.

The longitudinal and transverse velocities of an isotropic equivalent crystal can be calculated from the bulk modulus K_B and the Debye temperature T_D using the following equations:

$$K_B = \rho(V_l^2 - \tfrac{4}{3}V_t^2) \tag{2}$$

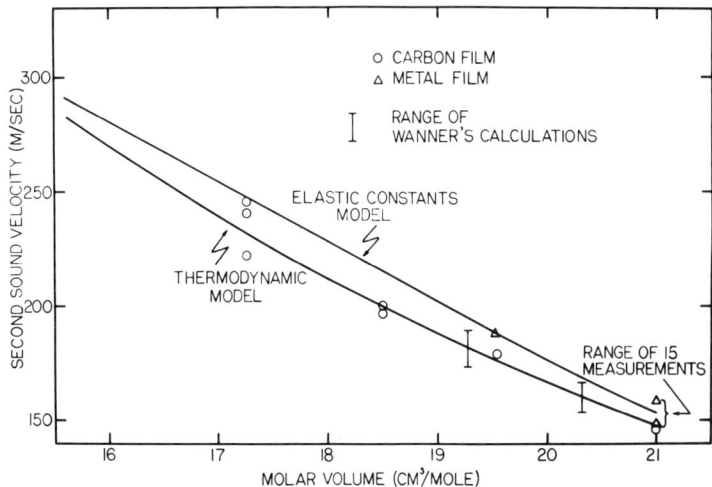

Fig. 3. Second-sound velocity in hcp ^{4}He as a function of molar volume. The solid lines are the results of calculations and the data points are labeled by the type of transducer used.

$$T_D = \frac{h}{k_B}\left(\frac{9N_0}{4\pi V_m}\right)^{1/3}(V_l^{-3} + 2V_t^{-3})^{-1/3} \qquad (3)$$

By using experimental values for the bulk modulus and Debye temperature, Wanner[12] has calculated V_l and V_t over a range of 16–21 cm^3/mole in molar volume.

Alternative values of V_l and V_t can be found from the elastic constants using the equations[13]

$$V_l^2 = (1/15\rho)(8C_{11} + 3C_{33} + 4C_{13} + 8C_{44}) \qquad (4)$$

$$V_t^2 = (1/15\rho)(C_{11} + 5C_{66} + 6C_{44} + C_{33} - 2C_{13}) \qquad (5)$$

The elastic constants of hcp ^{4}He have been determined from both neutron scattering measurements and first-sound measurements in oriented crystals. Using curves hand-fitted to the experimental values of the elastic constants vs. molar volume, one can obtain smooth curves for V_l and V_t vs. molar volume via Eqs. (4) and (5).

Values of V_{II} were calculated from Eq. (1), using the values of V_l and V_t calculated from Eqs. (2) and (3) or Eqs. (4) and (5), and compared as a function of molar volume in Fig. 3 with the directly measured values. The two computational models have a maximum difference of about 6% over this range of molar volume. Allowing for a spread in measured second-sound velocities due to different orientations, the experimental data are in good agreement with the calculated molar volume dependence. The metal film transducers could not be used for second-sound measurements at the highest pressures since the superconducting transition of the detector occurred at a temperature below that of pure second-sound propagation.

The results of this work clearly demonstrate an anisotropic second-sound velocity in hcp ^{4}He. In addition, they show that ballistic phonon data can be used to infer crystal orientation relative to the c axis. Isotropic models of the variation of second-sound velocity with molar volume are in good agreement with experiment over the range of our measurements.

References

1. T.F. McNelly, S.J. Rogers, D.J. Channin, R.J. Rollefson, W.M. Goubau, G.E. Schmidt, J.A. Krumhansl, and R.O. Pohl, *Phys. Rev. Lett.* **24**, 100 (1970); H.E. Jackson, C.T. Walker, and T.F. McNelly, *Phys. Rev. Lett.* **25**, 26 (1970).
2. V. Narayanamurti and R.C. Dynes, *Phys. Rev. Lett.* **28**, 1461 (1972).
3. C.C. Ackerman, B. Bertman, H.A. Fairbank, and R.A. Guyer, *Phys. Rev. Lett.* **16**, 789 (1966).
4. C.C. Ackerman and R.A. Guyer, *Ann. Phys. (N.Y.)* **50**, 128 (1968).
5. C.C. Ackerman and W.C. Overton, Jr., *Phys. Rev. Lett.* **22**, 764 (1969).
6. K.H. Mueller and H.A. Fairbank, in *Proc. 12th Intern. Conf. Low Temp. Phys., 1970,* Academic Press of Japan, Tokyo (1971), p. 135.
7. J.N. Fox, Thesis, Wesleyan University, 1971, unpublished.
8. R.H. Crepeau, O. Heybey, D.M. Lee, and S.A. Strauss, *Phys. Rev. A* **3**, 1162 (1971).
9. R. Wanner, Thesis, University of Alberta, 1970, unpublished.
10. R.H. Crepeau and D.M. Lee, *Phys. Rev. A* **6**, 516 (1972).
11. D. Trigg, *Rev. Sci. Instr.* **41**, 1298 (1970).
12. R. Wanner, private communication.
13. F.I. Fedorov, *Theory of Elastic Waves in Crystals,* Plenum Press, New York (1968).

Structure of Phase-Separated Solid Helium Mixtures*

A. E. Burgess and M. J. Crooks

Department of Physics, University of British Columbia
Vancouver, Canada

In a previous paper[1] the authors reported thermal conductivity measurements on solid helium mixtures in the temperature range between 0.15 and 0.6°K. The mixtures had ^{3}He concentrations $x_3 = 0.10$ and 0.90. The results suggested that below the phase separation temperature T_{ps} the thermal resistance was due mainly to boundary scattering of phonons by grains rich in the less abundant isotope embedded in a matrix rich in the more abundant isotope. The work described in this paper was undertaken to determine the temperature of the bcc to hcp transformation in the matrix and the change in conductivity when this transformation takes place. Preliminary kinetics of growth measurements are also presented.

It was found in the previous work that the phonon mean free path was temperature independent below T_{ps}. The phonon free path λ was calculated using the kinetic theory approximation $K = Cv\lambda/3$, where K is the thermal conductivity, C is the matrix phonon specific heat, and v is the velocity of sound in the matrix. The average grain radius a was then calculated using a simple model of phonon scattering by spherical grains that were large compared to the dominant phonon wavelengths. The average grain radius is given by the equation $a = 3\lambda ps/4$, where p is the fraction of the solid volume occupied by the grains and the scattering parameter s is the ratio of the phonon scattering cross section and the geometric cross section of the grains. The scattering parameter was estimated to be of order unity and the average grain radius was of the order of 1 μm. When the 10% ^{3}He phase-separated sample was warmed above T_{ps} and sufficient time had elapsed for the sample concentration to become isotropic, a boundary-scattering-like thermal conductivity was also observed; this was interpreted as being due to phonon scattering by bcc grains embedded in the hcp matrix. The mean free path for phonon scattering by these crystallographic "ghosts" of the phase-separated grains was not significantly different from the mean free path for scattering by the grains themselves. This result suggested that the phonon scattering was not mainly due to the large density difference between the grains and the matrix. We inferred from this and from the kinetics of growth that the grains scattered phonons like hard spheres.

The present conductivity experiments were done using samples of concentration $x_3 = 0.15$ and 0.21.

Homogeneous samples were cooled rapidly to a temperature between T_{ps} and 0.31°K and grains were grown isothermally in times typically between 10 and 60 min,

---

* Research supported by the National Research Council and the Unemployment Insurance Commission of Canada.

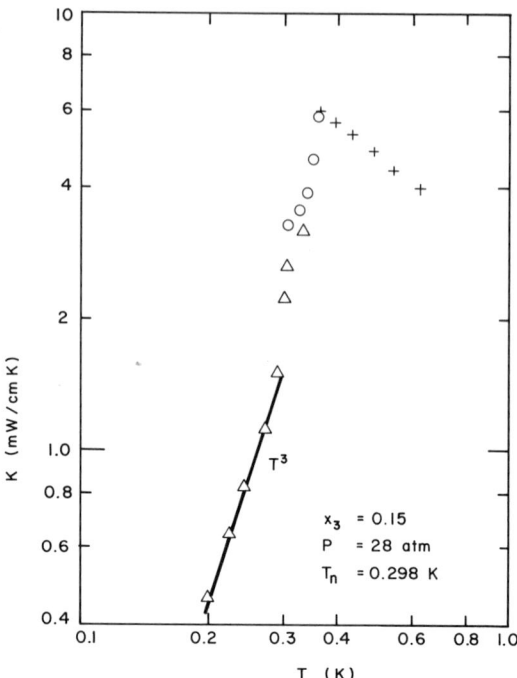

Fig. 1. The thermal conductivity of a 15% ^{3}He sample at 28.0 atm. The conductivity (crosses) was initially measured above T_{ps}. Grains were grown isothermally at 0.298°K and the conductivity (circles) was measured at increasing temperatures between 0.30 and 0.36°K. The sample was annealed, grains were once again grown at 0.298°K, and the conductivity (triangles) was measured at decreasing temperatures from 0.33 to 0.20°K. The matrix crystal structure is bcc above 0.31 and hcp below 0.29°K.

the longer times corresponding to higher temperatures. The conductivity was measured between 0.31°K and T_{ps}; then a series of measurements was made between 0.31 and 0.29°K, where a rapid decrease in conductivity took place, and finally measurements at lower temperatures were made. While making measurements above 0.31 and below 0.29°K we waited until the temperature gradient was constant to within 0.1 %/hr. Typical results are shown in Fig. 1. The phase separation takes place in the vicinity of 0.35°K (but T_{ps} cannot be accurately determined from these measurements). Extremely long relaxation times (in excess of 1 hr) for the conductivity measurements in the vicinity of 0.30°K, combined with the abrupt decrease in conductivity, suggest that a phase change is occurring in the sample. We conclude that a bcc–hcp crystallographic transformation is taking place in the matrix and that the matrix has a bcc crystal structure (corresponding to pure ^{3}He at low temperature and pressure) above 0.31°K and an hcp crystal structure (corresponding to pure ^{4}He at low temperature and pressure) below 0.29°K. The ratios of the bcc and hcp results extrapolated to 0.30°K and the extrapolated results for several samples are shown in Table I.

Table I

x_3	P, atm	$K(0.3°K)_{bcc}$, mW/cm·K	$K(0.3°K)_{hcp}$, mW/cm·K	K_{bcc}/K_{hcp}	$\lambda_{bcc}/\lambda_{hcp}$
0.21	28.5	2.7	1.5	1.8	1.25
0.15	28.0	2.5	1.5	1.6	1.1

The mean free path ratio was estimated by assuming Debye behavior for the solids and using 1.2 (from specific heat measurements of hcp and bcc ^{3}He) for the ratio of the Debye temperatures of the hcp and bcc matrices. The fraction of the sample volume in the grains and the average grain radius are expected to remain nearly constant during the transformation; therefore we conclude that the scattering parameter is nearly independent of matrix crystal structure. When this result is combined with the previous results for scattering by grains and "ghosts" in the 10% ^{3}He samples and the previously estimated scattering parameter one is lead to the following conclusions.

1. The scattering is neither solely due to the difference between the acoustical impedances of the grains and the matrix, nor solely due to differences in the grain and matrix crystal structures. The phonon scattering is possibly due to strain fields in the vicinity of the grains.

2. The grains act as hard-sphere phonon scatters.

3. All grain dimensions are large compared to the dominant phonon wavelengths.

We regard 0.31°K as a reproducible lower limit for the onset of the bcc–hcp transformation. It is possible that supercooling is present. Long-lived crystallographic "ghosts" were observed in a sample having $x_3 = 0.15$ and $P = 34.5$ atm, while "ghosts" were short lived in a sample having $x_3 = 0.15$ and $P = 28.0$ atm.

The grain size was found to depend on the temperature T_n at which isothermal growth took place. Results for samples at $x_3 = 0.15$ and $P = 28.0$ atm are shown in Table II. The three high-temperature growths were done on the same crystal, while the 0.259°K growth was done on a second crystal. The average grain radius is calculated assuming a pure ^{4}He matrix and a scattering parameter equal to one.

Preliminary isothermal measurements of the kinetics of grain growth have been performed. When grain growth takes place the sample pressure changes with time and one can determine the fraction of the sample that has been transformed, ζ, as a function of time. The sample pressure measurements were done using a strain gauge of 10^{-3} atm sensitivity. The results can be qualitatively interpreted using the Avrami equation[2] $\zeta = 1 - \exp(-Kt^n)$, where K is related to the diffusion coefficient of the

Table II

T_n, K	a, μm
0.298	1.9
0.298	1.7
0.284	1.2
0.258	0.9

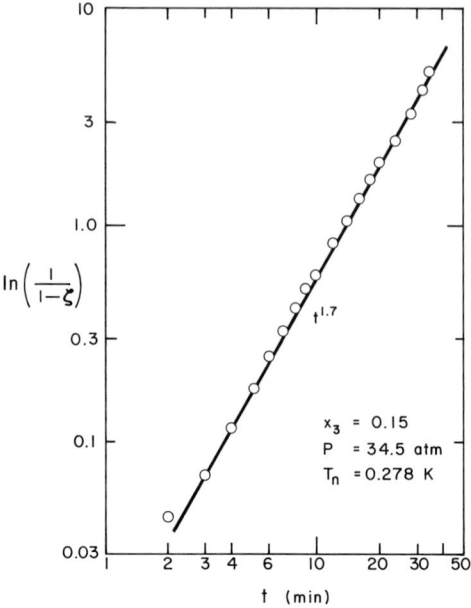

Fig. 2. The time dependence of the fraction of grain growth ζ.

matrix and the number of domains per unit volume, and the exponent n is determined by the nature of the transformation. Typical results are shown in Fig. 2. Transformations which are limited by the rate of boundary motion (e.g., crystals grown from supersaturated solution) typically have values of n between 3 and 4. Transformations in which diffusion of solute to a fixed number of sites is the limiting factor typically have values of n of 1.5 or less. In the case of phase separation in very dilute mixtures with a concentration-independent diffusion coefficient one would expect $n \simeq 1.5$ for ellipsoidal grains, $n \simeq 1.0$ for long, needlelike grains, and $n \simeq 0.5$ for platelike grains. For nondilute mixtures one would expect n to lie between 1 and 1.5 for ellipsoidal grains. The results of our measurements are given in Table III.

Table III

x_3	P, atm	T_n, °K	n	K, (min)$^{-n}$
0.15	34.5	0.278	1.7	1.2×10^{-3}
	28.0	0.258	1.3	6.5×10^{-3}
	28.0	0.259	1.7	3.5×10^{-3}
0.50	33.0	0.344	1.15	1.5×10^{-3}
	29.5	0.350	1.22	3.4×10^{-3}

It is not possible to reach an unambiguous conclusion without additional information. The results suggest that the grains are ellipsoidal in shape and grow in a diffusion-limited manner from a time-independent number of nucleation sites which are quickly saturated.

References

1. A.E. Burgess and M.J. Crooks, *Phys. Lett.* **39A**, 183 (1972).
2. J.W. Christian, *The Theory of Transformations in Metals and Alloys*, Pergamon Press, Oxford (1968).

Raman Scattering from Condensed Phases of ^{3}He and ^{4}He

C. M. Surko and R. E. Slusher

Bell Laboratories
Murray Hill, New Jersey

Inelastic light scattering has recently been used as a probe of the density fluctuations in both the liquid and solid phases of ^{4}He and ^{3}He.[1,2] In this paper we report the extension of these scattering observations to several new regions of pressure and temperature. First, normal liquid ^{4}He was studied at 3°K as a function of pressure up to the pressure of solidification. Second, the broad peak previously observed[2] in the spectrum of both liquids and solids was studied in ^{4}He as a function of temperature at constant density up to the Debye temperature of solid ^{4}He of the same density. Third, the scattered spectrum of the He II, including the two-roton peak, was investigated at 1.4°K as a function of pressure up to the pressure of solidification. Fourth, the spectrum of liquid ^{3}He, including the broad peak reported previously at higher temperatures,[2] was studied at 1.3°K as a function of pressure up to 13 atm. Finally, solid ^{4}He was investigated to temperatures as low as 1.3°K. This report includes the observed spectra and brief remarks on their interpretation. A more complete discussion of these results will be published elsewhere.

The experimental apparatus was similar to that used in previous experiments.[2] An Ar laser was used to produce from 100 to 300 mW of 5145-Å radiation in a focal volume 20 μm in diameter and 2 mm long. Laser powers greater than a few hundred milliwatts caused noticeable heating. At higher powers, heating at the cell windows produced large optical distortions in liquid ^{3}He at temperatures less than 1.5°K, where the thermal conductivity of the liquid is small. Better temperature control and lower temperatures were obtained using a Cu:Be pressure cell in place of the stainless steel cell used previously. The polarization of both incident and scattered light was distorted by the 0° sapphire windows used in these experiments,* so that an investigation of the polarization dependence of the scattering was not possible. Scattered light at 90° to the incident beam was collected with $f/4$ optics and analyzed with a double grating spectrometer. Stokes-scattered light was detected with a BX 754 channeltron photomultiplier tube and was digitally averaged with a small computer. Noise on the spectra shown in Figs. 1–3 is due to statistical fluctuations caused by the low counting rates.

A broad peak was observed previously[2] in the scattered spectrum of liquid ^{3}He. This peak shifted to higher frequencies with increased pressure and changed very

---

* To eliminate depolarization of the incident and scattered light by the windows, an attempt was made to use fused quartz in place of sapphire. The quartz was found to fracture at pressures greater than 50 atm and 0° sapphire was used for the cell windows for all experiments reported here.

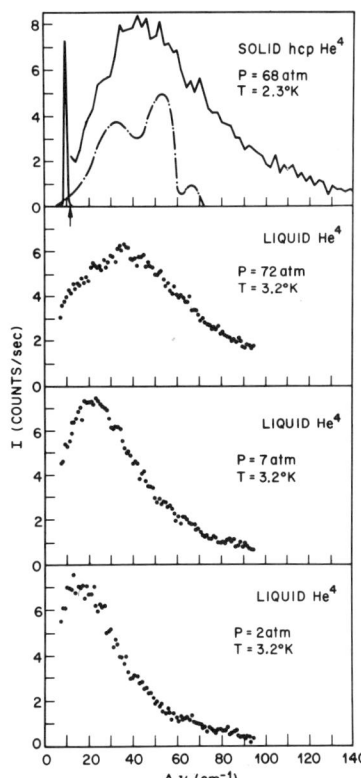

Fig. 1. The Raman-scattered intensity I of normal liquid ^{4}He as a function of frequency shift $\Delta \nu$ for several pressures up to the pressure of solidification. For comparison the previously published spectrum of hcp solid ^{4}He is shown (solid line), together with the predicted two-phonon spectrum for the hcp solid (dot-dash line); see Ref. 2 for details.

little upon solidification. As shown in Fig. 1, a similar peak has now been observed in normal liquid ^{4}He(He I). At low pressures these results agree with the spectra observed in He I by Pike and Vaughan.[3] It has been suggested that this peak is due to scattering due to excitations ("maxons") located near the maximum in the elementary excitation spectrum (the analog of scattering from the zone boundary phonons in the solid).[2,3] The fact that the peak shifts to higher frequencies with increased pressure is consistent with this conjecture. The large width of the peak is probably associated with damping of the elementary excitations and scattering involving creation of three or more excitations. This large width and the complexity of the possible scattering mechanisms make it difficult to obtain properties such as the "maxon–maxon" interaction from such data.

As noted previously,[2,3] the broad peak in both liquid ^{3}He and in normal liquid ^{4}He is not observed in simple classical liquids such as Ar and Kr where the scattered intensity decreases monotonically with increasing frequency shift. A major difference between He and other fluids is that it can be cooled far below the Debye temperature of the associated solid ($\sim 20°$K). It is of interest to compare the scattering spectra of liquid He near its "effective Debye temperature" to similar results for Ar and Kr. As shown in Fig. 2, the ^{4}He peak broadens near 20°K as compared to the results at the same density and lower temperatures. The intensity of the spectra at large frequency shifts $\Delta \nu$ is nearly an exponential function of $\Delta \nu$ as is also observed in liquid Ar and

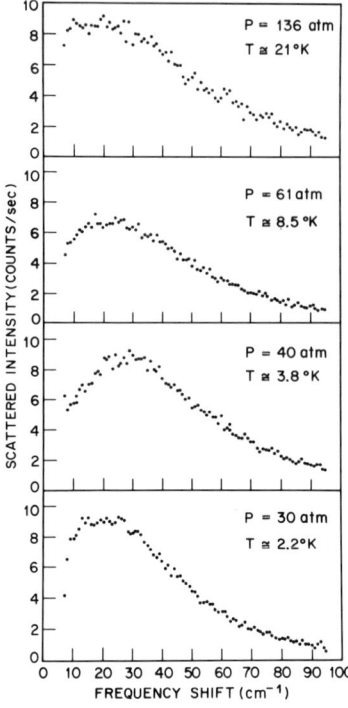

Fig. 2. The Raman spectrum of fluid ^{4}He at constant density at several temperatures between 3.2 and 21°K.

Kr. The difference in the spectra shown in Fig. 2 between 3.8 and 2.2°K for frequency shifts less than 20 cm^{-1} may be due to the onset of two-roton scattering.

The pressure dependence of the scattered spectrum from He II at 1.4°K is shown in Fig. 3. Qualitatively the most significant changes in the spectra occur in the roton portion of the spectrum (i.e., below 18 cm^{-1}), where the two-roton scattering peak decreases in intensity, broadens, and moves to lower frequency as the pressure is increased. Since the instrumental resolution is a major contribution to the observed width, we are unable to quote peak widths and shifts in peak position with any accuracy. However, the dominant contribution to the broadening appears to be due to the fact that the roton energy gap Δ is a decreasing function of the pressure P, and at a fixed temperature the rotons become more heavily damped due to roton–roton interactions. The largest contribution to Γ, the roton linewidth, is the Boltzmann factor $\exp[-\Delta(P)/T]$ and the pressure dependence of this factor accounts quite well for the increased damping which we observe.[4] This was verified by comparing the spectrum measured at an elevated temperature of 1.7°K and 1.7 atm with the spectrum measured at 1.4°K and 24.3 atm. Previous studies of the roton linewidth indicate that it should be the same at these two values of temperature and pressure.[4] We find that the spectra agree quite well when one shifts the curves along the frequency axis to account for the pressure dependence of Δ. If the two–roton interaction were to change sign as pressure is increased, one might expect an enhancement of the scattering due to excitations near the maximum in the elementary excitation spectrum. No such effect was apparent.

Two other changes in the spectra shown in Fig. 3 are apparent as the pressure is

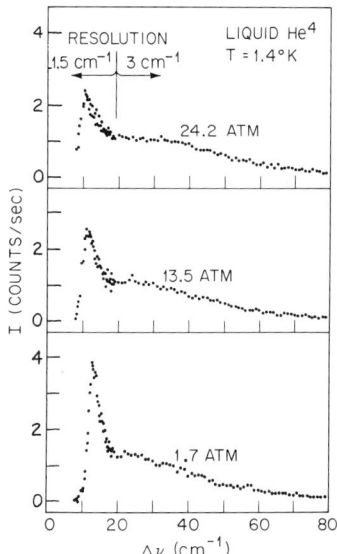

Fig. 3. The Raman spectrum of superfluid ^{4}He at several pressures.

varied. The slight peak in the spectrum near 24 cm^{-1} decreases as the pressure is increased, and the intensity of the spectrum at large frequency shifts (i.e., $\Delta v \geq 40$ cm^{-1}) increases slightly as the pressure is increased.

As shown in Fig. 3, there is considerable spectral weight at frequency shifts $\Delta v \geq 20$ cm^{-1}. We now comment on the possible origin of this part of the spectrum. The simplest scattering process which contributes to the Stokes-shifted Raman spectra in He II is the creation of two excitations with nearly equal and opposite wave vector. Neutron scattering results[5] indicate that the excitation spectrum of He II has a single-phonon part with phonons of energies up to 13 cm^{-1} and a multiphonon part which turns into a free-particle-like energy–momentum spectrum at high energies. There is very little weight in the single-phonon branch at energies greater than 11 cm^{-1} so that if the small peak at 24 cm^{-1} is due to the creation of two excitations from the upper end of the single phonon branch as has been suggested previously,[1] then this peak is on top of very large (and presumably structureless) background. The spectral weight between 20 and 40 cm^{-1} appears to be due to scattering involving both the single-phonon and multiphonon parts of the excitation spectrum. At larger frequency shifts ($\Delta v \geq 40$ cm^{-1}) the scattering can probably be interpreted as scattering from the single-particle-like multiphonon excitations. This would be the analog of "collision-induced scattering" observed in gases and in other rare gas liquids.

The Raman spectrum of liquid ^{3}He at pressures between 0 and 13 atm and at the significantly lower temperature of 1.3°K was measured, looking in particular for evidence of a "rotonlike" excitation. The spectra at 1.3°K were very similar to those measured previously[2] at 2.3°K, and no additional structure was observed.

The spectrum of solid hcp ^{4}He at 2.3°K was also published previously. We have now measured this spectrum at 1.3°K and 150 atm looking for possible structure in the spectrum.* None was observed, and the data at 1.3°K are very similar to those measured at higher temperatures.

The spectra of liquid ^{3}He and normal liquid ^{4}He as well as those of fluid ^{4}He at temperatures to 20°K have now been shown to have a broad peak, the position of which increases with increasing density. It seems likely that the peak in all of these spectra is due to scattering from more or less well-defined excitations near a maximum in the elementary excitation spectrum which should occur at a wave vector of approximately π/a, where a is the average interparticle spacing. The general nature of such short-wavelength excitations and in particular their existence in ^{3}He and normal liquid ^{4}He was predicted in 1965 by Pines.[7]

One feature has been found to be common to the Raman spectra of all phases of ^{3}He and ^{4}He, liquid, solid, and gas: The intensity at large frequency shifts Δv depends exponentially on Δv.[2,3] This dependence is characteristic of collision-induced scattering (CIS). At large Δv, CIS arises from the short-range interaction between atoms (see, e.g., Ref. 8) and it is therefore not surprising that the spectra of all phases display this characteristic spectral distribution.

Acknowledgments

We wish to thank D. Pines and N. R. Werthamer for helpful conversations.

References

1. T.J. Greytak and J. Yan, *Phys. Rev. Lett.* **22**, 987 (1969).
2. R.E. Slusher and C.M. Surko, *Phys. Rev. Lett.* **27**, 1699 (1971).
3. E.R. Pike and J.M. Vaughan, *J. Phys.* **4C**, L362 (1971).
4. O.W. Dietrich, E.H. Graf, C.H. Huang, and L. Passel, *Phys. Rev.* **5A**, 1377 (1972).
5. R.A. Cowley and A.D.B. Woods, *Can. J. Phys.* **49**, 177 (1971).
6. N.R. Werthamer, R.L. Gray, and T.R. Koehler, *Phys. Rev.* **4B**, 1324 (1971).
7. D. Pines, in *Quantum Liquids*, D.F. Brewer, ed., North-Holland, Amsterdam (1966), pp. 257–266.
8. J.I. Gersten, R.E. Slusher, and C.M. Surko, *Phys. Rev. Lett.* **25**, 1739 (1970), and references cited therein.

* The two-phonon Raman spectrum is calculated in Ref. 6.

Single-Particle Excitations in Solid Helium*

T. A. Kitchens and G. Shirane

Brookhaven National Laboratory
Upton, New York

V. J. Minkiewicz[†]

University of Maryland
College Park, Maryland

and
E. B. Osgood

Stevens Institute of Technology
Hoboken, New Jersey

During an extension of research on the inelastic neutron scattering from solid ^{4}He,[1,2] some observations of scattered neutron groups from the low-density phases of ^{4}He in the high-energy, high-momentum transfer (high-ω, high-Q) regime have been made. The observations reported here indicate that in yet another facet the solid phases of helium are liquidlike: In the low-density solid phases of helium, single-particle excitations appear to lie at energies as low as they do in the liquid. To our knowledge these observations are the first of single-particle excitations in any solid produced with thermal neutrons.

Most low-temperature physicists are familiar with the basis of the measurements, since the technique is nearly the same as that used in measuring the fraction of ^{4}He atoms in the zero momentum state in superfluid helium.[3] For large enough momentum transfer $\hbar Q$, such that scattering atoms may be considered to act independently, the scattering law for the energy transfer $\hbar\omega$ can be written as

$$S(Q, \omega) = \frac{\hbar}{N} \sum_{\mathbf{p}} n(\mathbf{p}) \delta\left(\hbar\omega - \frac{\hbar^2 Q^2}{2M_4} - \frac{\hbar^2 \mathbf{Q} \cdot \mathbf{p}}{M_4}\right) \quad (1)$$

where $n(\mathbf{p})$ is the number of particles with an initial momentum $\hbar\mathbf{p}$, and the delta function is due to the conservation of momentum and energy. This "Doppler approximation" was discussed earlier by Sjölander,[4] who found that the full-width is roughly 2.7 times the geometric mean of the mean kinetic energy of the atoms E_k and the mean recoil energy transferred to the atoms, $R = \hbar^2 Q^2 / 2M_4$.[‡]

As Q is reduced from large values, the scattering can no longer be characterized

---

* Work performed under the auspices of the U. S. Atomic Energy Commission.
† Supported in part by the Advanced Research Project Agency.
‡ Other work on multiphonon scattering is summarized in Ref. 5.

as single-particle scattering and is better considered as multiphonon scattering.[4] The difficulty in making the largely semantic distinction between multiphonon and single-particle scattering is that the average energy of the observed scattering should be exactly R for either case. This is due to the first moment sum rule,

$$\hbar^2 \int S(Q, \omega)\, \omega\, d\omega = R \qquad (2)$$

calculated first by Placzek,[6] and demonstrated for liquid ^{4}He by Cowley and Woods.[7] Therefore any distinction between multiphonon scattering and single-particle scattering must rely on the line shape. From Eq. (1) it is expected that the line shape of the single-particle scattering would be symmetric about R for any reasonable $n(\mathbf{p})$. The n-phonon scattering, on the other hand, depends on the nth convolution of the phonon density of states and thermal weighting factors. Only for quite low n does any of the residual character of the density of states appear in the n-phonon scattering; generally, the multiphonon scattering is quite diffuse, but it is not symmetric about R for low Q.[4] Cowley and Woods[7] have argued that for $Q > 3\text{Å}^{-1}$ the neutron groups are of single-particle scattering nature. Our measurements, taken in the same momentum transfer range used by Cowley and Woods, show that the characteristics of scattering from the low-density solid phases of ^{4}He are the same as those of the scattering from the liquid phase.

To our knowledge, single-particle-like dispersion, that is, a dispersion relation of the form $\hbar\omega = \hbar^2 Q^2/2M$, has not been observed for neutron scattering from a solid. This can be understood from the fact that most solids have much larger Debye–Waller factors $\exp(-2W) \equiv \exp[-2B(Q/4\pi)^2]$ (i.e., much smaller B) and binding energies than helium. Ambegaokar et al.[8] have shown that, even accounting for interference effects from multiphonon scattering, the first moment sum rule applied to the dynamic structure factor for *single*-phonon scattering, $S_1(Q, \omega)$, will be a fraction of the total Placzek sum rule, the fraction being equal to the Debye–Waller factor $\exp(-2W)$. Thus $\exp(-2W)$ must be small before much scattered intensity is available for the multiphonon or single-particle scattering. An uncertainty principle argument shows that the width of the observed neutron groups is consistent with the single particle state existing only until collision with a near neighbor.

This investigation was carried out on the Brookhaven High Flux Beam Reactor using neutrons with an initial energy of 44 meV on a three-crystal spectrometer operated in the constant-Q mode. The neutron beam resolution was 2 meV or better in these measurements and no corrections for resolution have been applied in the data shown here; the widths of the neutron groups are nearly an order of magnitude larger than the resolution width. Such corrections, if applied to the data, would tend to make the line shapes even more asymmetric but would leave the peak positions unaltered.

The bcc helium crystal was grown, as described before, in the cryostat used in our earlier investigations.[1,2] The hcp helium sample was grown "through" the bcc phase by holding the cell pressure at 27.7 atm, establishing some liquid at the top of the cell, and cooling the bottom of the cell to $\sim 1.56°$K, well below the hcp–bcc transition. When the cell top cooled from the bcc–liquid transition temperature the mass input indicated a molar volume of 20.8 cm^3. This molar volume is outside

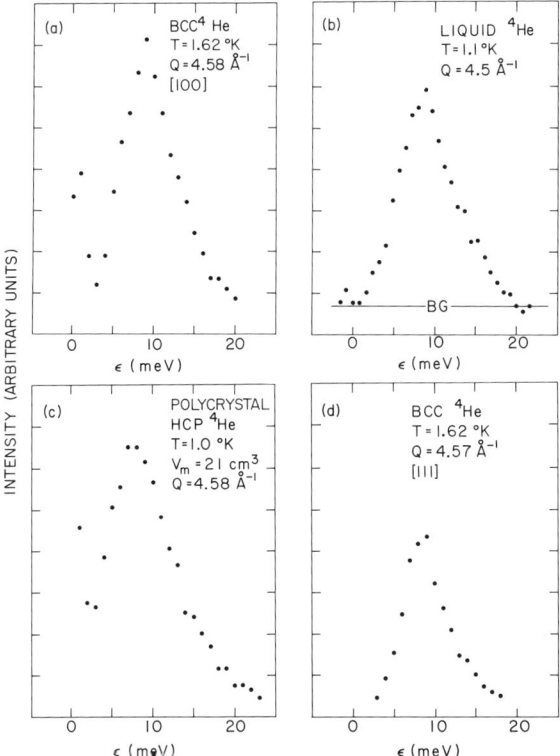

Fig. 1. Comparison of scattered neutron groups from the liquid and the low-density solid phases of ^{4}He. The data for the solid phases were obtained using a three-crystal spectrometer with 44-meV incident neutrons from the Brookhaven High Flux Beam Reactor. The background is negligible. The counting time in (d) was one-half that used for (a) and (c). The data for the liquid phase were taken from the work of Cowley and Woods.[7] The liquid was at low pressure and at low temperature, 1.1°K, and the data were taken on a three-crystal spectrometer with a final neutron energy of 62 meV.

the bcc phase, assuring us of an hcp sample (see, e.g., Ref. 9). Bragg scattering verified the phase to be hcp and that the sample was comprised of several crystallites, but not many crystallites.

Some typical neutron groups scattered from these samples are shown in Fig. 1. We have chosen to present data near $Q = 4.5$ Å^{-1} to facilitate comparison with the liquid scattering data collected in a similar way by Cowley and Woods.[7] These data are shown in part (b) of this figure. Quite clearly, the characteristics of the scattering observed in the liquid and in the solid phases are nearly identical. The figure also clearly demonstrates that the scattering appears to be the same for the bcc phase in the [100] and [111] directions. Measurements for other directions confirm that the scattering is isotropic. Data for the hcp sample composed of several crystals were also isotropic and, as seen from Fig. 1, of the same character as the bcc and the liquid data for the same momentum transfer Q.

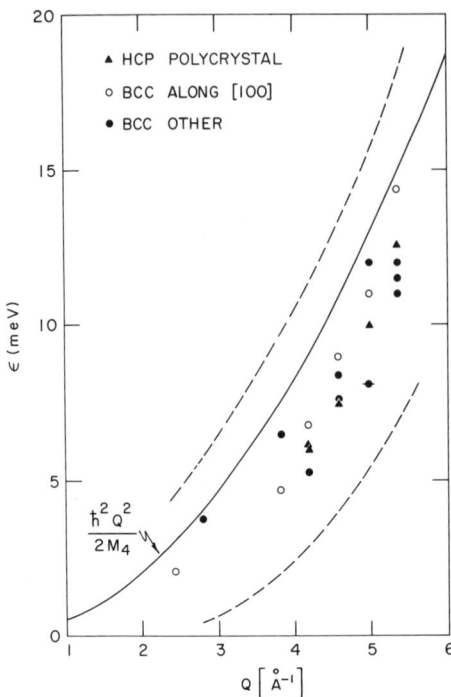

Fig. 2. The energy vs. transferred momentum relationship for the neutron groups scattered from the low-density solid phases of ^{4}He. The solid line corresponds to the expected relationship for single-particle scattering. The dashed lines show the approximate position of the intensities equal to one-half the peak of the observed neutron groups.

The dispersion relation for the data collected for both phases is shown in Fig. 2 along with the expected single-particle relation $R = \hbar^2 Q^2/2M_4$. The dashed lines are the approximate energies where the observed neutron groups are one-half as intense as they are at peak intensity. The effective mass for the least-squares fit to the peak intensities is $M^* = 1.27 M_4$. However, such deviations from the single-particle relation are quite consistent with the liquid data.

The data taken for $Q < 3$ Å^{-1} show weak single-phonon peaks, since the Debye–Waller factor $\exp(-2W) < 0.10$ for $Q > 2.4$. These groups also have more structure than those for $Q > 3$ Å^{-1} as might be expected from low-order multiphonon scattering. For $Q \gtrsim 3$ Å^{-1} the neutron groups have the same character as the single-particle-like groups shown in Fig. 1 and lie as low as 5–7 meV, about 1.5 times the energy of the most energetic phonon seen in the bcc phase.[1] The Doppler approximation width is somewhat greater than that observed. Although these preliminary data were not extensive enough to check the sum rule of Eq. (2), the integrations do suggest that the sum rule is complete; that is, there are no substantial contributions to the groups outside the distribution characterized in Fig. 1. These results also suggest that previous experimental results on the phonon spectra for hcp ^{4}He for medium and low densities[10] should be reexamined for effects of single-particle scattering.

In summary, we have presented yet more evidence of the similarity of the low-density solid phases of ^{4}He with the liquid phase.[11] The scattering from the hcp and bcc phases in the high-energy, high-momentum transfer regime is isotropic and independent of crystallographic phase and is nearly identical to that in the liquid. For $Q > 3$ Å^{-1} the scattering appears to be consistent with that from single-particle excitations. This work constitutes the first observation of such scattering from a solid phase.

Acknowledgments

We are grateful to M. Rosso and T. Oversluizen for invaluable technical assistance, and we acknowledge many beneficial discussions with our colleagues, in particular, J. D. Axe, M. Blume, R. A. Guyer, N. R. Werthamer, V. Korenman, and R. Prange.

References

1. E.B. Osgood, V.J. Minkiewicz, T.A. Kitchens, and G. Shirane, *Phys. Rev. A* **5**, 1537 (1972).
2. V.J. Minkiewicz, E.B. Osgood, G. Shirane, and T.A. Kitchens, (to be published).
3. P.C. Hohenberg and P.M. Platzman, *Phys. Rev.* **152**, 198 (1966).
4. A. Sjölander, *Ark. Fys.* **14**, 315 (1958).
5. W. Marshall and S.W. Lovesey, *Theory of Thermal Neutron Scattering*, Oxford Univ. Press (1971).
6. G. Placzek, *Phys. Rev.* **86**, 377 (1952).
7. R.A. Cowley and A.D.B. Woods, *Can. J. Phys.* **49**, 177 (1971).
8. V. Ambegaokar, J.M. Conway, and G. Baym, *Proc. Intern. Conf. Lattice Dynamics, J. Phys. Chem. Sol.* (Suppl. 1), 261 (1965).
9. J.K. Hoffer, Ph.D. Thesis, Univ. of California, Berkeley, 1968, unpublished.
10. R.A. Reese, S.K. Sinha, T.O. Brun, and C.R. Tilford, *Phys. Rev. A* **3**, 1688 (1971); V.J. Minkiewicz, T.A. Kitchens, F.P. Lipschultz, and G. Shirane, *Phys. Rev.* **174**, 267 (1968).
11. R.E. Slusher and C.M. Surko, *Phys. Rev. Lett.* **27**, 1699 (1971); N.R. Werthamer, *Phys. Rev. Lett.* **28**, 1102 (1972).

Single-Particle Density and Debye–Waller Factor for BCC ^{4}He

A. K. McMahan and R. A. Guyer

University of Massachusetts
Amherst, Massachusetts

Recently Osgood *et al.*[1] have reported results of inelastic neutron scattering from the bcc phase of solid ^{4}He. After subtracting the multiphonon background from their measured intensities and using the ACB sum rule[2] they obtained an effective Debye–Waller factor which exhibited remarkable oscillatory properties, not at all in agreement with the Gaussian behavior expected. This puzzle has stimulated considerable interest.[3-5] Horner[4] has suggested that Osgood *et al.* have *not*, in fact, determined the Debye–Waller factor, at least as defined by ACB, i.e., the square of the Fourier-transformed single-particle density. The purpose of this note is to support this observation in two ways. First, we find that all our attempts to fit the data with reasonable analytic curves have led, on taking the square root and Fourier transforming, to unphysical single-particle densities which go negative at some point.* Second, we report a calculation of the single-particle density using the Jastrow–Gaussian wave function, which has previously proved quite adequate in the description of quantum crystals. The resultant single-particle density is essentially Gaussian, as is its Fourier transform. There are no oscillatory features. Similar results have been independently found by Horner[4] and Sears and Khanna.[5] Our calculation was performed with a rather simple combination Monte Carlo and cluster approximation technique, which we describe.

Two examples of the analytic curves we have fitted to the experimental results are shown in Fig. 1. This presentation of the data is taken from Ref. 3. The analytic curves are of the form $\exp[-(Q^2/2A) - \lambda\Delta(Q)]$, where $A = 1.23$ Å^{-2}. The oscillatory curve (solid line) is obtained taking $\Delta(Q) = Q \sin(2\pi Q/3)$. The possibility that the data might represent a Gaussian-like Debye–Waller factor $\exp(-2W)$ except for a bump around $Q = 2.2$ Å^{-1} is represented by the dotted curve, taking $\Delta(Q) = -\sin^2(2Q)$ for $1.50 < Q < 3.07$ Å^{-1} and zero elsewhere. The harmonic or Debye prediction $\Delta(Q) = 0$ is also shown (dashed line) for comparison. The corresponding single-particle densities $\rho(\mathbf{r})$ shown in Fig. 2 are computed assuming $\exp(-2W)$ to be isotropic. The reason the single-particle density goes negative is clear. At values of $r \sim 2$ Å the Fourier transform of the Gaussian $\exp(-W)$ is positive, though extremely small, and results from a delicate balance of large positive and negative contributions from the regions $Q \lesssim$ and $Q \gtrsim 1.5$ Å^{-1}, respectively. The non-Gaussian $\exp(-W)$ curves are obtained by augmenting the Gaussian in the latter

---
* The group at Brookhaven has found similar results.[6]

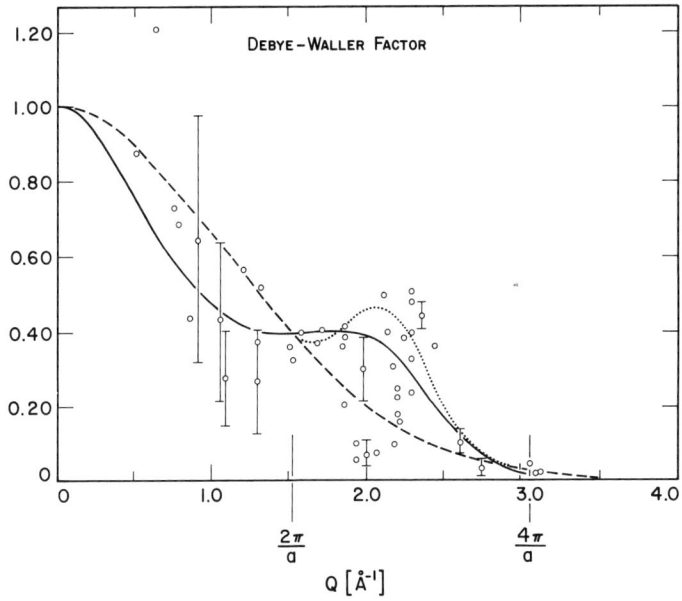

Fig. 1. Various analytic fits to the Debye–Waller factor data. The dashed line is the Gaussian appropriate for a harmonic solid. The solid line exhibits oscillatory behavior. The dotted curve is Gaussian except for the bump, as shown.

region, and perhaps depleting it in the former. Both effects shift this balance in the negative direction, and not much is needed to drive $\rho(\mathbf{r})$ negative. For $\exp(-2W)$ curves of the form we have considered the extent to which $\rho(\mathbf{r})$ goes negative decreases with decreasing λ. However, one does not get $\rho(\mathbf{r}) \geqq 0$ until λ is so small that these $\exp(-2W)$ curves would not be visually distinguishable from the Gaussian in Fig. 1. These comments are not conclusive by any means, although they do suggest that the data in Fig. 1 cannot correspond to the square of a Fourier-transformed single-particle density.

We have performed a calculation of the single-particle density taking a Jastrow–Gaussian wave function with the Nosanow[7] correlation function:

$$\psi = \psi_G \prod_{i<j} f(r_{ij}) \qquad (1)$$

$$\psi_G = \prod_i \phi_i(\mathbf{r}_i); \qquad \phi_i(\mathbf{r}_i) = \exp[-A(\mathbf{r}_i - \mathbf{R}_i)^2/2] \qquad (2)$$

$$f(r) = \exp\{-K[(\sigma/r)^{12} - (\sigma/r)^6]\}$$

$$K = 0.178, \qquad \sigma = 2.556 \text{ Å}$$

The calculation was performed for a density of $21.0 \text{ cm}^3/\text{mole}$, taking[1] $A = 1.23 \text{ Å}^{-2}$. The resultant $\rho(\mathbf{r})$ is found to be approximately Gaussian and mildly anisotropic. For $r \lesssim 2.5 \text{ Å}$ it is well described in the [100] and [011] directions and fairly well described (deviations $\lesssim 10\%$) in the [111] direction by the Gaussian parameters 1.49, 1.56, and 1.59 Å^{-2}, respectively. Beyond this region there are deviations from

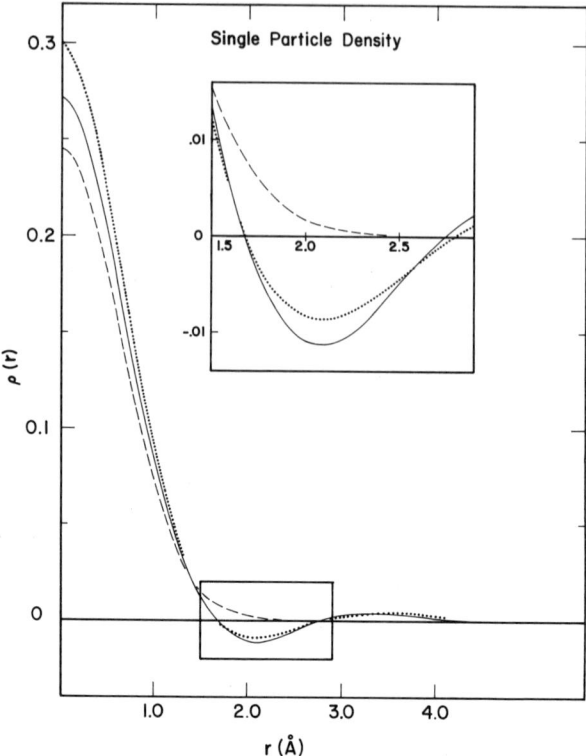

Fig. 2. The single-particle densities obtained from the three Debye–Waller factor curves in Fig. 1.

Gaussian behavior described by 10–20% changes in the appropriate Gaussian parameters. The major effect of the short-range correlations is to cause $\rho(\mathbf{r})$ to be a somewhat more compact Gaussian than the single-particle Gaussian used in ψ_G. Given the bcc geometry, it is quite understandable that $\rho(\mathbf{r})$ is most compact in the [111] and [011] directions, and least so in the [100] direction. An approximation to the Debye–Waller factor resulting from this $\rho(\mathbf{r})$ is shown (solid lines) in Fig. 3. These curves were calculated by treating $\rho(\mathbf{r})$ as spherically symmetric in performing the Fourier transformation, with the radial dependence taken to be that calculated for $\rho(\mathbf{r})$ in the various directions shown. In all three directions these $\exp(-2W)$ curves are seen to be Gaussian, to an excellent approximation, for $Q \lesssim 3.5\,\text{Å}^{-1}$.

In calculating the single-particle density, we have used n-body cluster approximations to the full integral

$$\rho_n(r_1) = \phi_1^2(r_1) \frac{\int d\mathbf{r}_2 \cdots d\mathbf{r}_n \prod_{i=2}^{n} \phi_i^2(\mathbf{r}_i)\, F_n^2}{\int \delta d\mathbf{r}_1 \cdots d\mathbf{r}_n \prod_{i=1}^{n} \phi_i^2(\mathbf{r}_i)\, F_n^2} \qquad (3)$$

$$F_n = \prod_{1 \leq i < j \leq n} f(r_{ij}) \qquad (4)$$

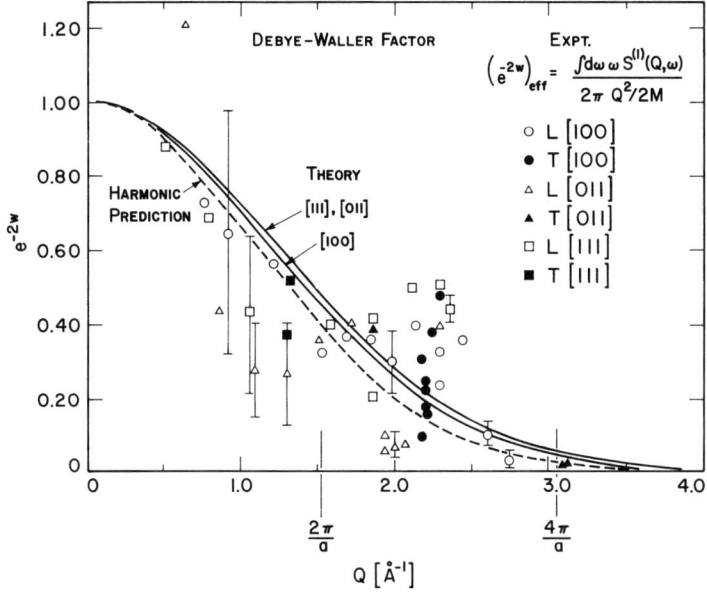

Fig. 3. Debye–Waller factor vs. momentum transfer Q. Comparison of the theoretical calculation (solid lines) to the experimental data. The dashed line gives the Debye–Waller factor for a harmonic solid.

where n specifies a group of atoms including atom 1 and all its first neighbors ($n = 9$), or, in addition, all second neighbors ($n = 15$), third neighbors ($n = 27$), and so on. Such approximations are motivated by the expectation that the short-range correlations involving atoms which are distant to atom 1 are relatively unimportant to the full integral, at least for reasonable deviations $\mathbf{r}_1 - \mathbf{R}_1$. One would expect rapid convergence of $\rho_n(\mathbf{r}_1)$ with n, and we find this to be the case. For example, in the [100] direction $\rho_{15}(\mathbf{r}_1) = \rho_{27}(\mathbf{r}_1)$ to within the 3% accuracy of our calculations for $r_1 \lesssim 2.0$ Å, in which region these functions fall off in size by nearly three orders of magnitude. We find differences between these two functions to be generally less than 25% beyond this region. We have used $n = 27, 27$, and 15 in obtaining the results shown in Fig. 3 for the [100], [011], and [111] directions, respectively.

The $3n$-dimensional integrals in both numerator and denominator of Eq. (3) were evaluated by the Monte Carlo integration technique. Each configuration of the n particles were generated according to the Gaussian part of the wave function by using $3n$ random numbers $w_{i\alpha}$, $0 \le w_{i\alpha} \le 1$, and interpolation of tabulated values for the function $g^{-1}(w)$:

$$r_{i\alpha} = g^{-1}(w_{i\alpha}) + R_{i\alpha} \qquad (5)$$

$$g(x) = (A/\pi)^{1/2} \int_{-\infty}^{x} dt \exp(-At^2) \qquad (6)$$

The denominator was determined by the average of F_n^2 over these configurations. For the numerator F_n^2 was averaged over this same set of configurations with the exception that r_1 was taken to be some prescribed value, the same for each configuration. Such averages were performed for 80 values of r_1 ranging from 0 to 8 Å. It is because we have generated the configurations by a function in which every particle is independent of every other particle that we may reuse the same set of configurations in this manner for each value of r_1 at which we wish to determine $\rho_n(\mathbf{r}_1)$. We generally found convergence by about 1000 configurations, and carried each calculation out to 2000 configurations. On the CDC 3800 computer and for $n = 27$ the running time was about 9 min per 1000 configurations. We tested the program for $n = 15$ by using the correlation function

$$f^{\text{test}}(r_{ij}) = \exp[-B_{ij}(\mathbf{r}_{ij} - \mathbf{R}_{ij})]$$

$$B_{ij} = \begin{cases} A/10 & i,j \text{ near neighbors} \\ 0 & \text{otherwise} \end{cases}$$

and taking $A = 1.23 \text{ Å}^{-2}$. Analytic evaluation of the integral gave $\rho_{15}(r) = c \exp(-Fr^2)$ with $F = 2.94 \text{ Å}^{-2}$. Monte Carlo evaluation in the [100] and [111] directions gave $F = 2.99 \pm 0.10$ and 2.90 ± 0.10, respectively. While the Monte Carlo evaluation with this test correlation function exhibited less obvious convergence and more scatter (10% as compared to 3%) than our calculation with the Nosanow correlation, we should point out that the ± 0.10 limits on the value of F represent some systematic deviation from the Gaussian behavior of the analytic result which is not within the 10% scatter. On the other hand, it is also apparent from the rather drastic effect of the test correlation functions on the single-particle density that the test calculation is numerically a good deal more difficult than the calculation with the Nosanow correlation functions.

We feel that the combination Monte Carlo and cluster approximation technique used in this work may be useful in the evaluation of other many-body integrals involving Jastrow–Gaussian wave functions. For example, exchange integrals and integrals of two-body potentials would also seem to be amenable to this technique. The recent criticism[8] of cluster approximations to such integrals is really directed at the two-body cluster approximations most often used. We suggest that cluster approximations which incorporate a more realistic number of neighboring atoms may be entirely adequate, and that Monte Carlo evaluation of such fairly low-dimensional integrals can be both relatively fast and accurate. We have in fact used this technique quite successfully in the evaluation of many-body exchange integrals (to be published). It should be noted, however, that in most of our calculations of this kind we have used the Nosanow correlation, a function which rises to unity well before the near-neighbor distance. In the case of correlation functions which level off well after the near-neighbor distance many of the advantages of this technique may be lost.

References

1. E.B. Osgood, V.J. Minkiewicz, T.A. Kitchens, and G. Shirane, *Phys. Rev. A* **5**, 1537 (1972).
2. V. Ambegaokar, J.M. Conway, and G. Baym, in *Lattice Dynamics*, R.F. Wallis, ed., Pergamon, London (1965).
3. N.R. Werthamer, *Phys. Rev. Lett.* **28**, 1102 (1972).
4. H. Horner, *J. Low Temp. Phys.* (to be published).
5. V.F. Sears and F.C. Khanna (to be published).
6. T.A. Kitchens, private communication.
7. L.H. Nosanow, *Phys. Rev.* **146**, 120 (1966).
8. N.R. Werthamer, Conference Summary, Banff Conf. Quantum Crystals, 1971, unpublished.

Thermal Defects in BCC ³He Crystals Determined by X-Ray Diffraction*

R. Balzer† and R. O. Simmons

*Department of Physics and Materials Research Laboratory
University of Illinois at Urbana—Champaign*

Information about thermally created defects in a solid can be obtained by comparing the change of macroscopic volume V and of X-ray cell volume Ω for different temperatures by means of the relation

$$\frac{\Delta V}{V} - \frac{\Delta \Omega}{\Omega} = c \tag{1}$$

where c is the net concentration of vacancy-type defects.[1] Solid bcc ³He can be investigated at essentially constant volume, thus reducing Eq. (1) to

$$-\Delta\Omega/\Omega = -3\,\Delta a/a = c \tag{2}$$

where a is the lattice parameter. Thus in a constant-volume experiment, observation of a single lattice parameter change with change of temperature can be directly related to the creation of vacancy-type defects.

An essential point to be emphasized is that such a determination is independent of microscopic models for the defect and of inferences from other experiments. When this point is understood it becomes apparent that results of direct structural measurements furnish evidence which puts useful constraints upon the interpretation of related experiments on solid helium. Some examples of related investigations of ³He and ⁴He at temperatures near the melting line are on heat capacity,[2,3] nuclear resonance and relaxation,[4] and ionic mobility.[5] Further advantages of X-ray diffraction are that its use permits direct verification of the crystalline structure of the specimen and permits direct determination of the X-ray cell volume Ω. These matters have frequently been uncertain, or subject to indirect inference, in previous investigations of solid helium.

Formidable problems arise in practical realization of the simple measuring principle, Eq. (2). They include (a) preparation of a single crystal appropriately oriented, (b) confinement of the crystal at pressures up to several hundred bars in a container sufficiently transparent to X rays that the weak coherent scattering by the helium can be measured accurately, and (c) control of specimen temperature and temperature gradients in an appropriate reversible manner which minimizes pressure

---

* Work supported in part by the U. S. Atomic Energy Commission.
† Present address: Technische Hochschule Darmstadt, Darmstadt, Germany.

gradients in the single crystal and which gives time for any change in defect content to equilibrate. The present paper is a preliminary report of work demonstrating that these problems can be surmounted for solid helium and that indeed effects of the type represented by Eq. (2) are present in bcc ^{3}He.

Helium crystals were grown from the liquid at constant pressure in a cylindrical Lucite pressure cell attached to a pot containing liquid ^{4}He. Constant pressure conditions were maintained by a special separator. The purified ^{3}He gas was pressurized by moving a piston inside the separator, driven by another gas, usually helium, from a high-pressure tank. The pressurizing gas and the pressurized ^{3}He gas were separated free of mutual contamination by flexible stainless steel bellows. The cell was progressively filled by solid helium, by cooling in a small temperature gradient. Progress of the solid–liquid interface could be monitored by X-ray diffraction. It was found that in the present cylindrical cell bcc ^{3}He tends to grow in a preferred [100] direction.

Visible access to the transparent cell also permitted inspection of the resulting solid helium for macroscopic flaws. When the cell was completely filled by crystal the helium was allowed to block the fill capillary. Subsequent experiments on the specimen crystal could therefore be carried out at essentially constant macroscopic volume. The experimental arrangement will be described in detail elsewhere.

To get information about the quality and orientation of the helium crystals, transmission Laue photographs were taken along the entire length of the specimen cell. A specimen was rejected if more than one single crystal was detected. For satisfactory crystals the angular coordinates of the lattice places were calculated from the positions of the Laue spots. A three-axis counter diffractometer specially designed for this purpose was then oriented about the crystal in order to carry out Bragg reflection measurements on (110) planes. Measurement of the Bragg angle θ yields the lattice parameter a and hence Ω. The first derivative of the Bragg equation leads to

$$-\Delta a/a = (\Delta\theta)\cot\theta \qquad (3)$$

Measurement of the angular displacement of the Bragg reflection $\Delta\theta$ thus yields, through Eqs. (2) and (3), information about changes in the net content of vacancy-type defects.

The formation of thermal defects at constant volume can be characterized by a free energy of formation, which, for the present brief discussion, is divided into the energy of formation w and the vibrational entropy of formation s. Then one has the relation

$$c = e^{s/k}e^{-w/T} \qquad (4)$$

leading to the relation

$$-3(\Delta a/a)_{\exp} - A = e^{s/k}e^{-w/T} \qquad (5)$$

A is the remaining net concentration of vacancy-type defects at the lowest temperature measured and $(\Delta a/a)_{\exp}$ is the experimentally obtained lattice parameter change compared to that at lowest temperature.

Figure 1 shows the relative change of the lattice parameter with change in temperature obtained from the shift of the (110) Bragg reflection. Measurements on three crystals prepared at molar volumes of 20.0, 20.5, and 20.9 ml, respectively, are

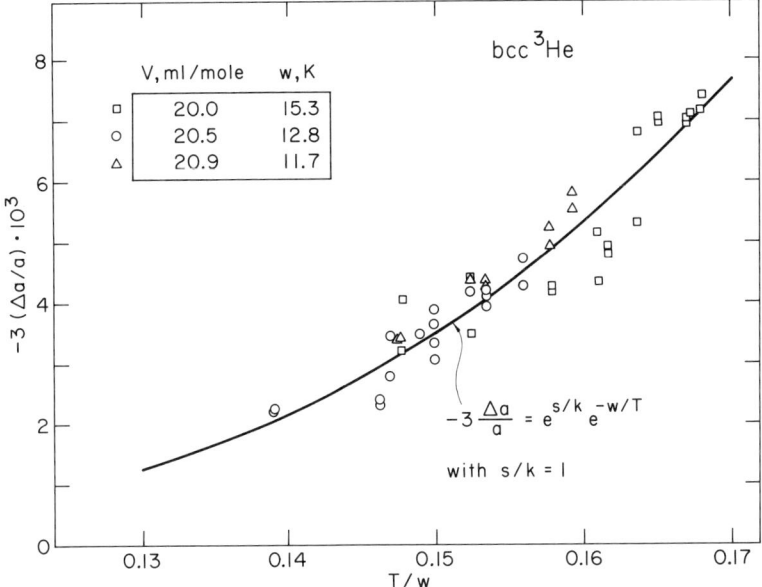

Fig. 1. Measured lattice parameter change of bcc ^{3}He vs. T/w at different molar volumes. (T = temperature, w = energy of formation; formation entropy s is assumed to be k.)

shown. In order to obtain a universal curve for the different molar volumes, a reduced temperature T/w is used. Further, the assumption of a constant entropy of formation ($s = k$) for different molar volumes has been made. Within the uncertainty of the measurements a reversible increase in lattice parameter with decreasing temperature was found, yielding the conclusion that a net concentration of vacancy-type defects is created with increasing temperature in bcc ^{3}He.

From Fig. 1 it is apparent that the values of A[Eq. (5)] in the present work are nonzero. It should also be pointed out that limitations in the cryostat have so far prevented X-ray diffraction measurements below 1.5°K from being accompanied by complete assurance of sufficient temperature homogeneity within the crystals. The difficulty arises in minimizing the necessary radiation and vacuum shields through which the soft X rays must penetrate. Nevertheless, the X-ray data, in spite of some scatter, can be analyzed in a manner which minimizes these uncertainties.

Another matter which requires investigation as the accuracy of X-ray diffraction work of the present type is refined is the possible production of electron-stabilized nonequilibrium vacancies or more complex defects by the X rays used in the Laue and Bragg measurements. Finally, it should be mentioned that the present lattice parameter work had to be kept apparently some significant increment away from the melting line, in order to avoid progressive changes in the specimen crystals.

The main results of the present work are: (a) demonstration of the possibility of precise X-ray Bragg diffractometry on single bcc ^{3}He crystals; (b) unambiguous identification of a net concentration of vacancy-type thermally generated defects in this substance; and (c) derivation of approximate and tentative values for the vacancy energy w. The remainder of this paper takes up the latter point.

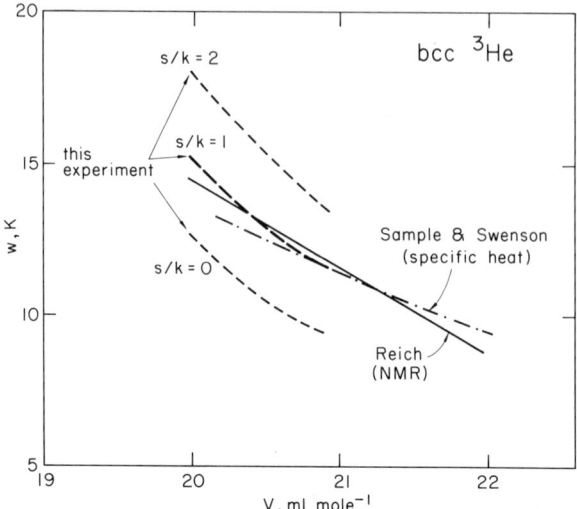

Fig. 2. Derived apparent energy of formation w for thermal defects in bcc ^{3}He at different molar volumes V compared with inferences from specific heat and NMR measurements.

The present data are insufficiently precise to demonstrate explicitly the Arrhenius behavior of Eq. (5). However, the solid curve in Fig. 1, of Arrhenius shape, is consistent with the data. Note that a different energy of defect formation w has been chosen for each molar volume, while the defect vibrational entropy s is assumed constant. Allowing for a small realistic variation of s with volume is not likely to change the derived values of w outside present experimental uncertainty. A more general and complete discussion of the appropriate defect free energy as a function of pressure, volume, and temperature will appear elsewhere.

Figure 2 shows the present derived energies w as a function of molar volume. For comparison, values for bcc ^{3}He of apparent defect activation energy inferred from earlier work on NMR[4] and on specific heat[2] are also shown. It should be noted that considerable uncertainty should be attached to these estimates. For example, although the high temperature heat capacity of bcc ^{3}He is well known,[2,6] the magnitude of the supposed "anomalous" or defect portion depends critically upon the value assumed for the unknown specific heat of a hypothetical defect-free crystal. For the latter a variety of differing theoretical values are available.[7] Within such uncertainties the general magnitude and qualitative volume dependence of the various derived energies are not in disagreement with one another.

Because X-ray measurements determine only the net vacancy content, it is not yet possible to determine whether the observed defects are created as single vacancies, as vacancy clusters, as nonlocalized vacancies, or as vacancy waves, as proposed and discussed in many theoretical papers.* Further measurements are necessary to obtain more detailed information about these defects.

---

* A review which gives some references in Ref. 8.

References

1. R.O. Simmons and R.W. Balluffi, *Phys. Rev.* **117**, 52 (1960).
2. H.H. Sample and C.A. Swenson, *Phys. Rev.* **158**, 188 (1967).
3. G.H. Ahlers, *Phys. Rev. A***2**, 1505 (1970).
4. H.A. Reich, *Phys. Rev.* **129**, 630 (1963).
5. G.A. Sai-Halasz and A.J. Dahm, *Phys. Rev. Lett.* **28**, 1244 (1972).
6. R.C. Pandorf and D.O. Edwards, *Phys. Rev.* **169**, 222 (1968).
7. H.R. Glyde and F.C. Khanna, *Can. J. Phys.* **49**, 2997 (1971).
8. R.A. Guyer, R.C. Richardson, and L.I. Zane, *Rev. Mod. Phys.* **43**, 532 (1971).

Possible Bound State of Two Vacancy Waves in Crystalline ^{4}He

William J. Mullin

Department of Physics and Astronomy
University of Massachusetts, Amherst, Massachusetts

Because of the large zero-point energy of quantum crystals, a particle next to a vacant site is able to tunnel into that site, so that the vacancy moves. This leads to the vacancy wave excitation in a quantum crystal.[1-4] The spectrum of a single vacancy lies in a band which is expected to have the form[1,3]

$$\varepsilon_{\mathbf{k}} = \phi + zt - t\sum_{\mathbf{a}} \cos \mathbf{k}\cdot\mathbf{a} \qquad (1)$$

where z is the number of near neighbors and $\mathbf{a}$ is a nearest-neighbor vector. The lowest energy in the band is ϕ, which is measured in specific heat[5] and NMR[2] experiments. The band width is $2zt$. (See Fig. 1.)

It has been shown[4,6] that two vacancies interact with one another by exchanging virtual phonons. This interaction is reasonably strong and can drastically alter the properties of the propagating vacancies. Indeed, it may make them diffuse rather than propagate.[4,6]

We have studied the spectrum of two interacting vacancy waves in ^{4}He, whose Hamiltonian is taken to be

$$H = \sum_{\mathbf{k}} \varepsilon_{\mathbf{k}} C_{\mathbf{k}}^+ C_{\mathbf{k}} + (-V) \sum_{\mathbf{k},\mathbf{p},\mathbf{q}} C_{\mathbf{p}+\mathbf{q}}^+ C_{\mathbf{k}-\mathbf{q}}^+ C_{\mathbf{k}} C_{\mathbf{p}}$$

We have used a contact interaction of strength $-V$ as a simplification. If one takes matrix elements of the Schrödinger equation in a basis of two vacancy-wave states,

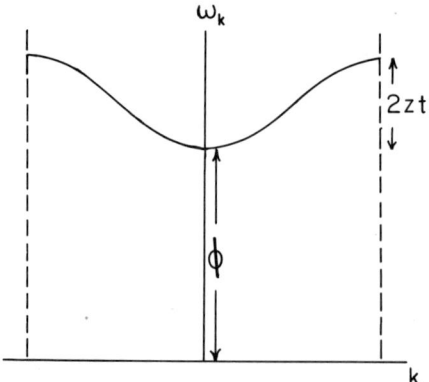

Fig. 1. Possible form of the vacancy wave excitation spectrum. The dashed lines represent the Brillouin zone boundaries.

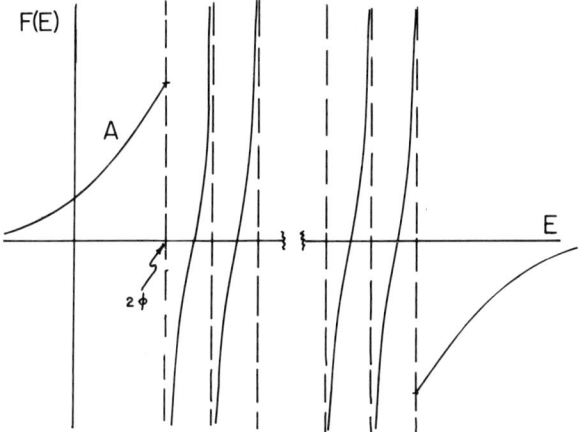

Fig. 2. The function $F(E)$ as defined in Eq. (3). Note that segment A has a finite intercept at $E = 2\phi$. A horizontal line drawn at $1/V$ gives the possible solutions for E.

one arrives at the following condition on the two-vacancy energy E:

$$1 = \frac{V}{\Omega}\sum_k \frac{1}{\varepsilon_{\mathbf{K}+\mathbf{k}} + \varepsilon_{\mathbf{K}-\mathbf{k}} - E} \quad (2)$$

where $\mathbf{K}$ is the center-of-mass momentum, $2\mathbf{k}$ is the relative momentum, and Ω is the volume of the system.

We consider the $\mathbf{K} = 0$ case. If we plot

$$F(E) = \frac{1}{\Omega}\sum_k \frac{1}{2\varepsilon_k - E} \quad (3)$$

we find the behavior shown in Fig. 2 for the particular case of ε_k given by Eq. (1). A horizontal line drawn at value $1/V$ gives the solutions for E. For an attractive potential there is a possible bound state below a continuum. Because of the nature of ε_k of Eq. (1) and its associated density of states, the part of the curve marked A in Fig. 2 approaches a finite limit at $E = 2\phi$. Thus the bound state disappears if the interaction is too weak. This problem is similar to one found in an investigation of electron–hole systems by Kohn.[7]

A rough solution to Eq. (2) can be obtained by making a quadratic approximation to ε_k:

$$\varepsilon_k \cong \begin{cases} \phi + 2zta^2k^2/12, & k < \alpha/a \\ 0 & k > \alpha/a \end{cases} \quad (4)$$

where α/a is some appropriate cutoff. The $\mathbf{k}$ integral can easily be carried out and a simple transcendental condition for the bound state results:

$$Ay = \tan^{-1}y \quad (5)$$

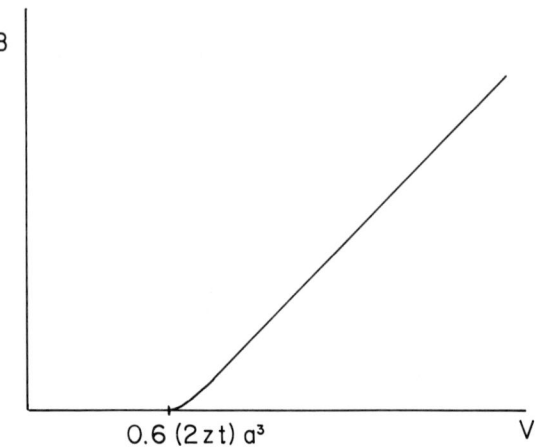

Fig. 3. The binding energy $B = 2\phi - E$ as a function of potential strength V.

where

$$y = \alpha(2zt/6B)^{1/2} \qquad (6)$$
$$A = 1 - \pi^2(2zt)/3V\alpha \qquad (7)$$

and

$$B = 2\phi - E \qquad (8)$$

B is the binding energy of the pair. The solution of Eq. (5) is shown qualitively in Fig. 3. This result is confirmed by an exact numerical calculation for a simple cubic lattice. For a bound state to exist it is necessary that

$$V > 0.6(2zt)\,a^3 \qquad (9)$$

For $\mathbf{K} \neq 0$ but still small, E solutions move to $E(\mathbf{K} = 0) + 2zt\,a^2 K^2/6$.

The interaction between vacancy waves involves virtual phonon exchange and so we expect[4,6]

$$V \simeq a^3 k\theta_D \alpha^2 \qquad (10)$$

where θ_D is the Debye temperature of ^{4}He and α is the relative volume distortion about a vacancy. Using $\alpha^2 \approx 10^{-2}$, we find for low pressures

$$V \simeq 0.3 a^3 \qquad (11)$$

Since the bandwidth is estimated from NMR experiments[2] to be $2zt \simeq 1°K$, we have the condition for a bound state as

$$V \gtrsim 0.6 a^3 \qquad (12)$$

Our approximate use of a contact interaction and our crude estimate of V leads to a large uncertainty in these numbers, so that a bound state is marginally possible. Further, as the density is increased, θ_D increases while $2zt$ remains nearly constant, so that the bound state becomes more likely.

Experiments involving the Rayleigh scattering of light from quantum crystals are planned for the near future.[8] Such experiments will measure the diffusion constant of vacancies, which is, in turn, a measure of vacancy–vacancy interactions[4,6] and should be strongly modified if some of the vacancies travel in pairs. It is also conceivable that Raman scattering experiments could show the two-vacancy spectrum, although this does not seem to have been observed in recent experiments.[9]

Celli and Ruvalds[10] have noted that hybridization of the one-roton and two-roton bound states in liquid ^{4}He may have an important effect on the stability of the one-roton state. Indeed, they claim that as the pressure is increased the one-roton state may become unstable, so that the transition to the crystalline state takes place. An analogous discussion could be given for crystalline ^{4}He. The renormalized one-vacancy gap ϕ_R would be

$$\phi_R = \tfrac{1}{2}\{\phi + B + [(\phi - B)^2 - 4g^2]^{1/2}\} \tag{13}$$

where ϕ is the unrenormalized gap, which is very pressure dependent, and g is the coupling constant between the one- and two-vacancy states. If at sufficiently low densities the square root becomes imaginary, the crystal will become unstable and melt. We doubt, however, that this mechanism provides a theory of melting for solid ^{4}He. Even assuming the validity of this Celli–Ruvalds type of analysis, we would expect B to become smaller as the pressure is decreased, rather than larger as would be needed in this theory.

The bound-state condition may also be solved for the rotons in liquid ^{4}He. If we take the spectrum to be

$$\varepsilon_k = \Delta + \hbar^2(k - k_0)^2/2\mu, \qquad k_0 - q \leq k \leq k_0 + q \tag{14}$$

where q is a cutoff, we again find a transcendental equation for the binding energy $B' = 2\Delta - E$:

$$A'/y' = \tan^{-1} y' \tag{15}$$

where

$$A' = \hbar^2 q \pi^2 / 2\mu V k_0^2 \tag{16}$$

and

$$y' = \hbar q/(2\mu B')^{1/2} \tag{17}$$

In contrast to Eq. (5), Eq. (15) has a bound state solution for *all* values of V. For small V we have

$$B' = 2\mu V^2 k_0^4/\hbar^2 \tag{18}$$

These roton results were previously found by Zawadowski *et al.*[11]

Acknowledgments

I wish to thank Harrell Broughton, Robert Guyer, and Michael Wortis for useful discussions. I would also like to thank J. A. Krumhansl for his hospitality this summer at the Laboratory of Atomic and Solid State Physics of Cornell University, where a portion of the manuscript was written.

References

1. J.H. Hetherington, *Phys. Rev.* **176**, 231 (1968).
2. R.A. Guyer, R.C. Richardson, and L.I. Zane, *Rev. Mod. Phys.* **43**, 532 (1971).
3. W.J. Mullin, *Phys. Rev. A* **4**, 1247 (1971).
4. R.A. Guyer (to be published).
5. H. Sample and C.A. Swenson, *Phys. Rev.* **158**, 188 (1967).
6. R.A. Guyer and W.J. Mullin (to be published).
7. W.A. Kohn, *Phys. Rev.* **133**, A171 (1964).
8. N.C. Ford, R.B. Hallock, and K.H. Langley, private communication.
9. C.M. Surko and R.E. Slusher, *Phys. Rev. Lett.* **27**, 1699 (1971).
10. V. Celli and J. Ruvalds, *Phys. Rev. Lett.* **28**, 539 (1972).
11. A. Zawadowski, J. Ruvalds, and J. Solana, *Phys. Rev. A* **5**, 399 (1972).

Multiphonon and Single-Particle Excitations in Quantum Crystals

Heinz Horner*

Institut für Festkörperforschung
Kernforschungsanlage Jülich, Jülich, West Germany

In a recent experiment Kitchens et al.[1] observed single-particle-like excitations in solid helium at momentum transfers $Q > 3$ Å^{-1}. Similar observations had previously been made on liquid helium[2] and neon.[3] The scattering in this regime in the solid, as well as in the liquid, is a broad peak centered at an energy slightly below the free-recoil energy $Q^2/2m$, where **Q** is the momentum transfer and m is the mass of helium or neon, respectively.

It can be shown[4] that even in a harmonic crystal the scattering will be free-particle-like, following the free-recoil energy, at large enough momentum transfer. In this case the scattering is just the superposition of multiphonon processes of all orders.

We wish to report here the results of a calculation of the dynamic scattering function $S(\mathbf{Q}, \omega)$ for solid helium and neon. This calculation includes all order multiphonon processes and the lowest order one–two phonon interference terms.

The calculation was done in the following way. One- and two-phonon processes, including their interference contributions, were calculated in a conventional way using a renormalized anharmonic theory[5] rather than the conventional anharmonic perturbation theory. To calculate higher order multiphonon processes, we use the expression for a harmonic crystal (see, e.g., Ref. 6)

$$S(\mathbf{Q}, \omega) = \int_{-\infty}^{\infty} dt \exp(i\omega t) \sum_i \exp[-2W(\mathbf{Q})]$$
$$\times \exp(-i\mathbf{Q} \cdot \mathbf{R}_i) \exp[\tfrac{1}{2}\mathbf{Q} \cdot \mathbf{D}_{0i}(0, t) \cdot \mathbf{Q}] \quad (1)$$

where $\exp[-2W(\mathbf{Q})]$ is the Debye–Waller factor with $W(\mathbf{Q}) = \tfrac{1}{2}\mathbf{Q} \cdot \mathbf{D}_{00}(0, 0) \cdot \mathbf{Q}$, $\mathbf{R}_i$ is a lattice vector, and $\mathbf{D}_{ij}(tt')$ is the phonon propagator given in harmonic theory by

$$\mathbf{D}_{ij}(tt') = \frac{1}{N} \sum_{\mathbf{k}\lambda} \{\exp[i\mathbf{k} \cdot (\mathbf{R}_i - \mathbf{R}_j)]\} \frac{\varepsilon_\lambda(\mathbf{k})\varepsilon_\lambda(\mathbf{k})}{2m\omega_\lambda(\mathbf{k})}$$
$$\times \left\{ \frac{\exp[-i\omega_\lambda(\mathbf{k})(t - t')]}{1 - \exp[-\beta\omega_\lambda(\mathbf{k})]} + \frac{\exp[i\omega_\lambda(\mathbf{k})(t - t')]}{\exp[\beta\omega_\lambda(\mathbf{k})] - 1} \right\} \quad (2)$$

---

* Partially supported by U.S. Atomic Energy Commission at Brookhaven National Laboratory, Summer 1972.

where $\omega_\lambda(\mathbf{k})$ and $\varepsilon_\lambda(\mathbf{k})$ are the harmonic frequencies and polarization vectors, respectively.

Bragg scattering and one- and two-phonon and higher multiphonon processes are obtained on expanding the last exponential in Eq. (1). This means that if we are interested in three-phonon and higher phonon processes only, we have to subtract the first three terms of this expansion from Eq. (1).

In the calculation for quantum crystals we have used this form but replaced the harmonic frequencies by the ones determined in an anharmonic self-consistent way.[7]

The results of this calculation for bcc ^{4}He are presented in Fig. 1 and for neon in Fig. 2. The solid lines give the frequency of the one-phonon peak and the position of the maximum of the multiphonon background which eventually develops into the free-particle scattering peak. The light solid lines give the frequency of the half-

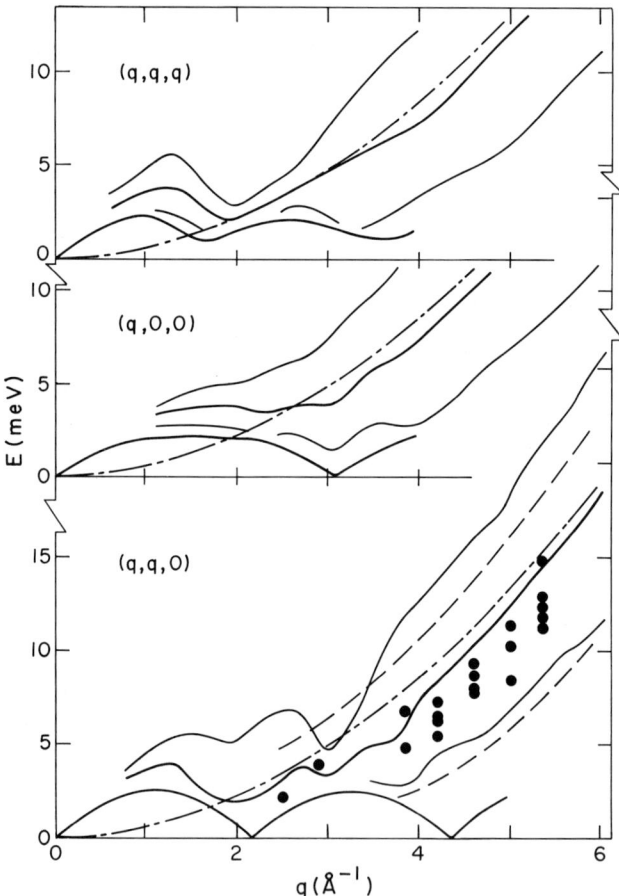

Fig. 1. Free-particle scattering in bcc ^{4}He for different directions: Solid lines give the frequency in the single-phonon and multiphonon scattering, respectively. Light lines give the frequencies of half-height of the multiphonon scattering. The free-recoil energy is shown dash-dotted. Experiments from Ref. 1.

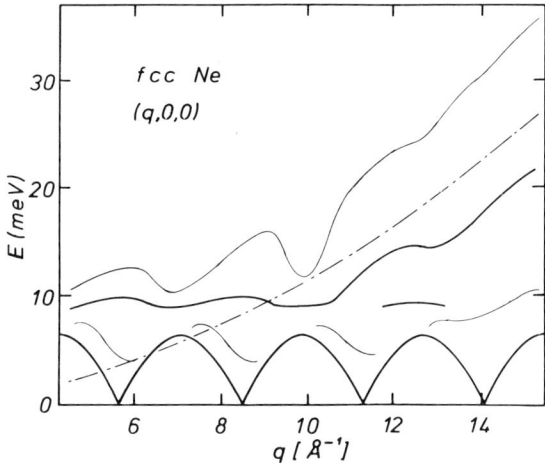

Fig. 2. Free-particle scattering in solid neon. (For explanation see caption of Fig. 1.)

height of this latter peak. The dash-dotted lines are the free-recoil energy; the experiments in solid helium are shown as dots for the maximum and as dashed lines for the half-height. It should be noted that the experiments are not corrected for changes in the instrumental resolution for changing energy. This correction would remove at least parts of the discrepancies.

In solid neon we see oscillations in the peak position as well as in the width following the periodicity in reciprocal space. Similar oscillations have actually been observed in the liquid.[3] The oscillations found in the calculation for solid helium are much weaker. Stronger oscillations might, however, result if higher-order anharmonicities and interference terms were included.

Acknowledgments

Discussions with T. A. Kitchens and V. J. Minkiewicz are acknowledged.

References

1. T.A. Kitchens, G. Shirane, V.J. Minkiewicz, and E.B. Osgood (to be published).
2. R.A. Cowley and A.D.B. Woods, *Can. J. Phys.* **49**, 177 (1971); O.K. Harling, *Phys. Rev. A* **3**, 1073 (1971).
3. W.J.L. Buyers, V.F. Sears, P.A. Lonngi, and D.A. Lonngi, Conf. Inelastic Neutron Scattering, Grenoble, 1972.
4. A. Sjölander, *Ark. Fys.* **14**, 315 (1958).
5. H. Horner, *Phys. Rev. Lett.* **29**, 556 (1972); also this volume.
6. W. Marshall and S.W. Lovesey, *Theory of Thermal Neutron Scattering,* Clarendon Press, Oxford (1971).
7. H. Horner, *J. Low Temp. Phys.* **9**, 547 (1972).

3
Helium-3 Nuclear Magnetism

What Is the Spin Hamiltonian of Solid BCC ^{3}He?*

L. I. Zane†

Physics Department, Colorado State University
Fort Collins, Colorado

Solid helium is a quantum crystal, i.e., a crystal in which the ratio of the zero-point energy to the binding energy is of order unity. This allows the atoms in the solid to have an unusually large mobility. This mobility of helium atoms becomes experimentally observable in solid ^{3}He because of its magnetic moment. Nuclear magnetic relaxation (NMR) experiments observe the hopping directly. The hopping leads to a spin ordering which effects the thermodynamics of solid ^{3}He at low temperatures. Hence a measurement of thermodynamic properties such as pressure, susceptibility, entropy, etc., allows us to garner information about the hopping helium atoms. The Heisenberg Hamiltonian,

$$\mathcal{H} = -2\hbar \sum_{i<j} J_2(ij)\, \boldsymbol{\sigma}_i \cdot \boldsymbol{\sigma}_j$$

has been used to describe solid ^{3}He, where $J_2(ij)$ is the exchange frequency between the pair of atoms at lattice sites i and j and $\boldsymbol{\sigma}_i$ is the spin of the atom on the ith lattice site. Normally only nearest neighbor (nn) interactions are important and the Hamiltonian is written as,

$$\mathcal{H} = -2\hbar J_2(\text{nn}) \sum_{i<j}^{\text{nn}} \boldsymbol{\sigma}_i \cdot \boldsymbol{\sigma}_j$$

where the nn above the summation restricts it to nearest neighbors. This Hamiltonian has been used to analyze the NMR, pressure, and susceptibility experiments and the result is that $J_2(\text{nn}) = -0.7\text{m}°\text{K}$ at a molar volume of 24 cm^3. The minus sign indicates that the interaction is antiferromagnetic.

Recently the Heisenberg Hamiltonian was generalized by the insertion of a term which represents triple exchange.[1] This alteration of the Hamiltonian is necessary in order to explain the pressure experiment of Kirk and Adams[2] (KA) which was performed on samples of solid ^{3}He immersed in high magnetic fields. The generalized Hamiltonian which is needed to understand the experiment of KA is[3]

$$\mathcal{H} = -2\hbar J(\text{nn}) \left(\sum_{i<j}^{\text{nn}} \boldsymbol{\sigma}_i \cdot \boldsymbol{\sigma}_j + A \sum_{i<j}^{\text{nnn}} \boldsymbol{\sigma}_i \cdot \boldsymbol{\sigma}_j \right)$$

where
$$A = J(\text{nnn})/J(\text{nn})$$
$$J(\text{nn}) = J_2(\text{nn}) + 6J_3 + 6J_4 + \cdots$$
$$J(\text{nnn}) = J_2(\text{nnn}) + 4J_3 + 4J_4 + \cdots$$

---

* Work supported by National Science Foundation Grant No. GP 22553
† Present address: Physics Department, University of Nevada, Las Vegas, Nevada.

and nnn is an abbreviation for next-nearest neighbor. Now $J(\text{nn})$ and $J(\text{nnn})$ include contributions from pairs, triples, quadruples, etc. The Heisenberg Hamiltonian only includes two-body interactions. Only the triple formed by two nn's and one nnn and nnn pair exchange are large enough to have an effect on the properties of the solid.

The experiment of KA allows us to determine the parameter A, which then allows us to calculate $J(\text{nnn})$. The value of A found is -0.22 and therefore $J(\text{nnn}) = 0.17$ m°K. It is important to notice that the sign of $J(\text{nnn})$ is positive, indicating a ferromagnetic interaction. Thouless[4] has shown that the cyclic permutation of an even number of atoms leads to an antiferromagnetic interaction, while the cyclic permutation of an odd number of atoms leads to a ferromagnetic interaction. In order for $J(\text{nnn}) > 0$, we must have

$$|J_3| > |J_2(\text{nnn})/4|$$

where we have ignored the small contribution due to J_4. The ferromagnetic interactions of the nnn's actually make the solid more antiferromagnetic and should contribute to raising the expected Néel temperature. For hcp ^{3}He, because there are triples of atoms which form triangles with three nn legs, there is only a renormalization of the Heisenberg exchange Hamiltonian. The Hamiltonian for the hcp solid is written as

$$\mathscr{H} = -2\hbar J(\text{nn}) \sum_{i<j}^{\text{nn}} \boldsymbol{\sigma}_i \cdot \boldsymbol{\sigma}_j$$

where

$$J(\text{nn}) = J_2(\text{nn}) + 4J_3$$

The important point is that the Hamiltonian for bcc ^{3}He has two exchange parameters, $J(\text{nn})$ and $J(\text{nnn})$, while the Hamiltonian for hcp ^{3}He has only one exchange parameter, $J(\text{nn})$.

In the above we have alluded to an expected increase in the expected Néel temperature for bcc ^{3}He. Now we will attempt to get a more quantitative idea of the magnitude of this increase. Callen and Callen[5] (CC) have done a two-particle cluster approximation for an antiferromagnetic system with nnn interactions. Baker et al.[6] (BGER) have offered a high-temperature expansion theory which is considered very reliable but have not extended it to the problem with nnn interactions. The Néel temperatures, not including nnn interactions, obtained by CC and BGER are 2.3 and 2.0 m°K, respectively. The Néel temperature obtained from the theory of CC with $A = -0.22$ is 2.9 m°K. If we assume that the two theories scale in a similar fashion, we would expect the high-temperature expansion theory to give a Néel temperature of around 2.5 m°K. Callen and Callen[7] have also used their basic theory to analyze ferromagnetic systems with nnn interactions and the results they obtain are in good agreement with experiment. If we assume that the antiferromagnetic theory of CC is as good as the ferromagnetic theory appears to be and that we can legitimately scale the theory of BGER, then we arrive at an expected Néel temperature for bcc ^{3}He of between 2.5 and 2.9 m°K.

The experiment of KA obliges us to revise the Hamiltonian for bcc ^{3}He. The generalized Hamiltonian given in this paper improves the agreement between theory and experiment. This agreement, though, is still qualitative in nature and more effort is required to make the theory and experiment agree quantitatively.

References

1. L.I. Zane, *Phys. Rev. Lett.* **28**, 420 (1972).
2. W.P. Kirk and E.D. Adams, *Phys. Rev. Lett.* **27**, 392 (1972).
3. L.I. Zane, *J. Low Temp.* **9**, 219 (1972).
4. D.J. Thouless, *Proc. Phys. Soc. (London)* **86**, 893 (1965).
5. H.B. Callen and E. Callen, *J. Phys. Soc. Japan* **20**, 1980 (1965).
6. G.A. Baker, Jr., H.E. Gilbert, J. Eve, and G.S. Rushbrooke, *Phys. Rev.* **164**, 800 (1967).
7. H.B. Callen and E. Callen, *Phys. Rev.* **136**, 1675A (1964).

Magnetic Properties of Liquid ^{3}He below 3 m°K*

D. D. Osheroff, W. J. Gully, R. C. Richardson, and D. M. Lee

*Laboratory of Atomic and Solid State Physics
Cornell University, Ithaca, New York*

Features observed along the ^{3}He melting curve in compressional cooling experiments have suggested that at least one and perhaps two phase changes take place in either the solid and/or the liquid ^{3}He within the cell at temperatures below 3 m°K.[1] In experiments in which the cell pressure was recorded as a function of time during periods in which the rate of compression dV/dt was held constant, two unexpected features were reproducibly obtained on the resulting pressurization curve which were labeled the A and B transitions, respectively, by Osheroff *et al.* The A transition is characterized by a sudden decrease in the rate of cell pressurization dP/dt upon cooling by roughly a factor of two in magnitude, all within a pressure interval of less than 3×10^{-4} atm. A nearly equal but opposite change in dP/dt is observed upon warming at precisely the same pressure. The pressure at which the A transition occurs in zero magnetic field is 1.70×10^{-1} atm above the cell pressure at 7 m°K, and $P(A)$ is lowered by the application of magnetic fields by an amount roughly equal to $-6.7 \times 10^{-5} H^2$ atm/kG^2. The pressure $P(A)$ corresponds to a temperature of about 2.7 m°K obtained by extrapolating the shifted melting curve of Johnson *et al.*[1,2] The B transition is characterized by a sudden drop in cell pressure by about 3×10^{-4} atm upon cooling and by a brief hesitation in the cell pressure as it decreases upon warming (B' transition). The pressure $P(B')$ is reproducibly 1.89×10^{-1} atm above the cell pressure at 7 m°K in zero magnetic field, and is shifted to higher pressures by the application of magnetic fields by an amount given roughly by $\Delta P = +2.02 \times 10^{-3} H^2$ atm/kG^2. The pressure $P(B)$ is always higher than $P(B')$ by an amount which varies from zero to 3×10^{-3} atm in zero field, depending upon the history of the compression, and by even larger pressure differences in magnetic fields of a few kG. We therefore conclude that the B transition (B and B') displays supercooling. The rather large magnetic field dependence of the B transition is considered of great interest, since it is much larger than is expected for magnetic ordering in the solid.

Since the initial publication presenting these results[1] similar phenomena have been observed in other compressional cooling experiments by Halperin and co-workers[3] and Wheatley and co-workers.[4] Although the pressure phenomena are clearly effects in the ^{3}He within the cell and are not strongly geometry dependent, an interpretation of these phenomena in terms of ordering in either the solid or

---

* Work supported in part by the National Science Foundation through Grant No. GP-24179 and the Advanced Research Projects Agency through the Cornell Materials Science Center.

liquid phase within the cell is uncertain. Such an interpretation requires that a large number of assumptions be made concerning the thermal behavior of the solid phase.

To obtain a fuller understanding of the changes manifested on the pressurization curve, we have undertaken an extensive survey of the magnetic properties of the solid and liquid phases existing within our cell, utilizing continuous wave nuclear magnetic resonance techniques. The results of this survey show changes in the magnetic behavior of the liquid phase which correlate exactly with the pressure phenomena. No changes in the magnetic properties of the solid phase were observed which correlated with the pressure phenomena, although thermal disequilibrium throughout the solid might mask such a change.

A specially designed NMR coil 0.48 cm in diameter and 2.54 cm long was oriented vertically within the cell,* extending to within 0.5 cm of the BeCu strain gauge used to monitor the cell pressure located at the base of the compression cell. This NMR coil replaced the ^{195}Pt thermometer used in previous experiments. The NMR coil was supported by an epoxy plug, but was open to the rest of the cell at both ends. A nearly linear gradient in the applied external field could be created parallel to the coil axis. In the gradient field only a thin layer of sample could satisfy the Larmor resonant condition at a given radio frequency. By sweeping the radio frequency, or equivalently H_0, the resonant layer could be moved up or down the length of coil, providing a means to obtain a profile of the resonant absorption of the sample as a function of distance along the coil length.

During compressions the solid formed within the volume occupied by the rf field of the coil tended to cluster at certain locations, with large portions of the coil length free from any solid whatsoever. This fortunate circumstance allowed us to observe separately the solid and liquid absorption signals. In the regions of the solid clusters the solid absorption signal at the lowest temperatures was 40–60 times larger than the nearly temperature-independent absorption due to the Fermi ordered liquid phase.

In Fig. 1 sets of absorption profiles, or position profiles, are shown schematically, showing the relative absorption signal of the helium as a function of position in the coil. The upper and lower ends of the coil are labeled, as is the direction of the field gradient. In both series the large solid absorption peaks have been truncated to permit better observation of the smaller liquid absorption signal. The pressure at which each profile was obtained is labeled with a reduced pressure notation:

$$p(t) = [P(t) - P(A)]/[2.04 \times 10^{-2} \text{ atm}]$$

where 2.04×10^{-2} atm is the value of $P(B') - P(A)$ in an 850-G magnetic field, the field in which these profiles were obtained. At $P(A)$, $p = 0$, and at $P(B')$, $p = 1$.

In the upper series of profiles, at cell pressures greater than $P(A)$, the right-hand edge of the liquid signal is seen to shift to the right, corresponding to a shift in the Larmor resonant frequency to higher and higher frequencies as the pressure increases, until at B the apparent frequency shift vanishes, and the resonant absorption signal of the liquid drops by nearly a factor of two in magnitude. This apparent drop in the liquid absorption at B is also evident in the lower series of

---

* For a schematic drawing of the cell the reader is referred to the paper by Lee elsewhere in these proceedings.

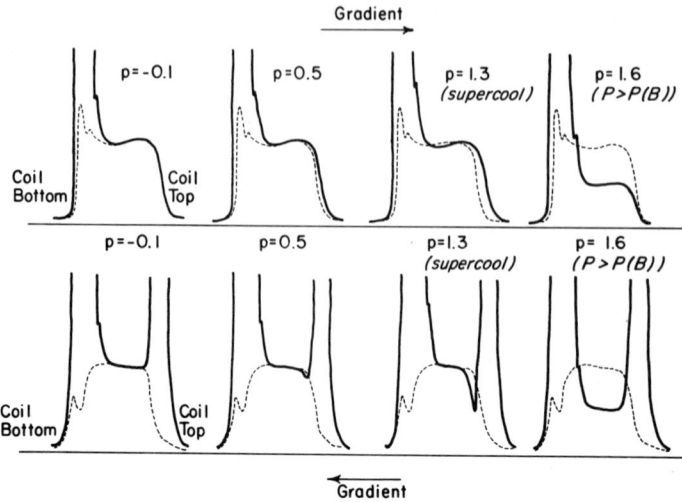

Fig. 1. NMR resonant profiles at various reduced pressures.

profiles. The "trough" seen to develop just below the upper solid peak in the lower series is totally explicable in terms of the apparent frequency shift of the liquid observed in the upper series.

By removing the external field gradient, the position profile obtained by sweeping the magnetic field was transformed into a frequency profile. This allowed the possibility that the liquid resonant frequency increases at pressures above $P(A)$ to be studied in greater detail. As the cell pressure was increased above $P(A)$ in this configuration a satellite line was indeed observed to split off the main resonant peak and to shift to higher frequencies as the pressure was increased to $P(B)$, at which point the satellite disappeared. As the cell pressure was lowered below $P(B')$ the satellite reappeared. In Fig. 2 a plot is shown of the frequency difference $\Delta v = v_{\text{solid}} - v_{\text{satellite}}$ as a function of p obtained at a constant resonant frequency of 1509 kHz. The frequency difference is independent of the history of the compression, and extrapolates to zero at $p = 0$.

There is strong evidence to suggest that the satellite line is the liquid portion of NMR signal, and that the Larmor resonant frequency of the liquid does increase at pressures above $P(A)$ as was suggested to explain the behavior of the liquid portions of the profiles in Fig. 1. The main evidences supporting this hypothesis are: (a) Δv is a unique function of cell pressure; (b) the magnitude of the satellite is nearly constant (provided the solid fraction within the NMR coil was not larger than about 40%), and nearly equivalent to the magnitude of the all-liquid absorption peak obtained prior to compression; (c) the shape of the satellite line closely resembles the all-liquid line shape, which was frequently a rather complex pattern; and (d) the satellite line shape does not distort or broaden as the splitting increases, even when Δv is as much as 15 times the satellite linewidth.

The frequency splitting was studied at a number of different resonant frequencies from 130 to over 2760 kHz. The splitting at all frequencies and pressures could be

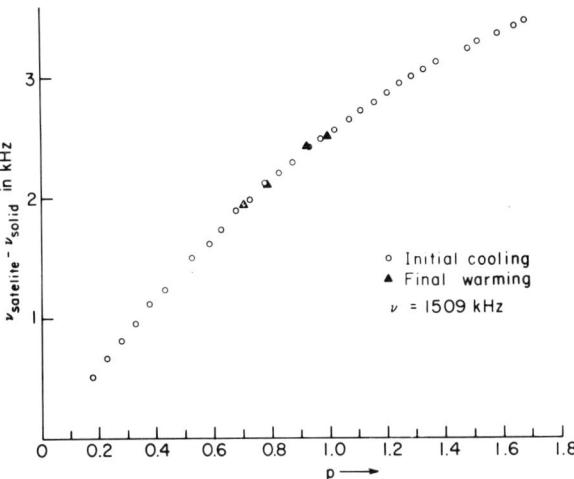

Fig. 2. Frequency difference of the satellite and solid peaks at 1509 kHz as a function of reduced pressure.

represented to $\pm 2\%$ by a single expression:

$$v^2_{\text{liquid}} - v^2_{\text{solid}} = (10.11p - 2.475p^2) \times 10^3 \quad \text{kHz}^2$$

By factoring, $v_l^2 - v_s^2 = (v_l - v_s)(v_l + v_s) \simeq 2(\Delta v) v_0$, one can see that the frequency shift in the liquid scales nearly as $1/v_0$.

The drop in the liquid absorption at B was studied in detail. It was observed that the transition to the lower absorption state, to the "B liquid," occurred as a wave which propagated from the bottom of the cell upward at B and from the top downward at B'. The velocity of this wave was directly proportional to the rate of change of pressure with time. The precise magnitude of the susceptibility drop at B appeared to vary from about 40% to 60%, although occasionally a small region in the B liquid appeared to have almost zero susceptibility. Such a region is shown in Fig. 3(a). This region appeared stable provided the cell pressure remained above $P(B')$.

The B liquid responded in a curious manner to the application of a small (25 ergs) heat pulse into a heater located at the bottom of the NMR coil. Upon the application of a heat pulse the entire B-liquid absorption profile appeared to drop nearly to zero susceptibility almost instantaneously, and relaxation to the normal B-liquid susceptibility was manifested as a wave traveling from the bottom of the coil upward at a velocity of about 1 cm/sec. Figure 3(b) shows the behavior of the absorption profile during this recovery period.

The mechanism which produces the magnetic features we observe in the liquid is not understood. It is not clear that these features are bulk properties of the liquid, independent of influences from the solid phase. Further, in the experiments of Halperin et al.[3] no drop in the static magnetization of the ^{3}He at $P(B)$ is observed using Squid detectors, and therefore the drop in resonant absorption at B should not be viewed as a drop in the overall susceptibility.

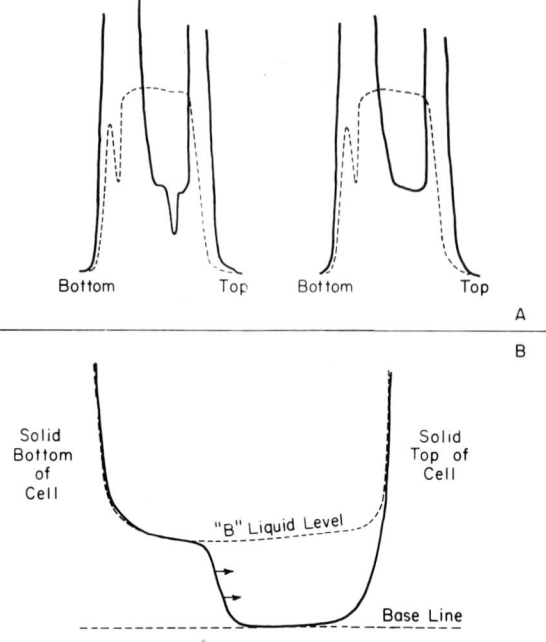

Fig. 3. (a) Anomalous susceptibility in the B liquid. (b) Recovery of the B liquid after the application of a heating pulse.

References

1. D.D. Osheroff, R.C. Richardson, and D.M. Lee, *Phys. Rev. Lett.* **28**, 855 (1972).
2. R.T. Johnson, O.V. Lounasmaa, R. Rosenbaum, O. G. Symko, and J.C. Wheatley, *J. Low Temp. Phys.* **2**, 403 (1970).
3. W.P. Halperin, R.A. Buhrman, W.W. Webb, and R.C. Richardson, this volume.
4. J.C. Wheatley, private communication.

Properties of ^{3}He on the Melting Curve*

W. P. Halperin, R. A. Buhrman, W. W. Webb, and R. C. Richardson

Department of Applied Physics
Cornell University, Ithaca, New York

We have made measurements of pressure, volume, and magnetization of a ^{3}He sample self-cooled by compressional cooling techniques. The method of this type of refrigeration is now well established (see, e.g., Ref. 1). In these experiments we investigated properties of both liquid and solid ^{3}He constrained to the melting curve at temperatures less than 3 m°K. This work was motivated by several interesting features of ^{3}He at low temperatures: the nuclear spin order anticipated in solid ^{3}He near 2 m°K, and the recent observations by Osheroff et al.[2] (OGRL) that at a very well-defined temperature below 2.65 m°K the NMR absorption signal of liquid ^{3}He appears to drop suddenly by approximately 50%.

Since most measurements made on self-cooled ^{3}He will have contributions from both liquid and solid phases, it is necessary to have a means for separating the effects from the two constituents. In order to solve this problem, we have designed a compression chamber in which the amount of solid can be accurately and continuously monitored and in which the compressional process itself is sufficiently adiabatic that more than 70% of the ^{3}He sample can be solidified isothermally at temperatures estimated to be less than 2 m°K.

The compression apparatus is constructed of beryllium–copper and is shown in Fig. 1. The ^{3}He sample (containing 20 ppm ^{4}He in these experiments) is contained in a pancake-shaped compartment, the top of which is a 0.074-cm-thick diaphragm 4 cm in diameter. A second diaphragm, 0.075 cm thick, 1.26 cm in diameter, at the bottom of the compartment serves as the active element in a capacitative pressure gauge.[3] The upper compartment in Fig. 1 contains ^{4}He, which is used as a hydraulic fluid. The position of the compression diaphragm can be accurately determined through measurement of the capacitance between the upper two plates shown in the figure. From this measurement changes in volume in the ^{3}He compartment can be determined. The ^{3}He pressure gauge has been calibrated to 0.05% using a Ruska Instrument Corp. quartz bourdon pressure gauge.

After the cell has been filled with 0.13 mole of ^{3}He at a pressure greater than 31 atm it is precooled by a dilution refrigerator operating continuously at 15 m°K. Typically, 36 hr is required to cool the sample from 1°K to 25 m°K. Starting the compression at 25 m°K, approximately 25% solid is formed in cooling the ^{3}He

---

* Work supported by the National Science Foundation under Grant No. GP-29682 and the Advanced Research Projects Agency through the Materials Science Center of Cornell University.

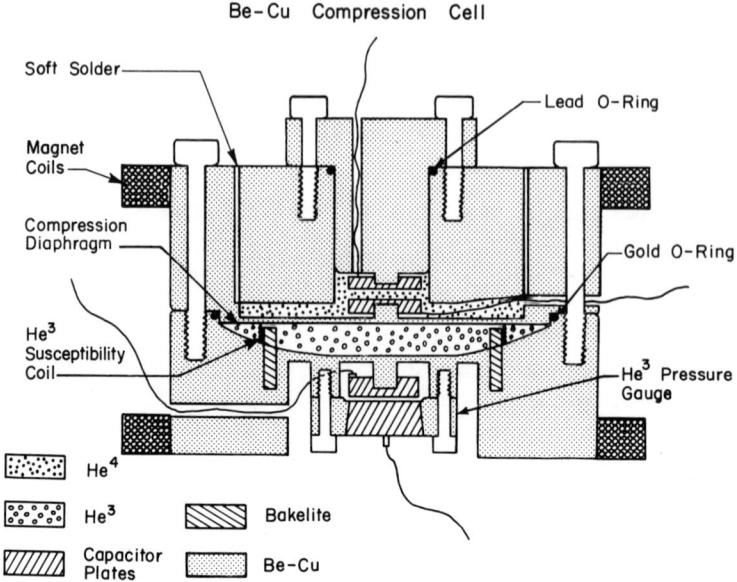

Fig. 1. The experimental Be–Cu compression chamber.

to the maximum melting pressures we have obtained, estimated to be at temperatures less than 2 m°K. Thereupon another 70% solid can be formed without any indication of warming until the compression is stopped. At the lowest temperatures the ambient heat leak into the cell is 4 ergs/min. In one experiment in which temperatures were measured by monitoring the magnetization of CMN powder with a superconducting magnetometer the formation of 90% solid was completed with T^* of the salt less than 1.95 m°K. The salt thermometer pill was contained in a right circular cylinder whose diameter was one-fourth of its height. The pill, the superconducting magnetometer coil, and a small solenoid magnet were contained in a hole 0.22 cm in diameter and 0.6 cm deep which was drilled vertically in the Bakelite form shown in Fig. 1. Unfortunately, this physical arrangement of the salt thermometer led to excessively long thermal equilibrium times (greater than 15 min) between the ^{3}He, as indicated by the ^{3}He melting pressure, and the salt magnetization. A comparison of the magnetization of the CMN and a 0.24-g copper sample immersed in the ^{3}He with 16-cm^2 surface area indicated a minimum copper magnetic temperature of 4.5 m°K. It appears that a much larger surface area is necessary for the copper to register the temperature of the ^{3}He down to lower temperatures.

Recent measurements in another apparatus in this laboratory by Osheroff et al.[4] (ORL) have indicated that the ^{3}He melting curve inferred from measurements of melting pressure as a function of time has two low temperature features. These are labeled as pressure A and B on cooling and A' and B', respectively, on warming. The feature A is reported by ORL to occur at 2.65 m°K as a sharp decrease in the rate of change of melting pressure with time. Under certain conditions this feature is reproduced in our cell at a pressure of 33.870 atm. We find, however, that if the cell contains more than 65% solid or if the volume is changed so slowly that the

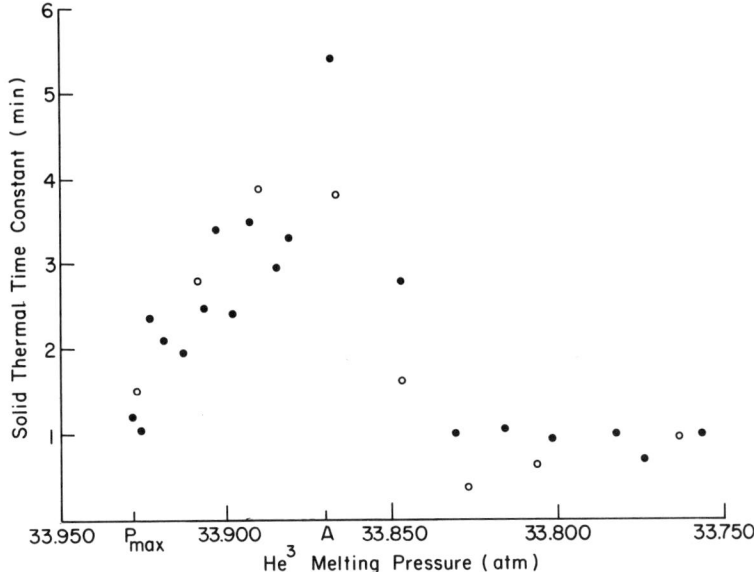

Fig. 2. Time constants for thermal equilibrium within the solid plotted as a function of the melting pressure. Open dots: run No. 31; closed dots: run No. 33.

melting pressure increases at a rate of less than 4×10^{-4} atm/min, the pressure in the cell changes smoothly as a function of time as the pressure passes through 33.870 atm.

We have made an analysis of the thermal time constant of the pressure response of the cell to heat applied to a small copper wire imbedded within the ^{3}He. The results of several compression experiments are shown in Fig. 2. In these experiments approximately 95% solid was created within the cell, cooled to the maximum melting pressure. Then small pulses of heat were applied to warm the ^{3}He. The time constants shown are those required for the rate of change of melting pressure to come into equilibrium with the drift rate due to external heat leak before the pulse was applied. The time constant passes through a maximum at the same pressure at which A occurs. Our heater occupies a volume of only 0.1 cm^3 in the 3.4-cm^3 compartment filled with 95% solid ^{3}He. Most likely, this heater is covered with solid. Putting heat into the heater at first produces some very local hot solid in response to which the liquid temperature increases (pressure decreases). We observe that the characteristic time for such response is less than 1 sec. After 60 sec the heat is removed. Then the liquid temperature immediately begins to relax back (pressure increases), cooling as thermal equilibrium is reestablished in the solid ^{3}He. This latter process has a characteristic time of the order of minutes and is shown in Fig. 2. At temperatures well above A the pressure does not "overshoot" but continues to decrease, approaching the ambient drift rate with an exponential time constant of order 1 min. During the experiment the volume changes by less than 5.5×10^{-2}%.

We conclude from these experiments that (1) a maximum in the thermal response of solid ^{3}He does occur at A; and (2) the initial interpretation of A given in Ref. 4

as a first-order transition of the solid is far too facile, since the thermal response of the cell is sufficiently rapid that the heat capacity of the solid already formed in compression will affect the rate of change of pressure.* (This contribution to the heat load of the cell was ignored in Ref. 4.) However, the time constants measured in this work most likely depend on the specific geometry of our cell.

The feature B appears as a sudden drop in pressure of the order of 1.0×10^{-4} atm and is accompanied by a simultaneous drop in volume of $1.3 \times 10^{-4}\%$. The observed volume change is larger than that compatible with the compressibility of either the solid or liquid measured at higher temperatures.[5] Further, these shifts have essentially the same magnitudes over a range of solid fractions from 25% to 65%. The feature B', which occurs at a lower pressure than B, appears as a flattening in melting pressure as a function of time. However, B' is frequently initiated by an abrupt drop in pressure and a simultaneous change in volume similar to that seen at B. From these observations it seems likely that the event at B is accompanied by a sudden conversion of a small amount ($< 0.04\%$) of liquid to solid. We measure a pressure difference between A (or A') and B' of $(2.04 \pm 0.07) \times 10^{-2}$ atm.

Osheroff et al.[2] have observed with NMR detection techniques a splitting of the ^{3}He line occurring between melting pressures of A and B, and between B' and A'. Their interpretation is that the small portion of the line which shifts to higher Larmor frequencies corresponds to the liquid signal. Further, they observe an abrupt change in the liquid signal, decreasing by 50% at B and increasing by the same amount at B'. In this work we have made static magnetization measurements of the ^{3}He sample when cooled through B and warmed through B'. For a sample which is 75% liquid in magnetic fields ranging from 5×10^{-3} to 10^2 Oe we observe no change in magnetization associated with B or B'.† Since the total magnetization M_z is related to the NMR absorption signal $A(\omega)$ at the frequency ω through the Kramers–Kronig relation $M_z = KP \int_{-\infty}^{\infty} [A(\omega)\,d\omega/\omega]$, where K is an instrumental constant and P is the Cauchy principle value of the integral at $\omega = 0$, there needs to be a component of $A(\omega)$ which appears at a frequency not measured by Osheroff et al.[2] in order for these two measurements to be consistent.

Finally, we wish to report that the maximum pressure on the melting curve we have obtained in these experiments is $(5.8 \pm 0.1) \times 10^{-2}$ atm greater than A. Using the melting curve measurements of Johnson et al.[7], the entropy of the solid at 2.65 m°K is $0.56R$. If the pressure difference between A and $T = 0$ is estimated by integration of the Clausius–Clapeyron equation $\Delta P = \int_{T_A}^{0} [(S_S - S_l)\,dT/\Delta V]$, assuming that the solid entropy decreases even linearly with temperature and that ΔV, the difference between the liquid and solid volumes, is essentially constant (consistent with direct volume measurements we have made here), then the pressure change we measure between A and the pressure maximum is too large by 25%. A substantial amount of the spin ordering in the solid must take place at temperatures well below A.

---

* See Ref. 5 for a more detailed discussion of this point.
† **Note added in proof:** With improved resolution we have now observed such a change (to be published in *Phys. Lett.*).

References

1. R.T. Johnson and J.C. Wheatley, *J. Low Temp. Phys.* **2**, 432 (1970).
2. D.D. Osheroff, W.J. Gully, R.C. Richardson, and D.M. Lee, this volume.
3. G.C. Straty and E.D. Adams, *Rev. Sci. Instr.* **40**, 1393 (1969).
4. D.D. Osheroff, R.C. Richardson, and D.M. Lee, *Phys. Rev. Lett.* **28**, 885 (1972).
5. H. Horner and L.H. Nosanow, *Phys. Rev. Lett.* **29**, 88 (1972).
6. E.R. Grilly, *J. Low Temp. Phys.* **4**, 615 (1971).
7. R.T. Johnson, O.V. Lounasmaa, R. Rosenbaum, O.G. Symko, and J.C. Wheatley, *J. Low Temp. Phys.* **2**, 403 (1970).

Low-Temperature Solid ^{3}He in Large Magnetic Fields*

R. T. Johnson, D. N. Paulson, R. P. Giffard,† and J. C. Wheatley

Physics Department, University of California, San Diego
La Jolla, California

Using continuous wave NMR, we have measured the bulk nuclear polarization of a liquid–solid ^{3}He sample compressionally cooled to below 5 m°K in a magnetic field of 54.5 kG (Larmor frequency of 177 MHz). Figure 1 shows the initial and maximum absorption lines for two compressional cooling runs. The lines were broadened to 155 G by applying a field gradient of 200 G/cm. This broadening permitted observation of the distribution of solid ^{3}He in the NMR coil. The lower absorption lines are the initial precompression signals, Fig. 1(a) representing all liquid and 1(b) representing some solid ^{3}He in the high-field end. The asymmetric all-liquid shape is caused by nonuniformities in the field gradient. The upper two absorption lines are the maxima obtained in the compression. In Fig. 1(b) we see a shape similar to that of the liquid, suggesting that the sample region is fairly uniformly filled with solid. In Fig. 1(a) the maximum signal is very unlike the liquid signal shape, suggesting that much liquid still remains in the sample region.

The absorption is proportional to the magnetization of the ^{3}He, while temperatures can be determined from a capacitive pressure transducer. The magnetization is calibrated using the liquid ^{3}He susceptibility and converted to polarization using the molar volumes of liquid and solid ^{3}He. Assuming the observation region is completely filled with solid ^{3}He, the largest peak polarization observed was 47% and the largest average polarization was 36%. The minimum temperature indicated by the pressure transducer was 3.1 m°K, which, using a mean field approximation,[1] would correspond to solid ^{3}He having a nuclear polarization of 64%. The presence of liquid ^{3}He will increase the experimentally deduced solid polarizations. For example, a 25% liquid–75% solid mixture would increase the above 47% to a 61% *solid* nuclear polarization. An interesting feature in the growth of the absorption lines during compression is that their basic shape did not change as more solid formed, suggesting that the solid ^{3}He forms on solid already present. This suggests using some means (e.g., a heater) initially to get the solid ^{3}He to form in the desired location.

To investigate the thermal relaxation time in solid ^{3}He, we studied the response of the magnetization to transient heating of two types. In one type we observed the response at one point on the absorption line to a several second pulse of rf energy at a frequency 3–4 MHz away from the resonant frequency. The absorption is decreased by the heating but then increases to a value slightly less than the initial value on a

---

* Supported by U.S. Atomic Energy Commission under Contract No. AT (04-3)-34, PA 143.
† Present address: Clarendon Laboratory, Oxford, England.

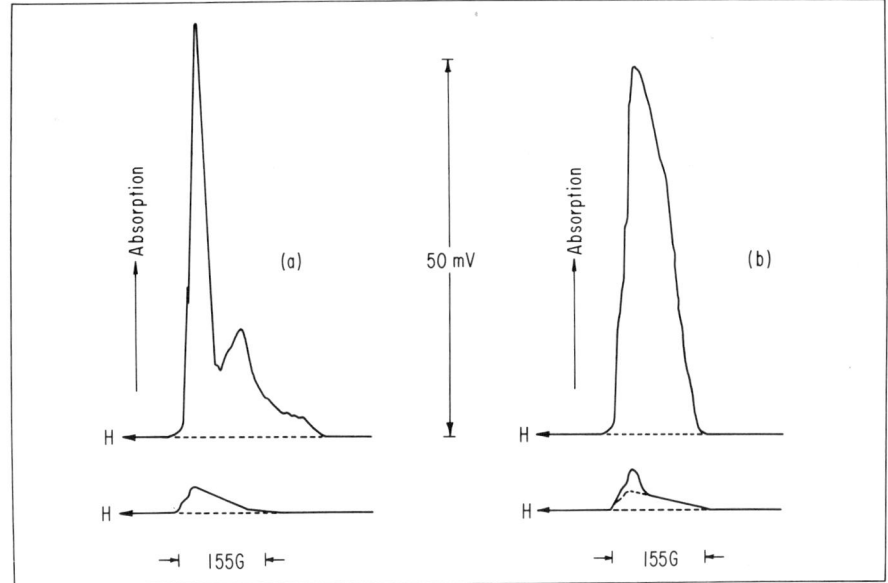

Fig. 1. Initial (bottom) and maximum absorption lines for two compressions.

time scale of several hundred seconds. The short time scale of the recoveries indicates the existence of both rapid, long-range mechanisms for removing the solid ^{3}He nuclear spin energy and a source of refrigeration for the heated spins. We believe there are relaxation zones in the solid ^{3}He, such as crystal defects as suggested by Giffard et al.,[2] which convert spin energy into phonons and vice versa. The energy diffusivity is 10^{-7} cm^2/sec, requiring characteristic lengths of 50 μm to explain the observed time scale of the recoveries. From the observed growth of the absorption line it seems unlikely that the solid ^{3}He formed in such small crystals. However, crystal defects on a 50-μm scale would not be unreasonable.[2] The refrigeration necessary for recovery is available from colder solid and liquid ^{3}He elsewhere in the cell (only 15% of the cell volume is in the NMR coil) and from conversion of liquid to solid ^{3}He. The quantitative description of the recoveries varied widely. Within scatter the recoveries were not correlated with either temperature (maximum about 45 m°K) or the observation position on the absorption line.

We also tried a localized heating of the solid ^{3}He nuclear spins using rf power at a frequency close to the resonant frequency—a resonant burn.[3] In a burn the spins are saturated locally on a surface of constant field and the recovery is thus sensitive to short-range phenomena like diffusion, as well as to long-range phenomena. The procedure was to sit at one location on the absorption line, apply heat for typically 2 sec, and then sweep the magnetic field back and forth through the burned region. Qualitatively, one would expect the burn to be a dip in the polarization as a function of field which would widen in time from spin diffusion and decrease in area because of thermal relaxation. The actual response to a burn is shown in Fig. 2 for both a low and a high temperature, with both upper limit estimates determined from the polarization assuming all solid ^{3}He at the observation point. Only a very small

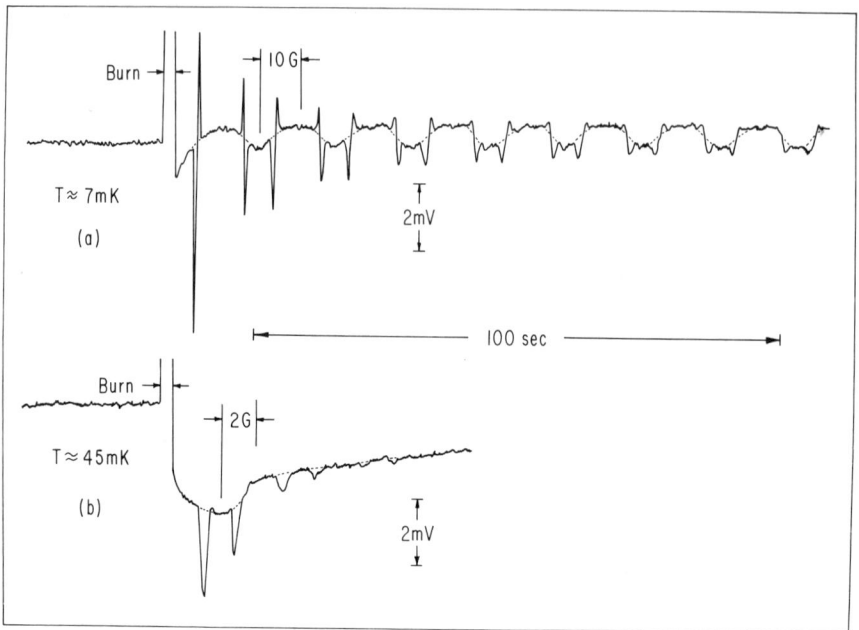

Fig. 2. Low- and high-temperature resonant burns.

section of the absorption line is represented here. The burn in no case wiped out the total absorption in the small region affected, and the changes in absorption shown are small compared with the total. The asymmetric burn shape, most prominent at low temperatures, is caused by the contribution of the solid ^{3}He nuclear magnetization to the local magnetic field. For unity polarization solid ^{3}He has a magnetization of $\frac{1}{4}$G, and a burn changes the magnetization on the scale of 25 μm. The resulting changes in the local field gradient are thus significant compared to the externally applied 200 G/cm. The absorption is proportional to the magnetization density per unit field or to the magnetization divided by the field gradient. A dip in the magnetization contributes an asymmetric dip plus rise to the field gradient, proportional to the gradient of the magnetization. The magnetization and field gradient thus combine to give the observed asymmetric shape. The solid ^{3}He magnetization decreases as the temperature increases and the observed asymmetry also decreases. For the high-temperature burn it is obvious that the net area is decreasing rapidly with time, indicating rapid thermal relaxation. For the low-temperature burn it is difficult to extract a net area, due to the presence of the large and persistent asymmetry. However, the burned region always became unmeasurably small on the scale of 100 sec or less, a result consistent with the nonresonant heating recoveries. No quantitative results could be deduced, although both types of heating suggest short thermal relaxation times for low-density solid ^{3}He in large magnetic fields and down to temperatures around 5 m°K.

We also present a reanalysis of previous measurements[4] of the ^{3}He melting curve at low temperatures and in high magnetic fields. We used a ^{54}Mn-in-iron γ-ray anisotropy thermometer in magnetic fields from 2.1 to 63.6 kG. The initial

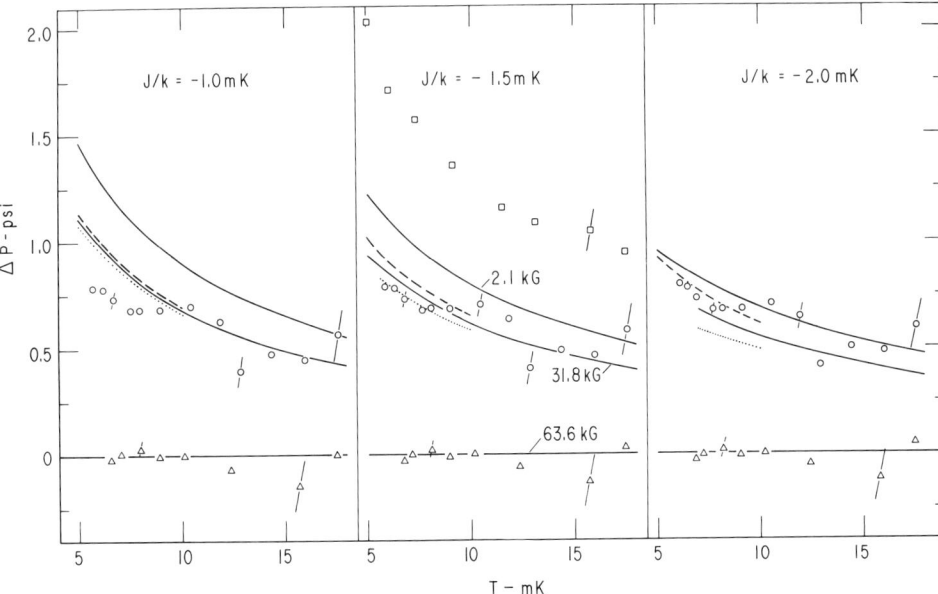

Fig. 3. Experimental melting pressure minus the theoretical melting pressure for 63.6 kG as a function of temperature for three values of the exchange constant J. Triangles, 63.6 kG; circles, 31.8 kG; and squares, 2.1 kG data. Representative error bars only indicate one standard deviation in the counting rate for the ^{54}Mn thermometer. The smooth curves are theoretical. The dotted and dashed curves below and above the 31.8-kG curve show the effect of an expansion vs. a mean field approximation to the solid ^{3}He entropy —the smooth curve is the average. (1 atm = 14.70 psi.)

analysis relied on the precision of the low-field measurements and indicated a much larger low-field dependence of the melting pressure on field than anticipated. Due to possible thermometric difficulties at low fields[5] and to the results of other measurements of the low-field dependence of the melting pressure on magnetic field which indicated the expected effect,[6] our reanalysis is based on the assumption that the highest-field theoretical melting curve is correct. Figure 3 shows our data and theoretical curves for the change in melting pressure with field for three values of the solid ^{3}He exchange constant J. The agreement between theory and experiment is reasonable between 31.8 and 63.6 kG. There is no major dependence of our analysis on J. Although the low-field measurements are not in agreement with theory, as expected, the low-field *calculated* melting curve is in quite good agreement with previous zero-field melting curve measurements made in our laboratory. Kirk and Adams[7] have measured the temperature derivative of the pressure at constant volume for solid ^{3}He as a function of field and found unexpected results at high fields. Their results have a much different theoretical basis than our melting curve measurements, and our results neither support nor refute their measurements.

References

1. R.T. Johnson and J.C. Wheatley, *J. Low Temp. Phys.* **2**, 423 (1970).
2. R.P. Giffard, W.S. Truscott, and J. Hatton, *J. Low Temp. Phys.* **4**, 153 (1971).
3. E.R. Hunt and J.R. Thompson, *Phys. Rev. Lett.* **20**, 249 (1968).
4. R.T. Johnson, R.E. Rapp, and J.C. Wheatley, *J. Low Temp. Phys.* **6**, 445 (1972).
5. N.J. Stone, private communication; R. Lines and N.J. Stone (to be published).
6. D.D. Osheroff, R.C. Richardson, and D.M. Lee, *Phys. Rev. Lett.* **28**, 885 (1972).
7. W.P. Kirk and E.D. Adams, *Phys. Rev. Lett.* **27**, 392 (1971).

High-Magnetic-Field Behavior of the Nuclear Spin Pressure of Solid ^{3}He*

W. P. Kirk and E. D. Adams

Physics Department, University of Florida
Gainesville, Florida

Much of the low-temperature behavior, particularly the ordering of the nuclear spins, in solid ^{3}He is determined by a sizable exchange energy J. Hence solid ^{3}He provides a unique system for studying the exchange energy and associated nuclear spin ordering effects. In order to gain more information about the nature of the exchange interaction and to make comparisons between theory and experiments, one has two options. The measurements can be extended down to the critical region $T_c \simeq 3|J|/k \simeq 2$ m°K, or one can, as suggested by Goldstein,[1] make measurements on the solid in the presence of strong magnetic fields at temperatures somewhat higher than T_c. The latter situation is in some ways more tractable experimentally.

In the comparison of experiment with a theoretical model, the initial attempt has been to use a spin-$\frac{1}{2}$ Heisenberg model with only pair exchange between the nearest neighbors. The appropriate Hamiltonian $\mathscr{H}$ is then of the form

$$\mathscr{H} = -2J_2 \sum_{i<j} \mathbf{I}_i \cdot \mathbf{I}_j - \sum_i \boldsymbol{\mu}_i \cdot \mathbf{H} \qquad (1)$$

where J_2 is the exchange energy for two-particle exchange, $\mathbf{I}_i$ is the nuclear spin operator for atom i, and the double summation sums all nearest-neighbor pairs $(i.j)$ once. Since the atoms have a magnetic moment μ and are in the presence of a magnetic field $\mathbf{H}$, then the Zeeman Hamiltonian is included as well by in the second term on the right of Eq. (1). From $\mathscr{H}$ given in Eq. (1) Baker et al.[2] have developed a high-temperature series expansion of the partition function in terms of $x = J_2/kT$ and $y = \mu H/kT$. The thermodynamic pressure, as studied in this work, is found by differentiation of the partition function with respect to volume V and is expressed by

$$P_V(T, H) = \frac{RT}{V} \frac{\partial \ln |J_2|}{\partial \ln V} \left[3x^2 - \frac{3x^3}{2} + \cdots + y^2(2x + 12x^2 + 52x^3 + \cdots) \right. \\ \left. + y^4(-1.33x - 23x^2 + \cdots) \right] \qquad (2)$$

We have measured $P_V(T, H)$ in solid ^{3}He in applied magnetic fields of about 0, 40, 60, and 70 kG for samples with molar volumes between $V \simeq 23.0$ and 24.0 cm^3/mole and over a temperature range of ~ 17–120 m°K. Some of the results of this work have been published elsewhere.[3] Figure 1 presents the results as the pressure

---
* Work supported in part by the National Science Foundation.

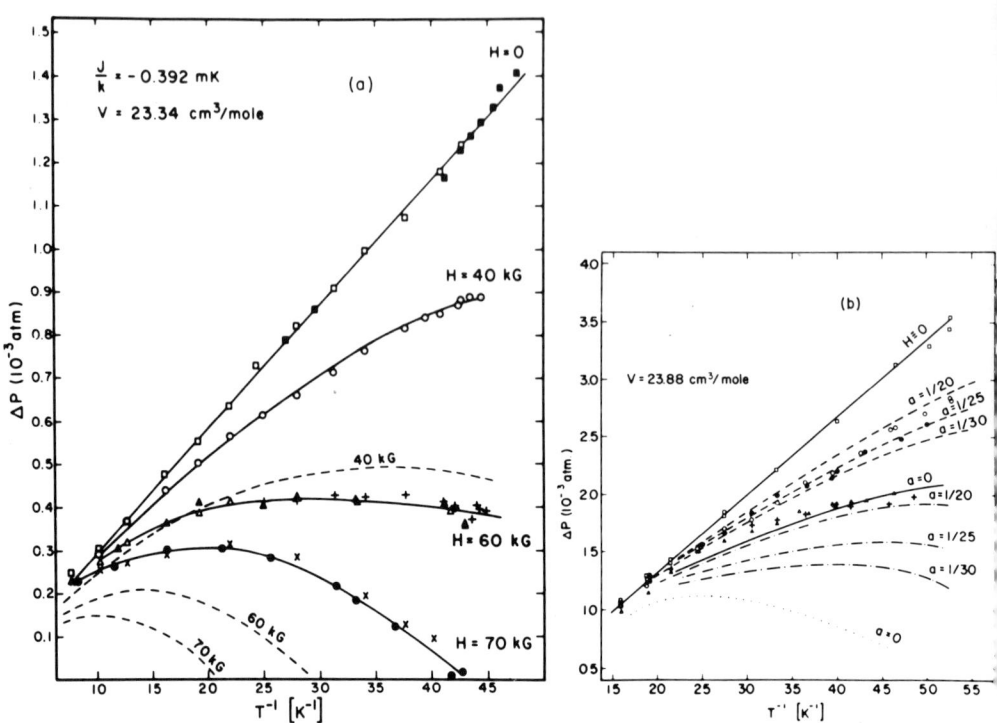

Fig. 1. (a) Pressure difference vs. T^{-1} for $V = 23.34$ cm^3/mole in fields of 70, 60, 40, and 0.5 ($H \simeq 0$) kG. Data taken on warming or cooling are shown by different symbols for a given H. Solid lines serve simply to connect points. Dashed curves are calculated behavior based on Eq. (2). (b) Pressure difference vs. T^{-1} for $V = 23.88$ cm^3/mole. Squares, circles, and triangles and crosses show data for fields at $H \simeq 0$, 40, and 60 kG, respectively. Calculated behavior, Eq. (2), shown by: solid line for $H \simeq 0$; solid line ($a = 0$) for 40 kG; and dotted line ($a = 0$) for 60 kG. Behavior calculated by Zane,[6] Eq. (3), is shown by dashed curves for $H = 40$ kG and dot-dashed curves for $H = 60$ kG (labeled $a = 1/20, 1/25, 1/30$).

change ΔP vs. T^{-1}. Data shown in Fig. 1(b) are for a sample with ~ 400 ppm ^{4}He impurity concentration. In Fig. 1(a) the ^{4}He concentration was ~ 2 ppm. The pronounced downward curvature of the pressure for increasing magnetic fields is a consequence of the effective exchange energy J being negative. This can be understood qualitatively from the spin-$\frac{1}{2}$ Heisenberg model. However, from our results it can also be shown by a general thermodynamic argument* that $J < 0$ without the necessity of using a Heisenberg model. In three earlier NMR susceptibility measurements[5] $J < 0$ has been found, and we consider the confirmation significant because of the completely different measuring methods.

In Fig. 1(a) the dashed lines show curves as computed from Eq. (2). Quantitative disagreement of nearly a factor of 2.5 is apparent. In Fig. 1(b) the calculated behavior using Eq. (2) is shown by the solid and dotted curves labeled with $a = 0$; here again the discrepancy is about a factor of two. We find in all our results that for a particular value of V the pressure data taken in various magnetic fields scale as H^2. This field dependence is as expected from thermodynamic arguments. Qualitative agreement

---

* More details are given in a review article by Trickey et al.[4]

with Eq. (2) is indicated by the fact that at a high enough field and small J the pressure is observed to go through a maximum as shown by the data in Fig. 1(a).

As suggested by us earlier,[3] perhaps the quantitative disagreement could be associated with the fact that only nearest-neighbor exchange was involved in the above Heisenberg model. Zane[6] has considered the possibility of higher-order exchange, in particular triple exchange, which is represented by a Hamiltonian of the form

$$\mathcal{H}^{(3)} = 2J_3 \sum_{i<j<k} (\mathbf{I}_i \cdot \mathbf{I}_j + \mathbf{I}_j \cdot \mathbf{I}_k + \mathbf{I}_k \cdot \mathbf{I}_i)$$

where J_3 is the exchange of three particles permuted cyclically at sites (i, j, k). The importance of the term $\mathcal{H}^{(3)}$ is that effectively it leads to ferromagnetic ordering between next-nearest neighbors. Thouless[7] first pointed out that in ^{3}He two-particle exchange should be antiferromagnetic, while three-particle exchange should be ferromagnetic. Adding $\mathcal{H}^{(3)}$ to Eq. (1), Zane obtained the following expression for the pressure:

$$P_V(T, H) = \frac{RT}{V} \frac{\partial \ln |J_2|}{\partial \ln V}$$

$$\times \{3x^2[1 - 6a(1+b) + 48a^2b] + \cdots + 2xy^2(1 - 9ab) + \cdots\} \qquad (3)$$

where $a = -J_3/J_2$, $x = J_2/kT$, $y = \mu H/kT$, and

$$b = \frac{d \ln |J_3|/d \ln V}{d \ln |J_2|/d \ln V}$$

Shown in Fig. 1(b) are some of our data compared with values of the pressure given by Eq. (3) using $b = 3/2$ and trial values of $a = 1/30, 1/25, 1/20$. As indicated by the small values of the parameter a, the theoretical fits to the data are improved by the addition of a small amount of triple exchange. This also indicates how sensitive the pressure measurements are to the more detailed aspects of the Heisenberg model. Unfortunately, a value for the parameter a in Zane's work is difficult to determine from a more fundamental theory of exchange.[8]

In our early work for $V \simeq 24.0$ cm^3/mole we found an upper limit for thermal equilibrium times of <1 min.[3] These short times are confirmed by the recent high-field susceptibility measurements of Johnson et al.[9] for solid ^{3}He on the melting curve at $T \gtrsim 5$ m°K. We now find for $V \simeq 23.0$ cm^3/mole that the equilibrium times in approximately zero- and 40-kG fields remain short; however, at 60 and 70 kG we observe somewhat longer equilibrium times. We were able to investigate these longer times by observing the behavior of the pressure as a function of time after a change in temperature.

In Fig. 2 we show a chart recording of the output from the pressure strain gauge with $H = 70$ kG. Upon warming from 24 to 27 m°K we first see the pressure decrease for about 1 min (about the time needed to establish a new, warmer equilibrium temperature for the dilution refrigerator) followed by a much slower increase. This behavior can be understood on the basis of a short exchange–lattice time constant and a longer Zeeman–exchange time constant. Thus as the exchange and lattice systems approach equilibrium with the Zeeman system relatively isolated, the

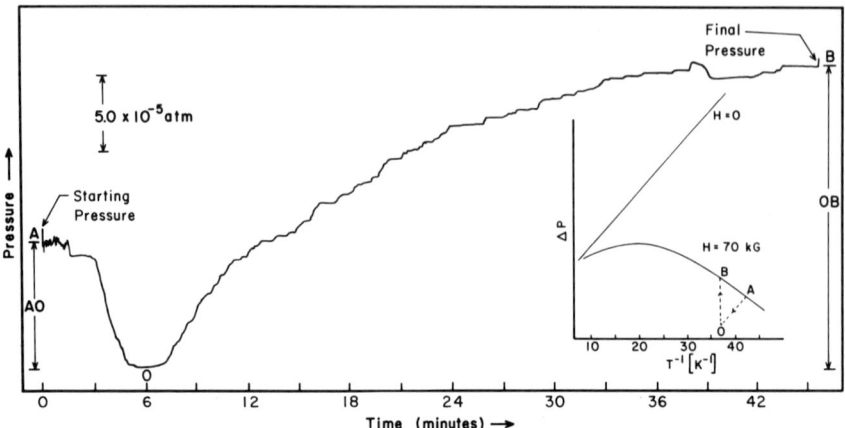

Fig. 2. Time dependence of the pressure recorded from the output of the capacitance strain gauge when T changed from 24 to 27 m°K. T was stabilized after ~ 1 min. Inset shows ΔP vs. T^{-1} schematically.

behavior is as in zero field. The magnitude of the initial pressure decrease for a certain ΔT corresponds reasonably well to the change one expects of the pressure at $H = 0$ for the same ΔT. This initial decrease is shown as the dashed line AO in the inset (plot of ΔP vs. T^{-1}) of Fig. 2 and it is drawn parallel to the $H = 0$ curve. At point O a turnaround takes place and the pressure begins to recover exponentially (shown in the inset along the dashed line OB) to a new value at point B. We interpret this recovery process as evidence for the Zeeman system coming to thermal equilibrium with both the exchange and lattice systems. The fact that the pressure (OB) recovered to a larger value than initially is easily understood by noticing that point B in the inset is higher than point A. Hence on warming from a low temperature a larger equilibrium pressure is expected. We find that the time constant τ for exponential recovery depends on the density of the sample. The values of τ were determined from the chart recordings by using a Mangelsdorf[10] analysis. The temperature dependence of τ was investigated in samples with three different density values and with ~ 2 ppm of ^{4}He impurity and the results are shown in Fig. 3.

We have analyzed the τ values and compared them with the work of Thomlinson et al.,[11] who studied T_1, the NMR spin-relaxation time, in the temperature region in which the ^{3}He exchange frequency modulation of the magnetic dipole field is the dominant relaxation mechanism. We find reasonable agreement with Thomlinson et al. which we consider as evidence that the behavior of the pressure recovery (OB, Fig. 2) is determined by a Zeeman–exchange relaxation mechanism. Furthermore this interpretation is substantiated by the temperature independence observed for τ (Fig. 3) since the NMR Zeeman–exchange relaxation mechanism is known to be temperature independent as well. More specifically, our τ values agree best with a T_1 analysis which employs a spectral density function of the form $J_1(\omega/\omega_e) = \text{const} \times \exp(-\omega/\omega_e)$, where ω_e is related to the exchange frequency (J_2) and ω is the Larmor frequency. Then T_1 is related to $J_1(\omega/\omega_e)$ by $T_1^{-1} = J_1(\omega/\omega_e) + 4J_1(2\omega/\omega_e)$. We consider our sample to be a powder specimen because of the many wires in the sample chamber and have accounted for this in the com-

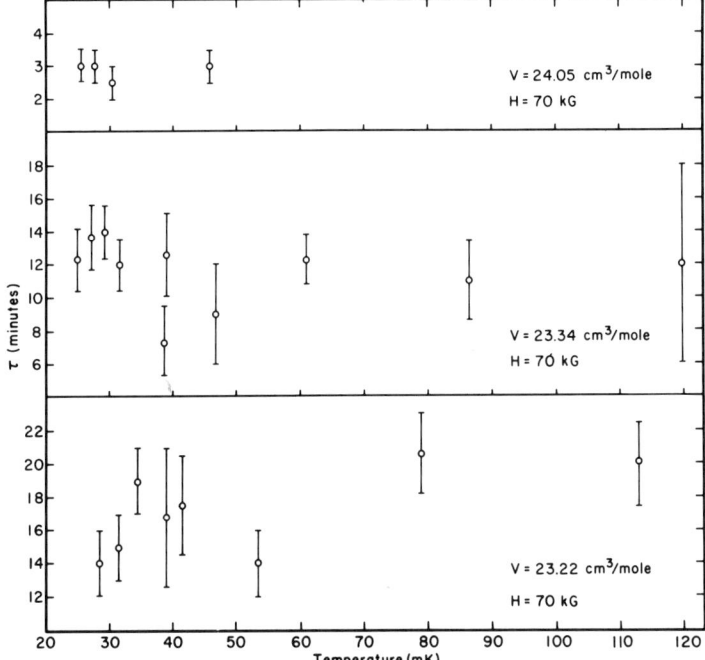

Fig. 3. Time constant τ vs. temperature for three molar volume samples in a 70-kG field.

parison with the work of Thomlinson et al., where the sample was more nearly a large crystal. In the work of Thomlinson et al. V (or ω_e) was held constant and ω varied, while in our case ω (or H) was constant and ω_e (or V) was varied. Comparison can be made at equal values of ω/ω_e in the two cases. Thomlinson et al. reported $T_1 = 0.8$ min for $\omega/\omega_e = 4.7$ as compared to our $\tau = 3$ min at $V = 24.05$ cm^3/mole; $T_1 = 14.4$ min for $\omega/\omega_e = 8.2$ compared to $\tau = 12$ min at $V = 23.34$; and $T_1 = 30.5$ min for $\omega/\omega_e = 9.1$ compared to $\tau = 18$ min at $V = 23.22$. The comparison at $V = 24.05$ cm^3/mole is not as good primarily because of the inherent limitation of our method to measure low values of τ. However, we consider the agreement sufficient to suggest that we have been able to identify and measure a nuclear-spin relaxation property in solid ^{3}He by a non-NMR means.

A brief description of the apparatus has been given elsewhere.[3] Several checks have been made so that spurious effects may be ruled out, including precautions to circumvent difficulties with magnetic field disturbance of the temperature sensors. We found that samples of the same molar volume but with different ^{4}He impurity concentration did not show different behavior of $P_V(T, H)$. The behavior of the empty pressure gauge at ~ 0 and 40 kG was observed and found to be temperature independent. Also the gauge (empty or filled) showed no indication of hysteresis in temperature or pressure and had a resolution of 2×10^{-6} atm. The magnetic field produced by the superconducting solenoid has been checked using a search coil. The current–field relation furnished by the manufacturer was verified to within the 5% accuracy of this method.

Thus we conclude that the observed departure from the behavior expected from the Heisenberg Hamiltonian when only two-body nearest-neighbor exchange is included is a real effect. Three-body exchange offers an interesting possibility for explanation of this discrepancy; however, further experimental and theoretical work is needed for a complete understanding of this problem.

Acknowledgments

We acknowledge the competent assistance of Brad Kummer and appreciate discussions with Dr. Allan Greenberg on aspects of this work.

References

1. L. Goldstein, *Phys. Rev.* **171**, 194 (1968).
2. G.A. Baker, Jr., H.E. Gilbert, J. Eve, and G.S. Rushbrooke, *Phys. Rev.* **164**, 800 (1967).
3. W.P. Kirk and E.D. Adams, *Phys. Rev. Lett.* **27**, 392 (1971).
4. S.B. Trickey, W.P. Kirk, and E.D. Adams, *Rev. Mod. Phys.* **44**, 668 (1972).
5. W.P. Kirk, E.B. Osgood, and M. Garber, *Phys. Rev. Lett.* **23**, 833 (1969); J.R. Sites, D.D. Osheroff, R.C. Richardson, and D.M. Lee, *Phys. Rev. Lett.* **23**, 836 (1969); P.B. Pipes and W.M. Fairbanks, *Phys. Rev. Lett.* **23**, 520 (1969).
6. L.I. Zane, *Phys. Rev. Lett.* **28**, 420 (1972).
7. D.J. Thouless, *Proc. Phys. Soc. (London)* **86**, 893 (1965).
8. A.K. McMahan, *Bull. Am. Phys. Soc.* **17**, 452 (1972).
9. R.T. Johnson, D.N. Paulson, R.P. Giffard, and J.C. Wheatley, *J. Low Temp. Phys.* **10**, 35 (1973).
10. P.C. Mangelsdorf, Jr., *J. Appl. Phys.* **30**, 442 (1959).
11. W.C. Thomlinson, J.F. Kelly, and R.C. Richardson, *Phys. Lett.* **38A**, 531 (1972).

The Interaction of Acoustic Phonons with Nuclear Spins in Solid ^{3}He*

K. L. Verosub†

Department of Physics, Stanford University
Stanford, California

During the past decade many conventional nuclear magnetic resonance experiments have been performed on solid ^{3}He (with inescapable ^{4}He impurities) at temperatures between 1.00 and 0.05°K. Guyer *et al.*[1] have synthesized the experimental data into a coherent theoretical framework. In their description the Zeeman energy is coupled to the lattice via three different particle motion excitations: vacancy waves, which result from the motion of vacancies in the lattice, tunneling excitation waves, which result from the pairwise interchange of two ^{3}He atoms on adjacent lattice sites, and mass fluctuation waves, which result from the pairwise interchange of a ^{3}He atom and a ^{4}He atom.

In the work reported here we perturb the solid helium by introducing acoustic phonons generated continuously by a piezoelectric quartz crystal resonating at 2.0 MHz. The ultrasonic energy creates a high phonon density over a narrow frequency range, with a corresponding temperature much higher than that of the lattice. By monitoring the continuous ^{3}He nuclear magnetic resonance signal we can study the ability of the phonons in the hot spot to transfer energy to the Zeeman system via the excitations. The experiments are performed at temperatures between 50 and 150 m°K. In this region the rate-determining interaction in the spin–lattice relaxation is that between the excitations and the phonons. The Zeeman system and the excitations come to equilibrium with each other in times on the order of tens of seconds, while the combined Zeeman–excitation system comes to equilibrium with the thermal phonons in times on the order of 300–3000 sec.[2]

The piezoelectric transducer is immersed in the solid helium in order to efficiently couple energy into the sample. Although all of the acoustic phonons are eventually thermalized, the corresponding heat leak is negligible. The lattice temperature of the solid helium is inferred from carbon resistance thermometry on a resistor mounted on the sample chamber and from nuclear magnetic resonance on a bundle of fine copper wires immersed in the solid helium. By changing the power into the temperature control heater, the thermal response of the carbon resistor, ^{3}He thermometer, and copper thermometer can be compared. The results are shown in Fig. 1. The error bars are typical for each thermometer. It is clear that the thermometers track each other very closely. The time lag between any pair is not more than

---

* Work supported by National Science Foundation and Advanced Research Projects Agency through Center for Materials Research, Stanford University.
† Present address: Amherst College, Amherst, Massachusetts.

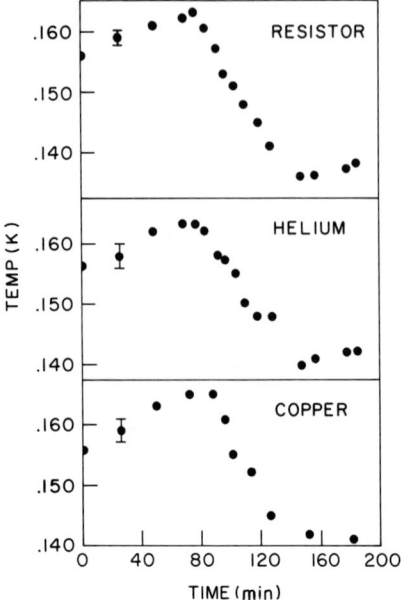

Fig. 1. Comparison of thermal response of thermometers.

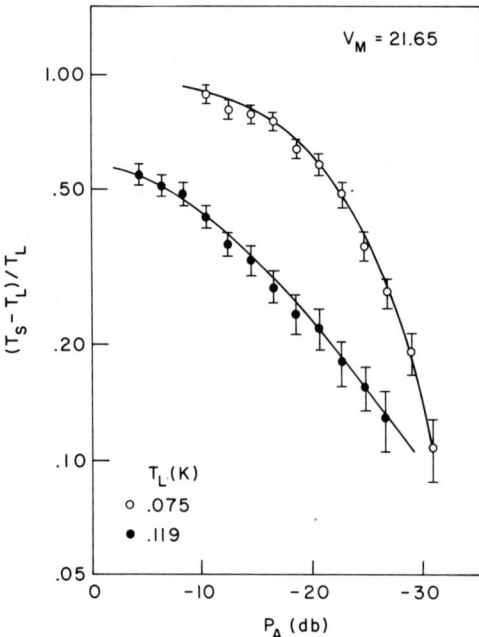

Fig. 2. Temperature difference between spins and lattice as a function of acoustic power.

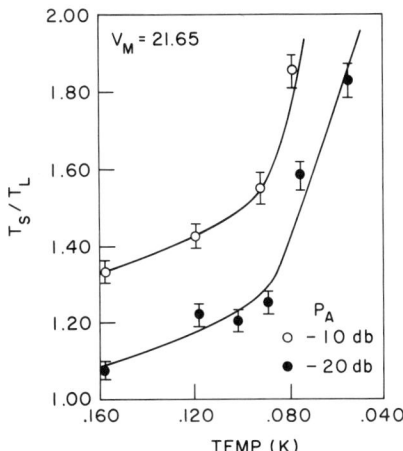

Fig. 3. Temperature difference between spins and lattice as a function of lattice temperature.

5 min. In addition each one is capable of changing at a rate of at least 1 m°K/min. Another experiment was performed at 50 m°K using only the resistor and the copper thermometer. These two thermometers bracket the lattice temperature of the solid helium. The results are the same as those at 150 m°K. Over a range of 150–50 m°K we are able to see changes in the solid ^{3}He lattice temperature of a few millikelvin in a few minutes.

After the sound has been turned on the ^{3}He spin temperature T_3 slowly rises while the lattice temperature of the solid helium T_L remains constant. Careful measurements indicate that the change in T_3 follows a $(1 - e^{-t/\tau})$ time dependence. The value of τ varies from about 8 min at 150 m°K to 12 min at 50 m°K, so that T_3 reaches its equilibrium value T_s in 30 min at 150 m°K to 50 min at 50 m°K. Thus we are able to heat preferentially the combined Zeeman–excitation system and to maintain a steady-state temperature difference between T_s and T_L. Values of T_s/T_L as high as 2.00 have been observed.

We have measured the steady-state value of T_s as a function of ultrasonic power at several temperatures between 150 and 50 m°K. The power is measured in terms of the dB attenuation of the 5-V peak-to-peak driver of the piezoelectric crystal. The data are conveniently presented as a semilog plot of $(T_s - T_L)/T_L$ vs. the power level. Fig. 2 shows the results at two lattice temperatures. The uncertainty in measuring T_s is about 2%. Since the uncertainty in T_L is less than 1%, the main error in $(T_s - T_L)/T_L$ is the error in T_s.

There are two qualitative features of the results. In the first place the heating effect is greatest at the lowest temperatures. In Fig. 3 we show the value of T_s/T_L as a function of T_L at two different power levels. The effect increases rapidly below about 100 m°K. Second, the value of $(T_s - T_L)/T_L$ appears to level off as the sound intensity is increased. This feature may be an artifice of the way the data are plotted.

It is unlikely that the known interactions between the three excitations and the lattice can be responsible for the heating effect. These interactions are all two-phonon Raman-type processes involving the entire phonon spectrum. They should be relatively insensitive to the presence of the hot spot if the energy density of the acoustic phonons E_A is small compared to the total thermal phonon energy density

E_T. In this experiment $E_A/E_T \approx 10^{-3}$. On the other hand, the rate at which a one-phonon process occurs is proportional to the phonon energy density over some frequency band. In this experiment $E_A/E_P \approx 10^6$, where E_P is the thermal phonon energy density for a 200-kHz bandwidth around 2.0 MHz. Thus the acoustic phonons will magnify a one-phonon process by a factor of one million. We suggest that we may be observing a direct one-phonon interaction probably between tunneling excitations and acoustic phonons.

These experiments were performed on a sample with a molar volume of 21.65 cm^3. Additional measurements have been made at three other molar volumes. We will analyze these data and publish them elsewhere shortly.

Acknowledgment

The author wishes to thank Dr. W. M. Fairbank for his encouragement and support for this experiment.

References

1. R.A. Guyer, R.C. Richardson, and L.I. Zane, *Rev. Mod. Phys.* **43**, 532 (1971).
2. R.P. Giffard, W.S. Truscott, and J. Hatton, *J. Low Temp. Phys.* **4**, 153 (1971).

4
Helium Monolayers

Motional Narrowing in Monolayer ^{3}He Film NMR*

R. J. Rollefson

Department of Physics, University of Washington
Seattle, Washington

Recent specific heat measurements[1-3] of monolayer helium films adsorbed on graphite have revealed striking new behavior. These measurements, as well as adsorption isotherms,[4] indicate that the graphite basal plane presents a particularly homogeneous surface for adsorption. The NMR measurements reported here were made with the intention of (1) continuing the study of monolayer helium film properties revealed by the heat capacity measurements using a different experimental technique, and (2) investigating the NMR in two dimensions for a monatomic, chemically inert system on a homogeneous substrate.

A wide-line spectrometer of standard Rollins design was used, operating at 20.5 MHz with lock-in detection. Graphitized carbon black† purified by an extended bakeout at 1000°C provided the substrate material, with a monolayer capacity of 3.17 cm^3 STP determined by a ^{3}He adsorption isotherm.

The observed linewidth as a function of temperature and fractional monolayer coverage x is shown in Fig. 1. The value quoted is the difference between the maximum and minimum of the derivative signal. At low temperatures the linewidth shows considerable coverage dependence, while at higher temperatures it attains a narrow, coverage-independent value, felt to be instrumentally determined. The change in linewidth with temperature is interpreted as motional narrowing[5] caused by an increased mobility of the helium atoms along the graphite surface. A broad line indicates a relatively lower adatom mobility.

On the basis of this interpretation the data can be divided into two coverage ranges showing distinctly different effects for changes in film density. For the higher-coverage films ($x > 0.7$, Fig. 1B) a low-mobility state exists for temperatures less than a temperature $T_{narrowing}$ which increases monotonically with coverage. On the other hand, for lower-coverage films (Fig. 1A) a low-mobility state is observed below 3°K for $x = 0.6$, while an *increase* and a *decrease* in the coverage each results in a *higher* low-temperature mobility.

The behavior observed for the higher-coverage films is very suggestive of melting, although it must be noted, as Resing[6] has pointed out, that a change in mobility as shown by motional narrowing does not necessarily indicate an abrupt transition. In the present case an abrupt transition has already been demonstrated

---

* Supported by the National Science Foundation.
† The graphitized carbon black was grade Sterling FT (similar to graphitized P-33) obtained from the Cabot Corp., Boston, Mass. Electron microscopy studies show the graphitized black to be in the form of small polyhedra ~ 0.2 μm in diameter, with the basal plane exposed.

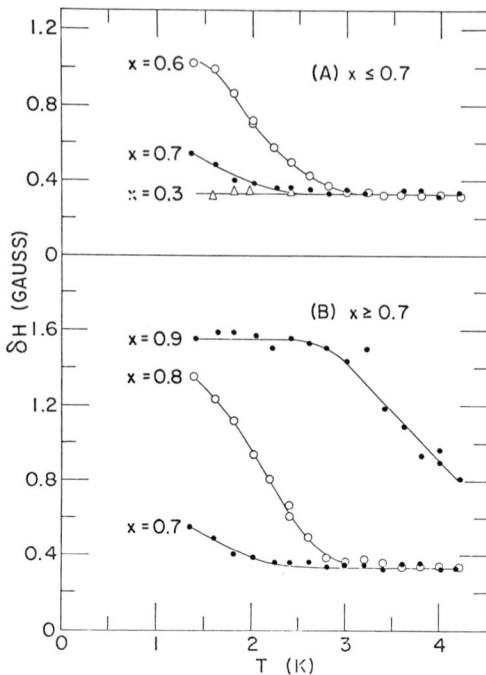

Fig. 1. Experimental linewidth as a function of temperature. x is the fractional ^{3}He monolayer coverage.

by the specific heat measurements,[2] in which sharp peaks in the heat capacity were observed. These peaks were interpreted as due to melting based on a correspondence in the density dependence of the melting and Debye temperatures between the films and bulk helium. The present measurements lend support to this interpretation by giving direct evidence of an increase in mobility at the melting temperature.

The data for coverages around $x = 0.6$ do not lend themselves to such an interpretation, for here an increase as well as a decrease in adatom density results in an increase in mobility. Again the specific heat[1] provides a clue: A sharp peak at 3°K was observed for $x = 0.6$ but was absent for both higher and lower coverages. It was proposed that this peak resulted from a transition to an ordered state in which the helium atoms occupy every third substrate potential minimum, a condition which exactly uses up the available helium atoms for 0.6 monolayer. The NMR results show that this ordered state is also one of low mobility, while for higher and lower coverages the lack of film–substrate registry allows an increased mobility.

The results presented here can be summarized in a phase diagram, originally proposed on the basis of the specific heat measurements,[7] as shown in Fig. 2.* On the right-hand side of the figure are schematic representations of the NMR linewidth measurements.

With the assumption that the low-temperature state for $x = 0.6$ consists of

---

* Only those specific heat results pertinent to an interpretation of the present NMR data are shown in Fig. 2.

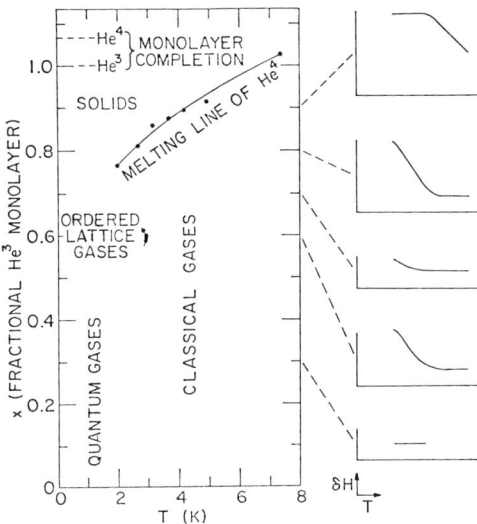

Fig. 2. Phase diagram for monolayer helium films (see Ref. 7). Data points denote peaks in the specific heat. The diagrams to the right are schematic representations of the data in Fig. 1. Although the melting line shown is for ^{4}He, the heat capacity of a single coverage of ^{3}He with $x \simeq 1$ was measured and showed similar behavior.

a regular triangular array ordered on the substrate potential, the second moment can be calculated. Since the signal-to-noise ratio was too small to allow a reliable measurement of the second moment, the linewidth was related to the second moment by assuming a Gaussian line shape. Using the lattice parameters of the graphite basal plane to determine an interatomic spacing for the helium film of 4.26 Å, a rigid lattice linewidth of 0.99 G was calculated, allowing for instrumental broadening. This is 3% smaller than the value measured at the lowest temperature, where the linewidth appears to have nearly reached its rigid lattice value, an agreement well within experimental uncertainty. If it is assumed that for coverages near one monolayer the film again orders in a regular triangular array, the second moment will scale as x^3. Although this seems a reasonable assumption, the lack of film–substrate registry makes it somewhat less certain than for $x = 0.6$. The linewidth calculated under this assumption is 9% larger than the lowest temperature point for $x = 0.8$ (the ultimate low-temperature width is probably somewhat greater than the value measured here, improving the agreement) and 12% larger than the low-temperature width for $x = 0.9$. Thus, as monolayer completion is approached the measured low-temperature width, i.e., the linewidth in the low-mobility state, becomes progressively less than that predicted by a second-moment analysis. It is known that in bulk solid helium the line is considerably narrowed by the exchange interaction. Perhaps for the higher-coverage films exchange is limiting the low-temperature linewidth, while for $x = 0.6$ the localization of the atoms by the substrate inhibits exchange. If this is so, it would appear that the effect here has the opposite density dependence to that observed in the bulk, though these preliminary data are rather scant. More extensive measurements near monolayer coverages will be required.

It is interesting to note the similarity between the motional narrowing observed here for two-dimensional films and that observed in many three-dimensional systems.[8] This similarity was not anticipated in the only paper[9] known to the author which attempts to calculate the effect of translation in two dimensions on the NMR linewidth. In this paper the authors assume that the film motion can be described by a diffusion equation. The spin correlation function, obtained numerically, is found to decrease rapidly to about one-half its initial value but to decrease very slowly thereafter—much more slowly than an exponential. In light of this slow decrease for large times and for the sake of mathematical tractability, the authors replace the correlation function by a constant, thus removing the correlation time from the expression for the linewidth. It is clear from the narrowing observed in helium films that the state of motion of the film has a substantial effect on the observed linewidth. It may be that by taking into account the initial rapid decrease in the correlation function, theory and experiment could be brought into better accord.

References

1. M. Bretz and J.G. Dash, *Phys. Rev. Lett.* **27**, 647 (1971).
2. M. Bretz, G.B. Huff, and J.G. Dash, *Phys. Rev. Lett.* **28**, 729 (1972).
3. D.C. Hickernell, E.O. McLean, and O.E. Vilches, *Phys. Rev. Lett.* **28**, 789 (1972).
4. A. Thomy and X. Duval, *J. Chim. Phys. (Paris)* **66**, 1966 (1969).
5. N. Bloembergen, E.M. Purcell, and R.V. Pound, *Phys. Rev.* **73**, 679 (1948).
6. H.A. Resing, *J. Chem. Phys.* **43**, 669 (1965).
7. M. Bretz, J.G. Dash, D.C. Hickernell, E.O. McLean, and O.E. Vilches (to be published).
8. A. Abragam, *The Principles of Nuclear Magnetism*, Oxford University Press (1961), Chapter 10.
9. A.A. Kokin and A.A. Izmestev, *Russ. J. Phys. Chem.* **39**, 309 (1965).

Neon Adsorbed on Graphite: Heat Capacity Determination of the Phase Diagram in the First Monolayer from 1 to 20°K*

G. B. Huff and J. G. Dash

Department of Physics, University of Washington
Seattle, Washington

Introduction

Using a cryostat equipped with a PDP-8/S on-line computer,[1] we have made high-resolution heat capacity measurements in the range 1–20°K on 14 submonolayer films of neon adsorbed on exfoliated graphite (Grafoil). Two previous studies[2,3] of the heat capacity of neon on graphite (graphitized carbon black powder) showed the existence of an anomaly in the range 10–20°K for 0.5 monolayer. Vapor pressure studies[4-7] of noble gases adsorbed on exfoliated graphite show this form of the material to be more homogeneous than graphitized carbon black. Heat capacity work on monolayer ^{3}He and ^{4}He films on Grafoil[8,9] gives evidence not only of surface homogeneity but also of good thermal conductivity.

The heat capacity cell was first used to independently confirm the findings of a lattice gas order–disorder transition at $x_g = 1/3$,[8,9] where x_g is the number of adatoms/number of graphite sites, for adsorbed ^{4}He, and in so doing, we measured the adsorption area 24 m^2/g from $N_{\text{critical}} = 60.12$ cm^3 STP and the area/site ratio of 5.24 Å^2. We have also completed a series of adsorption isotherms for neon from which isosteric heats of $\sim (450°\text{K}) k_B$ and monolayer completion at 115 cm^3 STP were determined.

Results

The heat capacity of a neon film at $x_g = 1/3$ is given in Fig. 1. The earlier reports[2,3] of an anomaly are substantiated. However, we obtain a considerably stronger and narrower peak than in either previous study: We attribute the sharpness to more uniform surfaces and better thermal equilibrium. Heatings of 5–50 m°K were used, and internal equilibrium was achieved in seconds. The experimental width (FWHM) is 75 m°K at 13.5°K. A finite difference integration of the C/Nk_BT vs. T curve gives a nearly vertical discontinuity in the entropy of $0.3 k_B$/atom, which at 13.5°K represents a latent heat of $\sim (4°\text{K}) k_B$ if this is a first-order transition. (The latent heat of fusion of bulk neon at the triple point is $(40°\text{K}) k_B$.[10]

A succession of coverages was studied of $0.06 < x < 0.84$ ($x = N_{\text{adsorbed}}/N_{\text{monolayer}}$). For $x < 0.22$ there is a single maximum in the heat capacity which sharpens and moves to lower temperatures with increasing coverage. At $x = 0.22$

---

* Research supported by the National Science Foundation.

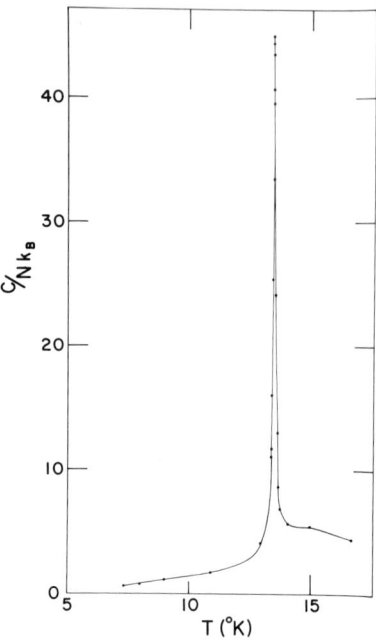

Fig. 1. C/Nk_B vs. T for $N = 60.30 \, \text{cm}^3$ STP.

the maximum splits into a sharp peak at $T = 13.5°\text{K}$ and a second maximum at 15.5°K. For $0.22 < x < 0.66$ the sharp peak increases rapidly in magnitude, remaining at 13.5°K; the second maximum remains at 15.5°K but diminishes in magnitude. For $0.66 < x < 0.84$ the sharp peak rapidly broadens, diminishes, and moves to higher temperatures; the second maximum has disappeared. These results are summarized in Fig. 2.

For all coverages the low-T heat capacity has an interesting feature: Below $\sim 2°\text{K}$ logarithmic plots give $C/Nk_B \propto \exp(-\theta_0/T)$, with θ_0 decreasing from 3.8 to 3.3°K with increasing coverage. For $3 < T < 5°\text{K}$ a 2D Debye solid T^2 dependence gives $\theta_D = 46°\text{K}$ at $x = 0.10$, increasing to $\theta_D = 52°\text{K}$ at $x = 0.84$.

It should be remarked that at 20°K the film signal is always considerably greater than one k_B/atom, the ideal 2D gas signature so ubiquitous in the ^{3}He and ^{4}He studies.[8,9]

Construction of the Phase Diagram

The principal features and regions of the first monolayer are given in Fig. 3. In comparison with helium films on graphite, which at a given coverage and temperature appear to be uniform, i.e., in a single phase, the neon films give evidence of phase equilibrium between patches of condensed film and a 2D vapor. A further comparison of the two systems is illuminating. The two adatoms are nearly the same size (2.749 Å diameter for neon compared to 2.556 Å for ^{4}He). A graphite hexagon is 1.42 Å on an edge, 2.84 Å across. Helium-4 adsorbed on graphite is a repulsive system; the adatoms see an essentially featureless substrate except at those coverages ($x_g = 1/4, 1/3$) where lattice gas ordering occurs. The ^{4}He zero-point energy,

Neon Adsorbed on Graphite: Phase Diagram in the First Monolayer

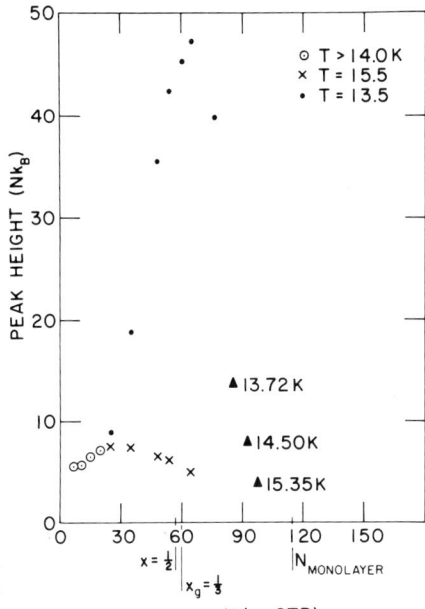

Fig. 2. Peak height vs. coverage: (◯) low-density peaks; (●) 13.5°K singularities; (▲) high-density transitions; (×) secondary maxima at 15.5°K.

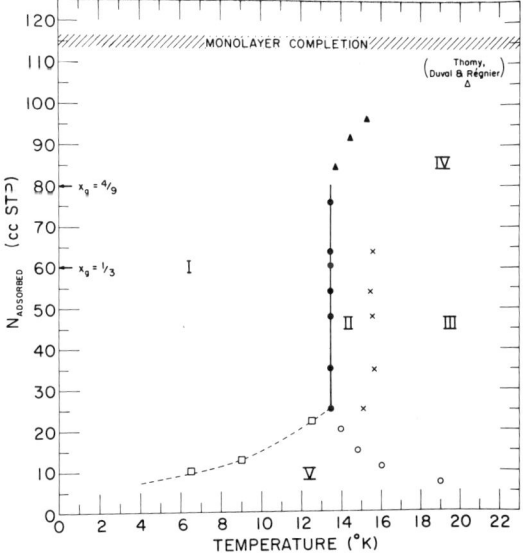

Fig. 3. Phase diagram of neon adsorbed on graphite: (◯) Low-density peaks; (●) 13.5°K singularities; (×) secondary maxima at 15.5°K; (▲) high-density transition; (△) Ref. 4; (- - □ - -) break in C_{total} vs. N isotherms. Tentative identification of thermodynamic regions: I—close packed solid + 2D vapor; II—commensurate liquid + 2D vapor; III—2D fluid; IV—incommensurate solid; V—low-density solid + 2D vapor.

~66°K,[11] dominates the attractive He–He potential, ~12°K, and the site-to-site barrier energy, ~17°K.[11] In comparison, neon adatoms are less active (zero point ~35°K[12]) and more attractive, ~35.6°K, and the site-to-site barrier is higher, ~25°K.[2] Whereas monolayer completion for neon is 115 cm³ STP in our sample, for ⁴He it is 108 cm³ STP. Clearly the neon film should be expected to exhibit interesting thermodynamic properties due to the delicate balance between energies and the graphite site–neon atom size correspondence.

Recently Ying[13] treated the case of classical adatoms in 2D periodic potentials and found that curious types of film phases, i.e., *commensurate* but not necessarily site-localized structures, were possible for propitious adatom-site size and energy ratios. One feature of these commensurate films is a low frequency cutoff to the phonon spectrum. This produces an exponentially decaying heat capacity signal at low T,[14] in agreement with our results. Also the characteristic temperature θ_0 can be related to the "repeat distance" of the Ying phase:

$$k_B \theta_0 = \hbar \omega_0 = \hbar c 2\pi / \lambda_{max} = \hbar c 2\pi / d_{repeat}$$

For $T > \theta_0$ a normal 2D Debye spectrum obtains with the Debye temperature θ_D representative of the high-frequency limit, which depends on the interatomic spacing:

$$k_B \theta_D = \hbar \omega_{max} = \hbar c 2\pi / \lambda_{min} = \hbar c 2\pi / d_{atomic}$$

Hence $\theta_D / \theta_0 = d_{repeat} / d_{atomic} \sim 48°K / 3.5°K \sim 15$. In Ying's terminology the film in region I is a 2D commensurate solid of order 15. Since order one is defined as having each adatom at the center of a site, it appears that the neon–neon forces have essentially overcome the localizing forces of the substrate.

Peierls[15] and Mermin[16] have shown theoretically that in a true 2D system there is no long-range order for $T > 0$, which precludes a first-order melting transition. Ying concludes that changes between commensurate phases would be first order. This implies a latent heat in agreement with the sharp step in the entropy observed for the 13.5°K singularity.

The rapid increase in the magnitude of the peak can be understood if we assume that sublimation occurs below the transition temperature so that some of the film is in a 2D vapor phase when 13.5°K is reached. As the coverage is increased a larger fraction of the film is in the solid phase, giving a larger total signal at the transition. The solid–vapor equilibrium keeps ϕ, the 2D vapor pressure, constant at 13.5°K, which produces the C_p-like signature.

However, it may be that no clear-cut distinctions between first- and higher-order transitions can be made for films. As high and narrow as the peak is, it is not infinite. Perhaps the sharpness of a singularity indicates the degree of commensurateness of the phases on either side of the anomaly. Indeed, the rapid decrease in the height of the peak and its broadening and shift to higher T for coverages above $x = 0.7$ may indicate that the film in region II is of lower order, i.e., more in registry with the substrate, than in region I. If so, we may have a system in which a "liquid" is more ordered than the solid. The density of the region II patches must be such that they fill all the adsorption area at $x = 0.7$, and when that coverage is surpassed the change in the order of the film at the transition becomes less and less pronounced, making region IV amorphous or incommensurate. The decreasing free area for $x > 0.7$ causes the 2D pressure to increase, forcing the transition to higher T. From a vapor pressure isotherm at 20.4°K Thomy et al.[4] report a phase transition in neon adsorbed on exfoliated graphite at $x \sim 0.9$. This datum fits nicely on the phase boundary indicated by the heat capacity results.

The line of secondary maxima, the broad anomalies at 15.5°K, is not as clearly understood since they disappear at about the $x_g = 1/3$ coverage, which has a Ying order of one. If the previous reasoning is correct, the transition should be as sharp as the one at 13.5°K. The peak has the same magnitude as the order–disorder transitions in helium films, but it should be much sharper at the critical coverage of 60 cm^3 STP. Region III is thought to be an incommensurate 2D fluid.

For coverages below $x = 0.22$ the 2D vapor pressure appears to remain below some critical value needed to create the phase of region II. Since at low T the film is assumed to form a solid, region V is composed of patches of low-density solid plus 2D vapor.

Summary

The properties of neon adsorbed on a homogeneous substrate such as Grafoil may be well described by Ying's treatment of surface phases. The standard terminology of 3D phases and phase transitions may be inappropriate for the nearly 2D world of monolayer films.

References

1. D.W. Princehouse, *J. Low Temp. Phys.* (to be published).
2. W.A. Steele and R. Karl, *J. Colloid. Interface Sci.* **28**, 397 (1968).
3. A.A. Antoniou, P. H. Scaife, and J.M. Peacock, *J. Chem. Phys.* **54**, 5403 (1971).
4. A. Thomy, X. Duval, and J. Regnier, *Compt. Rend.* **268C**, 1416 (1969).
5. A. Thomy and X. Duval, *J. Chim. Phys.* **66**, 1966 (1969).
6. A. Thomy and X. Duval, *J. Chim. Phys.* **67**, 286 (1970).
7. A. Thomy and X. Duval, *J. Chim. Phys.* **67**, 1101 (1970).
8. M. Bretz, G.B. Huff, and J.G. Dash, *Phys. Rev. Lett.* **28**, 729 (1972).
9. D.C. Hickernell, E.O. McLean, and O.E. Vilches, *Phys. Rev. Lett.* **28**, 789 (1972).
10. G.L. Pollock, *Rev. Mod. Phys.* **36**, 748 (1964).
11. D.E. Hagen, A.D. Novaco, and F.J. Milford, in *Proc. Intern. Symp. on Adsorption-Desorption Phenomena, Florence, 1971*, Academic Press, New York.
12. A.D. Crowell and R.B. Steele, *J. Chem. Phys.* **34**, 1347 (1961).
13. S.C. Ying, *Phys. Rev. B* **3**, 4160 (1971).
14. G.A. Stewart and J.G. Dash, *J. Low Temp. Phys.* **5**, 1 (1971).
15. R.E. Peierls, *Ann. Inst. Henri Poincaré* **5**, 177 (1935).
16. N.D. Mermin, *Phys. Rev.* **176**, 250 (1968).

NMR in ^{3}He Monolayers Adsorbed on Graphite below 4.2°K*

D. P. Grimmer and K. Luszczynski

Department of Physics, Washington University
St. Louis, Missouri

We have performed pulsed NMR experiments on ^{3}He monolayers adsorbed on Grafoil† pyrolytic graphite at temperatures between 4.2 and 0.3°K.

The motivation for the experiment is to determine the nature of the adsorbed ^{3}He system. On a suitable substrate one might expect adsorbed ^{3}He monolayers to behave like two-dimensional quantum systems. The substrate should possess uniform adsorption sites, with ^{3}He–substrate interaction strong enough to bind the absorbate to the surface, but weak enough to allow considerable ^{3}He–^{3}He interaction in the plane of the surface.

The adsorption substrate used in this experiment is Grafoil pyrolytic graphite sheet. Our adsorption isotherm surface area measurements, based on argon at 77°K (and ^{3}He at 4.2°K), agree with those of Bretz and Dash,[1] and indicate a uniform surface of specific area ~ 20 m^2/g. The substrate was constructed in the following manner. A strip of 0.010-in.-thick graphite sheet, together with a strip of Teflon sheet, was rolled into a circular cylinder around a sapphire rod 0.1 in. in diameter and slipped into a glass collar with an internal diameter of 0.53 in. The substrate was bonded to the sapphire rod with epoxy resin.‡ This configuration was used to reduce eddy currents in the Grafoil sheet caused by the radiofrequency pulses. The sapphire rod provided the thermal link to the ^{3}He refrigerator. The substrate was enclosed in a glass envelope which was attached to a copper plug by means of a housekeeper seal. A tapered collet on the plug held the sapphire rod and the plug was screwed to the bottom of the low-temperature probe. The radiofrequency coil was outside the glass envelope and was thermally isolated from it.

The sample chamber was heated to 130°C and evacuated for 70 hr. Subsequently a monolayer capacity of ^{3}He (about 2.3×10^{20} atoms) was admitted through a glass tube attached to the chamber and the tube was sealed off. The surface area of Grafoil was determined from our own measurements made with argon, together with measurements made by Bretz and Dash.[1] The area per atom is taken to be 12.8 Å^2 for argon and 9.95 Å^2 for ^{3}He.

Pulsed NMR techniques at 20 and 10 MHz were employed. The NMR equipment used is described elsewhere.[2] The rf heating in the Grafoil presented some experimental difficulties. Care was taken to achieve thermal equilibrium after

---

* Work supported in part by National Science Grants GU-1147 and GP-24572 and Navy Equipment Loan Contract No. NONR-1758(00).
† The Grafoil sheet, GTA grade, was obtained from Union Carbide, Carbon Products Division.
‡ Crest resin No. 7343/7139, Crest Products Company, Westminster, California.

pulses. Free induction decay (FID) times and spin–lattice (T_1) and spin–spin (T_2) relaxation times were measured at 20 and 10 MHz. Nuclear magnetic susceptibility and the spin diffusion coefficient were measured at 20 MHz.

We find that the FID (90° pulse) exponential decay time varies fairly linearly from 125 μsec at 4.2°K to 85 μsec at 0.3°K, for 20 MHz. At 10 MHz the FID decay time varies from 200 μsec at 4.2°K to 140 μsec at 0.3°K. This is considerably shorter than the magnetic field inhomogeneity decay time of 4 msec. However, the FID dephasing of spins is not random, since the spins can be rephased by a 180° pulse to produce an echo. Hence the effect is like a macroscopic magnetic field inhomogeneity of the order of 0.3 G/cm. The shape of the observed free induction decay corresponds to a Lorentzian line. This broadened Lorentzian line is believed to be caused by the macroscopic magnetization arising from the diamagnetism of the Grafoil.

The T_2 relaxation time varies from 1.25 msec at 4.2°K to 0.5 msec at 1.2°K, for 20 MHz. At 10 MHz the T_2 decay time varies from 2.25 msec at 4.2°K to 1.1 msec at 1.2°K. The T_2 decay was determined by applying 90°-180° and 90°-180°-180° pulse sequences, and was exponential out to about 2 msec. There is also a small component in the observed signal which has a much longer decay time. The T_2 decay may be due to the coupling of the ^{3}He to the free electrons in the substrate via a dipolar spin–spin interaction, rather than due to the ^{3}He–^{3}He dipolar coupling.

The T_1 relaxation time was measured at 4.2 and 1.2°K for frequencies of 20 and 10 MHz. In all cases the T_1 relaxation time cannot be characterized by a single exponential, but it is describable as a sum of two exponential decays. The time constants for the fast and slow decay components are respectively 115 (T'_1) and about 900 msec (T''_1) for 20 MHz, with the ratio of initial amplitudes of fast-decay to slow-decay components (M'_0/M''_0) given by ~0.36 at 4.2°K and ~1.10 at 1.2°K. For 10 MHz, $T'_1 \sim 80$ msec and $T''_1 \sim 540$ msec and (M'_0/M''_0) is given by ~1.08 at 4.2 and ~2.64 at 1.2°K. Qualitatively we can summarize the T_1 measurements as follows. For a fixed frequency the ratio (M'_0/M''_0) increases for lower temperature and the T'_1 and T''_1 times remain about the same. For a fixed temperature the ratio M'_0/M''_0 increases and the T'_1 and T''_1 times decrease for lower frequencies.

These results imply that there must be at least two types of regions in the substrate. Each region is characterized by different relaxation times.

We attempted to measure spin diffusion at 4.2 and 1.2°K for 20 MHz using a 90°-180°-180° (spin echo) pulse sequence. A linear gradient of up to 3 G/cm was produced by anti-Helmholtz coils fastened to the magnet pole faces. The standard theory of spin diffusion[3] predicts that the height of the second spin echo will follow an envelope given by

$$M(\tau, G) = M_0 \exp\left(-\frac{\tau}{T_2} - \frac{\gamma^2 G^2 D}{12}\tau^3\right)$$

where G is the gradient and τ is the time between the two spin echoes. However, a τ^3 behavior for the logarithm of the echo envelope was not observed, and in fact, the envelope decays were found to be exponential and independent of gradient, within experimental error. The diffusion data analysis is complicated by the fact that the diffusion of the ^{3}He is probably bounded in some manner by the finite size of the graphite flakes comprising the Grafoil sheet. A scanning electron mi-

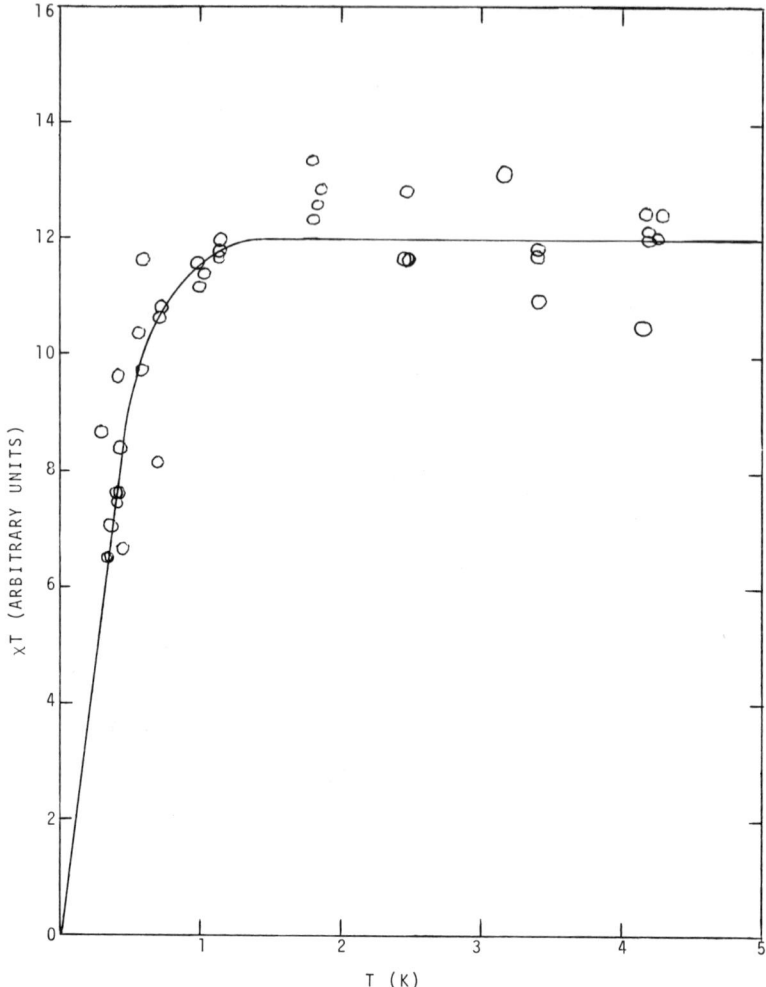

Fig. 1. Product of nuclear magnetic susceptibility and temperature vs. temperature.

croscope revealed that the flakes range in thickness from 500 to 1000 Å and in width from 1 to 4 μm. The theoretical treatment by Robertson[3] of bounded diffusion for the case where $\tau \gg 2a^2/\pi^2 D$ predicts an envelope exponential in τ given by

$$M(\tau, G) = M_0 \exp\left[-\left(\frac{\gamma^2 G^2 a^4}{120 D} + \frac{1}{T_2^{\text{observed}}}\right)\tau\right]$$

where a is the dimension of the bounded region. Since $M(\tau, G)$ is independent of gradient, then, following Robertson, we can write that

$$D \gg 2a^2/\pi^2 \tau \sim 3 \times 10^{-5} \text{ cm}^2/\text{sec} \quad \text{for} \quad a \sim 4 \times 10^{-4} \text{ cm}$$

Room-temperature NMR on ^{19}F in liquid Freon TF (trichlorotrifluoroethane, $C_2Cl_3F_3$), with the Freon saturating a Grafoil substrate, produced some of the

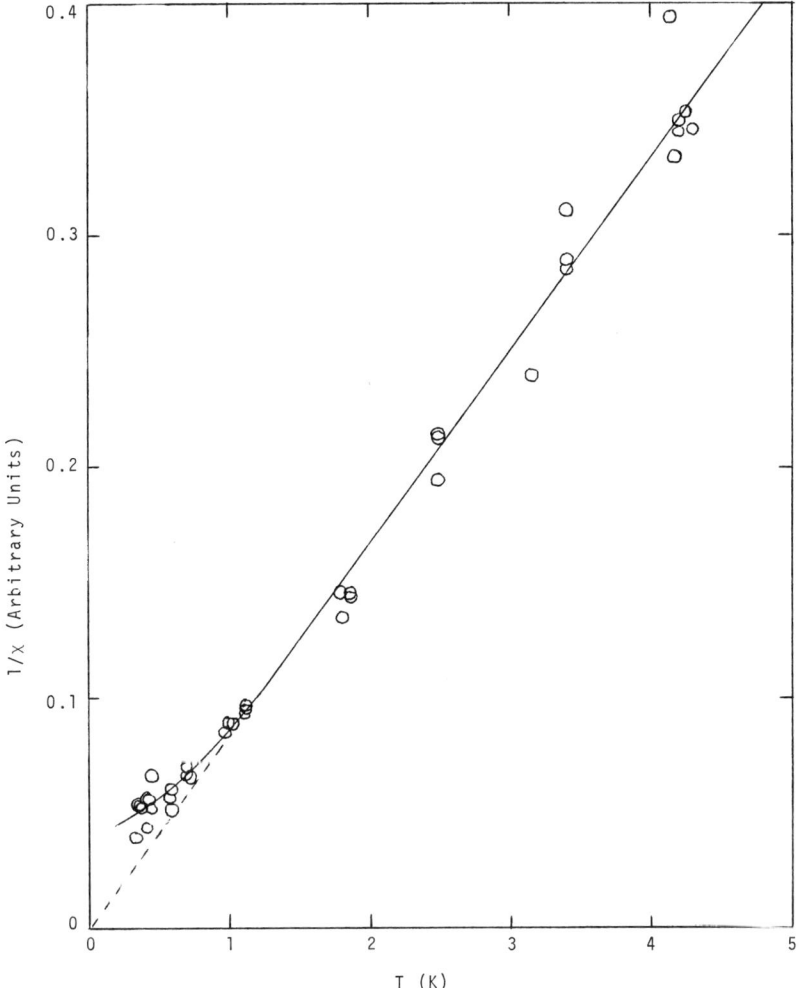

Fig. 2. Inverse nuclear magnetic susceptibility vs. temperature.

effects observed for the adsorbed ^{3}He: shortened FID's and multiple relaxation times. Also, the spin echo envelopes are independent of the field gradient. Since self-diffusion in bulk liquid Freon is large ($D \sim 10^{-5}$ cm^2/sec) and readily observable, even in saturated Freon films the effects of bounded diffusion play a dominant role in determining the dependence of the echo envelope on the field gradient.

The nuclear magnetic susceptibility χ of the adsorbed ^{3}He is found to obey Curie's law down to about 1°K; at lower temperatures, however, χ follows the susceptibility of a Fermi gas with a degeneracy temperature T_0 equal to 0.6°K (Figs. 1 and 2). It should be noted that in liquid ^{3}He T_0 varies from 0.45°K at saturated vapor pressure to 0.06°K as the solid density, or molar volume of 20 cm^3, is approached. The separation between the ^{3}He atoms in the adsorbed layer corresponds to that obtained in bulk solid ^{3}He with a molar volume of 20 cm^3. The observed behavior of nuclear magnetic susceptibility in the adsorbed monolayer of ^{3}He on Grafoil is consistent with a ^{3}He system of nonlocalized atoms in bounded regions.

References

1. M. Bretz and J.G. Dash, *Phys. Rev. Lett.* **26**, 963 (1971).
2. D.K. Biegelsen, Ph.D. Thesis, Washington University, St. Louis, 1970, unpublished; *Dissertation Abstracts* **31**, 6798–B (1971).
3. B. Robertson, *Phys. Rev.* **151**, 273 (1966).

Thermodynamic Functions for ^{4}He Submonolayers*

R. L. Elgin† and D. L. Goodstein

California Institute of Technology
Pasadena, California

Adsorption of gases on solids is generally studied by pressure isotherms, heat capacity isosteres, or heats of adsorption. By measuring these concurrently, the present study is able to calculate all other thermodynamic functions. The system of helium on Grafoil was chosen because it is exceptionally homogeneous, displaying mainly features of the adsorbate,[1] and because it is unusually reproducible.[2] This allows direct incorporation of data from other laboratories, which have shown heat capacity peaks attributed to Bose condensation, ordering transitions, and melting.[3-5]

The apparatus is similar to that used by Kierstead[6] to measure pressures near the critical point. The main differences are larger tubing and a capacitive manometer for higher accuracy at low pressures, manual temperature stabilization to allow heat capacity measurement, and Grafoil filling the central chamber (12.5 g in 15 cm^3). The preparation of the Grafoil was similar to the method of Bretz and Dash,[3] using a 22% larger sample.

The heat capacity C_N was measured at intervals of 10% or less in temperature and coverage ranges, these being $5 < T < 14°K$ and $0.01 < N < 1.5$ layers. Pressures were measured over the same region when in the 10^{-5}- to 800-Torr range of the manometer. At the highest temperatures the pressure was measurable to practically zero coverage. The use of pairs of isotherms and the Maxwell relation $(\partial S/\partial N)_T = -(\partial \mu/\partial T)_N$ gave the absolute entropy in terms of the measured chemical potential μ. These values were extended to lower temperatures from the heat capacity data by use of $(\partial S/\partial T)_N = C_N/T$. This procedure was repeated at higher coverages and lower temperatures, forming a band of accurate entropy values over the entire region for which both heat capacity and pressure measurements were accurate ($0.1 < P < 10$ Torr). The first relation was then applied alone to extend the entropy to higher coverages and the second to extend it to lower temperatures. The process was then reversed to generate the chemical potential and heat capacity in the regions not accessible to direct measurement.

The isosteric heat of adsorption Q_{st} was initially derived from pressure isotherms near 12.4°K. It was reduced to 5°K as explained above and then to practically 0°K by use of the data of Bretz et al.[3-5] (See Fig. 1.) The parameter x is the fraction of a monolayer coverage as defined by Bretz et al.[5] At 0°K, Q_{st} is identical to the partial molar energy. This energy is seen to be nearly constant at 144 ± 1 (all energies in

---

* Work supported in part by the Research Corporation.
† Supported by a National Science Foundation Fellowship.

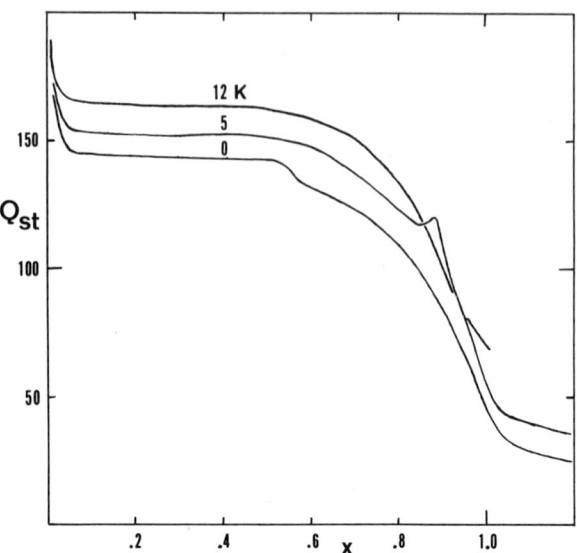

Fig. 1. Isosteric heats of adsorption. Coverage x in fraction of monolayer coverage, heat of adsorption Q_{st} in kelvins per atom. Curves shown for temperatures of 0, 5, and 12°K.

kelvins) for $0.1 < x < 0.5$, thus verifying the degree of homogeneity conjectured by Campbell.[7] However, at lower coverages the energy rises to at least 165 by $x = 0.02$. Above the ordering transition at $x = 0.57$ the energy falls rapidly due to mutual repulsion. The bump at 5°K and $x = 0.88$ indicates melting.

In three dimensions most thermodynamics is done at constant pressure. For systems adsorbed on solids the area is necessarily constant. This leads to unfamiliar results such as first-order phase transitions with the measured heat capacity everywhere finite. For purposes of orientation, the entropy of bulk helium is shown in Fig. 2 as though it were a two-dimensional system. The abscissa is the two-thirds power of the density relative to the monolayer. The ordinate is the areal entropy—numerically, the atomic entropy times the abscissa. The isotherms are spaced logarithmically from 1.5 to 13.6°K. At the bottom right is the solid with $C_v \propto T^3$. At the top left is the gas with C_v near 1.5. The dashed lines in the two-phase regions have been distorted by the dimensionality change but represent real equilibrium states. Near the center is the liquid and the lambda transition.

The entropy of ^{4}He adsorbed on Grafoil as found by this project is shown in Fig. 3. The data reduction is quasicontinuous except below 4.57°K, where the diamond points shown have been integrated downward from the heat capacity data of Bretz et al.[3-5] and the lines interpolated freehand.* The entropy appears accurate to 0.02 unit everywhere except where it is varying most rapidly. It integrates to zero at 0°K to within this uncertainty, showing that no low-temperature transitions have been missed.

---

* No other C_x data have been published except for one isostere at $x = 0.24$.[11] This extends to far lower temperatures but is not consistent with Ref. 3.

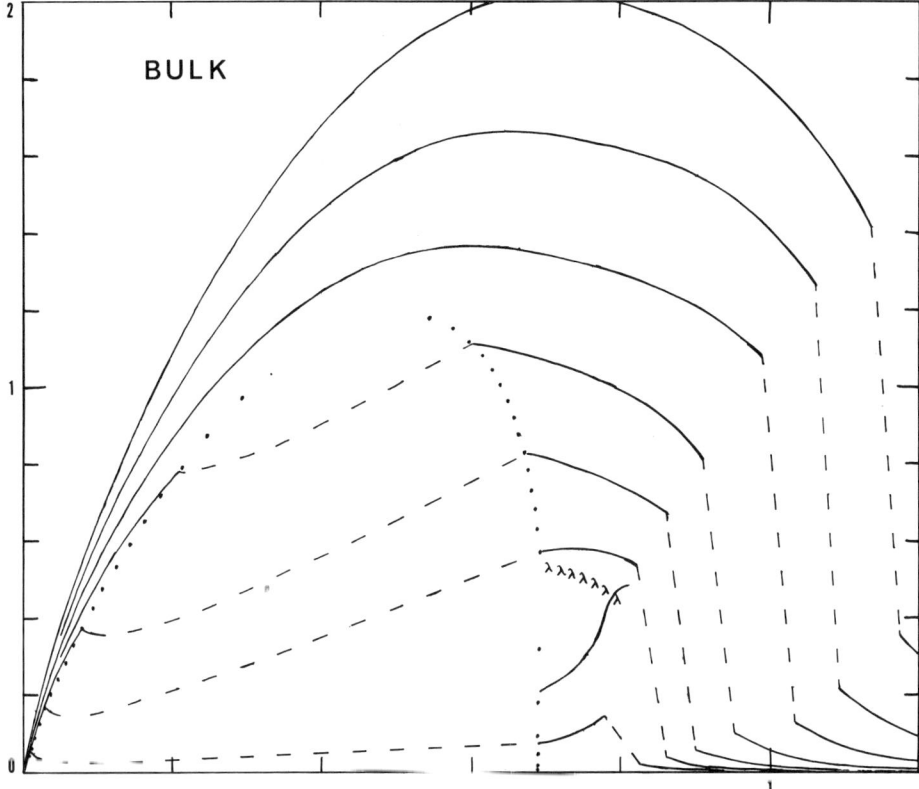

Fig. 2. Entropy isotherms for bulk ^{4}He. Axes as in Fig. 3. Temperatures of 1.53, (1.80), 2.20, 3.17, 4.57, 6.57, 9.44, and 13.6°K, from bottom to top. Gas data from Ref. 8, liquid data from Ref. 9, solid data from Ref. 10.

Nearly all the diagram is now thought to be understood. Above $x = 1$ the entropy rises rapidly as the second layer forms. The rest of the upper half of the figure gives heat capacities close to two-thirds that of the corresponding bulk point. Corresponding to the solid is a $C_N \propto T^2$ region. The entropy rise on melting has been attenuated about 50% relative to the bulk. The liquid region again has about two-thirds the heat capacity of the bulk except that there is no trace of the lambda transition. There may be a shadow of the bulk critical point at the same density ($x = 0.42$) and half the temperature, but the rest of the diagram appears unique to the film. The precipitous drop at $x = 0.57$ marks the ordering transition. Entropy far below ideal gas values for $x < 0.1$ can be explained by the high-energy sites shown in Fig. 1.

The tantalizing dip at the lowest temperature for $x = 0.34$ is less clear. The original suggestion was that a near-Bose condensation occurs due to small substrate inhomogeneities.[3,7] Widom and Sokoloff[12] proposed an independent check of this by use of a two-term virial expansion of low-pressure isotherms. More recently liquid condensation has been suggested by Novaco.[11] This should also show in the virial terms. Pressure isotherms were readily generated by the present study. For

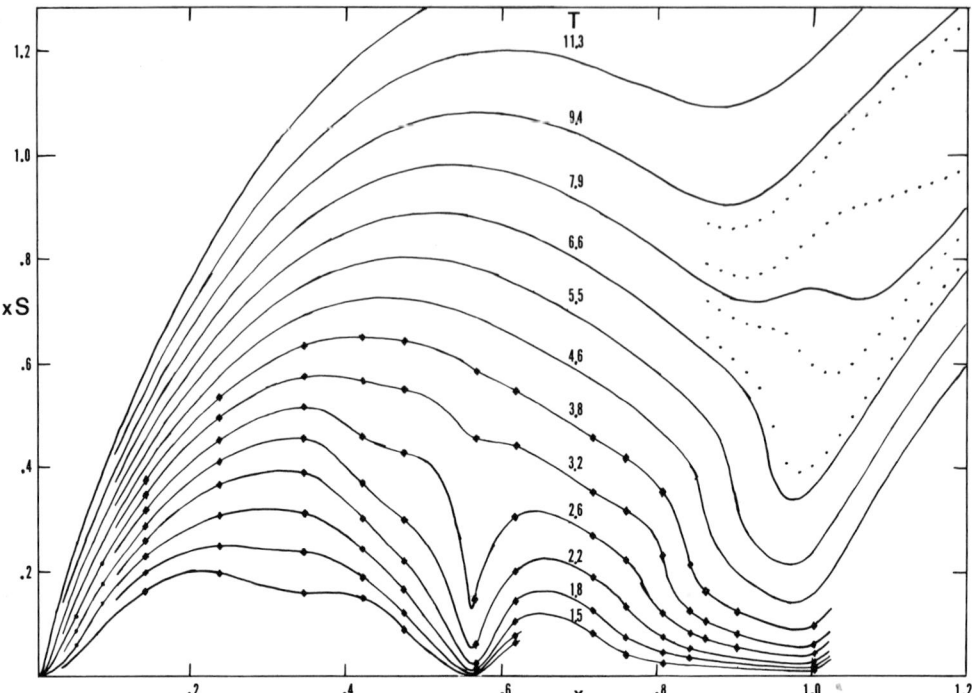

Fig. 3. Areal entropy for adsorbed ^{4}He. Coverage x in fractions of monolayer coverage. The dimensionless atomic entropy S has been multiplied by x to form the entropy per unit area. The precise values of the temperatures are 1.06, 1.27, 1.53, 1.83, 2.20, 2.64, 3.17, 3.80, 4.57, 5.48, 6.57, 7.87, 9.44, 11.33, and 13.62. (The dotted lines are for 6.87, 7.54, 8.23, 9.01.)

example, the pressure at the ordering peak is 1.3×10^{-17} Torr $\pm$ 20%. However, a plot of x/P vs. P, to fit the virial expansion, refused to fit straight lines. The trouble at low coverage was traced to the rapid rise of the binding energy. At coverages above $x = 0.1$ the two-term virial expansion ceases to be valid. For example, curvature is pronounced even using the Langmuir isotherm of an ideal substrate. As a last resort, tangents were drawn to the data at $x = 0.2$, but for both models the resulting virial terms predicted a transition well below 1°K.

The thermodynamic functions for the helium submonolayer are now more consistently known than they were for the bulk at the same temperatures and densities before 1958.* The data are sufficient to test complicated many-parameter, theories but not in the ideal dilute limit. On Grafoil, as on all other solid substrates, imperfections dominate at sufficiently low coverage.

---

* In 1958 Lounasmaa[14] finished his thesis on the dense liquid region. All the other functions can be found from the Helmholtz free energy, which is the integral of the μ data at constant T. Since its natural derivatives, μ and S, are the tabulated quantities, the requisite manipulations do not seriously degrade the data. Complete tabulations will be presented elsewhere.[15]

References

1. A.D. Novaco and F.J. Milford, *Phys. Rev. A* **5**, 783 (1972).
2. R.L. Elgin, 1972 Mid-Winter Solid State Conf., Irvine, Calif.
3. M. Bretz and J.G. Dash, *Phys. Rev. Lett.* **26**, 963 (1971).
4. M. Bretz and J.G. Dash, *Phys. Rev. Lett.* **27**, 647 (1971).
5. M. Bretz, G.B. Huff, and J.G. Dash, *Phys. Rev. Lett.* **28**, 729 (1972).
6. H.A. Kierstead, *Phys. Rev. A* **3**, 329 (1971).
7. C.E. Campbell, J.G. Dash, and M. Schick, *Phys. Rev. Lett.* **26**, 966 (1971).
8. R.D. McCarty, private communication.
9. O.V. Lounasmaa, *Cryogenics* **1**, 212 (1961).
10. J.S. Dugdale and J.P. Franck, *Phil. Trans. Roy. Soc.* **257**, 1 (1964).
11. D.C. Hickernell, E.O. McLean, and O.E. Vilches, *Phys. Rev. Lett.* **28**, 789 (1972).
12. A. Widom and J.B. Sokoloff, *Phys. Rev. A* **5**, 475 (1972).
13. A.D. Novaco (to be published).
14. O.V. Lounasmaa, Ph.D. Thesis, H.K. Onnes Laboratory, Leiden, Netherlands (1958).
15. R.L. Elgin, Ph.D. Thesis, California Institute of Technology, Pasadena, California (1973).

Elastic Properties of Solid ^{4}He Monolayers from 4.2°K Vapor Pressure Studies*

G. A. Stewart,† S. Siegel, and D. L. Goodstein

Department of Physics, California Institute of Technology
Pasadena, California

Heat capacity studies of monolayer ^{4}He on exfoliated graphite have revealed a striking variety of surface-phase equilibria.[1-4] In some cases the thermal behavior of these thin films has led to a direct comparison with the properties of the bulk. One of the most intriguing observations has been the near identity of the Debye temperatures at the same interatomic separation in the respective solid phases. This apparent coincidence has motivated the present study of those mechanical properties that determine the Debye θ at low temperature. In particular, how are the elastic constants for submonolayer solid helium related to those of the bulk? This question seems fundamental for eventual understanding of why the composite thermal behavior is analogous in two and three dimensions.

It is difficult to obtain information about the mechanical behavior of a homogeneous two-dimensional system from heat capacity studies alone. This is apparent from the functional dependence of the chemical potential:

$$d\mu = -\bar{S}\,dT - K_2^{-1}\,da \tag{1}$$

where $\bar{S}$ is the partial molar entropy, a is the atomic area, and K_2 is the two-dimensional isothermal compressibility,

$$K_2^{-1} = -a\left(\frac{\partial \varphi}{\partial a}\right)_T = -\left(\frac{\partial \mu}{\partial a}\right)_T \tag{2}$$

Here φ is the spreading pressure. Although the mechanical and thermal properties must be related from Eq. (1) by a Maxwell relation, it is not possible to generate an absolute compressibility solely from the heat capacity measurements. Even when combined with elementary Debye models that yield the correct temperature dependences for low-temperature heat capacities, it is not possible to detail the static lattice interactions and zero-point energies that may make appreciable contributions to thermodynamic potentials and derived compressibilities. However, the Debye temperatures alone are sufficient to determine a function of two required elastic constants in the simplest case of an isotropic solid. This may be seen in a two-

---

* Work supported in part by the Research Corporation.
† Present address: Department of Physics, University of Pittsburgh, Pittsburgh, Pennsylvania.

dimensional Debye model, where the Debye temperature is given by

$$\theta_2 = \frac{4\hbar}{k}\left(\frac{\pi}{m}\right)^{1/2} K_2^{-1/2} F_2(\sigma_2) \qquad (3)$$

$$F_2(\sigma_2) = \left[\frac{1-\sigma_2}{(1+\sigma_2)(3-\sigma_2)}\right]^{1/2}$$

where σ_2, Poisson's ratio, has been chosen as the second elastic constant.

Experimentally, measurement of the equilibrium vapor pressure above a homogeneous solid is a direct route to the compressibility. If the vapor is ideal, the chemical potential is readily determined from

$$\mu = kT \ln(p\lambda^3/kT) \qquad (4)$$

where λ is the thermal de Broglie wavelength. From Eq. (2)

$$K_2^{-1} = -kT\left(\frac{\partial \ln P}{\partial a}\right)_T \qquad (5)$$

Using the measured compressibility from Eq. (5) in conjunction with an elastic model and experimental θ values, Poisson's ratio for the solid may be determined. In this way the mechanical properties for an isotropic solid monolayer may be directly compared with the bulk just as the thermal properties have been compared using their implicit combination in the Debye temperature. It should be noticed in making any comparison that the range of σ and dimensionality of K differ for bulk and two-dimensional systems. Further, if Eq (5) is to be the compressibility of the solid, measurements must be made in a region where the surface system is homogeneous. Coverages and temperatures must be chosen where one is far from melting or its precursors as well as regions of probable multilayer formation. With the presence of these effects the chemical potential is no longer a probe of the pure solid even though most of the adsorbate may properly be considered in the solid phase.

Heat capacity studies are currently the most precise technique for determining the phase boundaries in the film. This fact, together with the above coverage requirements, almost necessitates that vapor pressure studies be done in combination with heat capacity measurements. Atomic areas corresponding to heat capacity melting peaks have been assigned by Bretz et al.[3] Thus to normalize coverages accurately in a vapor pressure cell for comparison with Ref. 3, it is convenient to pin the areas by measuring the temperature of the heat capacity maximum corresponding to a melting anomaly and then for the same coverage determine the vapor pressure at the appropriate temperature for the vapor pressure experiments.

Vapor pressure measurements in an adsorption cell containing Grafoil* have been made as a function of coverage for atomic areas between 8.8 and 11.8 $Å^2$ at 4.2°K. The basic apparatus and techniques have been described elsewhere.[5] In addition, partial molar entropies have been estimated from experimental isosteric heats of adsorption and chemical potentials at selected high coverages near the

---

* Union Carbide Corp., Carbon Products Division, New York.

monolayer. Atomic areas were normalized to 9.35 Å² in the manner indicated above, where this area corresponds to melting at 6.6°K. The normalization was performed using another Grafoil cell designed for heat capacity measurements[6] where the pressure at 4.2°K for 6.6°K melting coverage was 2.5×10^{-4} Torr. The total surface area consistent with this normalization is 105 m², and the uncertainty in normalization is estimated to be 1.5%.

The 4.2°K isotherm is shown in Fig. 1 and is approximately exponential over four orders of magnitude in pressure at the higher coverages. Figure 2 shows the chemical potential vs. inverse atomic area as generated from the isotherm and Eq. (5). The exponential character of the isotherm appears in Fig. 2 as the nearly linear dependence of $\mu(a^{-1})$ at small areas. The melting coverage at 4.2°K is indicated and the departure from linearity for atomic areas below 9 Å² is attributed to multi-layer formation. This interpretation is consistent with the full monolayer atomic area of 8.78 Å² from Ref. 3 and our isostere measurements, which indicate that

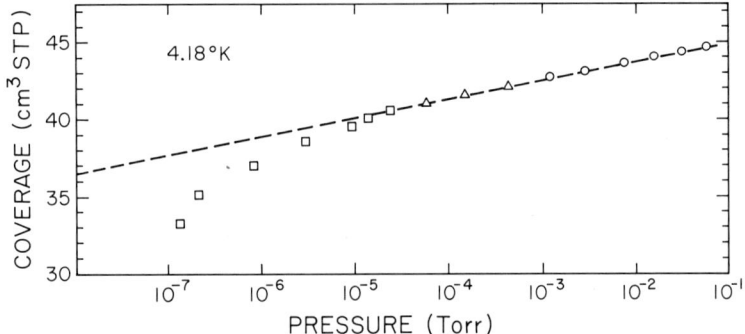

Fig. 1. 4.18°K ⁴He–Grafoil isotherm.

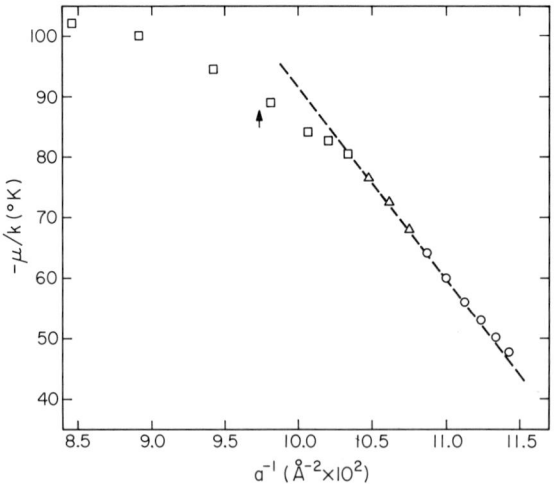

Fig. 2. 4.18°K chemical potential as a function of inverse atomic area.

Table I. Isosteric Heat of Adsorption, Chemical Potential, and Partial Molar Entropy near Monolayer Coverage at 4.18°K

a, Å^2	Q_{st}, °K	μ, °K	$\bar{S}/k$
8.79	41.8 ± 3	−47.8 ± 1	3.9 ± 1
8.90	53.1	−52.2	2.3
8.99	62.6	−56.2	0.9

the partial molar entropy becomes positive (Table I) and increases with coverage in the region of departure. Calculated values of $\bar{S}$ from heat capacity measurements and high-temperature vapor pressures[6] also show positive values of $\bar{S}$ near the monolayer. We restrict our discussion to the intermediate coverage region in Fig. 2, where the chemical potential appears linear and where the density of data points is sufficient to permit point-by-point numerical derivatives of μ. In this region, intermediate between melting and multilayer formation, $-d\mu/da$ is considered to be K_2^{-1}.

Derived values of K_2^{-1} are shown in Fig. 3(A), where the smooth curve is from a linear fit to $\mu(a^{-1})$ and the data points are numerical derivatives between experimental coverages with no smoothing. The values obtained are of the same order of magnitude as (but larger than) those that result from simple dimensional scaling of the zero-degree bulk compressibility[7,8], K_{03}, at the same atomic separation $\bar{d}$. For example, at 8.95 Å^2, $\bar{d}K_{03}^{-1}$ is 41 ergs/cm^2, approximately two-thirds the value of the film K_2^{-1}. It is interesting to note that $\frac{2}{3}K_2^{-1} = \bar{d}K_3^{-1}$ would be the exact scaling between the compressibilities if the variation of chemical potential with d were the same in film and bulk. The precision of the data is not sufficient to make detailed conclusions about the variation with coverage, but it does suggest that the compressibility of the film increases less rapidly with increasing $\bar{d}$ than in the scaled bulk: At 9.6 Å^2 the scaled bulk value of $\bar{d}K_{03}^{-1}$ is about 23 ergs/cm^2, approximately half that of the film from Fig. 3(A).

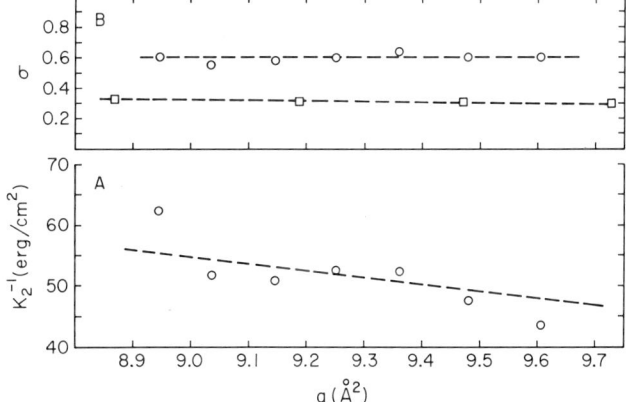

Fig. 3. (A) 4.18°K compressibility; the dashed line results from a linear approximation for molar areas 9.0–9.7 Å^2 in Fig. 2. (B) Poisson's ratio for the film (upper) and the bulk (lower).

The data of Fig. 3(A) may be combined in Eq. (3) with the Debye temperatures of Ref. 3 to estimate Poisson's ratio for the film. We neglect any temperature variation in the compressibility, by analogy with the bulk solid, where there is a maximum of 4% change from 0°K to the melting curve for molar volumes corresponding to the present experimental atomic separations. Over the coverage range 8.95–9.6 Å² the seven values of σ_2 have a mean value of 0.6 with about 4% scatter.

For a consistent comparison with the bulk solid we apply the isotropic model to a three-dimensional system, where the Debye temperature using the same elastic constants is

$$\theta_3 = \frac{2\pi h}{k} \left(\frac{9}{4\pi}\right)^{1/3} m^{-1/2} K_3^{-1/2} F_3(\sigma_3)$$

$$F_3(\sigma_3) = \left(\frac{1-\sigma_3}{1+\sigma_3}\right)^{1/2} \left[1 + 4(2)^{1/2} \left(\frac{1-\sigma_3}{1-2\sigma_3}\right)^{3/2}\right]^{-1/3}$$

(6)

Equation (6) may be combined with the θ_3 data of Ahlers[9] and the 0°K compressibilities of the bulk[7,8] to obtain σ_3 for the same atomic separations as in the film. Over this range σ_3, which is nearly constant at 0.31, is approximately one half of σ_2. Both σ_2 and σ_3 are shown in Fig. 3(B). As a check on the validity of the isotropic model applied to the bulk, we calculate the speed of sound and compare the result with the experimental values of Vignos and Fairbank[10] on polycrystalline hcp samples. At 9.3 Å, corresponding to a molar volume of 17.1 cm³ for $K_{03} = 11.6 \times 10^{-10}$ cm²/dyn with

$$C_{l3}^2 = 3(\rho_3 K_3)^{-1}(1-\sigma_3)(1+\sigma_3)^{-1}$$

(7)

the calculated speed of sound is 763 m/sec. The experimental longitudinal sound speed for this molar volume is 775 m/sec.

It is of interest to examine the significance of the result $\sigma_2 \approx 2\sigma_3$. Since σ is the ratio of fractional transverse contraction to fractional elongation for a pure longitudinal extension, mechanical stability allows the maximum values in two and three dimensions to differ by a factor of two. The maxima are 1.0 and 0.5, respectively. Thus the bulk and film possess approximately the same fractional value of the maximum allowed Poisson ratio. In addition, since the dilatation in volume and area corresponding to the same fractional elongation is given by

$$\frac{\delta V}{V} = \frac{(1-2\sigma_3)}{(1-\sigma_2)} \frac{\delta A}{A}$$

(8)

it follows that $\delta A/A \approx \delta V/V$. In this way the gross mechanical behavior of the film appears to be correlated with that of the bulk.

The experimental values of σ_2 and K_2 can be used to estimate the speed of sound in the monolayer from

$$C_{l2}^2 = 2(\rho_2 K_2)^{-1}(1+\sigma_2)$$
$$C_{t2}^2 = (\rho_2 K_2)^{-1}(1-\sigma_2)(1+\sigma_2)^{-1}$$

(9)

At 9.3 Å² for $K_2^{-1} = 5.3$ ergs/cm², $\sigma_2 = 0.6$, the longitudinal and transverse speeds are found to be 945 and 420 m/sec, respectively, where we have ignored distinctions between isothermal and adiabatic compressibilities. These estimates imply that for the film, as in the bulk, the low-temperature heat capacity is dominated by the transverse modes since in two dimensions

$$\theta_2 = \frac{2\pi h}{k}\left(\frac{2\rho}{\pi}\right)^{1/2}\left(\frac{1}{C_{2t}^2} + \frac{1}{C_{2l}^2}\right)^{-1/2} \tag{10}$$

References

1. M. Bretz and J.G. Dash, *Phys. Rev. Lett.* **26**, 963 (1971).
2. M. Bretz and J.G. Dash, *Phys. Rev. Lett.* **27**, 647 (1971).
3. M. Bretz, G.B. Huff, and J.G. Dash, *Phys. Rev. Lett.* **28**, 729 (1972).
4. D.C. Hickernell, E.O. McLean, and O.E. Vilches, *Phys. Rev. Lett.* **28**, 789 (1972).
5. J.L. Wallace and D.L. Goodstein, *J. Low Temp. Phys.* **3**, 283 (1970).
6. R.L. Elgin, this volume.
7. D.O. Edwards and R.C. Pandorf, *Phys. Rev.* **140**, A816 (1965).
8. J.S. Dugdale and J.P. Franck, *Phil. Trans. Roy. Soc.* **A257**, 1 (1964).
9. G. Ahlers, *Phys. Rev.* **A2**, 1505 (1970).
10. J.H. Vignos and H.A. Fairbank, *Phys. Rev.* **147**, 186 (1966); S.G. Eckstein, Y. Eckstein, and J.B. Ketterson, in *Physical Acoustics,* W.T. Mason, ed., Academic Press, New York (1970), Vol. VI, p. 360.

5
Molecular Solids

Sound Velocities in Solid hcp H_2 and D_2*

R. Wanner and H. Meyer

Department of Physics, Duke University
Durham, North Carolina

We report in summary fashion the results of a systematic study of transverse (v_t) and longitudinal (v_l) sound velocities in solid hcp H_2 and D_2. This research was carried out in the temperature range between 1.25 and 20°K on crystals grown with several orientations with respect to the sound propagation direction and at pressures up to about 200 bar. The crystals had various ortho–para ratios, and we have labeled them by the mole fraction X of molecules having a rotational angular momentum $J = 1$ (ortho-H_2 or para-D_2), the rest having $J = 0$. The sample chamber and the velocity measurement technique have been described previously.[1]

Crystals were grown under approximately constant pressure by slowly lowering the sample into a dewar containing liquid helium at the bottom. After all the sample had solidified it was annealed close to the melting temperature, the capillary was then blocked, and further cooling occurred at constant volume, accompanied by a drop in pressure. A melting pressure P_m of at least ~ 40 bar for H_2 and ~ 70 bar for D_2 was necessary to prevent differential thermal contraction from breaking the bond between quartz transducer and sample. These minimum melting pressures therefore correspond to approximately zero pressure at $T = 4.2°K$. It was found that two crystals grown consecutively from the liquid at the same pressure usually showed identical velocities, indicating that they were single crystals growing from the same nucleation point and having the same orientation with respect to the direction of sound propagation. In order to obtain a random selection of orientations, consecutive crystals were grown at widely different pressures. Some of the principal results are as follows.

v_l and v_t at 4.2°K As a Function of Melting Pressure for H_2 ($X = 0.75$) and D_2 ($X = 0.33$)

Figure 1 shows the results for about 300 crystals grown with a random orientation with respect to the sound propagation direction. The data fall between two parallel lines which represent the extreme values for a given melting pressure. In spite of the considerable number of crystals investigated, many more data points would be required to establish the limits with more accuracy. However, given these points and in the absence of other information, we have attempted to extract from them the

---

* This research has been funded by a grant from the Army Research Office (Durham) and a contract with the Office of Naval Research, Durham, N.C.

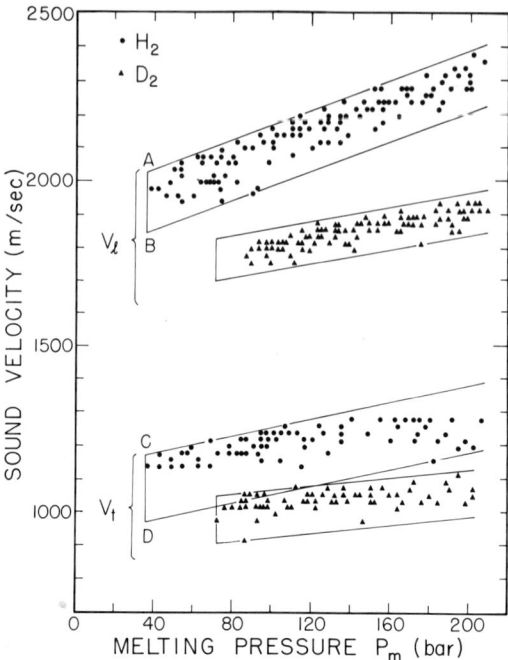

Fig. 1. Longitudinal (v_l) and transverse (v_t) sound velocities of H_2 ($X = 0.75$) and D_2 ($X = 0.33$) as a function of melting pressure.

five independent elastic constants of a hexagonal crystal. Taking advantage of the fact that our sample is very long and narrow, and from the qualitatively known dependence of the sound velocity on the propagation direction, we have concluded that this geometry prevents us from observing the fast transverse velocities.[2] Furthermore, the ratio c/a of the crystallographic axes is independent of the pressure in hcp D_2,* and is presumably so in hcp H_2, as in hcp ^{4}He, which leads to a reduction of the number of independent elastic constants from five to four.[4] By a variation technique, we have found a set of constants that yield the measured minimum and maximum velocities (denoted by A, B, C, and D in Fig. 1). For a melting pressure of 37 bar, which corresponds to a molar volume of about 22.9 cm³, the elastic constants for H_2 ($X = 0.75$) at 4.2°K, expressed in 10^6 m²/sec², are: $c_{11}/\rho = 4.1 \pm 0.1$, $c_{12}/\rho = 1.35 \pm 0.1$, $c_{13}/\rho = 0.46 \pm 0.1$, $c_{33}/\rho = 4.99 \pm 0.25$, and $c_{44}/\rho = 0.94 \pm 0.05$. For $P_m = 70$ bar for D_2, where we assume $V = 20.2$ cm³, $c_{11}/\rho = 3.34 \pm 0.1$, $c_{12}/\rho = 1.17 \pm 0.1$, $c_{13}/\rho = 0.61 \pm 0.1$, $c_{33}/\rho = 3.91 \pm 0.25$, and $c_{44}/\rho = 0.80 \pm 0.05$, where ρ is the density.

From these elastic constants we have also computed the Debye temperatures θ_D and the bulk modulus B_s for the same densities, and Table I presents our results together with those from other determinations.

---

* According to Nielsen[3] based on neutron scattering experiments under pressure.

Table I. Selected Constants for hcp H_2 ($X = 0.75$) and D_2 ($X = 0.33$) for $P \approx 0$ at 4.2°K

	H_2	D_2	Method	Ref.
Θ_D, °K	111 ± 6	105 ± 6	Sound velocities	This work
	115 ± 1	105 ± 1	Sound velocities (polycrystalline samples)	(*)
	122†, 116	106.5, 113.8 114‡	Specific heat Neutron scattering	8–10 11
B_s, (bar)	1740 ± 100	3370 ± 100	Sound velocities	This work
	2630, 2390†	4760	Sound velocities (polycrystalline samples)	(*)
	1440	2200	Density	12
	2000	3260	Density	13
		4000‡	Neutron Scattering	11
	1825†		Dielectric constant	14
γ_L	3.3, 2.5†	3.3	Sound velocities	This work
	2.1	2.0	$(\partial P/\partial T)_v$, C_v	15, 16
	2.2†	—	Specific heat C_v only	8

* H_2 ($X = 0.75$), Ref. 5. D_2 ($X = 0.33$), Ref. 6. H_2 ($X = 0$), Ref. 7.
† Data for H_2 ($X = 0$).
‡ Assuming $C_{13} = C_{11} + C_{12} - C_{33}$ and recalculating Θ_D and B_s.
The neutron data were taken for D_2 ($X = 0.02$).

Temperature Dependence of the Sound Velocity

We present precise measurements of the temperature dependence of both $v_l(T)$ and $v_t(T)$ for H_2 ($X \approx 0$) for crystals grown at a melting pressure $P_m \approx 70$ bar, corresponds to a molar volume $\sim 2.5\%$ smaller than grown under saturated vapor pressure. These data and the resulting temperature dependence of the bulk modulus, $B_s(T) - B_s(T = 0)$ are shown in Fig. 2. Assuming the Grüneisen equation of state, and assuming $d\gamma_L/dV = 0$, the relation between adiabatic modulus at constant volume and calorimetric data is given by

$$B_s(T) - B_s(0) = \frac{\gamma_L(\gamma_L + 1)}{V} \int_0^T C_v \, dT \qquad (1)$$

In addition to our own results, we have used the C_v data for H_2 ($X = 0$) due to Ahlers,[8] adjusted to correspond to the same molar volume. The resulting γ_L is shown in Fig. 2. Below about 7°K the sound velocity temperature dependence is so small as to introduce a large scatter in the calculation of γ_L. Nevertheless, the results can be estimated to extrapolate to about 2.5 for $T < 6°K$. This is somewhat lower than the value 3.3 quoted for H_2 ($X = 0.75$) from the pressure dependence of the sound velocities at 4.2°K, and somewhat higher than the results reported by Ahlers[8] from calorimetric data alone. The differences between γ_L obtained from the various methods are not considered unusual, however.

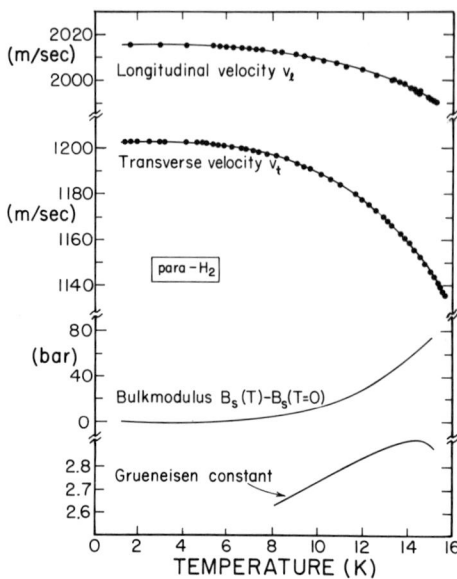

Fig. 2. Temperature dependence of v_l, v_t, the bulk modulus B_s, and the Grüneisen constant for H_2 ($X = 0$). The last quantity was obtained using Eq. (1).

Sound Velocity of H_2 at 4.2°K As a Function of X

We have measured the dependence on X of v_l and v_t in crystals of H_2 grown at a melting pressure $P_m \simeq 90$ bar. The original mole fraction when measurements started was $X = 0.70$ and it decreased by ortho–para conversion to $X = 0.35$, at which time the experiment was terminated. We have recorded an almost linear change of the velocities with X, but of different signs (Fig. 3), namely $1/v_l(\partial v_l/\partial X)_T = -7.7 \times 10^{-3}$ and $1/v_t(\partial v_t/\partial X)_T = 5.0 \times 10^{-3}$. Hence the bulk modulus change is $(\partial B_s/\partial X)_T = -78$ bar. This change is attributed to the decrease in the intermole-

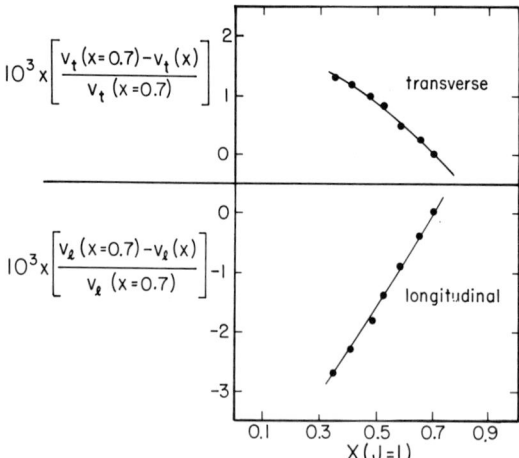

Fig. 3. Dependence of v_l and v_t on the ortho concentration X in H_2 at 4.2°K.

cular electric quadrupole–quadrupole interaction free energy as the number of $J = 1$ molecules decreases. A simple calculation based on the observed variation yields $(\partial P/\partial X)_{T,V} = -89$ bar for $0.35 < X < 0.75$, and the agreement is considered to be very good.

These results, together with qualitative data on the sound absorption and description of experiments in the cubic phase of D_2 ($X \simeq 0.9$), are described in more detail elsewhere.[17]

References

1. J.H. Vignos and H.A. Fairbank, *Phys. Rev.* **147**, 185 (1966).
2. R.H. Crepeau and D.M. Lee, *Phys. Rev. A***6**, 516 (1972).
3. M. Nielsen, private communication.
4. J.P. Franck and R. Wanner, *Phys. Rev. Lett.* **25**, 345 (1970).
5. P.A. Bezuglyi and R.Kh. Minyafaev, *Fiz. Tverd. Tela* **9**, 624 (1967) [*Soviet Phys.—Solid State* **9**, 480 (1967)].
6. P.A. Bezuglyi and R.Kh. Minyafaev, *Fiz. Tverd. Tela* **9**, 3622 (1967) [*Soviet Phys.—Solid State* **9**, 2854 (1968)].
7. P.A. Bezuglyi, R.O. Plakhotin, and L.M. Tarasenko, *Fiz. Tverd. Tela* **13**, 309 (1971) [*Soviet Phys.—Solid State* **13**, 250 (1971)].
8. G. Ahlers, *J. Chem. Phys.* **41**, 86 (1964).
9. R.J. Roberts and J.G. Daunt, *J. Low Temp. Phys.* **6**, 97 (1972).
10. R.W. Hill and O. Lounasmaa, *Phil. Mag.* **4**, 785 (1959).
11. M. Nielsen and H. Bjerrum Moller, *Phys. Rev. B* **3**, 4383 (1971).
12. H.D. Megaw, *Phil. Mag.* **28**, 129 (1933).
13. J.W. Stewart, *Phys. Rev.* **97**, 578 (1955).
14. B.G. Udovidchenko and V.G. Manzhelii, *J. Low Temp. Phys.* **3**, 429 (1970).
15. J.F. Jarvis, D. Ramm, and H. Meyer, *Phys Rev.* **178**, 1461 (1969).
16. D. Ramm, H. Meyer, and R.L. Mills, *Phys. Rev. B***1**, 2763 (1970).
17. R. Wanner and H. Meyer, *J. Low Temp. Phys.* **11**, 715 (1973).

Polarizability of Solid H_2 and D_2*

Barnie Wallace, Jr. and Horst Meyer

Department of Physics, Duke University
Durham, North Carolina

We have measured the dielectric constant ε of solid H_2 and D_2 at several densities ρ as a function of temperature and have calculated the polarizability α using the Clausius–Mosotti relation

$$\frac{\varepsilon - 1}{\varepsilon + 2} = \rho\alpha \frac{4\pi}{3} \frac{1}{M} \tag{1}$$

The aim of our experiments was to determine the effect on α caused by the orientational ordering of the H_2 and D_2 molecules with angular momentum $J = 1$, as predicted by Harris.[1] The measurements were expected to show a dependence on the mole fraction X of molecules with $J = 1$. As a control of our experimental method we have also measured ε in solid ^4He and Ne, where a priori very little temperature dependence of α is expected.

In solid H_2 and D_2 the molecules with $J = 1$ are coupled by an anisotropic electrostatic interaction between the electric quadrupole moments. In several respects these solids are similar to antiferromagnets. Just as the susceptibility in magnetic compounds is an indication of short-range order, we can imagine the electric polarizability to give the temperature dependence of the order parameter σ in H_2 and D_2. This parameter measures the average molecular orientational alignment along some crystalline axis, and is given by $\sigma = \langle 1 - 3\cos^2\Theta_{ns}\rangle_T$ where Θ_{ns} is the angle between the molecular axis and the symmetry axis. Harris[1] has shown that for a single crystal in an electric field E

$$\alpha(\Theta_{ES}, T) = \alpha(T = \infty) + K(\Theta_{ES})\sigma(T) \tag{2}$$

where $\alpha(T = \infty)$ is the power average over the molecular polarizabilities perpendicular and parallel to the molecular axis. The constant K is a function of the polarization anisotropy $(\alpha_\parallel - \alpha_\perp)$ and of the crystal configuration and depends on the angle Θ_{ES} between E and the crystalline symmetry axis. It is zero for a powder or for a single crystal with cubic symmetry. To provide the stimulus to develop the theoretical expression for $\alpha(T) - \alpha(T = \infty)$ and to check Harris' expectations,[1] we have taken precise data in solid hcp H_2 and D_2 which we report for the range $1.2 < T < 10°K$.

Our measurements of α were made at constant ρ by using a cylindrical beryllium–copper capacitor that was part of a resonant LC circuit. This circuit was powered

---

* Work supported by a grant from the U.S. Army Research Office (Durham) and by a contract with the Office of Naval Research.

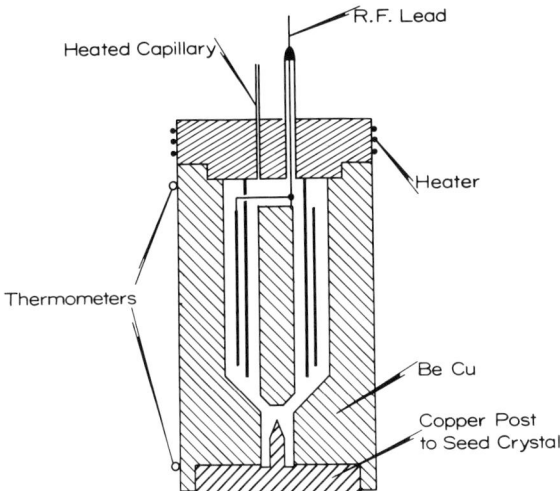

Fig. 1. Schematic diagram of the beryllium–copper capacitance cell; length 7 cm, outer diameter 2.6 cm. The concentric plate system, anchored mechanically, is to increase the capacitance. The lead O-rings and screws to tighten the top are not shown.

by a tunnel diode and operated at about 14 MHz. The sample container (Fig. 1) was designed to favor the growth of single crystals made necessary in order to obtain $K \neq 0$. The cell is cylindrical in shape, with 98% of the total cell capacitance along the symmetry axis. A cold tip at the bottom and a thermal gradient along the cell were used for slow growth of the crystal at constant pressure along the melting curve. By observing the change in dielectric constant, the crystal growth rate could be precisely regulated at any temperature. The crystals were annealed for several hours close to the melting curve before being cooled. The capillary to the cell was blocked once the growth was terminated and hence the density inside the cell stayed constant. From the observed resonance frequency ν of the tank circuit, ε was calculated. Account was taken of the temperature dependence of ν_0, the frequency of the empty capacitor. Small corrections were made for the temperature-dependent deformation of the BeCu cell caused by the pressure change $(\partial P/\partial T)_V$ of the sample.[2] The stability of the electronic system was such that changes in α of $1:10^5$ could be detected.

Measurements were made on a number of separate crystals of $H_2(X = 0.75)$, $H_2(X = 0)$, and $D_2(X = 0.33)$, where growth pressure and growth and annealing times were widely varied. Also, "powder" samples, where cooling from the fluid into the solid phase was carried out very rapidly, were prepared. In spite of careful annealing of the single crystals, some sudden jumps in the frequency were recorded at constant temperature above about 12°K, caused perhaps by adjustments of the crystal inside the cell. The reproducibility of the data upon successive warming and cooling cycles was best below 10°K, and for convenience we have normalized the data to those at 1.17°K, defining the ratio $R \equiv \alpha(T)/\alpha(T = 1.17°K)$. We will limit the presentation of the data to this range but mention that above 10°K the trend of the results was continued.

Figure 2 shows representative results for samples of H_2, D_2, ^4He, and Ne charac-

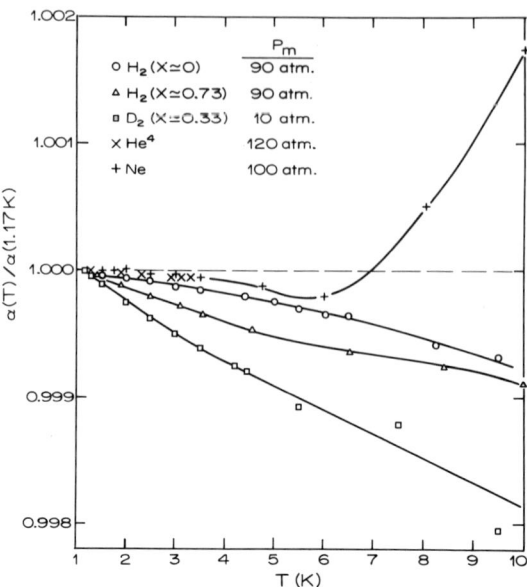

Fig. 2. The normalized polarizability $R = \alpha(T)/\alpha(1.17°K)$ for representative samples of solid H_2, D_2, ^4He, and Ne. These particular samples are characterized by their respective melting pressures P_m.

terized by their melting pressure P_m. For one type of solid the curves from various crystals and from the "powders," which we have not shown, agreed within $\pm 10\%$ of the deviation from $R = 1$. Solid ^{4}He up to the melting point at 3.7°K shows only a slight temperature variation of α and so does Ne up to 4°K.

The surprising results are the strong temperature dependence of Ne above 4°K and the smaller but nonnegligible one of $H_2(X = 0)$, both constituted of effectively spherical molecules. The temperature dependence of α for $H_2(0.75)$ is not very different from that of $H_2(X \simeq 0)$. Moreover, no discontinuous change in α was found at the transition from the hcp into the cubic phase near 1.5°K for $H_2(0.75)$. These results seem to indicate that orientational order has little if any effect on α. A further surprise is the still larger temperature dependence of α for $D_2(0.33)$.

There are no ready explanations for these results. A calculation by Harris[3] shows that a change with T of the axis ratio c/a for a monatomic crystal should result in a change of the observed α. This change, if any, should be dependent on the pressure change in the crystal and tend to zero approximately with T^4 as $T = 0$ is approached, contrary to the results we obtained.

Further results obtained during these experiments are as follows.

(1) The polarizabilities in the liquid and in the solid phase for $H_2(X \simeq 0)$ at the melting curve, calculated via Eq. (1) and using available density data for both phases,[4] were found to be the same within 0.2%. This latter result agrees with the recent work of Udovidchenko and Manzhelii,[5] but disagrees with the earlier work of Younglove,[4] who found α to be 0.45% larger in the solid phase.

(2) At a given temperature the melting pressure of $H_2(X \simeq 0)$ is 3.6 ± 0.2 atm

higher than that of $H_2(X \simeq 0.75)$. If we assume α is independent of X within 0.2%, we find that in the liquid for a given P and T close to the melting curve and for $P < 100$ atm, $\rho\{H_2(0.75)\} = 1.005\rho\{H_2(X \approx 0)\}$. This is consistent with the results quoted by Woolley et al.[6] for H_2 at saturated vapor pressure.

A more detailed report discussing all these results is being prepared and further experiments are planned down to lower temperatures and using purer Ne samples.*

Acknowledgements

We thank Professor A. Brooks Harris for many stimulating discussions and Dr. Dewey Lawson for his advice on single-crystal growth techniques for the capacitor cell.

References

1. A.B. Harris, *J. Appl. Phys.* **42**, 1574 (1971); and private communication.
2. J.F. Jarvis, H. Meyer, and D. Ramm, *Phys. Rev.* **170**, 320 (1968); **178**, 1461 (1969).
3. A.B. Harris, private communication.
4. B.A. Younglove, *J. Chem. Phys.* **48**, 4181 (1968), and references therein.
5. B.G. Udovidchenko and V.G. Manzhelii, *J. Low Temp. Phys.* **3**, 429 (1970).
6. H.W. Woolley, R.B. Scott, and F.G. Brickwedde, *J. Res. Nat. Bur. Std.* **41**, 374 (1948).

* **Note added in proof:** The nominal purity of the Ne used in this study was 99.7%. Thus, there is the possibility that the observed dielectric behavior might have been caused by impurities.

Scattering of Neutrons by Phonons and Librons in Solid o-Hydrogen

A. Bickermann, F. G. Mertens, and W. Biem

Institute for Theoretical Physics
University of Giessen, Giessen, West Germany

Recently Stein et al.[1] performed measurements of the incoherent cross section of the scattering of slow neutrons on solid o-hydrogen in the ordered $Pa3$ phase. The cross section of the librons seemed to be enhanced compared with the one calculated in the Bloch spin-wave approximation.[2] On the theoretical side calculations of phonons and librons and their coupling have been performed in the random phase approximation (RPA) by Mertens and Biem.[3] The librons alone have been considered, including their large anharmonicities, by Coll et al.[4]

In this paper we calculate the neutron scattering of phonons and librons in the RPA. In the theory of quantum crystals the short-range correlations play an important role and no dynamic question can be answered quantitatively without taking them into account. But since the structure of the crystal is relatively complicated and the number of the degrees of freedom is high (12 phonon modes + 8 libron modes), we neglect them in our calculation as a preliminary measure which has to be corrected in the future.

The RPA at $T=0$ can be formulated in short by introducing the creation and destruction operators $a_m^{i+}(\mu)$ and $a_m^i(\mu)$ of localized excitations[4] at the lattice site μ in cell m ($i = 1, 2, 3 = x, y, z$ means excitations of the center of mass motion; $i = 4, 5 = +1, -1$ means excitation of the orientational motion from the $Y_{1,0}$ state into the $Y_{1,\pm 1}$ states). Using sublattice coordinates in which these excitations are diagonal in a Hartree approximation, the Hamiltonian can be written

$$H = E_H^0 + H_2 + H' \qquad (1)$$

where E_H^0 is the Hartree energy, H_2 is the part quadratic in a^+ and a, and H' is the anharmonic part. The RPA then means the diagonalization of H_2 by a Bogoliubov transformation

$$a_m^i(\mu) = (N/s)^{-1/2} \sum_{k\lambda} \left[u_{km}^{\lambda i}(\mu) b_k^\lambda - v_{km}^{\lambda i}(\mu) b_k^{\lambda +} \right] \qquad (2)$$

neglecting H'; here N is the number of molecules and $s = 4$ is the number of molecules per cell.

With new operators b^+ and b the Hamiltonian reads

$$H = \sum_{k\lambda} \hbar \omega_k^\lambda (b_k^{\lambda +} b_k^\lambda + \tfrac{1}{2})$$

Scattering of Neutrons by Phonons and Librons in Solid o-Hydrogen

Since we are interested in the scattering of slow neutrons by o-H_2 without producing p-H_2, we need only to calculate the following part of the incoherent scattering function:

$$S_i(\kappa,\omega) = \tfrac{1}{2} \sum_{\mu m} \sum_f |<\psi_f | V_m^\mu | \psi_i >|^2 \delta(\omega - \omega_{fi})$$

where the spin sum has already been performed. $|\psi_i>$ is the initial state and $|\psi_f>$ is a final state. In our case we create single excitations from the ground state $|\phi_0>$ only. The scattering operator

$$V_m^\mu = \exp(i\kappa s_\mu^m) \cos \tfrac{1}{2}\kappa \rho_m^\mu$$

depends on the vector ρ_m^μ between the nuclei of the molecule (μ,m) and the displacement vector s_m^μ of this molecule. It is easy to express this operator by a^+ and a: for phonons we use

$$s_{m\mu}^i = (\hbar/2M\varepsilon_i)^{1/2} [a_m^{i+}(\mu) + a_m^i(\mu)]$$

where ε_i are the Hartree frequencies; for the librons we use the method of Raich and Etters[5] to introduce boson operators, which results in

$$\cos \kappa\rho/2 = c + l(a^+, a) + q(a^+, a)$$

where c, l, q mean terms constant, linear, and quadratic in a^+ and a.

The transformation of the operator V_m^μ as a function of (a^+, a) into another one as a function (b^+, b) can be performed and results in a lengthy expression containing the generalized polarization vectors e_μ^i which are connected with the u's and v's by

$$u_{km}^{\lambda i}(\mu) = \frac{\omega_k^\lambda + \varepsilon_i}{2\varepsilon_i} e_\mu^i(_k^\lambda) e^{ikR_m} \left(\frac{\varepsilon_i}{\omega_k^\lambda} \right)^{1/2}$$

$$\left. \begin{array}{l} v_{km}^{\lambda i}(\mu) \\ v_{km}^{\lambda j}(\mu) \end{array} \right\} = \frac{\omega_k^\lambda - \varepsilon_i}{2\varepsilon_i} e_\mu^i(_k^\lambda) \exp(ikR_m) \left(\frac{\varepsilon_i}{\omega_k^\lambda} \right)^{\frac{1}{2}} \begin{cases} \text{if } i = 1, 2, 3 \\ \text{if } i = 4, 5 \text{ (if } i = 4, j = 5 \\ \text{and vice versa)} \end{cases}$$

While u and v solve the RPA equations, the e_μ^i are the eigenvectors of the secular equation of the dynamic matrix which can be derived from the RPA equations.

We have to calculate $\langle b_k^{\lambda+} \phi_0 | V_m^\mu | \phi_0 \rangle$ to get $S_i(\kappa, \omega)$. This is a straightforward calculation.

Results for Single Crystals

For one-phonon scattering without exciting librons

$$S_{iP}(\kappa, \omega) = \left[j_0\left(\frac{\kappa\rho}{2}\right) - 2\frac{4\pi}{5} j_2\left(\frac{\kappa\rho}{2}\right) Y_{20}^\kappa \right]^2 \frac{1}{2} \frac{\hbar}{2M\omega}$$

$$\times \sum_{k_0 \lambda_0} \delta(\omega - \omega_{k_0}^{\lambda_0}) \sum_\mu |\kappa e_\mu(_{k_0}^{\lambda_0})|^2 e^{-2w_\mu}$$

with j_0 and j_2 spherical harmonics and w_μ the Debye–Waller factor,

$$2w_\mu = \frac{1}{N/s} \sum_{k\lambda} \frac{\hbar}{2M\omega_k^\lambda} |\kappa e_\mu(_k^\lambda)|^2 = (\kappa, A_\mu \kappa)$$

The Debye–Waller factor depends on the sublattice μ. Using coordinates with z axis pointing into the direction of the body diagonal of this sublattice, the tensor A_μ has diagonal form with two equal elements ($A_\mu^{xx} = A_\mu^{yy}$) and $A_\mu^{zz} \neq A_\mu^{xx}$, but the deviation from spherical symmetry may be small. The first factor in the formula is the form factor of the orientational order and the term with j_2 is small.

For one-libron scattering without exciting phonons,

$$S_{iL}(\kappa, \omega) = \frac{3}{10} j_2^2\left(\frac{\kappa\rho}{2}\right) \frac{\varepsilon}{\omega} \sum_{k_0 \lambda_0} \delta(\omega - \omega_{k_0}^{\lambda_0}) 4\pi \sum_\mu e^{-2w_\mu} F_\mu(\kappa)$$

with ε the excitation energy of one molecule from $Y_{1,0}$ into $Y_{1,\pm 1}$ and

$$F_\mu(\kappa) = \left| \sum_{i=4,5} e_\mu^i(_{k_0}^{\lambda_0}) Y_{2,i}^\kappa \right|^2$$

While $S_{iP} \approx \kappa^2$ for small κ, $S_{iL} \approx \kappa^4$ and therefore it is small compared with S_{iP}.

We will not write down the complicated formulas for the scattering amplitude or the scattering function in this case, because of their length.

Results for Polycrystals

To get the cross section here we have to integrate over all directions of κ relative to the crystal axes.

One-Phonon Scattering. (a) We get a simple and well known form of S_{iP} when we neglect the sublattice dependence of w_μ. If we neglect the small term j_2, too, we get the formula

$$S_{iP} = j_0^2\left(\frac{\kappa\rho}{2}\right) \frac{1}{2} \frac{\hbar}{2M\omega} \frac{\kappa^2}{3} e^{-2w} \sum_{k_0 \lambda_0} \delta(\omega - \omega_{k_0}^{\lambda_0})$$

with

$$2w = \frac{1}{N/s} \sum_{k\lambda} \frac{\hbar}{2M\omega_k^\lambda} \frac{\kappa^2}{3}$$

(b) More exact formulas need the calculation of sums with polarization vectors over the whole Brillouin zone. If the j_2 term is neglected here, too, we have to integrate

$$1/4\pi \int |\kappa e_\mu(_{k_0}^{\lambda_0})|^2 e^{-2w_\mu} d\Omega_\kappa$$

The integral $\int e^{-2w_\mu} d\Omega_\kappa$ can be performed with the tensor A_μ using the error function $\Phi(t)$:

$$\frac{1}{4\pi} \int (\exp - 2w_\mu) d\Omega_\kappa = (\exp - \kappa^2 A_\mu^{xx}) \frac{\sqrt{\pi}}{2} \frac{\Phi[\kappa(A_\mu^{xx} - A_\mu^{zz})^{1/2}]}{\kappa(A_\mu^{xx} - A_\mu^{zz})^{1/2}}$$

with

$$A_\mu^{ii} = \frac{1}{N/s} \sum_{\kappa\lambda} \frac{\hbar}{2M\omega_k^\lambda} e_\mu^i(_k^\lambda) e_\mu^{i\dagger}(_k^\lambda)$$

in the sublattice coordinates. Using this integral, the exact but unwieldy S_{iP} could be calculated.

As an approximation which takes into account the anisotropy in the sublattices to a certain extent, we can write

$$S_{iP} = j_0^2 \frac{1}{2} \frac{\hbar}{2M\omega} \frac{\kappa^2}{3} \sum_{k_0\lambda_0} \delta(\omega - \omega_{k_0}^{\lambda_0}) \sum_\mu |e_\mu(_{k_0}^{\lambda_0})|^2$$

$$\times (\exp - \kappa^2 A_\mu^{xx}) \frac{\sqrt{\pi}}{2} \frac{\Phi(\kappa[A_\mu^{xx} - A_\mu^{zz}]^{1/2})}{\kappa(A_\mu^{xx} - A_\mu^{zz})^{1/2}}$$

One-Libron Scattering. (a) Corresponding to the result for one-phonon scattering, we get

$$S_{iL} = \frac{3}{10} j_2^2 \left(\frac{\kappa\rho}{2}\right) \frac{\varepsilon}{\omega} \sum_{k_0\lambda_0} \delta(\omega - \omega_{k_0}^{\lambda_0}) e^{-2w}$$

where the only difference to the corresponding formula in the Bloch approximation[3] is the factor ε/ω, apart from an erroneous $1/2$ in the older paper. (b) Corresponding to the result for one-phonon scattering, we have

$$S_{iL} = \frac{3}{10} j_2^2 \left(\frac{\kappa\rho}{2}\right) \frac{\varepsilon}{\omega} \sum_{k_0\lambda_0} \delta(\omega - \omega_{k_0}^{\lambda_0}) 4\pi \sum_\mu \sum_{i=4,5} |e_\mu^i(_{k_0}^{\lambda_0})|^2$$

$$\times (\exp - \kappa^2 A_\mu^{xx}) \frac{\sqrt{\pi}}{2} \frac{\Phi(\kappa[A_\mu^{xx} - A_\mu^{zz}]^{1/2})}{\kappa(A_\mu^{xx} - A_\mu^{zz})^{1/2}}$$

In the case of *coupled phonons and librons* we can write down an approximative formula which is relatively exact in the energy region of interest, where the phonons and librons couple. In this coupling region we can write, assuming w independent of μ,

$$S_i = \frac{1}{2}\left(j_0^2 + \frac{4}{5} j_2^2\right) \frac{\hbar}{2M\omega} \frac{\kappa^2}{3} e^{-2w} \sum_{k_0\lambda_0} \delta(\omega - \omega_{k_0}^{\lambda_0}) |\alpha(_{k_0}^{\lambda_0})|^2$$

$$+ \frac{3}{10} j_2^2 \frac{\varepsilon}{\omega} e^{-2w} \sum_{k_0\lambda_0} \delta(\omega - \omega_{k_0}^{\lambda_0}) |\beta(_{k_0}^{\lambda_0})|^2$$

+ correction terms

Here we have introduced coefficients α and β which decompose the polarization vector $e_\mu(_k^\lambda)$ into its respective phononlike and libronlike parts:

$$e_\mu = \alpha e_\mu^P + \beta e_\mu^L$$

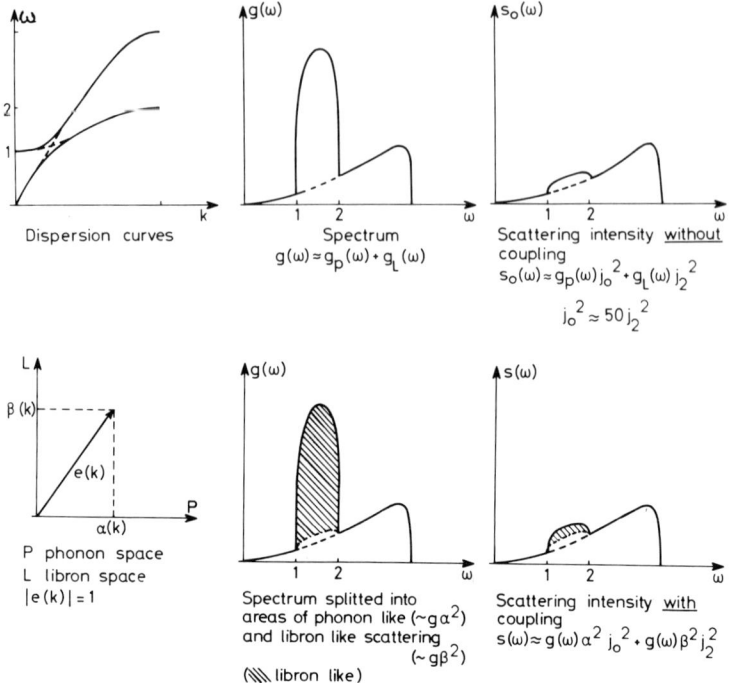

Fig. 1. The enhancement of the scattering function by coupling of phonons and librons in the region of the libron band (schematic).

where e_μ^P and e_μ^L are unit vectors in the P and L spaces. The correction terms can be neglected because they are proportional to j_2^2 times a factor V_{CR}/V_B, with V_B the volume of the Brillouin zone and V_{CR} the volume of the coupling region in the k space ($F \approx (1/3)^3 = 1/27$).

Outside of the coupling region other terms proportional to $(\varepsilon - \omega)^2$ influence the scattering and complicate the formula.

The above formula for S_i shows qualitatively the effect of enhanced scattering in the region of the libron energies, for the phononlike contribution of mixed modes to the scattering is proportional to $|\alpha|^2$ (see Fig. 1). Simple model calculations have resulted in an enhancement of the correct order of magnitude. We hope to calculate the effect for the real problem; much computer time is needed for such a program.

References

1. H. Stein, H. Stiller, and R. Stockmeyer, in *Proc. Conf. on Phonons*, Flammarion, Rennes (1971); H. Stein, Report of the Nuclear Research Center, Jülich, Germany, 838–FF (1972).
2. F.G. Mertens, W. Biem, and H. Hahn, *Z. Physik* **213**, 33 (1968); **220**, 1 (1969); F.G. Mertens, Report of the Nuclear Research Center, Jülich, Germany, 564–FN (1968).
3. F.G. Mertens and W. Biem, *Z. Physik* **250**, 273 (1972); F.G. Mertens, 734–FF (1971); W. Biem, 496–NP (1967).
4. C.F. Coll III, A.B. Harris, and A.J. Berlinski, *Phys. Rev. Lett.* **25**, 858 (1970).
5. J.C. Raich and R.D. Etters, *Phys. Rev.* **168**, 425 (1968); C.F. Coll III and A.B. Harris, *Phys. Rev.* B **4**, 2781 (1971).

Determination of the Crystal Structures of D_2 by Neutron Diffraction*

R. L. Mills, J. L. Yarnell, and A. F. Schuch

Los Alamos Scientific Laboratory, University of California
Los Alamos, New Mexico

Previous neutron diffraction studies[1,2] of solid D_2 have suffered from various difficulties but the basic problem has been one of preparing fine-grained powder specimens with no preferred orientation. The technique used in the present study was to inject D_2 gas, with para concentration[3,4] of 2, 29, and 96%, into boiling liquid He contained in a Ti-Zr cell as shown in Fig. 1. The cell material had a zero coherent cross section for neutron scattering and therefore contributed no peaks to the diffraction patterns. The filling procedure assured that D_2 bulk powder of random orientation would be formed in a structureless medium. The packing fraction of the D_2 was raised to about 0.5 by lightly tamping the powdered slurry in the cell with a screen plunger. Liquid He remained in the interstices of the sample and provided thermostatic cooling in the range 1.5–4°K without contributing significantly to the diffraction pattern. By boiling away the liquid He, the sample temperature could be raised to 10°K. Samples of H_2 could not be prepared by this technique since solid H_2 floats on the surface of liquid He.

The experiments were carried out on a neutron diffractometer located at the Los Alamos Omega West Reactor, which operated at a power of 6 MW. A monochromatic beam of neutrons of wavelength $1.28587 + 0.00014$ Å and free of second-order contamination was obtained from the (331) reflection from a germanium monochromator crystal. The incident beam flux was monitored by a fission counter. In a typical diffraction scan the number of scattered neutrons during the accumulation of 100,000 monitor counts was recorded with the scattering angle fixed at 2θ values ranging from 20° to 60° in 0.1° steps. The full-width at half-maximum of a diffraction peak varied slightly with angle from about 0.65° to 0.75° in 2θ.

Typical diffraction patterns for 96% paradeuterium (p-D_2) are shown in Fig. 2. Diffraction peaks were located using the correlation technique described by Black.[5] This technique provides an unbiased method of isolating weak peaks in the presence of statistical fluctuations in the background. Estimates of 2θ, integrated intensity, and their standard deviations for the peaks thus located were obtained from a least-squares fitting program.[15] In this program the diffraction peaks were represented by Gaussians and the slowly varying background by a cubic spline through nine points equally spaced over the interval $20° \leq 2\theta \leq 60°$. The integrated intensities

---

* Work performed under the auspices of the U. S. Atomic Energy Commission.

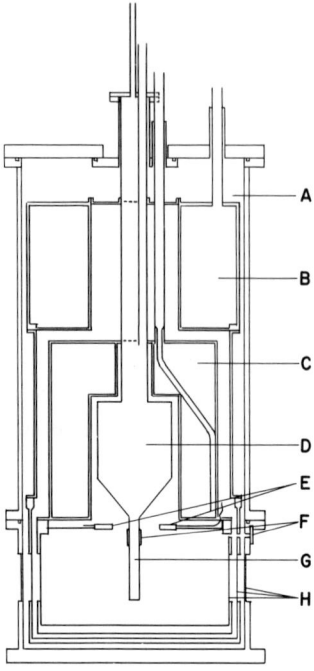

Fig. 1. Cryostat and cell for neutron diffraction studies. A, vacuum space; B, liquid N_2 pot; C, liquid He outer pot; D, liquid He inner pot; E, 40-mW lights; F, glass windows; G, Ti–Zr cell for solid D_2; H, Al thermal shields and windows for neutrons.

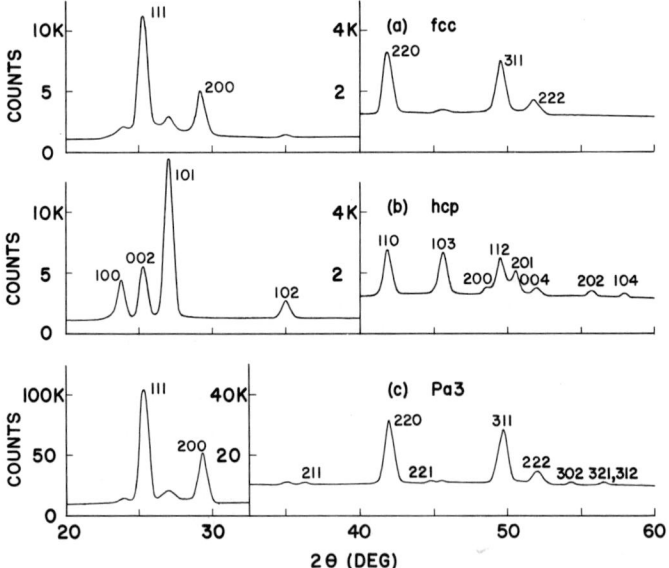

Fig. 2. Neutron diffraction patterns from the three crystal forms of 96% paradeuterium. (a) Disordered cubic solid, as-frozen at 4°K; (b) Disordered hexagonal solid at 4°K after warming to > 10°K; (c) ordered cubic solid at 1.5°K, summation of eight points at every 0.2° from three samples.

were used to calculate observed structure factors $|F_0|$ for each peak. Values of the temperature parameter B in the expression $\exp[-B(\sin^2\theta)/\lambda^2]$ and an overall scaling factor were refined for various trial structures by computer programs which minimized the disagreement parameter

$$R_{wF^2} = \left[\sum w(F_0^2 - F_c^2)^2\right]^{1/2} / \left[\sum wF_0^4\right]^{1/2}$$

where the F_c are the calculated structure factors and the weighting factors w are inversely proportional to the variance of the corresponding F_0^2.

Disordered Cubic

Deuterium gas at all para concentrations injected into liquid He at 4°K solidified predominantly in a structure which appeared to be fcc. A similar structure has been reported[6] when D_2 is frozen directly from the low-pressure gas on a cold, structureless surface at 4°K. When frozen from bulk liquid, D_2 at all para concentrations is known to form hcp solid.

As shown in Fig. 2(a) five sharp diffraction lines were observed which could be indexed as the (111), (200), (220), (311), and (222) reflections. In addition, weak, broad peaks from hcp D_2 were always present. The hcp intensities were assessed and were found to correspond to about 10% of the D_2 sample, independent of para concentration. The hcp component was then subtracted to give the final cubic pattern.

The refined structure factors for a model in which spherical molecules, or their equivalent, randomly oriented molecules, are located on an fcc lattice are given in Table I. With only five reflections the disagreement factor R_{wF^2} is quite small. Excellent fits were obtained at all para concentrations. However, there may be an indication that the structure is not strictly cubic. The apparent cell constant, as computed from each of the five reflections, showed patterns of variations which were the same for all para concentrations and which were up to ten times larger than the measuring errors. If distortions exist in the cubic cell, they are so small that the additional reflections arising from them remain undetected even in terms of peak broadening. It may be that the superimposed pattern from the 10% of material present as hcp caused the apparent shifts in 2θ. We conclude that in the as-frozen solid at 4°K

Table I. Observed and Calculated Structure Factors for As-Frozen Cubic D_2 at 4°K

		96% p-D_2		29% p-D_2		2% p-D_2	
		$|F_0|$	$|F_c|^a$	$|F_0|$	$|F_c|^a$	$|F_0|$	$|F_c|^a$
hkl:	111	5.822	5.818	5.861	5.864	5.836	5.817
	200	5.230	5.226	5.300	5.282	5.197	5.226
	220	3.328	3.383	3.393	3.455	3.335	3.382
	311	2.448	2.424	2.522	2.495	2.460	2.423
	222	2.334	2.166	2.385	2.236	2.311	2.165
a, Å		5.0868		5.1008		5.1030	
B, Å^2		7.290		7.077		7.366	
R_{wF^2}		0.014		0.015		0.015	

[a]Spherical molecules on fcc lattice.

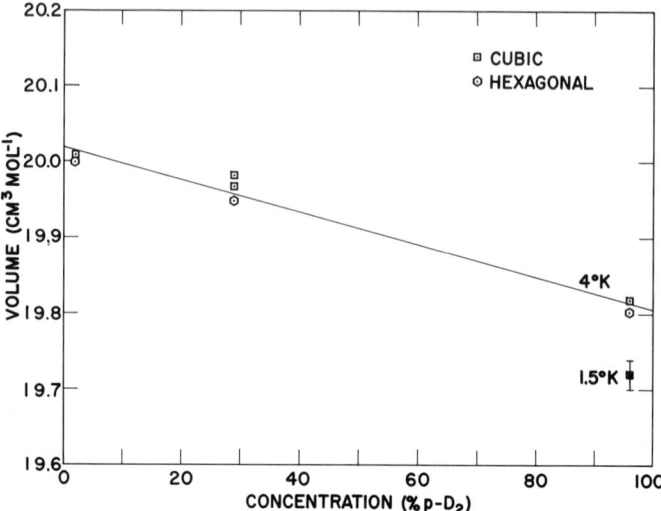

Fig. 3. Molar volume of solid deuterium as a function of para concentration for disordered cubic and hexagonal forms at 4°K. Solid square, ordered cubic at 1.5°K.

the molecules are orientationally disordered and are located on a lattice that is very nearly, if not exactly, fcc. This form of D_2 is probably the disoriented cubic structure that was indicated in earlier cycling experiments[7,8] on enriched material. Cubic specimens of 2% para content have a molar volume that is 1% larger than that of 96% para, as shown in Fig. 3.

Disordered Hexagonal

When the as-frozen samples were warmed to 10°K they transformed completely to the hexagonal structure. Samples at all para concentrations remained hexagonal when they were subsequently cooled back down to 4°K, as shown by the scan in Fig. 2(b). The 12 observed reflections could be indexed on the basis of an hcp structure as (100), (002), (101), (102), (110), (103), (200), (112), (201), (004), (202), and (104). Lattice constants were obtained by least-squares fitting. The c/a ratio was independent of the para content and was evaluated as 1.6317, compared to the ideal ratio of 1.6330. In Fig. 3 it can be seen that the molar volume and its variation with para content of the hcp samples at 4°K were quite similar to those of the as-frozen cubic solid.

Our trial structures for hexagonal D_2 were those in which the molecular centers are located at sites on an hcp lattice with the molecules disordered in the hexagonal planes and tilted from the c axis at various angles. This model is insensitive to whether the molecules precess or are stationary and are distributed randomly among the 24 general positions of space group $P6_3/mmc$. At an angle of about 55° the trial structure is essentially identical to one where the molecules are spherical or are freely rotating. In Table II are shown the structure factors at tilt angles giving the smallest disagreement factor for the various hcp samples at 4°K. The slight

Table II. Observed and Calculated Structure Factors for Hexagonal D_2 at 4°K

		96% p-D_2		29% p-D_2		2% p-D_2	
		$\|F_o\|$	$\|F_c\|^a$	$\|F_o\|$	$\|F_c\|^a$	$\|F_o\|$	$\|F_c\|^a$
hkl:	100	1.464	1.505	1.470	1.500	1.492	1.499
	002	3.021	2.953	3.207	2.966	3.179	3.009
	101	2.422	2.418	2.404	2.413	2.410	2.419
	102	1.059	1.111	1.002	1.112	1.068	1.128
	110	1.698	1.674	1.759	1.658	1.709	1.651
	103	1.284	1.304	1.274	1.315	1.348	1.359
	200	0.657	0.617	0.576	0.609	0.778	0.605
	112	1.327	1.236	1.316	1.230	1.257	1.242
	201	1.023	0.992	1.060	0.980	1.067	0.976
	004	1.345	1.135	1.322	1.162	1.286	1.242
	202	0.397	0.456	0.447	0.452	0.546	0.455
	104	0.350	0.427	0.324	0.436	0.410	0.465
a, Å		3.5967		3.6058		3.6089	
c, Å		5.8689		5.8833		5.8880	
B, Å^2		7.125		7.209		7.083	
R_{wF^2}		0.032		0.065		0.043	
Tilt, deg		57		58		60	

^{a}Molecules on hcp lattice tilted with respect to c axis.

deviations from the 55° or spherical-shell model are probably due to traces of preferred orientation in the powder samples following annealing at 10°K, which tends to enhance the (002) and (004) intensities.

Ordered Cubic

Only the hcp samples with the highest para concentration transformed to a cubic phase when cooled further to 1.5°K. The transition was never complete and from 7 to 14% of the material remained hcp. This behavior confirms earlier observations that the transition is martensitic.[9] To improve the counting statistics, eight points at every 0.2° were summed from scans on three different samples of 90–96% p-D_2 at 1.5°K. The results are plotted in Fig. 2(c). Again background and residual hcp intensities were assessed and were subtracted from the overall pattern. There remained nine cubic lines which were indexed as (111), (200), (211), (220), (221), (311), (222), (302), and (321, 312).

A quantum mechanical form factor was calculated, assuming all molecules to be in the $J = 1, m_J = 0$ state. Our trial structure was then one in which para molecules were located on an fcc lattice with their directions of quantization along the four cube diagonals in space group $Pa3$, as predicted theoretically.[10] The computer refinement based on this structure is shown in Table III, where the disagreement factor R_{wF^2} is 0.032. In earlier work[1] the weak cubic reflections were masked by coherent scattering from an Al cell. After making subtractions the authors reported that the relative intensities corresponded to diffraction from classical D_2 molecules

Table III. Observed and Calculated Structure Factors for Ordered Cubic D_2

		$\geq 90\%$ p-D_2	
		$|F_o|$	$|F_c|^a$
hkl:	111	6.026	5.931
	200	5.240	5.358
	210	—	0.306
	211	0.266	0.287
	220	3.536	3.558
	221	0.222	0.226
	311	2.646	2.599
	222	2.421	2.337
	302	0.419	0.488
	321, 312	0.415	0.475
a, Å		5.0760	
B, Å^2		6.514	
R_{wF^2}		0.032	

^{a}Quantum mechanical form factor for $J = 1$, $m_J = 0$ molecules in $Pa3$ space group.

fixed on an fcc lattice in space group $Pa3$. For this non-quantum mechanical model, we find the disagreement factor and the temperature parameter to be unacceptably large.

Ordered cubic p-D_2 at 1.5°K is 0.6% smaller in molar volume than the disordered solid at 4°K, as shown in Fig. 3. Only a portion of this decrease can be accounted for by published[11] thermal contraction data, so that the ordered cubic solid is more densely packed than the disordered solid. A similar result was inferred from earlier measurements[8] of the pressure change which accompanies the order–disorder transition.

For all three structures of solid D_2 the refined temperature parameters were reasonably consistent with values of Debye θ determined from heat capacity experiments.[12-14] Our values for 96% p-D_2 are: disordered cubic at 4°K, $\theta_D = 99$°K; disordered hexagonal at 4°K, $\theta_D = 101$°K; and ordered cubic at 1.5°K, $\theta_D = 110$°K.

Acknowledgments

We are indebted to Morris Klein, who prepared the computer codes for refining the trial structures. We also wish to thank Donald Clinton for fabricating the cryostat, and Robert MacFarlane for assisting with some of the scans. John Raich contributed helpful discussions during data analysis.

References

1. K.F. Mucker, P.M. Harris, D. White, and R.A. Erickson, *J. Chem. Phys.* **49**, 1922 (1968).
2. A.S. Bulatov and V.S. Kogan, *J. Exp. Theor. Phys. (USSR)* **54**, 390 (1968).
3. D.A. Depatie and R.L. Mills, *Rev. Sci. Instr.* **39**, 105 (1968).
4. E.R. Grilly, *Rev. Sci. Instr.* **24**, 72 (1953).

5. W.W. Black, *Nucl. Instr. Meth.* **71**, 317 (1969).
6. O. Bostanjoglo and R. Kleinschmidt, *J. Chem. Phys.* **46**, 2004 (1967).
7. A.F. Schuch, R.L. Mills, and D.A. Depatie, *Phys. Rev.* **165**, 1032 (1968).
8. D. Ramm, H. Meyer, and R.L. Mills, *Phys. Rev. B* **1**, 2763 (1970).
9. C.S. Barrett, L. Meyer, and J. Wasserman, *J. Chem. Phys.* **45**, 834 (1966).
10. H.M. James and J.C. Raich, *Phys. Rev.* **162**, 649 (1967).
11. H.D. Megaw, *Phil. Mag.* **28**, 129 (1939).
12. R.J. Roberts and J.G. Daunt, *J. Low Temp. Phys.* **6**, 97 (1972).
13. O.D. Gonzales, D. White, and H.L. Johnston, *J. Phys. Chem.* **61**, 773 (1957).
14. R.W. Hill and O.V. Lounasmaa, *Phil. Mag.* **4**, 785 (1959).
15. R.H. Moore and R.K. Ziegler, "Solution of the General Least Squares Problem with Special Reference to High Speed Computers," Los Alamos Scientific Laboratory Report LA 2367, October 15, 1959.

Velocity and Absorption of High-Frequency Sound Near the Lambda Transitions in Solid CD_4*

R. P. Wolf

Harvey Mudd College
Claremont, California

and

F. A. Stahl

California State University
Los Angeles, California

Deuterated methane CD_4 freezes at 89.6°K and in the solid phase undergoes two phase transitions characterized by lambda-type heat capacity maxima at 22.2 and 27.1°K.[1] At low pressures solid CH_4 exhibits clearly only one phase transition at 20.48°K.[1] The object of our study was to determine the acoustical properties associated with the phase transitions in CD_4 and to compare the results with prior work[2] on CH_4. We have measured velocity and attenuation of 4- and 12-MHz longitudinal sound in CD_4 at temperatures between 2 and 77°K and have observed discontinuities in the slope of the velocity vs. temperature curve and attenuation peaks in the vicinity of the transition temperatures.

We grew solid CD_4 samples by a procedure reported earlier[3] for CH_4, at rates of about 1 mm/hr. Using a conventional two-transducer pulse system, we measured propagation times as well as output amplitudes for fixed input. Temperature was measured with a germanium resistance thermometer in a low-frequency ac bridge circuit, with resolution of $\pm\,0.02\%$. Accuracy was about 0.1% when compared with selected known triple-point temperatures.

Samples were allowed to warm from 4°K at a rate not exceeding 6 K°/hr, by boiling liquid from the dewar and using the large enthalpy of helium adsorbed on a cold, saturated molecular sieve to provide the controlled slow rise in temperature. Following all warmup runs on a given methane sample absolute values of velocity v and attenuation α were measured for 77°K equilibrium conditions by melting the driving transducer into the methane in approximately 1/4-mm steps. Propagation time and amplitude were recorded after equilibrium was restored at each step. Displacements of the transducer were measured with a traveling microscope, and the total change in displacement was corrected for thermal contraction of the stainless steel connecting line. Typical velocity meltdown data for one sample are shown in Fig. 1. The average 77°K velocity for four samples is 2135 $\pm$ 36 m/sec at both 4 and 12 MHz.

* Work supported by the National Science Foundation.

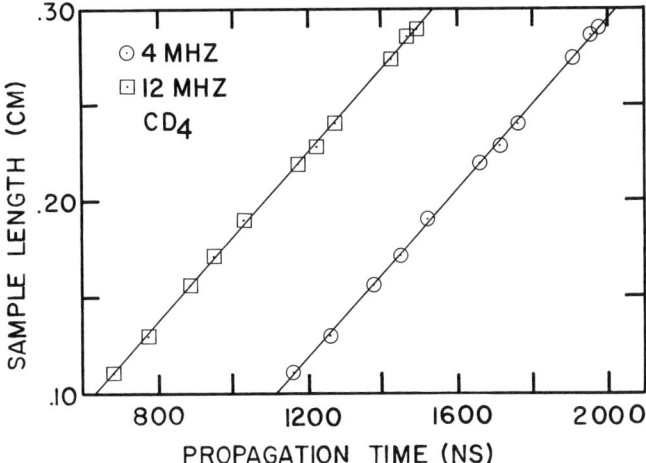

Fig. 1. Sample length vs. propagation time for longitudinal sound in one CD_4 crystal at 4 and 12 MHz, 77°K. The 4-MHz slope is 2162 m/sec, the 12-MHz slope is 2174 m/sec.

Values of v and α determined for 77°K equilibrium conditions were used in calculating $v(T)$ and $\alpha(T)$ at other temperatures from corresponding propagation times $t(T)$ and amplitudes $A(T)$. The effective length of CD_4 at 77°K was calculated for each run from the product of equilibrium velocity at 77°K and propagation time corrected for delay in the electronics. We made a total of ten warmup runs on three CD_4 samples. Figure 2 shows results for average velocity, corrected for thermal contraction in CD_4, along with uncorrected 4-MHz data for one typical run and, for comparison, the average velocity of 4-MHz longitudinal sound in CH_4. The total thermal contraction[4] in CD_4 from 77 to 2°K is about $2\frac{1}{2}\%$, but is only 0.4% from 20 to 30°K, whereas the velocity changes 7.0% from 20 to 30°K. The average 2°K velocity is 2750 ± 75 m/sec at both 4 and 12 MHz.

Dispersion in the velocity between 4 and 12 MHz was not found at either 2 or 77°K within experimental error. At intermediate temperatures and especially in the vicinity of the λ-transitions the dispersion, if any, was less than the scatter in our data—about 0.05%.

There were some indications that the sound velocity increased on the order of 0.01% or less as temperature rose from 2 to 8°K. The changes were at the limit of resolution of our apparatus. Measurements of the coefficient of thermal expansion[5] of CH_4 indicate a fractional decrease in length $\Delta L/L$ equal to 2.3×10^{-5} as T increases from 4.2 to 8.7°K. A similar behavior might be expected in CD_4, and might account for the increase in velocity.

The velocity of sound in CD_4, as shown in Fig. 2, exhibits discontinuities in slope at both 4 and 12 MHz on the average at 22.3 ± 0.16 and 26.8 ± 0.14°K, whereas the heat capacity maxima reported by Colwell et al.[1] occur at 22.2 and 27.1°K. Our measurements of the CD_4 crystal temperature are accurate to ± 0.1 K° near the transition temperatures. Thus the difference observed for the upper transition is probably not due to experimental error; the discontinuity in slope of $v(T)$ vs. T occurs below the heat capacity maximum for the upper transition in CD_4. In CH_4 the velocity

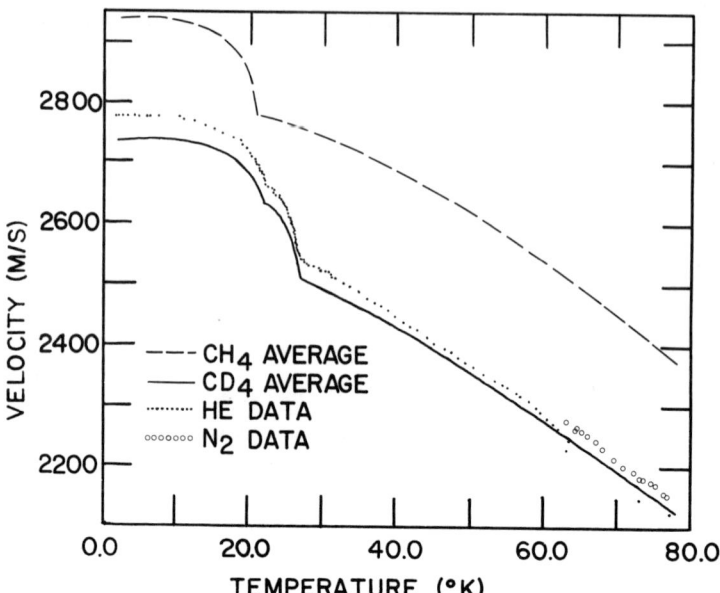

Fig. 2. Average velocity of 4-MHz longitudinal sound in CD_4 vs. temperature; data from one liquid helium and one liquid nitrogen warmup run (not corrected for thermal expansion) and average 4-MHz longitudinal sound velocity in CH_4 are also given for comparison.

kink occurs 0.35 K° above the heat capacity maximum and is appreciably more abrupt than either of the discontinuities associated with the CD_4 transitions.

In measurements of attenuation we find average values at 77°K of 2.38 ± 0.20 cm^{-1} at 4 MHz and 4.84 ± 0.50 cm^{-1} at 12 MHz. The variation of attenuation with frequency measured for CD_4 agrees reasonably well with the variation observed for CH_4 at this temperature[2] and may be represented as proportional to $\omega^{0.65 \pm 0.2}$.

The upper limit for attenuation due to thermal conductivity is estimated to be 0.005 cm^{-1}. This is very much less than the normal attenuation or the excess attenuation measured near the transitions, about 1.0 cm^{-1}; hence thermal conductivity should make a minor contribution to the attenuation. However, the attenuation due to thermal conduction could be much larger if our samples were composed of crystallites much smaller than the sound wavelength (0.2 mm) with large thermal gradients across them. In that case the attenuation is proportional to $\sqrt{\omega}$.[6]

The attenuation is observed to rise to a maximum in the vicinity of each transition temperature with peaks occurring on the average at 21.4 ± 0.2 and 26.3 ± 0.2°K, as shown in Fig. 3. These values are 0.8 K° below the heat capacity maxima for each transition, and 0.9 and 0.5 K° below the kinks in the velocity curve, respectively. There is no systematic difference between the amplitude of the attenuation peaks at 4 and at 12 MHz, although the 12-MHz attenuation is slightly greater than the 4-MHz attenuation.

In CH_4 two attenuation peaks are associated with the upper lambda transition; the lower transition is not evident in velocity or attenuation data. One peak is sharp

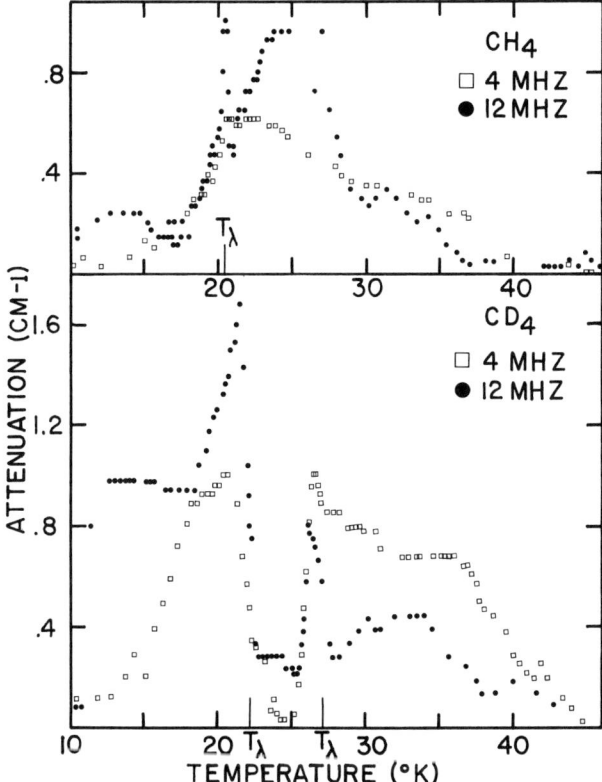

Fig. 3. Attenuation of 4- and 12-MHz longitudinal sound in CD_4 vs. temperature in transition region; attenuation vs. temperature of 4- and 12-MHz longitudinal sound in CH_4 is also shown for comparison. Values of T_λ are from the heat capacity anomalies.

(less than 1 K°), occurs at T_λ, and shows no temperature shift with changing frequency. We believe it is due to a thermal conduction contribution caused by large temperature gradients across many small crystals. The other peak is broad (about 10 K°), occurs above T_λ, and shifts to higher temperature with higher frequency. It can be described by a simple relaxation process with a relaxation time τ given by $\tau_0 e^{E/kT}$ where $E/k \approx 200°K$.[2,7] Figure 3 shows a comparison of our present attenuation data for one CD_4 sample with data for one CH_4 sample at 4 and 12 MHz. By comparison we attribute the CD_4 attenuation peaks predominantly to thermal conduction contributions to the attenuation, and conclude that our samples contain an appreciable number of very small crystallites. There is evidence of a smaller, broad peak above the upper transition in CD_4 which shifts to a higher temperature at 12 MHz. The attenuation data show considerable scatter and variation between runs. We believe this is due to fluctuations in the sample bonding to the quartz transducers caused by rapid expansion of the sample at the temperatures of the thermal expansion anomalies.

In CD_4 we observe the broad peak to vary with temperature less rapidly than

ω^1. Hence we conclude that the attenuation peak is due to a simple relaxation process and occurs when $\omega\tau = 1$. We observe a shift to higher temperature for higher frequency and conclude that τ is a decreasing function of temperature, probably of the form $\tau_0 e^{E/kT}$. Scatter in these amplitude data makes determination of τ_0 or E impossible.

The other phenomenological theory that can be discussed is that developed by Landau and Khalatnikov and others, in which the relaxation time is given by $C(T - T_\lambda)^{-b}$, where b is usually 1.0 and C may be different above and below the transition. Although we do not expect the sharp attenuation peaks to be proportional to the relaxation time for the reasons stated earlier, we have fitted the attenuation to the formula $C/(T - T_\lambda)^{-b}$ and we find reasonable agreement for both transitions if $b = 1.1 \pm 0.4$ and $C(T > T_\lambda)/C(T < T_\lambda) = 2.0 \pm 0.5$.

Acoustical behavior associated with the lower-temperature transition in CD_4 is qualitatively much the same as that associated with the upper transition. Hence we conclude that a lower transition in CH_4 ought to have been apparent in acoustical behavior if one analogous to the lower CD_4 transition occurs at normal pressure. That no such transition was apparent for CH_4 indicates a qualitative difference between the two isotopes. The broad heat capacity anomaly at $8°K$ in CH_4 is more likely associated with nuclear spin species conversion than with an ordering of molecular orientations, as presumably occurs in CD_4 and at the upper transition in CH_4.

References

1. J.H. Colwell, E.K. Gill, and J.A. Morrison, *J. Chem. Phys.* **39**, 635 (1963).
2. R.P. Wolf, F.A. Stahl, and J.A. Watrous, *J. Chem. Phys.* **59**, 115 (1973).
3. F.A. Stahl, R.P. Wolf, and M.B. Simmonds, *Phys. Lett.* **27A**, 482 (1968).
4. V.G. Manzhelii, A.M. Tolkachev, and V.G. Gavrilko, *J. Phys. Chem. Solids* **30**, 2759 (1969).
5. D.C. Heberlein and E.D. Adams, *J. Low Temp. Phys.* **3**, 111 (1970).
6. L.D. Landau and E.M. Lifshitz, *Theory of Elasticity*, 2nd ed., Pergamon Press, London (1970), Section 35.
7. A.A. Thiele, W.M. Whitney, and C.E. Chase, in *Proc. 9th Intern. Conf. Low Temp. Phys., 1964*, Plenum Press, New York (1965), p. 1122.

Growth and Neutron Diffraction Experiments on Single Crystals of Deuterium

J. Votano,* R. A. Erickson, and P. M. Harris

Department of Chemistry, The Ohio State University
Columbus, Ohio

Neutron and X-ray diffraction studies[1-3] of crystalline solid solutions of orthodeuterium and paradeuterium using powder-sample techniques have clearly demonstrated the existence of a low-temperature cubic phase for solutions of high (> 65%) paradeuterium content. The neutron diffraction studies[2,3] indicate the cubic phase to be of space group $Pa3$ with four molecules per unit cell, whereas the X-ray diffraction[1] indicates an fcc structure. To further elucidate the low-temperature structure of enriched paradeuterium samples, we have undertaken a neutron diffraction study of single crystals in the temperature interval 1–18°K. We report here some preliminary experiments to establish techniques for preparing and cooling single-crystal samples. Four cell designs were tested in these experiments. All cells were cylindrical and of nominal size 3 mm diameter × 4 cm long.

The first cell was of aluminum. The filling and cooling were performed through the top. This cell did not produce any large crystals or show any preferential direction of crystal growth.

The second cell was of thin-wall stainless steel, filled and cooled from the bottom. A Teflon plug with collimating aperature ± 5° was inserted into the bottom of the cell. Excellent large crystals were produced by a slow cooling through the triple point of D_2. In general the crystals grew with the hexagonal c axis parallel to a cell diameter and an a axis approximately parallel to the cell axis. Nine Laue Bragg (LB) reflections were obtained for a converted (3% para) sample at 17.5°K. The data were quite consistent with the hcp structure and parameters obtained from powder diffraction. The molecules have random (spherical) orientation and the thermal vibration is isotropic. This crystal was cooled to 2°K and showed both time- and temperature-dependent LB line broadening and fragmentation.

The third cell was of stainless steel, filled from the top and cooled from below. The Teflon collimator plug was used. An external helium gas system was added to pressurize, via the fill line, up to 100 atm. A sample of normal (33% para) deuterium, n-D_2(I), was crystallized at 18.5°K under 30 atm He pressure. Twelve LB reflections in the ($h0l$) set were measured and again the results were consistent with the powder analysis. This crystal was cooled very slowly (10 hr) to 13.8°K with the (closed off) helium pressure following the line $dP/dT = 1.3$ atm°K^{-1} and *no* change in lattice constants (cell volume) or linewidth was observed. At 13.8°K the (002) line broadened

---

* Present address: Department of Biology, Massachusetts Institute of Technology, Cambridge, Massachusetts.

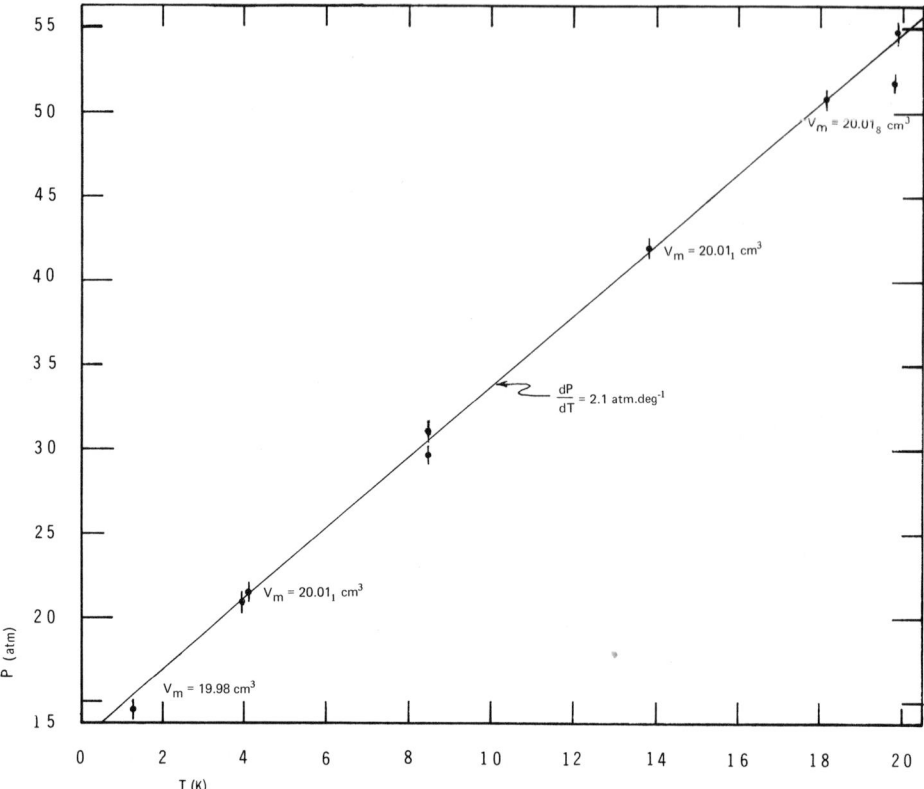

Fig. 1. The observed helium pressure vs. temperature in cell 4 with $n\text{-}D_2$ crystal. The indicated molar volumes were deduced from measured parameters of the hcp lattice.

from 0.5 to 0.6° and shifted (contracted) by 0.015°. We attributed the nonequilibrium cooling to adhesion at the side wall. In an attempt to pressurize this sample to 100 atm at 13.8°K, it was broadened and fragmented beyond further use.

The fourth cell was the same as the third except for the addition of a thin-wall Teflon liner in the main cell and the incorporation of a finer ($\pm 1.5°$) collimator plug at the bottom. A crystal of $n\text{-}D_2(\text{II})$ was grown at 18.2°K, 50 atm He, and cooled over a period of ~ 60 hr to 1.3°K. The initial (002) line width (0.6°) increased uniformly to 0.87° at 13.8°K and then remained constant to 1.3°K. No fragmentation was observed. The helium pressure was observed to decrease along the line $dP/dT = 2.2$ atm K^{-1}, as shown in Fig. 1. The molar volumes indicated in the figure were computed from measurements of the lattice parameters. It is to be noted that except for a small ($\sim 0.15\%$) decrease at 1.3°K *no* volume change was observed. After the crystal had been at 1.3°K for several hours we attempted to return to 4.2°K to make some compressibility measurements. Though the warmup was as slow as the prior cooldown, the (002) line rapidly broadened and the crystal fragmented. We interpret this result (and perhaps the small crystal contraction at 1.3°K) as evidence for the onset of some orientation ordering at 1.3°K.

It was expected that during cooling the (002) and (100) LB line positions would be shifted in some uniform manner that would allow a measure of the volumetric expansion coefficient, after making the standard correction for pressure change using the compressibility data of Megaw.[4] If we assume $K \sim 3 \times 10^{-4}$ atm^{-1}, independent of temperature, and that the measured helium pressure is the equilibrium pressure on the D_2 at 13.8°K in cell 3 (after the volume relaxed), then the average volumetric expansion coefficient is $\bar{\alpha} \simeq 13 \times 10^{-4}$ K^{-1}. If we assume thermodynamic equilibrium in crystal II (cell 4), then for 18–4.2°K, $\bar{\alpha} \simeq 7 \times 10^{-4}$ K^{-1} and for 4.2–1.3°K, $\bar{\alpha} \simeq 13 \times 10^{-4}$ K^{-1}. The indicated expansion coefficients are smaller than the estimates of Megaw based on C_p–C_v data, but our suggested values are much larger than those corresponding to the precise measurements of Ramm et al.[5] From their measurement of $(\partial P/\partial T)_v$, $\alpha \simeq 3 \times 10^{-4}$ K^{-1}, 1.3–4.2°K. The discrepancy cannot be accounted for by a simple change of compressibility. For our results to be made consistent with those of Ramm et al. we must assume that our crystal II is not in equilibrium with the He pressure, but that the side walls exert sufficient force, without plastic flow, on the horizontal layers to hold the ($h0l$) reflections in fixed positions (except at 1.3°K). But if the required compression occurs without plastic flow, then the pressure cell used by Ramm et al. appears to be unsuitable for measuring thermodynamic pressure in D_2. The results are clearly ambiguous.

We propose an additional experiment with n-D_2 to independently measure the compressibility and expansion in the range 2–13°K, and to study line intensities at 1.3°K for direct evidence of rotational ordering.

With minor modifications to the apparatus suggested by the present studies extension of the investigation to samples of enriched paradeuterium is planned.

References

1. A.F. Schuch and R.L. Mills, *Phys. Rev. Lett.* **16**, 616 (1966); A.F. Schuch, R.L. Mills, and D.A. Depatie, *Phys. Rev.* **165**, 1032 (1968).
2. K.F. Mucker, S. Talhouk, P.M. Harris, D. White, and R.A. Erickson, *Phys. Rev. Lett.* **16**, 799 (1966); K.F. Mucker, P.M. Harris, D. White, and R.A. Erickson, *J. Chem. Phys.* **49**, 1922 (1968).
3. R.L. Mills, J. Yarnell, and A.F. Schuch, this volume.
4. H.D. Megaw, *Phil. Mag.* **28**, 129 (1939).
5. D. Ramm, H. Meyer, and R.L. Mills, *Phys. Rev. B* **1**, 2763 (1970).

Proton Resonance in Highly Polarized HD*

H. M. Bozler and E. H. Graf

Department of Physics, State University of New York at Stony Brook
Stony Brook, New York

Introduction

The prospect of producing a statically polarized proton target for elementary particle research has been of interest for some time and has become technically feasible with the advent of dilution refrigerators and > 100-kG superconducting magnets. Protons in a 100-kG field can, in principle, be polarized to 76% at temperatures of 10 m°K.

Solid H_2 would be the most obvious choice for a proton target were it not for the fact that H_2 is polarizable only in its ortho state and that the heat (172°K per molecule) associated with the slow but unavoidable ortho-to-para conversion at low temperatures makes the cooling of o-H_2 to the millidegree range impractical. Solid HD, where this problem does not arise since the nuclei in the molecule are distinguishable, thus becomes the most suitable possibility.

A property most important to the production and maintenance of nuclear polarization is the spin–lattice relaxation time T_1, which in HD is affected by small amounts of o-H_2 impurity $c_{o\text{-}H_2}$, temperature T, and magnetic field H_0.[1-5] In this communication, we present results of our studies of proton relaxation times in HD as a function of each of these variables.

Apparatus

The HD samples were condensed into a thin-walled stainless steel chamber housed within the mixing chamber of a dilution refrigerator. Magnetic fields of up to 100 kG were produced by a superconducting magnet that surrounded the mixing chamber and lower heat exchangers of the refrigerator. Temperature was measured by three thermometers: (1) a calibrated 100-Ω Speer carbon resistor, (2) a powdered copper nuclear resonance thermometer, and (3) the proton spins of the HD sample itself. The NMR thermometers were calibrated absolutely against ^{3}He vapor pressure at temperatures between 0.7 and 2.2°K. The resistor and the copper thermometers were in direct contact with the mixing chamber liquid.

Proton resonances in HD were observed with a CW NMR spectrometer, basically a high-frequency Rollin circuit, shown in Fig. 1. A length of transmission line driven by a constant-current rf source through a directional coupler was termi-

---

* Work supported in part by the National Science Foundation and the U. S. Atomic Energy Commission.

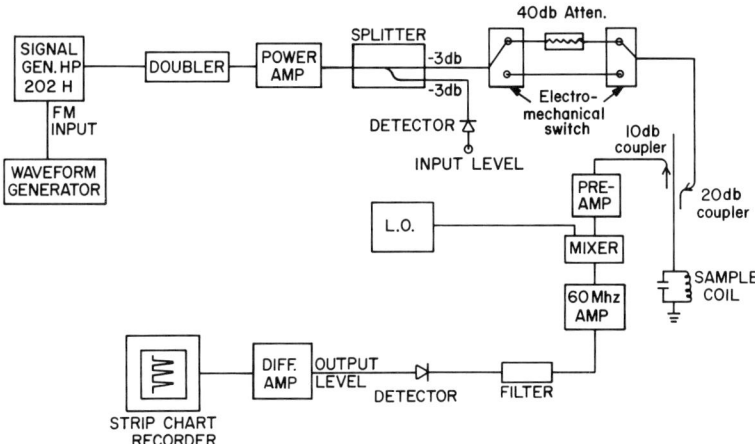

Fig. 1. Block diagram of 250–420-MHz NMR spectrometer. Electronics for lower frequencies are similar.

nated at one end by a coil containing the sample and left open-circuited at the opposite end. The frequency was tuned to a circuit resonance corresponding to a voltage standing-wave maximum, as observed through a second directional coupler. Circuit resonances occurred at regular intervals in the driving frequency, and these could be shifted by changing the position of the open-circuit termination. The Q of each circuit resonance was ~ 40. With this arrangement it was possible to cover the entire range of magnetic fields between 14.1 and 98.9 kG by tuning in a circuit resonance (between 60 and 421 MHz) and then adjusting the magnetic field to the corresponding nuclear resonance. The NMR line was observed by modulating the driving frequency, holding the static field constant to avoid possible eddy-current heating. The NMR line appeared as an amplitude modulation of the CW carrier. The depth of modulation under conditions of high polarization could be as high as 30%, requiring corrections for nonlinearity and admixture of dispersive mode. The details of these corrections will be described in a later communication.

Results

The T_1 measurements were made by observing the regrowth of signal amplitude following saturation. The relaxation sometimes exhibited a small deviation from exponential form, as has been observed by others,[1,2] which was probably due to nonuniform distribution of o-H_2. Nevertheless T_1 was taken as the time constant of the best least-squares exponential fit to our data. Points were taken for $14.1 \leq H_0 \leq 98.9$ kG. Measured values of T_1 ranged from 5 to 10^5 sec. Care was taken to ensure good thermal contact between the sample and the dilution refrigerator by introducing a dilute liquid mixture of ^{3}He in ^{4}He or pure liquid ^{3}He as a heat exchange medium between the sample and the walls of the sample chamber.

The samples used in our work were 97.5% HD with 1.8% H_2 and 0.7% D_2 impurities. We varied $c_{o\text{-}H_2}$ by storing the HD samples at 4.2°K, allowing the o-H_2 to convert to p-H_2. We have determined the ortho to para conversion rate to be 0.57 ± 0.03%/hr. in reasonable agreement with earlier work.[1,2]

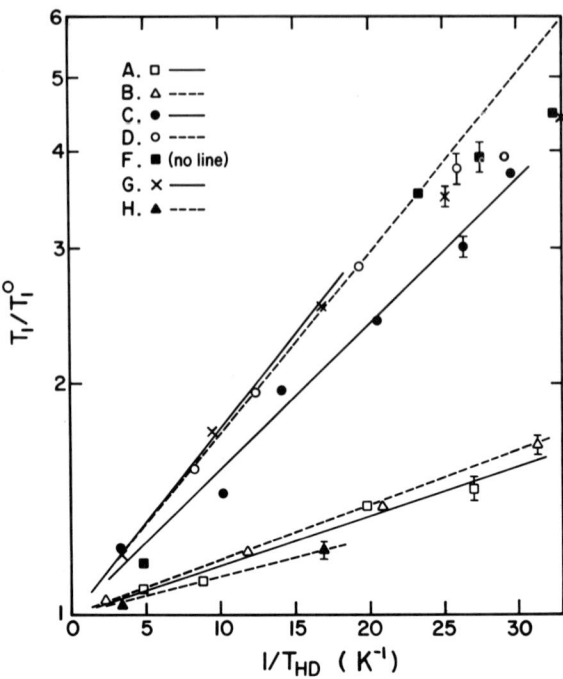

Fig. 2. Temperature dependence of T_1. Values of T_1^0, α [see Eq. (1)], H_0, and c_{o-H_2} are given below. Typical error bars are shown on the figure.

	A	B	C	D	F	G	H
T_1^0, sec	91.4	136	73.1	86.4	16.8	333	827
α, m°K	15	16	44	54	51	57	10
H_0, kG	14.1	14.1	63.4	98.9	94.1	98.9	14.1
$c_{o-H_2} \times 10^4$	2.9	2.0	3.6	4.8	7.5	2.9	1.1

Figures 2 and 3 are plots of our results. Figure 2 shows the temperature dependence of T_1 for several values of H_0 and c_{o-H_2}. Except at the highest fields and lowest temperatures, the data can be fitted to the form

$$T_1 = T_1^0 e^{\alpha/T} \qquad (1)$$

Values of α vary considerably with H_0, from about 16 m°K at 14.1 kG to 54 m°K at 98.9 kG. The dependence of α on c_{o-H_2}, however, is seen to be quite weak for $10^{-4} \lesssim c_{o-H_2} \lesssim 5 \times 10^{-4}$.

Figure 3 shows the dependence of T_1 on c_{o-H_2} and H_0. The points follow a simple power law $T_1 \propto c_{o-H_2}^{-3}$ for $c_{o-H_2} < 2 \times 10^{-4}$ in samples with $H_0 = 14.1$ kG, and for all measured concentrations at higher fields. When $c_{o-H_2} > 2 \times 10^{-4}$ the 14.1-kG data deviate from the power law. Previous studies have shown that a minimum in T_1 occurs at an $o-H_2$ concentration $c_{min} (\simeq 10^{-2})$ where the molecular reorientation correlation times become equal to the Larmor precession period.[1] We would thus expect c_{min} to increase with increasing H_0. Our data are consistent with this expectation.

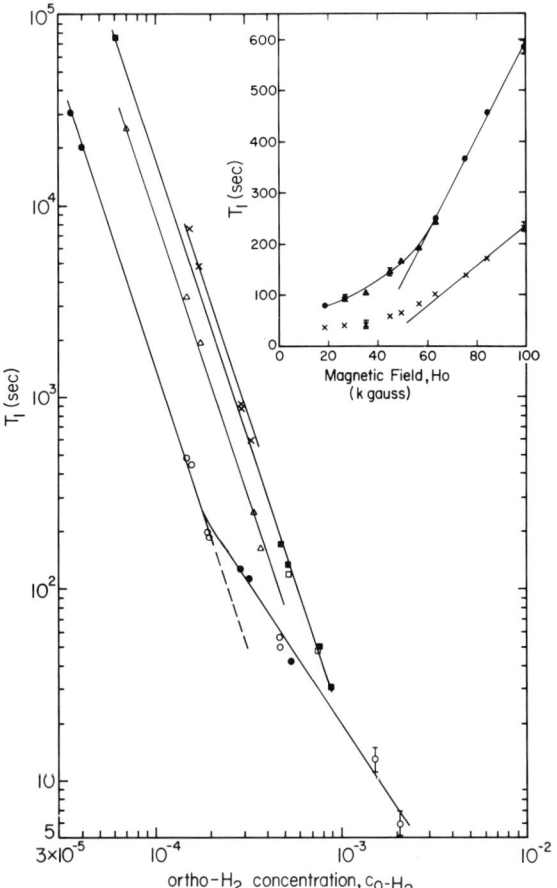

Fig. 3. Dependence of T_1 on o-H_2 concentration and magnetic field at $T = 50$ m°K. On the concentration plot (main figure) the curves, from bottom to top, were taken at $H_0 = 14.1, 63.4, 94.1,$ and 98.9 kG. A few representative error bars for T_1 are shown at high values of $c_{o\text{-}H_2}$. At low concentrations the error is about the size of the symbols. Field dependence curves (inset) were taken at $c_{o\text{-}H_2} = 4.7 \times 10^{-4}$ (bottom) and 3.3×10^{-4} (top).

The T_1 vs. $c_{o\text{-}H_2}$ data of Fig. 3 (taken at constant H_0) show a marked field dependence. We have also measured this field dependence directly, holding $c_{o\text{-}H_2}$ constant. Two plots of such data are shown in the inset of Fig. 3. The breaks in the curves are associated with the deviation of T_1 from the $c_{o\text{-}H_2}^{-3}$ power law. At high fields T_1 appears to be linear in H_0.

Discussion

HD proton relaxation in the presence of small o-H_2 impurity concentrations has been treated theoretically by Sung.[6] He calculated the o-H_2 relaxation due to the intramolecular dipole–dipole interaction in the presence of a strong intermolecular electric quadrupole–quadrupole interaction[7] and a crystal field. The HD protons are in turn strongly coupled to the o-H_2 by an energy-conserving cross-relaxation.[8] Sung's calculated temperature dependence of T_1 does not seem applicable to our system in the ranges of T and $c_{o\text{-}H_2}$ that we have covered. However, his calculation of the T_1 dependence on $c_{o\text{-}H_2}$ and H_0 does describe in a qualitative way the general characteristics of our results. For example, his theory gives $T_1 \propto c_{o\text{-}H_2}^{-2.25}$ for $c_{o\text{-}H_2} < 10^{-4}$, versus our experimental $c_{o\text{-}H_2}^{-3}$. The calculated field dependence is $T_1 \propto (aH_0 + E_0)^{7/4}$, where a and E_0 are constants. We find $T_1 \propto H_0$ at high fields. A more complete theory will have to include a temperature-dependent relaxation process valid in the millidegree region, which, moreover, becomes more strongly dependent on T as H_0 increases. The effects of the p-H_2 and D_2 impurities (which we have neglected) also bear study.

Polarized HD Targets

The problems in constructing a usable polarized HD target appear to be, in the main, technical. Our direct susceptibility measurements at ~ 100 kG and 30 m°K have given polarizations of 33%, with T_1 varying from ~ 20 sec to more than one day. Relaxation times can be kept reasonably short even in high fields if $c_{o\text{-}H_2}$ is not too small. Loss of thermal equilibrium between the dilution refrigerator and HD samples due to ortho–para conversion heating does not seem to occur (in the presence of heat exchange liquid) as long as $c_{o\text{-}H_2} < 2 \times 10^{-4}$, even at the lowest temperatures reached in this experiment.

Acknowledgments

We wish to thank Professors M. Good and J. Kirz for their interest in and encouragement of this project.

References

1. W.N. Hardy and J.R. Gaines, *Phys. Rev. Lett.* **17**, 1278 (1966).
2. R. Rubins, A. Feldman, and A. Honig, *Phys. Rev.* **169**, 299 (1968).
3. J.H. Constable and J.R. Gaines, "High Field NMR Thermometry below 1°K Using HD" (to be published).
4. A. Honig, in *Proc. II Intern. Conf. on Polarized Targets*, G. Shapiro, ed., Berkeley (1971), p. 99.
5. H.M. Bozler and E.H. Graf, in *Proc. II Intern. Conf. on Polarized Targets*, G. Shapiro, ed., Berkeley (1971), p. 103.
6. C.C. Sung, *Phys. Rev.* **177**, 435 (1969).
7. T. Moriya and K. Motizuki, *Prog. Theoret. Phys.* **18**, 183 (1957).
8. M. Bloom, *Physica* **23**, 767 (1957).

Thermal Conductivity of Solid HD Containing Isotopic Impurities*

J. H. Constable and J. R. Gaines

Department of Physics, The Ohio State University
Columbus, Ohio

The thermal conductivity of a sample of solid HD containing fixed H_2 and D_2 isotopic impurities has been measured over a period of 271 hr in the temperature range 4–0.2°K. During this time a substantial number of o-H_2 impurities convert to p-H_2, enabling one to determine in the same crystal the o-H_2 dependence of the thermal conductivity. The o-H_2 dependence is compared to the theory of Ebner and Sung[1] and found to give reasonable agreement for temperatures above 0.6°K. Isotopic scattering was found to dominate the high-temperature data. The low-temperature results were characteristic of boundary scattering, from which it was inferred that the sample was polycrystalline.

The thermal conductivity measurements were made in the sample cell shown in Fig. 1. While the cell was filled with ^{4}He the two 220-Ω Speer resistance thermometers inside the cell were calibrated against the 220-Ω Speer resistor mounted on the outside of the mixing chamber of the dilution refrigerator. The resistor on the mixing chamber had been calibrated from the NMR susceptibility of platinum in earlier experiments. The calibration has been stable to within ± 5% over a period of three years.[2] The NMR calibration was shifted to zero magnetic field by repeating the last set of thermal conductivity measurements at the NMR field. Under the assumption that the thermal conductivity is independent of magnetic field, corrected temperatures were found with the largest correction being 2%.

The spiral of copper foil (thickness 0.012 cm) in the top of the cell has a surface area of 162 cm² and was used to make thermal contact to the solid HD. The surface spacing of the foil spiral was about 0.019 cm, which was larger than the phonon mean free path found for the HD. Thus the HD thermal conductivity was not appreciably limited in the channels between the foil. The temperature differential ΔT between the copper and the HD was measured using the top two thermometers in Fig. 1. The thermal resistance R to the heat flow Q between the copper and the HD was found to approximately fit the following equation

$$R = \Delta T/Q = (4/T + 0.1/T^3) \quad W^{-1} {}^\circ K \tag{1}$$

A reasonable assumption is to assign the term inversely proportional to T to the thermal resistance through the various copper pieces and the term inversely proportional to the cube of the temperature to the HD–Cu contact resistance. Using

---

* Work supported in part by the National Science Foundation under grant No. GH-33011X.

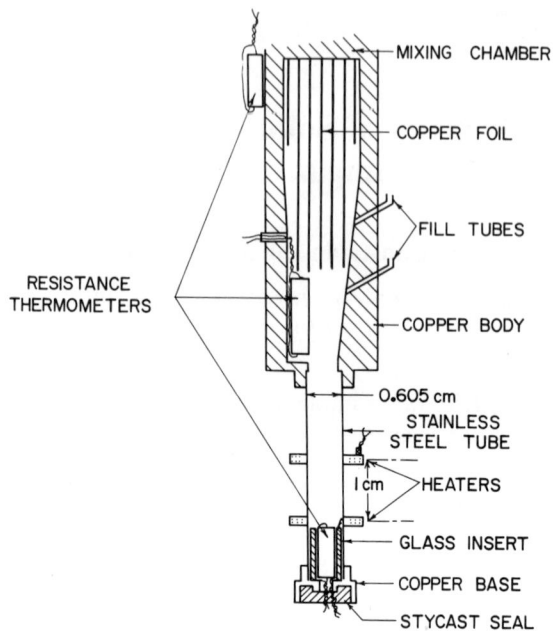

Fig. 1. Thermal conductivity sample cell.

the area of the copper foil, one obtains a contact resistance of $16.2T^{-3}$ (cm^2°K^4/W), which is higher than the Kapitza resistance expected for a solid-to-solid boundary. The high boundary resistance found between HD and copper, or liquid helium,[2] is probably due to mechanical fractures in the solid HD.

The inability to effectively remove heat from the HD combined with the heat generated in the sample from ortho–para conversion of the o-H$_2$ impurities severely limits the ultimate temperature to which an HD sample can be cooled. Taking the rate of ortho–para conversion to be 7.09×10^{-3} per hour,[3] one finds that an HD sample containing as little as 0.1% o-H$_2$ liberates 0.13 μW/cm^3.

Although the conversion heat was sometimes comparable to the measuring heat, the use of the one-thermometer, two-heater technique for measuring the thermal conductivity eliminated the effect of this extra heat flux.[4] The two matched 10,000-Ω heaters shown in Fig. 1 were made from 390 cm of Evanohm wire which was loosely coiled in the donut-shaped recess of the sample tube (height and outside diameter of the recess were 0.8 mm and 1.24 cm, respectively). This construction was chosen to provide a large HD-to-wire surface area and to remove from the main column of HD any local heating of the HD due to poor wire-to-HD contact. It should be noted that no correction due to the thermal conductivity of the empty cell has been used, since its thermal conductivity is less than 1% that of the HD even at the lowest temperatures.

The initial o-H$_2$ concentration in the HD was determined from NMR relaxation measurements[5] to be 0.97%. During the course of the experiment ortho–para conversion reduced this concentration to 0.14%. Figure 2 shows the data taken near the beginning of the experiment (curve i), at the end (curve f), and a third curve (d).

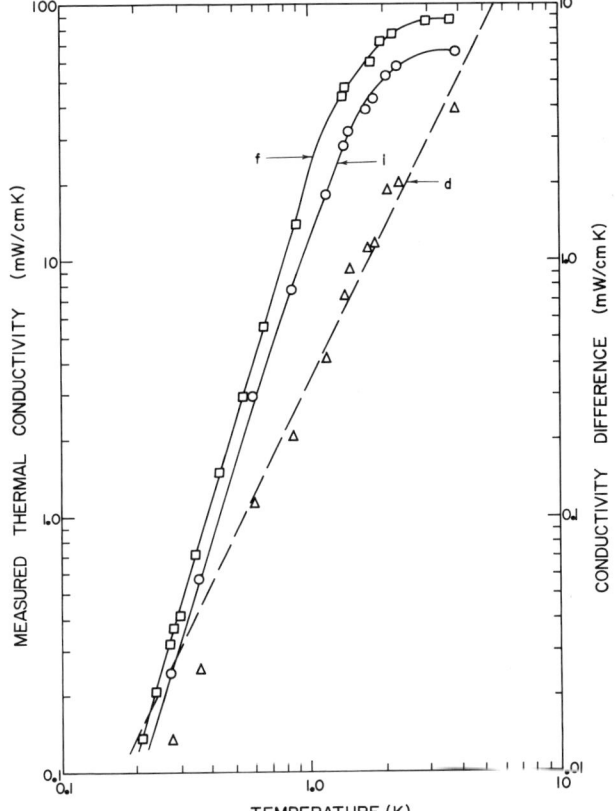

Fig. 2. HD thermal conductivity data; (○) initial data; (□) final data; (△) conductivity difference.

Curve d represents the o-H_2-dependent conductivity obtained by calculating:

$$K_d = (1/K_i - 1/K_f)^{-1} \qquad (2)$$

where K_i and K_f are the initial and final thermal conductivities, respectively.

The above equation represents a valid method for obtaining the o-H_2 dependent contribution to the thermal conductivity if Matthiessen's rule holds–namely that thermal resistances are additive. Matthiessen's rule is generally correct in cases where the normal phonon processes proceed so rapidly that they dominate the distribution functions, making it necessary to take the harmonic average over the phonon distribution[6] to obtain the mean free path. Since the lattice forces in the solid hydrogens have large anharmonic characteristics, we will assume that it is proper to add thermal resistances.

Assuming the Umklapp processes in H_2 and HD are similiar, so that the estimated contribution to the thermal conductivity in H_2 by Bohn and Mate[4] can be used for HD, the limitation on the thermal conductivity in this experiment at high temperatures probably arises from scattering of phonons by the isotopic impurities in the HD. Analysis of the data based on this assumption and use of the harmonic average gives an impurity concentration of 1.4%,[6] which compares favorably to the 1.3% H_2 known to be present from the NMR data. The alternative approach to taking the harmonic average and using Matthiessen's rule is to calculate the average over the phonon distribution of the sum of the reciprocal relaxation times.[7] When applied to the data of this experiment the latter approach predicts an anomalously high impurity concentration. This would seem to be an experimental justification of the assumption regarding the additivity of thermal resistances.

The low-temperature data show the T^3 temperature dependence characteristic of boundary scattering with a temperature-independent mean free path for phonons. Using a Debye temperature of 112°K, we obtain a mean free path of 70 μm. From this we infer that the sample is polycrystalline with a typical crystallite size of the order of the mean free path.

The contribution to the phonon scattering due to o-H_2 impurities can be inferred by subtracting from the thermal resistance values measured earlier in the experiment the thermal resistance measured after most of the o-H_2 molecules have converted to p-H_2 molecules in the same sample. This has been done by means of Eq. (2) and yields the points in Fig. 2 indicated by curve d. The temperature dependence is seen to be basically that predicted by Ebner and Sung, namely T^2, as shown by the dashed line. Similar curves corresponding to different o-H_2 concentrations scale inversely as the square of the concentration. It is because these measurements were all performed on the same sample that the o-H_2 concentration-independent parts could be subtracted out, displaying the T^2 dependence heretofore unseen.[4]

Since the experiments of Bohn and Mate were performed on different samples, it was impossible for them to subtract out the concentration-independent scattering unless they had single crystals for each experiment. Because of the similarity of their data and ours, it seems unreasonable to attribute all the phonon scattering at low temperatures in their data to o-H_2 molecules. This would imply that their samples were also polycrystalline, in spite of their appearance. The analog of the experiment we have performed on HD is virtually impossible in H_2, due to pairing of the o-H_2 molecules, an effect which is essentially negligible in HD.

References

1. C. Ebner and C.C. Sung, *Phys. Rev. B* **2**, 2115 (1970).
2. J.H. Constable and J.R. Gaines, in *Temperature, Its Measurement and Control in Science and Industry*, Vol. IV, ISA, Pittsburgh, Pa. (1972).
3. J.H. Constable and J.R. Gaines (to be published).
4. R.G. Bohn and C.F. Mate, *Phys. Rev. B* **2**, 2121 (1970).
5. W.N. Hardy and J.R. Gaines, *Phys. Rev. Lett.* **17**, 1278 (1966).
6. J.M. Ziman, *Electrons and Phonons*, Clarendon Press, Oxford (1960).
7. P. Carruthers, *Rev. Mod. Phys.* **33**, 92 (1961).

Self-Consistent Calculations of the Lattice Modes of Solid Nitrogen

J. C. Raich*

*Department of Physics, Colorado State University
Fort Collins, Colorado*

and

N. S. Gillis†

*Sandia Laboratories
Albuquerque, New Mexico*

Recently considerable interest has been shown in the lattice modes of solid nitrogen. The librational and translational modes at $\mathbf{k} = 0$ have been measured by Raman and infrared scattering.[1,2] Solid nitrogen is observed in three phases: the α-phase, with an fcc molecular lattice, the hcp β-phase, and the body-centered tetragonal γ-phase. The space group for α-N_2 is $P2_13$; however, the distortions from a $Pa3$ structure are very small.[3] The α–β transition at zero pressure occurs at 35.6°K. This transition is accompanied by considerable anharmonicity in the librational modes.[3,4] In the α phase all lines broaden substantially with increasing temperature. The frequency of the most intense line at about 32 cm^{-1} is found to decrease above about 20°K and to soften markedly near $T_{\alpha\beta}$. Similar behavior is observed for the translational modes.

Calculations of the phonon and librational spectrum have been based on classical lattice dynamics[5] as well as variational methods.[6,7] The intermolecular potential for solid nitrogen is not known accurately. Various forms of the intermolecular potential have been examined: (i) the superposition of a spherically symmetric Lennard-Jones potential with an anisotropic part due to either (a) electrostatic quadrupole–quadrupole (EQQ) interactions or (b) EQQ interactions plus directional corrections due to attractive dispersive and repulsive forces; and (ii) an atom–atom Lennard-Jones potential. None of these potentials has been entirely successful in explaining the observed properties of the solid.

We have calculated the temperature dependence of the librational frequencies assuming a rigid lattice model and a $Pa3$ structure. The random phase approximation (RPA) of Fredkin and Werthamer[8] is used. Best agreement with experiment is found when only EQQ interactions, with a molecular quadrupole moment of $eQ = -1.52 \times 10^{-26}$ esu, are used. The single-particle wave functions are taken to be linear combinations of spherical harmonics, the free rotator eigenfunctions,

---

* Work supported by National Science Foundation Grant GP-22553.
† Work performed under the auspices of the U. S. Atomic Energy Commission.

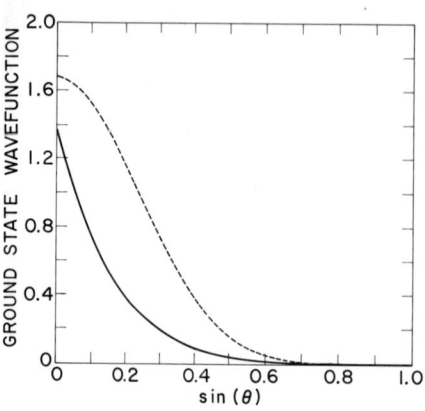

Fig. 1. The ground-state wave function ψ_{00} vs. sin θ is shown by the solid line. The ground-state wave function of Jacobi and Schnepp is shown by the dashed line.

$$\psi_{ml}(\theta, \phi) = \sum_j a_j^{ml} Y_{jm}(\theta, \phi) \quad (1)$$

Here m and l label the librational states and (θ, ϕ) specify the orientation of the molecule in the crystal. The ground-state wave function is shown in Fig. 1. For comparison, the variational ground-state wave function of Jacobi and Schnepp[6] is also shown. Figure 2 shows the temperature dependence of the librational modes at $\mathbf{k} = 0$ calculated using the RPA. The energy levels calculated from the molecular field approximation[7] and the temperature dependence of the nuclear quadrupole resonance frequency, normalized to the molecular field results, are shown in the same figure. An order–disorder transition in the molecular orientations is predicted by the molecular field approximation at 54.4°K. For the present calculations the

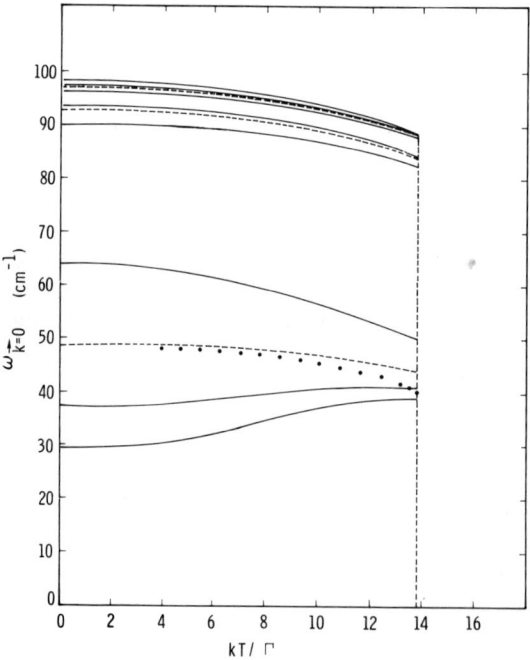

Fig. 2. The temperature dependence of the librational frequencies, as calculated using the RPA,[8] are shown by the solid lines. The dashed lines are the molecular field results. The temperature dependence of the nuclear quadrupole resonance frequency,[4] normalized to the molecular field results, is shown by the dots.

second excited states are included. Inclusion of the higher excited states is found to lower the librational frequencies by about 2 cm^{-1}. Although no detailed measurements of the temperature dependence of the librational spectrum at **k** = 0 have been made, it is clear that the RPA does not provide a satisfactory description of the temperature dependence of the librational frequencies for solid α-N$_2$. EQQ interactions with $eQ = -1.51 \times 10^{-26}$ esu, however, seem to give a reasonably accurate description of the librational frequencies at zero temperature.

One of the deficiencies of the RPA is that it is not a self-consistent calculation. The self-consistent phonon approximation[9] and its refinements have been used successfully in recent years to treat lattice anharmonicity in rare gas crystals up to near the melting points. It is seen from Fig. 2 that the molecular orientations for α-N$_2$ are well localized. One would therefore expect that oscillator wave functions would serve as an adequate basis for a description of both the translational and librational molecular motions. This assumption is supported by the success of variational calculations using oscillator-type wave functions.[6] In addition, an accurate description of the α–β transition requires a treatment of the phonons as well as the librational excitations.

The most straightforward extension of the self-consistent phonon approximation is to treat solid α-N$_2$ as an assembly of atoms instead of molecules. The advantage of such a description is the simplicity of the self-consistent formalism in rectangular atomic coordinates. A disadvantage is the problem of accurately treating the strong correlations between bonded atoms in a molecule. The intermolecular interactions between the atoms of two different molecules are represented by

$$v(r) = \varepsilon_{12}(\sigma/r)^{12} - \varepsilon_6(\sigma/r)^6 \qquad (2)$$

The parameters ε_{12} and ε_6 are chosen such that the interaction between two molecules, given by

$$V = \sum_{n=1}^{4} v(r_n) \qquad (3)$$

when averaged over the orientations of the two molecules, will yield a Lennard-Jones potential of the form

$$\langle V \rangle = 4\varepsilon[(\sigma/r)^{12} - (\sigma/r)^6] \qquad (4)$$

In Eq. (3) the r_n are the four different seperations between nonbonded atoms and in Eq. (4) $\varepsilon = 1.18 \times 10^{-14}$ erg and $\sigma = 3.71$ Å. This requires $\varepsilon_{12} = 0.495 \times 10^{-14}$ erg and $\varepsilon_6 = 0.967 \times 10^{-14}$ erg. The intramolecular interactions are represented by a harmonic potential so chosen that there is no net force on each atom in the crystal. The interatomic distance is taken to be 1.1 Å.

It is then straightforward to extend the methods of Ref. 9 to the present problem, where one has eight atoms per unit cell. As in Ref. 9, we introduce a short-range Jastrow correlation function to take asymmetric atomic motions into account. The Jastrow function was chosen to be of the form

$$f(r) = \exp\left[-\tfrac{1}{2}(B\sigma/r)^5\right]$$

with the parameter B chosen to minimize the free energy of the system.

The self-consistent frequencies at **k** = 0 calculated in this manner are compared

Table I. Calculated and Observed Librational and Translational Frequencies at k = 0 for α-N$_2$.*

	Experiment	Classical†	This work
E_g	32[1]	26.5	27.9
T_g	36.5[1]	29.6	32.5
T_g	60[1]	35.7	41.4
A_u	Inactive	47.5	41.8
T_u	48.8[2]	51.3	45.2
E_u	Inactive	54.9	49.3
T_u	70[2]	75.7	67.1
A_g	2326[1]	—	2329.2
T_g	2327[1]	—	2329.5

*An atom–atom intermolecular potential of the form $V = \sum_{n=1}^{4} [\varepsilon_{12}(\sigma/r_n)^{12} - \varepsilon_6(\sigma/r_n)^6]$ is used. The potential parameters are $\sigma = 3.71$ Å, $\varepsilon_{12} = 0.495 \times 10^{-14}$ erg, $\varepsilon_6 = 0.967 \times 10^{-14}$ erg, the interatomic spacing is 1.1 Å, and the short-range correlation parameter is $B = 0.81$.
†Ref. 5, potential model III.

with the corresponding experimental values at zero temperature in Table I. For comparison the harmonic frequencies calculated using the potential model III of Ref. 5, a 6–12 atom–atom potential, are also given. The advantage of the present intermolecular potential is that it has fewer adjustable parameters, being largely determined by the form of the interaction as well as the better known spherically averaged potential. We find small changes of the self-consistent frequencies from their corresponding harmonic values, varying from about 15% for the librational frequencies to about 3% for the phonon frequencies. The frequency shifts due to the cubic anharmonic terms were recently estimated as $[\Delta\omega^{(3)}/\omega] \approx 0.02$ at zero temperature.[10] We therefore feel that the present self-consistent method shows promise as an accurate description of the lattice modes of diatomic molecular crystals. These calculations do, however, point out the need for more accurate intermolecular potentials. Calculations of the temperature dependence of the **k** = 0 frequencies and their relationship to the α–β transition are currently in progress.

References

1. A. Anderson, T.S. Sun, and M.C.A. Donkersloot, *Can. J. Phys.* **48**, 2265 (1970).
2. R.V. St. Louis and O. Schnepp, *J. Chem. Phys.* **50**, 5177 (1969).
3. D.C. Heberlein, E.D. Adams, and T.A. Scott, *J. Low Temp. Phys.* **2**, 449 (1970).
4. J.R. Brookeman, M.M. McEnnan, and T.A. Scott, *Phys. Rev. B* **4**, 3661 (1971).
5. T.S. Kuan, A. Warshel, and O. Schnepp, *J. Chem. Phys.* **52**, 3012 (1970).
6. N. Jacobi and O. Schnepp, *Chem. Phys. Lett.* **13**, 344 (1972).
7. J.C. Raich and R.D. Etters, *J. Low Temp. Phys.* **7**, 449 (1972).
8. D.R. Fredkin and N.R. Werthamer, *Phys. Rev.* **138**, A1527 (1965).
9. N.S. Gillis, T.R. Koehler, and N.R. Werthamer, *Phys. Rev.* **175**, 1110 (1968).
10. A.B. Harris and C.F. Coll III, *Sol. St. Commun.* (to be published).

6
Other Topics in Quantum Crystals

Ionic Mobilities in Solid Helium*

G. A. Sai-Halasz and A. J. Dahm

Department of Physics, Case Western Reserve University
Cleveland, Ohio

In his pioneering work in the area of ion motion in solid helium Shal'nikov[1] demonstrated that ions could be used to probe excitations in solid helium if the crystals were of sufficiently high quality. We have measured the mobilities of ionic carriers in solid ^{4}He using a space-charge-limited current technique. Here we wish to report the results of these measurements, (the details of which are reported elsewhere),[2,3] and the conclusions drawn from these data concerning both the nature of the carriers and the fundamental nature of the transport process involved, and we also report on the use of positive ions to monitor the equilibrium vacancy concentration.

The temperature dependences of the mobilities are presented in Fig. 1. The data presented are from individual runs with a 0.121-cm diode spacing. Relative errors for a given crystal are less than the width of the symbols on the graph except where error bars are shown. We estimate an absolute error of $\pm 50\%$ due to the variation of mobility from crystal to crystal. The drastic temperature dependences displayed in Fig. 1 suggest thermally activated mechanisms for the motion of the carriers near the melting curve. In analyzing our data we relate the mobility to the diffusion coefficient of an ion with the Einstein relation $\mu = (e/kT)D$, and assume D to obey an Arrhenius equation of the form

$$D_\pm = D_{0\pm} \exp(-\Delta_\pm/kT) \tag{1}$$

Values of $\Delta_\pm$ and $D_{0\pm}$ are extracted from plots of $\ln(\mu T)$ vs. T^{-1}. The values of Δ/k for the positive and negative ions are, respectively, 40 ± 1 and $62 \pm 1°$K at a molar volume of 17.9 cm^3/mole and 20 ± 2 and $25 \pm 2°$K at 20.6 cm^3/mole.

The mechanism for positive-ion motion near the melting curve can immediately be identified as vacancy diffusion by comparing the activation energies associated with positive ion and spin diffusion in hcp solid helium. This comparison is made in Fig. 2 with a plot of the activation energies in hcp solid ^{3}He and ^{4}He vs. molar volume. The ionic activation energies are taken from our data and the work of Ifft et al.[6] It is clear from this comparison that the same mechanism is involved in the diffusion of these two entities, namely vacancy diffusion. The activation energy represents the enthalpy of vacancy creation. The preexponential factors associated with positive-ion motion in ^{4}He and spin diffusion[5] in dilute mixtures of ^{3}He in ^{4}He at 20.6 cm^3/mole agree to within the experimental errors. The absence of ob-

---

* Work partially supported by the U. S. Atomic Energy Commission.

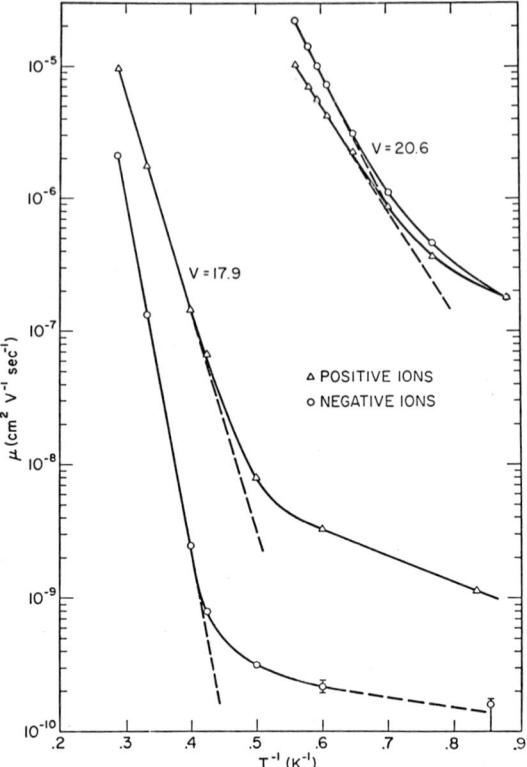

Fig. 1. Ionic mobility μ vs. inverse temperature. Molar volumes in cm^3/mole.

servable effects due to the electrostriction of the lattice surrounding the charge is not understood.

The activation energies associated with the motion of the two carriers near the melting curve differ, indicating that the negative ion differs in character from the positive ion. Our data are consistent with a model of the negative ion in which the electron is trapped in a void[7] which moves by the diffusion of adsorbed atoms along its surface. The free-electron model as well as void motion by other mechanisms can be excluded by our data.[2,3] An expression[8] relating the negative ion mobility μ_- to the adatom surface-diffusion coefficient D_s is

$$\mu_s = (3ea^4/2\pi kTR^4)\, D_s \qquad (2)$$

where a is the interatomic spacing, $R \simeq 9$ Å is the radius of the void, and D_s is the surface-diffusion coefficient, which may be written[9] as

$$D_s = \alpha d^2 v\, e^{-\delta G/kT} \qquad (3)$$

Here α is a constant of order unity, d is the average jump distance along the surface, v is taken as the Debye frequency, and δG is the change in the Gibbs free energy required to create and move a diffusable atom. We rewrite this expression as $D_s =$

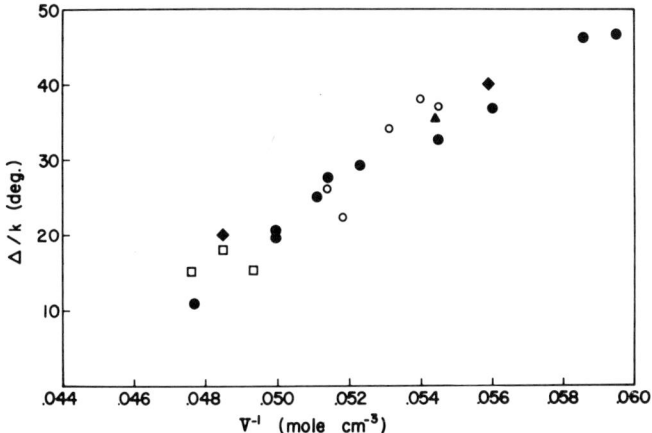

Fig. 2. Activation energies from positive ion mobility μ_+ and spin diffusion D_s measurements in hcp solid helium vs. inverse molar volume. (◆) ^{4}He, μ_+, this work; (●) ^{4}He, (▲) ^{3}He, μ_+, Ifft et al.[6]; (○) ^{3}He, D_s, Reich[4]; (□) 1.94% ^{3}He in ^{4}He, D_s Miyoski et al.[5]

$D_{0s} \exp(-\delta H/kT)$, with

$$D_{0s} = \alpha d^2 v \, e^{(\delta S/k)} \qquad (4)$$

Here δH and δS are the respective changes in the enthalpy and entropy. Our experimental values of D_{0s} obtained with the use of Eq. (2) are on the order of 1 cm^2/sec. This may be compared with a value of $\alpha a^2 v \simeq 10^{-3}$ cm^2/sec, and if the phenomenological prefactors are considered to be quantitatively correct, a large jump distance and entropy change are implied. These comparatively large values of D_{0s} are consistent with experimental values for adatom diffusion on metal surfaces and provide evidence for void motion by adatom diffusion. We note that the negative ion provides us with a unique opportunity to study the free surface of solid helium. Ionic mobilities at temperatures less than seven-tenths of the melting curve are due to other processes.

Positive ions can serve as useful probes to monitor the vacancy concentration since the mobility is proportional to it. The time response of the space-charge-limited positive ion current to a small sudden temperature change is expected to yield the relaxation time τ_v for vacancies to attain a new thermal equilibrium concentration. We report here an upper limit for this relaxation time in our crystals.

The responses of the crystal temperature (measured with a carbon resistor) and of the positive-ion current to step function changes in the heat input to the crystal chamber are presented in Fig. 3 for a 2% concentration of ^{3}He in ^{4}He. Both parameters as well as the negative-ion current, which depends only on the crystal temperature, relax with a time constant of 13 sec. With an electronically simulated setup we determined that a 2-sec vacancy relaxation time would have resulted in a detectable difference in the response times for the two curves shown. We quote upper limits on τ_v of 2 sec in crystals of pure ^{4}He and 2% ^{3}He in ^{4}He and of 4 sec for crystals of pure ^{3}He and 5% ^{3}He in ^{4}He. The first two crystals met our criterion for good crystals, which is that the mobility is field dependent, while in the latter two the mobilities were slightly field dependent.

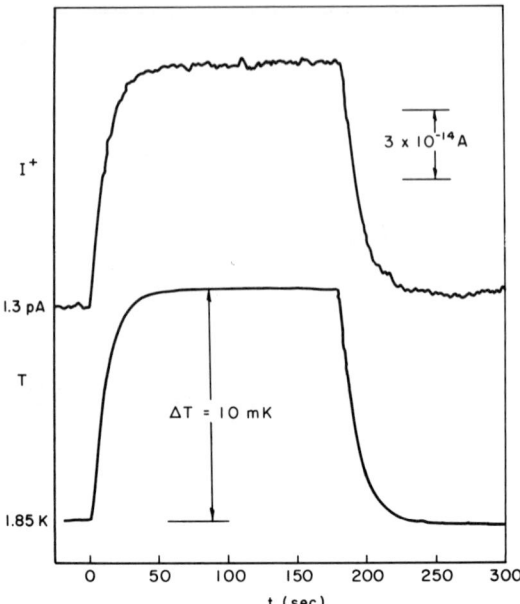

Fig. 3. Crystal temperature and positive space-charge-limited current vs. time following sudden changes in heat input to a crystal with a 2% mixture of ^{3}He in ^{4}He at 20 cm^3/mole.

The most likely sources and sinks of vacancies in crystals are crystal surfaces, grain boundaries, and dislocations. If we assume that the efficiency of these sinks for the annihilation of vacancies is unity, then the relaxation times τ_v for these processes are given, respectively, by $\tau_{cs} = L^2/\pi^2 D_v$,[10] $\tau_{gb} = R^2/\pi^2 D_v$,[11] and $\tau_d = b^2 \ln(b/a)/2D_v$;[12] here D_v is the vacancy diffusion coefficient, $L = 0.458$ cm is the diode spacing, R is the radius of a spherical crystallite, a is the effective radius of the dislocation core, and $b = (\pi N_d)^{-1/2}$, where N_d is the density of dislocations. The small value of τ rules out diffusion of vacancies to the crystal surface in the case of ^{4}He, where the vacancy diffusion coefficient for vacancy waves may be large.[13] In ^{3}He,[4] $D_v \simeq 5 \times 10^{-5}$ cm^2/sec. With this value of D_v these measured relaxation times can be accounted for by crystallites of 1 mm dimensions or with a dislocation density of 10^4 cm^{-2}. We believe that crystallites of this dimension are not present in our sample, but we have little quantitative information on dislocation densities.

Ionic currents are very sensitive to crystal quality. In crystals strained by cooling at rates greater than 1 m°K/sec the mobilities increase with the applied electric field. This phenomenon can be explained by the space charge of ions trapped on crystal imperfections, presumably dislocations. These trapped ions are then extracted by the application of larger electric fields. In more severely strained crystals the time to establish a steady-state current following a voltage change increases. In even poorer quality crystals the current vanishes altogether. However, our data are not extensive enough to determine either the depth of the trapping potential or the density of imperfections. Until more extensive measurements are taken we must conclude that the measured relaxation times can be accounted for by dislocations as vacancy sources and sinks.

In conclusion, our data provide evidence for vacancy diffusion as the mechanism for positive-ion motion near the melting curve. Experimental evidence supporting the void model of the negative ion have been presented, and the void motion is consistent with a surface-diffusion mechanism. The vacancy equilibrium relaxation time is less than 2 sec and can be explained by the creation and annihilation of vacancies at dislocations.

Acknowledgments

We wish to thank R. Guyer and P. Zilsel for some very helpful conversations.

References

1. A.I. Shal'nikov, *Zh. Eksperim. i Teor. Fiz.* **47**, 1727 (1964), [*Soviet Phys.—JETP* **20**, 1161 (1965)].
2. G.A. Sai-Halasz and A.J. Dahm, *Phys. Rev. Lett.* **28**, 1244 (1972).
3. G.A. Sai-Halasz, Ph.D. Thesis, 1972, University Microfilms, Ann Arbor, Michigan.
4. Haskel A. Reich, *Phys. Rev.* **129**, 630 (1963).
5. D.S. Miyoski, R.M. Cotts, A.S. Greenberg, and R.C. Richardson, *Phys. Rev. A* **2**, 870 (1970).
6. E. Ifft, L. Mezhov-Deglin, and A. Shal'nikov, in *Proc. 10th Intern. Conf. Low Temp. Phys., 1966*, VINITI, Moscow (1967), Vol. I, p. 224.
7. M.H. Cohen and J. Jortner, *Phys. Rev.* **180**, 238 (1969).
8. R. Kelly, *Phys. Stat. Sol.* **21**, 451 (1967).
9. H.P. Bonzel and N.A. Gjostein, in *Molecular Processes on Solid Surfaces*, E. Drauglis, R.D. Gretz, and R.I. Jaffee, eds., McGraw-Hill, New York (1969), p. 533.
10. P.G. Shewmon, *Diffusion in Solids*, McGraw-Hill, New York (1952), p. 16.
11. R.W. Balluffi and D.N. Seidman, *J. Appl. Phys.* **36**, 2708 (1965).
12. J. Crank, *The Mathematics of Diffusion*, Oxford University Press, London (1956), p. 85.
13. R.A. Guyer, R.C. Richardson, and L.I. Zane, *Rev. Mod. Phys.* **43**, 532 (1971), Appendix C.

Vapor Pressure Ratios of the Neon Isotopes

G. T. McConville

*Monsanto Research Corporation, Mound Laboratory**
Miamisburg, Ohio

The vapor pressure ratio of ^{20}Ne to ^{22}Ne has been measured in the temperature region from 13 to 24.5°K. Preliminary measurements have also been made of the ^{21}Ne to ^{22}Ne ratio. Previous measurements of the ^{20}Ne to ^{22}Ne ratio over the solid phase are those of Bigeleisen and Roth[1] down to 16.5°K and those of Keesom and Haantjes[2] over a smaller temperature range. In the region of overlap the former measurements are in good agreement with the data presented here.

The ^{20}Ne and ^{22}Ne samples were at least 99.99% isotopically pure. Mass spectrometric analysis showed the starting material for the samples to have < 30 ppm H_2 or He. The ^{21}Ne sample was 150 standard cm^3 of 91.01% ^{21}Ne, 3.45% ^{20}Ne, and 5.54% ^{22}Ne. The starting material had 2.5% H_2 in it. This sample was cleaned by passing the gas back and forth through the charcoal trap (see Fig. 1) from the cooled storage chamber to the vapor pressure bulb. The trap was warmed and pumped into an external vessel after each pass. After two passes each way there was 105 cm^3 of gas left in the sample. The ^{20}Ne and ^{22}Ne samples were passed from the cooled storage chamber through the trap into the vapor pressure bulb. More material was condensed and frozen than needed. Then the vapor was transferred back to the storage container. The remaining solid was melted, the liquid and vapor were equilibrated, and the process was repeated. Each time the H_2 content of the sample was reduced by an order of magnitude.

The measurements were made in a shielded high-purity copper vapor pressure bulb. A vacuum of 10^{-6} Torr was maintained around the vapor pressure tubes and bulb. The temperature of the bulb was maintained using a germanium thermometer in a feedback loop with a controller and heater to within 1 m°K between 12 and 25°K. The temperature could be maintained constant for 1 hr or more. The temperature was measured using a calibrated platinum thermometer which was independent of the feedback loop. Independently determined fixed points of the triple point and boiling point of ^{20}Ne were found to be consistent with the calibration to 0.02°K. Thus the temperature measurement in this work is independent of the neon vapor pressure scale.

The calibrations of the manufacturer for the 3-, 30-, and 100-Torr Baratron pressure-measuring heads were found to be consistent with each other but not with the 1000-Torr head. This instrument was recalibrated in our standards lab using a Ruska air dead weight tester and *in situ* using a wide-bore mercury manometer.

---

* Operated for the U.S. Atomic Energy Commission under Contract No. AT-33-1-GEN-53.

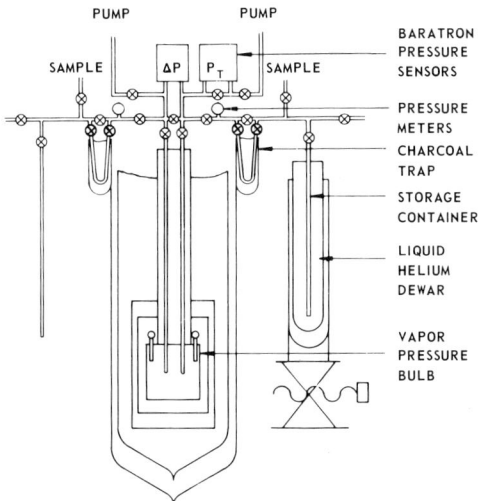

Fig. 1. Diagram of vapor pressure measurement system.

The consistency of the two calibrations and a measurement against the 30-Torr head are shown in Fig. 2.

The results for ^{20}Ne and ^{22}Ne are shown in Fig. 3, where $\ln(P_{20}/P_{22})$ is plotted vs. the measured temperature. The measurements were made with two pairs of Baratron pressure heads and were started at 15.5°K with the 30-Torr head measuring $P_{20} - P_{22}$ and the 1000-Torr head measuring P_{22}. The points between 15.5 and 24.5°K were measured in three days, with overlap at 19.5 and 21°K. The samples were maintained at 15°K during the nights. Then the Baratrons were changed to a 3-Torr head measuring ΔP and the 30-Torr head measuring P_{22} and the heads were outgassed for four days while the samples were maintained at 4.2°K. Then the data between 13.5 and 16.5°K were taken. The bending over of the data at low temperatures is caused by the thermomolecular pressure making P_{22} appear too large. This can be seen in that $\ln(P_{20}/P_{22}) \approx (P_{20} - P_{22})/P_{22}$.

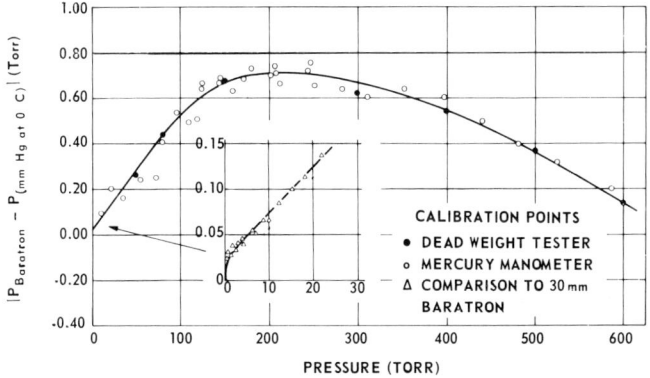

Fig. 2. Calibration of 1000-Torr Baratron head.

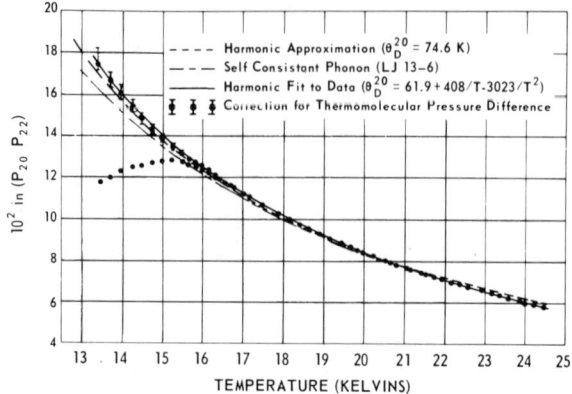

Fig. 3. Vapor pressure ratio of ^{20}Ne and ^{22}Ne. Error bars indicate data corrected for thermomolecular pressure; (- - -) harmonic approximation ($\theta_D^{20} = 74.6$); (-- --) self-consistent phonon (LJ–13); (——) harmonic fit to data ($\theta_D^{20} = 61.9 + 408/T - 3023/T^2$).

The thermomolecular pressure correction originally applied was deduced from our previous work on thermomolecular pressure,[3] where it was found that the correction for He and Ar could be determined from the same parameters in the Weber–Schmidt equation provided the temperature dependence of the viscosity of the specific gas was used. The equation was integrated using the viscosity $\eta = \eta_0(T/T_0)^{n+(1/2)}$, which is a good representation for He and Ar. For Ne in the temperature range between 20 and 300°K, n is clearly not constant. Calculations for limiting values of n gave values for the thermomolecular pressure correction which produce points in Fig. 3 from the center to the top of the error bars. The pressure correction was then measured for Ne between 20 and 300°K in a tube similar to the vapor pressure tubes. A considerably smaller correction was measured. Since the tubes in the vapor pressure apparatus had been pumped on for many months, the correction measurement was repeated after the tube had been pumped on at room temperature at 5×10^{-6} Torr for three weeks; a larger correction resulted, but still not as large as the calculated values. This correction is represented in Fig. 3 by the bottom of the error bars as a lower limit.

In Fig. 3 the corrected data are compared to theoretical models. The harmonic model with Bigeleisen's value of $\theta_D = 74.6$ for ^{20}Ne is high near the triple point and on the lower limit of the error bars at low pressures. The curve for the improved self-consistent phonon theory[4] using a Lennard-Jones 13–6 potential falls even lower at low pressures. The data between 16 and 24.5°K, where there is no thermomolecular correction, were fit to seven terms of the harmonic expansion with a Debye θ of the form of $A + B/T + C/T^2$ with a maximum deviation of $<0.5\%$. When extended to lower temperatures this curve agrees well with points corrected for the thermomolecular pressure difference. The larger effective θ_D than is found in Ne heat capacity measurements reflects the fact that only even anharmonic terms contribute a large effect to the vapor pressure ratio.[5]

A preliminary vapor ratio measurement of ^{21}Ne and ^{22}Ne has been made. We had the ^{21}Ne sample for a very limited time and unfortunately all of the H_2 was not removed from the sample. The vapor pressure ratio was measured between 15 and 24.5°K and the points showed a rapid increase with decreasing temperature of about 50% between 17.5 and 15°K. This increase is consistent with ~ 50 ppm H_2 in the ^{21}Ne sample as analyzed by an Atlas CH-4 mass spectrometer. Subtracting the contribution of the H_2, the ^{21}Ne to ^{22}Ne ratio falls about 2% below the calculated harmonic curve with a ^{20}Ne $\theta_D = 74.6$°K. The reason for this difference is not yet understood. The analysis of the impurity content in the ^{21}Ne sample leads to an estimate of only a few ppm or less impurities in the ^{20}Ne to ^{22}Ne ratio measurement, which is below our limit of detectability.

References

1. J. Bigeleisen and E. Roth, *J. Chem. Phys.* **35**, 68 (1961).
2. W.H. Keesom and J. Haantjes, *Physica* **2**, 986 (1935).
3. G.T. McConville, W.L. Taylor, and R.A. Watkins, *J. Chem. Phys.* **53**, 912 (1970).
4. M.L. Klein, W. Blizard, and V.V. Goldman, *J. Chem. Phys.* **52**, 1633 (1970).
5. M.L. Klein and J.A. Reissland, *J. Chem. Phys.* **41**, 2773 (1964).

Investigation of Phonon Radiation Temperature of Metal Films on Dielectric Substrates

J. D. N. Cheeke, B. Hebral,* and C. Martinon

Centre de Recherches sur les Très Basses Températures, CNRS
Grenoble-Cedex, France

We present experimental results and a comparison with theoretical models for the temperature of pulse-heated metal films deposited on dielectric substrates. The interest of this work is twofold. First, the experimental situation corresponds closely to that pertaining to heat pulse propagation in perfect dielectric solids at low temperatures and a knowledge of the phonon-emitting temperature is useful in analyzing the results of such experiments. Second, and of most interest in the present context, the experiments can be used to verify the theoretical formulation of the thermal boundary resistance between solids at low temperatures due to phonon transmission.[1]

The experimental techniques have been described in detail elsewhere.[2,3] Thin metal films were vacuum-deposited on clean, optically polished surfaces of quartz or sapphire substrates. Eighty-nanosecond electrical pulses of variable amplitude were applied to a 50-Ω transmission line terminated by the film. Because of its $R(T)$ characteristics, the resistance of the film is generally different from 50 Ω. The thermal response time of the film is very short (several nanoseconds), so that when the substrate is held at 4°K the film quickly heats up during the pulse duration to a steady temperature determined by the thermal contact resistance between film and substrate. From measurements of the amplitudes of incident and reflected pulses one can determine both the power absorbed by the film and its final temperature, which can then be compared directly with that expected from the theory.

Representative results are shown in Figs. 1–3 for Pb–Al$_2$O$_3$, Sn–Al$_2$O$_3$, and Sn–SiO$_2$, respectively. The ensemble of all experimental results obtained thus far has been given elsewhere.[4] The sound velocities used were estimated from values of Poisson's ratio and the 0°K Debye temperature.[5] In the case of quartz, which is quite anisotropic, we used average values of the calculated sound velocities over a 60° cone[6] normal to the face used (Z cut).

Two curves have been drawn in each figure to represent the theoretical expectations. The upper one (AM) corresponds to the radiation temperature expected on the basis of Little's acoustic mismatch model.[1] The lower one corresponds to the so-called "perfect match" (PM) model, which represents an upper limit on the heat transfer and where the phonon transmission coefficient for incidence within the critical cone is artificially put equal to unity. Both of these curves have been calculated numerically,[3] using a Debye model.

* Present address: Department of Physics, University of Ottawa, Ottawa, Ontario.

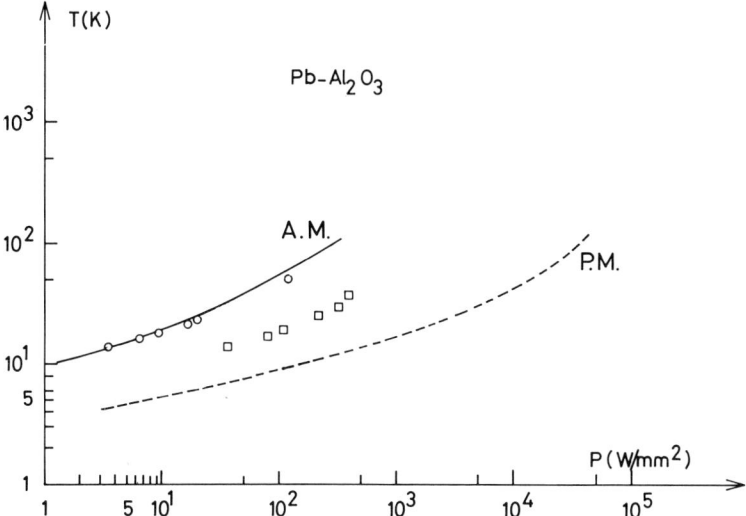

Fig. 1. Radiation temperature of Pb–Al$_2$O$_3$. (□) Film deposited at 300°K. (○) Film deposited at 77°K.

The results of Fig. 1 for Pb–Al$_2$O$_3$ are in good agreement with the original results of Herth and Weis.[2] We observe, as did they, that for a film deposited without unusual precautions the data points fall significantly below the AM curve. Herth and Weis found that agreement with the AM curve could be obtained after thorough glow discharge cleaning of the surface and film deposition at 77°K, while we have found that the latter was sufficient. A similar effect was observed for tin, as seen in Fig. 2. We conclude that evaporating on a cooled substrate prevents the formation of a matching layer between the two media, as suggested in Ref. 2.

Other characteristics of the experiment are seen in all three figures. The data points approach the Little curve asymptotically at the lower temperatures, with an

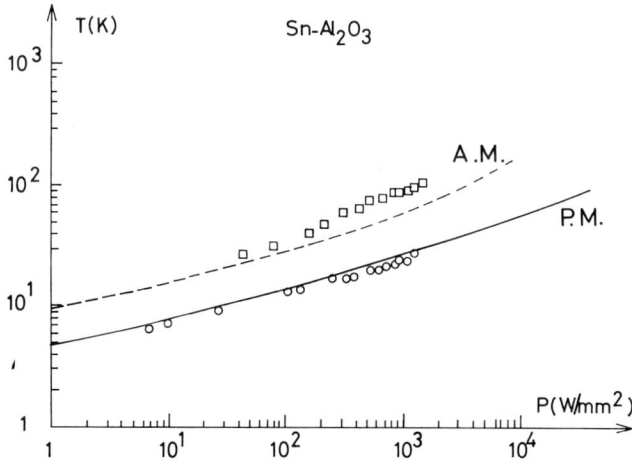

Fig. 2. Radiation temperature of Sn–Al$_2$O$_3$. (○) Film deposited at 300°K. (□) Film deposited at 77°K.

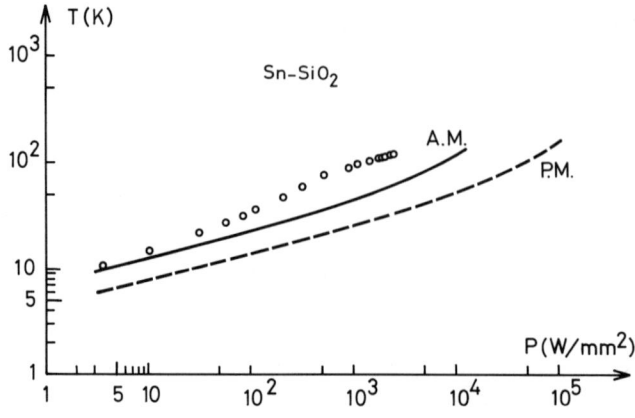

Fig. 3. Radiation temperature of Sn–SiO$_2$.

upward deviation in the case of tin at the higher temperatures. We have often observed this upward curvature in our other results, as have Herth and Weis, but as seen here, it does not seem to occur systematically. It has been suggested that it may be due to phonon dispersion in the film[2] or to a limitation of the phonon mean free path in the substrate at higher temperatures.[4] Further work is needed on this point.

The fact that good agreement with the theory is obtained in the low-temperature limit for a quasiisotropic substrate (Al$_2$O$_3$) and a quite anisotropic substrate (quartz) suggests that effects of anisotropy tend to be washed out in such experiments, which is not unexpected. Likewise it has been shown recently[7] that the electrical resistance of such films (and hence their temperature) is quite insensitive to the very large deviations which are expected from a Bose–Einstein distribution function in the film.

In summary, the present and similar results confirm the validity of the acoustic mismatch model for the solid–solid thermal resistance. At the same time, we feel that many fine details of the problem are inaccessible using thermal phonons, and it will be interesting to conduct such experiments using high-frequency monochromatic phonons of variable incidence angle.

References

1. W.A. Little, *Can. J. Phys.* **37**, 334 (1959).
2. P. Herth and O. Weis, *Z. Physik* **29**, 101 (1970).
3. J.D.N. Cheeke, B. Hebral, and C. Martinon, *J. de Physique* **34**, 257 (1973).
4. J.D.N. Cheeke, B. Hebral, and C. Martinon, *Compt. Rend.* **274**, B–1237 (1972).
5. J.D.N. Cheeke, B. Hebral, and C. Martinon, CNRS Colloquium at Ste. Maxime, France on the Physics of Very High Frequency Phonons, June 1972, *J. de Physique Colloque* **C4** C4–57 (1972).
6. R.J. Von Gutfeld, *Physical Acoustics V*, W.P. Mason, ed., Academic Press, New York (1968).
7. N. Perrin and H. Budd, *Phys. Rev. Lett.* **28**, 1701 (1972).

Surface Thermal Expansion in Noble Gas Solids

V. E. Kenner and R. E. Allen

Department of Physics, Texas A&M University
College Station, Texas

In this paper we present the results of the first realistic calculations of surface thermal expansion.* Our method, which represents a modification of the one that we proposed earlier,[2] is based on the quasiharmonic approximation, in which the vibrational Helmholtz free energy is given by

$$F_{\text{vib}} = k_B T \sum_\omega \ln \left[2 \sinh \left(\hbar\omega / 2 k_B T \right) \right] \tag{1}$$

where k_B is the Boltzmann constant, T is the temperature, and the summation is over all the normal mode frequencies ω of the solid. We assume that the stresses are small enough to be neglected so that the Gibbs free energy is equal to the Helmholtz free energy F. If Φ is the static energy of the solid, then

$$F = \Phi + F_{\text{vib}} \tag{2}$$

We define a_0 to be an interplanar spacing (i.e., spacing between two planes parallel to the surface) in the bulk and A_i to be the ith interplanar spacing near the surface (with $i = 1$ at the surface), and we let

$$A_i = a_0 + a_i, \quad i = 1, 2, \ldots \tag{3}$$

Then the condition for thermal equilibrium is

$$\partial F / \partial a_i = 0, \quad i = 0, 1, 2, \ldots \tag{4}$$

We now use a Taylor series expansion of F_{vib} about $a_i = a_i^{(0)}$ and Φ about $a_i = a_i^{(1)}$. (In Ref. 2 we expanded both about $a_i = a_i^{(0)}$.) We assume that the $a_i^{(0)}$ and $a_i^{(1)}$ are chosen in such a way that we can neglect all terms beyond the second. We define

$$\Delta a_i = a_i - a_i^{(0)}, \quad \delta a_i = a_i^{(1)} - a_i^{(0)} \tag{5}$$

---

* A calculation for iron has been reported by Dobrzynski and Maradudin.[1] This calculation cannot be considered quantitative, however, since the results for the bulk thermal expansion coefficient are in error by about a factor of three. As can be seen in Fig. 1, the results of the present paper are in good agreement with the experimental values for the bulk. Also, the method of Dobrzynski and Maradudin is based on an expansion about the positions of static equilibrium in the bulk crystal, with only the cubic anharmonic terms retained. This method will not be very accurate at high temperatures, and in fact leads to the physically incorrect prediction that the thermal expansion coefficients are constant at high temperatures. As can be seen in Figs. 1 and 2, the present method leads to changes in the thermal expansion coefficients at high temperatures that are in agreement with experimental data for the bulk.

Then
$$\partial\Phi/\partial a_i \approx \Phi_i + \sum_j \Phi_{ij}(\Delta a_j - \delta a_j) \tag{6}$$

where

$$\Phi_i = \left(\frac{\partial\Phi}{\partial a_i}\right)_1, \quad \Phi_{ij} = \left(\frac{\partial^2\Phi}{\partial a_i \partial a_j}\right)_1 \tag{7}$$

We will use the subscript 1 to represent quantities evaluated at $a_i = a_i^{(1)}$ and the subscript or superscript 0 to represent quantities evaluated at $a_i = a_i^{(0)}$. We also have

$$\frac{\partial F_{\text{vib}}}{\partial a_i} \approx \left(\frac{\partial F_{\text{vib}}}{\partial a_i}\right)_0 + \sum_j \left(\frac{\partial^2 F_{\text{vib}}}{\partial a_i \partial a_j}\right)_0 \Delta a_j \tag{8}$$

From Eqs. (2), (4), (6), and (8) we get

$$\sum_j \left[\Phi_{ij} + \left(\frac{\partial^2 F_{\text{vib}}}{\partial a_i \partial a_j}\right)_0\right] \Delta a_j = -\left[\Phi_i - \sum_j \Phi_{ij} \delta a_j + \left(\frac{\partial F_{\text{vib}}}{\partial a_i}\right)_0\right] \tag{9}$$

If we represent the terms in square brackets on the left-hand and right-hand sides of Eq. (9) by K_{ij} and F_i, respectively, Eq. (9) implies that

$$\Delta a_i = -\sum_j (\mathbf{K}^{-1})_{ij} F_j \tag{10}$$

If we use Eq. (1) to obtain the derivatives of F_{vib}, we get

$$F_i = \Phi_i - \sum_j \Phi_{ij} \delta a_j - \frac{\hbar}{2A_i^{(0)}} \sum_{\omega_0} \omega_0 \gamma_i(\omega_0) \coth\left(\frac{\hbar\omega_0}{2k_B T}\right) \tag{11}$$

where $\gamma_i(\omega) = -A_i \partial(\ln\omega)/\partial a_i$, and

$$K_{ij} = \Phi_{ij} + \frac{\hbar}{2A_i^{(0)} A_j^{(0)}} \sum_{\omega_0} \omega_0 \gamma_i(\omega_0) \gamma_j(\omega_0)$$

$$\times \left[\coth\left(\frac{\hbar\omega_0}{2k_B T}\right) - \left(\frac{\hbar\omega_0}{2k_B T}\right) \operatorname{csch}^2\left(\frac{\hbar\omega_0}{2k_B T}\right)\right] \tag{12}$$

We have assumed in Eq. (12) that $\partial^2(\ln\omega)/\partial a_i \partial a_j \approx 0$. From inspecting typical results for $\partial^2(\ln\omega)/\partial a_i \partial a_j$ in our calculations, we estimate that this assumption will lead to errors of a few percent.

Using the high-temperature versions of Eqs. (11) and (12) (with $\delta a_j = 0$), we previously calculated the surface thermal expansion coefficients in the high-temperature limit for a model Lennard-Jones solid.[3] The results were in good agreement with the prediction[2]

$$\frac{\alpha_{\text{surface}}}{\alpha_{\text{bulk}}} \approx \frac{3}{4} \frac{\langle u_z^2 \rangle_{\text{surface}}}{\langle u_z^2 \rangle_{\text{bulk}}}, \quad T \text{ large} \tag{13}$$

where $\alpha_{\text{surface}} = \alpha_1$ and $\alpha_{\text{bulk}} = \alpha_0$, with

$$\alpha_i = \frac{1}{A_i} \frac{\partial A_i}{\partial T} \tag{14}$$

(and $A_0 = a_0$). Here $\langle u_z^2 \rangle$ is the mean-square amplitude in the direction perpendicular to the surface.

In performing the present calculations for the noble gas solids, we have assumed, as usual, a Lennard-Jones 12-6 potential

$$\phi(r) = 4\varepsilon[(\sigma/r)^{12} - (\sigma/r)^6] \tag{15}$$

We have performed our calculations for an 11-layer slab with two surfaces and have determined the frequencies ω in a manner described previously.[4] We could have determined the derivatives $(\partial\omega/\partial a_i)_0$ using first-order perturbation theory, but found it easier just to calculate the frequencies at two values of a_i close to, and on either side of, $a_i^{(0)}$. We used a sample mesh of two-dimensional wave vectors[4] $\bar{q}$ which contains eight and 12 values of $\bar{q}$ in the irreducible element of the two-dimensional Brillouin zone for the (111) and (100) surfaces, respectively. There are thus $3 \times 11 \times 8 = 264$ independent sample frequencies for the (111) surface and $3 \times 11 \times 12 = 396$ for the (100) surface which are used in approximating the summation over ω (which contains an infinite number of values of ω for a slab which is infinite in the directions parallel to the surface). Finally, we assumed that the third and deeper interplanar spacings expand in proportion to the bulk interplanar spacing—i.e., we took $\Delta a_i = 0$ for $i \geq 3$, which is a good approximation.

In test calculations we found that the variation of the static energy (and its derivatives) with the interplanar spacings is more important than the variation of the frequencies (and their derivatives). In our final calculations we therefore used only one set of $a_i^{(0)}$—namely the values corresponding to the positions of static equilibrium—but several values of δa_i, determined in the way described below to make the Taylor series expansion for $\partial\Phi/\partial a_i$ in Eq. (6) reasonably accurate at all temperatures. In other words, we expanded $\partial F_{\text{vib}}/\partial a_i$ about the positions of static equilibrium in Eq. (8) for all temperatures, but we expanded $\partial\Phi/\partial a_i$ about a number of configurations corresponding to different temperature ranges.

For the Lennard-Jones potential of Eq (15) we define a dimensionless temperature T^* and a dimensionless thermal expansion coefficient α^* by

$$T^* = \frac{k_B}{\hbar}\left(\frac{M\sigma^2}{\varepsilon}\right)^{1/2} T \tag{16}$$

$$\alpha_i^* = \frac{1}{A_i}\frac{dA_i}{dT^*} \tag{17}$$

where M is the atomic mass.

Our procedure consisted of two steps. To obtain a first estimate of the interplanar spacings, we expanded both $\partial F_{\text{vib}}/\partial a_i$ and $\partial\Phi/\partial a_i$ about the static equilibrium configuration. This first estimate gave the results for the bulk indicated by the solid line in Fig. 1; notice that the curve for α_{bulk}^* unphysically levels off to a constant value at high temperatures. (We do not show the results for the first estimate of α_1^* and α_2^*.) Then, using the values of the interplanar spacings obtained in this estimate at several values of the temperature (e.g., $T^* = 20, \ldots, 70$ for Xe), we took the δa_i of Eqs. (5) and (6) to have the corresponding values in our final calculation of the Δa_i at nearby temperatures. That is, in finally determining the thermal expansion coefficients α_i^* at a given temperature, we expanded $\partial\Phi/\partial a_i$ about the configuration

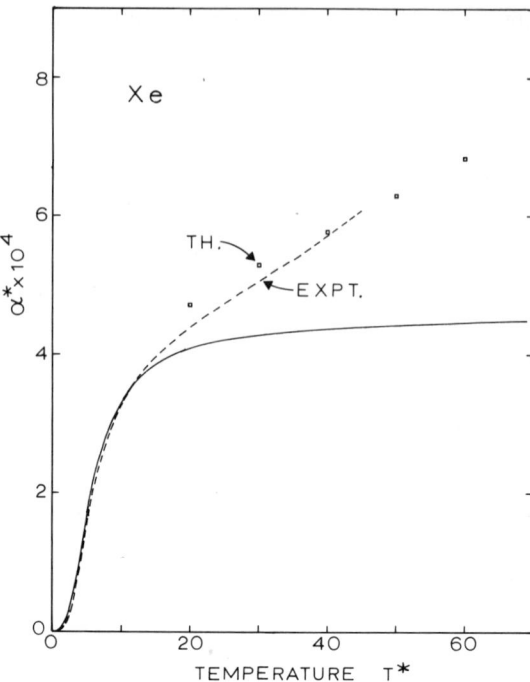

Fig. 1. Results for bulk thermal expansion. The solid line represents the "first estimate" described in the text, which is obtained from a Taylor series expansion of both $\partial\Phi/\partial a_i$ and $\partial F_{\text{vib}}/\partial a_i$ about the positions of static equilibrium. The squares represent the final calculated values of α_{bulk} with $\partial F_{\text{vib}}/\partial a_i$ expanded about the static equilibrium positions and $\partial\Phi/\partial a_i$ expanded about the configuration at the given temperature (e.g., $T^* = 30$) which includes the thermal expansion obtained in the "first estimate." The dashed line represents the experimental data of Ref. 5.

which was predicted in our first estimate for this temperature. However, as mentioned above, we always expanded $\partial F_{\text{vib}}/\partial a_i$ about the configuration of static equilibrium.

In the case of α_{bulk} for Xe the results obtained from the above procedure are indicated by the squares at $T^* = 20, 30, \ldots, 60$ in Fig. 1. The experimental results[5] are indicated by the dashed line. We thus find that our method yields calculated values for the bulk thermal expansion coefficient of Xe which are in remarkably good agreement with experiment. We find the same agreement for Ar and Kr (for which reliable experimental data exist up to the melting point.[5-7] This agreement with the bulk data provides some confidence in our results for the surface thermal expansion coefficients.

We mention that our results for α_{bulk} at high T are considerably better than those obtained from much more elaborate calculations based on the quasiharmonic approximation,[8] because of a cancellation of errors in our method. As can be seen in Fig. 1, our first estimate of the interplanar spacing is well below the correct value at high T. As a result our final calculated values of α_{bulk} will be too small. On the

Surface Thermal Expansion in Noble Gas Solids

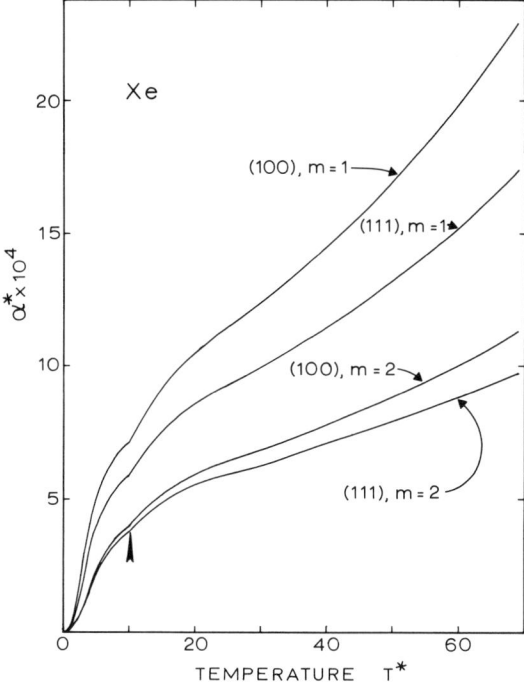

Fig. 2. Thermal expansion coefficients for the first ($m = 1$) and second ($m = 2$) interplanar spacings beneath the surface. These results are for the (111) and (100) surfaces of Xe. To the left of the vertical arrow ($T^* \leq 10$) the results were obtained by expanding $\partial\Phi/\partial a_i$ about the positions of static equilibrium, and to the right ($T^* > 10$) by the interpolation scheme described in the test.

other hand, the quasiharmonic approximation predicts values of α_{bulk} which are too large.[8] The errors involved in our method and in the quasiharmonic approximation thus tend to cancel, so that our results are more accurate than those obtained in more rigorous calculations based on the quasiharmonic approximation.

In order to obtain values of the surface thermal expansion coefficients between those determined at a small number of sample temperatures ($T^* = 20, \ldots, 70$ for Xe) in the way described above, we have simply interpolated between the sample temperatures using a cubic interpolation formula.* The results obtained in this way for the (111) and (100) surfaces of Xe are shown in Fig. 2. Below $T^* = 10$ the results are those obtained in the first estimate (i.e., by expanding $\partial\Phi/\partial a_i$ about the positions of static equilibrium), and above $T^* = 10$ the results were obtained by interpolation.

In Fig. 3 we plot the ratios $\alpha_{\text{surface}}/\alpha_{\text{bulk}}$ and $\alpha_2/\alpha_{\text{bulk}}$ for the (111) and (100) surfaces of Xe. Two points are noteworthy: First, Eq. (13) is approximately valid for $T \gtrsim \frac{1}{2}\theta_D$, where θ_D is the bulk Debye temperature. [For Xe $(k_B/\hbar)(M\sigma^2/\varepsilon)^{1/2}\theta_D = 27.4$ if θ_D is taken to be the calorimetric Debye temperature at $T \rightarrow 0°\text{K}$.][9] Second,

---

* The value at $T^* = 10$ used in the interpolation for Xe was that obtained in the "first estimate," in order to make the values for $T^* > 10$ and $T^* \leq 10$ match up.

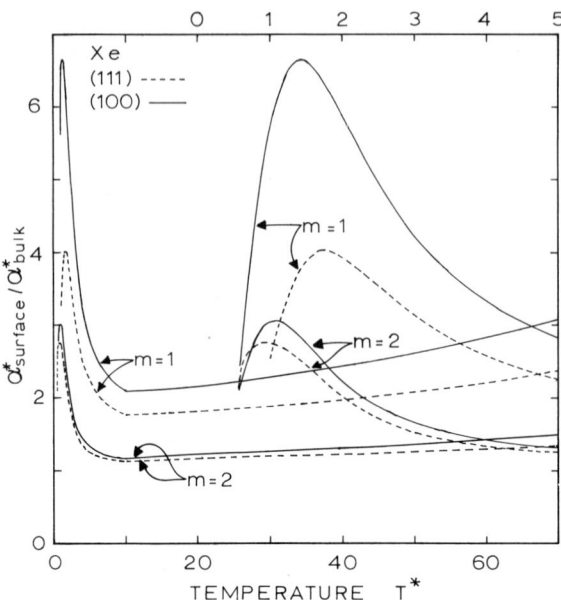

Fig. 3. Ratio of α for first ($m = 1$) and second ($m = 2$) interplanar spacings to α_{bulk}. These results are again for Xe(111) and Xe(100). On the right-hand side we show the behavior at low temperatures ($0 < T^* \leq 5$) on the expanded scale at the top.

as $T \to 0$, $\alpha_{surface}/\alpha_{bulk}$ undergoes a large increase, passes through a peak (at roughly 6% of the bulk Debye temperature), and then decreases again. This behavior will be explained in detail elsewhere, but it can basically be attributed to the presence of low-frequency surface modes. For the (100) surface there are more surface modes with lower frequencies than for the (111) surface, and the peak will therefore be higher and will occur at a slightly lower temperature for the (100) surface.

The qualitative behavior discussed in the preceding paragraph may occur generally, e.g., in metals (except at extremely low temperatures, where the electronic contribution to the thermal expansion will become important). Experimental tests of this behavior would be of interest.

References

1. L. Dobrzynski and A.A. Maradudin, *Bull. Am. Phys. Soc.* **17**, 133 (1972).
2. R.E. Allen, *J. Vac. Sci. Tech.* **9**, 934 (1972).
3. V.E. Kenner and R.E. Allen, *Phys. Lett.* **39A**, 245 (1972).
4. R.E. Allen and F.W. de Wette, *Phys. Rev.* **179**, 873 (1969).
5. C.R. Tilford and C.A. Swenson, *Phys. Rev.* **B5**, 719 (1972).
6. O.G. Peterson, D.N. Batchelder, and R.O. Simmons, *Phys. Rev.* **150**, 703 (1966).
7. D.L. Losee and R.O. Simmons, *Phys. Rev.* **172**, 944 (1968).
8. M.L. Klein, G.K. Horton, and J.L. Feldman, *Phys. Rev.* **184**, 968 (1969).
9. R.E. Allen and F.W. de Wette, *Phys. Rev.* **188**, 1320 (1969).

Pair Potentials for van der Waals Solids from a One-Electron Model*

S. B. Trickey, F. W. Averill,† and F. R. Green, Jr.

Quantum Theory Project, Department of Physics and Astronomy
University of Florida, Gainesville, Florida

Recent years have seen much attention paid to the details of the interatomic potentials used to calculate the lattice properties of the rare gas crystals. Thus, for example, we have the extensive work of Klein and co-workers[1] on argon. They determined a multiparameter potential by fitting to beam scattering and second virial coefficient data, and then examined that potential by stringent comparisons of calculated liquid and solid properties to empirical values. As this example might suggest, it is not far from correct to say that, with the exception of solid He, the essential problem for the theory of the ground state of rare gas crystals has been to test various empirical and semiempirical potentials. Such a situation is somewhat unsatisfying, for it would be preferable to have a theory of rare gas crystals which incorporated an *ab initio* treatment of the electronic structure of the solid into its description of lattice properties. The present work is motivated by that desire.

There exist a number of energy band calculations on solid Ar using the OPW (with local exchange),[2] APW (with local exchange),[3] relativistic KKR,[4] pseudopotential,[5] OPW (with local exchange, then Hartree–Fock exchange, and then an approximate treatment of correlation),[6] and APW (with Hartree–Fock exchange, then a simple treatment of correlation)[7] methods. In all these cases the calculations were non-self-consistent, i.e., the energy bands were obtained for some assumed crystal potential—usually one constructed from an atomic superposition. Furthermore, all these calculations had as their purpose the study of the classical questions of band theory, namely the size of the direct band gap, the width of the valence bands, the values of exciton binding energies, etc. None of the calculations referenced involved any attempt to utilize energy band theory to describe the lattice cohesive properties. This fact is not surprising in view of the relative newness of a form of energy band theory which is well-adapted to that task. Though there is only one energy band calculation for Xe extant,[8] comments of similar nature apply.

As a result of several rather different calculations on alkali metals,[9] magnetized vanadium,[10] aluminum,[11] etc., it is now known that the augmented plane wave, statistical exchange ("APW-$X\alpha$") energy band model[12] yields at least semiquantitative static lattice energies and PV relations (at $T = 0°K$) for metals. Our work represents the first application of that model to rare gas crystals, with good results. Slater[13]

---

* Supported in part by the National Science Foundation.
† Present Address: Department of Materials Science, Northwestern University, Evanston, Illinois.

has argued that, although the good results are at first sight surprising, there is adequate reason to believe that the model is capable of handling a van der Waals system.

The fundamental assumption[12] of the $X\alpha$ method is that the total energy of a system may be written as

$$\langle E_{x\alpha} \rangle = \langle T_{x\alpha} \rangle + \langle W_{x\alpha} \rangle \tag{1}$$

where (in Rydberg units)

$$\langle T_{x\alpha} \rangle = -\sum_i n_i \int u_i^*(1) \nabla_1^2 u_i(1) \, d\mathbf{r}_1 \tag{2a}$$

$$\langle W_{x\alpha} \rangle = \sum_v \int \rho(1) g_{1v} \, d\mathbf{r}_1 + \tfrac{1}{2} \sum_{\mu\nu}' g_{\mu\nu}$$

$$+ \tfrac{1}{2} \int \rho(1) \rho(2) g_{12} \, d\mathbf{r}_1 \, d\mathbf{r}_2 + \tfrac{1}{2} \int \rho(1) U_{x\alpha}(1) \, d\mathbf{r}_1 \tag{2b}$$

in which the charge density is

$$\rho(1) \equiv \rho(\mathbf{r}_1) = \sum_i n_i u_i^*(1) u_i(1) \tag{3}$$

Here the u_i are one-electron spin orbitals, each with occupancy n_i; the Roman subscripts refer to electrons, the Greek subscripts to nuclei, and the various g's are the usual Coulomb operators. The exchange correlation potential, which is the essence of the model, is

$$U_{x\alpha}(1) = -9\alpha [(3/8\pi) \rho(1)]^{1/3} \tag{4}$$

with α a parameter such that $2/3 \leq \alpha \leq 1$. Application of the variation principle to Eq. (1) gives a one-electron effective Schrödinger equation which we solve self-consistently in the irreducible wedge of the first Brillouin zone using the APW method.

The static lattice energy is found by computing $\langle E_{x\alpha} \rangle$ for the solid at a given density and subtracting from that the total energy of the isolated atom, also found via Eq. (1). The same value of α is used in both calculations. The choice of α has been much discussed;[12] for brevity we note that we used what is commonly known[14] as α_{VT}. This value of α is guaranteed, by construction, to yield an atomic total energy from Eq. (1) which is equal to the expectation value of the original many-electron Hamiltonian (for the atom) with respect to a single determinant constructed from the $X\alpha$ orbitals. Thus $\alpha_{Ar} = 0.72131$ and $\alpha_{Xe} = 0.69962$, from Ref. 14 and similar calculations.

To extract an interatomic potential, we choose some analytic form, then perform a lattice sum at each solid density for which we have an APW-$X\alpha$ static lattice energy. The potential parameters are adjusted until the lattice sum for the static lattice energy is fitted best (in the sense of least squares) to the APW-$X\alpha$ static lattice energy over the available density range. Because of the exploratory nature of the present work, we have restricted ourselves to the Lennard-Jones form of the potential, in spite of its known deficiencies.

We have applied the technique just summarized to solid Ar and Xe in the fcc structure. In the case of Ar we have computed static lattice energies at seven lattice constants, ranging from 4.76 to 10.58 Å, while in Xe we have made six calculations in the range 5.292–6.77 Å. Our computed results for Ar are: static lattice energy $E = -1041.1$ K; equilibrium lattice constant $a = 5.08$ Å; Lennard-Jones well depth ε and radius for zero of the potential σ, 118.4 K and 3.303 Å, respectively. These are to be compared with empirical values[15] of $E = -1033.2$ K, $a = 5.312$ Å, $\varepsilon = 119.3$ K, and $\sigma = 3.448$ Å. The corresponding computed values for Xe and $E = -1261.2$ K, $a = 6.184$ Å, $\varepsilon = 143.8$ K, and $\sigma = 4.052$ Å, as opposed to empirical values[15] of $E = -1926.5$ K, $a = 6.131$ Å, $\varepsilon = 228$ K, and $\sigma = 3.973$ Å.

The discrepancies in E and ε for Xe may well stem from truncation error problems, at least in part. That is, the calculated atomic total energy for Ar is of the order of 1053 Ry, while the static lattice energy is -0.0066 Ry, or about one part in 10^5 of the total energy, while in Xe the atomic total energy is around 14,464 Ry and the static lattice energy is about -0.0126 Ry, about one part in 10^6. Thus we believe our Ar cohesive energy uncertainty to be ± 0.0002 Ry compared to a Xe value of ± 0.002 Ry. It would appear, therefore, that subtraction of two large total energies from one another to find the static lattice energy is at least part of our problem. It may also be that the quoted uncertainty in the experimental static lattice energy for Xe is artificially small, since the value given ($E = -1261.2$ K) for $T = 0°K$ is based on an extrapolation from the latent heat of vaporization of liquid Xe at its boiling temperature ($\sim 165°K$). We note that our calculated E vs. a curve for Xe has a physically acceptable shape, for if we rigidly shift all our energies down by 662.5 K ($= 0.0042$ Ry) in order to match the empirical equilibrium static lattice energy, we find $\sigma = 3.965$ Å, $\varepsilon = 222.88$ K, and a significant reduction in the standard deviation of our least-squares fit.

Because of the size of the uncertainty in the calculated cohesive energies, our method cannot at this point be used to study the old question of the relative stability of the hcp and fcc lattice structures for rare gas crystals. In Ar, for example, the fcc lattice structure is stable with respect to the hcp by about 0.1% of the binding energy,[5] compared to an uncertainty in the calculated cohesive energy of about 3%. It is not obvious that even with improved calculational techniques (e.g., multiple-precision arithmetic in place of the double-precision arithmetic presently employed) the model is sufficiently sensitive to give an adequate account of the stability problem, though it may be. With the computers presently available to us (IBM 360/65, 370/165) we see no hope of making a direct attack on the question.

Computational limitations also determine the extent to which we can use a more realistic form for the pair potential than the Lennard-Jones expression. The problem is mainly one of economics. Since we must calculate static lattice energies at more lattice constants than there are adjustable parameters in the pair potential, a many-parameter potential form (such as that met in Ref. 1) is beyond our reach. A more subtle difficulty in potentials with a large number of parameters is the near-linear dependence of some of those parameters. Though this feature of such potentials helps in making fine adjustments of their behavior, it poses a serious problem for almost any feasible fitting routine. We are currently exploring a four-parameter potential, as well as extending the calculation to Kr.

Acknowledgments

We are grateful to Professors J. B. Conklin, Jr., J. W. D. Connolly, and J. C. Slater of the Quantum Theory Project for several valuable conversations. Support of a substantial portion of the computing costs by the University of Florida Computing Center is gratefully acknowledged.

References

1. M.L. Klein, J.A. Barker, and T.R. Koehler, *Phys. Rev. B* **4**, 1983 (1971), and references therein.
2. R.S. Knox and F. Bassani, *Phys. Rev.* **124**, 652 (1961).
3. L.F. Mattheiss, *Phys. Rev.* **133**, A1399 (1964).
4. U. Rössler, *Phys. Stat. Sol.* **42**, 345 (1970).
5. R. Ramirez and L.M. Falicov, *Phys. Rev. B* **1**, 3464 (1970).
6. N.O. Lipari and W.B. Fowler, *Phys. Rev. B* **2**, 3354 (1970).
7. L. Dagens and F. Perrot, *Phys. Rev. B* **5**, 641 (1972).
8. M.H. Reilly, *J. Phys. Chem. Sol.* **28**, 2067 (1967).
9. F.W. Averill, *Phys. Rev. B* **4**, 3315 (1971), and references therein.
10. T.M. Hattox, J.B. Conklin, Jr., J.C. Slater, and S.B. Trickey, *J. Phys. Chem. Sol.* **34**, 1627 (1973).
11. M. Ross and K.W. Johnson, *Phys. Rev. B* **2**, 4709 (1970).
12. J.C. Slater and K.H. Johnson, *Phys. Rev. B* **5**, 844 (1972).
13. J.C. Slater, *J. Chem. Phys.* **57**, 2389 (1972).
14. K. Schwarz, *Phys. Rev. B* **5**, 2466 (1972), and references therein.
15. G.L. Pollack, *Rev. Mod. Phys.* **36**, 748 (1964).

MAGNETISM
7
Plenary Topics

Magnetic Interaction Effects in Dilute Alloys

R. F. Tournier

*Centre de Recherches sur les Très Basses Températures, CNRS
Grenoble-Cedex, France*

We are interested here in the evolution of dilute alloys from nonmagnetic behavior at very low temperatures to magnetic behavior when the concentration increases.

Above T_{Kf}, which is called the Kondo temperature[1,2] or the fluctuation temperature,[2] a single impurity follows a Curie–Weiss law over a large range of temperature[3,4,9] (from $\frac{1}{3}T_{Kf}$ to $10T_{Kf}$). Well below T_{Kf} it tends to a finite susceptibility according to a T^2 law.[8] Magnetization is induced by the external field more easily for lower T_{Kf}. The resistivity[5-7] decreases as T^2 from its residual value at $0°K$ and then varies as $\log T$ when the temperature increases[10]; for $T < T_{Kf}$ the specific heat increases linearly with T, goes through a maximum, and then slowly decreases.[11]

To explain these properties, two types of theory exist. The first is based on the Kondo Hamiltonian[1,15] ($-2J_{sd}\mathbf{S}\cdot\mathbf{s}$), which couples the spin S of the impurity to the spin s of the conduction electrons through an antiferromagnetic exchange coupling. It is clear in this case that the impurity spin exists and also that the Blandin–Friedel–Anderson condition[12-14] for the magnetism of the impurity, $1-\frac{1}{2}u_{\text{eff}}\rho_d(E_F) < 0$, is respected, where u_{eff} is an effective intraatomic interaction dominated by the Coulomb interaction and $\rho_d(E_F)$ is the density of states at the Fermi level for the two spin directions per orbital. With the aid of this Hamiltonian the logarithmic anomaly of the resistivity and its minimum in the alloys[15] have been explained. The explanation has opened up a very large field of work[2,10,11] attempting to approach the ground state of the impurity at $0°K$. Schrieffer[16] has suggested that such a model might be applied to all impurities in noble metals to describe their magnetic properties above T_{Kf} (the $\log T$ term in the resistivity, the Curie–Weiss law for the susceptibility). The theoretical attempts to describe the apparent nonmagnetic situation below T_{Kf} have not been totally convincing, although some results are promising.[17,18]

The second type of theory, local spin fluctuation (LSF) theory, assumes that the ground state of the impurity is nonmagnetic and thus $1-\frac{1}{2}u_{\text{eff}}\rho_d(E_F) > 0$ for all impurities. u_{eff} is not actually accessible to calculation. When the temperature increases we have local spin fluctuations[19] on the impurity site, as was initially shown by Lederer and Mills in the Pd–Ni system. It has been suggested by Caplin and Rizzuto[6] and Rivier and Zuckermann[20] that this model can be extended to all impurities. The fluctuations attain their maximum amplitude around T_{Kf}. Their lifetime is equal to $\hbar/kT_{Kf}$. Above this temperature the local fluctuation cannot easily be distinguished from the thermal fluctuations of a true paramagnetic spin.[20] Magnetic behavior may be observed in the resistivity and susceptibility,[21-23] as has been shown by Suhl and co-workers. The behavior of transition impurities in transition metals has been analyzed at temperatures higher than T_{Kf} by Kaiser and Doniach.[24]

The Interactions between Nearly Magnetic Impurities

Density-of-States Effects. We use here the LSF model because we are essentially interested in the modification of properties under the influence of interactions in the region well below T_{Kf}, where the LSF theory seems to be particularly successful. We also use the critical condition for magnetism because at very low temperatures it is possible to separate the properties of magnetic impurities and those of nonmagnetic impurities. Here we call "magnetic" the impurities that create a magnetic ordering at low temperatures through RKKY interactions.

A single impurity has an enhanced Pauli-like susceptibility

$$\chi_1 = \eta \mu_B^2 (2l + 1) \rho_d(E_F) \quad \text{with} \quad \eta = 1/[1 - \tfrac{1}{2} u_{\text{eff}} \rho_d(E_F)] \tag{1}$$

In this scheme the density of states at the impurity site is a Lorentzian function[14] of the energy of the conduction electrons. If the density of states $\rho(E_F)$ of the matrix varies slowly at the Fermi level, its width[12-14] Δ is proportional to $\rho(E_F)$. Thus a local change $\delta\rho(E_F)$ in the density of states will produce a change $d\Delta$ in the width Δ of the virtual bound state $d\Delta/\Delta = d\rho(E_F)/\rho(E_F)$. This change may be produced by another impurity situated at the distance r as follows.

It is well known that the screening charge $Z(E)$ inside a sphere of radius r around an impurity[12] is an oscillating and decreasing function of kr for electrons having an energy less than $E = \hbar^2 k^2/2m$. We observe Friedel oscillations at the distance r. The electronic density is dZ/dv, where dv is the element of volume $4\pi r^2\, dr$. Then the density of states at the Fermi level[29] created at the distance r at $T = 0°K$ is

$$\left(\frac{dZ}{dv\, dE}\right)_{E=E_F} = \delta\rho(E_F, r)$$

$$\frac{\delta\rho(E_F, r)}{\rho(E_F)} = (2l + 1) \sin \varphi_l \frac{\sin(2k_F r + \varphi_l)}{(k_F r)^2}$$

$$\varphi_l = \frac{\pi Z}{2(2l + 1)}$$

Z is the difference of charge introduced by the impurity in the metal. φ_l is given by the preceding formula when only d electrons are responsible for the screening charge ($l = 2$).

Daniel[29] has successfully analyzed the influence of the density of states on the physical properties of the matrix and especially the influence on the Knight shift. Caroli[25] and Kim[28] have shown, using an Anderson model, that two impurities at a distance r modify their energy position in the Fermi sea and the width of their virtual bound states.

Caroli obtains the effect per orbit:

$$\frac{\delta\rho_d(E_F)}{\rho_d(E_F)} = \sin\varphi_l \frac{\sin(2k_F r + 3\varphi_l)}{(k_F r)^2}$$

With the same φ_l this quantity will be positive or negative as r varies.

The Different Aspects of the Pair Interaction. Considering a pair of impurities separated by a distance r, the response χ to a local field is given by Eq. (1) in which we have $\rho_d(E_F) + \delta\rho_d(E_F)$ taking the place of $\rho_d(E_F)$. We will have the following different situations according to the sign and the value of $\delta\rho_d(E_F)$.

(a) $\delta\rho_d(E_F) < 0$; consequently, $\chi < \chi_1$; the susceptibility is decreased compared to the susceptibility χ_1 of the single impurity; each impurity is less magnetic.

(b) $0 < \delta\rho_d(E_F) < \delta\rho_c(E_F)$; $\chi > \chi_1$; the susceptibility is increased; each impurity is more magnetic. $\delta\rho_c(E_F)$ is the critical value to obtain magnetism:

$$1 - \tfrac{1}{2}u_{\text{eff}}[\rho_d(E_F) + \delta\rho_c(E_F)] = 0$$

$$\frac{\delta\rho_c(E_F)}{\rho_d(E_F)} = \frac{1 - \tfrac{1}{2}u_{\text{eff}}\rho_d(E_F)}{\tfrac{1}{2}u_{\text{eff}}\rho_d(E_F)} \simeq \frac{1}{\eta}$$

When the exchange enhancement factor η is large the change of density of states necessary to create magnetism is small. Thus long-distance interactions can create magnetism.

(c) $\delta\rho_d(E_F) > \delta\rho_c(E_F)$; magnetism is created. We may obtain within the critical distance the same number of ferromagnetic pairs and antiferromagnetic pairs because of the oscillating character of the RKKY interaction. The possibility of the appearance of magnetism under the influence of interactions was first proposed by Blandin and Friedel[13] to explain the c^2 variation of the magnetization of Au–Co and Cu Co alloys [30,31]

Statistics of the Interaction Effects. We consider here that an impurity on one site receives all the changes of the density of states induced by all other impurities. If M_2 and M_4 are, respectively, the second and fourth moments of the distribution of the density of states calculated for a cfc lattice, we have $M_4/M_2^2 \sim 1/c$, and then the distribution is probably a truncated Lorentzian[32] distribution having a width $\Delta\rho$:

$$\Delta\rho = \frac{\pi}{2\sqrt{3}} M_2 \left(\frac{M_2}{M_4}\right)^{1/2} = \frac{41.6c}{(k_F d)^2} \rho_d(E_F) \sin\varphi_l$$

This distribution is centered on $\delta\rho_d(E_F) = 0$. In a very dilute alloy an equal number of impurities "see" an increase and a decrease in the density of states $\rho_d(E_F)$. Consequently, there are the same number of impurities with an increased susceptibility as with a decreased susceptibility.

Statistics of Local Susceptibilities. The distribution of $P(X)$ of the local susceptibilities is also Lorentzian as a consequence of the Lorentzian density of states distribution:

$$P(X) = X_0/\pi[X_0^2 + (X - A)^2]$$

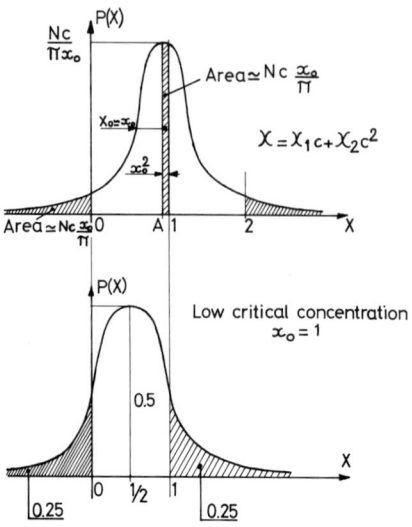

Fig. 1. The distribution function $P(X) = X_0/[\pi(X_0^2 + (X-A)^2)]$ plotted vs. the reduced susceptibility $X = X/X_1$. At low concentrations the width X_0 is equal to x_0 and so is proportional to c. The shift $(1 - A)$ is x_0^2. At the concentration for which $x_0 = 1$ ($c = 0.125\%$ in Cu–Fe alloys) the magnetic impurities ($X < 0$) represent a fraction equal to one-fourth of the total number of impurities.

and X is the reduced susceptibility $X = \chi/\chi_1$, X_0 is the width, given by $X_0 = x_0/(1 + x_0^2)$, and A is the position of the center of the distribution, $A = 1/(1 + x_0^2)$, with $x_0 = \eta \Delta\rho/\rho_d(E_F)$. x_0 is proportional to c because $\Delta\rho \sim c$.

X_0 and A have been obtained after neglecting some small terms and after assuming the following approximation: $\delta\rho_c(E_F)/\rho_d(E_F) = 1/\eta$. In Fig. 1 we have represented a distribution for the case where $\delta\rho_d(E_F)$ can easily be larger than $\delta\rho_c(E_F)$ (Cu–Fe, for example). When the impurity is magnetic it is represented by $X < 0$ because the denominator of χ is negative when $\delta\rho_d(E_F) > \delta\rho_c(E_F)$. The distribution is not centered at $X = 1$ due to the fundamental property that the number of sites with increased susceptibility is equal to the number of sites with decreased susceptibility. Thus the three hatched areas for $X < 0$, $A < X < 1$, and $X > 2$ are equal to Ncx_0/π. Then $1 - A \simeq x_0^2$.

We now examine some consequences of the existence of this local susceptibility distribution; we do not take into account the molecular field (RKKY oscillations) created by all the small magnetic moments induced by the external field. The discussion in zero field is probably realistic; the discussion of the properties in an external field should be seen as an attempt to understand the behavior of dilute alloys.

The Bulk Susceptibility Is Equal to $N\chi_1 c + N\chi_2 c^2$. The number of impurities for which $X < 0$ and $X > 2$ is $2Ncx_0/\pi$; this number varies as c^2; χ_2 does not depend on c; it is a mean value. We see also (Fig. 1) that the same number of magnetic impurities ($X < 0$) and of nearly magnetic impurities ($X > 2$) contribute to the c^2 term. This point is verified with the Cu–Fe system.[62]

The number of impurities for which $0 < X < 2$ is proportional to c. Their mean reduced susceptibility is $A = 1 - x_0^2$. At low concentrations x_0^2 is negligible. Then we see that the isolated impurities detected[4] in the Cu–Fe system have a mean susceptibility which is equal to the single-impurity susceptibility χ_1. Their Kondo temperatures[39] extend between $T_{Kf}/2$ and infinity.

The Nuclear Magnetic Resonance of the Impurity. χ_0 varies as c at low concentrations, then goes through a maximum, and thereafter progressively decreases; the shift $(1 - A)$ increases with c^2. The properties of such a distribution would have to be directly observed with the NMR of the impurity because the external field induces on each impurity site a hyperfine field proportional to its susceptibility. The resonance line is the distribution of the local susceptibilities if we neglect the influence of the RKKY interactions. In this model $X_0 = W/2(\%)/K_d(\%)$, where $W(\%)$ is the relative peak-to-peak distance in the derivative NMR line and $K_d(\%)$ is the Knight shift of the single impurity. X_0 follows exactly the law $X_0 = x_0/(1 + x_0^2)$, with $x_0 = 78c$ for the Au–V system, according to Narath and Gossard's results[33,35] (Fig. 2). The relative variation of the Knight shift is directly obtained knowing x_0. There is a good agreement between the calculation and the experiment (Fig. 2) when $x_0 < 1$. From Al–Mn results of Alloul et al.[34] we also obtain a relative variation of the Knight shift which can be calculated from the linewidth. Here $(1 - A) \simeq x_0^2 \sim c^2$.

The Specific Heat of the Isolated Impurities. Such a distribution of the local susceptibility or of the inverse of the Kondo temperature may be seen in the specific heat. Its maximum should be shifted progressively toward higher temperatures when the concentration increases, but the precision of the experimental results is not sufficient or the concentration low enough to observe this behavior, if it exists.

The Residual Magnetism. At very low temperatures the magnetic moments freeze because of the RKKY interactions between them at a temperature T_{Rk} proportional to the number of magnetic impurities Ncx_0/π. Thus T_{Rk} is proportional to c^2.[13,40] This ordering temperature will be observed well below T_{Kf}.

In the model of Blandin, Friedel, and Anderson when the magnetism condition is verified the moment is at a maximum.[13] This point is verified for Cu–Fe and Au–Co alloys.

The Magnetic and Nonmagnetic Sites. At the concentration for which $x_0 = 1$ the width X_0 goes through a maximum; one-fourth of the impurities carry a magnetic moment $(X < 0)$ and three-fourths are nonmagnetic $(X > 0)$. We define here a critical concentration; well below it the c^2 term can be observed; above it the ordering temperature T_{Rk} increases progressively. The magnetic sites induce an RKKY polarization and consequently a magnetic moment on the nonmagnetic sites. Then the alloy contains *two* types of moments: the maximum moments of the magnetic impurities and the induced magnetic moments which progressively increase with the concentration.

The alloy will be a magnetic alloy when the induced moments attain the saturation value. This will happen at a concentration for which $T_{Rk} \geq T_{Kf}$.

The Cu–Fe Alloy

Separation of the c and c^2 Contributions to the Magnetization. The magnetization below 30°K of the alloy Cu–Fe can be divided in two parts,[4] $M(H) = M_1(H)c + M_2(H)c^2$, when $c < 0.03\%$. $M_2(H)$ is easily saturated at 40 kOe. The slope dM/dH

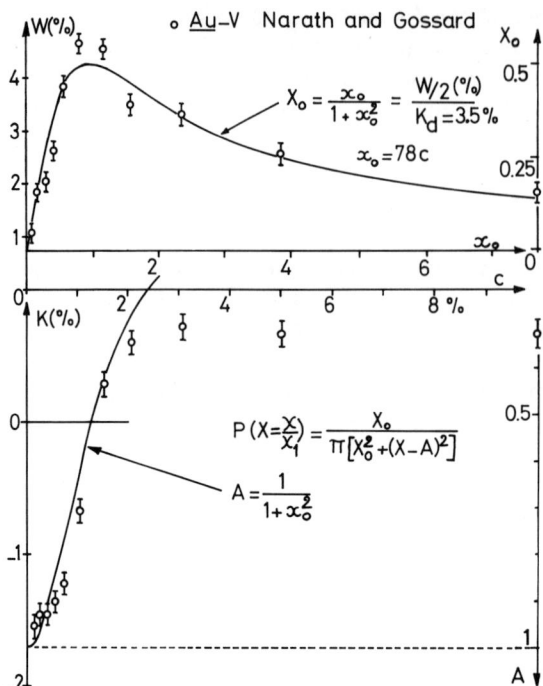

Fig. 2. The experimental NMR results of Narath and Gossard[33] on Au–V alloys plotted vs. $x_0 = 78c$ and c. $W(\%)$ is the peak-to-peak distance of the derivative of the absorption curve; $K_d(\%)$ is taken to be 3.5%.[33] The maximum of X_0 is obtained for $x_0 = 1$. In this case $A = 1/2$. A is calculated only from the parameter x_0 determined from the width X_0.

for $H > 40$ kOe is proportional to c. The specific heat measured in a 40-kOe field[38] is linear with T and proportional to c.

The initial susceptibility is also composed of two terms $\chi = \chi_1 c + \chi_2 c^2$, where $\chi_1 = C_1/(T + 29)$, $\chi_2 \simeq C_2/T$. Knowing C_2 and the saturation of M_2, we determine the spin $S = 2.7$ of the carrier and the concentration of carriers $Zc^2 = 130c^2$. Then the spin S corresponds to the spin of a ferromagnetic pair of iron impurities carrying a moment per atom of $2.7\mu_B$.

The Residual Magnetism. If the preceding analysis is correct, the c^2 term is bound to dominate the susceptibility at very low temperatures, $\chi_1 c$ being negligible compared to $\chi_2 c^2$.

Hirschkoff et al.[37] have observed that the bulk susceptibility varies as c^2. An ordering temperature T_{Rk} proportional to c^2 has been observed down to $0.01°K$ ($T_{Rk} = 3.52 \times 10^5 c^2$) or down to temperatures as low as $T_{Kf}/3000$. For a concentration of $130c^2$ carriers we deduce from T_{Rk} an ordering temperature of $27°K$ per percent impurities rather as for the Cu–Mn system ($20°K$ per percent).

In Fig. 3 we have plotted the susceptibility well below T_{Rk} and the T term[38] of the specific heat due to magnetic impurities vs. c^2. A term independent of c^2, obtained in both cases, characterizes the RKKY nature of the magnetic interaction between pairs of impurities. Using the ratio γ/χ calculated in an Ising model[40] which has given a good spin for the carrier for Cu–Mn and Au–Fe, we obtain $S = 2.35 \pm 0.3$, in agreement with the value obtained earlier.

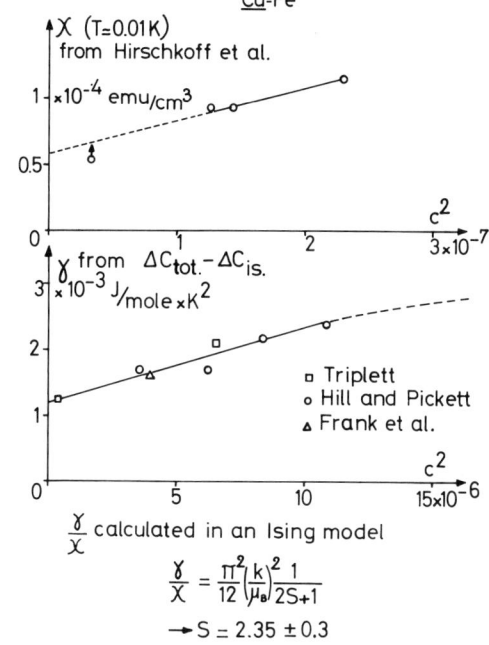

Fig. 3. Plotted at the top of the figure is the susceptibility measured by Hirschkoff et al.[37] in the magnetic ordering of Fe pairs. For $c^2 = 0$ we obtain a value of χ independent of the concentration as for Cu–Mn alloys.[40] The linear term of the specific heat due to magnetic impurities well below the ordering temperature T_{Rk} has been evaluated by Tholence[62] from Ref. 66.

Analysis of the c^2 Term in the Magnetization. Hirschkoff et al.[37] plotted their susceptibility χ vs. $T^{-2/3}$. But we write it as follows:

$$\chi\left(\frac{\text{emu}}{\text{cm}^3}\right) = \frac{5 \times 10^{-7}}{T + 0.009} + 15.5 \times 10^{-7}$$

for $c = 112$ ppm and $0.01 < T < 0.4°$K. The first term is the magnetic part; the second term is the nonmagnetic part of the initial susceptibility. The Curie constant evaluated here is $C^2/2$. Then only one-half of the c^2 term contributes to the residual magnetic ordering.

Ferromagnetic pairs carrying a magnetic moment of $5.4\mu_B$ should be saturated easily in an external field at very low temperatures. Daybell and Steyert[41] have shown the existence of a magnetization term which is easily saturated in a 1-kOe field. In Fig. 4 in low field at $0.07°$K this saturation effect is clearly seen in a magnetization curve obtained by Tholence. Its saturation magnetization represents 30–40% of the total saturation of the c^2 term; the slope dM/dH at $H = 2$ kOe is equal to that part of the initial susceptibility independent of T as determined for $c = 170$ ppm from the results of Hirschkoff.

Thus we can distinguish in the thermal variation of χ and in the magnetization curve two terms varying as c^2: The first corresponds to the behavior of the magnetic part, the second is only saturated in a 40-kOe field and its magnetic moment is apparently not spontaneous but is caused by the external field; its temperature T_{Kf} is apparently of the order of several degrees Kelvin.

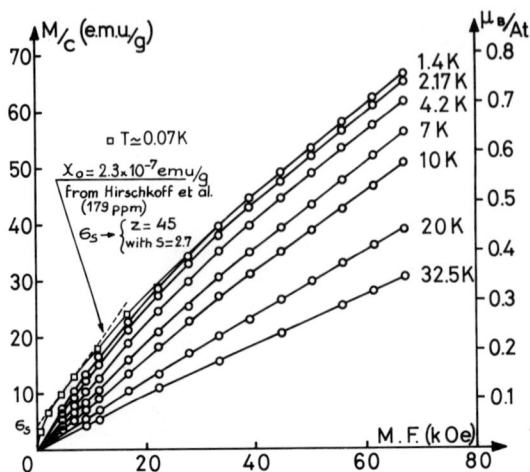

Fig. 4. The magnetization of a Cu–Fe alloy containing 195 ppm impurities measured at 0.07°K and at different temperatures. The position of the magnetic part corresponds approximately to $45c^2$ with $S = 2.7$ per pair. Thus the nonmagnetic part varying as c^2 corresponds to $85c^2$. The slope dM/dH is equal to the nonmagnetic susceptibility deduced from Hirschkoff's work[37] in a very low field.

Analysis of the c^2 Term in the Resistivity. Star[5] has measured the Cu–Fe resistivity below 1°K. It can be written as

$$\rho = \rho_0 + aT^2 + b \ln T$$

The last term is a magnetic term which varies as c^2 (Fig. 5). Using the formula of Hamann, Star obtained the number of pairs as less than $130c^2$. When the magnetoresistance is measured at $H = 2$ kOe the log T term disappears. Only the T^2 term remains. The term aT^2 contains a c term and a c^2 term (Fig. 5). The c^2 part has a temperature T_{Kf} of the order of several degrees Kelvin if we use a concentration of $260c^2$ for the number of nearly magnetic impurities in pairs ($2 \times 65c^2$ for the nearly ferromagnetic pairs and $2 \times 65c^2$ for the nearly antiferromagnetic pairs). This value is in agreement with the fact that a 40-kOe field is necessary to saturate the nearly ferromagnetic pairs.

The Nuclear Magnetic Resonance of Copper. Heeger et al.[42] by studying the NMR linewidth (ΔH) of ^{63}Cu in Cu–Fe and comparing the bulk susceptibility χ to the local Fe susceptibility deduced from the Mössbauer effect,[44,45] found evidence for an extra contribution to χ and ΔH due to a negative polarization cloud around the impurity. In Fig. 6 it is clear that only the magnetization of isolated impurities is seen by the Mössbauer effect,[46] so that the hyperfine field is proportional to $M_1(H)$. It is also known that this hyperfine field at high temperatures varies as the susceptibility $\chi_1 \sim c/(T + 30)$. Thus the extra contribution seen by NMR is also, as apparently for the magnetization, due to the c^2 term and not to a negative polarization cloud.

We analyze here the NMR results of Potts and Welsh.[47] We determine at high temperatures the constant of proportionality between ΔH and χ using different results from the literature. A good proportionality is obtained above 30°K for each concentration. Using the scaling constant $M(emu/g) \times 1.9 \times 10^3 = \Delta H$, a good proportionality is observed above 40 kOe at 1.3°K as shown in Fig. 7. The final slope represents in each case the one-impurity effect. Thus the saturation in $\Delta H(H)$ is also due to pairs.

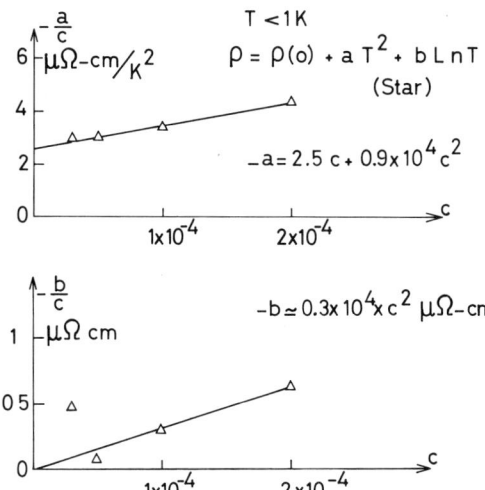

Fig. 5. The resistivity[5] of Cu–Fe alloys below 1°K expressed as $\rho = \rho_0 + aT^2 + b \ln T$. The c^2 variation of b and the c^2 contribution to a are observed.

Two types of pairs are observed with ΔH. The first, which represents 30–50% of the total saturation, is easily saturated and corresponds to the magnetic part. The second part is only saturated in higher fields.

The fact that the saturation value of ΔH is not proportional to c^2 above 0.03% is related to the problem of what happens to the NMR of copper[63] when the number of magnetic impurities increases beyond a certain point. Looking at the results on Cu–Mn alloys in the literature, we observe a linewidth which ceases to be proportional to the concentration of Mn in the same range of c and in the region of magnetic ordering.

Thus it is clear from resistivity, magnetization, and NMR results that two types of sites contribute to the c^2 term. This is in good agreement with the general ideas that we developed earlier.

The Magnetic Sites and the Sites Carrying an Induced Magnetic Moment. By means of the Mössbauer effect Window[48] has studied the hyperfine distribution in Cu–Fe alloys for $0.2\% < c < 1\%$. Three types of sites can be distinguished:

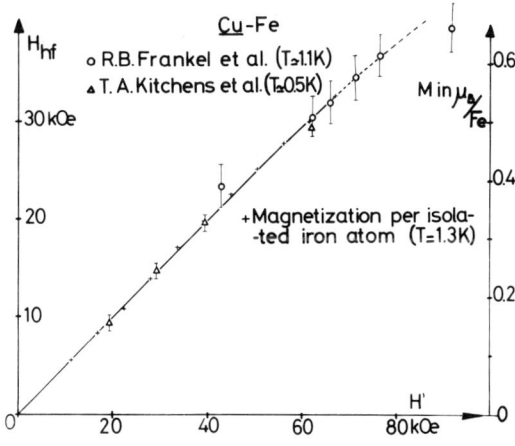

Fig. 6. The hyperfine field deduced from the Mössbauer effect[44,45] plotted against H. The magnetization of the isolated impurities[4] is also plotted against H. A good proportionality is observed with a scaling factor of 60 kOe per Bohr magneton.

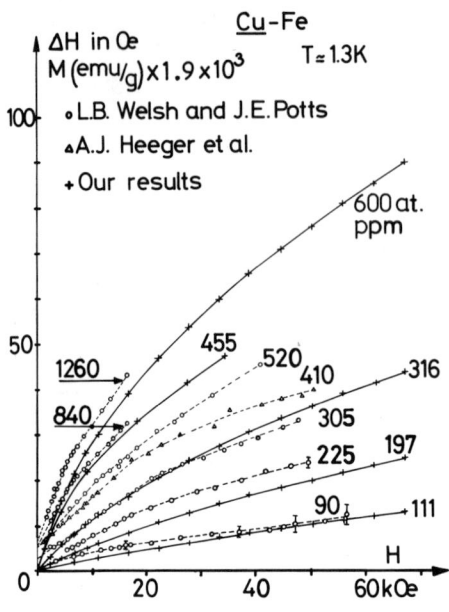

Fig. 7. The NMR linewidth ΔH of $Cu^{42,47}$ in Cu–Fe and the total magnetization of Fe plotted vs. H using a constant of proportionality M(emu/g) $\times 1.9 \times 10^3 = \Delta H$ (Oe) determined by comparing ΔH and χ^{43} at high temperatures.

first, the isolated sites, which represent 75% of the total number of sites, having a hyperfine field which slowly increases with c ($H \simeq 80$ kOe); second, sites having a well-defined hyperfine field equal to 160 kOe; third, sites with an upper hyperfine field corresponding to first-neighbor pairs and groups of Fe atoms, which are more numerous after annealing.

Using the constant of proportionality of 60 kOe/μ_B determined in Fig. 6, we calculate a magnetic moment of 2.6μ_B for the sites carrying well-defined magnetic moments and a moment of 1.3μ_B for the isolated atoms in Cu–Fe 1%.

Thus the magnetic impurities in pairs carry the maximum moment of 2.6μ_B as determined with magnetic measurements. The isolated sites are nonmagnetic and carry moments induced by the magnetic sites in the magnetic ordering state. The amplitude of the induced moments increases with c. The proportion of nonmagnetic sites (75%) compared to the proportion of magnetic sites (25%) is in agreement with the values deduced from our model.

First-Neighbor Interactions, Giant Moments, Nearly Magnetic Clusters

Au–V, Au–Co, Cu–Co. As for the long-range interactions, two opposite effects can be produced by interaction between two first-neighbor 3*d* atoms. The susceptibility can be increased, as is the case for cobalt pairs in gold and in copper; or the susceptibility can be decreased, as is the case for V pairs in gold.[9,49] Calculating the *d–d* interaction, Morya has shown, in agreement with the preceding results, that if the *d* state is less than half full, the interactions move the impurities away from magnetism and if the *d* state is more than half full, the impurities go toward magnetism.

In *Au–V* two distinct resonance lines[33,35] have been observed corresponding,

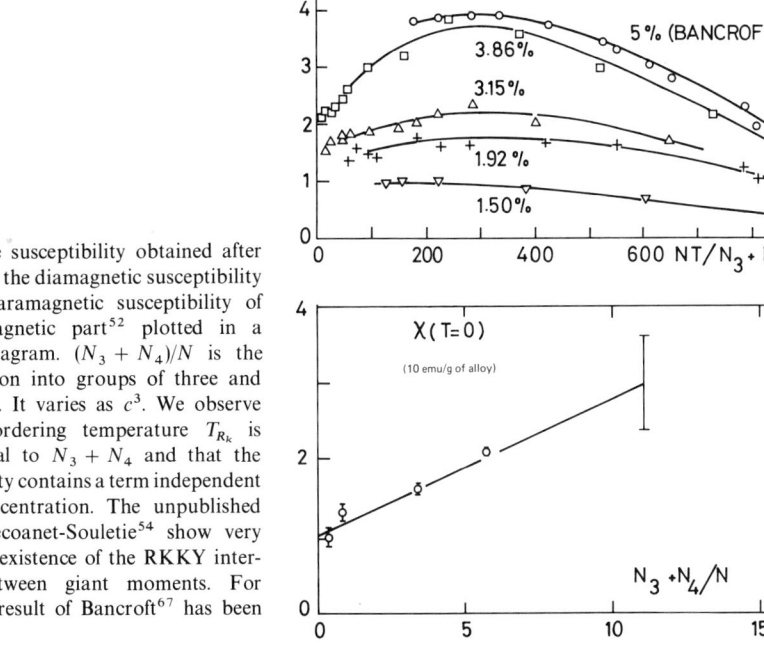

Fig. 8. The susceptibility obtained after subtracting the diamagnetic susceptibility and the paramagnetic susceptibility of the nonmagnetic part[52] plotted in a reduced diagram. $(N_3 + N_4)/N$ is the concentration into groups of three and four atoms. It varies as c^3. We observe that the ordering temperature T_{R_k} is proportional to $N_3 + N_4$ and that the susceptibility contains a term independent of the concentration. The unpublished data of Lecoanet-Souletie[54] show very clearly the existence of the RKKY interactions between giant moments. For $c = 5\%$ a result of Bancroft[67] has been used.

respectively, to the V pairs and to the isolated V atoms. The susceptibility and the electronic specific heat[49] decrease when the concentration increases because the number of V atoms having first neighbors is increased and their temperature T_{Kf} is increased.

In Au–Co[52-54] the properties can be explained by three types of sites: isolated impurities ($T_{Kf_1} = 250°K$), pairs of first neighbors ($T_{Kf_2} = 30°K$), and groups of three atoms or more, which carry a moment of $(1.2 \pm 0.2)\,\mu_B$ per atom and have a hyperfine field on each nucleus equal to 200 kOe. In Cu–Co we have the same situation[51] but the tendency to clustering increases the dimensions of the magnetic groups.

In these last two cases the properties can be analyzed as due to two types of impurities: magnetic impurities and nonmagnetic impurities. The magnetic impurities are groups of cobalt atoms building giant moments which order at low temperatures. The ordering temperature T_{Rk} is proportional to c^3 as a consequence of the RKKY interaction between giant moments (Fig. 8). The nonmagnetic part is due to isolated atoms and to pairs; at low temperatures the specific heat and the susceptibility are dominated by the c^2 term. The superparamagnetic behavior of nearly magnetic pairs above T_{Kf} has not been verified with Au–Co or Cu–Co alloys as it has been with the nearly magnetic pairs of iron in Cu–Fe.[4]

Giant Moments in Cu–Ni, Rh–Ni. We examine the properties of concentrated alloys studied near the critical concentration in the region where their magnetism is small. Kouvel and Comly[59] have shown by neutron diffraction the existence of giant moments of $10\mu_B$ in Cu–Ni extending to a distance of 10 Å around a magnetic nickel atom. To describe the appearance of magnetism on nickel atoms, Robbins et al.[56] and Perrier et al.[57] used a local environment model analogous to those used by Jaccarino and Walker to describe the appearance of magnetism in the Nb–Mo alloys; the magnetism appears if the $3d$ atom has a minimum number m of first neighbors of a certain nature ($m = 8$ Ni in Cu–Ni, $m = 7$ Mo for Fe in Nb–Mo).

Perrier et al.[57] showed that the existence of giant moments is a direct consequence of the existence of a critical condition for magnetism and of the structure of the lattice, and that it is not necessary to have chemical clustering to obtain giant moments. The nonmagnetic part of the magnetization of Cu–Ni alloys has been described in terms of a simple local environment model.[58] Let us imagine, as Perrier did for $n > m$, that a nickel atom has $n = m - 1$ first neighbors of nickel atoms. It is nonmagnetic but has a very large susceptibility; the probability for a nickel atom to have $m - 1$ first neighbors on the first four shells will be enhanced because these different shells have a certain number of sites in common with the first shell which have a probability $(m - 1)/12$ of being occupied by nickel atoms. Thus in this situation we will have a very nearly magnetic cloud. In an external field this cloud will be easily polarized and a giant moment will be built by the external field.

These clouds can exist near the critical concentration of all nonmagnetic–ferromagnetic transitions of disordered alloys. It is necessary to note that the nearly magnetic atoms are much more numerous than the magnetic atoms near the critical concentration.

Houghton et al.,[61] by resistivity measurements, and Muellner and Kouvel,[60] by magnetization measurements on Rh–Ni alloys, have apparently observed the properties of these clouds. Kouvel and Muellner separated the magnetization of the alloy into two parts: a nonmagnetic part approximately constant with temperature and proportional to the external field, and a part σ_0, easily saturated, which has a Curie–Weiss-like susceptibility with a paramagnetic Curie temperature varying from -2 to $-54°K$ for 58–62% of nickel. A moment of $25\mu_B$ is calculated from the Curie constants and from σ_0. The authors have concluded that there is an antiferromagnetic coupling at long distances between giant moments to explain this behavior.

We suggest that this apparent magnetic part is mainly due to nearly magnetic clouds. The proof is given by the resistivity measured by Houghton et al.,[61] which varies as T^2 between 1 and $4°K$; the coefficient of T^2 varies as σ_0 with the concentration. This T^2 behavior is the sign of near magnetism; σ_0 is the saturation magnetization of these nearly magnetic clouds induced by the external field. Above their fluctuation temperature T_{Kf} these clouds would have a superparamagnetic behavior ($u_{\text{eff}} = 25\mu_B$), each impurity fluctuating in phase with the other impurities of the cloud.

Conclusion and Perspectives

We have shown that the residual magnetism characterized by an ordering temperature $T_{Rk} \sim c^n$, $n = 2, 3, \ldots$, is observed at temperatures as low as $10^{-4}T_{Kf}$. We have described the low-temperature properties as composed of two parts: a magnetic part and a nonmagnetic part. The condition for the existence of magnetism seems a useful tool for explaining physical properties. The traditional model that we have used with success does not solve the fundamental problem of the behavior of the isolated impurity or of isolated groups of impurities at 0°K. Attempts to calculate the Kondo temperature of pairs in the Kondo model[64] fail for such large effects of interaction. The model is good only in the "magnetic" region of the alloys. Now the fundamental question for an experimentalist is to determine if the magnetism of pairs will disappear at very low temperatures when the concentration in "magnetic" pairs is very small. Some authors (Beck, Star, etc., ...) have suggested that chemical clustering plays a leading role in the existence of this residual magnetism. In the model that we have used the properties studied are in agreement with perfectly disordered solid solutions. Thus we think it is a parasitic phenomenon and not fundamental. However, the appearance of magnetism can be a useful tool for studying the first stage of precipitation.

The situation in concentrated alloys is in agreement with the same analysis; here the condition for the existence of magnetism has also a physical sense: It is the first-neighbor interaction which is dominant; when the number of first neighbors is known the local magnetism can be described. But the residual magnetism remains to be studied; the ordering temperature should vary as c^n, n being larger than two. Here also the fundamental problem is to determine if the magnetism of the giant moments will disappear when their number is very small.

We have shown the existence of nearly ferromagnetic pairs having a superparamagnetic behavior above their T_{Kf} temperature. By analogy we have been led to ask the question concerning the existence in concentrated alloys of nearly magnetic clouds with a superparamagnetic behavior above a certain temperature T_{Kf}.

We have also suggested the existence of antiferromagnetic pairs in very dilute alloys. The experimental proof of their existence does not exist, but the fact that a dilute alloy has a sort of antiferromagnetic behavior in the magnetic state is good evidence in favor of this assumption. We also suggest by analogy that the existence of antiferromagnetic regions is possible in the nonmagnetic–antiferromagnetic transition; very low-temperature studies should permit them to be detected.

We have also shown that over a large range in concentration the magnetic sites induce moments on the nonmagnetic sites. Numerous studies of the hyperfine field distribution and of the moment distribution are necessary to understand the transition to the magnetism. Another problem appears in "magnetic" dilute alloys such as Cu–Mn, Au–Mn, Ag–Mn, and Au–Fe alloys. Some nonmagnetic sites perhaps exist at very low temperatures well below the ordering temperature T_{Rk}. It is well known that at low temperatures some sites are submitted to very low molecular fields. These impurities can be submitted to a large positive or negative electronic density created by all other impurities and thus to a large supplement of density of states. Their Kondo temperatures can be greatly increased or decreased. We expect the existence of nonmagnetic sites well below T_{Rk} but around or above T_{Kf} in these dilute magnetic alloys. This type of phenomenon apparently has been observed in the resistivity of Cu–Mn alloy.[65]

Finally, it is necessary to study the one-impurity behavior and the effect of interactions when they are small. We have begun to do this, analyzing the effect of the density of states on the isolated impurities. New specific heat measurements at very low concentrations around T_{Kf} are necessary. New hyperfine measurements like Mössbauer effect studies are also necessary to observe the evolution of the susceptibility of the isolated impurities. There seems to be a promising future for very low-temperature studies in the field of magnetism in dilute and concentrated alloys.

Acknowledgments

In this paper I have used some unpublished data of Tholence on the Cu–Fe alloy. I would like to thank him and also Dr. Gilchrist for reading the manuscript.

References

1. J. Kondo, in *Solid State Physics,* F. Seitz, D. Turnbull, and H. Ehrenreich, eds., Academic Press, New York (1969), Vol. 23, p. 184.
2. A.J. Heeger, in *Solid State Physics,* F. Seitz, D. Turnbull, and H. Ehrenreich, eds., Academic Press, New York (1969), Vol. 23, p. 284.
3. C.M. Hurd, *Phys. Rev. Lett.* **18**, 1127 (1967).
4. J.L. Tholence and R.F. Tournier, *Phys. Rev. Lett.* **25**, 867 (1970).
5. W.M. Star, F.B. Basters, G.M. Nap, E. De Vroede, and C. Van Baarle, *Physica* **58**, 585 (1972).
6. A.D. Caplin and C. Rizzuto, *Phys. Rev. Lett.* **21**, 746 (1968).
7. W.M. Star and G.T. Nieuwenhuys, *Phys. Lett.* **30A**, 22 (1969).
8. J.E. Van Dam, *Phys. Lett.* **38A**, 19 (1972).
9. L. Creveling and H.L. Luo, *Phys. Rev.* **176**, 614 (1968).
10. M.D. Daybell and W.A. Steyert, *Rev. Mod. Phys.* **40**, 380 (1968).
11. J.E. Van Dam and G.J. Van Den Berg, *Phys. Stat. Sol.* **3**, 11 (1970).
12. J. Friedel, *Can. J. Phys.* **34**, 1190 (1956); *Nuovo Cimento* (Suppl.) **7**, 287 (1958).
13. A. Blandin and J. Friedel, *J. Phys. Radium* **20**, 160 (1959).
14. P.W. Anderson, *Phys. Rev.* **124**, 41 (1961).
15. J. Kondo, *Progr. Theoret. Phys.* **32**, 37 (1964).
16. J.R. Schrieffer, *J. Appl. Phys.* **38**, 1143 (1967).
17. P.W. Anderson, G. Yuval, and D.R. Hamann, *Phys. Rev. B* **1**, 4464 (1970).
18. K.D. Schotte and U. Schotte, *Phys. Rev. B* **4**, 2228 (1971).
19. P. Lederer and D.L. Mills, *Solid State Commun.* **5**, 131 (1967); *Phys. Rev.* **165**, 837 (1968); *Phys. Rev. Lett.* **19**, 1036 (1968).
20. N. Rivier and M. Zuckermann, *Phys. Rev. Lett.* **21**, 904 (1968).
21. M.J. Levine, T.V. Ramakrishnan and R.A. Weiner, *Phys. Rev. Lett.* **20**, 1370 (1968).
22. M.J. Levine and H. Suhl, *Phys. Rev.* **171**, 567 (1968).
23. H. Suhl, *Phys. Rev. Lett.* **19**, 442 (1967).
24. A.B. Kaiser and S. Doniach, *Inter. J. Magnetism* **1**, 11 (1970).
25. B. Caroli, *J. Phys. Chem. Solids* **28**, 1427 (1967).
26. P.W. Anderson, G. Yuval and D.R. Hamann, *Phys. Rev. B* **1**, 4464 (1970).
27. K.D. Schotte and U. Schotte, *Phys. Rev. B* **4**, 2228 (1971).
28. D.J. Kim, *Phys. Rev. B* **1**, 3725 (1970).
29. E. Daniel, *J. du Phys. et le Radium* **20**, 51 (1959); **20**, 849 (1959).
30. R. Tournier and L. Weil, *J. Phys. Radium* **23**, 522 (1962).
31. R. Tournier, Thesis, Université de Grenoble, 1965 (unpublished).
32. A. Abragam, *Les principes du magnétisme nucléaire,* Bibliothèque des Sciences et Techniques Nucléaires 114 (1961); J. Friedel, *J. Phys. Radium,* **23**, 692 (1962).
33. A. Narath and A.C. Gossard, *Phys. Rev.* **183**, 391 (1969).
34. H. Alloul, P. Bernier, H. Launois, and J.P. Pouget, *J. Phys. Soc. Japan* **30**, 101 (1971).
35. A. Narath, in *Proc. 12th Intern. Conf. Low Temp. Phys., 1970,* Academic Press of Japan, Tokyo (1971).
36. J.L. Tholence and R. Tournier, *J. Phys.* (Colloque C1) **32**, (C1), 211 (1971).

37. E.C. Hirschkoff, M.R. Shanabarger, O.G. Symko, and J.C. Wheatley, *Phys. Lett.* **34A**, 341 (1971); *J. Low Temp. Phys.* **5**, 545 (1971).
38. B.B. Triplett and N.E. Philipps, in *Proc. 12th Intern. Conf. Low Temp. Phys., 1970*, Academic Press of Japan, Tokyo (1971), p. 747.
39. J. Souletie and R. Tournier, *J. Phys.* **32**, (Colloque C1), 172 (1971).
40. A. Blandin, Thesis, Université de Paris, 1961 (unpublished); J. Souletie and R. Tournier, *J. Low Temp. Phys.* **1**, 95 (1969).
41. M.D. Daybell and W.A. Steyert, *Phys. Rev. Lett.* **20**, 195 (1968).
42. A.J. Heeger, L.B. Welsh, M.A. Jensen, and G. Gladstone, *Phys. Rev.* **172**, 302 (1968).
43. H.E. Ekström and H.P. Myers, Preprint, 1971.
44. R.B. Frankel, N.A. Blum, B.B. Schwartz, and D.J. Kim, *Phys. Rev. Lett.* **18**, 1050 (1967).
45. T.A. Kitchens, W.A. Steyert, and R.D. Taylor, *Phys. Rev.* **138**, A467 (1965).
46. S. Hüfner, *Z. Physik* **247**, 46 (1971).
47. J.E. Potts and L.B. Welsh, *Phys. Rev. B* **5**, 3421 (1972).
48. B. Window, *J. Phys. C: Metal Phys. Suppl.* **3**, S 323 (1970).
49. M. Saint-Paul, J. Souletie, D. Thoulouze, and B. Tissier, *J. Low Temp. Phys.* **7**, 129 (1972).
50. M. Inoue and T. Moriya, *Progr. Theoret. Phys. (Kyoto)* **38**, 41 (1967).
51. R. Tournier and A. Blandin, *Phys. Rev. Lett.* **24**, 397 (1970).
52. E. Boucai, B. Lecoanet, J. Pilon, J.L. Tholence, and R. Tournier, *Phys. Rev. B* **3**, 3834 (1971).
53. P. Costa-Ribeiro, J. Souletie, and D. Thoulouze, *Phys. Rev. Lett.* **24**, 900 (1970).
54. B. Lecoanet-Souletie, Thesis, Université de Grenoble, 1971 (unpublished).
55. V. Jaccarino and L.R. Walker, *Phys. Rev. Lett.* **15**, 258 (1965).
56. C.G. Robbins, H. Klaus, and P.A. Beck, *Phys. Rev. Lett.* **22**, 1037 (1969).
57. J.P. Perrier, B. Tissier, and R. Tournier, *Phys. Rev. Lett.* **24**, 313 (1970).
58. B. Cornut, J.P. Perrier, B. Tissier, and R. Tournier, *J. Phys.* **32** (Colloque C1), 746 (1971).
59. J.S. Kouvel and J.B. Comly, *Phys. Rev. Lett.* **24**, 598 (1970).
60. W.C. Muellner and J.S. Kouvel, Presented at the Conf. on Magnetism and Magnetic Materials, Chicago, Nov. 16–19, 1971.
61. R.W. Houghton, M.P. Sarachik, and J.S. Kouvel, *Sol. State Commun.* **10**, 369 (1972).
62. J.L. Tholence and R. Tournier (to be published).
63. H. Nagasawa and W.A. Steyert, *J. Phys. Soc. Japan* **28**, 1202 (1970).
64. K. Matho and M.T. Beal-Monod, *Phys. Rev. B* **5**, 1899 (1972).
65. O. Laborde (to be published).
66. R.W. and G.R. Pickett, in *Proc. 10th Intern. Conf. Low Temp. Phys., 1966*, VINITI, Moscow (1967), Vol. 4, p. 300; J.P. Franck, F.D. Manchester, and D.L. Martin, *Proc. Roy. Soc. (London)* **263A**, 494 (1961).
67. M.H. Bancroft, *Phys. Rev. B* **2**, 2597 (1970).

Magnetic Phase Transitions at Low Temperatures

W. J. Huiskamp

Kamerlingh Onnes Laboratory
Leiden, The Netherlands

Introduction

Magnetism is associated with the degree of freedom of electronic and nuclear spins. We shall defer the subject of nuclear magnetism to the end. Electronic magnetism in crystal lattices may be due to itinerant electrons, such as in metals and alloys, or to localized magnetic moments, such as in insulators. Unpaired electron spins at fixed lattice points may interact through direct exchange, superexchange, and dipolar interactions. In metals magnetic moments may interact over relatively long distances via the conduction electrons, i.e., the RKKY interaction. Do these interactions lead to phase transitions? It is experimentally known that phase transitions occur from the microkelvin to kilokelvin range, i.e., nine decades in the temperature scale.

What does one understand theoretically about magnetic phase transitions? For instance, why do they occur at a finite temperature $(T \neq 0)$?

Theory

For the Ising model in two dimensions (and assuming $H = 0$) one can demonstrate easily that at sufficiently low temperatures, i.e., low entropy, a magnetization spontaneously develops. In Fig. 1 an island of minus spins in an environment of plus spins will tend to convert into plus spins, since the boundary energy gain is $2LJ$, where L is the length of the boundary and $2J$ is the energy of a pair of spins interacting across the boundary. This energy should be sufficient to compensate for the entropy times kT, which amounts to $kT \ln 3^L$, since at every lattice point there are three possibilities for continuation of the boundary. Hence (Fig. 1) at low T spontaneous magnetization will develop by removing boundaries. It can be shown more rigorously that at low T the magnetization has a lower bound in the limit of vanishing field. It is well known that Onsager's solution for the 2D Ising model bears this out in more detail, in particular $T_c = 2J/\ln(1 + \sqrt{2})$. (It should be noted that the boundary entropy, as estimated above, does not include translational degrees of freedom. The corresponding entropy contribution can be neglected in translationally invariant systems if one counts the entropy per lattice site. This is not valid for a ferromagnet split up in domains).

The above argument cannot be repeated for the Heisenberg model, in which spins are assumed to be isotropic, i.e., all three spatial components participating equally in the interaction. The greater degree of freedom of spins can, in a two-

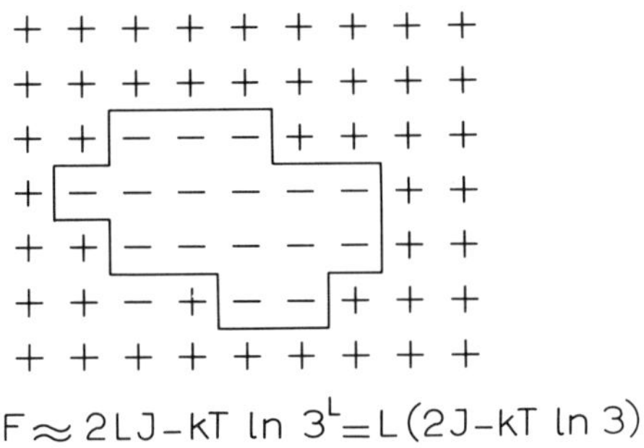

```
+ + + + + + + + +
+ + + + + + + + +
+ +|- - -|+ + + +
+|- - - - - -|+ +
+ +|- - - - -|+ +
+ + - +|- -|+ + +
+ + + + + + + + +
```

$$\Delta F \approx 2LJ - kT \ln 3^L = L(2J - kT \ln 3)$$

Fig. 1. Free energy of a boundary in a 2D Ising spin system.

dimensional lattice, be utilized to change the spin direction gradually between regions of opposite spin alignment. This reduces the energy. Hence even at low T the magnetization can go continuously to zero when the magnetic field is reduced to zero. It is rigorously shown that the ideal 2D Heisenberg system does not have a spontaneous magnetization.

Two-dimensional lattice models have also been invoked to explain phase transitions in ferroelectrics and antiferroelectrics. The reason, of course, is that these two-dimensional models are amenable to exact solutions, one-dimensional models usually fail to show phase transitions, and three-dimensional models are too difficult to be solved exactly.

Recently it has been shown that the 2D Ising square lattice model is mathematically related to the ferroelectric models. Furthermore, a relatively modest generalization of the 2D Ising model leads to phase transitions for which the heat capacity does not have a logarithmic singularity. It may be concluded that the logarithmic singularity in the specific heat of the Onsager solution of the simple square Ising model is *not* a common feature of 2D phase transitions in general, since slightly more complicated 2D Ising systems have a different kind of singularity. Neither should two-dimensional systems, although exactly solvable and very interesting by themselves, be taken as a reliable guide to the three-dimensional reality.

Quite recently the key importance of lattice dimensionality in the *critical* properties of magnetic systems has been elucidated.[1-3] It is well known that in the study of critical phenomena the concept of scaling has been very fruitful. We shall discuss shortly a scaling model of Baker (earlier known as the hierarchal model of Dyson). Let us consider an array of spins $S_1, S_2, \ldots$ and as a start we consider a one-dimensional chain of Ising spins. There is no phase transition for the 1D system, but the model is easily extended to arbitrary dimensions. Now these spins are supposed to interact in pairs by the exchange coupling $-2JS_1 S_2$ (where the subscript z has been dropped for convenience), $-2JS_3 S_4$, etc. In addition there is a weak interaction between pairs of all pairs, and further between pairs of all pairs of pairs, etc. (Fig. 2). The Hamiltonian representing the sum of all interactions has the scaling properly built in. Namely, if one approaches the critical point, the correlations among spins increase and in most cases nearest-neighbor (nn) spins in a pair will be aligned. Hence we consider $\hat{S}_1 = (S_1 + S_2)/\sqrt{2}$, and $\hat{S}_2 = (S_3 + S_4)/\sqrt{2}$ as new variables, "block spins." Still closer to the critical point the correlations will

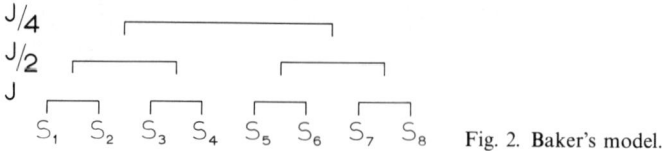

Fig. 2. Baker's model.

extend over four spins, so that we use $\hat{S}_1 = (\hat{S}_1 + \hat{S}_2)/\sqrt{2}$, etc., as new variables. At every stage the Hamiltonian is written as

$$-2JS_1S_2 = +J[(S_1 - S_2)^2 - S_1^2 - S_2^2]$$
$$-2J\hat{S}_1\hat{S}_2 = +J[(\hat{S}_1 - \hat{S}_2)^2 - \hat{S}_1^2 - \hat{S}_2^2]$$

Since these spin variables become large and quasicontinuously distributed in space, one may approximate this expression by

$$\mathscr{H} = J[(\nabla S)^2 + Q(S^2)]$$

where $Q(S^2) = -S^2$, but may also be a more general function of S^2. For instance, by adding an S^4 term one may give Q two minima at spin values $S = \pm 1$, the values they would exclusively have in the Ising model (Fig. 3). The above expression is familiar as the Landau–Ginzburg Hamiltonian. The gradient expression represents the energy of nonuniform magnetization, as already discussed, whereas the Q term prevents S from becoming small. Now on the basis of the Baker Hamiltonian and using methods developed by Wilson and Fisher, one has calculated exactly, although not yet very accurately, the critical indices for various lattice dimensionalities. For instance, the critical index γ for the 3D Ising model susceptibility $\chi = (T - T_c)^{-\gamma}$ is calculated as $\gamma = 1 + (1/6)\varepsilon + (25/324)\varepsilon^2$, where $\varepsilon = d - 4$, d being the lattice dimensionality. The result for $d = 3$ is then $\gamma = 1.244$, very close to the best numerical estimate $\gamma = 1.250 \pm 0.003$, whereas for hypothetical 4D and higher-dimensionality lattices the molecular field prediction $\gamma = 1$, $\chi = (T - T_c)^{-1}$, i.e., the Curie–Weiss law is obtained. Furthermore, for $T - T_c$ quite small many details of the Hamiltonian can be varied without affecting the critical exponent if only the Landau–Ginzburg-type Hamiltonian is retained at every level of scaling upward. In particular, the lattice structure should have little influence on the critical behavior if the scaling concept is valid.

It would be outside the scope of this review to discuss critical phenomena further. However, it should be emphasized that the extensive computation of terms in series expansions for the partition function, susceptibility, etc., for various crystal

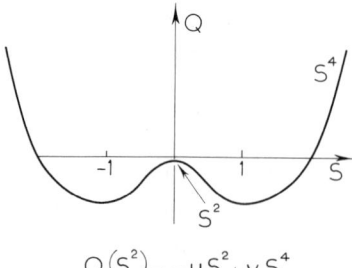

$Q(S^2) = -uS^2 + vS^4$

Fig. 3. Second term in the Landau–Ginzburg Hamiltonian.

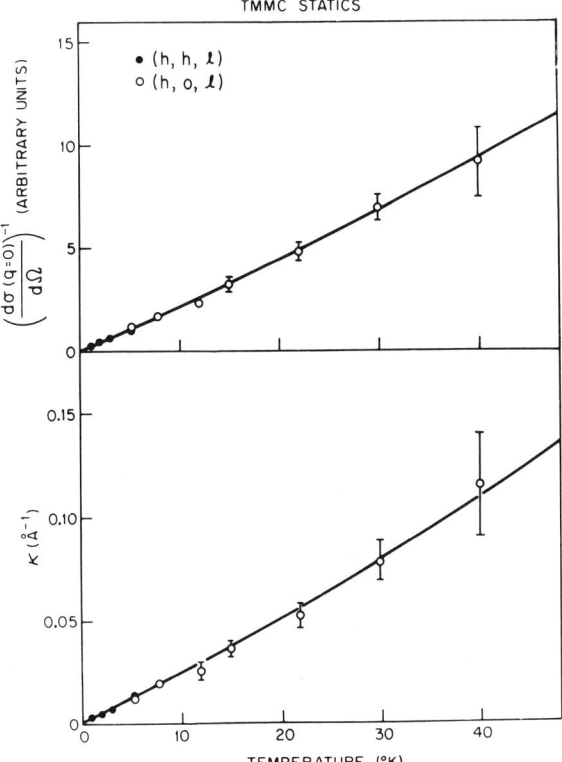

Fig. 4. Upper graph shows the inverse of neutron scattering differential cross section as a function of temperature while lower graph gives inverse of correlation length in tetramethylammonium manganese chloride (TMMC) as a function of temperature.

structures and Hamiltonians (Heisenberg, Ising, XY, and intermediate forms) has produced very useful results in the last few years.

Comparison with Experiment

In order to illustrate the influence of dimensionality, we shall give some examples of 1D and 2D spin systems, and then concentrate on 3D Ising systems in the remainder of this paper.

One-Dimensional Systems

Correlation Length. Do correlations really grow when the critical point is approached, as was assumed in the scaling predictions? Neutron scattering is a fairly direct method for determining the correlation length in magnetic compounds. The quasielastic scattering intensity distribution measured at temperatures slightly above the critical point is rather simply related to the correlation length.

For a linear chain of classical spins having $T_c \approx 0$ as its critical point the correlation length ξ should increase as $\xi \propto (T - T_c)^{-1} = T^{-1}$. Figure 4 shows that for

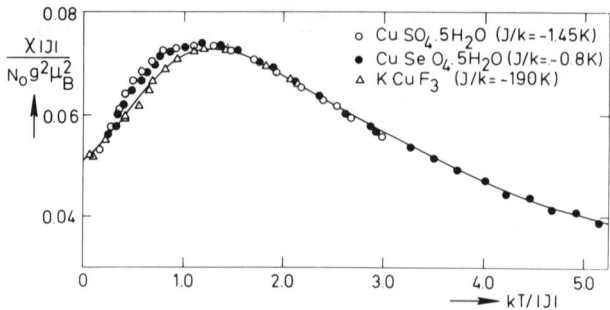

Fig. 5. Susceptibility χ of three linear chains in reduced units, χJ vs. kT/J, where J is the exchange constant.

$(CD_3)_4$ NMnCl$_3$ (or TMMC) this is experimentally verified with fairly good precision over a wide temperature region.[4]

Magnetic Susceptibility. For linear chains of spins, whose interactions can be reasonably well described by the Heisenberg model, the susceptibility as a function of temperature has the behavior depicted in Fig. 5. For two compounds, namely CuSO$_4$ · 5H$_2$O and CuSeO$_4$ · 5H$_2$O, Miedema and Haseda[30] measured the specific heat and found that these compounds have Heisenberg linear chain characteristics. The susceptibility data are derived from proton NMR at low temperatures (by Wittekoek[31] and Hirakawa[32]) and the line drawn in Fig. 5 represents the theoretical calculations of Bonner and Fisher.[33] These calculations are not exact, however.

Two-Dimensional Systems

2D Ising. We shall first mention a few experimental results on 2D Ising systems. First we show the susceptibility data on K$_2$CoF$_4$ (Fig. 6) of Breed et al.[34] In K$_2$CoF$_4$ the χ data had to be corrected for a rather large Van Vleck susceptibility and furthermore the Ising model may be only approximately realized. Figure 7 shows the experimental points for the heat capacity of CoCs$_3$Br$_5$ compared to the exact calculation of Onsager (solid curve). However, the 2D character of CoCs$_3$Br$_5$ is question-

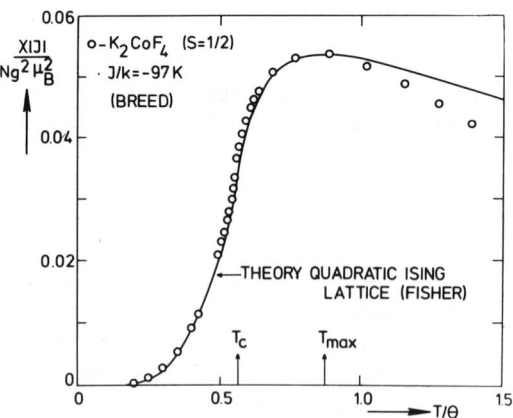

Fig. 6. Susceptibility of two-dimensional Ising-type antiferromagnet compared to Fisher's calculated curve.

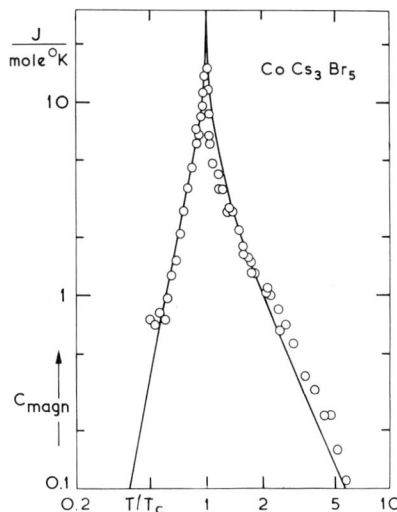

Fig. 7. Heat capacity of $CoCs_3Br_5$ as compared to Onsager's theory (solid curve).

able in view of conflicting evidence on measured exchange constants. Similar results have been reported on Co formate by Takeda et al.[35] We conclude that good 2D Ising systems have yet to be found.

2D Heisenberg. As to 2D Heisenberg spin systems, we mention in particular the work of Miedema, De Jongh, and co-workers[5] on ferromagnetic layers of Cu ions, separated by organic molecules of variable length.[5] The crystal structure (Fig. 8) favors the 2D character, and the interchain interaction can be made virtually negligible. Before discussing results obtained on these compounds, a few remarks should be made on the theoretical knowledge on 2D Heisenberg systems. There is

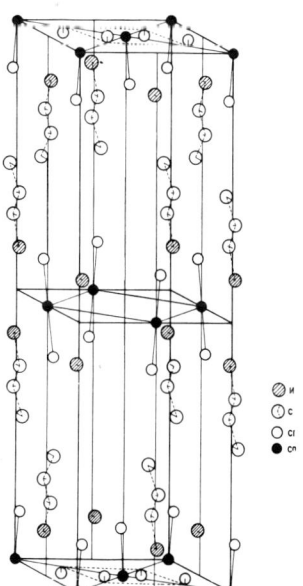

Fig. 8. Crystal structure of $(C_2H_5NH_3)_nCuCl_4$, showing magnetic layers interspaced by layers of organic molecules, of which the length can be varied ($n = 2$–10).

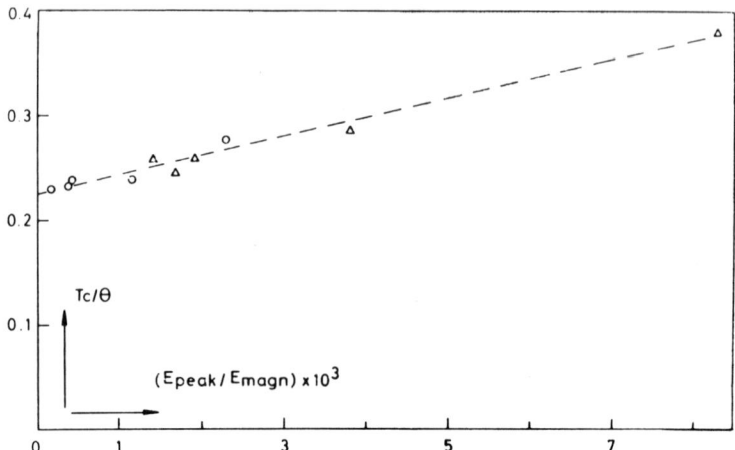

Fig. 9. Transition temperature T_c of 2D Heisenberg ferromagnet vs. anisotropy parameter. At vanishing anisotropy a finite Stanley–Kaplan transition temperature may be inferred from the data. T_c is divided by the Curie–Weiss constant θ.

the proof of Mermin and Wagner[6] that the 2D Heisenberg system does not sustain long-range order above $T = 0$. However, the susceptibility series diverges (as for a 3D ferromagnet) at a finite temperature. This has led to the suggestion by Stanley and Kaplan[7] that a new type of phase transition would occur at T_c such that the susceptibility becomes infinite but the magnetization remains zero. Below T_c the correlation length among the spins would be rather large but not infinite. Under such conditions it is plausible that a small amount of anisotropy introduced into the 2D Heisenberg model may trigger the phase transition into a truly long-ranged ordered state. (This is comparable to the situation encountered with TMMC. It is known that TMMC shows an extremely sharp phase transition of a truly 3D character at $T = 0.84°K$.) Such anisotropy arises, by example, from dipolar interactions between the layers, which can be varied by increasing the interlayer distance. It is found experimentally that the critical temperature of this series of compounds does not go to zero if the anisotropy parameter is made very small. The Miedema group concludes from their data (Fig. 9)[36] that in the limit of very large interplane separations there is still a finite transition temperature, to be identified with Stanley–

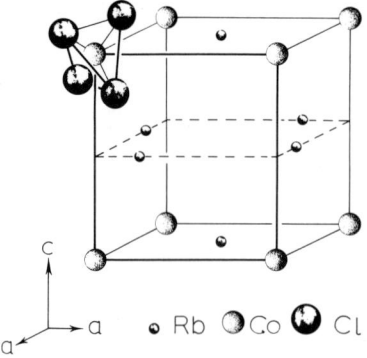

Fig. 10. Crystal structure of $CoRb_3Cl_5$.

Kaplan transitions. At large interplane separations it is found from the heat capacity data that practically only short-range ordering occurs.

Three-Dimensional Ising Systems. We shall first discuss why Ising systems are found in real crystals, and we take $CoCs_3Cl_5$ as an example. Cobalt has spin 3/2 in the cubic crystalline-field potential arising from a tetrahedral Cl environment (Fig. 10). A small distortion of the tetrahedron creates an axial crystalline field, which causes the $S_z = \pm 3/2$ states to be energetically lower than the $S_z = \pm 1/2$ states. The energy difference is found from EPR, optical, and specific heat measurements to be about $\Delta/k = 12.4$ K. Hence at sufficiently low temperature only $S_z = \pm 3/2$ is populated. These two states can be described by an effective spin 1/2 having very anisotropic g values, $g_z = 7.2$ and $g_\perp \approx 0$. (All Co ions are magnetically equivalent.) Consequently, the magnetic exchange interaction can only occur among the axial z components of the magnetic moments; hence $\mathcal{H} = -2JS_{1z}S_{2z}$. The neglect of $J(S_{1x}S_{2x} + S_{1y}S_{2y})$ is valid provided $S_z = \pm 1/2$ states are not mixed into the $S_z = \pm 3/2$ states. This requires $J \ll \Delta$ and $T \ll \Delta$, which is realized in the compounds just mentioned. The Co ions are arranged at the corners of a tetragonal unit cell. ESR on pairs of Co ions in the isomorphous Zn compound has shown that the exchange interaction $J_\perp$ among nearest neighbors in the aa plane is somewhat smaller than the interaction J_c of neighbors along the c axis. Adding the nn dipolar interaction to the exchange makes $J_\perp \approx J_c$. This corresponds reasonably well to the simple cubic structure, i.e., the interaction is equally strong among six nearest neighbors lying at mutually orthogonal directions. However, there may be further than nearest-neighbor interactions.

The experimental data on $CoCs_3Cl_5$[8] and on $CoRb_3Cl_5$[9] plotted on a reduced temperature scale (Fig. 11) agree rather well with the calculated results for a simple cubic lattice. The $CoCs_3Cl_5$ results show some deviation at both the low- and high-temperature sides. The long-range dipolar interaction can spoil the agreement between the theoretical nearest-neighbor Ising model and experiment and this would be more noticeable in the Cs compound than in the Rb compound, since the latter has a higher T_N.

More recently an even better 3D Ising system has been found in $DyPO_4$. The Dy ion has a nearly pure $J_z = \pm 15/2$ ground doublet, which is describable in terms of effective spin 1/2, and $g_\| = 19.4$ and $g_\perp = 0.51$. The antiferromagnetic interactions lead to $T_N = 3.39°$K, which is quite low compared to the next higher doublet, which lies at $\Delta E/k \approx 100$ K above the ground state. The Dy ions are arranged in a nearly tetrahedral lattice (Fig. 12). The specific heat has been measured by Colwell et al.[10] and is given in Fig. 13. Also given in Fig. 13 are the theoretical heat capacity c/R curves for $T > T_c$ based on series expansion coefficients of Sykes,[37] and the low-temperature ($T < T_c$) c/R as computed from a series expansion given by Baker[38] for the tetrahedral lattice. It is seen that the agreement between theory and experiment is quite good. Although the general shape of the curve for a simple cubic lattice and that of the tetrahedral (diamond) lattice are the same, on closer inspection noticeable differences exist. Hence one can differentiate between the results for various lattice structures, so that a satisfactory state of affairs is reached for the heat capacity of simple Ising systems, both theoretically and experimentally. Other properties of $DyPO_4$, for example the susceptibility, also agree with Ising model

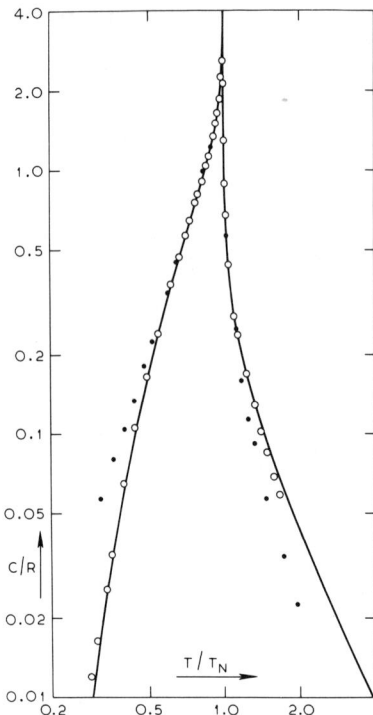

Fig. 11. Heat capacity of $CoRb_3Cl_5$ (circles) and $CoCs_3Cl_5$ (dots). The curve is calculated from a series expansion for the simple cubic Ising model.

predictions (see Koonce et al.[11] and Wright et al.[12]). One can conclude that the Ising model is realized in typical quantum systems (i.e., discrete spin value) and that to reveal its properties low temperatures are required for the purpose of bringing the system into a single quantum state. Furthermore, experimentally, low-temperature measurements obviate the need for subtraction of lattice heat capacities.

Phase Boundary

Antiferromagnetic compounds in the presence of magnetic fields often exhibit properties, represented by phase diagrams, which can be related to other subjects in low-temperature physics, e.g., mixtures of ^{3}He–^{4}He.

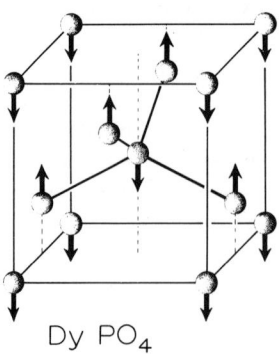

Fig. 12. Crystal and magnetic structure of $DyPO_4$.

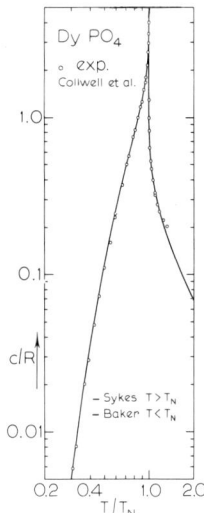

Fig. 13. Heat capacity of DyPO$_4$ compared to series expansion results for the diamond lattice.

Ising antiferromagnets ideally should have only the antiferromagnetic-to-paramagnetic phase boundary. They do not possess the intermediate spin-flop phase usually encountered in Heisenberg systems. In Heisenberg antiferromagnets anisotropy in single-ion properties or exchange may be expressed as an anisotropy field H_A which is small compared to the exchange field. In the Ising model the anisotropy field is considered to be infinite. Of course, in reality many intermediate cases occur, such as FeCl$_2$ and Ni(NO$_3$)$_2 \cdot$ 2H$_2$O, and these are given the name metamagnets. The phase boundary may be determined by measuring the magnetization as a function of field, by magnetocaloric measurements, etc. More recently ultrasonic attenuation and resistivity measurements have been successfully utilized; Fig. 14 gives a beautiful example of the present capabilities for tracing the phase boundary accurately in high fields and at low temperatures. Both techniques rely on the difference in the scattering of phonons and electrons in the paramagnetic and antiferromagnetic phase.[13]

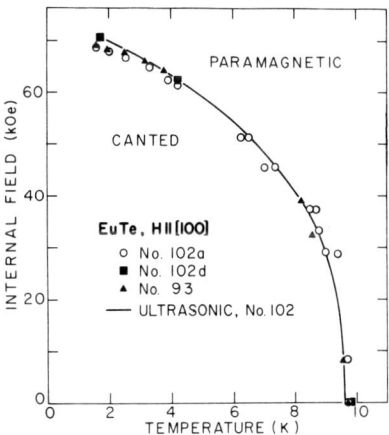

Fig. 14. Phase boundary determined by ultrasonic and resistivity measurements (data points) on EuTe.[13]

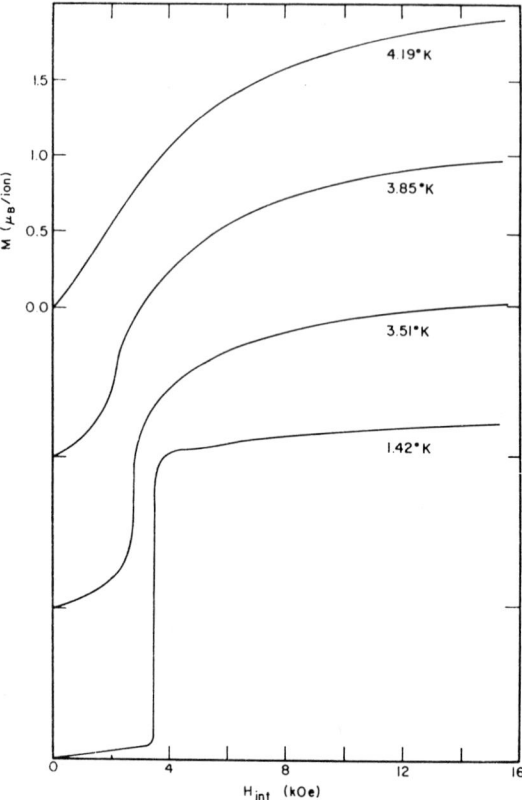

Fig. 15. Magnetization vs. temperature in $Ni(NO_3)_2 \cdot 2H_2O$. Below about 3.85°K a discontinuity in M is observed, corresponding to a first-order phase transition.

Recently interest in the phase boundary has been stimulated by discussions regarding the tricritical point. Let us first recall that Néel, Gorter, Kanamori, and others had predicted that the phase boundary in metamagnets may be first order for $T > 0$ but changes into a second-order phase transition (continuous magnetization) for T approaching T_N. Figure 15 illustrates this with results on $Ni(NO_3)_2 \cdot 2H_2O$.[14] At low T one sees a discontinuous rise of the magnetization, but at $T > 3.85°K$ there is only an inflection point in the m vs. T curve.

Now let us see what the theory has to say about a spin-1 system. Suppose there is an anisotropy effect, expressed as a ΔS_z^2 term, which causes an energy difference between (in first order) a nonmagnetic singlet and a $S_z = \pm 1$ doublet. Suppose interactions occur among the magnetic $S_z = \pm 1$ nearest neighbors. This interaction is necessarily of the Ising type $-JS_i^z S_j^z$ if $J \ll \Delta$, but we assume it is more generally valid. Now it was shown by Capel[39] that for certain values of the Δ/J ratio the phase transition is first order, i.e., different in character from that to be expected for the Ising model only, i.e., $\Delta = -\infty$. The reason, of course, is that the system may condense into the magnetically ordered state if the temperature is sufficiently low to make the molecular field on the $S_z = \pm 1$ states so large that, for instance, the

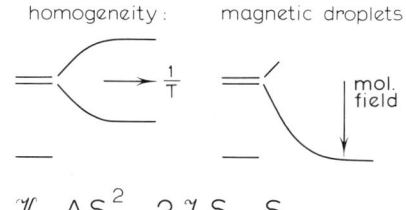

Fig. 16a. Formation of magnetic droplets energetically favored.

$$\mathcal{H} = \Delta S_z^2 - 2J S_{z1} S_{z2}$$

$S_z = -1$ state falls below the $S_z = 0$ state (Fig. 16a). This could be associated with the formation of magnetic droplets, since ions in $S_z = \pm 1$ states would tend to be surrounded by $S_z = \pm 1$ ions rather than $S_z = 0$ ions in order to increase the molecular field. Hence we obtain a phase boundary in the Δ–T plane (Fig. 16b). Now the left part of the phase boundary represents the second-order transition from paramagnetic to antiferromagnetic phase, while the right part represents transition to the coexistence region (droplet formation).

Recently Blume et al.[15] applied this model to the ^{3}He–^{4}He system. One replaces the $S_z = \pm 1$ ions by ^{4}He atoms, the $S_z = 0$ ions by ^{3}He atoms, the parameter Δ by the Gibbs free energy difference between ^{3}He and ^{4}He atoms, and the exchange interaction J by the attractive potential for ^{4}He particles to condense from the normal phase ($S_z = \pm 1$) into the superfluid phase ($S_z = -1$). Below a finite temperature T_{tric} a first-order transition occurs which leads to a separation into two phases.

The series expansion procedures which have proved quite successful in predicting critical temperatures in simple Ising models may also be utilized in the aforementioned, somewhat more complicated case. Oitmaa[16] has calculated part of the phase boundary by means of a series expansion for the susceptibility of such an Ising system. The divergence of the susceptibility at a certain temperature for arbitrary Δ defines the phase boundary. He did not get information about the first-order part. Undoubtedly, further results will be forthcoming, such as calculations for critical exponents which can be defined around the tricritical point.

Finally, it should be pointed out that similar behavior of the phase boundary is to be expected for antiferromagnets, in which the ions have ferromagnetic nnn interaction in addition to the antiferromagnetic nn interaction. Experimental results of Landau et al.[17] are shown in Fig. 17.

In this section we have discussed some attempts to approach the study of phase

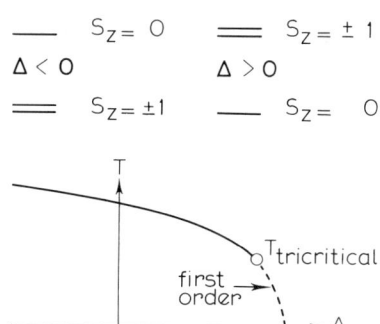

Fig. 16b. Phase boundary for $S = 1$ system having anisotropy parameter Δ.

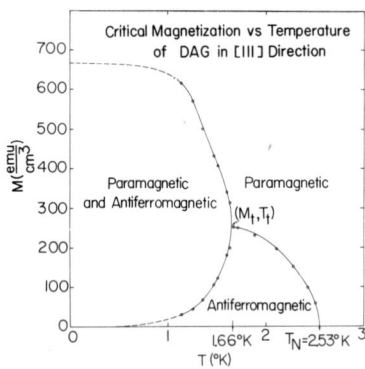

Fig. 17. Phase diagram for DAG, M vs. T.

boundaries following procedures which have been successfully employed for the Ising model calculations of critical phenomena. We have also seen that there are only a few detailed experimental investigations. These have to be carried out at low temperatures if one wants to restrict oneself to magnetic fields in the 100-T range or below. Clearly, however, we have only made a few steps beyond the ground covered by the work of, for instance, Ehrenfest, Landau, and Tisza (see Ref. 40).

Other Dichotomic Systems

Nearly Degenerate Singlets. The Ising model is characterized by a variable which can take two values, a dichotomic variable or pseudospin variable. This can be utilized in a variety of systems, as discussed earlier, and would also be appropriate, for instance, in the order–disorder transition of NH_4Cl. We shall discuss some other examples studied at low temperatures beginning with the phase transition in non-Kramers compounds having a near-degenerate doublet ground state consisting of two singlets, e.g., Pr, Tb, and Tm salts or intermetallic compounds. Suppose we consider Tm^{3+} ions in an axial crystalline field, in which the $J_z = |\pm 6\rangle$ state would be lowest, represented by effective $S_z = \pm 1/2$. Then, by a lowering of the crystal symmetry, two nonmagnetic singlets are formed from the $S_z = \pm 1/2$ states by the familiar linear combinations $S_z = |1/2\rangle + |-1/2\rangle$ and $S_z = |1/2\rangle - |-1/2\rangle$. This can be formally accomplished by adding a $\Gamma S_x = (\Gamma/2)(S_+ + S_-)$ term to the Hamiltonian, where Γ is the singlet–singlet splitting. This is formally equivalent to adding a Zeeman term to the Hamiltonian, which arises from a transverse (pseudo) field, i.e., transverse to the axis of preferential alignment of the spins. The original states, $J_z = \pm 6$, presumably would have nearest-neighbor exchange interaction, or in any case dipolar interaction. Clearly these interactions would occur among the axial components of the magnetic moments only, and therefore the Ising model $-JS_i^z S_j^z$ would be applicable. This would lead at low T to magnetic ordering. Now the ΓS_x term, on the other hand, tends to make the system nonmagnetic and therefore we have a competition between the magnetic coupling J and the effect of the "transverse" crystal field Γ. Hence the critical temperature decreases with Γ (Fig. 18). Series expansions of the susceptibility of such a system are able to mark the critical points as a function of the Γ parameter. This has been computed[18,19] for simple cubic and simple square lattices (3D and 2D). The resulting line may be considered

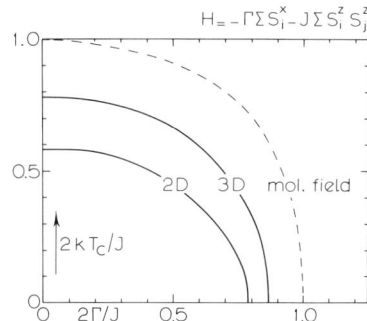

Fig. 18. Computed phase boundary for Ising system having transverse field term, e.g., for a (nearly degenerate) two-singlet antiferromagnet.

as a phase boundary: for low T and Γ one is in the magnetically ordered state, for high Γ or high T there is no ordering and, for example, in the heat capacity only a Schottky anomaly for two energy levels would remain.

Let us look at one experimental result[20] on such a system, $Tm_2(SO_4)_3 \cdot 8H_2O$ (Fig. 19). One sees a competition between short-range ordering (Schottky anomaly for two singlets) and long-range ordering arising from the magnetic interaction. It might be of interest to study such systems in the presence of real magnetic fields, thereby adding a $g\beta HS_z$ term to the spin Hamiltonian.

At this stage our models may look somewhat artificial, since it is argued that we study phase transitions in a "transverse" field, which cannot be easily varied. Let us therefore look at a phase transition occurring in a real magnetic field where a pseudospin variable is also at first sight arbitrarily introduced. We refer to measurements of Van Tol et al.[21] on the heat capacity of $Cu(NO_3)_2 \cdot 2\tfrac{1}{2}H_2O$. This compound has dimers of Cu ions, which couple antiferromagnetically into a $S = 0$ ground state and a $S = 1$ higher state. A magnetic field of about 36 kOe causes a crossing of the $S_z = -1$ and $S = 0$ levels (Fig. 20). At the crossing point we have a pseudo spin-1/2 system, in which interactions occur predominantly along linear chains. Specific heat measurements in a 35.7-kOe field show a perfectly sharp phase transi-

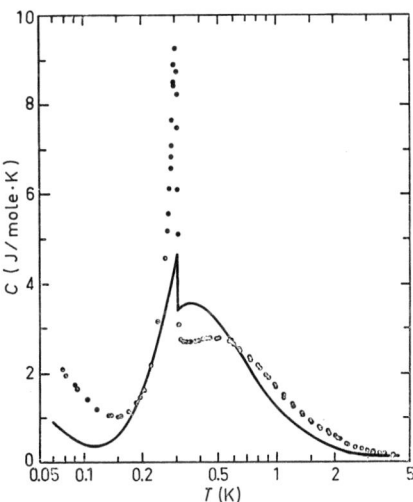

Fig. 19. Heat capacity of $Tm(SO_4)_3 \cdot 8H_2O$, which is a two-singlet compound.

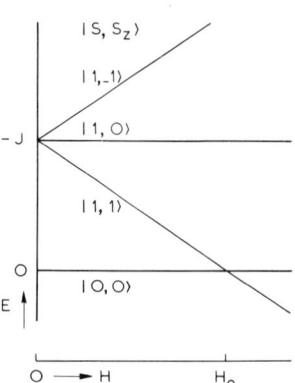

Fig. 20. Zeeman splittings for Cu dimers in $Cu(NO_3)_2 \cdot 2\frac{1}{2}H_2O$. The crossing field $H_0 = 35.7$ kOe.

tion (Fig. 21), which gradually vanishes for smaller or larger fields. The entropy change corresponds to $S = 1/2$, for a pair.

Jahn–Teller Distortion. Recently heat capacity measurements on crystals such as $DyVO_4$, $TbVO_4$, $TmVO_4$, have shown that phase transitions occur which are not due to magnetic interactions but arise from Jahn–Teller distortions. This has been shown by Cooke and co-workers[22,23] for $TmVO_4$ (Fig. 22), in which there is no magnetic phase transition at all, and in the case of $DyVO_4$, where in addition to the Jahn–Teller phase transition a magnetic phase transition also occurs (Fig. 23).

Let us follow the Gehring et al.[24] treatment of $DyVO_4$ as an example. The crystal has tetragonal symmetry and crystal fields are such as to make the aa plane preferential for the Dy spins at low temperatures. Under the influence of external stress the crystal rather easily develops an orthorhombic distortion such that either of two states are preferred, namely $J_a = |\pm 15/2\rangle$, or $J_b = |\pm 15/2\rangle$. In the absence of strain the J_a and J_b states are equivalent. The possibility of tunneling from one state to another at high temperatures produces a JT splitting $\Gamma = \hbar\omega = 9$ cm^{-1}, where ω is the tunneling frequency. This situation is formally described by $S_z =$

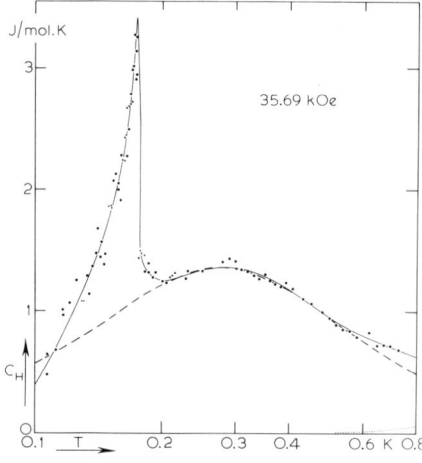

Fig. 21. Heat capacity of $Cu(NO_3)_2 \cdot 2\frac{1}{2}H_2O$ magnetic field H_0.

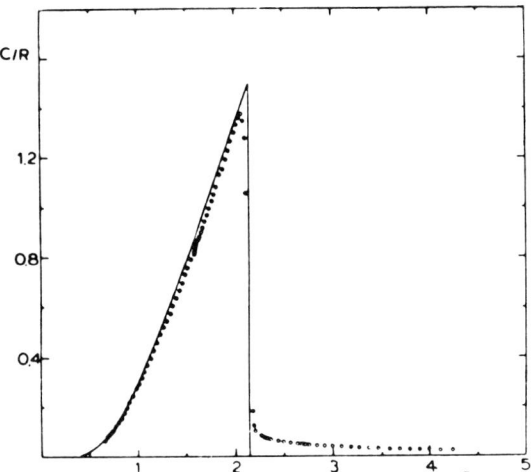

Fig. 22. Cooperative Jahn–Teller transition in TmVO$_4$.

$-1/2$ representing the $J_a = |\pm 15/2\rangle$ states and $S_z = \pm 1/2$ representing the J_b states. The tunneling term ΓS_x produces mixing of the two states. The question we are concerned with here is: Why does a crystallographic phase transition into orthorhombic symmetry occur when cooling the crystal, i.e., why does the crystal order into, for instance, the $S_z = -1/2$ state?

It should be realized that phonons may be operative to create local distortions at a particular lattice site i, which then couples to a distortion at a neighboring site j in such a way as to make these distortions energetically most favorable, e.g., both along the a axis, or equivalently both along the b axis. The local distortion would, in the pseudo spin formalism, be represented by a S_i^z term (for a uniform strain in the crystal the index i can be dropped) and the coupling can be represented by $JS_i^z S_j^z$, just as in the Ising model.

This interaction would, if J is sufficiently larger than Γ, cause phase transitions. The nn approximation suggested here would lead to appreciable short-range ordering. The heat capacity data do indeed show some short-range ordering. On the other

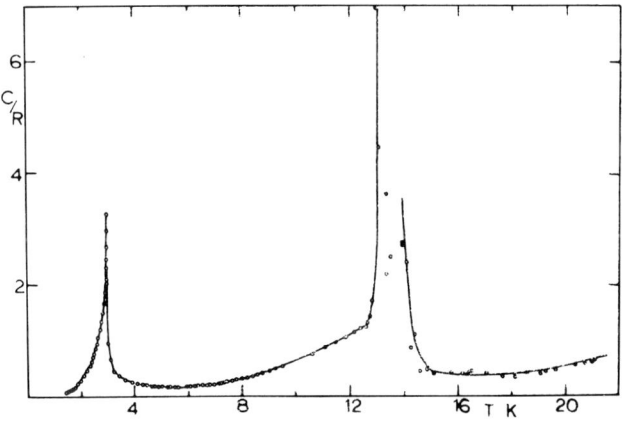

Fig. 23. Jahn–Teller transition and magnetic phase transition in DyVO$_4$.

hand, it is not surprising that JT distortions have longer range than purely magnetic interactions. Hence the molecular field approximation, as shown for $TmVO_4$, is in many respects quite sufficient. In $DyVO_4$, however, the analysis of Gehring et al.[24] showed that the phase transition temperature is appreciably lower than the molecular field estimate.

We shall not treat this subject further, except for mentioning that (1) external stress and magnetic fields may influence the transition and their interplay lead to pseudospin-flop phenomena at phase boundaries; and (2) the influence of stress can be experimentally well determined because these JT transitions occur at low temperatures, corresponding to relatively small JT splittings.

We conclude that measurements at low temperatures have disclosed an interesting field of cooperative JT distortions and that the accompanying phase transitions can now be quantitatively studied along the same theoretical lines as and on the basis of the experience gained in magnetic transitions.

Phase Transitions in Dipole–Dipole-Coupled Spin Systems

Electronic Moments. In quite a number of rare earth ionic compounds the exchange interactions are quite small, so that dipolar interactions dominate. It has been questioned whether such systems would sustain long-range order and whether they tend to become ferromagnetic or antiferromagnetic. Experiments have to be carried out at quite low temperatures since dipolar couplings are rather weak and hence transition temperatures usually fall below 1°K.

Let us first consider cerium magnesium nitrate (CMN). It is known that exchange interactions in CMN are negligible or, at most, quite small compared to dipolar interactions. The susceptibility obeys Curie's law approximately but at a temperature of about 2 m°K there is a maximum corresponding to a constant minimum in $1/\chi$. That minimum value depends on the crystal shape and equals the demagnetizing factor N for an ellipsoidal sample. This agrees with what one expects for a ferromagnetically ordered sample split in long domains. Furthermore, the susceptibility as a function of field shows a rapid decrease for the elongated ellipsoidal sample, whereas for the spherically shaped sample the decrease occurs at approximately $(4\pi/3) M$, i.e., the demagnetizing field (Fig. 24). The decrease, of course, is due to the approach of the saturation magnetization at low T.

Other systems are known to behave in the same way, e.g., recently the specific heats of $DyCl_3 \cdot 6H_2O$ (Fig. 25) and $ErCl_3 \cdot 6H_2O$ were investigated by Lagendijk et al.[25] On the basis of the heat capacity data one can determine the magnetic energy at $T = 0$ rather accurately for $ErCl_3 \cdot 6H_2O : E/R = -0.272$ K.

Now it is theoretically possible to obtain a rather reliable estimate of the energy at $T = 0$ by the so-called Luttinger–Tisza method, later used by Daniels and Felsteiner, Meyer, Niemeyer, and others (see Ref. 41). It is assumed that the magnetic structure is described by a Bravais lattice. Helical spin structures, for instance, cannot be treated by this method. Hence once the directions of the spins, e.g., at the corners of a simple cubic Bravais lattice, are known, translational operations define the directions of all spins in the entire lattice. In particular for crystals having single ion anisotropy, e.g., all spins pointing either in $+z$ and $-z$ direction (Ising type), this method should be very close to reality. However, such calculations prac-

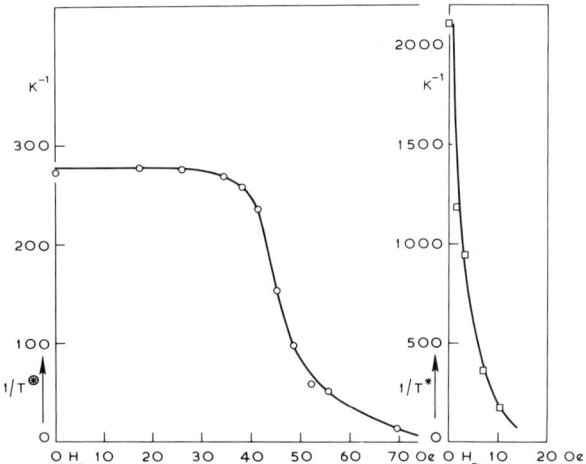

Fig. 24. Susceptibility of CMN vs. magnetic field for a spherical sample (left) and an elongated ellipsoid (right).

tically always lead to the result that the lowest-energy state corresponds to antiferromagnetic spin structures. Only by including the possibility of domain formation, i.e., by subtracting the demagnetization energy $\frac{1}{2}NM^2$ (where M is the saturation magnetization) are ferromagnetic states competitive. In fact, in $ErCl_3 \cdot 6H_2O$ the calculations show a ferromagnetic state and an antiferromagnetic structure to have

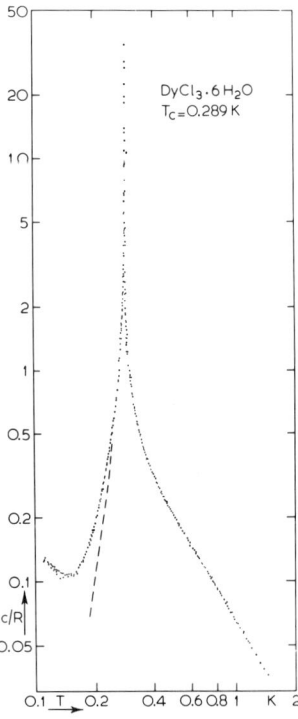

Fig. 25. Specific heat of $DyCl_3 \cdot 6H_2O$.

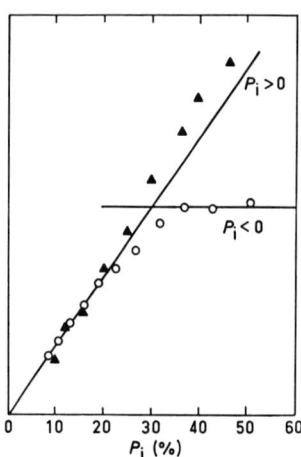

Fig. 26. Nuclear magnetic susceptibility vs. nuclear polarization in CaF_2 in the microdegree range. The circles correspond to negative temperatures and show a maximum in the susceptibility.

nearly equal energies, $E/R = -0.246$ K and -0.244 K, respectively. Experiments on $ErCl_3 \cdot 6H_2O$ indicate that ferromagnetism is indeed realized. In such cases even a minute exchange contribution may be decisive. Since dipolar interactions have scaling properties, further investigation of dipolar interactions would be of interest in connection with remarks made in the theory section on critical behavior for Baker-type Hamiltonians.

Nuclear Moments in Diamagnetic Crystals. Spin systems which are perfectly dipolar may be found among nuclear magnetic moments in insulators. Here, of course, the pioneering experiment of Abragam et al.[26] should be mentioned. They showed that the nuclear magnetic susceptibility of CaF_2 at exceedingly low temperatures reaches a maximum provided that the experiment is started by polarizing the nuclear moments in a direction opposite to the external field (Fig. 26). That is, they prepare their spin system in a negative-temperature state. If their polarization is large enough, they see indications of magnetic ordering. Again on the basis of Luttinger–Tisza calculations, they propose for the field along the cubic crystal [001] axis two magnetic structures, both antiferromagnetic, one having the lowest energy of all the structures and the other having the highest energy. The latter state will obviously be populated at negative temperatures. For a field along the [110] axis there is a different antiferromagnetic structure which has lowest energy, etc. (Fig. 27). It is not yet entirely clear whether domain formation occurs in these systems.

Nuclear Moments in Van Vleck Paramagnetic Substances. Various intermetallic compounds of rare earth metals have the property that the rare earth ground state is a nonmagnetic singlet which exhibits an appreciable Van Vleck paramagnetism due to the rather close proximity of higher electronic states. In such cases the singlet can be substantially polarized by an external field or by coupling to other magnetic moments. We shall consider in particular the hyperfine coupling of the nuclear magnetic moment, which causes some polarization of the singlet, or in other words, creates an electronic magnetic moment. In some intermetallic compounds Andres, Bucher and co-workers[27] found that the nuclear magnetic moment is enhanced

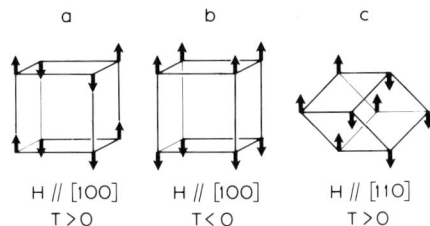

Fig. 27. Lowest-energy configurations of fluorine nuclear magnetic moments in a vertical external field.

by as much as a factor of ten by the induced electronic moment. Even if these moments interacted only by dipolar interactions, phase transitions would occur at a much higher temperature than, e.g., in CaF_2. However, in addition to this dipolar coupling, one also has RKKY interaction of the induced electronic moments, which gives appreciable long-range coupling and can cause phase transitions. For the interesting results obtained so far we refer to the original publications, and we shall mention here only one result, the nuclear ordering in $PrCu_2$ (Fig. 28).

Conclusion

Magnetic phase transition studies at low temperature are a fruitful field since results on various spin systems can be utilized to study various exactly solvable models of statistical mechanics. This is because (a) low temperatures select a single quantum ground state, which sometimes can be shown to obey the Ising Hamiltonian, for which exact results are most extensively available; (b) low-temperature phase transitions can be influenced relatively strongly by applied fields, stress, etc.; and (c) at low temperatures the influence of lattice vibrations can in many cases be neglected (in contrast, e.g., to ferroelectric systems) so that statistical theories for fixed lattices are applicable.

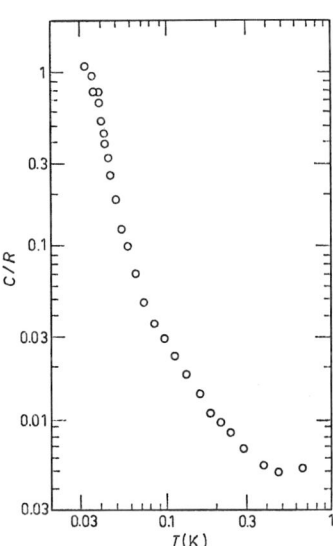

Fig. 28. Nuclear magnetic moments of Pr ordering in $PrCu_2$ at 54 m°K by cooperative interaction via the RKKY interaction and hyperfine enhancement of the nuclear magnetic moments.

References

1. K.G. Wilson, *Phys. Rev. B* **4**, 3174, 3184 (1971).
2. K.G. Wilson and M.E. Fisher, *Phys. Rev. Lett.* **28**, 240 (1972).
3. G.A. Baker, *Phys. Rev. B* **5**, 2622 (1972).
4. R.J. Birgeneau, R. Dingle, M.T. Hutchings, G. Shirane, and S.L. Holt, *Phys. Rev. Lett.* **26**, 718 (1971).
5. J.L. De Jongh and A.R. Miedema, *Adv. Phys.* (1973) (to appear).
6. N.D. Mermin and H. Wagner, *Phys. Rev. Lett.* **17**, 1133 (1966).
7. H.E. Stanley and T.A. Kaplan, *Phys. Rev. Lett.* **17**, 913 (1966).
8. R.F. Wielinga, H.W.J. Blöte, J.A. Roest, and W.J. Huiskamp, *Physica* **34**, 223 (1967).
9. H.J.W. Blöte and W.J. Huiskamp, *Phys. Lett.* **29A**, 304 (1969).
10. J.H. Colwell, B.W. Mangum, D.D. Thornton, J.C. Wright, and H.W. Moos, *Phys. Rev. Lett.* **23**, 1245 (1969).
11. C.S. Koonce, B.W. Mangum, and D.D. Thornton, *Phys. Rev. B* **4**, 4054 (1971).
12. J.C. Wright, H.W. Moos, J.H. Colwell, B.W. Mangum, and D.D. Thornton, *Phys. Rev. B* **3**, 843 (1971).
13. Y. Shapira, S. Foner, N.F. Oliveira, and T.B. Reed, *Phys. Rev. B* **5**, 2647 (1972).
14. V.A. Schmidt and S.A. Friedberg, *Phys. Rev. B* **1**, 2250 (1970).
15. M. Blume, V.J. Emery, and R.B. Griffiths, *Phys. Rev. A* **4**, 1071 (1971).
16. J. Oitmaa, *J. Phys. C* **4**, 2466 (1971); *J. Phys. C* **5**, 435 (1972).
17. D.P. Landau, B.E. Keen, B. Schneider, and W.P. Wolf, *Phys. Rev. B* **3**, 2310 (1971); *Phys. Rev. B* **5**, 4472 (1972).
18. R.J. Elliott and C. Wood, *J. Phys. C* **4**, 2359 (1971).
19. P. Pfeuty and R.J. Elliott, *J. Phys. C* **4**, 2370 (1971).
20. T.E. Katila, N.E. Phillips, M.C. Veuzo, and B.B. Triplett (to be published).
21. M.W. Van Tol, K.M. Diederix, and N.J. Poulis, *Physica* **64**, 363 (1973).
22. A.H. Cooke, S.J. Swithenby, and M.R. Wells, *Sol. State Comm.* **10**, 265 (1972).
23. A.H. Cooke, C.J. Ellis, K.A. Gehring, M.J.M. Leask, D.M. Martin, B.M. Wanklyn, M.R. Wells, and R.L. White, *Sol. State Comm.* **8**, 689 (1970).
24. G.A. Gehring, A.P. Malozemoff, W. Staude, and R.N. Tyte, *J. Phys. Chem. Sol.* **33**, 1487, 1499 (1972).
25. E. Lagendijk and W.J. Huiskamp, *Physica* **65**, 118 (1973).
26. M. Chapellier, M. Goldman, V.H. Chau, and A. Abraham, *J. Appl. Phys.* **41**, 849 (1970).
27. K. Andres and E. Bucher, *Phys. Rev. Lett.* **28**, 1652 (1972).
28. K.W. Mess, J. Lubbers, L. Niesen, and W.J. Huiskamp, *Physica* **41**, 260 (1969).
29. J. Stephenson and D.D. Betts, *Phys. Rev. B* **2**, 2702 (1970).
30. A.R. Miedema, H. van Kempen, T. Haseda, and W.J. Huiskamp, *Physica* **28**, 119 (1962).
31. S. Wittekoek, T.O. Klaassen, and N.J. Poulis, *Physica* **39**, 293 (1968).
32. K. Hirakawa, I. Yamada, and Y. Kurogi, *J. Phys. (France)* **32**, C1-890 (1971).
33. J.C. Bonner and M.E. Fisher, *Phys. Rev.* **135**, A·640 (1964).
34. D.J. Breed, K. Gilijamse, and A.R. Miedema, *Physica* **45**, 205 (1969).
35. K. Takeda, T. Haseda, and M. Matsuura, *Physica* **52**, 225 (1971).
36. P. Bloembergen, K.G. Tan, F.H.J. Lefevre, and A.H.M. Bleyendaal, *J. Phys. (France)* **32**, C1-878 (1971).
37. M.F. Sykes, D.L. Hunter, D.S. McKenzie, and B.R. Heap, *J. Phys. A* **5**, 667 (1972).
38. G.A. Baker, *Phys. Rev.* **129**, 99 (1963) and private communication.
39. H.W. Capel, *Physica* **37**, 423 (1967).
40. L. Tisza, *Ann. Phys.* **13**, 1 (1961).
41. Th. Niemeyer, *Physica* **57**, 281 (1972).

8
Phase Transitions

Magnetic Equations of State in the Critical Region*

S. Milošević

Institute of Physics, University of Belgrade
Belgrade, Yugoslavia

and
D. Karo, R. Krasnow, and H. E. Stanley

Physics Department, Massachusetts Institute of Technology
Cambridge, Massachusetts

Introduction

Two of the main advances in the theory of phase transitions were the van der Waals equation of state for gases[1] and Onsager's solution of the two-dimensional Ising model.[2] The scaling hypothesis[3-10] may not be as significant an advance as these two, but it has had a wide impact—and, if nothing else, at least we did not have to wait for the new step forward as long as it took between the thesis of van der Waals (1873) and Onsager's masterpiece (1944)!

The scaling hypothesis makes the prediction that data everywhere in the critical region should collapse onto a single curve, called the *scaling function*. Although this prediction has been confirmed by experimental measurements on a wide variety of magnetic materials,[11-18] there has yet to be a single calculation of the scaling function directly from the Heisenberg Hamiltonian. In this work we first present a general method for calculating scaling functions for theoretical model systems using high-temperature series expansions. We then apply our method and calculate the Griffiths[7] scaling function $h(x)$ for both the Ising and the Heisenberg models. A principal motivation for this work is that the only previous calculation[19] of $h(x)$ was for the Ising model, and the method used for this system cannot be extended to the case of the Heisenberg model (since it relies upon the availability of low-temperature series). Indeed, many experimental measurements of $h(x)$ are on materials whose interactions are not described by the Ising model, and these measurements do *not* agree with the calculated Ising model $h(x)$.

We then consider some specific properties of the scaling function $h(x)$ implied by our results. We also consider the "universality question"[20,21]† by comparing scaling functions calculated from different Hamiltonians on different lattices. We conclude by comparing our calculated scaling functions with experimental data from a wide variety of magnetic materials, and we discuss some forms of presenting

---

* Supported by the National Science Foundation under Grant GP-15428. Invited paper.
† A promising effort at justifying the scaling and universality hypothesis is provided by the renormalization group approach.[21]

data. In particular, the parametric representation of Josephson and Schofield is treated.[22]

Outline of Calculational Method

Our calculational method has two major steps. First we must find a valid approximation for an equation of state in the critical region, and second we must use an appropriate scaling hypothesis to modify the equation of state and produce a scaling function. We will consider in detail the problem of calculating the Griffiths scaling function $h(x)$, but it is important to point out that our method is *not* restricted to the calculation of $h(x)$. If we accept a general scaling hypothesis for thermodynamic functions,[10] then the method can be used to calculate scaling functions for the appropriate part of any thermodynamic quantity when high-temperature series information is available. Furthermore, if we accept a general scaling hypothesis for the static two-spin correlation function[10] as well, then by the same method one can calculate scaling functions for additional quantities related to the correlation function. Similar statements are true for the dynamic two-spin correlation function, and for "scaling with a parameter." Work along these extensions is in progress.

We now present our procedure for calculating the Griffiths scaling function[7]

$$h(x) \equiv H/M^\delta \qquad (1)$$

where

$$x \equiv \varepsilon/M^{1/\beta} \qquad (2)$$

and H, ε, and M denote, respectively, the singular parts of the magnetic field divided by kT_c/μ, the reduced temperature $(T - T_c)/T_c$, and the magnetization divided by its saturation value. The critical point exponents δ and β are defined by the asymptotic relations

$$H \sim M^\delta [\varepsilon = 0] \qquad (3)$$

and

$$M \sim (-\varepsilon)^\beta [H = 0] \qquad (4)$$

We find it sufficient for the Ising and Heisenberg models to use *only* information from the available high-temperature series expansions for the Gibbs potential $G(z, H)$, where H is the magnetic field and z is proportional to the inverse temperature. We require

$$M(z, H) = -(\partial G/\partial H)_z \qquad (5)$$

The series expansion is in the form of a double power series

$$M(z, H) = \sum_{n=0}^{\infty} \sum_{m=0}^{\infty} a_{nm} H^m z^n \qquad (6)$$

where we only know the coefficients for reasonably small orders n and m. We then need a reversion from $M(z, H)$ to a series for $H = H(z, M)$. During the reversion we may keep z and M independent and we find

$$H(z, M) = \sum_{n=0}^{\infty} \sum_{m=0}^{\infty} b_{nm} M^m z^n \qquad (7)$$

Alternatively, we may impose a "constraint" $y \equiv y(z, M)$ and write

$$H(z, y) = \sum_{n=0}^{\infty} \sum_{m=0}^{\infty} c_{nm} y^m z^n \qquad (8)$$

In general, we prefer a reversion that allows us to evaluate exactly coefficients of powers of z. This usually means we choose (7) or (8) with $y = M/z$.

Thus H, the magnetic field, is now a double power series, and we know only the low-order terms. In order to describe the behavior of H in the critical region, we must find a way to extrapolate beyond the low-order terms. To utilize the knowledge that H is zero on the phase boundary (i.e., the spontaneous magnetization curve), we fix $M = c =$ const in (7) or $y = a =$ const in (8), so that the resulting path of constant M or y intersects the phase boundary (see Fig. 1 for paths corresponding to $M = c$ and $y = M/z = c$, the two paths used for all series treated here). We make the rather general assumption that when the phase boundary is approached along the path under consideration H goes to zero according to a simple power law,

$$H(z, \text{const}) = (1 - z/z_0)^q \phi(z) \qquad (9)$$

where z_0 corresponds to the temperature of the path's intersection point with the phase boundary, q is the power, and $\phi(z)$ contains the remaining temperature dependence of H along the path. We find z_0 and q from appropriate Padé approximants or by studying $H^{-1/q}$ with ratio methods, etc. We next find $\phi(z)$ from Padé approximants to $(1 - z/z_0)^{-q} H(z, \text{const})$. This seems adequate because H appears not to have other real singularities in the one-phase critical region.

We now have an expression for $H(z, M)$ calculated along a path in the critical region. We proceed to present a scaling hypothesis that instructs us how to transform $H(z, M)$ into a scaling function $h(x)$. Our method of calculating the scaling function $h(x)$ was discovered after consideration of a recent formulation of the scaling hypothesis that involves the concept of generalized homogeneous functions.[10,23-25] By definition, a function $f(x, y)$ is a generalized homogeneous function if there exist two numbers a and b such that for all positive values of λ the relation

$$f(\lambda^a x, \lambda^b y) = \lambda^p f(x, y) \qquad (10)$$

is satisfied, where the exponent p is called the "scaling power" of the function $f(x, y)$ (one can choose $p = 1$ by redefining $a \to a/p$ and $b \to b/p$). The reader can verify by inspection that the scaling hypothesis in the form of Eq. (1) implies that $H(\varepsilon, M)$ is a generalized homogeneous function in the critical region, since there exist two numbers $a\ [= 1/\beta(\delta + 1)]$ and $b\ [= 1/(\delta + 1)]$ such that for all positive λ one has

$$H(\lambda^{1/\beta(\delta+1)}\varepsilon, \lambda^{1/(\delta+1)}M) = \lambda^{\delta/(\delta+1)} H(\varepsilon, M) \qquad (11)$$

Strictly speaking, Eq. (11) is expected to be true asymptotically for the singular part of the magnetic field H.* However, given equation (11) and choosing λ to be $(c/M)^{\delta+1}$, where c is assumed to be a very small constant, (11) results in

$$c^\delta H(\varepsilon, M)/M^\delta = H(\varepsilon c^{1/\beta}/M^{1/\beta}, c) \qquad (12)$$

---

* While intuitively we identify the singular part of a thermodynamic function with a dominant, changing portion, an analytic definition requires care. See, e.g., Ref. 26 or Ref. 10.

and thereby one may further argue that

$$h(x) = H(xc^{1/\beta}, c)/c^\delta \tag{13}$$

The constant c has been incorporated in order to ensure that arguments of the H function in Eq. (4) are small quantities, since values of H fall in the critical region only if its arguments are small. From Eq. (13) it follows that if we knew the function $H(\varepsilon, M)$, then $h(x)$ could be obtained by replacing the variables ε and M by $xc^{1/\beta}$ and c, respectively. We recall that there may be a constraint between the variables ε and μ and this constraint must be satisfied by the new variables $xc^{1/\beta}$ and c. We will see that a suitable constraint can act to keep the arguments in (13) small even for very large x and thus allow us to represent $h(x)$ by a single expression. In other instances we will see that we cannot keep the arguments in (13) for large x, and to find $h(x)$ valid for large x, we will be forced to make an additional calculation using critical amplitudes.

Calculation of Scaling Function for Ising Model

As mentioned earlier, Gaunt and Domb[19] have calculated the scaling function $h(x)$ for the two- and three-dimensional ($d = 2, 3$) Ising model utilizing both high- and low-temperature series expansions. In this section we obtain an expression for $h(x)$ that requires for its calculation only high-temperature series expansions. The appropriate high-temperature expansions have been obtained by Gaunt and Baker[27]* in connection with their calculation of $M(T, H = 0)$, the spontaneous magnetization or "phase boundary." These expansions have the form

$$H(\varepsilon, M) = (\varepsilon + 1)\tanh^{-1}[M\tau(M, v)] \tag{14}$$

where

$$\tau(M, v) \equiv \sum_{n=0}^{\infty} \psi_n(M) v^n \cong \sum_{n=0}^{L} \psi_n(M) v^n \tag{15}$$

Here $\psi_n(M)$ are polynomials in M^2 of degree n,

$$v \equiv \tanh(J/kT) \equiv \tanh[K_c/(\varepsilon + 1)] \tag{16}$$

$K_c \equiv J/kT_c$, k is the Boltzmann constant, and J is the exchange parameter in the Ising Hamiltonian ($J > 0$). Only a limited number L of polynomials $\psi_n(M)$ have been calculated,[27] the number L being equal to 8, 12, and 12 for the fcc, bcc, and sc lattices respectively. Hence Eq. (14) truncated at order L is not expected to describe the behavior of $H(\varepsilon, M)$ in the critical region unless it can be approximated by some closed-form expression that represents an extrapolation beyond order L.

According to Eq. (13), to obtain the scaling function $h(x)$, we must set $M = c$ in Eqs. (14)–(16), where c is a small positive constant. Formally, this procedure is similar to a problem encountered by Gaunt and Baker in a different context,[27] and we shall therefore follow their approach here. Specifically, we shall assume that the function $\tau(M = c, v)$ in (15) vanishes at the phase boundary with the power law form

$$\tau(c, v) \cong \sum_{n=0}^{L} \psi_n(c) v^n \cong \left(1 - \frac{v}{v_0}\right)^q f(v) \tag{17}$$

---

* We generally use the same notation as Ref. 27, but we call $q \equiv 1/\iota$.

where v_0, q, and $f(v)$ are to be estimated by the method of Padé approximants (PA). Thus one first must find v_0 and q by considering PA's to $(d/dv)[\ln \tau(c, v)]$, and afterward $f(v)$ can be determined by studying the product $[1 - (v/v_0)]^{-q}\tau(c,v)$.

Gaunt and Baker[27] noticed that the series (17) was not sufficiently lengthy for reliable estimates for v_0 and q to be obtained unless c was inside the interval $0.6 \lesssim c \lesssim 0.975$. Since the smaller are the values of c, the larger is the region of x where the relation (13) is satisfied, we shall choose $c = 0.6$ for our further analysis. For the Ising model analysis we shall consider first the bcc lattice. Using the Padé method, we estimated that $q = 1.076$ and $v_0 = 0.1658$.[24]* Then we form PA's to the function $f(v)$ of Eq. (17); these PA's were found to be consistent up to five decimal places. We therefore almost arbitrarily chose the [4,4] PA, whence Eq. (17) becomes

$$\tau(0.6, v) = \left(1 - \frac{v}{0.1658}\right)^{1.076} \frac{1 - 4.566v + 5.406v^2 + 5.842v^3 + 0.3907v^4}{1 - 5.936v + 17.603v^2 - 37.098v^3 + 25.811v^4} \quad (18)$$

If we now combine Eqs. (18), (17), and (15), we can obtain from (13) an approximate expression for $h(x)$ that depends upon the exponents β and δ. If we use the generally accepted estimates $\beta \cong 5/16$ and $\delta \cong 5$, then we finally obtain for $h(x)$ the small-x expression

$$h_1(x) = [(0.195x + 1)/0.07776] \tanh^{-1}[0.6\tau(0.6, \tilde{v})] \quad (19)$$

where $\tau(0.6, \tilde{v})$ is given by (18) and $v \equiv \tanh[0.15743/(0.195x + 1)]$.

It is important to emphasize that the expression of Eq. (19) for $h_1(x)$ is not expected to be accurate for very large values of x.[24] This is the reason we use the subscript 1 in Eq. (19), reserving the notation $h_2(x)$ for an expression for large x.

Since $x \equiv \varepsilon/M^{1/\beta}$ [cf. Eq. (1)], positive x corresponds to $T > T_c$ and it turns out that we can also calculate $h(x)$ for *large* x directly from high-temperature expansions. The method has, in fact, been carefully explained by Gaunt and Domb.[19] Thus we find the following expression for $h(x)$, valid at large x,†

$$h_2(x) = x^\gamma \frac{1.0097 + 1.0189x^{-2\beta} + 0.4945x^{-4\beta} + 0.1696x^{-6\beta}}{1 + 0.4388x^{-2\beta}} \quad (20)$$

where $\gamma = 1.25$ and $\beta = 5/16$. This expression differs only very slightly from the fifth expression of Ref. 19, and the difference is probably caused by the rounding off of results at different stages of the calculation.

At $x \cong 1$ the large-x expression of Eq. (20) for $h_2(x)$ overlaps the small-x expression of Eq. (19) for $h_1(x)$ (within an accuracy of about 1%). Hence by using high-temperature expansions exclusively we have obtained two expressions that represent the scaling function of the Ising model in its whole region of definition: $h(x) \equiv h_1(x)$ for $x \lesssim 1$, while $h(x) \equiv h_2(x)$ for $x \gtrsim 1$ [Eqs. (19) and (20), respectively].

We mentioned before that Gaunt and Domb[19] have calculated the scaling function $h(x)$ for the Ising model. They derived five different expressions for five different domains of x. Their first four expressions were obtained using "low-temperature" expansions,[29] and they cover the same domain of x as does Eq. (19) for

---

* The possibility $q \neq 1$ corresponds to an infinite susceptibility on the phase boundary. This problem will be discussed later in this paper.
† In deriving expression (20) we have used results of Essam and Hunter[28] concerning the amplitudes for higher-order derivatives of magnetization with respect to field.

$h_1(x)$, while their fifth expression for large x coincides with our Eq. (20) for $h_2(x)$. Therefore we made a comparison[24] only between our $h_1(x)$ and the four "low-temperature" expressions of Ref. 19, and found that the discrepancy was at most several percent.

Calculation of Scaling Function for $S = 1/2$ Heisenberg Model

We have utilized the technique illustrated for the Ising model in the previous section to calculate for the first time the scaling function of the $S = 1/2$ Heisenberg model, with Hamiltonian

$$\mathscr{H} = -\frac{1}{2}J \sum_{\langle ij \rangle} \boldsymbol{\sigma}^{(i)} \cdot \boldsymbol{\sigma}^{(j)} - kT_c H \sum_{i=1}^{N} \sigma_3^{(i)} \quad (21)$$

Here $\boldsymbol{\sigma}^{(i)}$ and $\boldsymbol{\sigma}^{(j)}$ are the Pauli spin operators at the nearest-neighbor pair of sites $\langle i,j \rangle$, H is the external magnetic field divided by kT_c/μ (to make it dimensionless), and σ_3 is the component of $\boldsymbol{\sigma}$ parallel to the field $\mathbf{H}$.

Baker et al.[30] have calculated high-temperature series expansions for the Hamiltonian (21) that are analogous to the Ising model series of Eq. (15):

$$H(\varepsilon, M) = (\varepsilon + 1) \tanh^{-1}[Mg(M, z)] \quad (22)$$

where

$$g(M, z) \equiv \sum_{n=0}^{\infty} (2^{-n}/n!) P_n(M) z^n \cong \sum_{n=0}^{L} (2^{-n}/n!) P_n(M) z^n \quad (23)$$

Here $P_n(M)$ are polynomials in M^2 of degree n,

$$z \equiv K_c/(\varepsilon + 1) \quad (24)$$

$K_c = J/kT_c$, and $L = 8$ for all three lattices (fcc, bcc, and sc).

Again we assume that the one-phase region is analytic and in analogy with (17) we assume that the function $g(M, z)$ for fixed $M = c$ vanishes at the phase boundary as

$$g(c, z) \cong 1 + \sum_{n=1}^{L} (2^{-n}/n!) P_n(c) z^n \cong \left(1 - \frac{z}{z_0}\right)^q \phi(z) \quad (25)$$

The first lattice considered is the fcc lattice. The Padé approximant analysis of $(d/dz) \ln g(c, z)$ provides relatively reliable results for z_0 and q providing c is in the range $0.4 \leq c \leq 0.85$.[30] For the reasons discussed in the section outlining the calculational method, we will choose the smallest possible value, $c = 0.4$. The PA's for z_0 and q are somewhat less consistent than in the Ising case. Our estimates are essentially the same as those of Ref. 30.

The Padé approximants to the function $\phi(z)$ of Eq. (25) were next formed. They were found to be quite consistent,[24] and we chose the [3, 3] approximant as representative. The corresponding closed-form expression for $g(0.4, z)$ is, from (25),

$$g(0.4, z) = \left(1 - \frac{z}{0.25526}\right)^{1.29} \frac{1 + 3.789z + 1.671z^2 + 3.612z^3}{1 + 3.775z + 4.622z^2 + 14.397z^3} \quad (26)$$

Substituting Eq. (26) into Eq. (22) and using Eq. (13), we obtain the following expression for the scaling function $h(x)$ of the $S = 1/2$ Heisenberg model:

$$h_1(x) = \frac{(0.4)^{1/\beta} x + 1}{(0.4)^\delta} \tanh^{-1}[0.4g(0.4, \tilde{z})] \qquad (27)$$

where

$$\tilde{z} \equiv 0.2492/[(0.4)^{1/\beta} x + 1] \qquad (28)$$

Again we must emphasize that Eq. (27) is valid for small x. In order to obtain an expression for large x we used the Gaunt–Domb method, which is quite sensitive to the estimated values of the critical point exponents. We considered two possible choices,[24] either $\beta = 0.35$ and $\delta = 5$ or $\beta = 0.385$ and $\delta = 4.71$. We found that the second set of exponents gave more reliable estimates of the amplitudes for higher-order field derivatives of magnetization and hence we decided to prefer this set in deriving $h_2(x)$ for the Heisenberg model. [We have to stress that in the expression (27) one can put in *either* set of exponents since none of these were used in its derivation]. We thereby obtained the following expression* for $h_2(x)$:

$$h_2(x) = x^{\beta(\delta-1)} \frac{0.9328 + 0.2805 x^{-2\beta}}{1 - 0.2472 x^{-2\beta}} \qquad (29)$$

The large-x expression (29) matches perfectly with the small-x expression (27); the region of overlap extends from $x = 1.2$ to $x = 1.5$, within which the discrepancy between the two expressions is never larger than 1%. Therefore we have found that in the case of the $S = 1/2$ Heisenberg model, just as in the case of the Ising model, *two* expressions are sufficient to represent the scaling function $h(x)$ over its entire range of definition.

Calculation of Scaling Function for Classical Heisenberg Model ($S \to \infty$)

The classical ($S = \infty$) Heisenberg model is no less interesting than the $S = 1/2$ Heisenberg model, either from an experimental or from a theoretical point of view. In fact, according to the universality hypothesis,[20,21] critical point exponents are *independent* of spin quantum number S. However, even for those properties that may depend on spin, the $S = \infty$ Heisenberg model is probably no worse an approximation than the $S = 1/2$ Heisenberg model[23,31] for the Heisenberg magnets CrBr$_3$ ($S = 3/2$) and EuO ($S = 7/2$).† Therefore in this section we calculate the scaling function $h(x)$ for the classical Heisenberg model, described by the Hamiltonian

$$\mathcal{H} = -2J \sum_{\langle ij \rangle} \mathbf{S}^{(i)} \cdot \mathbf{S}^{(j)} - kT_c H \sum_{i=1}^{N} S_3^{(i)} \qquad (30)$$

where $\mathbf{S}^{(i)}$ and $\mathbf{S}^{(j)}$ are unit vectors ("classical spins") at the nearest-neighbor pair of sites $\langle ij \rangle$ of the lattice, while the other variables are defined following Eq. (21).

For the purpose of calculating the magnetic phase boundary for the classical

---

* Reference 24 contains expressions for $h_2(x)$ calculated for $\beta = 0.35$ and $\delta = 5$.
† See, e.g., the comparison of the data for EuO and EuS with the $S = \infty$ Heisenberg model in Ref. 32.

Heisenberg model, Stephenson and Wood[33,34] generated the following high-temperature series expansion for the fcc lattice:

$$H(\varepsilon, M) \cong M(\varepsilon + 1) \sum_{n=0}^{9} B_n(y) [K_c/(\varepsilon + 1)]^n \qquad (31)$$

where B_n is a polynomial of order n in the variable

$$y \equiv M(\varepsilon + 1)/K_c \qquad (32)$$

In this case it is convenient[34] to fix $y = a = $ const, thereby obtaining *two* coupled equations

$$H(\varepsilon, M) \cong M(\varepsilon + 1) \sum_{n=0}^{9} B_n(a) [K_c/(\varepsilon + 1)]^n \qquad (33)$$

and

$$M(\varepsilon + 1)/K_c = a \qquad (34)$$

To obtain the scaling function $h(x)$ we *begin* as in the previous sections. First we assume that the summation of Eq. (33) vanishes at the phase boundary in the form [cf. Eqs. (17) and (25)]

$$\sum_{n=0}^{9} B_n(a) z^n \cong \left(1 - \frac{z}{z_0}\right)^q \phi_\infty(z) \qquad (35)$$

where $z \equiv K_c/(\varepsilon + 1) \equiv J/kT$ as before. We must choose the constant a in the range $0.8 < a < 3$. (For $a > 3$, Padé approximants do not give consistent predictions[34] for z_0 and q, while for $a < 0.8$, z_0 is of the order of K_c, and thus available techniques are not sufficient for a precise estimation of z_0.) Since we wish both ε and M to be in the critical region, it follows from (34) that we should choose a as small as possible, and accordingly we choose $a = 0.8$.

Next we seek to estimate the numbers z_0 and q that appear in Eq. (35) by studying respectively the poles and residues of Padé approximants to the logarithmic derivative of (35),

$$(d/dz) \ln \left[\sum_{n=1}^{9} B_n(0.8) z^n \right] \qquad (36)$$

From the corresponding Padé tables[24] we infer that $z_0 = 0.1577$ and $q = 1.33$. We then formed Padé approximants to the function $(1 - z/z_0)^{-q} \sum_{n=1}^{9} B_n(0.8) z^n$ [cf. Eq. (35)], and choosing the [3, 3] PA as representative, we obtain, on substituting Eq. (35) into (33),

$$H(\varepsilon, M) = M(\varepsilon + 1) \left(1 - \frac{z}{0.1577}\right)^{1.33} \frac{3 - 22.169z + 23.71z^2 + 9.512z^3}{1 - 7.823z + 10.745z^2 + 1.697z^3} \qquad (37)$$

where ε and M are restricted to those values that satisfy (34) with $a = 0.8$,

$$M(\varepsilon + 1) = 0.8 K_c \qquad (38)$$

Next we apply Eq. (13) and we obtain from Eqs. (37) and (38) an expression for the scaling function $h(x)$ that is in the form of two coupled equations,

$$h(x) = \frac{0.8K_c}{c^\delta}\left(1 - \frac{\tilde{z}}{0.1577}\right)^{1.33} \frac{3 - 22.169\tilde{z} + 23.71\tilde{z}^2 + 9.512\tilde{z}^3}{1 - 7.823\tilde{z} + 10.745\tilde{z}^2 + 1.697\tilde{z}^3} \tag{39}$$

and

$$c(xc^{1/\beta} + 1) = 0.8K_c \tag{40}$$

where $\tilde{z}$ denotes $K_c/(xc^{1/\beta} + 1)$ as in the $S = 1/2$ case.

The coupling between Eqs. (39) and (40) essentially means that in order to calculate $h(x)$ at some particular value of x, one first has to find the corresponding value of c from Eq. (40). It follows from (40) that large values of x imply small values of c. But according to the discussion that followed Eq. (13), it is very important to have c small; indeed, the fact that c decreases when x increases leads to Eq. (39) being a valid expression for large x as well as for small x, and this is why we do not place a subscript 1 on $h(x)$ in (39).

The meaning of the last sentence is graphically illustrated in Fig. 1, which shows the two paths of approach to the phase boundary utilized by the high-temperature expansions used in this work. In *this* work, for the Ising and $S = 1/2$ Heisenberg models the phase boundary is approached along paths of the type labeled "path 1" in Fig. 1 (c fixed). In the case of the classical Heisenberg model we approach the phase boundary along the hyperbolic path labeled "path 2" (a fixed). When x increases c remains constant for path 1 and decreases for path 2. Therefore, although expressions (19) and (27) for the Ising and $S = 1/2$ Heisenberg models cease to be valid at large x, we might expect that the corresponding expression (39) for the $S = \infty$ Heisenberg model will be valid for large x. In fact, the results from calculating the large-x scaling function using amplitudes for the $S = \infty$ Heisenberg model show good agreement with (39) until extremely large x is reached.

We remark that, as in the case of the $S = 1/2$ Heisenberg model, there is more

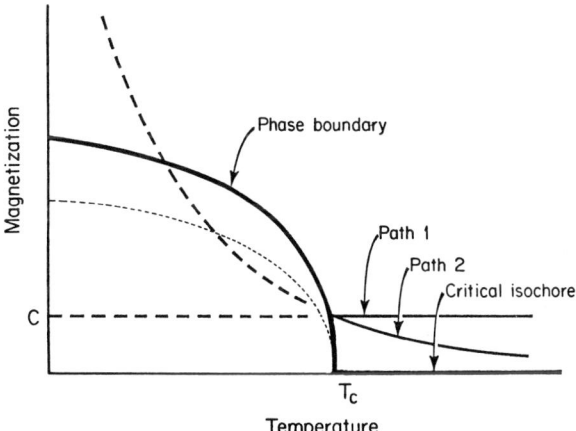

Fig. 1. Two ways of approaching the phase boundary by the high-temperature expansion of the magnetic field $H(\varepsilon, M)$. The path labeled 1 is used in the Ising and $S = 1/2$ Heisenberg model cases, whereas path 2 is used in the case of the $S = \infty$ model. The dotted line represents a possible spinodal curve.

Behavior of the Scaling Function $h(x)$ Near the Phase Boundary

From Eqs. (17), (25), and (35) it can be shown that the scaling function $h(x)$ vanishes at $x = -x_0$, the phase boundary, as

$$h(x) = (x + x_0)^q \bar{\phi}(x) \qquad (41)$$

where the function $\bar{\phi}(x)$ is regular and nonzero at $-x_0$. Hence the first derivative $h'(x)$ will also vanish at $x = -x_0$ if $q > 1$, and the susceptibility

$$\chi = (\partial M/\partial H)_{H=0} = M^{1-\delta}[\delta h(x) - xh'/\beta]^{-1} \qquad (42)$$

will then be infinite on the coexistence curve.

The locus of infinite susceptibility is usually called the "spinodal," and it represents the limit of thermodynamic stability.[35] Note that there is no general argument requiring the existence of such a curve.

While the coincidence of the spinodal with the "binodal," as the coexistence curve is also called,[35] has never been physically observed, it is not forbidden by any thermodynamic or statistical mechanical principle known to us. It in fact occurs in certain model systems, namely the droplet model[36] and the spherical model,[37] for which $q = 2$.

Thus while the results $q \neq 1$ may possibly be artifacts of the calculational method, they are not ruled out by presently known physical requirements. Gaunt and Baker,[27] in determining the phase boundary of the Ising model from high-temperature series, defined the exponent $\iota = 1/q$, and argued that the value $\iota = 1$ was the correct one for this model. They also discussed the effect of the finiteness of the series on the determination of q, stating that values of q different from unity would be obtained on any path (except $H = 0$) crossing the phase boundary if the series under study were not sufficiently long.

Of course, we explored the possibility of using $q = 1$. Table I provides values of $h(0)$ obtained from expressions fully analogous to (18) only constructed with different Padé approximants to the function $f(v)$ of (17). Column A is constructed using $q = 1.076$, column B uses $q = 1.08$, while column C uses the prediction $q = 1$. One can notice in column A the rather striking consistency between different Padé approximants [as is typical for other values of $h(x)$ as well]. It is noticeable from column B that this consistency is weakened when one takes a sort of average estimate[24] $q = 1.08$. From column C we see that the consistency gets still worse when one takes $q = 1$. The consistency of the Padé approximant results led us to use $q = 1.076$ in (18) instead of the presumably correct value $q = 1$. However, this may be merely a way to achieve a consistent approximation but not an attempt to disprove the possibility of q being equal to unity.

For the Heisenberg model low-temperature (far from the critical region) spin-wave calculations[38] support an equation of state which also shows a spinodal–binodal coincidence:

$$M(H, T) \approx M_0(T) + X(T) H^{1/2} + \cdots \qquad (43)$$

Table I. Values of $h(0)$ of the Ising Model (the bcc lattice) Obtained by Using Different PA's (different rows of the table) to the Function $f(v)$ [Eq. (17)] and By Choosing Different Estimates (different columns) of the Exponent q

Padé [N, D]	A ($q = 1.076$)	B ($q = 1.08$)	C ($q = 1$)
[3, 3]	0.3812	0.3807	0.3911
[4, 3]	0.3819	0.3812	0.3906
[5, 3]	0.3819	0.3812	0.3889
[6, 3]	0.3819	0.3813	0.3874
[3, 4]	0.3819	0.3812	0.3906
[4, 4]	0.3819	0.3813	0.3914
[5, 4]	0.3819	0.3813	0.3769
[6, 4]	0.3819	0.3811	0.3856
[3, 5]	0.3819	0.3813	0.3899
[4, 5]	0.3819	0.3814	0.3812
[5, 5]	0.3819	0.3812	0.3868
[6, 5]	0.3830	0.3820	0.3842
[3, 6]	0.3819	0.3813	0.3888
[4, 6]	0.3819	0.3812	0.3863
[5, 6]	0.3819	0.3801	0.3846
[6, 6]	0.3819	0.3817	0.3830

An interesting observation concerning the coincidence of the spinodal with the binodal comes from an argument by Chu et al.[39]* If we define an exponent β_s and an amplitude M_s to characterize the shape of a spinodal (see Fig. 1)

$$M = M_s |\varepsilon|^{\beta_s}, \quad \chi^{-1} = 0 \tag{44}$$

then the scaling hypothesis, Eq. (1), predicts $\beta_s = \beta$.[27] Clearly $\beta_s = \beta$ if the spinodal and the binodal are the *same* curve near the critical point, as is the case when $q > 1$.

Chu et al.[39] argue that if a spinodal is to exist, its exponent β_s must be an integer multiple of $1/2$, and most likely is $1/2$. Thus $\beta_s \neq \beta$ unless $\beta = 1/2$. Their argument has no relation to the scaling hypothesis; it hinges on (i) the physically reasonable assumption that on the $M = 0, \varepsilon > 0$ line the equation of state $H(M, \varepsilon)$ should be analytic, and (ii) a particular choice for the analytic form of χ^{-1} near the phase boundary. They propose that along a line $M = M_1 = $ const the existence of the spinodal should manifest itself *outside* the phase boundary in the behavior of the susceptibility, in the following manner:

$$\chi^{-1} \sim [\varepsilon - \varepsilon_s(M)]^\gamma \tag{45}$$

where

$$\varepsilon_s(M) \equiv (M/M_s)^{1/\beta_s} \tag{46}$$

---

* This work is stated for a binary mixture.

is the equation for the spinodal line, from (44). In other words, along a line $M = M_1$ the susceptibility will seem to diverge at $\varepsilon_s(M_1)$, a locus inside the phase boundary. But Eq. (45) means that the equation of state is not analytic on the $M = 0, \varepsilon > 0$ line unless $\beta_s = 1/2n$, with n an integer. This violates the physically reasonable requirement (i). Experimentally, however, the susceptibility often appears to diverge at a locus such as (46), with $\beta_s = \beta \neq 1/2$. This led Chu et al. to discard the form (45) postulated for χ^{-1}.

Their argument is not without worth, though, for it points out the possibilities open to the form of the susceptibility in the event that the spinodal exists. If a spinodal is present, then the susceptibility will appear to diverge at that locus, and unless the divergence is of an essential nature, an exponent γ^* should be definable,

$$\chi^{-1} \sim [\varepsilon - \varepsilon_s(M)]^{\gamma^*} \qquad (47)$$

where γ^* need not equal γ. The form (47) allows for what is often experimentally observed, although one is usually interested in obtaining β_s and not γ^*. The choice of $\gamma^* \neq \gamma$ does not change the Chu et al. argument. What they do not allow for, however, is that forms (47) [and even (45)] may be true, but only sufficiently near the phase boundary. Therefore trying to expand them near the $M = 0$ line is an invalid procedure. This is the case for the spherical model as well as for the model systems we have studied (according to the obtained numerical results). For the spherical model[37] the equation of state is

$$H = (M^2 + \varepsilon)^2 \qquad (48)$$

and thus

$$\chi^{-1} = (M^2 + \varepsilon)(5M + \varepsilon) \qquad (49)$$

Clearly both the phase boundary and the spinodal are given by $M^2 = -\varepsilon$, so that $\beta_s = \beta = 1/2$. Sufficiently near the phase boundary for a fixed M only the first factor in (49) contributes to the divergence of χ, and thus $\gamma^* = 1$. (Of course, the fact that $\beta_s = 1/2$ guarantees the analyticity of the equation of state at $M = 0$ anyway.)

For the Heisenberg and Ising models studied here the form of the equation of state near the phase boundary is Eq. (1) with $h(x)$ given by (41). From (41)-(42) the susceptibility near the phase boundary is given by

$$\chi^{-1} \sim [\varepsilon + x_0 M^{1/\beta}]^{q-1} \qquad (50)$$

so that $\gamma^* = q - 1$. In order to evaluate the susceptibility near the $M = 0$ line, a *different* form for $h(x)$ must be used, namely $h_2(x)$. This form has already been constructed to satisfy the analyticity requirement.

In conclusion, we have seen that, without violating analyticity requirements, it is possible to have a spinodal (i) with $\beta_s = \beta \neq 1/2$, and (ii) with a limiting form, Eq. (47), for the susceptibility near the phase boundary.

Comparative Study of Scaling Functions for $S = 1/2$ Ising, $S = 1/2$ Heisenberg, and $S = \infty$ Heisenberg Models

Universality of the Critical Properties. There have been many "conjectures"[20-21] about the extent to which $h(x)$ depends upon parameters of the system such as type

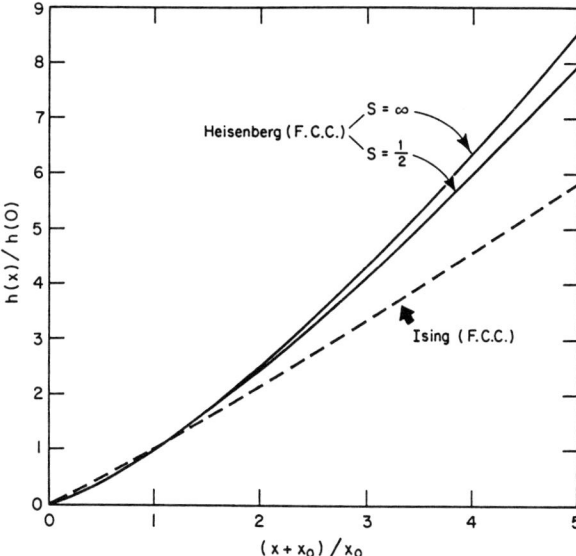

Fig. 2. Comparison of the normalized scaling functions $h(x)/h(0)$ plotted vs. $(x + x_0)/x_0$ for the three ferromagnetic models: the Ising model and $S = 1/2$ and $S = \infty$ Heisenberg models.

Table II. Comparison of the Normalized Ising Model Scaling Functions for the Three Cubic Lattices

	$h_1(x)/h_1(0)$		
$(x + x_0)/x_0$	fcc	bcc	sc
0.25	0.224	0.224	0.226
0.50	0.473	0.473	0.474
0.75	0.733	0.733	0.733
1.25	1.27	1.27	1.27
1.50	1.55	1.55	1.55
1.75	1.83	1.83	1.83
2.00	2.11	2.12	2.12
2.25	2.40	2.41	2.40
2.50	2.69	2.70	2.69
2.75	2.98	3.00	2.99
3.00	3.27	3.29	3.28
3.25	3.57	3.59	3.58
3.50	3.87	3.89	3.88
3.75	4.16	4.20	4.18
4.00	4.46	4.50	4.48
4.25	4.76	4.81	4.78
4.50	5.07	5.11	5.09

of lattice and magnitude of the spin quantum number, *if* in fact $h(x)$ depends at all upon these parameters!

In particular, there has been almost no detailed evidence concerning what features of a system the scaling function $h(x)$ might depend upon, since the scaling function has hitherto been calculated only for the $d = 2$ and $d = 3$ Ising models.[19] Gaunt and Domb observed that $h(x)$ differed drastically for the two different lattice dimensionalities $d = 2$ and $d = 3$. Gaunt and Domb also calculated $h(x)$ for the bcc, sc, and fcc lattices (though they do not display the results of their calculation), and they state that the scaling functions agree to within the accuracy of their calculation.

We have varied the parameters "spin space dimensionality" D (with $D = 1$ and $D = 3$ corresponding, respectively, to the Ising and Heisenberg models), lattice structure, and spin quantum number S. From Fig. 2 it is clear that our calculated scaling functions depend strongly upon "D" even when the normalized plots $h(x)/h(0)$ vs. $(x + x_0)/x_0$ are compared. In the remainder of this section we examine in detail the question of whether our calculated functions $h(x)$ depend upon lattice structure and spin quantum number.

Independence of the Normalized Scaling Function with Respect to Lattice Structure.

Ising Model. The *normalized* Ising model scaling functions $h(x)/h(0)$ are compared as functions of $(x + x_0)/x_0$ in Table II for the bcc, fcc, and sc lattices. The disagreement is at most 2% and we conclude that our calculations support the conjecture that normalized plots of $h_1(x)/h_1(0)$ vs. $(x + x_0)/x_0$ are independent of lattice structure for the Ising model. For large x we rely on Gaunt and Domb's claim that the scaling functions agree to within the accuracy of their calculations.[19] Our calculations of scaling functions along a curved path (cf. Fig. 1), while not of greater accuracy, do support the conjecture of lattice independence.

Heisenberg Model. For the $S = 1/2$ Heisenberg model we compare in Table III normalized plots of $h_1(x)/h_1(0)$ for the fcc and bcc lattices. It appears that the normalized scaling function is lattice independent, at least for small arguments. For the $S = \infty$ Heisenberg model we have reliable scaling functions only for the fcc lattice. Our calculations on the bcc and sc lattices, while not sufficiently accurate for detailed comparison, do not suggest any evidence of lattice dependence.

Possible Independence of the Normalized Scaling Function with Respect to Spin Quantum Number S. As was made clear in the preceding two sections, the critical point exponents for the $S = 1/2$ and $S = \infty$ Heisenberg models are not yet firmly established, but it is at least *possible* that these exponents are independent of spin quantum number. Therefore in this section we consider the question of whether the corresponding scaling functions $h(x)$ are spin independent.

Figure 2 compares the normalized scaling functions $h(x)/h(0)$ for the $S = 1/2$ and $S = \infty$ Heisenberg models; also shown for comparison is the Ising model scaling function.[19] One observes that for all negative values of x [i.e., for all values of the abscissa $(x + x_0)/x_0 < 1$] corresponding to $T < T_c$, and for small positive values of x, there is almost no difference between $S = 1/2$ and $S = \infty$ Heisenberg model cases (cf. also Table IV). Hence we are led to conclude that for $T < T_c$ any

Table III. Comparison of the Normalized Scaling Functions of the $S = 1/2$ Heisenberg Model for the fcc and bcc Lattices

	$h_1(x)/h_1(0)$	
$(x + x_0)/x_0$	fcc	bcc
0.25	0.167	0.164
0.50	0.409	0.405
0.75	0.690	0.688
1.25	1.33	1.34
1.50	1.69	1.69
1.75	2.06	2.06
2.00	2.44	2.45
2.25	2.84	2.85
2.50	3.26	3.26
2.75	3.68	3.68
3.00	4.12	4.11
3.25	4.57	4.55
3.50	5.02	5.00
3.75	5.46	5.46
4.00	5.96	5.92
4.25	6.45	6.38
4.50	6.94	6.86

spin dependence of the scaling function is sufficiently small ($\lesssim 2\%$) that it is within the accuracy of our calculation of $h(x)$.

On the other hand, for larger positive values of x the discrepancy slowly increases. However, these comparisons were made using $h(x)$ for the $S = \infty$ model, which has an incorrect asymptotic large-x dependence. The large-x calculation produces a scaling function in substantially better agreement with the $S = 1/2$ results. The agreement is within the uncertainty produced by differing estimates of T_c and the critical point exponents.

In summary, then, we believe that there is no firm evidence for doubting spin independence of the normalized Heisenberg scaling function.

Comparison of Calculated Scaling Functions with Experimental Results

Plots of Scaled Magnetization $M/H^{1/\delta}$ vs. Scaled Temperature $\varepsilon/H^{1/\beta\delta}$. In the section outlining calculational method we assumed that the magnetic field $H(\varepsilon, M)$ is a generalized homogeneous function in the critical region [cf. Eq. (11)].

Now if one formulates the scaling hypothesis by means of the essentially equivalent hypothesis that a thermodynamic potential (e.g., the Gibbs potential G) is a generalized homogeneous function, then one can straightforwardly show[10] that all its Legendre transforms and derivatives are *also* generalized homogeneous functions. In particular, the magnetization $M(\varepsilon, H) = -(\partial G/\partial H)_T$ is also a generalized

Table IV. Comparison of the Normalized Scaling Functions of the $S = 1/2$ and $S = \infty$ Heisenberg Models (fcc lattice)

$(x + x_0)/x_0$	$h_1(x)/h_1(0)$, $S = 1/2$	$h(x)/h(0)$, $S = \infty$
0.125	0.07	0.06
0.250	0.17	0.16
0.375	0.28	0.27
0.500	0.41	0.40
0.625	0.55	0.53
0.750	0.69	0.68
0.875	0.84	0.84
1.125	1.16	1.17
1.250	1.33	1.35
1.375	1.51	1.53
1.500	1.69	1.72
1.625	1.87	1.92
1.750	2.06	2.12
1.875	2.25	2.32
2.00	2.44	2.53

homogeneous function—i.e., there exist two numbers a and b such that for all positive values of the number λ

$$M(\lambda^a \varepsilon, \lambda^b H) = \lambda M(\varepsilon, H) \tag{51}$$

The "scaling parameters" a and b in Eq. (51) are unspecified, but they may be related to critical point exponents. For example, by setting $\lambda = |\varepsilon|^{-1/a}$ and choosing $H = 0$ in (51), we find that $M(\varepsilon, 0) \propto |\varepsilon|^{1/a}$ and $a = 1/\beta$; similarly by setting $\lambda = |H|^{-1/b}$ and choosing $\varepsilon = 0$ in (51), we find that $M(0, H) \propto H^{1/b}$ and $b = \delta$. Equation (51) thus becomes

$$M(\lambda^{1/\beta} \varepsilon, \lambda^\delta H) = \lambda M(\varepsilon, H) \tag{52}$$

If we follow the argument leading up to Eq. (12), we argue that (52) must be valid for all values of λ and hence in particular for the choice $\lambda \equiv (c/H)^{1/\delta}$, whence (52) becomes

$$M(\varepsilon, H)/H^{1/\delta} = M(c^{1/\beta\delta} \varepsilon/H^{1/\beta\delta}, c)/c^\delta \tag{53}$$

Equation (53) says that plots of "scaled magnetization" $\tilde{M} \equiv M(\varepsilon, H)/H^{1/\delta}$ vs. "scaled temperature" $\tilde{\varepsilon} \equiv \varepsilon/H^{1/\beta\delta}$ should be described by the function $M(c^{1/\beta\delta} \varepsilon, c)/c^\delta$, which is essentially the magnetization function for a constant but small (if c is small) value of the magnetic field. In other words Eq. (53) states that the scaled magnetization $M/H^{1/\delta}$ is a function of only one variable, the scaled temperature $\varepsilon/H^{1/\beta\delta}$.

It turns out[10,25] that plots of the experimental data in the form $M/H^{1/\delta}$ vs. $\varepsilon/H^{1/\beta\delta}$ have some advantages over the original presentations of experimental results, namely all experimental data can be captured within one curve for both regions, $T < T_c$ and $T > T_c$, without breaking the curve into two parts and without applying a log-log plot. This advantage arises from the fact that with current equipment it is difficult to make measurements for H arbitrarily close to zero and hence the ordinate $M/H^{1/\delta}$ and the abscissa $\varepsilon/H^{1/\beta\delta}$ never become extremely large. Furthermore, data do not appear clustered in a certain region, nor does the behavior of the corresponding curve become dominated by the value of only one critical point exponent.

Therefore we will compare our calculations of the scaling function with the experimental results in the form of plots $M/H^{1/\delta} (\equiv 1/h^{1/\delta})$ against $\varepsilon/H^{1/\beta\delta} (\equiv x/h^{1/\beta\delta})$. In doing this, we will deal only with the normalized quantities $h(x)/h(0)$ and x/x_0.

Chromium Tribromide (CrBr$_3$). Since CrBr$_3$ was the first *insulating* magnet for which accurate measurements germane to the scaling law equation of state were made,[14] we treat this material first. Now chromium has $S = 3/2$, so we compared the experimental data for the scaling function with both the $S = 1/2$ and the $S = \infty$ Heisenberg calculations, and also with the $S = 1/2$ Ising model calculation. The data disagreed strongly with the Ising model calculation, and disagreed with the $S = \infty$ Heisenberg model in those regions where the $S = 1/2$ and $S = \infty$ Heisenberg calculations are slightly distinguishable. We show in Fig. 3 the $S = 1/2$ Heisenberg model calculation compared with the data.

It is important to emphasize that there are no adjustable parameters whatsoever used in the Heisenberg model calculation, so we found the agreement to be rather gratifying and even perhaps somewhat surprising when one remembers that the critical point exponents of CrBr$_3$ ($\beta = 0.368$, $\gamma = 1.215$, and $\delta = 4.28$) are not at all close to the values used in our calculations for the $S = 1/2$ Heisenberg model ($\beta = 0.385$, $\gamma = 1.43$, and $\delta = 4.71$), where γ is the critical point exponent of the initial susceptibility.[23] This difference in exponents is likely the reason for the slight discrepancy between the "tails" of the experimental and theoretical curves in Fig. 3.

Figure 4 shows an enlarged region of Fig. 3 centered about $x = 0$ (corresponding

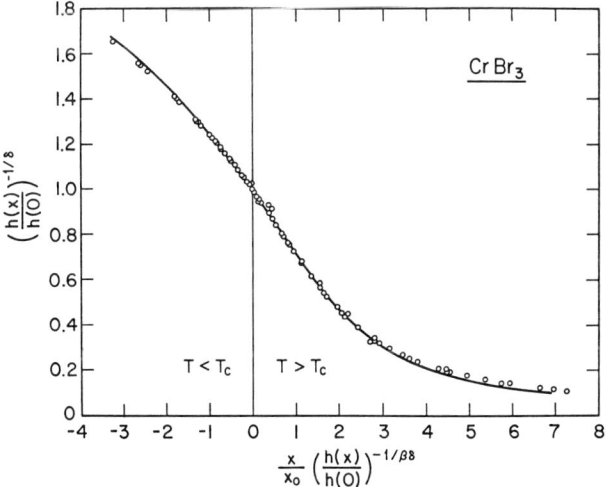

Fig. 3. Comparison between Heisenberg model scaling functions and experimental data[14] on CrBr$_3$.

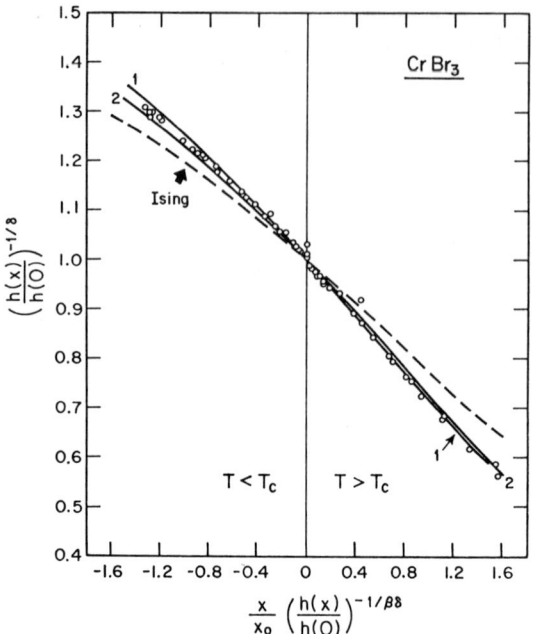

Fig. 4. Enlarged portion of Fig. 3. The curve labeled 1 is the same curve as that in the previous figure, i.e., it corresponds to the $S = 1/2$ Heisenberg model with $\beta = 0.385$ and $\delta = 4.71$; the curve numbered 2 corresponds to the same model but the estimates $\beta = 0.35$ and $\delta = 5$ were used in its construction. The broken curve corresponds to the Ising model scaling function.

to $T = T_c$). The curve labeled 1 in Fig. 4 is the $S = 1/2$ Heisenberg calculation from Fig. 3, while the curve labeled 2 in Fig. 4 is the same functional form, Eq. (27) as curve 1 except that we have used $\beta = 0.35$ and $\delta = 5$ rather than $\beta = 0.385$ and $\delta = 4.71$ (cf. the discussion in the section on the calculation of the scaling function for the $S = 1/2$ Heisenberg model). We note that the two curves are barely distinguishable in this region, though the discrepancy between them is somewhat larger in the region $T < T_c$ ($x < 0$) than in the region $T > T_c$ ($x > 0$). Also shown in Fig. 4 is the scaling function calculated for the Ising model, and one sees rather dramatically how much more closely the data are fit by the Heisenberg model than by the Ising model.

According to our understanding of the universality hypothesis,[20,21] the fact that $CrBr_3$ has a rather large amount of "lattice anisotropy" (different coupling strengths in different directions, the coupling in the z direction, J_z, being about 17 times weaker than in the xy plane, J_{xy}) does not mean that the scaling function should be any different from the "isotropic" case ($J_z = J_{xy}$). Since we cannot easily calculate the scaling function for the case of arbitrary lattice anisotropy, we cannot test this hypothesis *theoretically*, but the fact that our isotropic calculation agrees as well as it does with the experimental data for $CrBr_3$ (for which $J_z/J_{xy} = 1/17$)

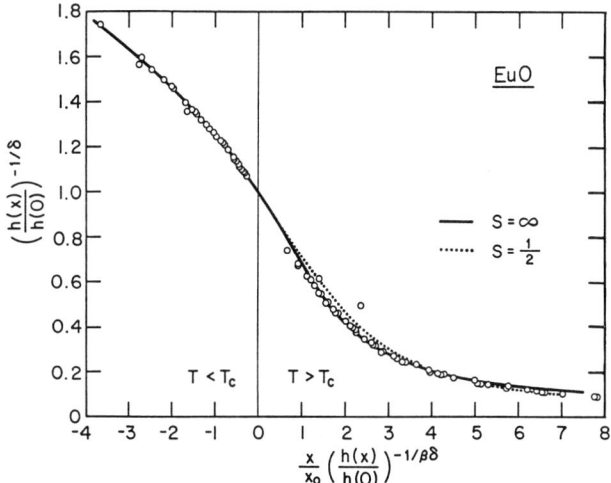

Fig. 5. Comparison between experimental data[16] for EuO and the Heisenberg model calculations.

suggests that *if* there is a dependence of the scaling function upon lattice anisotropy, this dependence is very weak.

Europium Oxide (EuO). A second ferromagnet for which accurate experimental data have very recently been obtained is EuO.[16] Europium oxide has spin quantum number $S = 7/2$, and hence in Fig. 5 we compare the data with both the $S = 1/2$ and the $S = \infty$ scaling functions.

It might be somewhat surprising that the agreement between measured and calculated scaling functions is as good as it is for EuO, since the measured exponents were $\beta = 0.368$ and $\delta = 4.46$, while the exponents used in the calculation were $\beta = 0.385$ and $\delta = 4.71$ for the $S = 1/2$ case and $\beta = 0.38$ and $\delta = 4.63$ for the $S = \infty$ case. Besides, EuO has next-nearest-neighbor interactions that are of magnitude almost comparable to the nearest-neighbor interactions,[16] while our calculations were for the nearest-neighbor Hamiltonians. Considering this point, perhaps we should remark that it is at least possible that the scaling function $h(x)$ is independent of the strength of second-neighbor interactions since a necessary condition for this to be so, the invariance of the critical point exponents, has been established.*

Nickel (Ni). We will next compare our calculated scaling functions with experimental results[13] for Ni, the first ferromagnet for which the scaling law equation of state was tested.[11] Our calculations have been done for models which presumably describe *insulating* ferromagnets, whereas Ni is a *metal*, and hence it might seem unrealistic to expect agreement between the theoretical and experimental results.

---

* In particular, model calculations have been done where independence of critical properties with respect to lattice structure and spin magnitude was checked. Recently Paul and Stanley[40] checked, for various models, independence of critical exponents with respect to range of interaction (second-neighbor interactions) and type of bonds between magnetic moments ("lattice anisotropy").[41]

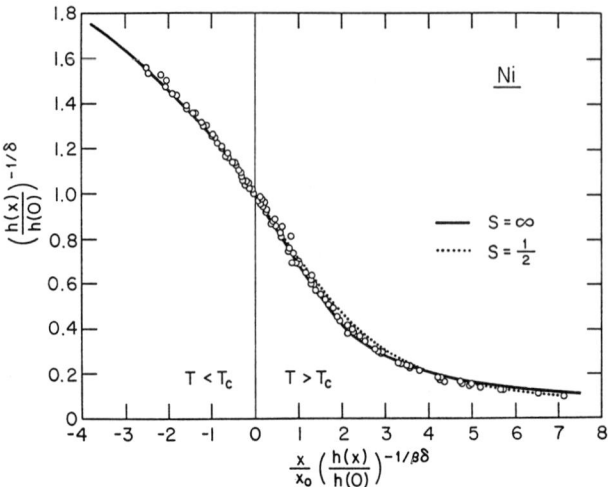

Fig. 6. Comparison between experimental data[13] for Ni and the Heisenberg model calculations.

However, Fig. 6 reveals indeed a good agreement of the Ni data with calculated scaling functions for both the spin-1/2 and spin-∞ Heisenberg models.

The agreement is somewhat better with the classical Heisenberg model, which may be due to the fact that the critical exponents[13] of Ni ($\beta = 0.378$ and $\delta = 4.58$) are closer to the values for $S = \infty$ ($\beta = 0.38$ and $\delta = 4.63$) than to those for $S = 1/2$ ($\beta = 0.385$ and $\delta = 4.71$).

We find that comparison with experimental results for Ni suggests very strongly that insulating or conducting properties of a material are not important near the critical point as regards the normalized scaling function $h(x)$.

Palladium–Iron Alloy (Pd$_3$Fe). Figure 7 shows rather good agreement between experimental data[17] for the conducting alloy Pd$_3$Fe in its *disordered* state with our calculated scaling function for the $S = 1/2$ Heisenberg model. By contrast, the data for the ordered alloy shown in Fig. 8 are in rather poor agreement with the Heisenberg model calculation.

What is the interpretation of these results? One possible interpretation is that the interactions in the *random* alloy may be rather short range,[17] and indeed the measured critical point exponents $\beta = 0.364$ and $\delta = 4.61$ are rather close to those for the nearest-neighbor Heisenberg model, Eq. (21). On the other hand, the interactions in the *ordered* alloy might be quite long range,[17] and in fact the measured critical point exponents $\beta = 0.444$ and $\delta = 3.64$ are considerably closer in magnitude to the predictions $\beta = 1/2$ and $\delta = 3$ of the Curie–Weiss or "mean-field theory," which corresponds to each spin interacting with every other spin with a force of equal magnitude. It is precisely because of this case of "infinite-range interactions" that, according to our understanding, the universality hypothesis[20,21] would predict different behavior. Thus the agreement of experiment and Heisenberg model in

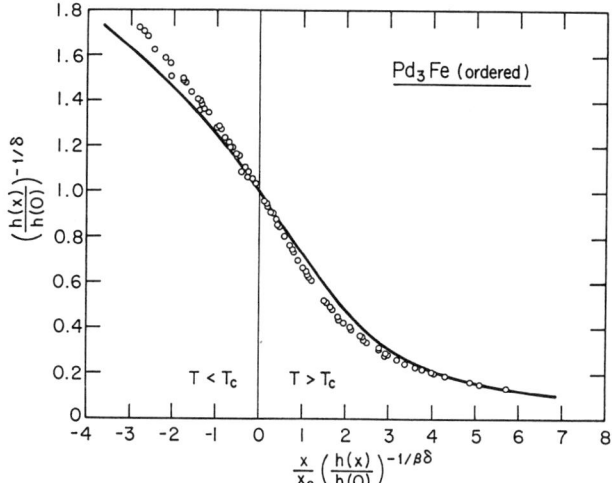

Fig. 7. Comparison between experimental data[17] for Pd_3Fe (disordered) and the Heisenberg model calculations.

Fig. 7 and the disagreement in Fig. 8 might in fact be consistent with the universality prediction that systems with short-range interactions have different critical properties from systems whose interactions are infinite in range.[20,21]

Yttrium Iron Garnet (YIG). Recently experimental data for YIG have become available.* YIG is a cubic ferrimagnet insulator and has $S = 5/2$. The measured critical point exponents are $\beta = 0.370$ and $\delta = 4.65$. Figure 9 shows good agreement between the YIG data and the Heisenberg model. Experimental data for YIG are available for very small values of the applied magnetic field. We do not display the resulting larger range of abscissa and ordinate in Fig. 9. The uncertainty in the Heisenberg model exponents is sufficient to allow for good agreement with the YIG data.

In summary, then, we have seen that the nearest-neighbor isotropic Heisenberg model equation of state appears to be adequate for a wide range of physical systems: (i) $CrBr_3$ (a rhombohedral two-sublattice ferromagnet with $S = 3/2$ and $J_z/J_{xy} = 1/17$); (ii) EuO (an $S = 7/2$ semiconducting magnet with next-nearest-neighbor interactions); (iii) Ni (a metallic ferromagnet); (iv) disordered Pd_3Fe alloy; and (v) $Y_3Fe_5O_{12}$ (a ferrimagnet).

Thus we see that the scaled equation of state, when properly normalized, appears to be independent of many specific physical parameters. The one case of disagreement, ordered Pd_3Fe alloy, is thought to correspond to an example of infinite-range interactions for which case the universality hypothesis would indeed predict that the nearest-neighbor models considered in this work would be inadequate.

* We wish to thank Berkner and Litster[18] for communicating to us their data on YIG in tabular form prior to publication.

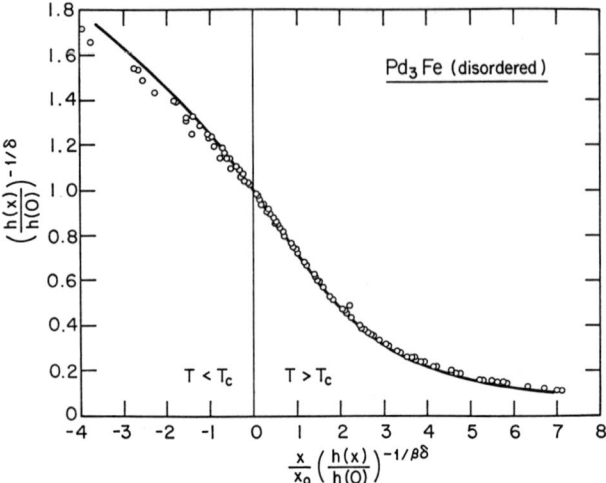

Fig. 8. Comparison between experimental data[17] for Pd_3Fe (ordered) and the Heisenberg model calculations.

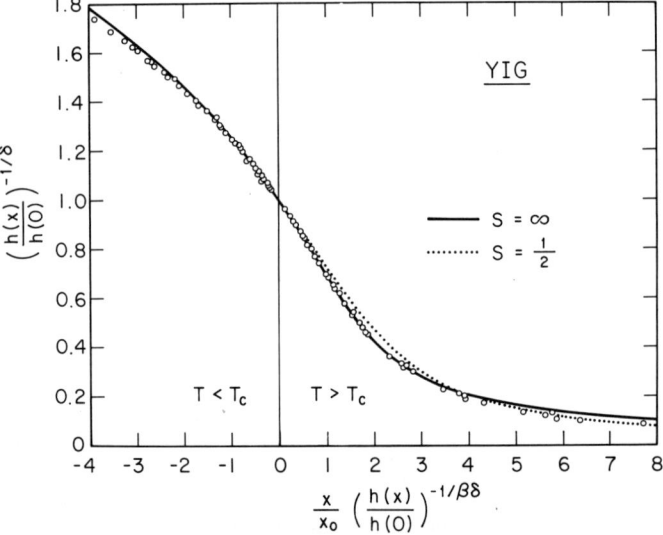

Fig. 9. Comparison between experimental data[18] for YIG and the Heisenberg model calculations.

The Parametric Representation and the Linear Models

An increasingly popular way of presenting experimental data near the critical point is through the parametric representation[22] of the equation of state. This representation is defined through the transformations

$$H = r^{\beta\delta} h_p(\theta) \tag{54}$$

$$\varepsilon = r t(\theta) \tag{55}$$

which, in principle, define r and θ as functions of H and ε. The equation of state is then given by

$$M = r^\beta m(\theta) \tag{56}$$

The functions $h_p(\theta)$ and $t(\theta)$ are usually chosen to suit some general smoothness criteria; $m(\theta)$ is then determined by comparing form (56) with actual data.

The purpose of this representation is to encompass all the singularities in the r variable, which in some sense is a "distance" away from the critical point. It is then straightforward to show that this representation is equivalent to form (1), and therefore to $H(\varepsilon, M)$ being a generalized homogeneous function (GHF).

Schofield's and Josephson's choices for $h_p(\theta)$ and $t(\theta)$ are[22]

$$h_p(\theta) = a\theta(1 - \theta^2) \tag{57}$$

$$t(\theta) = (1 - b^2\theta^2), \quad b > 1 \tag{58}$$

with a and b adjustable.

It is immediate from the definitions (57) and (58) that the special loci are as follows.

Coexistence curve $(H = 0, M \neq 0), \theta = \pm 1$.
Critical isotherm $(\varepsilon = 0), \theta = \pm 1/b$.
Critical isochore $(H = 0, M = 0), \theta = 0$.

These are irrespective of the form of $m(\theta)$.

It is experimentally verified that for a number of systems $m(\theta)$ is nearly linear with θ,[42]

$$m(\theta) \cong q\theta, \quad q = \text{const (near unity)} \tag{59}$$

This led Schofield et al.[42] to postulate the "linear model," where (59) is taken to be the form of the equation of state. This leaves a, b^2, and g as adjustable parameters. Schofield et al. go further and show that under certain conditions the choice of b^2 that makes the data "most linear" is

$$b^2 = \frac{\delta - 3}{\delta - 1} \frac{1}{1 - 2\beta} \tag{60}$$

Results for Heisenberg and Ising Models. From the scaling functions $h(x)$ we have calculated the functions $m(\theta)$ for the $S = 1/2$ and $S = \infty$ Heisenberg models considered above. These are displayed in Figs. 10 and 11. The parameters a and b^2 in Eqs. (57) and (58) were chosen so as to make the functions $m(\theta)$ the most linear near $\theta = 0$. It is obvious that even with the "best" choices of a and b^2 it is not possible to ascribe a linear-model behavior to the $m(\theta)$ functions. The same conclusion was deduced by Gaunt and Domb[19] from their Ising model scaling function calculation.

That the function $m(\theta)$ should be linear is, of course, merely a conjecture. For several systems it is verified that the linearity holds up to about the value $\theta = 1/b$, i.e., the critical isotherm, after which point the function begins to curve downward.[42] Real systems, however, do not exhibit the negative infinite slope at $\theta = 1$, which may be noticed in Figs. 10 and 11 in the case of the Heisenberg models as well as in

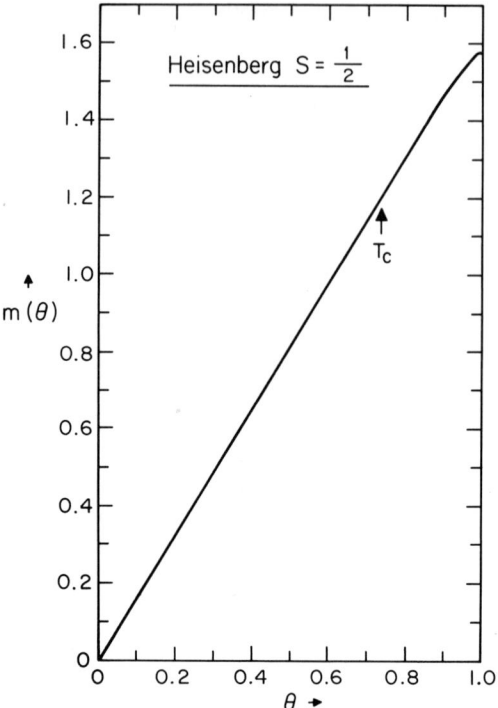

Fig. 10. Parametric representation for $S = 1/2$ Heisenberg model; $a = 1.489, b = 1.356$.

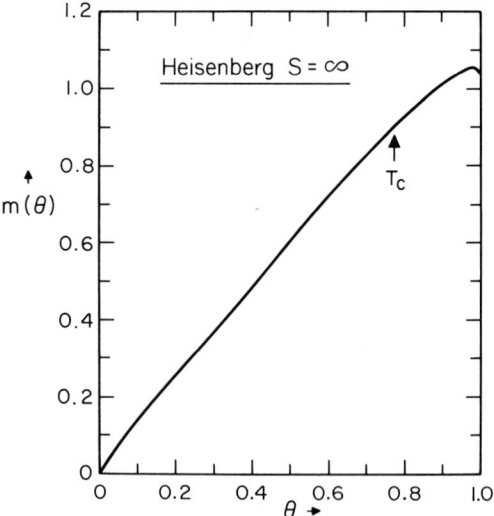

Fig. 11. Parametric representation for $S = \infty$ Heisenberg model $(D = 3)$; $a = 4.349, b = 1.288$.

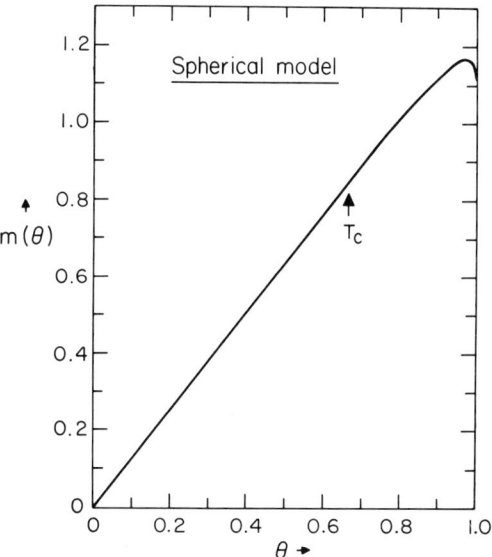

Fig. 12. Parametric representation for the spherical model; $a = 1.20$, $b = 1.50$.

Fig. 12 in the case of the spherical model (the only exactly soluble three-dimensional model).

The negative infinite slope at $\theta = 1$ exhibited by the models represents the coincidence of the spinodal and the coexistence curve discussed previously. To see this explicitly, we will next consider the susceptibility in the parametric form. It follows from Eqs. (54)–(57) that for the Schofield–Josephson parametrization, Eqs. (58) and (59),

$$\chi_T = \left(\frac{\partial M}{\partial H}\right)_T = \frac{r^{-\gamma}}{a} f(\theta) \tag{61}$$

where

$$f(\theta) \equiv \frac{2b^2 \beta \theta m(\theta) + (1 - b^2\theta^2) m'(\theta)}{D(\theta)} \tag{62}$$

with

$$D(\theta) \equiv 1 + (2b^2\beta\delta - b^2 - 3)\theta^2 - b^2(2\beta\delta - 3)\theta^4 \tag{63}$$

From the above one can see that

$$\chi_T(H = 0, T > T_c) = \frac{r^{-\gamma}}{a} f(0) = \frac{r^{-\gamma}}{a} m'(0) \tag{64}$$

and

$$\chi_T(H = 0, T < T_c) = \frac{r^{-\gamma}}{a} f(1) = \frac{r^{-\gamma}}{a} \frac{2b^2\beta m(1) + (1 - b^2) m'(1)}{2(b^2 - 1)} \tag{65}$$

For the simple linear model (59), Eq. (61) becomes

$$\chi_T = \frac{g}{a} r^{-\gamma} \frac{1 - b^2\theta^2(1 - 2\beta)}{D(\theta)} \tag{66}$$

If condition (60) is invoked as well, Eq. (66) simplifies considerably to

$$\chi_T = \frac{g}{a} r^{-\gamma} \frac{1}{1 + [(2\beta\delta - 3)/(1 - 2\beta)]\theta^2} \qquad (67)$$

The distinction between invoking or not invoking Eq. (60) is important for the mean-field and spherical models, in which b^2, as given by Eq. (60), is indeterminate. Schofield et al.[42] use $b^2 = 3/2$ for the mean-field theory and then use (66) to fit the susceptibility.

It can be seen immediately from Eq. (65) that the only way one can have a spinodal *at the coexistence curve*, $|\theta| = 1$, is if $m'(1)$ is negative infinite, since the denominator cannot vanish and $m(1)$ is known to be finite on account of finite M. This rules out the linear model altogether, since $m' = g$ for all θ, irrespective of the choice of b^2. Thus the linear model cannot represent such model systems as the spherical model,[37] the droplet model,[36] and possibly the Heisenberg and Ising models (according to the present calculations), since they all have a spinodal coinciding with the coexistence curve. [The curving over of $m(\theta)$ does not pose any thermodynamic problems; it can curve over and χ_T can still be > 0, as well as monotonic with M for fixed T.]

Thus the nonlinearity of $m(\theta)$ at $\theta = 1$ for the Heisenberg models corresponds to the spinodal coinciding with the phase boundary. One is then tempted to speculate on the nonlinearity of other systems that do not have a negative *infinity* in the slope at $\theta = 1$, but merely curve over, such as YIG.[18] The possibility that the nonlinearity in the $\theta < 1$ region is a precursor of a singularity in the $\theta > 1$ region is currently being investigated.

Summary

In this work we have presented a method for calculating directly from high-temperature series expansions the scaling function $h(x) = H/M^\delta$, where $x \equiv \varepsilon/M^{1/\beta}$. Our method is made possible by directly utilizing the assumed property of the magnetic field $H(\varepsilon, M)$ being a generalized homogeneous function.[10] We hope that our calculated expressions for $h(x)$ will prove useful to those wishing to compare experimental data with model systems.

Acknowledgments

We wish to thank several workers for supplying us with their experimental data in tabular form: John T. Ho and J. David Litster (CrBr$_3$), Norman Menyuk (EuO), James Kouvel and J. Comly (Pd$_3$Fe and Ni), and D. Berkner and J. D. Litster (YIG). One of us (S.M.) is grateful for the hospitality and support provided at the Massachusetts Institute of Technology during the time (August 1972) when this work was put in its final form.

References

1. J.D. van der Waals, Ph.D. Thesis, University of Leiden, 1873.
2. L. Onsager, *Phys. Rev.* **65**, 117 (1944).
3. B. Widom, *J. Chem. Phys.* **43**, 3892, 3898 (1965).
4. C. Domb and D.L. Hunter, *Proc. Phys. Soc.* **86**, 1147 (1965).

5. L.P. Kadanoff, *Physics* **2**, 263 (1966).
6. A.Z. Patashinskii and V.L. Pokrovskii, *Zh. Eksperim. i Teor. Fiz.* **50**, 439 (1966) [*Soviet Phys.—JETP* **23**, 292 (1966)].
7. R.B. Griffiths, *Phys. Rev.* **158**, 176 (1967).
8. B.I. Halperin and P.C. Hohenberg, *Phys. Rev. Lett.* **19**, 700 (1967); *Phys. Rev.* **177**, 952 (1969).
9. E. Riedel and F. Wegner, *Z. Physik* **225**, 195 (1969).
10. A. Hankey and H.E. Stanley, *Phys. Rev. B* **6**, 3515 (1972).
11. J.S. Kouvel and D.S. Rodbell, *Phys. Rev. Lett.* **18**, 215 (1967).
12. A. Arrott and J.E. Noakes, *Phys. Rev. Lett.* **19**, 786 (1967).
13. J.S. Kouvel and J.B. Comly, *Phys. Rev. Lett.* **20**, 1237 (1968).
14. J.T. Ho and J.D. Litster, *Phys. Rev. Lett.* **22**, 603 (1969); *Phys. Rev. B* **2**, 4523 (1970).
15. K. Miyatani, *J. Phys. Soc. Japan* **28**, 259 (1970).
16. N. Menyuk, K. Dwight, and T.B. Reed, *Phys. Rev. B* **3**, 1689 (1971).
17. J.S. Kouvel and J.B. Comly, in *Critical Phenomena in Alloys, Magnets, and Superconductors*, R.E. Mills, E. Ascher, and R.I. Jaffee, eds., McGraw-Hill, New York (1971).
18. D. Berkner and J.D. Litster, *AIP Conference Proceedings*, **10**, 894 (1973).
19. D.S. Gaunt and C. Domb, *J. Phys. C* **3**, 1442 (1970).
20. S. Fisk and B. Widom, *J. Chem. Phys.* **50**, 3219 (1969); P.G. Watson, *J. Phys. C* **2**, 1883, 2158 (1969); L.P. Kadanoff, Newport Beach Conference, Jan. 1970; and in *Proc. Enrico Fermi Summer School of Physics, Varenna, 1970*, M.S. Green, ed., Academic Press, New York (1972); C. Domb, in *Statistical Mechanics at the Turn of the Decade*, E.G.D. Cohen, ed., Marcel Dekker, New York (1971); R.B. Griffiths, *Phys. Rev. Lett.* **24**, 1479 (1970).
21. K.G. Wilson, *Phys. Rev. B* **4**, 3174, 3184 (1971); K.G. Wilson and M.E. Fisher, *Phys. Rev. Lett.* **28**, 248 (1972); K.G. Wilson, *Phys. Rev. Lett.* **28**, 540 (1972); and F.J. Wegner, *Phys. Rev. B* **5**, 4529 (1972).
22. P. Schofield, *Phys. Rev. Lett.* **22**, 606 (1969); B.D. Josephson, *J. Phys. C* **2**, 1113 (1969).
23. H.E. Stanley, *Introduction to Phase Transitions and Critical Phenomena*, Oxford Univ. Press, London (1971).
24. S. Milošević and H.E. Stanley, *Phys. Rev. B* **6**, 986 (1972).
25. S. Milošević and H. E. Stanley, *Phys. Rev. B* **6**, 1002 (1972).
26. A. Hankey, Ph.D. Thesis, Massachusetts Institute of Technology, 1972.
27. D.S. Gaunt and G.A. Baker, *Phys. Rev. B* **1**, 1184 (1970).
28. J.W. Essam and D.L. Hunter, *J. Phys. C* **2**, 392 (1968).
29. M.F. Sykes, J.W. Essam, and D.S. Gaunt, *J. Math. Phys.* **6**, 283 (1965).
30. G.A. Baker, Jr., J. Eve, and G.S. Rushbrooke, *Phys. Rev. B* **2**, 706 (1970).
31. R.F. Wielinga, in *Progress in Low Temperature Physics*, C.J. Gorter, ed., North-Holland, Amsterdam (1970), Vol. VI, pp. 333–373.
32. J. Als-Nielsen, O.W. Dietrich, W. Kunnmann, and L. Passell, *Phys. Rev. Lett.* **27**, 741 (1971).
33. R.L. Stephenson and P.J. Wood, *Phys. Rev.* **173**, 475 (1968).
34. R.L. Stephenson and P.J. Wood, *J. Phys. C* **3**, 90 (1970).
35. L. Tisza, *Generalized Thermodynamics*, MIT Press, Cambridge, Mass. (1966).
36. M.E. Fisher, *Physics* **3**, 255 (1967).
37. G. Joyce, in *Phase Transitions and Critical Phenomena Vol. 2*, C. Domb and M.S. Green, eds., Academic Press, London (1972).
38. M.E. Fisher, *J. Appl. Phys.* **38**, 981 (1967).
39. B. Chu, F.J. Schoenes, and M.E. Fisher, *Phys. Rev.* **185**, 219 (1969).
40. G. Paul and H.E. Stanley, *Phys. Rev. B* **5**, 3715 (1972).
41. G. Paul and H.E. Stanley, *Phys. Lett.* **37A**, 347 (1971); *Phys. Rev. B* **5**, 2578 (1972).
42. P. Schofield, J.D. Litster, and J.T. Ho, *Phys. Rev. Lett.* **23**, 1098 (1969), and references contained therein.

Properties of PrPb₃ in Relation to Other $L1_2$ Phases of Pr

E. Bucher, K. Andres, A. C. Gossard, and J. P. Maita

Bell Laboratories
Murray Hill, New Jersey

The low-temperature electronic properties of $PrPb_3$ have recently received considerable attention due to the potential of this compound for nuclear adiabatic cooling. As will be discussed below, a consistent interpretation of the available data requires that the Pb ligands carry a negative (or possibly close to zero) effective point charge and that an appreciable degree of antiferromagnetic exchange coupling exists. A review of the properties of isomorphous PrX_3 compounds indicates that the effective point charge may be correlated with the electronegativity of the ligands.

There exist five PrX_3 compounds having the cubic $L1_2$ structure in which the Pr^{3+} ion has $z = 12$ nearest neighbors in dodecahedral symmetry. The energy levels of Pr^{3+} in these compounds (X = In, Sn, Tl, Pb, or Pd) should therefore closely resemble those found in rocksalt structure compounds (octahedral nearest neighbors). In the notation of Lea et al.[1] the fourth- and sixth-order point-charge potentials in the first-neighbor approximation are*

$$B_4 = -\frac{7}{32}\frac{Ze^2}{R^5}\langle r^4 \rangle \beta \qquad (1)$$

$$B_6 = -\frac{39}{256}\frac{Ze^2}{R^7}\langle r^6 \rangle \gamma \qquad (2)$$

where B_4 differs by a factor $-1/2$ and B_6 by $-13/4$ with respect to octahedral coordination. Using the level diagram of Ref. 1 for $J = 4$, $x < 0$, and $W < 0$, we see that the crystal-field ground state may be either Γ_3 or Γ_5 for negatively charged (point charge) neighbors, but must be a Γ_1 singlet for positively charged neighbors.

Figure 1 shows the inverse molar susceptibility χ_m^{-1} vs. temperature T of $PrPb_3$ and the molar magnetic moment σ_m in fields of up to 50 kOe at 1°K, both indicating a nonmagnetic crystal-field ground state. Figure 2 further proves that this nonmagnetic crystal-field ground state of Pr^{3+} holds throughout the whole system of $La_{1-x}Pr_xPb_3$, as indicated by the slow decrease of the superconducting transition temperature T_s vs. x for $LaPb_3$[3] and the Van Vleck susceptibility $\chi_m(0)$. The concentration dependence of $\chi_m(0)$ is rather large. The strong increase of $\chi_m(0)$ from $PrPb_3$ to the dilute limit in $LaPb_3$ is predominantly due to the antiferromagnetic exchange forces in the concentrated material. Judging from the Néel temperature $T_N = 2.7°K$ in $NdPb_3$, we estimate that in $PrPb_3$ one is very close to the critical exchange leading to an induced moment system. From our specific heat results

---
* The calculation follows Ref. 2.

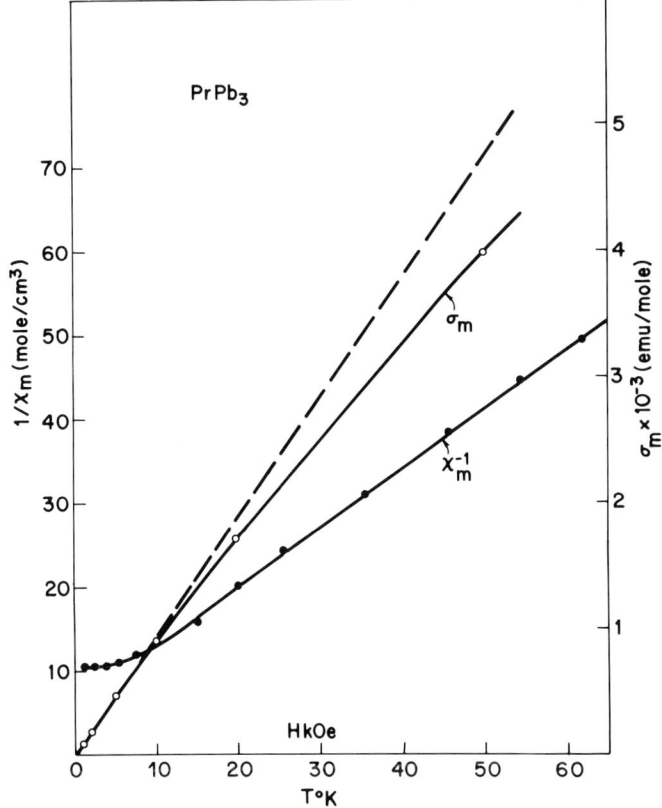

Fig. 1. Inverse molar susceptibility χ_m^{-1} vs. T and molar magnetic moment σ_m vs. magnetic field H at 1 K.

there is no evidence for a detectable shift in the crystal-field splitting between the lowest two levels in $(La_{0.90}Pr_{0.10})Pb_3$[3] and $PrPb_3$,[4] which might be invoked to explain the change in Van Vleck susceptibility.

Hyperfine-enhanced NMR could not be detected in $PrPb_3$, presumably because of the line broadening due to the almost critical exchange. Usually this was easy to observe in many other Pr intermetallic compounds where Pr is in a cubic point symmetry and in a singlet ground state.[5] Likewise, attempts to cool $PrPb_3$ by nuclear adiabatic cooling were not very successful. Some cooling was observed. Starting from about 15 kOe and 130 m°K about 90 m°K was reached. The experiment indicated a large specific heat load when cooling the material down. Subsequent specific heat measurements of $PrPb_3$ between 0.04 and 1.0°K indeed indicated a specific heat anomaly with a maximum occurring at 0.35°K, as shown in Fig. 3.

The only consistent interpretation of these data is the assumption of a non-Kramers (nonmagnetic) ground-state doublet Γ_3. The integral $\int_{0.04}^{1°K}(C_m/T)\,dT$ is indeed found to be close to $R \ln 2$. Within the point-charge approximation given by Lea et al.[1] a Γ_3 non-Kramers doublet in $PrPb_3$ could be understood if the Pb ligand carries a negative charge and if the sixth-order crystal-field potential is relatively strong. This would be consistent with the properties of the isoelectronic

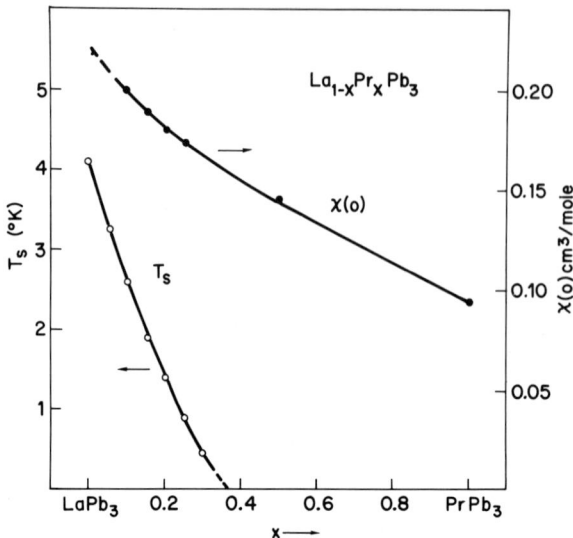

Fig. 2. Superconducting transition temperature T_s and Van Vleck susceptibility $\chi_m(0)$ vs. praseodymium concentration x in $La_{1-x}Pr_xPb_3$.

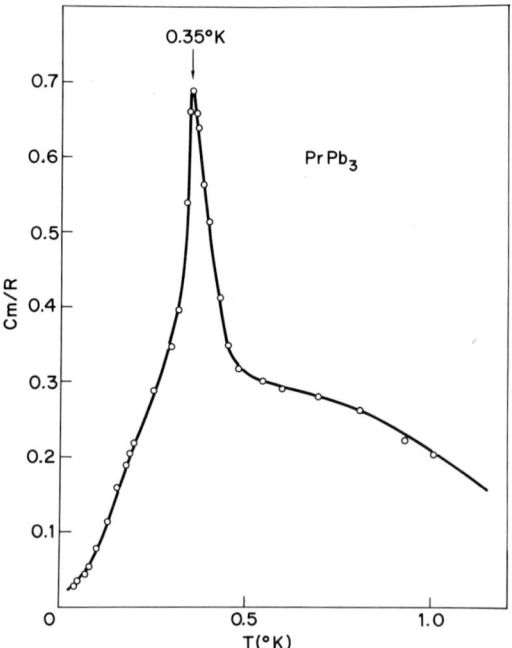

Fig. 3. Molar specific heat C_m/R of $PrPb_3$ between 0.04 and 1.0°K.

Table I. Lattice Constant a, Néel Temperature T_N, Van Vleck Susceptibility $\chi_m(0)$, and Crystal-Field Splittings between the Lowest Two Levels for All Five PrX$_3$ Phases with L1$_2$(Cu$_3$Au) Structure*

Compound	$a(L1_2)$ (this work)	T_N, °K	$\chi_m(0)$, cm^3/mole	Point-charge crystal-field ground state	Splitting between lowest CF levels, °K
PrIn$_3$	4.67$_0$	—	0.025	Γ_1	124 ± 5
PrSn$_3$	4.71$_5$†	7.92	—	Γ_5	Not determined
PrTl$_3$	4.74$_5$	—	0.096	Γ_1	37 ± 2
PrPb$_3$	4.86$_5$	—	0.095	Γ_3	18 ± 2
PrPd$_3$	4.13$_6$‡	1.05	—	Γ_5	Not determined

* Parameters determined from specific heat data. The electronic and lattice parts were subtracted from the corresponding LaX$_3$ compounds.
† In agreement with Ref. 9.
‡ In agreement with Ref. 10.

system PrSn$_3$, where the ground state is a magnetic triplet (Γ_5 if Sn is assigned a negative point charge). This conclusion can be drawn again by evaluating the entropy under the λ-type anomaly occurring at 7.92°K in PrSn$_3$. It is found to be close to $R \ln 3$.

It is interesting at this point to consider the magnetic properties of all PrX$_3$ compounds together, particularly as far as their crystal-field ground state is concerned. All their magnetic properties are summarized in Table I. In PrIn$_3$ and PrTl$_3$ it can be shown unambiguously by susceptibility and specific heat measurements that the ground state is a singlet followed by a triplet Γ_4. Therefore one has to assign In and Tl a *positive* point charge. Moreover, the crystal-field splitting of Γ_1–Γ_4 in PrTl$_3$ can be quantitatively calculated from the lattice constant and a point charge of $Z = +3$ (Tl can be monovalent or trivalent). We calculate 35°K, as compared with the experimentally observed value of 37 ± 2°K.[4] In the case of PrPd$_3$ the ground state again is magnetic, as concluded from the magnetic properties and specific heat in the Pr$_{1-x}$La$_x$Pd$_3$ system. Therefore one has to assume, as in the case of PrSn$_3$ and PrPb$_3$, that Pd also carries a negative charge. Recent Mössbauer effect studies[6] in YbPd$_3$ also suggest from the observed level sequence Γ_7-Γ_8-Γ_6 (with Γ_7 lowest) to assign Pd zero or a weakly negative point charge.

If the influence of the six second-nearest Pr neighbors is taken into account, one has to replace $Z (\equiv Z_1)$ in (1) and (2) by $Z_1 - (1/2\sqrt{2}) Z_2$ and $Z_1 - (1/26\sqrt{2}) Z_2$, respectively. One can also explain the ground states in PrSn$_3$, PrPb$_3$, and PrPd$_3$ assuming $|Z_1| \sim 0$ and $Z \gg 1$. However, this is not very reasonable and there is no evidence of such a strong influence of second nearest neighbors in similar compounds.[7]

We will not consider here the accuracy of the point-charge model; the good agreement with the point-charge calculation in the case of PrTl$_3$ might be considered as a mere coincidence. It is more important here to point out that in intermetallic compounds both charge signs do exist and that the sign of the total effective point charge will, in general, determine the nature of the crystal-field ground state.

One might try to go one step further and consider all PrX$_3$ phases as one case in the totality of rare earth intermetallic phases. The question at hand is whether it is possible to unravel empirical parameters which could predict at least the sign of the point charge with reasonable reliability (e.g., irrespective of crystal structure).

Intuitively, a correlation of the ligand point charge sign and strength with the electronegativity ε is expected. A preliminary statistical survey recently made[7] in the case of all analyzed Pr compounds indeed reveals a change of the point-charge sign occurring at a value of $\varepsilon \sim 1.7$ (the ε values were taken from Gordy and Thomas' table[8]); Pr itself has an ε value of 1.1. However, the scope of this speculation goes beyond the present paper. A detailed study will be published elsewhere; in particular, it will include transition metal compounds in which the role of the d-electron contribution to the crystal-field splitting is still obscure.

Acknowledgments

We would like to thank Mrs. A.S. Cooper for X-ray analysis and W.M. Walsh, Jr. for constructive comments on this work.

References

1. K.R. Lea, M.J.M. Leask, and W.P. Wolf, *J. Phys. Chem. Solids* **23**, 1381 (1962).
2. M.T. Hutchings, *Solid State Physics*, F. Seitz and D. Turnbull eds., Academic Press, New York (1964), Vol. 16.
3. E. Bucher, K. Andres, J.P. Maita, and G.W. Hull, Jr., *Helv. Phys. Acta* **41**, 723 (1968).
4. E. Bucher, K. Andres, J.P. Maita, A.S. Cooper, and L.D. Longinotti, *J. de Physique* **32** (Cl), 114 (1971).
5. E. Bucher, A.C. Gossard, K. Andres, J.P. Maita, and A.S. Cooper, in *Proc. 8th Rare Earth Res. Conf.*, U.S. GPO, Washington D.C. (1970) p. 74.
6. I. Nowik, B.D. Dunlap, and G.M. Kalvius, *Phys. Rev. B* **6**, 1048 (1972).
7. E. Bucher and J.P. Maita, *Sol. State Commun.* **13**, 215 (1973).
8. W. Gordy and W.J.O. Thomas, *J. Chem. Phys.* **24**, 439 (1956).
9. A. Iandelli, in *Constitution of Binary Alloys*, R.P. Elliott, ed., McGraw-Hill, New York (1965).
10. I.R. Harris and G.V. Raynor, *J. Less Common Metals* **9**, 263 (1965).

On Nuclear Magnetic Ordering Phenomena in Van Vleck Paramagnetic Materials

K. Andres

Bell Laboratories
Murray Hill, New Jersey

Magnetic interactions between nuclei of Van Vleck paramagnetic ions are greatly enhanced in the presence of magnetic exchange interactions between the ions. This leads to relatively large and easily observable nuclear magnetic susceptibilities in the temperature range below 1°K and to cooperative nuclear magnetic ordering phenomena in the millikelvin range. If the exchange interactions exceed a critical value, the singlet ground state becomes unstable and electronic order occurs at low temperatures. For exchange interactions close to the critical value (but somewhat below it) a mixed electronic–nuclear phase transition can be expected. This critical region is examined in the molecular field approximation and the results are discussed and compared with calculations by Murao.[1]

Reference

1. T. Murao, *J. Phys. Soc. Japan* **31**, 683 (1971); and **33**, 33 (1972).

Electronic Magnetic Ordering Induced by Hyperfine Interactions in Terbium and Holmium Gallium Garnets

J. Hammann and P. Manneville

Commissariat à l'Energie Atomique, CEN
Saclay, Gif-sur-Yvette, France

Introduction

Previous susceptibility measurements on the terbium and the holmium gallium garnets (TbGaG and HoGaG) by Wolf et al.[1] down to 1°K and by Cooke et al.[2] down to 0.6°K indicated no magnetic ordering for these compounds. The results have been used in the case of TbGaG by Ayant et al.[3] to evaluate the crystal-field splitting of the Tb^{3+} ion and they found two low-lying singlet energy levels well isolated from the upper ones with a separation of $\Delta = 4.2°K$ and a magnetic moment at saturation m_s of $7.35\mu_B$. In the case of HoGaG, Onn and Meyer published specific heat measurements[4] which are consistent with the same two-singlet scheme with $\Delta = 7.4°K$.

In order to gain more information on the energy levels of these rare earth ions and to study the possibility of a magnetic ordering in these garnets, we performed magnetization, anisotropy, and specific heat measurements down to 0.36°K. The results, which will be published elsewhere, did not show any transition down to this temperature and in both cases fit well a two-singlet model with parameters given in Table I. The only nonvanishing component of the magnetic moment of each ion is found to lie along one of the local twofold symmetry axes, namely [001] for the ions located at $\pm [0, \frac{1}{4}, \frac{1}{8}]$, [100] for those at $\pm [\frac{1}{8}, 0, \frac{1}{4}]$, and [010] for those at $\pm [\frac{1}{4}, \frac{1}{8}, 0]$.

Transition Condition

From the crystal-field point of view the situation of the garnets considered here is therefore the same as for the terbium aluminum garnet (TbAlG), which has been studied in detail (see, for instance, Ref. 5). Within the model which has been developed for TbAlG the molecular field transition temperature is given by $x = Th\Delta/2T_N$, where $x = \Delta/2\beta m_s^2$ and β is the molecular field constant, which we shall assume to be mainly due to dipolar interactions. The constant β depends on the particular magnetic structure considered; it has been shown in Ref. 6 that the value of β that minimizes the dipolar magnetic energy corresponds to the AFA structure, which has already been found for TbAlG[7] and which consists of six antiferromagnetic sublattices with magnetic moments lying along $\pm[001]$, $\pm[100]$, and $\pm[010]$.

This structure leads to the value of β given in the table below along with the

Table I

	HoGaG	TbGaG
Δ, °K	7.40	2.87
m_s, μ_B	7.69	6.68
β, G/μ_B	467.8	469.9
x_{dip}	1.955	1.019
a_J,[10] 10^{-2} °K	3.90	2.54
T_N(dip)	0.103	0.345
T_N(exp)	0.19	0.25
x_{exp}	1.45	1.03

corresponding value of x. From these values it appears that no purely electronic magnetic ordering should be possible. This is first supported by the fact that no transition has been observed down to 0.36°K which would already lead to values of x very close to one, and second by the fact that the molecular field transition condition allows a much higher limiting value of x than more elaborated theories.[8]

This model neglects the hyperfine interactions, which may be expected to become very important at the lower temperatures since the Tb and the Ho nuclei have nuclear spins of 3/2 and 7/2, respectively. Murao[9] has already studied such systems with coupled nuclei and electrons and showed that the hyperfine interactions could induce electronic magnetic ordering.

In our present case we write the one-ion Hamiltonian as

$$\mathcal{H} = V_c - g_J\mu_B \mathbf{J} \cdot \mathbf{H} - a_J \mathbf{J} \cdot \mathbf{I}$$

where V_c represents the crystal-field Hamiltonian, which leads to the splitting Δ of the two lowest-lying singlets; the electric quadrupole part of the hyperfine interactions is much smaller than the magnetic dipolar part $a_J \mathbf{J} \cdot \mathbf{I}$, and has been neglected.

As has already been stated, the effects related to the excited energy levels which are much higher than the two lowest-lying ones may be neglected and this allows a formal exact treatment of the above Hamiltonian within the molecular field approximation. In particular, one obtains for the electronic magnetic transition temperature an exact expression which leads to the following expression when expanded to the second order in α: with

$$\alpha = 2 \frac{m_s}{g_J\mu_B} \frac{a_J}{\Delta} \left[\frac{I(I+1)}{3}\right]^{1/2}$$

$$x = Th\frac{\Delta}{2T_N} + \alpha^2 \left[\frac{\Delta}{4T_N}\left(3 - th^2\frac{\Delta}{2T_N}\right) - \frac{3}{2}th\frac{\Delta}{2T_N}\right]$$

From that expression, and in contrast to the case of a purely electronic system, it appears that the magnetic ordering is always possible and, further, the transition temperature is simply given by

$$T_N = \frac{\Delta}{2}\frac{\alpha^2}{x-1} \quad \text{when} \quad T_N \ll \Delta$$

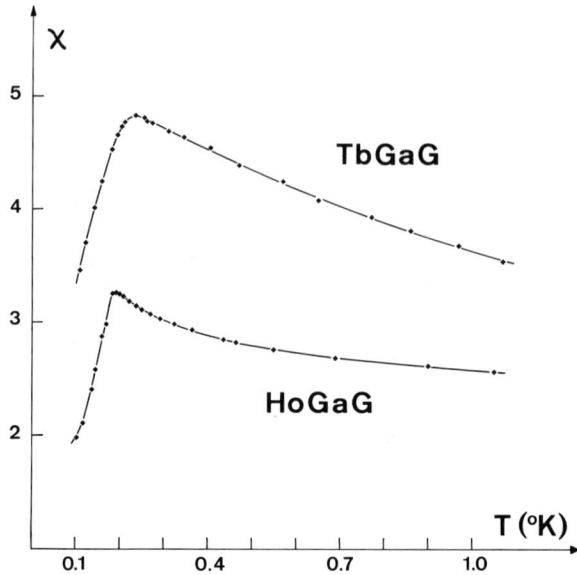
Fig. 1. Susceptibility (in emu) of TbGaG and HoGaG.

In Table I values of T_N calculated with the exact expression have been reported together with the experimental values we obtained from susceptibility and specific heat measurements.

Susceptibility Measurements

The susceptibilities were measured with a mutual inductance bridge working with an alternative field of an amplitude of 3 G and a frequency of 20 Hz. The sample consisted of approximately 1 g of powder obtained by grinding a monocrystal and the cooling was obtained with an adiabatic demagnetization device using potassium chromium alum.

The determination of the temperature was done by use of a 220-Ω Speer carbon resistor the resistance of which was measured with an alternative Wheatstone bridge dissipating 2×10^{-10} W in 10 kΩ. The carbon resistor was calibrated against susceptibility measurements on potassium chromium alum and on CMN. Unfortunately, we found that this calibration did not remain valid at the lowest temperatures from one run to another, despite the fact that the resistor had been cycled from room temperature to 4.2°K many times. The estimated error in the absolute temperature determination was approximately 5% between 0.1 and 0.3°K. This could not be improved because of the lack of experimental room to measure simultaneously the susceptibilities of the sample and of CMN.

Within these restrictions Fig. 1 shows the temperature dependence of the susceptibilities of TbGaG and HoGaG. The two curves display a sharp maximum and decrease very quickly at the lowest temperatures. Their behavior is thus in agreement with the theoretical statement of an antiferromagnetic ordering. The

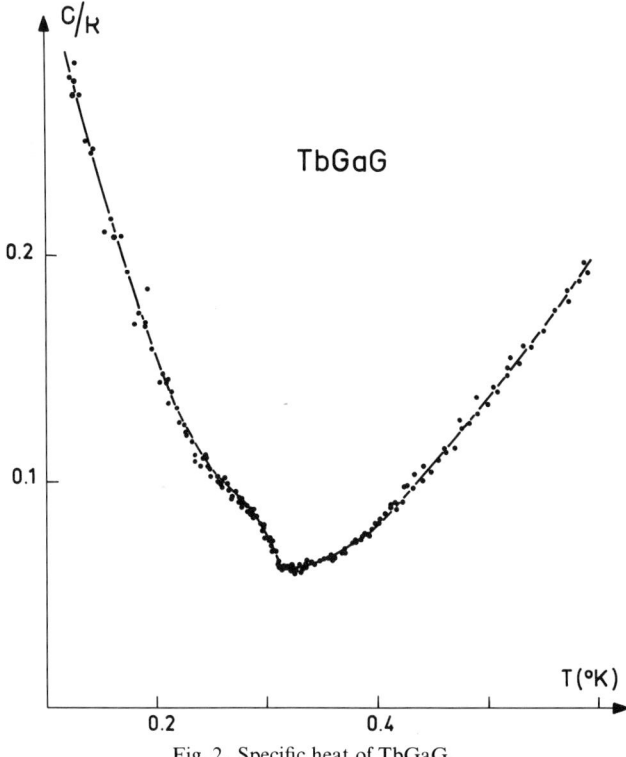

Fig. 2. Specific heat of TbGaG.

transition temperatures given in Table I have been taken as corresponding to the maximum of the susceptibilities.

Specific Heat Measurements

The specific heat measurements were performed with the same cryogenic device; the only difference was that the sample was now coupled to the demagnetization stage through a superconductive thermal switch consisting of a tin wire which was located on the fringe of the superconductive coil. The measurements were made by the usual heat pulse method, heat being supplied to the sample through a resistance which was wound inside the mixture of the ground crystal and silicon grease charged with tiny pieces of copper. The temperature was measured with the same 220-Ω Speer carbon resistor.

The results are shown in Figs. 2 and 3. They confirm clearly the results of the susceptibility measurements, since the observed singularities coincide, within the temperature calibration errors. However, though these singularities do exist, their shape could not be well defined. This was due to the high-temperature increase during every heat pulse which was necessary to have a correct measurement and could be comparable to the width of the singularity, and to the thermal contact among the heater, the sample, and the thermometer, which was not sufficient at the lowest temperatures, where the specific heat increased quickly. This was par-

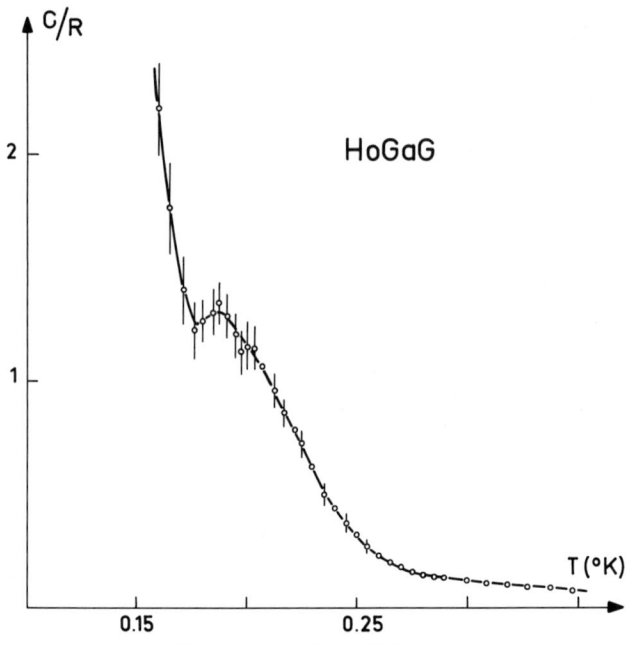

Fig. 3. Specific heat of HoGaG.

ticularly true in the case of HoGaG, the specific heat of which is much greater than that of TbGaG, and where the relative dispersion of the great number of experimental points led us to report only average values with their corresponding maximum error limits.

In TbGaG the electronic Schottky anomaly became preponderant above 0.3°K, whereas the nuclear contribution increased very sharply below the singularity. In HoGaG the electronic Schottky anomaly shifted toward much higher temperatures and the nuclear contribution was much greater due to greater value of the nuclear spin; the transition singularity was located within the nuclear Schottky anomaly.

In both cases if there were no polarization of the electronic magnetic moments at temperatures lower than that of the singularity, the nuclear specific heat would be much smaller, even taking into account the perturbation due to higher excited levels. This is again in agreement with the presence of an electronic magnetic ordering.

Conclusion

From the above discussion it appears that TbGaG and HoGaG undergo a magnetic transition to an antiferromagnetic structure which can be predicted as being of the AFA type as in TbAlG and HoAlG.[7] This transition can be understood as induced by hyperfine interactions. Further experiments are needed in order to check the magnetic structure and to fully describe the magnetic behavior. In particular, it would be very interesting to measure the temperature dependence of the electronic and the nuclear magnetic moments, since their theoretical behavior shows a very unusual shape which has been found by Murao[11] for similar systems.

References

1. W.P. Wolf, M. Ball, M.T. Hutchings, M.J.M. Leask, and A.F.G. Wyast, *J. Phys. Soc. Jap.* **17** (suppl. B1), 443 (1962).
2. A.H. Cooke, T.L. Thorp, and M.R. Wells, *Proc. Phys. Soc.* **92**, 400 (1967).
3. Y. Ayant, A. Belorizky, and J. Rosset, *Czech. J. Phys.* **B17**, 361 (1967).
4. D.G. Onn and H. Meyer, *Phys. Rev.* **156**, 663 (1967).
5. A. Gavignet-Tillard and J. Hammann, in *Proc. 12th Intern. Conf. Low Temp. Phys.* Academic Press of Japan, Tokyo 1971, p. 697.
6. H.W. Capel, R. Bidaux, P. Carrara, and B. Vivet, *Phys. Lett.* **22**, 400 (1966).
7. J. Hammann, *Acta Cryst.* **B25**, 1853 (1969).
8. Y-L. Wang and B.R. Cooper, *Phys. Rev.* **172**, 539 (1968).
9. T. Murao, *J. Phys. Soc. Japan* **31**, 683 (1971).
10. A. Abragam and B. Bleaney, *EPR of Transition Ions*, Clarendon Press, Oxford (1970).
11. T. Murao (to be published).

Low-Field Magnetic Properties of $DyVO_4$ and $TbPO_4$

H. Suzuki and T. Ohtsuka

Department of Physics, Faculty of Science
Tohoku University, Sendai, Japan

and

T. Yamadaya

Matsushita Research Institute
Kawasaki, Japan

Introduction

$DyVO_4$ and $TbPO_4$ belong to the family of rare earth phosphates, arsenates, and vanadates, which have the tetragonal zircon structure. In particular, $DyVO_4$ has attracted great interest[1-15] since it shows a cooperative Jahn–Teller distortion to orthorhombic symmetry at $T_D \cong 14°K$ accompanied by a change in magnetic properties. In the paramagnetic region below T_D the principal g values become very anisotropic, with $g_a = 19 \gg g_b, g_c$, where the c axis corresponds to the tetragonal axis. An interesting feature is that the g_a axis can be rotated by application of a magnetic field. At $T_N \cong 3°K$, $DyVO_4$ becomes antiferromagnetic. Due to the large g-value anisotropy, the state can be regarded as an Ising antiferromagnet.

One of the objects of the present work is to utilize the high sensitivity of a SQUID (superconducting quantum interference detector) to measure in detail the dc magnetic susceptibility in very low fields, particularly near T_D, to shed light on the relation between the magnetic properties and the crystal distortion.

In contrast to $DyVO_4$, there appears to be no crystal lattice transition in $TbPO_4$ right down to the magnetic transition temperature $T_N \cong 2.2°K$. Work by Lee, Moos, and co-workers[16-18] shows that the transition is into a canted antiferromagnet with two collinear sublattices 40° off the tetragonal c axis, and that a Jahn–Teller distortion accompanies the magnetic transitions. Detailed magnetic measurements by a SQUID have been made on this substance near T_N.

Experiment

Single crystals of $DyVO_4$ were grown by a flux method. The crystals used in this study were about $1 \times 2 \times 4$ mm in size with the tetragonal axis lying in the long direction. Single crystals of $TbPO_4$ were grown from $Pb_2P_2O_7$ flux. Two samples in the form of parallelepipeds were used, with dimensions of about $0.3 \times 0.9 \times 10$ mm and $0.3 \times 2 \times 6$ mm, respectively, with the tetragonal axis lying in the long directions. The SQUID used was a double-point contact type made from niobium, similar to that described by Zimmerman and Silver.[19] With an appropriate feedback circuit, changes in magnetic fields of the order of 10^{-8} G could be detected.

Results and Discussion

The susceptibility χ of $DyVO_4$ was measured from 1.7 to 70°K in fields ranging from 0.5 to 50 Oe applied parallel to the a axis. Figure 1 shows the overall temperature dependence of χ. An anomaly at $T_D \cong 14°K$ and an antiferromagnetic transition at $T_N \cong 3°K$ is noted in accordance with previous results. In the field range measured no dependence of χ on the applied field was detected. Therefore χ may be interpreted as equivalent to the zero-field dc susceptibility. The susceptibility for $T > T_D$ followed Curie's law very well, whereas for the paramagnetic region $T_N < T < T_D$ it appeared to follow the Curie–Weiss law with a Weiss constant $\theta = -3°K$. This may be indicative of a change in interaction upon crystal distortion. In the vicinity of the Jahn–Teller transition temperature T_D a fairly sharp change in the susceptibility was noted. Figure 2 shows a recording of the output of the SQUID when the temperature was drifted slowly through T_D, where it is seen that the change takes place within a temperature region of about 0.1°K. This is in contrast with the previous report by Hudson and Mangum,[8] which shows a much broader transition. The reason for this discrepancy is not clear, although it may be due to the difference in the samples used. Since the transition was fairly sharp and the sensitivity of the detector was sufficient, we looked into the critical behavior of this transition. A

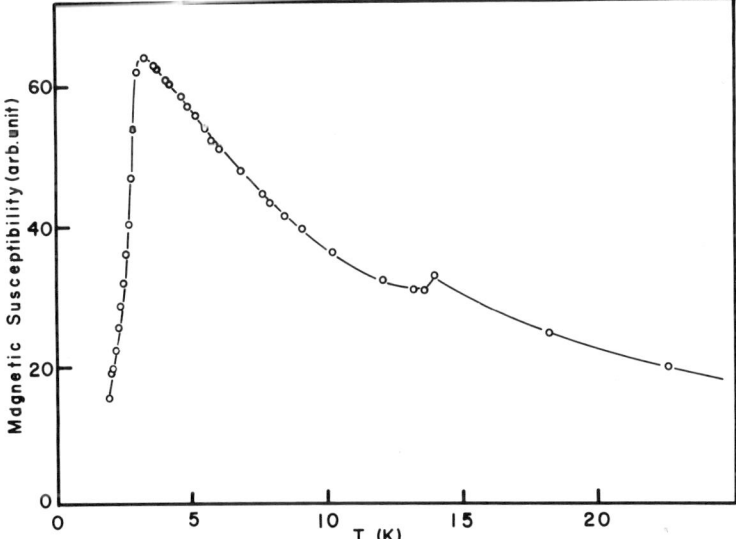

Fig. 1. Magnetic susceptibility of $DyVO_4$. The open circles represent the data obtained from the continuous recording of the output of the SQUID in the temperature range 1.7–70°K. A field of ~15 Oe was applied parallel to the a axis. In the field range measured (0.5–50 Oe) no dependence of χ on the applied field was obtained.

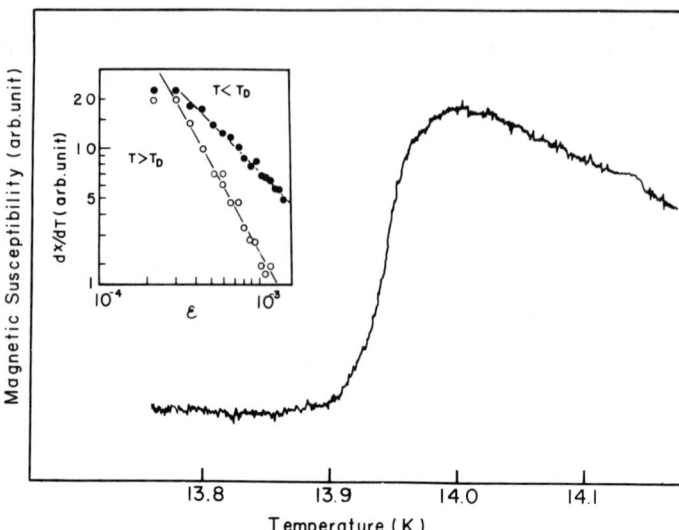

Fig. 2. A recording of the output of the SQUID in the vicinity of T_D. A field of 25 Oe was applied parallel to the a axis. The inset shows a plot of $d\chi/dT$ vs. $\varepsilon \equiv |1 - T/T_D|$ obtained from the figure. Here T_D was taken as the temperature at which $d\chi/dT$ takes a maximum. For $T < T_D$ we obtain $d\chi/dT \propto |1 - T/T_D|^{-1}$, and for $T > T_D$, $d\chi/dT \propto |1 - T/T_D|^{-1.73}$.

detailed examination shows that the susceptibility may be fitted to the equation $\chi = A \log |1 - T/T_D| + B$ for $T < T_D$ within the temperature region $2 \times 10^{-3} < |1 - T/T_D| < 2 \times 10^{-2}$; and to $\chi = A'|1 - T/T_D|^{-0.73} + B'$ for $T > T_D$ within the region $3 \times 10^{-3} < |1 - T/T_D| < 1.5 \times 10^{-2}$. Here T_D was taken as the temperature at which $d\chi/dT$ takes a maximum. A plot of $d\chi/dT$ versus $\varepsilon \equiv |1 - T/T_D|$ is shown in the inset of Fig. 2. The significance of this critical behavior is not clear and we present it as an observation to be clarified in the future. Experimentally speaking, more work is necessary with different samples. We remark that within the low applied fields used no field dependence of the susceptibility near T_D was noted and that the susceptibility at $T < T_D$ pertains to a multidomain configuration.

The overall temperature dependence of the susceptibility in TbPO$_4$, observed with the field ($\sim$ 19 Oe) applied parallel to the c axis, is shown in Fig. 3. An antiferromagnetic-like transition at $T_N \cong 2.1°K$ is seen, except for the appearance of an anomalous peak directly below T_N. No anomaly which might correspond to a lattice transition was visible above T_N and the susceptibility followed the Curie–Weiss law with the Weiss constant $\theta = -6°K$. The anomalous peak at T_N is very sensitive to applied fields, as is shown in the inset of Fig. 3, where it is seen that a field of 25 Oe depresses the peak appreciably. It is also noted that substantial hysteresis is present. The results for the second TbPO$_4$ specimen were similar, except for the behavior of the anomalous peak. In contrast to the first specimen, the peak was already depressed at a field of only 5 Oe. The origin of this peak is not clear, but it may be related to the canted antiferromagnetic structure reported by Lee, Moos, co-workers.[16,17] The difference of the behavior of the peak between the two specimens and the presence of hysteresis may be related to the simultaneous occurrence of crystal distortion, which may give rise to different domain structures, depending on the treatment. More work is necessary to clarify this point.

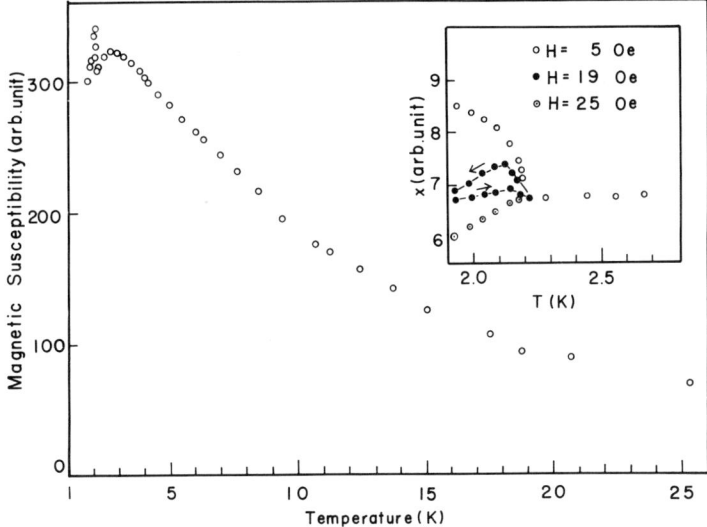

Fig. 3. Magnetic susceptibility of TbPO$_4$. Open circles represent the data obtained from the continuous measurements. A field of about 19 Oe was applied parallel to the c axis. The inset shows the enlarged portions of transition temperature region in some different applied fields parallel to the c axis. A typical hysteresis is also shown in the inset.

Acknowledgment

We wish to thank Dr. T. Fujita, who constructed part of the SQUID circuitry used in this work.

References

1. P.J. Becker, G. Dummer, H.G. Kahle, L. Klein, G. Müller-Vogt, and H.C. Schopper, *Phys. Lett.* **31A**, 499 (1970).
2. A.H. Cooke, C.J. Ellis, K.A. Gehring, M.J.M. Leask, D.M. Martin, B.M. Wanklyn, M.R. Wells, and R.L. White, *Sol. State Comm.* **8**, 689 (1970).
3. A.H. Cooke, D.M. Martin, and M.R. Wells, *J. Phys. (Paris)* **32** (C1), 488.
4. C.J. Ellis, K.A. Gehring, M.J.M. Leask, and R.L. White, *J. Phys. (Paris)* **32** (C1), 1024.
5. K.W.H. Stevens, *J. Phys. C: Sol. State Phys.* **4**, 2297 (1971).
6. F. Sayetat, J.X. Boucherle, M. Belakhovsky, A. Kallel, F. Tcheou, and H. Fuess, *Phys. Lett.* **34A**, 361 (1971).
7. J.C. Wright and H.W. Moos, *Phys. Rev. B* **4**, 163 (1971).
8. R.P. Hudson and B.W. Mangum, *Phys. Lett.* **36A**, 157 (1971).
9. J.B. Forsyth and C.F. Sampson, *Phys. Lett.* **36A**, 223 (1971).
10. E. Pytte and K.W.H. Stevens, *Phys. Rev. Lett.* **27**, 862 (1971).
11. R.T. Harley, W. Hayes, and S.R.P. Smith, *Sol. State Comm.* **9**, 515 (1971).
12. A.H. Cooke, D.M. Martin, and M.R. Wells, *Sol. State Comm.* **9**, 519 (1971).
13. G. Gorodetsky, B. Lüthi, and B.M. Wanklyn, *Sol. State Comm.* **9**, 2157 (1971).
14. R.J. Elliott, A.P. Young, and S.R.P. Smith, *J. Phys. C: Proc. Phys. Soc., London* **4L**, 317 (1971).
15. R.L. Melcher and B.A. Scott, *Phys. Rev. Lett.* **28**, 607 (1972).
16. J.N. Lee, H.W. Moos, and B.W. Mangum, *Sol. State Comm.* **9**, 1139 (1971).
17. S. Spooner, J.N. Lee, and H.W. Moos, *Sol. State Comm.* **9**, 1143 (1971).
18. J.N. Lee and H.W. Moos, *Phys. Rev. B* **5**, 3645 (1972).
19. J.E. Zimmerman and A.H. Silver, *Phys. Rev.* **141**, 367 (1966).

Magnetic Ordering of the Delocalized Electron Spin in DPPH

S. Saito and T. Sato

The Research Institute for Iron, Steel and Other Metals
Tohoku University, Sendai, Japan

It is well known that the magnetism of itinerant electrons in metals or alloys remains a fundamental problem to be clarified. This problem also exists in quite a different kind of substance, solid aromatic free radicals having delocalized electrons. In contrast with metals or alloys, however, there are few investigations about the spatial distribution of electron spin in solid aromatic free radicals and about the spin ordering mechanism in these substances at low temperatures.

In addition to shedding light on the magnetism, the data on spin distributions in solid free radicals at low temperatures may be helpful in establishing effective proton nuclear polarization mechanisms, and in developing a thermometer for the ultralow-temperature region. For the latter purpose it is suggested that magnetic transition temperatures of solid free radicals could be lowered by the dilution of the electron spin concentration with corresponding isomorphous derivatives which have no unpaired electrons.

Aromatic free radicals are characteristic in that the spin distributions on their molecular frameworks can be estimated from NMR shifts of the protons bonded to the ring carbons (see, e.g., Ref. 1). Furthermore, by observing the shifts of the proton NMR in the region above and below the magnetic transition temperature, we can estimate the change of electron spin density during the transition.

The stable free radical DPPH is used as a calibration standard for EPR investigations. One of its widely used forms is that obtained by recrystallization from benzene solution, in which case a benzene molecule is incorporated into the crystal structure for each molecule of DPPH. In this crystal every DPPH molecule has one unpaired electron,[2] which is partially delocalized on the molecular framework. The paramagnetic susceptibility[3] of DPPH–benzene crystal obeys the Curie–Weiss law in the range from the ambient temperature down to liquid helium temperature, but a slight deviation from the Curie–Weiss law begins below about 4°K. It has been reported by Prokhorov and Fedorov[4] that the EPR line of this substance disappears between 0.2 and 0.3°K.

In this report we investigate electron spin magnetism in the DPPH–benzene crystal through magnetic susceptibility and proton NMR. Single or powdered DPPH–benzene crystals were cooled by the adiabatic demagnetization method or with a dilution refrigerator. The magnetic susceptibility was measured by a mutual inductance bridge in the temperature region down to 30 m°K. The proton NMR measurements were done with a Robinson-type marginal oscillator detector with

Magnetic Ordering of the Delocalized Electron Spin in DPPH

the sweep of the magnetic field in the temperature region down to 70 m°K. In the NMR measurements with the dilution refrigerator a pair of small modulation coils were used for the small-amplitude magnetic field modulation at 80 Hz in order to minimize the eddy current heating by the modulated field. The modulation coils were attached to the wall of the glass vacuum jacket and near the glass mixing chamber containing the sample. Temperatures of the sample were determined from resistances of a calibrated carbon resistor inserted in the mixing chamber. During the cooling by the adiabatic demagnetization, temperatures were determined from the magnetic susceptibility of the cooling agent, CrK alum or Mn Tutton salt. From both magnetic susceptibility χ and NMR measurements the antiferromagnetic transition of DPPH–benzene crystals was observed at temperatures between 0.18 and 0.25°K. Based on the obtained NMR spectra and χ data we will discuss the spin magnetism of the delocalized electron in the paramagnetic and the ordered state. We have taken an additional step in the exploration of a thermometer working in the ultralow temperature region.

Magnetic Susceptibility χ

The measured χ for the polycrystalline sample of DPPH–benzene in the region down to 30 mK is shown in Fig. 2 (solid curve). The χ measurement for the single crystal was not of sufficient sensitivity, because of the small sample. As is clearly seen, the solid curve in Fig. 2 has a broad shoulder in the region between about 2 and about 0.5°K. This shoulder may be attributed to the short-range order effect near the ordering.

The dilution of the electron spin concentration was tentatively tested. The measured χ for the sample diluted by 9.8% DPP–hydrazine, whose molecular form is shown in Fig. 1, is shown in Fig. 2 by the broken curve: The χ peak observed for

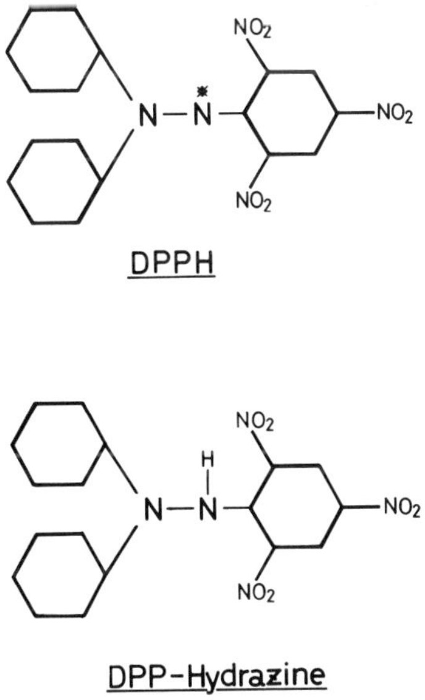

Fig. 1. Structures of DPPH (11-diphenyl-2-picryl-hydrazyl) and DPP–hydrazine (11-diphenyl-2-picrylhydrazine).

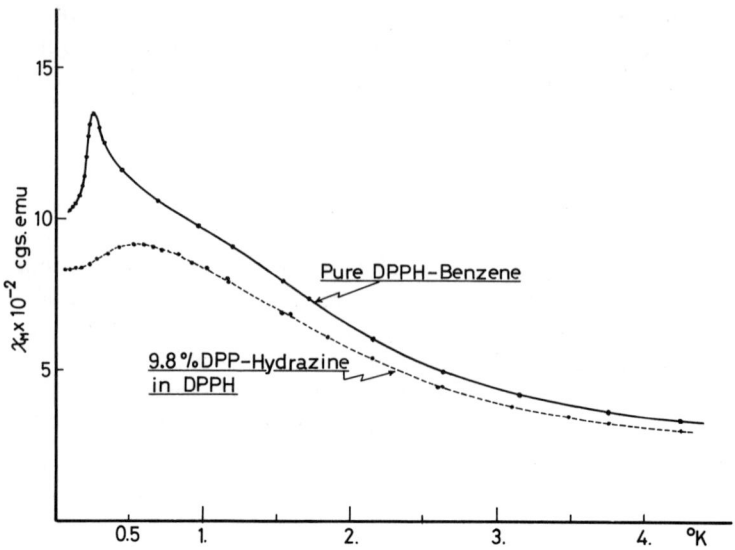

Fig. 2. Temperature dependences of the molar magnetic susceptibility for DPPH–benzene crystal and DPPH(diluted by 9.8% DPP–hydrazine)–benzene crystal.

pure DPPH–benzene crystal near 0.25°K disappeared in the case of the sample diluted by 9.8% DPP–hydrazine, and a broad maximum probably caused by the exaggerated short-range order effect was observed. This broad maximum may correspond to the shoulder observed in the solid curve between 2 and 0.5°K. By further investigation, it may be possible to make use of the diluted substance as a thermometer working in the ultralow-temperature region.

Proton NMR

The proton NMR spectra for the powdered specimen of DPPH–benzene above the liquid helium temperature have been reported by several authors.[5,6] Our experimental results for the powdered and single crystals in the paramagnetic region are shown in Fig. 3 (a, c). The result for the powdered specimen nearly agrees with the results in Ref. 6. As for the spectrum for the single crystal, the number of peaks are more numerous than that for the powdered specimen, as is shown in Fig. 3 (c).

The electron spin density in aromatic free radicals can be calculated with the MO or VB method.[7] Recently Hegyhati[8] calculated correctively the spin density of free DPPH, based on the assumptions that the molecule is planar, that the bonding of two central nitrogens involve sp^2 hybrid orbitals, and that the pi-electronic struc-structure is affected by the NO_2 groups of the picryl ring primarily through the inductive effect. If we apply McConnell's empirical equation $a_H = Q\rho_i$ to this problem, where a_H is the hf splitting of the EPR (which is also evaluated from the NMR) shift from the frequency of the proton, ρ_i is the pi-electron spin density of ith ring carbon,

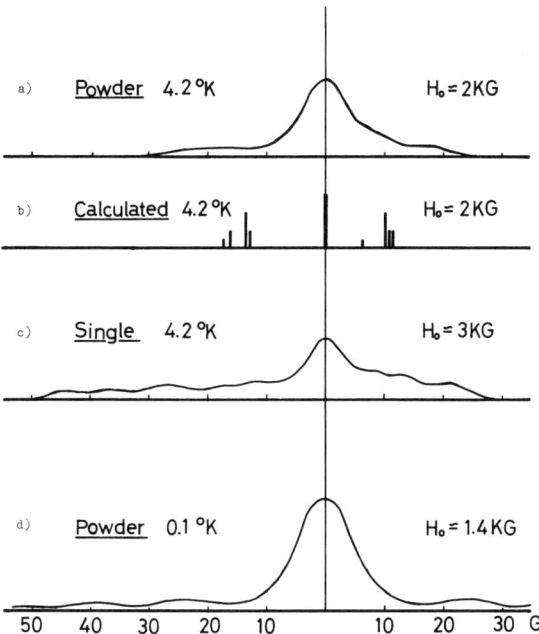

Fig. 3. Proton NMR absorption of DPPH–benzene crystals in (a–c) paramagnetic and (d) antiferromagnetic states.

and Q is empirical constant which is estimated to -22.5 G for the typical aromatic benzene ring, we can compare the calculated spin densities with the NMR shifts. The calculated spectrum is shown in Fig. 3 (b), which was obtained by the use of Hegyhati's spin densities and the empirical Q value mentioned above. There are several disagreements between the calculated and observed spectra. These disagreements may be attributed to the distortion of the molecule. Indeed, the molecule is not planar as assumed in the calculation but is distorted in the crystalline state.[9]

As shown in Fig. 3 (a–c), the NMR line shape is not symmetric around the central intense line, which is just at the resonance frequency of the free proton. When the temperature was lowered, passing through T_N, the line shape became nearly symmetric. The line shape in the temperature region far below T_N is shown in Fig. 3 (d). In the region below T_N the line shapes of the single crystal and of the powdered specimen were similar, as was true in the paramagnetic state. The intensity of the central intense line did not change drastically with lowering of the temperature, but it decreased gradually, as did the linewidth. From these results we conclude that the delocalization of the electron spin remains in the region, below T_N. This conclusion leads to an interesting suggestion when we consider the antiferromagnetic spin structure in the crystalline lattice. That is, if we suppose two sublattices in the region below T_N, it is suggested from the observed spectrum that in half of the DPPH molecules in the crystal the sign of the spin density at each position on the molecular framework should be inverted with saturating sublattice magnetization.

References

1. H.M. McConnell and D.B. Chesnut, *J. Chem. Phys.* **28**, 107, (1958) and references therein.
2. D.J.E. Ingram, *Free Radicals as Studied by Electron Spin Resonance*, Academic Press, New York and London (1958); S.A. Al'tshuler and B.M. Kozyrev, *Electron Paramagnetic Resonance*, Academic Press, New York (1964), Chapter VII.
3. H.J. Gerristen, R. Okkes, H.M. Gisman and J. Van den Handel, *Physica* **20**, 13 (1954).
4. A.M. Prokhorov and V.B. Fedorov, *Soviet Phys.—JETP* **16**, 1489 (1963).
5. H.S. Gutowsky, H. Kusumoto, T.H. Brown, and D.H. Anderson, *J. Chem. Phys.* **30**, 860 (1959).
6. Yu.S. Karimov and I.F. Shchegolev, *Soviet Phys.–JETP* **13**, 1 (1961).
7. T.H. Brown, D.H. Anderson, and H.S. Gutowsky, *J. Chem. Phys.* **33**, 720 (1960).
8. M.M. Hegyhati, *J. Chem. Phys.* **50**, 3123 (1969).
9. D.E. Williams, *J. Am. Chem. Soc.* **88**, 5665 (1966).

Tricritical Susceptibility in Ising Model Metamagnets and Some Remarks Concerning Tricritical Point Scaling*

F. Harbus,[†] H. E. Stanley, and T. S. Chang[‡]

Physics Department, Massachusetts Institute of Technology
Cambridge, Massachusetts

There are now several "metamagnetic" materials whose phase diagrams in the magnetic field–temperature ($H-T$) plane have been studied and are known to exhibit both first- and second-order phase transitions. Examples of such systems are dysprosium aluminum garnet (DAG),[1,2] $FeCl_2$[2,3] § and $Ni(NO_3)_2 \cdot 2H_2O$,[5] with Néel temperatures of 2.53, 23.55, and 4.2°K, respectively. These are basically antiferromagnetic systems with extreme anisotropy so that they may be well represented by an Ising-like model. No spin flopping occurs at low temperatures—rather, the critical field induces an abrupt, discontinuous transition from the low-moment antiferromagnetic phase (where the order parameter is nonzero) to a high-moment paramagnetic phase (where the order parameter is zero).

A schematic phase diagram of a metamagnet in the $H-T$ plane is shown in Fig. 1. The point where the first-order and second-order (or lambda) lines meet is called the tricritical point (TCP), and is of special theoretical interest, since it is an example of where the hypothesis of universality[6] or "smoothness"[7] may be expected to break down. In particular, critical point exponents at the TCP will differ from those characterizing the second-order line. At the TCP two ordering processes occur simultaneously. There are critical fluctuations not only in the order parameter M_{st}, the staggered magnetization, but also in the nonordering density M, the direct magnetization.

In ideal two-sublattice Ising antiferromagnets with nearest-neighbor exchange only and with exchange isotropic with respect to lattice direction, the phase transition would never be expected to become first order except at absolute zero. Thus real metamagnets and models designed to simulate them must incorporate more complex sorts of interactions to produce tricritical behavior. In both $FeCl_2$ and $Ni(NO_3)_2 \cdot 2H_2O$ there exist layers of ferromagnetically coupled spins, with adjacent layers themselves coupled antiferromagnetically. Thus the perfectly ordered state at $T = 0°K$ would consist of planes of aligned spins, one layer up, the next down, and so on. The situation in DAG is more complicated, and it appears that further-neighbor interactions may be responsible for the metamagnetic transition. Only in DAG have

---

* Supported by NSF, ONR, AFOSR, and NASA.
† National Science Foundation Predoctoral Fellow.
‡ Permanent address: Riddick Laboratories, North Carolina State University, Raleigh, North Carolina.
§ See also the $H = 0$ work on $FeCl_2$ of Birgeneau and co-workers.[4]

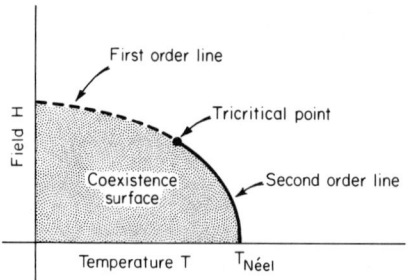

Fig. 1. Schematic $H - T$ phase diagram of metamagnetic.

some data on TCP exponents been reported, although investigations in $FeCl_2$ are now also underway.[8]

We were thus motivated to study, by the method of high-temperature series expansions, the critical behavior of two Ising model Hamiltonians which depart from the nearest-neighbor isotropic case by the inclusion of mixed interactions, i.e., the presence of both ferromagnetic interactions within a sublattice and antiferromagnetic interactions between sublattices. The ferromagnetic "paste" may be pictured as causing spins on one sublattice to become locked together at low temperatures and to flip *en masse* under the influence of a sufficiently large external critical field, giving rise to the first-order phase transition behavior. Of course, the lower the temperature, the greater the discontinuity in magnetization between the antiferromagnetic and paramagnetic phases. The theory for such models has hitherto been considered only from the molecular field approach[8] (see, e.g., Ref. 9) and recently by Monte Carlo methods.[10] Extrapolations from series expansions have proven extremely useful in the study of critical phenomena in the past, and often yield results markedly different from the predictions of "classical" theory.

The specific models considered will be referred to as the "meta" model and the "nnn" model. The Hamiltonian for the *meta model* is

$$\mathcal{H} = -J_{xy} \sum_{\langle ij \rangle}^{xy} s_i s_j - J_z \sum_{\langle ij \rangle}^{z} s_i s_j - \mu H \sum_i s_i \qquad (1)$$

Here each moment $s_i = \pm 1$, the first sum is over nearest-neighbor (nn) spins in an xy plane, the second sum is over nn spins coupled in the z direction, and H is the external magnetic field. To simulate a material like $FeCl_2$, with in-plane ferromagnetism and between-plane antiferromagnetism, we choose $J_{xy} > 0$ and $J_z < 0$. The high-temperature series for the susceptibility χ and staggered susceptibility χ_{st} are obtained to eighth order in inverse temperature on the sc lattice. Further, the series are *exact* in the external field, a feature which is extremely advantageous in studying the critical properties.

The second model studied, the *nnn model*, is an isotropic nearest-neighbor antiferromagnet with the addition of next nearest-neighbor *ferromagnetic* interactions. In an obvious notation, the Hamiltonian is

$$\mathcal{H} = -J_1 \sum_{\langle ij \rangle}^{nn} s_i s_j - J_2 \sum_{\langle ij \rangle}^{nnn} s_i s_j - \mu H \sum_i s_i \qquad (2)$$

with $J_1 < 0$ and $J_2 > 0$. This model was shown[10] by Monte Carlo techniques to have a tricritical point, and perhaps is a better representation of the interactions important in DAG. We emphasize, however, the intrinsic theoretical interest of such tricritical models apart from their experimental applicability.

The function χ_{st} is the strongly divergent quantity along the second-order part of the phase boundary, while χ should diverge only weakly, and experimentally, in fact, appears finite.[11] However, in DAG as the TCP is approached from above in temperature the susceptibility on the phase boundary gets larger and larger, until the magnetization vs. internal field isotherms show an infinite gradient at T_t, the tricritical temperature. To understand the basis for such behavior, it is useful to consider the complete three-dimensional metamagnetic phase diagram, where the third field variable is a staggered magnetic field H_{st} acting in one sense on one sublattice and oppositely on the other. Then the possibility of a strongly divergent susceptibility χ at the TCP arises because of the special features of the geometry of the phase diagram at the TCP, where three critical lines and three coexistence surfaces intersect. It should be noted that mean-field theory predicts no divergence in χ as $T \to T_t$ from the high-temperature side.

The exponent characterizing the behavior of χ at the TCP is thus a new exponent $\bar{\gamma}$, with $\chi \sim (T - T_t)^{-\bar{\gamma}}$. One of the results of the present work is an estimate for the $\bar{\gamma}$ of the models of Eqs. (1) and (2) (denoted by $\bar{\gamma}_{meta}$ and $\bar{\gamma}_{nnn}$ respectively). These are determined from the consistency criterion that the χ and χ_{st} series produce coincident singularities along the $h \equiv \mu H/k_B T =$ const path corresponding to the TCP. Although the χ series were irregular and not easily analyzable, it was expected that at the TCP the χ series raised to the inverse of a power close to the correct $\bar{\gamma}$ would show a convergent Padé approximant table with a pole matching that from the χ_{st} series along the same "h" path. Shown in Tables I and II are Padé tables obtained for the nnn model with $J_1 = -1$ and $J_2 = +1/2$, along the path $h = 0.84$. The first table is for $(\chi_{st})^{4/5}$ and the second is for χ^4 along the same path. The good convergence of each table and agreement between the critical temperatures predicted by these two completely different series lead us to the estimate of $\bar{\gamma}_{nnn} \sim 1/4$. Similar calculations for the meta model provide evidence for $\bar{\gamma}_{meta} \sim 1/2$.

In the remainder of this paper we discuss some implications and applications of a scaling law hypothesis at the TCP. According to Hankey et al.,[12]* the hypothesis for TCP scaling can be made by assuming the singular part of the Gibbs potential to be asymptotically a generalized homogeneous function,[14]

$$G(\lambda^{\bar{a}_1}\bar{x}_1, \lambda^{\bar{a}_2}\bar{x}_2, \lambda^{\bar{a}_3}\bar{x}_3) = \lambda G(\bar{x}_1, \bar{x}_2, \bar{x}_3) \qquad (3)$$

Table I. Singularities Given by Padé Approximants to the Series $(\chi_{st})^{4/5}$ along the Path $h = 0.84$ for the nnn Model with $J_1 = -1$, $J_2 = +1/2$†

D	N = 2	3	4	5	6
2	6.25	6.29	6.35	6.36	6.37
3	6.27	6.45	6.36	6.37	
4	6.43	6.35	6.37		
5	6.36	6.37			
6	6.37				

* See also a recent preprint by Griffiths[13] which follows this approach.
† We estimate the critical temperature to be $k_B T_c = 6.37 \pm 0.01$.

Table II. Singularities Given by Padé Approximants to the Series χ^4 along Path Identical to That in Table I*

D	N = 2	3	4	5	6
2	5.91	6.89	6.57	6.35	6.42
3	6.37	6.39	6.38	6.39	
4	6.39	6.38	6.39		
5	6.38	6.39			
6	6.39				

* The critical temperature is estimated to be $k_B T_c = 6.39 \pm 0.01$, in excellent agreement with the value from the staggered susceptibility series.

where $\lambda(>0)$ is an arbitrary parameter, $\bar{x}_1$ is in the direction of H_{st} coming out of the H–T plane, $\bar{x}_2$ is any direction in the H–T plane not parallel to the critical line, and $\bar{x}_3$ is the direction parallel to the critical line. The $\bar{a}_i$ are called scaling powers. It has been shown[12] that the shape of the critical line is determined by Eq. (3) and has the form

$$\bar{x}_2 = -k\bar{x}_3^{1/\varphi} \qquad (4)$$

where k is a constant and $\varphi \equiv \bar{a}_3/\bar{a}_2$ is defined as the "crossover" exponent.

All TCP exponents may be obtained by appropriate differentiation of Eq. (3).[12-15] Assuming the first-order line is asymptotically parallel to the critical line at the TCP and that the phase boundary at the TCP is parallel to neither the T nor the H axes, we find from Eq. (3) the discontinuity in magnetization for $T < T_t$ to be given by

$$\Delta M \sim (T_t - T)^{\bar{\beta}} \quad \text{as} \quad T \to T_t^- \qquad (5)$$

where $\bar{\beta} = (1 - \bar{a}_2)/\bar{a}_3$. The denominator here is $\bar{a}_3$ (not $\bar{a}_2$) because the path of approach to the TCP in determining ΔM must be along the first-order line, and this is the "independent direction" $\bar{x}_3$. Similarly, the staggered magnetization is found to vary as

$$M_{st} \sim (T_t - T)^{\bar{\beta}_{st}} \qquad (6)$$

where $\bar{\beta}_{st} = (1 - \bar{a}_1)/\bar{a}_2$, and the denominator is now $\bar{a}_2$ since the path of approach generally will *not* be asymptotically parallel to the critical line or the first-order line.*

For the susceptibility we have

$$\chi \equiv \left(\frac{\partial M}{\partial H}\right)_{H_{st}=0} \sim |T - T_t|^{(1-2\bar{a}_2)/\bar{a}_i} \qquad (7)$$

where $i = 2$ if the path of approach is not parallel to the critical line and $i = 3$ if the path is parallel to the critical line. We denote these exponents by $-\bar{\gamma}^{(2)}$ and $-\bar{\gamma}^{(3)}$, respectively, and note that they are related by $\bar{\gamma}^{(2)} = \bar{\gamma}^{(3)}\varphi$.

* We consider only those paths of approach to the TCP lying in the physically accessible H–T plane.

Tricritical Susceptibility in Ising Model Metamagnets

For the *staggered* magnetic susceptibility we have, similarly,

$$\chi_{st} \equiv \left(\frac{\partial M_{st}}{\partial H_{st}}\right)_{H=H_t} \sim |T - T_t|^{(1-2\bar{a}_1)/\bar{a}_t} \tag{8}$$

with

$$\bar{\gamma}_{st}^{(2)} = \bar{\gamma}_{st}^{(3)} \varphi = -(1 - 2\bar{a}_1)/\bar{a}_2.$$

Unfortunately, experimental data on the TCP properties of metamagnetic materials are rather sparse, but values for $\bar{\beta}$ and $\bar{\gamma}$ have been estimated for DAG.² These recent experimental results give $\bar{\beta} = 0.65 \pm 0.05$ and $\bar{\gamma}^{(3)} = 1.3 \pm 0.1$, with superscript (3) because the data for determining $\bar{\gamma}$ appear to be taken from a path following the phase boundary. Now, if scaling holds, these values predict that $\bar{a}_2 \simeq 0.75$, $\bar{a}_3 \simeq 0.38$, and $\varphi^{-1} \simeq 1.95$ (or a nearly analytic phase boundary at the TCP). Clearly, more detailed measurements of the phase boundary near the TCP would be very desirable (although undoubtedly experimentally extremely difficult), since these would provide a test of the scaling prediction for φ. Scaling also predicts a value for $\bar{\delta}$ in $M - M_t \sim (H - H_t)^{1/\bar{\delta}}$ of $\bar{\delta} = (1 - \bar{a}_2)/\bar{a}_2 \simeq 1/3$, which might also be experimentally measurable.

We now discuss how, in principle, one may determine *all three* TCP exponents $\bar{a}_1$, $\bar{a}_2$, and $\bar{a}_3$ from a rather minimal knowledge of the behavior of the system when the TCP is approached along paths in the physical plane only. We can obtain $\bar{a}_2$ from the tricritical susceptibility exponent alone, since $\bar{\gamma}^{(2)} = -(1 - 2\bar{a}_2)/\bar{a}_2$. Next $\bar{a}_3$ can be obtained from the shape of the phase boundary at the TCP, since $\varphi \equiv \bar{a}_3/\bar{a}_2$.

Finally, $\bar{a}_1$ may be obtained using the "double power law" prediction

$$\chi_{st} \sim \bar{x}_3^{(\gamma_{st} - \bar{\gamma}_{st}^{(2)})/\varphi}(\bar{x}_2 + k\bar{x}_3^{1/\varphi})^{-\gamma_{st}} \tag{9}$$

Equation (9) was obtained by making two scaling hypotheses, one about the TCP and the other about the critical line. Details of the derivation of (9) are presented elsewhere.¹⁶ In (9), γ_{st} denotes the staggered susceptibility exponent for the critical line; the universality hypothesis predicts $\gamma_{st} = 5/4$. Then if χ_{st} is written in the form

$$\chi_{st} \sim A(T_c)[T - T_c(H)]^{-\gamma_{st}} \tag{10}$$

for various field values; the factor $T - T_c(H)$ in (10) corresponds to the factor $(\bar{x}_2 + k\bar{x}_3^{1/\varphi})$ in (9). The amplitude function $A(T_c)$ gives information about the "mixed" exponent $\gamma_{st} - \bar{\gamma}_{st}^{(2)}$ in (9). Since γ_{st} is known and $\bar{\gamma}_{st}^{(2)} = -(1 - 2\bar{a}_1)/\bar{a}_2$, we can determine $\bar{a}_1$.*

Using the systematic approach outlined above, we have made preliminary estimates of the scaling powers $(\bar{a}_1, \bar{a}_2, \bar{a}_3)$ for the nnn and the meta models. For the nnn model these numbers are very roughly in the ratio 15:10:6. This can be compared with the results of Ref. 12 on the Blume–Emery–Griffiths mean field calculation.¹⁷,¹⁸

---

* Note that a divergent amplitude indicates $\bar{\gamma}_{st}^{(2)} > \bar{\gamma}_{st}$, while if $\bar{\gamma}_{st}^{(2)} < \gamma_{st}$, the amplitude approaches zero.

References

1. D.P. Landau, B.E. Keen, B. Schneider, and W.P. Wolf, *Phys. Rev. B* **3**, 2310 (1971).
2. W.P. Wolf, B. Schneider, D.P. Landau, and B.E. Keen, *Phys. Rev. B* **5**, 4472 (1972).
3. I.S. Jacobs and P.E. Lawrence, *Phys. Rev.* **164**, 866 (1967).
4. R.B. Birgeneau, W.B. Yelon, E. Cohen, and J. Makovsky, *Phys. Rev. B* **5**, 2607 (1972); W.B. Yelon and R.B. Birgeneau, *Phys. Rev. B* **5**, 2615 (1972).
5. V.A. Schmidt and S.A. Friedberg, *Phys. Rev. B* **1**, 2250 (1970).
6. R.B. Griffiths, *Phys. Rev. Lett.* **24**, 715 (1970).
7. R.B. Griffiths, in *Critical Phenomena in Alloys, Magnets, and Superconductors*, R.E. Mills, E. Ascher, and R.I. Jaffee, eds., McGraw-Hill, New York (1971), pp. 377–391.
8. B. Birgeneau and G. Shirane, private communication.
9. R. Bidaux, P. Carrara, and B. Vivet, *J. Phys. Chem. Solids* **28**, 2453 (1967).
10. D.P. Landau, *Phys. Rev. Lett.* **28**, 449 (1972).
11. R.B. Griffiths and J.C. Wheeler, *Phys. Rev. A* **2**, 1047 (1970).
12. A. Hankey, H.E. Stanley, and T.S. Chang, *Phys. Rev. Lett.* **29**, 278 (1972).
13. R.B. Griffiths, Preprint.
14. A. Hankey and H.E. Stanley, *Phys. Rev. B* **6**, 3515 (1972).
15. E.K. Riedel, *Phys. Rev. Lett.* **28**, 675 (1972).
16. T.S. Chang, A. Hankey, and H.E. Stanley, this volume.
17. M. Blume, V.J. Emery, and R.B. Griffiths, *Phys. Rev. A* **4**, 1071 (1971).
18. M. Blume, *Phys. Rev.* **141**, 517 (1966); H.W. Capel, *Physica (Utrecht)* **32**, 966 (1966).

Double-Power Scaling Functions Near Tricritical Points*

T. S. Chang, A. Hankey, and H. E. Stanley

*Physics Department, Massachusetts Institute of Technology
Cambridge, Massachusetts*

Introduction

There have been extensive experimental measurements[1] near tricritical points (TCP) in metamagnets, NH_4Cl, and $^3He-^4He$ mixtures. These measurements indicate that a TCP is a special symmetry point characterized by its own set of critical exponents.

In a recent paper[2] we discussed how the geometry of curves of singularities near a TCP in the three-dimensional space (spanned by the temperature T, ordering field η, and nonordering field g) is determined by the homogeneity scaling hypothesis. In this paper we extend these ideas and present a new three-dimensional scaling prediction using certain invariant theorems of continuous groups and generalized homogeneous functions[3] (GHF). We demonstrate that near a TCP and a critical line thermodynamic functions may be represented asymptotically in terms of expressions involving "double-power" scaled variables. Such an expression will allow data, from regions where it is valid, to collapse from a *volume* onto a *line*.

Selection of Scaling Directions

Before we can proceed to make the scaling hypothesis for a TCP, it will be necessary to determine the relevant directions for scaling. The three thermodynamic fields (T, η, g) constitute an affine space in which directions may be defined by parallelism only. A TCP is a point of intersection[4] of three critical lines in this three-dimensional affine space (Fig. 1). At each point P on a critical line three different types of directions can be established. The first direction $x_1(P)$ is a direction locally not parallel to the coexistence surface, and the second $x_2(P)$ is locally parallel to the coexistence surface but not parallel to the critical line. These are "strong" and "weak" directions of Griffiths and Wheeler.[5] The third direction $x_3(P)$ is locally parallel to the critical line and we call this an "independent direction."

As the point P moves toward the TCP these directions attain limiting orientations. Since there are three critical lines terminating at a TCP, three "rival" sets of directions of this type exist at the TCP. It has been shown[2] that if scaling holds at a TCP, these three sets of directions are equivalent. Thus we choose the relevant

---

* Supported by Laboratory of Nuclear Science (MIT), NSF, ONR, AFOSR, and NASA. Portions of this work were presented at the March 1972 meeting of the American Physical Society.

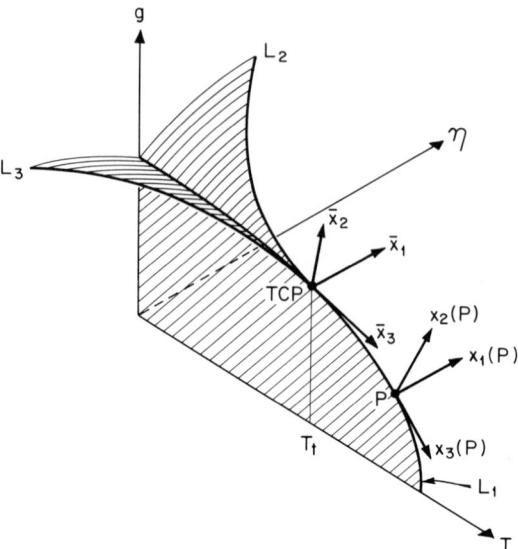

Fig. 1. Schematic phase diagram showing a TCP (at $(T = T_t)$. Shaded areas are coexistence surfaces. At a point P on L_1, a triad of directions $x_i(P)$ is shown. This triad becomes $\bar{x}_i$ at TCP.

directions for scaling at a TCP as $\bar{x}_i \equiv \lim[P \to \text{TCP}]x_i(P)$, where P is a point on the critical line L_1 (see Fig. 1).

Scaling Hypothesis for TCP

Having ascertained the relevant scaling directions $\bar{x}_i$ for TCP, we now introduce a scaling parameter $\lambda (> 0)$ and make the homogeneity hypothesis that, the singular part of the Gibbs potential is asymptotically a GHF of $\bar{x}_i$,

$$G(\lambda^{\bar{a}_1}\bar{x}_1, \lambda^{\bar{a}_2}\bar{x}_2, \lambda^{\bar{a}_3}\bar{x}_3) = \lambda G(\bar{x}_1, \bar{x}_2, \bar{x}_3) \tag{1}$$

where the $\bar{a}_i$ are the scaling powers. Equation (1) is equivalent to the statement that

$$G = F_3(\bar{x}_1, \bar{x}_2, \bar{x}_3) \tag{2}$$

is an invariant equation under the one-parameter (λ) group $(\mathcal{G}_3)$ of transformations

$$G' = \lambda G, \qquad \bar{x}'_i = \lambda^{\bar{a}_i}\bar{x}_i, \qquad i = 1, 2, 3 \tag{3}$$

In other words, under the transformations Eq. (2) becomes $G' = F_3(\bar{x}'_1, \bar{x}'_2, \bar{x}'_3)$.

It can be shown[3] that the group $\mathcal{G}_3$ admits a basis set of three (3) functionally independent absolute invariants, $y_i(G', \bar{x}'_1, \bar{x}'_2, \bar{x}'_3) = y_i(G, \bar{x}_1, \bar{x}_2, \bar{x}_3)$, such that all other absolute invariants are expressible in terms of these. One such basis set is

$$y_0 \equiv G/\bar{x}_3^{1/\bar{a}_3}, \qquad y_1 \equiv \bar{x}_1/\bar{x}_3^{\bar{a}_1/\bar{a}_3}, \qquad y_2 \equiv \bar{x}_2/\bar{x}_3^{\bar{a}_2/\bar{a}_3} \tag{4}$$

The scaling hypothesis, Eq. (1), requires Eq. (2) to be expressible in terms of the basis set as follows[3]:

$$y_0 = \bar{F}_2(y_1, y_2) \tag{5}$$

which states that G (and other thermodynamic functions), when appropriately

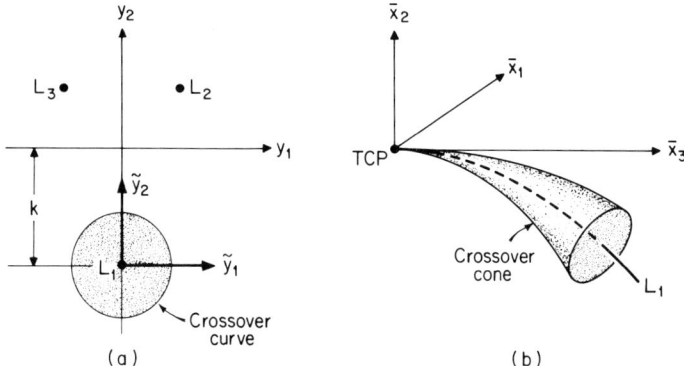

Fig. 2. (a) The invariant (y_1, y_2) plane. The strong and weak directions for L_1 are y_1 and y_2. (b) The principal points of interest of (a) in the $(\bar{x}_1, \bar{x}_2, \bar{x}_3)$ space.

scaled, are functions of the invariants (y_1, y_2) alone. This result allows data near a TCP to collapse from a *volume* onto a *surface*.

We remark that using Eq. (1), it is possible to determine all exponent relations and scaling laws for a TCP. These results have been tabulated elsewhere.[3,4,6]

The Geometry of Surfaces and Curves Near a TCP

Since the quantities y_1 and y_2 defined in Eq. (4) form a basis set of functionally independent absolute invariants of $\bar{x}_i$ under the group of transformations ($\bar{x}'_i = \lambda^{\bar{a}_i} \bar{x}_i$), points in the invariant (y_1, y_2) plane give rise to invariant curves in the $(\bar{x}_1, \bar{x}_2, \bar{x}_3)$ space. We have seen that the scaling hypothesis requires scaled thermodynamic functions near a TCP to depend on y_1 and y_2 only. This implies that each of the three critical lines near the TCP can be expressed as a point $y_i = k_i$ in the (y_1, y_2) plane, where the k_i are constants. This result was used previously[2] by the authors to demonstrate that the scaling directions defined for one critical line are always consistent with the directions defined in terms of the other two lines.

Usually, for systems exhibiting a TCP one of the critical lines is a planar curve lying entirely in the (g, T) plane. This is the curve denoted by L_1 in Fig. 1. Such a line is given by $(y_1, y_2) = (0, -k)$ in the invariant plane, while the coexistence plane bounded by L_1 maps into a line along the vertical axis $y_1 = 0$ (Fig. 2a). Near L_1 it is expected that the symmetry property of the critical line will also influence the asymptotic form of the thermodynamic functions.* The region of influence is bounded by some "crossover" curve $f_c(y_1, y_2) = 0$ (Fig. 2a), or

$$f_c(\bar{x}_1/\bar{x}_3^{\bar{a}_1/\bar{a}_3}, \bar{x}_2/\bar{x}_3^{\bar{a}_2/\bar{a}_3}) = 0 \qquad (6)$$

which is a conical surface surrounding L_1 in the $(\bar{x}_1, \bar{x}_2, \bar{x}_3)$ space (Fig. 2b). Scaling will not tell us the actual shape of the curve in the (y_1, y_2) plane, but it does limit the shape of the conical "crossover" surface in the $(\bar{x}_1, \bar{x}_2, \bar{x}_3)$ space, since all points in the (y_1, y_2) plane give rise to curves approaching the TCP along the $\bar{x}_3$ axis (corresponding to the minimum $\bar{a}_i$).†

---

* The scaling ideas for $L_{2,3}$ require more detailed discussions and are given in Ref. 3.
† In this paper we assume $\bar{a}_1 \neq \bar{a}_2 \neq \bar{a}_3$ (see Ref. 2 for discussion of other cases).

Double-Power Scaling Functions for L_1

We now proceed to deduce the restriction on the asymptotic form of the thermodynamic functions near a TCP adjacent to the critical line[6] L_1 (see Ref. 3 for $L_{2,3}$). Along L_1 the conventional scaling hypothesis is normally stated in terms of a GHF equation

$$G(\mu^{a_1}x_1, \mu^{a_2}x_2; x_3) = \mu G(x_1, x_2; x_3) \tag{7}$$

where $\mu (>0)$ is an arbitrary parameter, (a_1, a_2) are the scaling powers for L_1, and x_3 is an inactive variable which does not scale. Near the TCP, however, the value of $y_0 \equiv G/\bar{x}_3^{1/\bar{a}_3}$ changes only if the values y_1 and/or y_2 change. It is therefore more appropriate to make a scaling hypothesis about L_1 near the TCP using y_1 and y_2.

Since the coexistence surface bounded by L_1 maps into the vertical axis of the (y_1, y_2) plane, the direction y_1 is strong and y_2 is weak. Thus we deduce that the proper scaling variables for L_1 near the TCP are

$$\tilde{y}_1 \equiv y_1 \qquad \tilde{y}_2 \equiv y_2 + k \tag{8}$$

which vanish at the line $(y_1, y_2) = (0, -k)$.

We now hypothesize that along L_1 near the TCP, $\tilde{y}_0 \equiv y_0$ is a GHF of $(\tilde{y}_1, \tilde{y}_2)$

$$\tilde{y}_0(\mu^{a_1}\tilde{y}_1, \mu^{a_2}\tilde{y}_2) = \mu \tilde{y}_0(\tilde{y}_1, \tilde{y}_2) \tag{9}$$

In other words, we postulate that

$$\tilde{y}_0 = F_2(\tilde{y}_1, \tilde{y}_2) \tag{10}$$

is an invariant equation under the group $(\mathcal{G}_2)$ of transformations

$$(\tilde{y}_0' = \mu \tilde{y}_0; \tilde{y}_1' = \mu^{a_1}\tilde{y}_1, \tilde{y}_2' = \mu^{a_2}\tilde{y}_2).$$

As before, this scaling hypothesis requires Eq. (10) to be expressible as

$$z_0 = F_1(z_1) \tag{11}$$

where $z_0 \equiv \tilde{y}_0/\tilde{y}_2^{1/a_2}$ and $z_1 \equiv \tilde{y}_1/\tilde{y}_2^{a_1/a_2}$ form a set of functionally independent absolute invariants of $\mathcal{G}_2$ and therefore of the variables (G, x_1, x_2, x_3) under the direct-product group $\mathcal{G}_2 \otimes \mathcal{G}_3$.

Reexpressing Eq. (11) in terms of the original variables (G, x_1, x_2, x_3), we obtain the "double-power" form of

$$G/x_3^{1/\bar{a}_3}(\bar{x}_2/\bar{x}_3^{\bar{a}_2/\bar{a}_3} + k)^{1/a_2} = F_1[\bar{x}_1/\bar{x}_3^{1/\bar{a}_3}(\bar{x}_2/\bar{x}_3^{\bar{a}_2/\bar{a}_3} + k)^{a_1 a_2}] \tag{12}$$

Equation (12) predicts that near the TCP and L_1 data will collapse from a *volume* onto a *line*. Clearly, this happens only within the "crossover" cone of Eq. (6).

In the plane $\bar{x}_1 = 0$ (i.e., the $g-T$ plane), Eq. (12) requires

$$G \sim \bar{x}_3^{1/\bar{a}_3}(\bar{x}_2/\bar{x}_3^{\bar{a}_2/\bar{a}_3} + k)^{1/a_2} \tag{13}$$

and the conical surface of Eq. (6) becomes two "crossover" lines, $\bar{x}_2 = C_{1,2}\bar{x}_3^{\bar{a}_2/\bar{a}_3}$ or $y_2 = C_{1,2}$ (Fig. 3). The crossover exponent $\varphi \equiv \bar{a}_3/\bar{a}_2$ which determines the shape of the crossover lines can be obtained directly from the shape of L_1.

The entire discussion in this paper may be extended to the scaling of any ther-

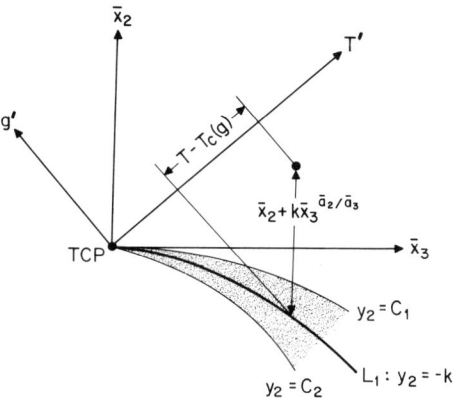

Fig. 3. Fig. 2(b) sliced in the $\bar{x}_1 = 0$ plane. The crossover lines are labeled $y_2 = C_{1,2}$. The projection of $\bar{x}_2 + k\bar{x}_3^{\bar{a}_2/\bar{a}_3}$ along the T axis is $T - T_c(g)$ and $(T', g') = (T - T_t, g - g_t)$.

modynamic function f. For example, for the staggered magnetic susceptibility $\chi_{st} \equiv \partial^2 G/\partial \eta^2 \propto \partial^2 G/\partial \bar{x}_1^2$ (or $\partial^2 G/\partial x_1^2$), and the magnetic susceptibility $\chi \equiv \partial^2 G/\partial g^2 \propto \partial^2 G/\partial \bar{x}_2^2$ (or $\partial^2 G/\partial x_2^2$) for a metamagnet, the analogous expressions of Eq. (13) are

$$\bar{x}_3^{(1-2\bar{a}_i)\bar{a}_3}(\bar{x}_2/\bar{x}_3^{\bar{a}_2/\bar{a}_3} + k)^{(1-2a_i)/a_2} \tag{14}$$

with $i = 1, 2$ respectively. We note that these expressions have the appropriate divergence properties at the critical line and the TCP.

Finally, we make a few remarks about the exponents and directions of approach toward the TCP and L_1. Using the staggered susceptibility as an example, we note that if we approach the TCP along a curve $\bar{x}_2/\bar{x}_3^{\bar{a}_2/\bar{a}_3} = $ const, the scaling exponent is $-\bar{\gamma}_{st}\varphi^{-1} = (1 - 2\bar{a}_1)/\bar{a}_3$. If we approach the critical line L_1 along a line $\bar{x}_3 = $ const, the scaling exponent is $-\gamma_{st} = (1 - 2a_1)/a_2$, as expected. On the other hand, if the TCP is approached along a path outside the crossover lines, χ_{st} scales with an exponent $-\bar{\gamma}_{st} = (1 - 2\bar{a}_1)/\bar{a}_2$. Similar remarks may be made with respect to the three-dimensional "double-power" scaling functions.

Expression (14) may be cast in "mixed-exponent" form; e.g., for $i = 1$

$$\chi_{st} \sim C[T - T_c(H)]^{-\gamma_{st}} \tag{15}$$

in which $(\bar{x}_2 + k\bar{x}_3^{\bar{a}_2/\bar{a}_3})$ has been replaced by its projection along the T axis, and

$$C \sim \bar{x}_3^{(\gamma_{st} - \bar{\gamma}_{st})/\varphi} \tag{16}$$

is the asymptotic amplitude. Thus, depending on the relative magnitudes of γ_{st} and $\bar{\gamma}_{st}$, χ_{st} may diverge, vanish, or stay essentially constant as the TCP is approached within the crossover cone.

Acknowledgments

We are grateful to F. Harbus, R. B. Griffiths, E. K. Riedel, and J. C. Wheeler for discussions and for sending us preprints prior to their publication.

References

1. D.P. Landau, B.E. Keen, B. Schneider, and W.P. Wolf, *Phys. Rev. B* **3**, 2310 (1971); C.W. Garland and B.B. Weiner, *Phys. Rev. B* **3**, 1634 (1971); E.H. Graf, D.M. Lee, and J.D. Reppy, *Phys. Rev. Lett.* **19**, 417 (1967).
2. A. Hankey, H.E. Stanley, and T.S. Chang, *Phys. Rev. Lett.* **29**, 278 (1972).
3. T.S. Chang, A. Hankey, and H.E. Stanley, *Phys. Rev. B* **7**, 4263 (1973); *Phys. Rev. B* **8**, 346 (1973); A. Hankey, T.S. Chang, and H.E. Stanley, *Phys. Rev. B* **8**, 1178 (1973); A. Hankey and H.E. Stanley, *Phys. Rev. B* **6**, 3515 (1972).
4. R.B. Griffiths, to be published; also *Phys. Rev. Lett.* **24**, 715 (1970).
5. R.B. Griffiths and J.C. Wheeler, *Phys. Rev. A* **2**, 1047 (1970).
6. E.K. Riedel, *Phys. Rev. Lett.* **28**, 675 (1972).

Low-Temperature Spontaneous Magnetization of Two Ferromagnetic Insulators: $CuRb_2Br_4 \cdot 2H_2O$ and $CuK_2Cl_4 \cdot 2H_2O$

C. Dupas, J.-P. Renard, and E. Velu

*Institut d'Electronique Fondamentale,**
Université de Paris—Orsay, France

Introduction

We have performed accurate measurements of the low-temperature spontaneous magnetization in some of the isomorphous compounds $CuM_2X_4 \cdot 2H_2O$. The crystal structure of these salts is tetragonal but close to body centered cubic (bcc).[1] The ferromagnetic interactions between the Cu^{2+} ions are nearly isotropic.[2]

The theoretical spontaneous magnetization of the bcc Heisenberg ferromagnet has already been calculated using the self-consistently renormalized spin-wave theory,[3] or Green's function formalism.[4] At low temperature our experimental data do not fit these calculations. We have thus been led to take into account a weak axial exchange anisotropy.

Experiment

The transition temperatures of $CuRb_2Br_4 \cdot 2H_2O$ and $CuK_2Cl_4 \cdot 2H_2O$ are respectively 1.876 and 0.893°K. Below T_c the NMR frequency of the protons in zero magnetic field is proportional to the local field at the site of the protons and thus to the spontaneous magnetization of the crystal. NMR signals have been studied with a frequency-modulated and frequency-swept Robinson oscillator. At low temperature the PMR frequencies are close to 6.5 MHz for $CuRb_2Br_4 \cdot 2H_2O$, and 3.7 MHz for $CuK_2Cl_4 \cdot 2H_2O$.

Cryogenic Apparatus

(a) $0.4 < T < 1°K$: The crystal is immersed in a pumped ^3He bath, the temperature for $T > 0.6°K$ being derived from the ^3He vapor pressure (1962 scale), and temperatures below this from the value of a 10-Ω, 1/8-W Allen-Bradley resistor first calibrated against the ^3He thermometer.

(b) In order to obtain temperatures below 0.4°K, we use adiabatic demagnetization of a 20-g CrK alum single crystal (Fig. 1). Alum and sample are precooled at 0.6°K by a ^3He-pumped bath in an applied 10-kOe field; all of this device is placed in a vacuum jacket, ensuring insulation from the main ^4He bath at 1.2°K.

---

* Laboratory associated with CNRS.

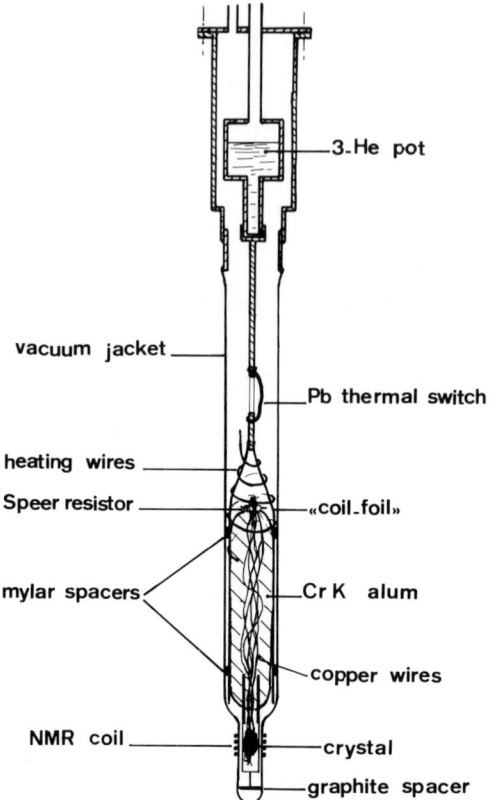

Fig. 1. Schematic diagram of the demagnetization apparatus.

Helium-4 exchange gas at low pressure (2×10^{-2} mm Hg at 4.2°K) is used to rapidly precool the apparatus to 1.2°K; it is then cryopumped by the ^{3}He stage. A Pb superconducting thermal switch actuated by the fringe field of the magnet ensures thermal insulation between alum and ^{3}He bath in zero magnetic field. Thermal contact between CrK alum and ^{3}He first stage under the switch is achieved by means of a copper wire "coil-foil"[5] into which the alum is glued with GE 7031 varnish. The alum crystal was grown on a bundle of 200 enamelled copper wires 0.1 mm in diameter into which we have glued the sample and a 220-Ω, 1/2-W Speer carbon resistor with GE varnish, in order to obtain a tight thermal link.

With this apparatus we were able to cool single crystals of $CuM_2X_4 \cdot 2H_2O$ from 77 to 0.09°K in 5 h. Temperatures from 0.09 to 0.6°K were easily obtained by using heating constantan wires wound around the "coil-foil"; thermal equilibrium was attained in a few minutes in all cases. The Speer resistor, measured with a low-dissipation Wheatstone bridge (10^{-13} W), was first calibrated against ^{4}He and ^{3}He pressure measurements. The fit of the experimental points by a Clement–Quinnell law[6] was made by mean square calculations on a Univac 1108 computer. An extrapolation of this law gives a determination of the sample temperature within a few percent accuracy for $T < 0.6$°K.

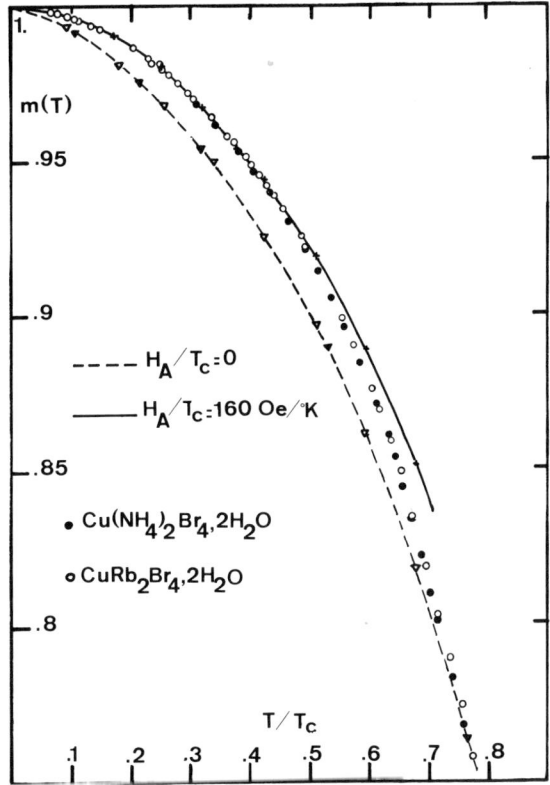

Fig. 2. Reduced magnetization vs. T/T_c in $Cu(NH_4)_2Br_4 \cdot 2H_2O$ and $CuRb_2Br_4 \cdot 2H_2O$. The dashed line corresponds to the theoretical curve calculated for a Heisenberg ferromagnet; the solid curve is deduced from our calculations.

In addition, the PMR spectra in $CuRb_2Br_4 \cdot 2H_2O$ are relatively simple, and the intensity of the lines varies as $1/T$; this characteristic gave us a direct approach to the temperature of the bromide sample. This is not the case for $CuK_2Cl_4 \cdot 2H_2O$, in which the complicated PMR spectra have not been as yet completely explained.

Results

Experimental values of the reduced magnetization vs. temperature in the two copper compounds are shown in Figs. 2 and 3. It is clear that in these two cases the theoretical curves calculated with the hypothesis of an isotropic Heisenberg interaction do not fit the experimental results.

We have thus introduced into the Hamiltonian a weak uniaxial exchange interaction between first neighbors; at low temperature this term can be represented by an anisotropy field H_A parallel to the c axis. The spontaneous magnetization has been calculated using the self-consistent method of spin-wave renormalization introduced by Bloch.[7] Detailed information about our calculations can be found in Ref. 8.

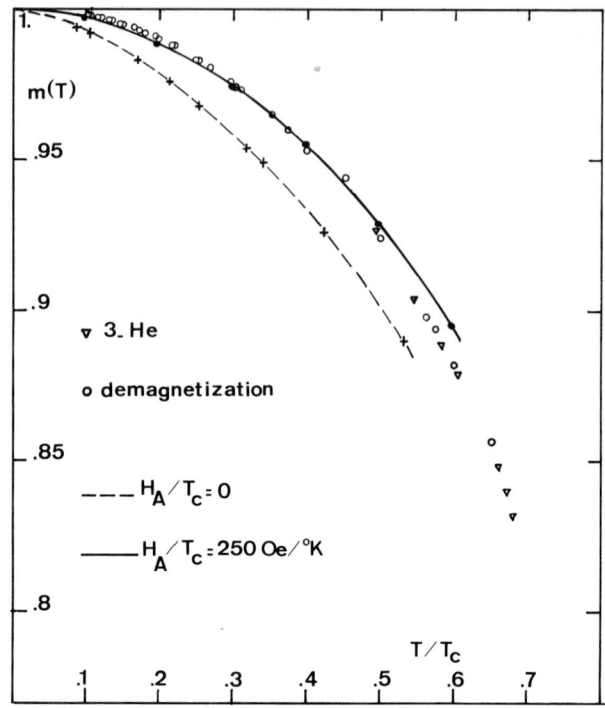

Fig. 3. Reduced magnetization vs. T/T_c in $CuK_2Cl_4 \cdot 2H_2O$.

The only parameter to be determined is H_A. The exchange integrals J_1 and J_2 between first and second neighbors are given by the relation

$$2k_B T_c / J_1 = 5.11 + 4.95 J_2 / J_1$$

arising from the high-temperature series expansions of Dalton and Wood.[9] We have checked that varying J_2/J_1 for given H_A and T_c does not change the calculated curve.

In $CuRb_2Br_4 \cdot 2H_2O$ the agreement between theoretical and experimental data is excellent with $H_A/T_c = 160$ Oe/°K, or $H_A = (300 \pm 25)$ Oe. This salt is a uniaxial ferromagnet with spin easy axis along c, just as in $Cu(NH_4)_2 Br_4 \cdot 2H_2O$.[10] In these compounds the simple model of an axial anisotropy proves very satisfactory.

In $CuK_2Cl_4 \cdot 2H_2O$ the experimental points are very close to the theoretical curve drawn with $H_A/T_c = 250$ Oe/°K, or $H_A = (225 \pm 25)$ Oe. It does not seem possible to explain the small observed differences in terms of the few percent accuracy of the temperature measurements. Our preliminary studies of this salt suggest a more complex magnetic structure with two possible spin directions [110] and [1$\bar{1}$0], and the anisotropy may be more complex than that which we used in the calculations.

In the next step we shall perform new NMR experiments using $CuRb_2Br_4 \cdot 2H_2O$ as secondary thermometer, the PMR frequency in this salt giving a measure of the temperature. The signals in the bromide are very intense, the signal-to-noise ratio being of about 10^2. The linewidth is of the order of 10 kHz, which allows us to point out the mean PMR frequency with an uncertainty δf of about 0.1 kHz. The resolution δT of this thermometer is then deduced from the slope of the magnetization curve, to which $\delta f/\delta T$ is proportional. Temperatures would be measured with an accuracy better than 1 m°K betwen 0.6 and 0.2°K, and better than 2 m°K at 0.09°K.

References

1. R.W.G. Wyckoff, *Crystal Structures*, Interscience, New York (1948), Vol. III.
2. A.R. Miedema, R.F. Wielinga, and W.J. Huiskamp, *Physica* **31**, 1585 (1965).
3. P.D. Loly, *J. Appl. Phys.* **39**, 1109 (1968).
4. J.F. Cooke, *Phys. Rev. B* **2**, 220 (1970).
5. A.C. Anderson, G.L. Salinger, and J.C. Wheatley, *Rev. Sci. Instr.* **32**, 1110 (1961).
6. J.R. Clement and E.H. Quinnell, *Rev. Sci. Instr.* **23**, 213 (1952).
7. M. Bloch, *Phys. Rev. Lett.* **9**, 286 (1962); *J. Appl. Phys.* **34**, 1151 (1963).
8. E. Velu, J-P. Renard, and C. Dupas, *Sol. St. Comm.* **11**, 1 (1972).
9. N.W. Dalton and D.W. Wood, *Phys. Rev.* **138**, A779 (1965).
10. J.P. Renard and E. Velu, *J. de Physique* **32** (C1), 1154 (1971).

Heat Capacity Measurements on KMnF$_3$ at the Soft Mode and Magnetic Phase Transitions

W. D. McCormick and K. I. Trappe

Department of Physics, University of Texas
Austin, Texas

Introduction

We made high-resolution heat capacity measurements on a 0.792-g single crystal of KMnF$_3$ between 80 and 200°K. We found four specific heat anomalies in this temperature range, two associated with soft mode structural phase transitions and two with magnetic transitions. These will all be discussed in the following paragraphs, with special emphasis on a detailed study of the 187°K soft mode transition.

Experimental Technique

We used the standard discontinuous heating method in these experiments. The sample was attached with Nonaq stopcock grease to a small gold-plated copper disk. A miniature platinum resistance thermometer (PRT) and a resistance heater were imbedded in the disk with their leads attached to heat sinks also imbedded in the disk. At 200°K this calorimeter, which had been designed for measurements on a larger crystal of SrTiO$_3$, constituted about 75% of the measured heat capacity. The temperature of the calorimeter was monitored by a 94-Hz resistance bridge. The Smith four-lead bridge circuit was used to reduce the effect of lead resistance and the off-balance signal from the bridge was displayed on a strip-chart recorder. Noise and short-term stability of the temperature measurement was better than 50 μ°K. Based on the precision claimed by the manufacturer of the PRT we believe that temperatures we quote are accurate to $\pm$ 0.2°K. The reproducibility from run to run seems to be at least an order of magnitude better. The calorimeter was surrounded by an isothermal radiation shield the temperature of which was measured by a PRT like the one in the calorimeter. A dc bridge and commercial temperature controller operated the shield. Sample and shield were in a vacuum jacket surrounded by double dewars filled with liquid nitrogen.

Previous Measurements

Rao et al.[1] found specific heat anomalies at 84.3 $\pm$ 0.2 and 179.5 $\pm$ 0.2°K, attributing them respectively to an antiferromagnetic and a structural transition. Moruzzi and Teaney[2] found peaks at 87.6 and 186°K and observed that there was about 0.3°K hysteresis at the lower transition. Hirakawa and Furukawa[3] examined the upper transition in a single crystal using a temperature drift technique. They found sharp peaks at 186.75°K on heating and 186.55°K on cooling, the hysteresis indicating a first-order transition.

Experimental Results

Figure 1 shows the heat capacity over a 5°K temperature interval in the neighborhood of the upper structural phase transition. High-resolution measurements of the same transition with the temperature scale expanded by a factor of 20 are shown in Fig. 2. Figure 3 shows the heat capacity over the low-temperature range, where the other three anomalies are found. The inset here shows a second run through the 92.3°K anomaly, which is very weak. Arrows indicate the transition temperatures at 82.5, 87.8, and 92.3°K. The transitions at 82.5, 92.3, and 186.6°K all show the hysteresis and long time constants typical of first-order transitions. The 87.8°K transition, however, exhibits no observable hysteresis and seems to be a typical second-order λ anomaly. Note that at the 187°K transition in Fig. 2 and the 82.5°K transition in Fig. 3 there are points indicated by arrows which are off scale.

Discussion

The specific heat anomaly at 87.8°K is typical of second-order antiferromagnetic transitions and we identify this as the Néel temperature. The magnetic properties of $KMnF_3$ near this transition have been extensively studied.[4,5] Probably because of strains caused by the structural transitions, the specific heat anomaly at 87.8°K is rounded. It is not, therefore, as good a candidate for the study of critical phenomena as, for example, $RbMnF_3$, which retains its cubic perovskite structure through the magnetic transition.

Static magnetic susceptibility measurements[4] indicated a second magnetic

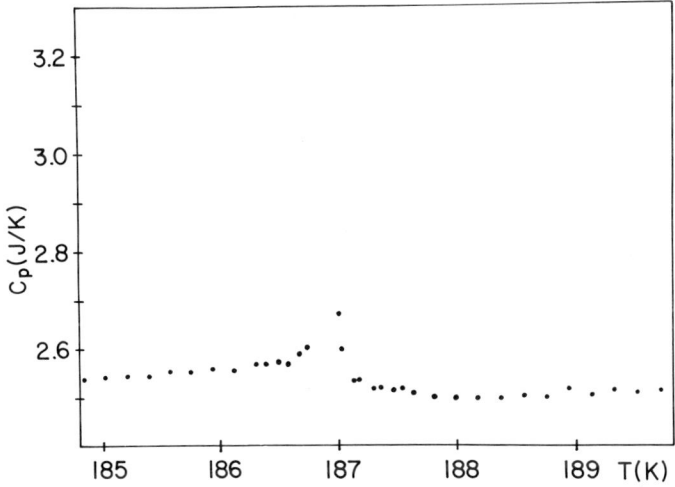

Fig. 1. Heat capacity of $KMnF_3$ crystal plus sample holder, $\Delta T \simeq 180$ m°K.

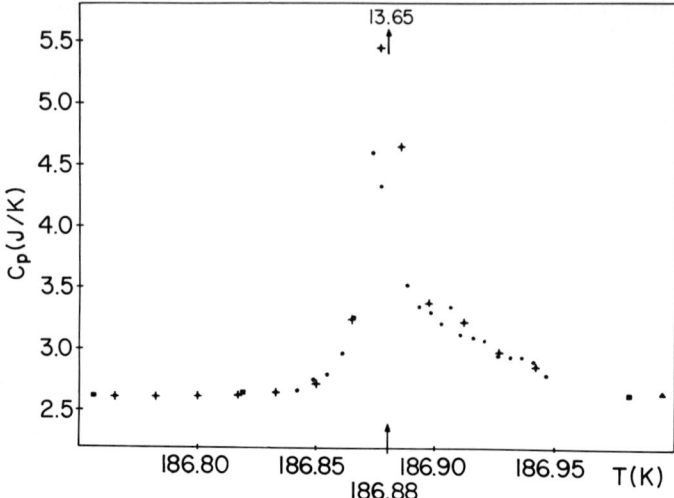

Fig. 2. Heat capacity of $KMnF_3$ crystal plus sample holder. ΔT (in m°K): Triangles, 180; squares, 60; crosses, 18; dots, 6.

transition in $KMnF_3$ at 81.5°K and it was identified as a transition to a canted, weakly ferromagnetic state. This is doubtless the first-order transition which we see at 82.5°K. We estimate a value of 17.7 J/mole for the latent heat. After our measurements at this transition we found that our sample and calorimeter had a longer thermal time constant than before, indicating that the thermal bond between the sample and calorimeter was worse. This probably indicates a considerable structural change at this transition.

It has recently been shown, mainly by neutron scattering experiments,[6-9] that $KMnF_3$ has two structural phase transitions resulting from instabilities of

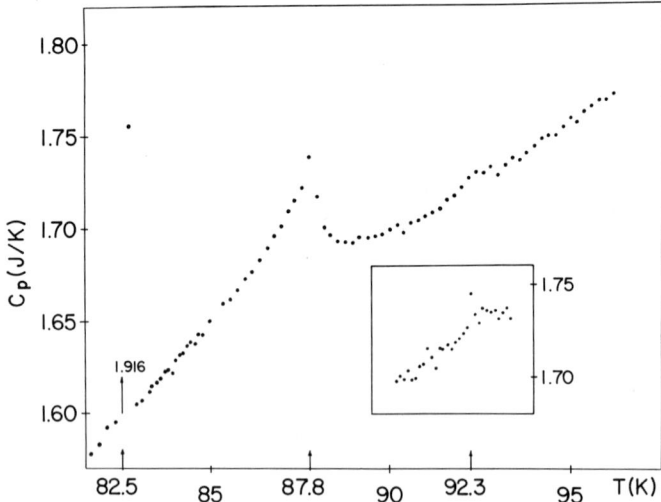

Fig. 3. Heat capacity of $KMnF_3$ crystal plus sample holder. Displaced insert shows additional run near $T = 92°K$.

low-energy transverse acoustic phonon branches at the zone boundary. The transition at 187°K has been identified with the condensation of the mode with Γ_{25} symmetry at the [111] Brillouin zone corner (R point) of the cubic structure. The lower-temperature structural transition, originally confused with the antiferromagnetic transition, was found[8] to be due to the M_3 mode condensing at the [110] zone boundary. The neutron scattering experiments[8] suggest a transition temperature of 91.5°K; we take this to be the transition that we find near 92.3°K. On each run through the transition we saw one or more points with the long thermal equilibrium times or hysteresis typical of first-order transitions. The small latent heat, however, was smeared out over a range of 1–2°K, probably because of strains introduced by the domain structure due to the 187°K transition. If we assume that the anomalous heat capacity is all due to smeared-out latent heat, we may place an an upper limit on this quantity of 1.2 J/mole.

The specific heat anomaly at the soft mode transition just below 187°K shows some particularly interesting features. Approaching the transition from below, the specific heat rises sharply very close to the transition. Above the transition, however, the specific heat falls much more slowly, taking about 1 deg to recover to the apparent baseline. The question is: Is the rounding of the transition an intrinsic property of $KMnF_3$, or is it caused by impurities or strains in the crystal? Clearly this cannot be answered without investigating other samples.* Ignoring the specific heat rise close to the transition, we may estimate a specific heat discontinuity, a feature typical of both first- and second-order transitions. From Fig. 1 we find the discontinuity in the specific heat at the 187°K transition to be $\Delta C_p = 17.2$ J/mole-°K.

Structurally, the Γ_{25} mode transition in $KMnF_3$ is identical to the soft mode transition in the perovskite compound $SrTiO_3$ at 105°K.[10] This transition has been very extensively studied both experimentally and theoretically,[11] and it is interesting to compare the two cases. In $SrTiO_3$ the transition is second order and the specific heat discontinuity is much smaller. We have found a value[12] of $\Delta C_p = 0.17$ J/mole-°K. According to a recent theory of Feder,[13] this fact that the discontinuity is larger for $KMnF_3$ is connected to the much greater prominence in $KMnF_3$ of the so-called "central peak" in the neutron scattering cross section.

The much larger values for both the tetragonal strain and the rotation angle for the fluorine (or oxygen) octahedron below the transition temperature indicate that the coupling between the soft Γ_{25} mode and the elastic strain in the crystal is much larger for $KMnF_3$. According to Pytte,[14] the first-order nature of the transition in $KMnF_3$ indicates a strong third-order coupling to the strain.†

Summarizing the results, we find the following transitions:

1. $T = 186.6°K$, first-order structural transition, soft Γ_{25} mode, specific heat tail much wider above transition, superimposed $\Delta C_p = 17.2$ J/mole-°K.
2. $T = 92.3°K$, first-order structural transition, soft M_3 mode, latent heat < 1.2 J/mole, broadened probably because of strains produced by twinning at the 187°K transition.
3. $T = 87.8°K$, second-order λ type, antiferromagnetic transition.

---

* Our sample, a single crystal of 0.8 g, was very kindly loaned to us by Dr. Brage Golding, Bell Telephone Laboratories.
† However, see Ref. 15.

4. $T = 82.5°K$, first-order transition, canting of manganese spins, latent heat about 17.7 J/mole.

References

1. R.V.G. Rao, C.D. Das, H.V. Keer, and A.B. Biswas, *Proc. Phys. Soc. (London)* **81**, 191 (1963).
2. V.L. Moruzzi and D.T. Teaney, *Bull. Am. Phys. Soc.* II **9**, 225 (1964); D.T. Teaney, *Critical Phenomena*, M.S. Green, ed.; J.V. Sengers, NBS Misc. Publ. 273, 50, 1965.
3. K. Hirakawa and M. Furukawa, *Japan J. Appl. Phys.* **9**, 971 (1970); M. Furukawa, Y. Fujimori, and K. Hirakawa, *J. Phys. Soc. (Japan)* **29**, 1528 (1970).
4. A.J. Heeger, O. Beckman, and A.M. Portis, *Phys. Rev.* **123**, 1652 (1961), and references therein.
5. M.J. Cooper and R. Nathans, *J. Appl. Phys.* **37**, 1041 (1966).
6. V.J. Minkiewicz and G. Shirane, *J. Phys. Soc. (Japan)* **26**, 674 (1969).
7. W.J.L. Buyers, R.A. Cowley, and G.L. Paul, *J. Phys. Soc. (Japan Suppl.)* **28**, 242 (1970).
8. G. Shirane, V.J. Minkiewicz, and A. Linz, *Sol. St. Comm.* **8**, 1941 (1970).
9. K. Gesi, J.D. Axe, G. Shirane, and A. Linz, *Phys. Rev. B* **5**, 1933 (1972).
10. G. Shirane and Y. Yamada, *Phys. Rev.* **177**, 858 (1969).
11. J. Feder and E. Pytte, *Phys. Rev. B* **1**, 4803 (1970), and references therein.
12. W.D. McCormick and K.I. Trappe, *Bull. Am. Phys. Soc.* II **8**, 850 (1971).
13. J. Feder, *Sol. St. Comm.* **9**, 2021 (1971).
14. E. Pytte, *Phys. Rev. Lett.* **28**, 895 (1972).
15. N.S. Gillis and T.R. Koehler, *Phys. Rev. Lett.* **29**, 369 (1972).

Magnetic and Magnetooptical Effects at the Phase Transition in Antiferromagnetic Ferrous Carbonate

V. V. Eremenko, K. L. Dudko, Yu. G. Litvinenko, V. M. Naumenko, and N. F. Kharchenko

Physico-Technical Institute of Low Temperatures
Ukrainian Academy of Sciences, Kharkov, USSR

The optical Faraday effect, magnetization, and light absorption spectrum in antiferromagnetic ferrous carbonate have been studied in detail in pulsed magnetic fields up to 300 kOe at 4.2–40°K.

The metamagnetic transition in $FeCO_3$ from an antiferromagnetic state to a ferromagnetic state ($H \parallel C_3$) has been found to occur in the field range of $\sim$ 150–180 kOe. The magnetization curve shows sharp jumps at the beginning and the end of the transition; between the jumps the magnetic moment changes smoothly. The jumps are seen as spikes on the differential magnetic susceptibility curve, which is presented in the oscillogram of Fig. 1. The magnitudes of both the magnetic moment jumps and differential susceptibility spikes depend on the initial sample temperature. They decrease with the temperature increase and completely vanish at $T \geq 22°K$. The deviation of the magnetic field orientation from the axis C_3 by angles up to 20° does not affect the magnetization curve essentially.

The studies of the Faraday light rotation within the region of the metamagnetic transition revealed a characteristic continuous smooth growth of the rotation angle for the polarization plane and a sharp decrease in the intensity of the linearly polarized light component. Near the transition end a jumplike increase in the rotation

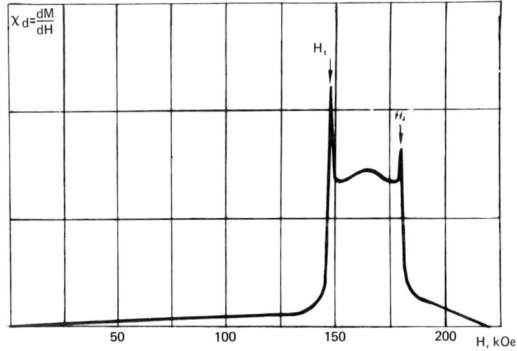

Fig. 1. Oscillogram of differential magnetic susceptibility in $FeCO_3$, $T = 4.2°K$.

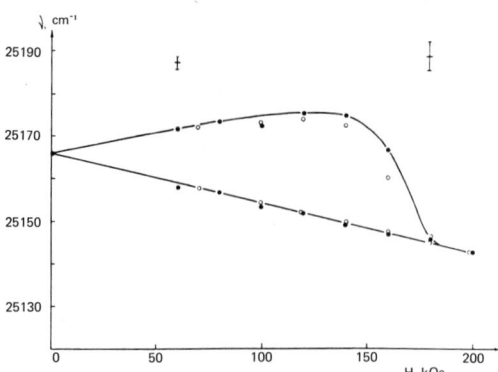

Fig. 2. Dependence of sublattice splitting of the band $v = 25{,}166$ cm^{-1} on the magnetic field strength.

becomes well pronounced. In general, the Faraday effect displays a monotonic change in the crystal magnetization within the metamagnetic transition region. The behavior of pure exciton light absorption in $FeCO_3$ in a magnetic field permitted an immediate observation of the motion of the magnetic sublattices. The nature of a sublattice splitting of one of these bands (Fig. 2) shows that the metamagnetic transition in $FeCO_3$ to a ferromagnetic configuration is realized through an intermediate homogeneous noncollinear magnetic structure. In this structure one of the sublattices remains fixed along the C_3 axis and the other smoothly rotates with the growing external field. Thus, unlike laminar antiferromagnets of $FeCl_2$ type, in which the metamagnetic transition is the first-order transition at low temperatures, no magnetic phase layering in the transition region occurs in $FeCO_3$.

Since the metamagnetic transition is usually studied in pulsed magnetic fields of small duration, i.e., practically under the adiabatic conditions, investigations were carried out of the magnetocaloric effect which accompanies the transition under consideration. The effect was studied by the magnetooptical method (Fig. 3). The value of the magnetocaloric heating under magnetization appears to be anomalously large (ΔT up to 23°K) and not equal to the cooling value under the sub-

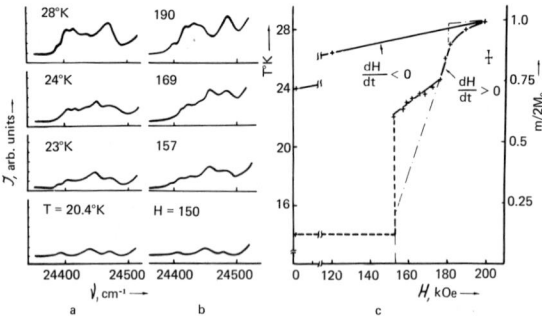

Fig. 3. Microphotograms of $FeCO_3$ absorption spectrum at various (a) temperatures and (b) magnetic fields (kOe); (c) the magnetocaloric effect procedure (broken curve shows the magnetic moment change).

sequent demagnetization. This temperature hysteresis is in good agreement with the magnetic hysteresis.

All the observed effects are well described on the assumption that the anisotropy energy in the crystal considerably exceeds the energy of both the antiferromagnetic and ferromagnetic exchange interactions.

9
Low-Dimensional Systems

Fluctuations in One-Dimensional Magnets: Low Temperatures and Long Wavelengths

R. J. Birgeneau

Bell Laboratories
Murray Hill, New Jersey

and

G. Shirane and T. A. Kitchens

Brookhaven National Laboratory
Upton, New York

The near-ideal one-dimensionality of the linear-chain antiferromagnet $(CH_3)_4NMnCl_3$ (TMMC) was first demonstrated by Dingle et al.[1] in their study of its susceptibility and optical properties. They showed that the susceptibility could be accurately fitted from room temperature down to about 10°K using Fisher's[2] exact solution for the one-dimensional Heisenberg antiferromagnet. They found in addition that there was no three-dimensional ordering down to 1.5°K although an apparent anomaly in $d\chi/dT$ at 0.84°K could have indicated the onset of long-range order. The molecular field ordering temperature is $\sim 76°K$. The one-dimensionality of the magnetism in TMMC was most directly demonstrated by Birgeneau et al.[3] in their quasielastic neutron scattering experiments. The scattering cross section is given simply by

$$\frac{d\sigma}{d\Omega_f}(\mathbf{Q}) \propto \mathscr{S}(\mathbf{Q}) = \sum_{\mathbf{r}} (\exp i\mathbf{Q} \cdot \mathbf{r}) \langle \mathbf{S}_o \cdot \mathbf{S}_r \rangle$$

so that the neutron scattering measures directly the static correlations $\langle \mathbf{S}_o \cdot \mathbf{S}_r \rangle$ in Fourier transform. Birgeneau et al. found that the scattering took the form of planes perpendicular to the chain axis with a Lorentzian cross section peaking at $Q = (2n + 1)\pi/a$, where a is the nearest-neighbor Mn^{2+} separation. The planar scattering necessitated that the correlations $\langle \mathbf{S}_o \cdot \mathbf{S}_r \rangle$ were purely one dimensional. They found in addition that the measured $d\sigma/d\Omega_f$ could be accurately fitted at all temperatures between 1.1 and 40°K using Fisher's exact solution for the correlations $\langle \mathbf{S}_o \cdot \mathbf{S}_r \rangle$ for the classical linear-chain antiferromagnet. Equivalently, Fisher's solution for the classical nearest-neighbor Heisenberg model correctly predicted both the spatial and thermal variation of the instantaneous correlations in TMMC at all temperatures between 40 and 1.1°K.

The dynamics of TMMC were studied by Hutchings et al.[4] with the inelastic neutron scattering technique. They found that there were well-defined spin waves at low temperatures for $q > \kappa$, the inverse correlation length. Furthermore, the spin waves were found to follow a simple sine wave dispersion curve over the range measured: $0.5\pi/a \leq Q \leq 0.95\pi/a$ at 1.9°K with a slope in agreement with that deduced from the susceptibility. As the temperature was raised from 1.9 to 40°K the "spin waves" typically weakened in intensity and broadened asymetrically, with the scattering increasing on the low-energy side.

We have extended the measurements of Birgeneau et al. and Hutchings et al. both to a wider Q range and to lower temperatures. We find that at 0.841°K there is a transition to three-dimensional long-range order. The transition occurs abruptly with little precursive effects either in the form of three-dimensional critical scattering or in a marked change in the planar scattering. The planar scattering is found to drop uniformly in intensity in the ordered phase, mirroring the corresponding rise in the three-dimensional Bragg scattering. The transition appears to be second order with the order parameter critical exponent $\beta = 0.26$ for $0.001 < 1 - T/0.841 < 0.1$. We have also studied the spin waves over the Q range $0.1\pi/a \leq Q \leq 0.985\pi/a$ at 1.45°K. The spin-wave dispersion relation is found to have the form of a double sine wave, thus exhibiting the full symmetry of the paramagnetic Brillouin zone. Also, measurements of a few spin waves down to 0.6°K show that the three-dimensional ordering has no observable effect on the spin-wave energies or intensities. Finally, we studied the excitations at $q^* = Q - (\pi/a) = 0.015\pi/a$ to $0.04\pi/a$ with high energy and q resolution over the temperature range 1.45–9°K; over this same temperature range κ, the inverse correlation length, changes from $0.0034\pi/a$ to $0.022\pi/a$. From these measurements we can deduce detailed information about the dynamic response function as q^* is varied from greater to less than κ; in particular, the data show clearly the evolution from overdamped to spin-wave-type behavior.

Full details of these experiments together with an extensive theoretical analysis will be published at a later date.

References

1. R. Dingle, M.E. Lines, and S.L. Holt, *Phys. Rev.* **187**, 643 (1969).
2. M.E. Fisher, *Am. J. Phys.* **32**, 343 (1964).
3. R.J. Birgeneau, R. Dingle, M.T. Hutchings, G. Shirane, and S.L. Holt, *Phys. Rev. Lett.* **26**, 718 (1971).
4. M.T. Hutchings, G. Shirane, R.J. Birgeneau, and S.L. Holt, *Phys. Rev. B* **5**, 1999 (1972).

Nuclear Relaxation in a Spin-$\frac{1}{2}$ One-Dimensional Antiferromagnet*

E. F. Rybaczewski, E. Ehrenfreund, A. F. Garito, and A. J. Heeger

Department of Physics and Laboratory for Research on the Structure of Matter
University of Pennsylvania, Philadelphia, Pennsylvania

and

P. Pincus †

Department of Physics, University of California at Los Angeles
Los Angeles, California

The spin dynamics in a spin–1/2, one-dimensional (1D) antiferromagnet (AF) at low temperatures is probed by nuclear spin–lattice relaxation measurements. The absence of a long-range order and a true phase transition even at 0°K results in large-amplitude spin fluctuations which dominate the relaxation even at very low temperatures and give rise to a finite relaxation rate at $T \to 0°K$.

To calculate the nuclear relaxation we use the fermion representation of the $s = 1/2$ 1D Heisenberg Hamiltonian.[1]

$$\mathcal{H} = J \sum_k (\cos k - 1) c_k^+ c_k + \frac{J}{N} \sum_{k_1, k_2, q} (\cos q) c_{k_1+q}^+ c_{k_2-q}^+ c_{k_2} c_{k_1} \tag{1}$$

where c_k^+ and c_k are, respectively, creation and annihilation operators for spinless fermions. The total number of fermions is not conserved. The dispersion relation has been calculated using the Hartree–Fock approximation,

$$\varepsilon(k) = Jp \cos k \tag{2}$$

where at low temperatures $p \approx 1.637[1 - 0.240(k_B T/J)^2 + \ldots]$. The ground state in zero magnetic field consists of $N/2$ occupied states and $N/2$ empty states. There is no gap in the excitation spectrum even in finite magnetic field as long as $\mu_B H \ll J$. Bulaevskii[1] has also shown that the ground-state energy in this approximation is within 5% of the exact result.

The relaxation rate, for a hyperfine interaction $A\mathbf{I}\cdot\mathbf{S}$, is calculated using the relation (for spin-1/2 nuclei)[2]

$$T_1^{-1} = A^2/(4\hbar^2 N) \sum_j \left\{ \langle S_j^+ S_j^- \rangle_{\omega_N} + \langle S_j^- S_j^+ \rangle_{\omega_N} \right\} \tag{3}$$

---

* Supported in part by the National Science Foundation and in part by ARPA through the LRSM.
† Supported by the National Science Foundation and the U. S. Office of Naval Research.

where $\langle S_j^+ S_j^- \rangle_{\omega_N}$ is the Fourier transform (at the nuclear Zeeman frequency ω_N) of the spin–spin time correlation function. Expressing the spin operators in the fermion representation and using the Green's function formalism, one can show that

$$T_1^{-1} = (A^2/4\,\hbar^2 N) \sum_k \rho^+(k, \omega_N) \tag{4}$$

where $\rho^+(k, \omega)$ is the anticommutator spectral weight function of the operators c_k^+ and c_k.

The simplest process contributing to the relaxation is the direct one in which a nuclear spin flip is accompanied by emission or absorption of an elementary excitation. For this process T_1^{-1} is calculated using the random phase approximation (RPA), where $\rho^+(k, \omega) = 2\pi\delta(\omega - \varepsilon(k)/\hbar)$. Thus T_1^{-1} for $\hbar\omega \ll J$ is

$$T_1^{-1} = A^2/2\hbar J p \tag{5}$$

This expression for T_1^{-1} is expected to be valid only at low temperatures, since it does not take into account the thermal broadening of the excitation spectrum. The finite width of the spectrum would give rise to multiexcitation emission (and absorption) and will be much more temperature dependent than that expected from Eq. (5).

Dealing with 1D systems requires special attention to the finite contribution of the uniform mode ($q = 0$) to the relaxation, since, unlike higher-dimensional systems, the density of states at $q = 0$ is not negligible. Given the uniform mode with susceptibility whose imaginary part is $\chi''_{\alpha\beta}(\omega)$, one can use the fluctuation-dissipation theorem to get the contribution to the spin–spin correlation function,

$$\langle S^\alpha S^\beta \rangle_\omega = (2k_B T/g^2\mu_B^2\omega)\,\chi''_{\alpha\beta}(\omega) \tag{6}$$

where α, β are spin component indices.

Consider now the anisotropic nuclear–electron dipolar interaction in order to calculate the contribution of the uniform mode. This interaction includes a term proportional to $I^+ S^z$ and involves a nuclear spin flip unaccompanied by an electron spin flip. Its contribution to the nuclear relaxation is given by[3] $T_1^{-1}|_{\text{dip}} = \alpha_D^2 \langle S^z S^z \rangle_{\omega_N}$, where $\alpha_D^2 = (3/5)g^2\mu_B^2\gamma_N^2 r^{-6}$ is the appropriate dipolar coupling constant averaged over all angles. χ''_{zz}, the response function in the z direction to an alternating field in the z direction, is given by the Bloch equations as $\chi''_{zz} = \chi_0 \omega \tau/(1 + \omega^2\tau^2)$, where χ_0 is the uniform static susceptibility and τ is the electronic relaxation time. $T_1^{-1}|_{\text{dip}}$ thus becomes

$$T_1^{-1}|_{\text{dip}} = \frac{2\alpha_D^2 k_B T}{g^2\mu_B^2} \frac{\chi_0 \tau}{1 + \omega_N^2\tau^2} \tag{7}$$

Other terms in the dipolar interaction involve simultaneous nuclear–electron spin flips which require much more energy, and are thus neglected.[3]

These two mechanisms, the direct process and the uniform mode contribution, predict at low temperatures a frequency-independent term with finite zero-temperature extrapolation [Eq. (5)] and a frequency-dependent term whose con-

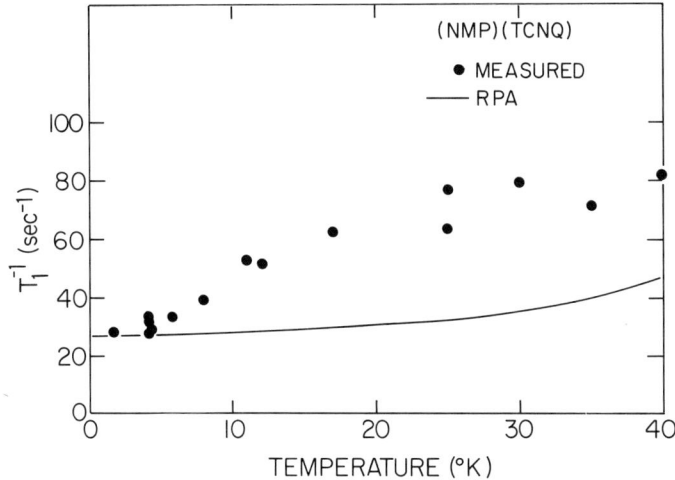

Fig. 1. Temperature dependence of the proton spin–lattice relaxation rate in (NMP) (TCNQ). These measured rates are frequency independent in the range 20–40 MHz. The solid curve represents the weak temperature dependence predicted by the Hartree–Fock theory.

tribution becomes less important at high frequencies and low temperatures [provided χ_0 and τ remain nondivergent as $T \to 0$; Eq. (7)].

Proton spin–lattice relaxation times were measured in polycrystaline samples of the the spin–1/2, one-dimensional AF (NMP) (TCNQ).* The recovery of the free induction magnetization after saturation was found to be exponential in time at all temperatures.

The main features of the results are as follows: (a) The relaxation rate extrapolated to zero temperature is, to within the experimental accuracy, frequency independent from 4 to 40 MHz with a value $T_1^{-1}(0) = 29 \pm 2$ sec^{-1}; (b) above 20 MHz T_1^{-1} is frequency independent and possesses a weak temperature dependence, as shown in Fig. 1, whereas at low frequencies a somewhat stronger temperature dependence is observed, with the relaxation becoming considerably slower at the lowest temperatures; (c) the frequency dependence at 4.2°K is as shown in Fig. 2; and (d) the measured rates were insensitive to the magnetic impurity content of the sample.

The frequency-independent zero-temperature extrapolation of T_1^{-1} indicates that the direct process discussed above, Eq. (5), is the main relaxation mechanism at low temperatures and high frequencies. Using the values $A/g\mu_B = 1.57$ G and $J = 5.2 \times 10^{-3}$ eV,[4,5] one finds $T_1^{-1}(0) = A^2/3.7\ \hbar J$, in good agreement with the value $A^2/3.28\ \hbar J$ deduced from Eq. (5). The experimental rate is much more temperature dependent than that predicted by the RPA (Fig. 2). We attribute this to the thermal broadening of the excitation spectrum, which has not been taken into account in the RPA.[6] The contribution of the longitudinal fluctuations of the uniform mode [Eq. (7)] is expected to be pronounced at low frequencies. The frequency dependence

---

* For a review of the properties of N = methylphenazinium tetracyanoquinodimethan (NMP) (TCNQ) see Epstein et al.[4]

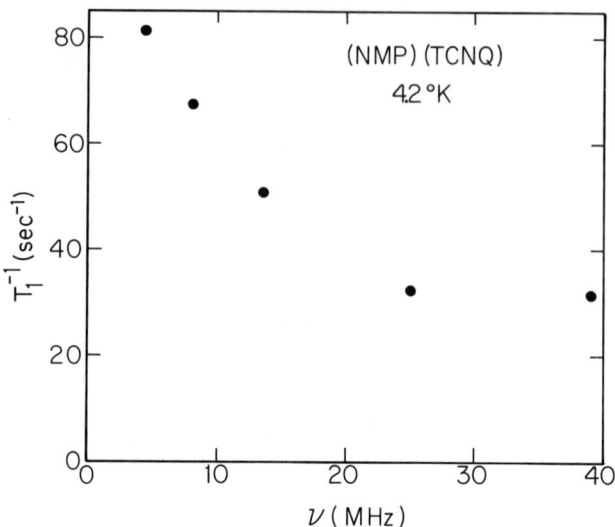

Fig. 2. Frequency dependence of the proton spin–lattice relaxation rate in (NMP) (TCNQ) at 4.2°K.

of the low-frequency data yields $\tau \approx 4 \times 10^{-8}$ sec, which is about ten times shorter than the inverse of the measured exchanged narrowed ESR linewidth.[4] However, since the nuclear relaxation mechanism measures $\tau(\omega = \omega_N)$ i.e., far off resonance $(\omega_N/\omega_{ESR} \approx 10^{-3})$, it might not be surprising to find a discrepancy when compared with the simple Bloch equation, which predicts strictly exponential relaxation and Lorenzian linewidth.

We conclude that this example of spin-1/2, 1D AF is, to a reasonable accuracy, described by fermionlike excitations, that the electronic correlation time is of the order of $\hbar/J$ throughout the AF regime due to the substantial spin fluctuations, and that the contribution of longitudinal fluctuations of the uniform mode has to be taken into account in nuclear relaxation measurements in 1D systems.

References

1. L.N. Bulaevskii, *Soviet Phys.—JETP* **16**, 685 (1963).
2. T. Moriya, *Prog. Theoret. Phys. (Kyoto)* **16**, 23 (1956).
3. A. Abragam, *Nuclear Magnetism,* Oxford (1961), p. 380.
4. A.J. Epstein, S. Etemad, A.F. Garito, and A.J. Heeger, *Phys. Rev. B* **5**, 952 (1972).
5. E. Ehrenfreund, E.F. Rybaczewski, A.F. Garito, and A.J. Heeger, *Phys. Rev. Lett.* **28**, 873 (1972).
6. D. Beeman and P. Pincus, *Phys. Rev.* **166**, 359 (1968).

Low-Dimensional Magnetic Behavior of $Cu(NH_3)_2 \cdot Ni(CN)_4 \cdot 2C_6H_6$

H. Kitaguchi, S. Nagata, Y. Miyako, and T. Watanabe

*Department of Physics, Hokkaido University
Sapporo, Japan*

The crystal symmetry of $Cu(NH_3)_2 \cdot Ni(CN)_4 \cdot 2C_6H_6$ is tetragonal and the space group is P_4/m, as shown in Fig. 1.[1,2] The magnetic susceptibility was measured by the ac Hartshon bridge method at 80 Hz. The temperature range for the present experimental measurement is between 0.2 and 20.4°K. The temperature range in the specific heat measurement is between 1.2 and 20.4°K.[3,4]

The experimental results on the magnetic susceptibility are given as open circles in Fig. 2. It has broad maximum near 3°K and decreases with decreasing temperature down to 0.2°K. For the analysis of this result two procedures were introduced: the high-temperature series expansion method and the Bonner–Fisher procedure for the one-dimensional antiferromagnet.[6] Resulting values for the exchange interaction

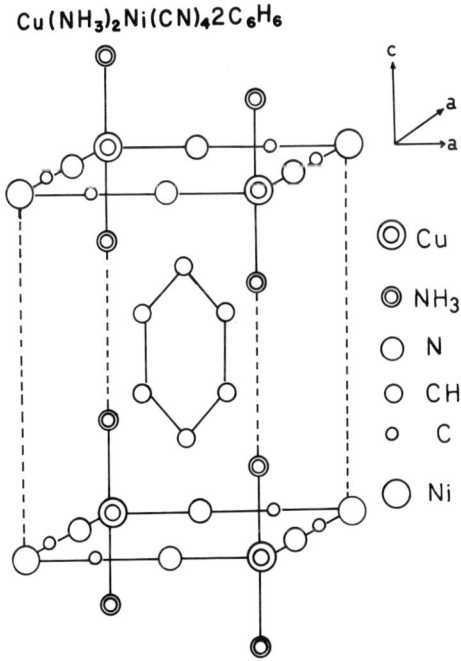

Fig. 1. Unit cell of $Cu(NH_3)_2 \cdot Ni(CN)_4 \cdot 2C_6H_6$.
($a = 7.39$ Å, $c = 8.24$ Å)

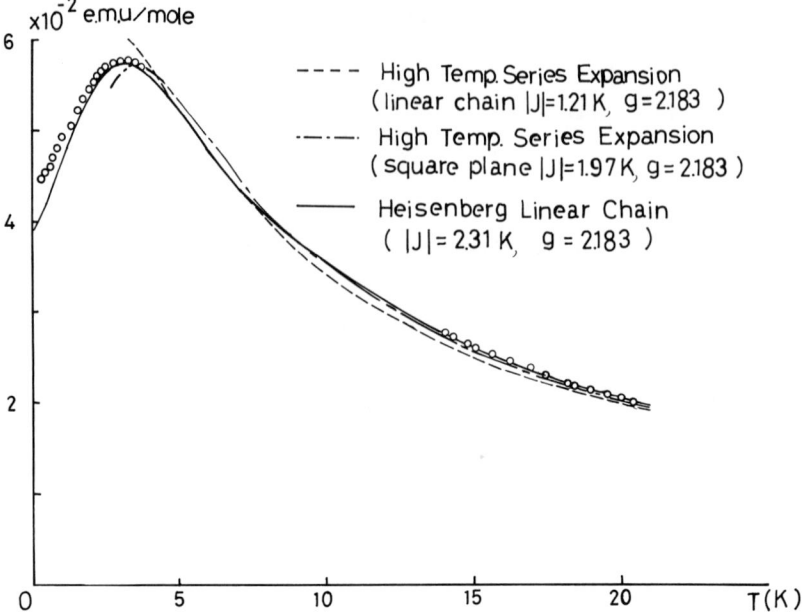

Fig. 2. Experimental susceptibility (circles) vs. temperature.

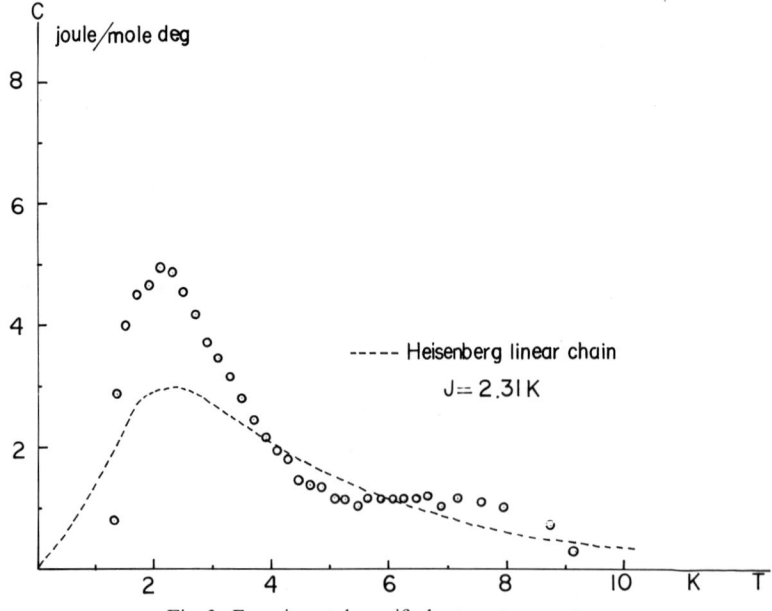

Fig. 3. Experimental specific heat vs. temperature.

are $|J| = 1.97°K$ for linear chain and $|J| = 1.21°K$ for plane square lattice, respectively, in the case of the high-temperature series expansion method. The linear-chain model ($|J| = 1.97°K$) seems to be better in this case. On the other hand, the value obtained from the Bonner–Fisher method is $|J| = 2.31°K$. The agreement with the experimental result is very clear in these cases, as given in Fig. 2. From these analyses we can conclude that $Cu(NH_3)_2 \cdot Ni(CN)_4 \cdot 2C_6H_6$ shows one-dimensional antiferromagnetic characteristics for the Heisenberg spin system of $S = 1/2$.

The experimental result of specific heat measurements for $Cu(NH_3)_2 \cdot Ni(CN)_4 \cdot 2C_6H_6$ gives a broad maximum at about $3°K$ and decreases rapidly below $3°K$, as shown in Fig. 3. The calculated entropy change ΔS equals 17.40 (J/mole-°K) from the experimental data. On the other hand, the entropy change expected for this spin system ($S = 1/2$) is $\Delta S_{mag} = R \ln(2S + 1) = 5.76$ J/mole-°K. Therefore $Cu(NH_3)_2 \cdot Ni(CN)_4 \cdot 2C_6H_6$ contains large excess entropy in the liquid helium temperature region; that is $\Delta S_{ex} = \Delta S - \Delta S_{mag} = 11.64$ J/mole-°K. This entropy change can be ascribed to the hindered rotation of NH_3 group.[3,4] The entropy change for such a hindered rotation is $\Delta S_{sh} = 2R \ln 2 = 11.52$ J/mole-°K for the present system. That is, the present excess entropy change ΔS_{ex} agrees very well with the above result $S\Delta_{sh}$. Using the value of $J = 2.31°K$, the values of C_{max} and T_{max} obtained from the Bonner–Fisher theory are $C_{max} = 2.91$ J/mole-°K and $T_{max} = 2.23°K$, respectively.[6] Using the above three values, the analytical result for the one-dimensional antiferromagnet is as shown in Fig. 3.[6]

References

1. T. Iwamoto, T. Miyoshi, T. Miyamoto, Y. Sasaki, and S. Fujiwara, *Bull. Chem. Soc. (Japan)* **40**, 1174 (1967).
2. J.H. Rayner and H.M. Powell, *J. Chem. Soc. (London)* **1952**, 319.
3. S. Takayanagi, S. Nagata, and T. Watanabe, in *Proc. 12th Intern. Conf. Low Temp. Phys., 1970* Academic Press of Japan, Tokyo (1971), p. 791.
4. S. Nagata, T. Maruyamauchi, and T. Watanabe, *J. Phys. Soc. (Japan)* **30**, 1054 (1971).
5. G.S. Rushbrooke and P.J. Wood, *Mol. Phys.* **1**, 168 (1958).
6. J.C. Bonner and M.E. Fisher, *Phys. Rev.* **135**, 640 (1964).

Electron Paramagnetic Resonance in Cu(NO$_3$)$_2$ · 2.5H$_2$O*

Y. Ajiro,† N. S. VanderVen, and S. A. Friedberg

Carnegie–Mellon University
Pittsburgh, Pennsylvania

The low-field susceptibility,[1] specific heat,[2] proton resonance,[3] and magnetization isotherms[4] of Cu(NO$_3$)$_2$ · 2.5H$_2$O at low temperatures can be described approximately by a model in which the cupric ions are antiferromagnetically coupled in pairs. The members of a pair ($S_1 = S_2 = \frac{1}{2}$) are assumed to interact via isotropic exchange $J_1 \mathbf{S}_1 \cdot \mathbf{S}_2$ with $J_1/k = 5.2°$K. Quantitative discrepancies between this simple model and the data were removed initially by including a weaker antiferromagnetic interpair coupling in a molecular-field approximation.[4]

A full determination of the structure[5] of this salt placed it in the space group I2/a, and revealed the presence of regular chains of cupric ions linked by NO$_3^-$ groups along the b axis. Translational invariance evidently requires that the dominant pairing interaction occur between Cu^{2+} ions on adjacent chains. However, there are three different interchain separations in this structure of comparable magnitude and at least three multiatom superexchange paths which could mediate the pairing interaction. Thus the basic pair could not be clearly identified, although it appeared likely that it was formed by Cu^{2+} ions 5.53 Å apart on nearest-neighbor chains. A center of inversion lies between these two Cu^{2+} ions.

It has proved possible[6] to account quantitatively for all of the available low-field data on Cu(NO$_3$)$_2$ · 2.5H$_2$O by assuming the pairs to be coupled in extended one-dimensional arrays. However, two different models yield essentially identical fits of the data. In the first the pairs Cu-J_1-Cu form the rungs of a ladder rising in the b direction, each Cu coupled by an interaction J_2 to Cu's above and below it. In the second, staggered, alternating chains of the form -Cu-J_1-Cu-J_2-Cu-J_1-Cu- intersect the b axis. Both models are compatible with the structure. They also explain[6,7] the unusual magnetocaloric behavior[8] of this salt in fields of $\sim$ 36 kOe at temperatures near 0.3°K as arising from pronounced short-range order in a one-dimensional system. The observation[9] of long-range spin order below $\sim$ 0.16°K in fields of $\sim$ 36 kOe indicates, of course, that these arrays are not completely isolated from one another.

We have studied the EPR in single crystal Cu(NO$_3$)$_2$ · 2.5H$_2$O between 1.4 and 300°K in an effort to: (1) locate the dominant spin pair in the structure; (2) verify our assumption that interpair coupling is predominantly one dimensional and identify the actual one-dimensional array of pairs; and (3) detect if possible the interaction among these arrays which makes possible the propagation of long-range

---

* Work supported by National Science Foundation and Office of Naval Research.
† Present address: Kyoto University, Kyoto, Japan.

spin order at sufficiently low temperature and high fields. All measurements were carried out at X band (9.3 GHz).

At all but the very lowest temperatures a single exchange-narrowed resonance line was observed for all orientations of the static field H_0 relative to the crystalline axes. The angular dependence of the resonance locates the principal axes of susceptibility K_1, K_2, K_3. It also provides the components of the g tensor ($g_\parallel = 2.40$, $g_\perp = 2.07$) whose symmetry and principal axes are consistent with the tetragonal pyramidal coordination of Cu^{2+} in this salt.

Below liquid H_2 temperatures the intensity of the line follows the relation $I \propto (1/T)[4/(3 + e^{J_i/kT})]$ with $J_1/k = 5.2°K$. This is shown in Fig. 1, where the integrated intensity (solid dots) is compared with the theoretical curve (solid line). Thus the resonance occurs essentially in a triplet state lying 5.2°K above a singlet ground state, in good agreement with inference from other observations. Unlike the resonance seen[10] in the well-known Cu^{2+} pairs in $Cu(CH_3COO)_2 \cdot H_2O$, however, this one shows no resolvable hyperfine structure. This fact indicates the presence of significant interpair coupling and can be interpreted in terms of the transfer triplet spin excitons (see, e.g., Ref. 11). The absence also of resolved fine structure such as might arise from anisotropic intrapair interactions may be a further consequence of such interpair coupling.

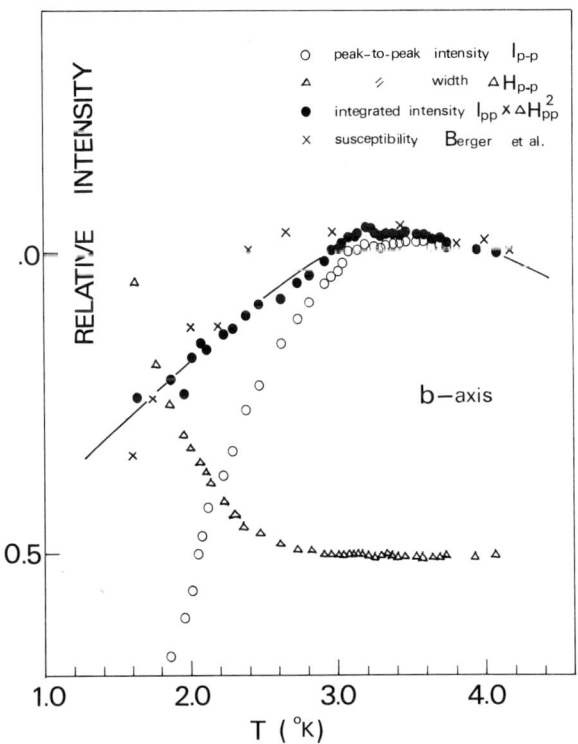

Fig. 1. The temperature variation of the intensity and linewidth of the resonance line in $Cu(NO_3)^2 \cdot 2.5H_2O$ with $H_0 // b$.

The resonance linewidth is unusual in its anisotropy and its temperature dependence. It has proved possible to account for the observations in a reasonable way by extending the linewidth theory for Frenkel spin excitons.[12] The contributions which have been considered include the exchange-narrowed (1) intrapair dipolar interaction, (2) interpair dipolar interaction between similar ions, (3) interpair dipolar interactions between dissimilar ions; as well as (4) amalgamation of lines from dissimilar ions due to exchange interaction, and (5) hyperfine interaction. The overall anisotropy at low temperatures is found to arise from contributions 1 and 5. Thus from the low-temperature data it is possible to locate the intrapair axis. It is found to be the axis of the pair separated by 5.53 Å and linked by two parallel bridges of the form $Cu-NO_3-H_2O-Cu$, as previously hypothesized. At high temperatures the overall anisotropy is dominated by contributions 2 and 3 above. From the magnitude and angular variation of the linewidth it is possible to identify $J_2/k \approx 1°K$ with exchange between similar ions along the b axis. Thus the ladder model is selected as the realistic one. Finally, it is possible to associate differences in linewidth between certain orientations of H_0 with contribution 4 listed above and thus to establish the exchange interaction between dissimilar ions, J_3/k, as $\sim 0.1°K$. This accounts for the dominant interaction between adjacent spin ladders.

The full experimental results and details of the linewidth calculations will be published elsewhere. To illustrate the rather good agreement obtained between theory and experiment, however, we show in Fig. 2 the angular variation of the linewidth at 77 and 1.5°K with H_0 in the ac (xy) plane. (The directions x, y, and z form a convenient set of orthogonal axes with $z \parallel b$ and y along the intrapair axis.) Also shown are the calculated curves for the relevant contributions in this plane together with their sums. The data for 77°K were fitted using $J_2/k = 0.84°K$ and $J_3/k = 0.17°K$ as well as $J_1/k = 5.2°K$. No further adjustment of parameters was required to give the curves for 1.5°K. In Fig. 3 the observed and calculated temperature

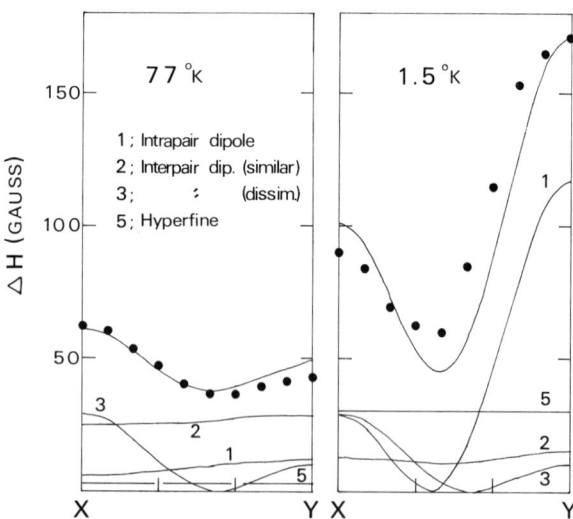

Fig. 2. The observed and calculated linewidth at 77 and 1.5°K with H_0 in the xy (ac) plane.

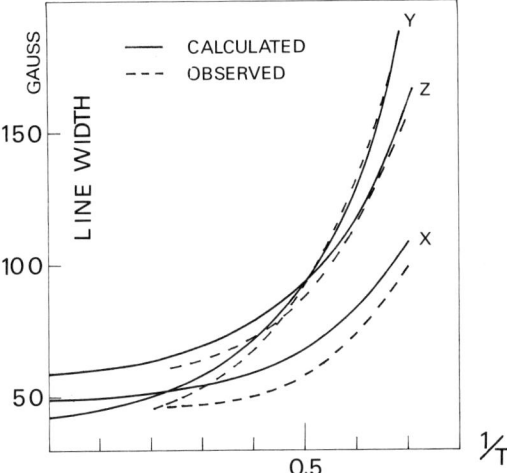

Fig. 3. The observed and calculated temperature dependence of the linewidth with H_0 in the x, y, and z directions.

variations of the linewidth are shown for the helium region with H_0 along the x, y, and z (b) axes. Again the parameters are those fitted at 77°K. The fact that the linewidth is largest at the lowest temperatures for H_0 parallel to the y axis is evidence that this is actually the direction of the intrapair axis.

References

1. L. Berger, S.A. Friedberg, and J.T. Schriempf, *Phys. Rev.* **132**, 1057 (1963).
2. S.A. Friedberg and C.A. Raquet, *J. Appl. Phys.* **39**, 1132 (1968).
3. S. Wittekoek and N.J. Poulis, *J. Appl. Phys.* **39**, 1017 (1968).
4. B.E. Myers, L. Berger, and S.A. Friedberg, *J. Appl Phys* **41**, 1149 (1969).
5. B. Morosin, *Acta Cryst.* **B26**, 1203 (1970).
6. J.C. Bonner, S.A. Friedberg, H. Kobayashi, and B.E. Myers, in *Proc. 12th Intern. Conf. Low Temp. Phys.*, 1970, Academic Press of Japan, Tokyo (1971), p. 691.
7. M. Tachiki, T. Yamada, and S. Maekawa, *J. Phys. Soc. Japan* **29**, 663 (1970).
8. T. Haseda, Y. Tokunaga, R. Yamada, Y. Kuramitsu, S. Sakatsume, and K. Amaya, in *Proc. 12th Intern. Conf. Low Temp. Phys.*, 1970, Academic Press of Japan, Tokyo (1971), p. 685.
9. M.W. van Tol, L. Henkens, and N.J. Poulis, *Phys. Rev. Lett.* **27**, 739 (1971).
10. B. Bleaney and K.D. Bowers, *Proc. Roy. Soc.* **A214**, 451 (1952).
11. P.L. Nordio, Z.G. Soos, and H.M. McConnell, *Ann. Rev. Phys. Chem.* **17**, 237 (1966).
12. Z.G. Soos, *J. Chem. Phys.* **46**, 4284 (1967).

One-Magnon Raman Scattering in the Two-Dimensional Antiferromagnet K_2NiF_4

D. J. Toms and W. J. O'Sullivan

Department of Physics and Astrophysics
University of Colorado, Boulder, Colorado

and

H. J. Guggenheim

Bell Telephone Laboratories
Murray Hill, New Jersey

The properties of magnetic excitations in the class of planar antiferromagnets typified by K_2NiF_4 have been of great interest, primarily because of their two-dimensional (2D) character. In particular, the prototype system K_2NiF_4 seems to behave as a nearly isotropic 2D Heisenberg system with respect to its $q \neq 0$ excitation spectrum. K_2NiF_4 consists of simple cubic NiF_2 and KF planes identical to those in the 3D antiferromagnet $KNiF_3$. However, in K_2NiF_4 each successive NiF_2 plane is separated by two KF planes and the nearest-neighbor interplane separation between Ni^{2+} ions is approximately twice the intraplane separation. Neutron diffraction measurements[1] of the spin-wave spectrum in K_2NiF_4 have shown that the intraplane exchange interaction is greater than 270 times the interplane exchange. Antiferromagnetic resonance measurements[2] have demonstrated the existence of a very small anisotropy field ($H_A/H_{exch.} \simeq 1/500$) the presence of which is essential to the establishment of 3D long-range order below the Néel temperature T_N of 97.1°K, but which seems to have no detectable effect on the 2D character of the $q \neq 0$ excitation spectrum.[3]

The neutron diffraction[1,3] studies have provided the essential information regarding the properties of magnetic excitations in the planar antiferromagnets. With improvements in both energy and wavenumber resolution it has been possible to study the dynamic critical behavior of magnons over the entire Brillouin zone in K_2NiF_4 and to definitely establish the 2D nature of the excitations for $q \neq 0$. This work has shown that long-range 2D transverse spin correlations continue well above the 3D transition temperature T_N and that renormalization effects are evidenced almost completely in the imaginary rather than the real part of the magnon energy, up to the highest temperature measured.

In addition, the temperature behavior of the $q \simeq 0$ magnon in K_2NiF_4 has been investigated with neutron scattering.[3] At $q \simeq 0$ the magnon energy renormalizes in typical 3D fashion, although the fact that it renormalizes as the sublattice magnetization is surprising. The $q \simeq 0$ neutron results are in essential agreement with those of the AFMR measurements of Birgeneau et al.,[2] except that the zone center magnon was observed to within 0.1°K of T_N in the neutron studies, while the AFMR signal could be distinguished only to within about 10°K of T_N. The onset of line broadening of the AFMR signal occurred well below T_N. No linewidth variation was observed up to T_N in the neutron results for $q \simeq 0$. Although the intrinsic instrumental resolution of about 1.3 cm^{-1} is significantly less than the AFMR linewidth above 50°K, the lack of observable line broadening probably follows from the finite q-space sampling ($\Delta q \simeq 0.005$ Å^{-1}) in the neutron scattering measurements.

To date inelastic light scattering has been applied in K$_2$NiF$_4$ only to the study of Brillouin-zone-boundary magnon pair modes.[4,5]

In this paper we report preliminary results of the first observations of single-magnon Raman transitions in K$_2$NiF$_4$. The motivation for this work follows from the recognition that by studying the first-order Raman spectrum as a function of scattering angle, one can sample the momentum range $10^{-6} q_{max}$ (forward scattering) $\lesssim q \lesssim 10^{-3} q_{max}$ (backward scattering), where $q_{max} = \pi/a$ and we assume an incident light wavelength of about 5000 Å. The effective momentum transfer corresponding to an AFMR experiment is about $10^{-6} q_{max}$ and, assuming a resolution of 0.005 Å^{-1}, a neutron diffraction study of the $q \simeq 0$ excitation samples a wavenumber regime somewhat larger than the full range available in a first-order Raman experiment. Thus if there exists a property of the magnetic excitations which is highly momentum sensitive in the $q \simeq 0$ region, it seems possible that Raman scattering studies can contribute information which is complementary to that derived from AFMR and neutron diffraction experiments.

The first suggestive factor that such interest may exist is the observation that the renormalization process for finite q magnons in K$_2$NiF$_4$ seems to be a lifetime effect, and that, while the AFMR resonance was observed to broaden drastically well below T_N, no linewidth broadening of the $q \simeq 0$ magnon was observed in the neutron data up to T_N. The second follows from the observed persistence of a $\lambda \simeq 110$ Å magnon to K$_2$NiF$_4$ to temperatures well above T_N.[1] The magnon wavelength corresponding to backscattering with an argon-ion laser source is about 2500 Å. Thus it seems possible that with its intrinsically finer q resolution and by taking advantage of the full range of available momenta about $q = 0$, first-order Raman scattering from magnons in K$_2$NiF$_4$ may be used to study the properties of magnons in the transition region between 3D and 2D behavior. The results presented here are not only preliminary, but also correspond entirely to scattering at a single angle, 90°.

The sample used in this work was a single crystal of K$_2$NiF$_4$ approximately 2 mm wide × 5 mm wide × 6 mm long. Although the crystal reveals internal cracks and contains numerous small bubblelike inclusions, it is of generally higher optical quality than the crystal used by Fleury and Guggenheim[4] and with proper care it was possible to produce a fairly high-quality surface finish on the sample. With few exceptions the techniques used in these measurements were standard. All measurements were made with a Spectra Physics Model 165-03 laser operating in a single longitudinal mode at 5145 Å. An I$_2$ vapor filter was used to reduce the elastic scattered light background. Photon counting was employed and the Spex Model 1401 spectrometer was operated in a stepping mode with a step size of 0.5 cm^{-1}. To assure reproducibility of frequency results the laser line was included in each sweep. When the laser profile had been recorded the stepping rate was reduced and the magnon

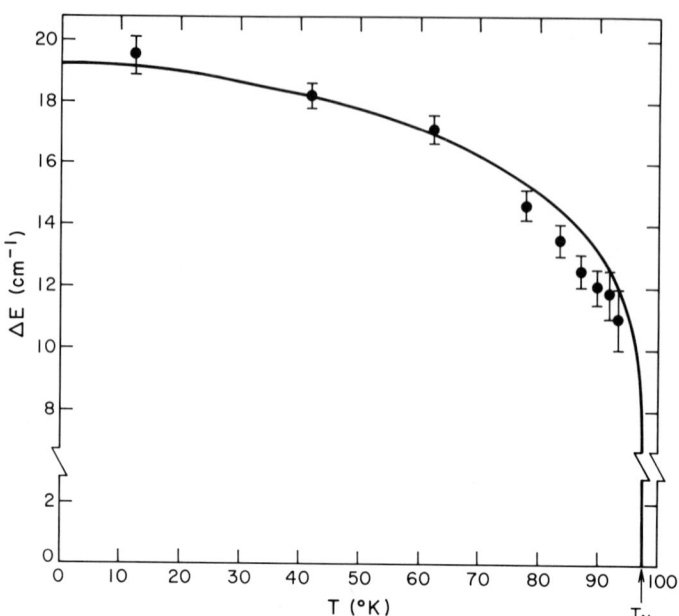

Fig. 1. Magnon frequencies in K_2NiF_4 vs. temperature. The solid line is the magnetization normalized to 19.1 cm^{-1} at $T = 0°K$.

spectrum taken. To assure that our system throughput remained constant during the long runs that were required, the optical phonon at 375 cm^{-1} was recorded immediately before and after each run. To eliminate the possibility that sample heating effects could modify the observations, particularly near T_N, spectra were taken at fixed temperatures with both 400 and 200 mW incident laser power. The temperature of the heavy copper block within which the sample was mounted could be held stable to 0.01°K. Absolute temperatures were determined by a Pt resistance thermometer calibrated to ± 0.02°K which was mounted in the sample block.

Figure 1 is a plot of the magnon frequency in K_2NiF_4 as a function of temperature for right-angle scattering. The highest temperature at which a well-defined excitation could be resolved was 93.5°K, although a diffusive component was observed to persist for a few degrees above T_N. Our frequency results are indistinguishable from the $q \simeq 0$ neutron results. In Fig. 2 we plot data representing several single-magnon Raman spectra for temperatures above 77°K. In each case the measured linewidth exceeds the instrumental width and estimates of the lifetime variation with temperature can be extracted. Signal levels corresponding to the data in Fig. 2 range from a peak of about eight counts/sec at 77.5°K to about one count/sec at 93.5°K. Although improvements will follow from the use of digital signal-averaging techniques, with our present system we are required to count for long periods (250 sec per data point in Fig. 2) in order to gain the necessary statistics. Although nothing different has appeared in the magnon frequency vs. temperature results for 90° scattering, there appears to be a difference between the lifetime vs. temperature behavior of the magnons we are observing and the behavior noted in the AFMR and the $q \simeq 0$ neutron scattering measurements. In Fig. 3 we plot Γ_{magnon} the full magnon linewidth

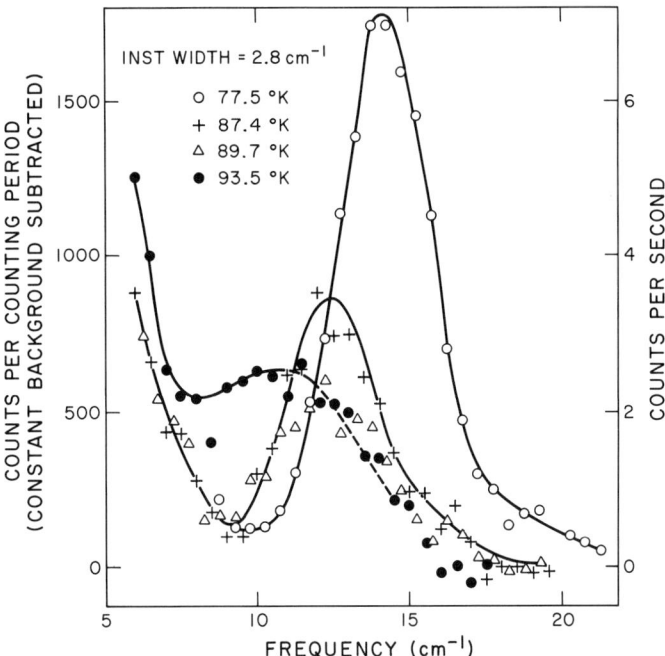

Fig. 2. Raman magnon spectra at several temperatures.

at half-maximum extracted from our measurements, as a function of temperature. The instrumental width was determined from the measured laser line profile. We assumed the deconvolution results, valid for Gaussian distributions,

$$\Gamma^2_{magnon} = \Gamma^2_{obs} - \Gamma^2_{instr}$$

The corresponding AFMR linewidths are also plotted in Fig. 3. These were simply extracted with a prejudiced eye from a figure appearing in Ref. 2. The entire exercise cannot be considered legitimate enough to warrant the inclusion of error bars. It does appear, however, that line broadening in our data seems to be delayed until higher temperatures than that experienced in the AFMR data.

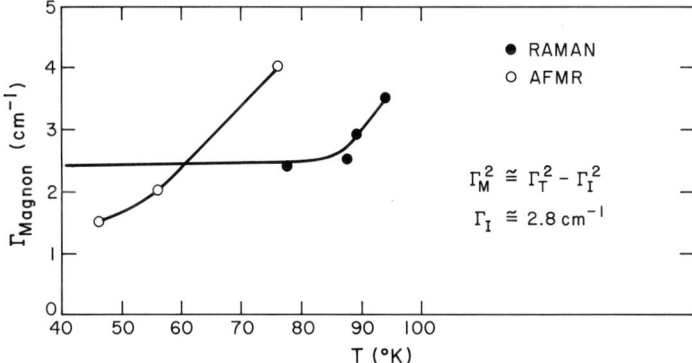

Fig. 3. Estimated Raman and AFMR magnon linewidths vs. temperature.

In any case, the magnon linewidth as a function of temperature must be measured with greater care and for at least one other scattering angle, probably $\theta_{\text{scat}} = 180°$, before conclusions can be drawn regarding the significance of our results.

References

1. J.S. Skalyo, Jr., G. Shirane, R.J. Birgeneau, and H.J. Guggenheim, *Phys. Rev. Lett.* **23**, 1394 (1969).
2. R.J. Birgeneau, F. DeRosa, and H.J. Guggenheim, *Sol. St. Comm.* **8**, 13 (1970).
3. R.J. Birgeneau, J.S. Skalyo, Jr., and G. Shirane, *Phys. Rev. B* **3**, 1936 (1971).
4. P.A. Fleury and H.J. Guggenheim, *Phys. Rev. Lett.* **24**, 1346 (1970).
5. S.R. Chinn, H.J. Zeiger, and J.R. O'Connor, *Phys. Rev. B* **3**, 1709 (1971).

NMR Study of a Two-Dimensional Weak Ferromagnet, Cu(HCOO)$_2$ · 4D$_2$O, in Magnetic Fields up to 10 kOe

A. Dupas and J.-P. Renard

*Institut d'Electronique Fondamentale**
Université de Paris—Orsay, France

Introduction

Copper formate tetrahydrate crystallizes in the monoclinic system. It has a definite layer structure with alternate sheets of copper formate and of water molecules parallel to the (001) plane. At room temperature its space group is $P2_{1/a}$.[1] Below 234.1°K the unit cell is doubled along the c axis.[2] According to recent X-ray studies the low-temperature space group is $P2_1$.[3] Within a copper formate layer each copper ion is coupled to four others by a large superexchange interaction integral J of about 71.5°K.[4] On the other hand, interlayer couplings are of the order of 0.02°K.[5] At about 16.6°K there is a sharp transition to an antiferromagnetic state with a weak parasitic ferromagnetism arising from a Dzyaloshinsky–Moriya antisymmetric exchange interaction.[6] The purpose of the present work is to investigate the magnetic phase diagram of copper formate tetrahydrate in applied fields up to 10 kOe.

Magnetic Structure in Zero Applied Field

We have reported in a previous paper the NMR determination of the proton internal fields in Cu(HCOO)$_2$ · 4H$_2$O.[7] Assuming negligable displacements of the atoms in a copper formate layer at the doubling of the unit cell, the magnetic arrangement I in Fig. 1(a) fits the proton internal fields well, as was shown elsewhere.[8] In a copper formate layer the magnetic moments are antiferromagnetically arranged in the ac plane at about 10° from a axis, and have a 2.2% ferromagnetic component along the b axis; two copper ions are connected by a c-vector translation have opposed antiferromagnetic components and parallel ferromagnetic components. With this two-sublattice structure there must be a net magnetization along b, in contradiction with the results of Kobayashi and Haseda,[6] unless "weak ferromagnetic domains" exist with opposed magnetizations. If there are significant displacements of the atoms in copper formate layers, the proton internal fields may be explained by the four sublattice structure II in Fig. 1(b), which differs from structure I in that two copper ions connected by a c-vector translation have completely antiparallel magnetic moments. However, with structure II it is not possible to determine the canting of the moments without knowing the displacements of the atoms in copper formate layers.

---

* Laboratory associated with CNRS.

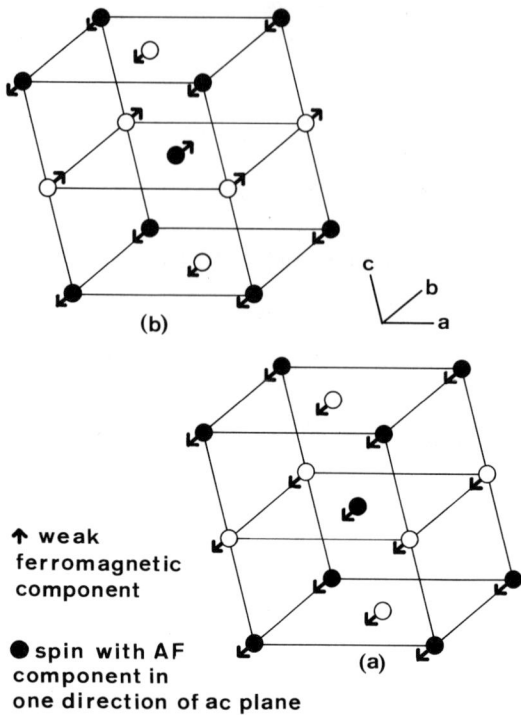

↑ weak
ferromagnetic
component

● spin with AF
component in
one direction of ac plane

○ spin with opposed AF component

Fig. 1. The two proposed magnetic arrangements in copper formate tetrahydrate: (a) two-sublattice structure I; (b) four-sublattice structure II.

Experiments

Proton NMR is performed on single crystals of $Cu(HCOO)_2 \cdot 4D_2O$ at 4.2°K. During a set of measurements the frequency is kept constant, the magnetic field being successively applied in various directions of the ab and ac planes; for each direction of the external field the resonance field values are measured.

Experimental results

(1) External field in the ac plane. We observe four proton resonance lines. When the applied field is rotated in the plane the position of the lines changes regularly in a continuous manner with a 360° periodicity. There is no dramatic change of the diagrams when the frequency is increased up to 40 MHz. The only noticeable effect is a progressive distortion.

(2) External field H in the ab plane. (a) For H lower than $H_c \simeq 5.2$ kOe we observe eight proton resonance lines whose positions change regularly with a 360° periodicity when the applied field is rotated (Fig. 2). On increasing the magnitude of the magnetic field, the intensities of four lines are reduced while the intensities of

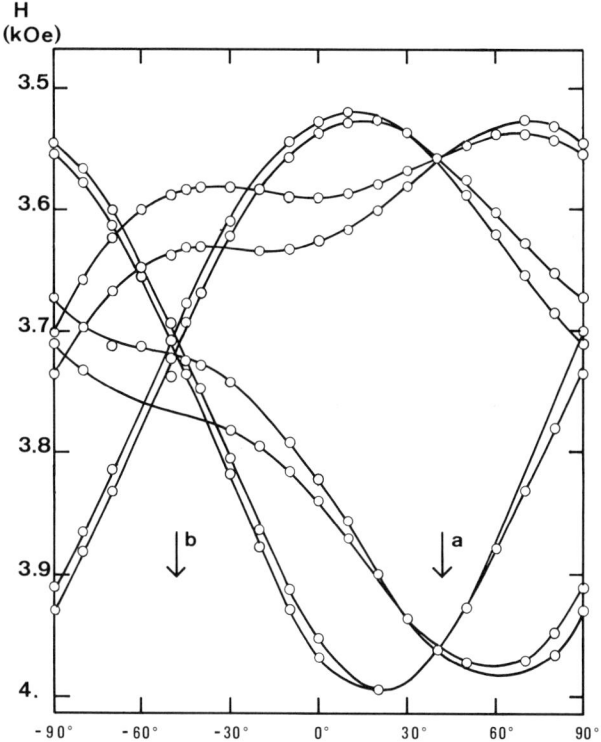

Fig. 2. Resonance diagram of Cu(HCOO)$_2 \cdot$ 4D$_2$O when the field H is rotated in the ab plane. The frequency is 16.04 MHz.

the four others are increased. (b) For H higher than H_c we observe no more than four proton resonance lines. When the field is rotated in the plane there are discontinuities in the positions of all the lines for an angle θ_c between the field and the b axis (Fig. 3). As the field is raised the product $H \cos \theta_c$ remains constant and equal to H_c.

Phase Diagram

Figure 4(a) shows the magnetic phase diagram we infer from our experimental results. In the A phase the magnetic moment arrangement derives continuously from the zero-field arrangement as the external field is increased. If structure I is assumed, the A' phase would be similar to A but without "weak ferromagnetic domains"; if structure II is considered, the A' phase would be a two-sublattice structure II', where every copper formate plane has the same magnetic arrangement. The B structure is a two-sublattice structure which derives from the I or II' structure by a sudden important rotation of the magnetic moments when the component of the applied magnetic field along the b axis reaches the critical value H_c. By increasing the magnitude of the external field from zero to 10 kOe in one direction of the ab plane, the three phases are successively encountered, as is shown by Fig. 4(b). The 5.2-kOe value of H_c is in accordance with results of Seehra and Castner.[9]

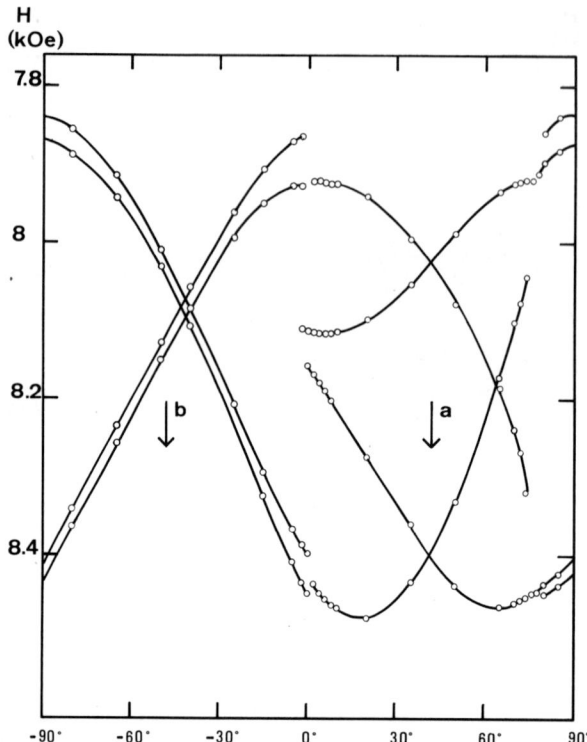

Fig. 3. Resonance diagram of $Cu(HCOO)_2 \cdot 4D_2O$ when the field H is rotated in the ab plane. The frequency is 35 MHz.

A–A' Transition

If the zero applied field structure I is assumed, the transition from A to A' appears as the complete disappearance of the "weak ferromagnetic domains." If structure II is considered, the A–A' transition occurs from a four-sublattice structure to a two-sublattice structure. The second case has been discussed by some authors.[5,10] The first hypothesis is more likely, due to the fact that the transition is not sharp and depends only on the magnitude of the external field.

A'–B Transition

This transition is a very peculiar spin flop in which the applied field is perpendicular to the spin direction, and thus is very interesting. The plane magnetic arrangement is the same for the two structures I and II'. We can thus discuss these two cases in the same manner if we neglect the small interlayer interaction. The energy of the system may be written in the molecular field approximation as

$$W = NzJS^2 \{ \mathbf{u}_1 \cdot \mathbf{u}_2 - \eta u_{1Y}u_{2Y} - \zeta u_{1Z}u_{2Z} + \mathbf{DM} \cdot \mathbf{u}_1 \wedge \mathbf{u}_2 \\ + \mathbf{DE} \cdot ([g_1]\mathbf{u}_1 + [g_2]\mathbf{u}_2) \}$$

NMR Study of a Two-Dimensional Weak Ferromagnet

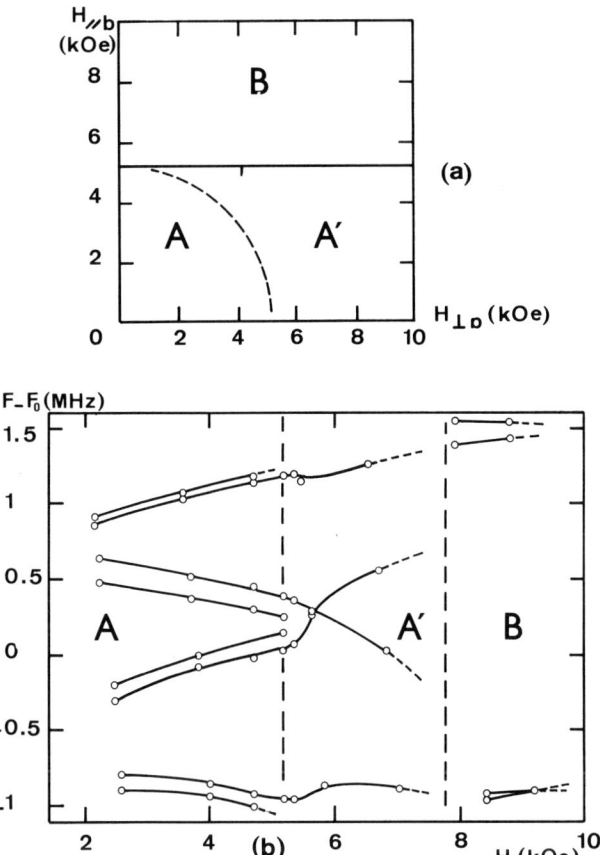

Fig. 4 (a) Magnetic phase diagram of copper formate tetrahydrate in fields up to 10 kOe. (b) $F - F_0$ is the difference between the resonance frequencies of proton in $Cu(HCOO)_2 \cdot 4D_2O$ and the free proton resonance frequency. H is applied at 39° off the a axis in the ab plane.

using $\mathbf{DM} = \mathbf{D}/J$ and $\mathbf{DE} = \mu_B \cdot \mathbf{H}/zJS$. Here $\mathbf{u}_1$ and $\mathbf{u}_2$ are the unit vectors of the spin direction in the two sublattices, z is the number of first neighbors, N is the number of spins in a sublattice, η and ξ are parameters which characterize the symmetric exchange anisotropy, $\mathbf{D}$ is the Dzyaloshinsky–Moriya vector, and $OXYZ$ are the principal axes of the symmetric exchange interaction ($OY = b$). The g tensor of the two nonequivalent copper ions are symmetry related and, assuming they are symmetric, they may be written as

$$[g_1] = \begin{bmatrix} g_{xx} & g_{xy} & g_{xz} \\ g_{xy} & g_{yy} & g_{yz} \\ g_{xz} & g_{yz} & g_{zz} \end{bmatrix}, \quad [g_2] = \begin{bmatrix} g_{xx} & -g_{xy} & g_{xz} \\ -g_{xy} & g_{yy} & -g_{yz} \\ g_{xz} & -g_{yz} & g_{zz} \end{bmatrix}$$

In order to ensure that the zero-applied-field state coincides with the experimental one, we choose OX along the antiferromagnetic direction, and $\mathbf{DM}$ along

OZ. However, other choices may be possible. With an applied field along the OY axis the energy is then written

$$W = NzJS^2[-1 + \xi \sin^2 \delta + 2DE_y(g_{xy} \cos \delta + g_{yz} \sin \delta) + 2\alpha(g_{yy}DE_y + DM_z \cos \delta) + 2\alpha^2]$$

where δ is the angle between a axis and the antiferromagnetic component of the spins that lies in the ac plane and α is the canting angle. With $g_{yz} = 0$ one finds two exact solutions for the equilibrium conditions, with the following values for δ:

$$\sin \delta_1 = 0 \qquad (1)$$
$$\cos \delta_2 = (2g_{xy} - g_{yy}DM_z) DE_y/(2\xi + DM_z^2) \qquad (2)$$

Solution (1) is energetically the most interesting if $2\xi + DM_z^2 > 0$. We have solved the problem by means of a computer for $g_{yz} \neq 0$. We have thus shown that when $2\xi + DM_z^2 \gtrsim 0$ and for suitable values of g_{xy}, g_{yz}, and DM_z there may be a jump at a critical field from a solution with $\delta \simeq 0$ to a solution with $\delta \simeq \delta_2 \simeq 90°$. We think that the A'–B transition corresponds to this type of spin flop.

Conclusion

We have determined the phase diagram of copper formate tetrahydrate in fields up to 10 kOe and established the occurrence of a very peculiar spin flop when the field is applied along the b axis. Further work is needed to discriminate between the two proposed structures, and investigations at higher fields would be interesting.

References

1. R. Kiriyama, H. Ibamoto, and K. Matshuo, *Acta Cryst.* **7**, 482 (1954).
2. K.C. Tuberfield, *Sol. St. Comm.* **5**, 887 (1967).
3. M.J. Bird and T.R. Lomer, *Acta Cryst.* **B27**, 859 (1971).
4. M.S. Seehra, *Phys. Lett.* **28A**, 754 (1969).
5. Y. Ajiro, and N. Terata, in *Proc. 12th Intern. Conf. Low Temp. Phys. 1970*, Academic Press of Japan, Tokyo (1971), p. 813.
6. H. Kobayashi and T. Haseda, *J. Phys. Soc. Japan* **18**, 541 (1963).
7. A. Dupas and J.-P. Renard, *Compt. Rend.* **271B**, 154 (1970).
8. A. Dupas, 3eme cycle Thesis, Université Paris (1971), unpublished.
9. M.S. Seehra and T.G. Castner, *Phys. Rev.* **1B**, 2289 (1970).
10. K. Yamagata, M. Hayama, and T. Odaka, *J. Phys. Soc. Japan* **31**, 1279 (1971).

10
Ferromagnetism

Heisenberg Model for Dilute Alloys in the Molecular Field Approximation

Michael W. Klein*

Department of Physics, Bar Ilan University
Ramat Gan, Israel

Introduction

In the 1950's several experiments were performed in which dilute concentrations of magnetic impurities were dissolved in a nonmagnetic metal host. These experiments were done prior to the discovery of the Kondo[1] effect and their purpose was to study the magnetic properties of materials in which the probability of having a near neighbor is not appreciable. Therefore any cooperative magnetic phenomena which these systems exhibit necessarily arise from the long-range magnetic interaction. That long-range magnetic phenomena existed was shown by Owen et al.[2] and Schmitt and Jacobs[3] for Cu–Mn and by Lutes and Schmit[4] for Au–Fe. This long-range interaction can be obtained theoretically by using an s–d Hamiltonian and calculating the effective Hamiltonian in second-order perturbation theory. This calculation was done by Ruderman and Kittel,[5] Kasuya,[6] and Yosida,[7] and will be denoted in this paper as the RKKY interaction. The effective Hamiltonian $\mathscr{H}$ for the interaction between the magnetic impurities is[7]

$$\mathscr{H} = \left(\frac{3n}{N_o}\right)^2 \frac{2\pi}{E_F} J(0)^2 \sum_{i,j} F(2k_F r_{ij}) \mathbf{S}_i \cdot \mathbf{S}_j = \sum_{i,j} v_{ij} \mathbf{S}_i \cdot \mathbf{S}_j \qquad (1)$$

where $F(x) = (x \cos x - \sin x)/x^4$, n/N_0 is the number of conduction electrons per atom, E_F is the Fermi energy of the host metal, and $J(0)$ is the s–d exchange interaction and is assumed to be a constant. $\mathbf{S}_i$ and $\mathbf{S}_j$ are the spin operators of the impurities located at sites r_i and r_j, respectively, and k_F is the Fermi wave vector. The second form of Eq. (1) defines the potential v_{ij}.

Later, Zimmerman and Hoare[8] measured the low-temperature specific heat of dilute Cu–Mn and found the remarkable phenomenon that the system exhibits a large excess of low-temperature specific heat which is linear in T and independent of c. To explain this phenomenon, Marshall[9] and Herring[10] independently proposed that the excess specific heat arose from the magnetic disordering of the impurities which interact via the RKKY potential. Marshall[9] worked out the low-temperature specific heat in the molecular field approximation, and found a qualitative agreement with experiment in the Ising model.† Later, Klein and Brout[12] considered the statis-

---

* Part of this work supported by AFOSR contract No. F44620-71-C-0013 at Belfer Graduate School of Science, Yeshiva University, and part by the Bar Ilan University Research Committee.
† An alternate mechanism was proposed by Overhauser.[11]

tical mechanics of the random alloy system using a cluster expansion of the Ising model partition function in combination with the molecular field approximation. They found that there are only short-range spin–spin correlations between the magnetic impurities, and obtained a low-temperature specific heat in agreement with that of Marshall[9] and with experiment.[8]

There are several major difficulties with the Marshall–Klein–Brout[9,12] (MKB) calculations: (1) The calculation was done in the Ising model, rather than in the more correct Heisenberg Hamiltonian given in Eq. (1). (2) The calculations give the properties of the system near $T = 0$ only, and there is no way to obtain the specific heat, the magnetic susceptibility, and other thermodynamic properties of the system at temperatures far from zero. The latter difficulty has been overcome in Ref. 13, where I derived the probability distribution of the internal fields using a self-consistent mean random molecular field (MRF) approximation for all temperatures. Only a single parameter γ, which characterizes the strength of the RKKY interaction, enters the MRF calculation. Once this parameter is determined the theory gives the magnetic susceptibility, the specific heat, and the magnetization for all temperatures. It was recently shown by Klein and Shen[14] that the high-temperature magnetic susceptibility predicted by the theory is in reasonable qualitative agreement with experiment. We also found[14] that the parameter γ obtained from the high-temperature experiments, gives the low-temperature magnetic susceptibility in reasonable agreement with experiment. This fact enhanced our confidence in the self-consistent method, in spite of the disturbing inconsistency of using an Ising model instead of the more correct Heisenberg Hamiltonian given in Eq. (1).

The rather good agreement between the Ising model calculation and experiment was quite puzzling, particularly so, since it was argued* that in the Heisenberg model the internal field is a three-dimensional vector quantity and the low-temperature specific heat should be proportional to T^3 rather than T. Thus we were faced with the dilemma in which an incorrect theory (the Ising model) gave very good agreement with experiment, whereas the correct Heisenberg model presumably did not.

Previous Results and the Limitations of the Derivations

Before we present our derivation of the Heisenberg model probability distribution it is useful to describe the previous work on the Ising model as well as the previous conjectures on the Heisenberg model. It is also necessary to discuss the limitations under which the work presented here is valid.

In the treatment of the dilute magnetic alloy system it is convenient to discuss two different regions of temperature T and impurity concentration c: (1) A region of very low concentration c, where c is sufficiently low that the system is characterized by the Kondo effect,[1] and the impurity–impurity interaction given in Eq. (1) can be neglected, and (2) a region of somewhat higher concentration where the behavior of the system is characterized by the impurity–impurity interaction. The latter case is expected to be valid when the average impurity–impurity interaction energy as well as the temperature under consideration are much greater than the Kondo

---

* In Ref. 15 and by many other researchers in private communication with the author.

temperature T_K. In general, both of the above effects coexist, as was discussed by Suhl,[16] Harrison and Klein,[17] Tholence and Tournier,[18] Tsay and Klein,[19] and Béal Monod.[20] However, there are some ideal situations where the properties of the system are determined by *only* one or the other of the above effects. For example, Cu–Mn has a Kondo temperature of about 0.05°K and the average impurity–impurity interaction energy (as determined from the maximum in the magnetic susceptibility) is much greater than T_K for $c > 0.05\%$. It is therefore typical of a system whose behavior is dominated by the impurity–impurity interaction. The discussion in this paper addresses itself to an ideal system where Kondo-like effects can be completely neglected, disregarding the derivations which show that this condition is almost never satisfied.[19]

Let us now consider a brief sketch of the previous derivations. The magnetic impurities are each assumed to be randomly and uniformly distributed throughout the volume V of the solid with a probability V^{-1}. The effective internal (Ising) field $H_I(o)$, acting on a spin at site o is, in the Ising model,

$$H_I(o) = \sum_j v_{oj} \langle \mu_j \rangle \tag{2}$$

where $\langle \mu_j \rangle = \tanh \beta H_I(j)$, where $H_I(j)$, [which is also a random variable], is the effective (Ising) field at site j, and v_{oj} is defined in Eq. (1); $\beta = (k_\beta T)^{-1}$, where k_B is the Boltzmann constant. Thus the spins are assumed to be oriented in the z direction of the (Ising) field. The probability distribution $P(H_I)$ of the random Ising field H_I is obtained,[13] and the magnetic specific heat C_M is evaluated using the approximate expression[9,12]*

$$C_M \approx \frac{N_0 c}{2 k_B T^2} \int_{-\infty}^{\infty} P(H_I) H_I^2 \operatorname{sech}^2 \beta H_I \, dH_I \approx \frac{N_0 c P(0)}{2 k_B T^2} \int_{-\infty}^{\infty} H_I^2 \operatorname{sech}^2 \beta H_I \, dH_I \tag{3}$$

where $N_0 c$ is the number of impurities and $P(0)$ is the probability for zero field. In the second form of Eq. (3) it is assumed that near $T \to 0$ the value of $P(0)$ can be taken outside the integral sign. Changing variables in Eq. (3) and letting $x = \beta H_I$ gives

$$C_M \approx \tfrac{1}{2} N_0 c k_B^2 P(0) \, T \int_{-\infty}^{\infty} x^2 \operatorname{sech}^2 x \, dx \tag{4}$$

The arguments that were presented for the temperature dependence of C_M in the Heisenberg model (HM) are as follows. Since in the HM the spins are quantized not in the z component but in the direction of the total field **h**, the volume of integration in Eq. (2) is $4\pi h^2 dh$ [rather dH as in Eq. (3)], where h is the magnitude of the vector field **h**. When we substitute this into Eq. (2) we obtain

$$C_M \approx \frac{N_0 c}{2 k_B T^2} \int_0^\infty P(h) \, h^2 \operatorname{sech}^2 \beta h \, \{4\pi h^2 \, dh\} \tag{5}$$

Again removing $P(0)$ from under the integral sign as in Eq. (3) and integrating the

---

* A more convenient method is to evaluate the specific heat from the entropy as was done in Ref. 13.

three-dimensional field gives

$$C_M \propto T^3 \tag{6}$$

in contrast to the linear dependence upon T found in Eq. (4).

Derivation of the Probability Distribution in the Heisenberg Model

We now proceed to derive the expression for the probability distribution of the field in the Heisenberg model using the statistical model of Margenau,[21] and show that even though the spins are quantized in the magnitude (and along the direction) of the total field **h**, the low-temperature specific heat is still linear in T and independent of c.

We let the magnetic impurities be randomly distributed in the nonmagnetic host matrix with the interaction Hamiltonian given by Eq. (1). We are concerned with the case of very low impurity concentrations, such that the average interimpurity distance is very large. For Cu–Mn, $2k_F r_{nn} \approx 7$, where r_{nn} is the near-neighbor distance. Thus we neglect the r^{-4} term compared to the r^{-3} term in Eq. (1). Also, since the average distance between the impurities is much greater than the period of oscillation of $\cos 2k_F r_{ij}$, if we place any one impurity in a random position, it is, on the average, just as likely to experience a positive potential as a negative one. For this reason we approximate the interaction potential v_{ij} in Eq. (1) between the magnetic impurities at sites i and j by the relation

$$v_{ij} = \pm a/r_{ij}^3 \tag{7}$$

with probability of $\frac{1}{2}$ for each sign, where a is the strength of the interaction between two impurities at a distance of one lattice constant. The approximation made in Eq. (7) will enable us to evaluate our resulting integrals exactly without changing the physics of the problem.

In the presence of an external magnetic field H_{ext} the Hamiltonian $\mathscr{H}$ for the system is

$$\mathscr{H} = -\sum_{i,j} v_{ij} \mathbf{S}_i \cdot \mathbf{S}_j - \mathbf{H}_{\text{ext}} \cdot \overline{\sum} \mathbf{S}_i \tag{8}$$

We now define the vector molecular field $\mathbf{h}_o$ at an arbitrary site o by the relation

$$\mathbf{h}_o = \sum_j v_{oj} \langle \mathbf{S}_j \rangle + \mathbf{H}_{\text{ext}} \equiv \mathbf{H}_o + \mathbf{H}_{\text{ext}} \tag{9}$$

where the spin $\langle \mathbf{S}_j \rangle$ is the thermal average of the spin at site j. The spin $\mathbf{S}_j$ is quantized in the direction of the local vector field (and in the magnitude of the field) at site j, $\mathbf{h}_j$. Thus if we let h_j be the magnitude of the vector field at site j, i.e., $h_j = |\mathbf{h}_j|$, then

$$\langle \mathbf{S}_j \rangle = \mathbf{u}_j \sum_{M=-s}^{s} M \exp[\beta h_j M] \bigg/ \sum_{M=-s}^{s} \exp[\beta h_j M] \equiv \mathbf{u}_j m_j(h_j) \tag{10a}$$

where $\mathbf{u}_j$ is a unit vector in the direction of $\mathbf{h}_j$, and

$$m_j(h_j) = \frac{2s+1}{2} \coth\left(\frac{2s+1}{2}\right) \beta h_j - \frac{1}{2} \coth\left(\frac{\beta h_a}{2}\right) \tag{10b}$$

is the well-known Brillouin function. Thus the spin S_j is quantized in the direction of $\mathbf{h}_j$. From Eq. (9) we get

$$h_o = (H_o + H_{ext}^2 + 2H_{ext}H_o \cos\theta)^{1/2} \quad (11)$$

where θ is the angle between $\mathbf{H}_{ext}$ and the internal field $\mathbf{H}_o$. For simplicity we let $H_{ext} \to 0$ in all our calculations. (Note: In calculating the magnetization or magnetic susceptibility one must let $H_{ext} \to 0$ only after the appropriate derivatives are taken.)

The energy E_0 of the spin located at the origin o is from Eq. (9) (in the limit as $H_{ext} \to 0$)

$$E_o = -\mathbf{h}_o \cdot \langle \mathbf{S}_0 \rangle = -(h_o \mathbf{u}_o) \cdot (m_o \mathbf{u}_o) = -h_o m_o \quad (12a)$$

also

$$E_o = -\sum_j v_{oj} m_j \mathbf{u}_j \cdot (m_o \mathbf{u}_o) = -\left|\sum_j v_{oj} m_j (\mathbf{u}_j \cdot \mathbf{u}_o)\right| m_o \quad (12b)$$

Equating Eqs. (12a) and (12b) gives

$$h_o = \left|\sum_j v_{oj} m_j (\mathbf{u}_j \cdot \mathbf{u}_o)\right| \quad (13)$$

where

$$\mathbf{h}_j = \sum_k v_{jk} \langle \mathbf{S}_k \rangle \quad (14)$$

Since the magnetic impurities are randomly distributed in the solid, v_{oj} in Eq. (13) is a random variable, and so is h_o. We wish to obtain the probability distribution of the random variable h_o, $P(h_o)$, allowing $\mathbf{u}_j$ to take any direction of orientation. We can formally write the expression for the magnitude of the field at site o, h_o, by using the statistical model of Margenau[21] as follows[13]:

$$P(h_o) = \sum_{v_{ij} = \pm a/r_{ij}^3} \int_{r_N} \delta\left(h_o - \left|\sum_j v_{oj} m_j (\mathbf{u}_o \cdot \mathbf{u}_j)\right|\right) P(r_1, r_2, \ldots, r_N) d^3 r_N \quad (15)$$

where the first sum is a summation over all $v_{ij} = \pm a/r_{ij}^3$ as given in Eq. (7), $P(r_1, r_2, \ldots, r_N)$ is the probability that particle 1 is in volume $d^3 r_1$ at r_1, particle 2 in volume $d^3 r_2$ at r_2, etc., and the integral represents a product over a $3N$-dimensional differential volume.

We now let the positions of the particles be *independent* random variables. Let V be the volume of the solid; then

$$P(r_1, r_2, \ldots, r_N) = V^{-N} \quad (16)$$

This approximation allows multiple occupancy of the sites; however, the error made is of order c^2, and is negligible in the limit as $c \to 0$.

For calculational purposes we let

$$x_o = \sum_j v_{oj} m_j (\mathbf{u}_o \cdot \mathbf{u}_j) \quad (17)$$

in Eqs. (13) and (15). We then calculate $P(x_o)$ using Eq. (15) with the absolute value signs dropped from the second sum. We obtain

$$P(h_o) = P(x_o) + P(-x_o) \quad (18)$$

Using Eqs. (17) and (16) in Eq. (15) gives

$$P(x_o) = (2\pi)^{-1} \sum_{v_{ij} = \pm a/r_{ij}^3} \int_{-\infty}^{\infty} d\rho (\exp[i\rho x_o])$$

$$\times \prod_{j=1}^{N} \int \exp[-i\rho v_{oj} m_j (\mathbf{u}_o \cdot \mathbf{u}_j)] \frac{d^3 r_{oj}}{V} \qquad (19)$$

We consider now the product in Eq. (19). If $m_j(\mathbf{u}_o \cdot \mathbf{u}_j)$ were independent of the site j under consideration, we could evaluate one of the integrals, say I_1, and then the product over N would simply become I_1^N. However, $m_j(\mathbf{u}_o \cdot \mathbf{u}_j)$ depends upon all sites of the system, as is exhibited in Eq. (14). Thus in order to evaluate Eq. (19), an approximation will have to be made, and we will be guided by the physics of the problem to make what we consider a reasonable approximation.

We remark that x_o is a random variable because the impurities are randomly distributed in the system. But then x_j is similarly a random variable which has a well-defined probability distribution. We therefore make the approximation that when we calculate $P(x_o)$ at site o, functions of the random variables at other sites are replaced by their average values. We denote this approximation as the mean random molecular field (MRF) approximation. Thus in the MRF approximation

$$\exp[-i\rho v_{oj} m_j(x_j)(\mathbf{u}_o \cdot \mathbf{u}_j)] \xrightarrow{\text{MRF}} \frac{1}{2} \int_0^\pi d\Omega_j \int_{-\infty}^{\infty} P(x_j) dx_j \{\exp[-i\rho v_{oj} m_j(x_j)(\mathbf{u}_o \cdot \mathbf{u}_j)]\} \qquad (20)$$

where the factor of $\frac{1}{2}$ on the right-hand side arises from $[\int d\Omega_j]^{-1}$, where $d\Omega_j = \sin \theta_j \cdot d\theta_j$. Substituting Eq. (20) into Eq. (19) and summing over each $v_{ij} = \pm a/r_{ij}^3$, we obtain

$$P(x_o) = \frac{1}{2\pi} \int_{-\infty}^{\infty} d\rho e^{i\rho x_o} \prod_{j=1}^{N} \left\{ \frac{1}{2V} \int_{-\infty}^{\infty} P(x_j) dx_j \right.$$

$$\left. \times \int_0^\pi d\Omega_j \int d^3 r_{oj} \cos\left[\frac{\rho a m_j \cos \phi_j}{v_{oj}^3}\right] \right\} \qquad (21)$$

Now as a self-consistency condition we require that $P(x_j)$ be independent of site j. Then Eq. (21) gives an integral equation for $P(x)$. Except for the trivial integration over θ, Eq. (21) is now the same as Eq. (2.12) of Ref. 13. We thus follow the steps in Ref. 13 to obtain

$$P(x) = \frac{1}{2\pi} \int_{-\infty}^{\infty} \exp(i\rho x) \exp(-\gamma c \|m\| |\rho|) d\rho \qquad (22)$$

where c is the impurity concentration and

$$\gamma = \tfrac{2}{3} \pi^2 n_0 |a| \qquad (23)$$

where n_0 is the number of sites per unit cell ($n_0 = 4$ for an fcc lattice) and

$$\|m\| = \int_{-\infty}^{\infty} |m(x)| P(x) \, dx \tag{24}$$

where m is given in Eq. (10b). Finally, using Eq. (18), we obtain

$$P(h) = \frac{2}{\pi} \frac{\Delta}{\Delta^2 + h^2} \tag{25}$$

where

$$\Delta = \gamma c \|m\| \tag{26}$$

with

$$\|m\| = \int_0^{\infty} P(h) m(h) \, dh \tag{27}$$

Equation (25) is the final result of our calculation. Equation (25) in combination with Eq. (27) gives an integral equation for $P(h)$ which can be solved self-consistently for all temperatures.

We now make the very important observation that Eq. (25), which gives the probability distribution of the magnitude of the field, is normalized using the differential field element dh [as in Eq. (3)] and not with the volume element $4\pi h^2 \, dh$ as given in Eq. (5). This shows that the probability distribution for the magnitude of the field is one dimensional. Substituting Eq. (25) into Eq. (3) immediately shows that the specific heat is linear in T and independent of c, in agreement with experiment and with the Ising model calculation. This completes our proof.

The Magnetization and Magnetic Susceptibility

Since we set $H_{ext} = 0$ in our derivation, the angle θ in Eq. (11) does not enter our calculations. However, in the expression for the magnetization M we obtain

$$M(H_{ext} \to 0) = N_0 c \int_0^{\infty} P(h) m(h) \, dh \, \langle \cos \theta \rangle \tag{28}$$

where $\langle \cos \theta \rangle = (\int \cos \theta \, d\Omega) / \int d\Omega$, where $d\Omega = 2\pi \sin \theta \, d\theta$. Averaging Eq. (28) over all angles, we obtain that

$$M(H_{ext} \to 0) = 0$$

The physical interpretation of this result is that even though each individual spin is quantized in the direction of the total field h, the direction of quantization (in zero external field only) is uniformly distributed over all solid angles, and thus the magnetization vanishes in zero external field. A detailed calculation of the magnetic susceptibility in this model will be presented elsewhere. We remark, however, that preliminary calculations show that the very low-temperature magnetic susceptibility is, except possibly for a multiplicative constant, the same as for the Ising model of Ref. 13.

It should also be remarked that a similar argument to the one presented here may justify the recent use of the "Ising like" model for glasses by Anderson et al.[22] This model also gives a specific heat linear in T.

Conclusion

It is shown that in the molecular field approximation the Heisenberg model gives a specific heat which is linear in T and independent of c. This shows, for the first time, that at least in the molecular field approximation the low-temperature specific heat arising from the RKKY potential is in agreement with experiment and with previous calculations using the Ising model.

Acknowledgment

I wish to thank Professor Marshall Luban for some useful discussions.

References

1. J. Kondo, *Progr. Theor. Phys. (Kyoto)* **32**, 37 (1964).
2. J. Owen, M. Browne, V. Arp, and A. F. Kip, *J. Phys. Chem. Sol.* **2**, 85 (1957).
3. I.S. Jacobs and R.W. Schmitt, *Phys. Rev.* **113**, 459 (1959).
4. O.S. Lutes and J.L. Schmit, *Phys. Rev.* **134**, A676 (1964).
5. M.A. Ruderman and C. Kittel, *Phys. Rev.* **96**, 99 (1954).
6. T. Kasuya, *Prog. Theor. Phys. (Kyoto)* **16**, 45 (1956).
7. K. Yosida, *Phys. Rev.* **106**, 893 (1957).
8. J.E. Zimmerman and F.E. Hoare, *J. Phys. Chem. Sol.* **17**, 52 (1960).
9. W. Marshall, *Phys. Rev.* **118**, 1520 (1960).
10. C. Herring, unpublished.
11. A.W. Overhauser, *Phys. Rev. Lett.* **3**, 414 (1959).
12. M.W. Klein and R. Brout, *Phys. Rev.* **132**, 2412 (1963).
13. M.W. Klein, *Phys. Rev.* **173**, 552 (1968); **188**, 933 (1969).
14. M.W. Klein and L. Shen, *Phys. Rev.* **135**, 1174 (1972).
15. W. Marshall in *Proc. 8th Intern. Conf. Low Temp. Phys.,* 1962, Butterworth, Washington, D.C. (1963).
16. H. Suhl, *Phys. Rev. Lett.* **20**, 656 (1968).
17. R.J. Harrison and M.W. Klein, *Phys. Rev.* **154**, 540 (1967); **B1**, 940 (1970).
18. J.L. Tholence and R. Tournier, *Phys. Rev. Lett.* **25**, 867 (1970).
19. Y.C. Tsay and M.W. Klein, in *Proc. 1971 Conf. Magnetism and Magnetic Materials,* American Institute of Physics (1972); p. 1145, *Phys. Rev.* (to be published); this volume.
20. M.T. Béal Monod, *Phys. Rev.* **178**, 874 (1969).
21. H. Margenau, *Phys. Rev.* **48**, 755 (1935).
22. P.W. Anderson, B.I. Halpern, and C.M. Varma, *Phil. Mag.* **25**, 1 (1972).

Spin Polarization of Electrons Tunneling from Thin Ferromagnetic Films

R. Meservey and P.M. Tedrow

Francis Bitter National Magnet Laboratory*
Massachusetts Institute of Technology
Cambridge, Massachusetts

We have measured the conductance $\sigma = dI/dV$ of tunnel junctions between ferromagnetic thin films and very thin superconducting Al films[1] at a temperature of 0.5°K and in magnetic fields H up to 55 kOe. From plots of σ vs. the applied bias voltage V we can determine the spin polarization $P \equiv (n\uparrow - n\downarrow)/(n\uparrow + n\downarrow)$ of the tunneling electrons, where $n\uparrow$ and $n\downarrow$ are the numbers of electrons with magnetic moment parallel and antiparallel to H, respectively. The energy of the electrons involved is within about 1 meV of the Fermi energy E_F. This method depends on the fact[2] that the density of states of very thin Al films is split into spin-up and spin-down components by an applied magnetic field. Thus by choosing the bias voltage appropriately, one can control whether spin-up or spin-down electrons can tunnel.

The samples were made by first exposing a 50-Å-thick evaporated Al film to air, thereby forming an insulating oxide layer, and then evaporating a cross-strip of a ferromagnetic metal. The conductance dI/dV of the resulting tunnel junction was measured using a standard ac technique. Figure 1 shows plots of σ vs. V for several values of H for an Al–Al$_2$O$_3$–Co junction. The degree of asymmetry about V = 0 is a measure of the spin polarization of the tunneling electrons. Similar measurements[2] on Al–Al$_2$O$_3$–Ag junctions, for example, yielded completely symmetric curves.

The conductance curves in Fig. 1 can be understood with the help of the tunneling model represented in Fig. 2. The tunneling conductance at a given voltage V is proportional to an integral over all energies of a product of two terms.[3] The first term is the density of states of the superconductor, shown in Fig. 2(a). The usual BCS density of states is split by energy $2\mu H$ into spin-up (dotted) and spin-down (dashed) parts. The second term is a sharply peaked function shown in Fig. 2(b), the width of which is temperature dependent. At $T = 0$ it reduces to a δ-function. We assume that this function can be thought of as consisting of a spin-up and a spin-down part. The unequal size of these two parts is the result of the polarization due to the ferromagnet. Changing the bias voltage in effect moves this function horizontally relative to the superconductor density of states. Assuming there is no spin mixing during tunneling, the conductance curves of Fig. 2(c) are obtained. The spin-up electrons contribute the dotted curve, while the spin-down electrons contribute the dashed curve. The total conductance is the sum of these two, the solid curve in Fig. 2(c). The similarity between this curve and those of Fig. 1 is clear.

* Supported by the National Science Foundation.

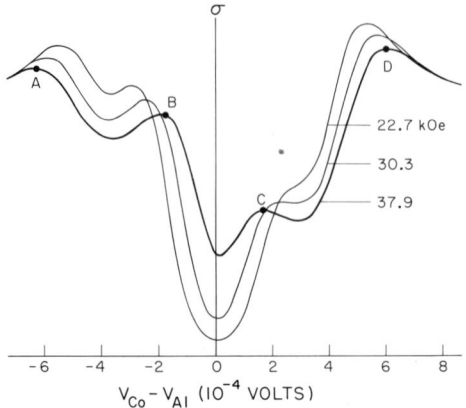

Fig. 1. Tunneling conductance σ vs. voltage for various values of H. The points $A, B, C,$ and D are defined following Eq. (1).

To extract a value for P from curves such as those in Fig. 1, we first assume that no spin scattering occurs during tunneling and that the tunneling probability for a given spin direction is independent of energy. The absence of spin scattering has been demonstrated experimentally[2] in tunneling between two Al films. The lack of dependence on H of the derived values of P seems to justify the assumption of energy independence of the tunneling probability. Then

$$P = \left\{ \frac{2[\sigma(-V_m) - \sigma(V_m - 2\mu H/e)]}{\sigma(-V_m) - \sigma(V_m - 2\mu H/e) + \sigma(V_m) - \sigma(-V_m + 2\mu H/e)} \right\} - 1 \quad (1)$$

where $\sigma(-V_m)$, $\sigma(-V_m + 2\mu H/e)$, $\sigma(V_m - 2\mu H/e)$, and $\sigma(V_m)$ are, for example, points A, B, C, and D, respectively, in Fig. 1. Here μ is the electron magnetic moment and e is the magnitude of the electronic charge. The voltage V_m could be chosen to be any value, but picking it to be near the right-hand maximum of the conductance curve makes the calculation of P relatively insensitive to errors in the measurement of σ and V. Values of P calculated from experimental curves similar to those in Fig. 1 are shown in Fig. 3 as a function of H. The metals studied were Fe, Co, Ni, and Gd.

Two important results are shown in Fig. 3. First, in all cases the sign of P is positive; that is, the tunneling electrons have their magnetic moments preferentially aligned parallel to the applied field. Second, the magnitude of P appears to be correlated with the magnetic moment per atom of the metals. These results are somewhat unexpected if the measured values of P are characteristic of electrons inside the ferromagnet, since according to accepted band models the density of states for spin-down electrons should be larger than for spin-up electrons at the Fermi surface.

A similar result[4] for electrons 0.4 and 0.8 eV below the Fermi surface was obtained by Bänninger et al. using a photoemission technique. The spin polarization of photoelectrons emitted from ferromagnetic films was measured as a function

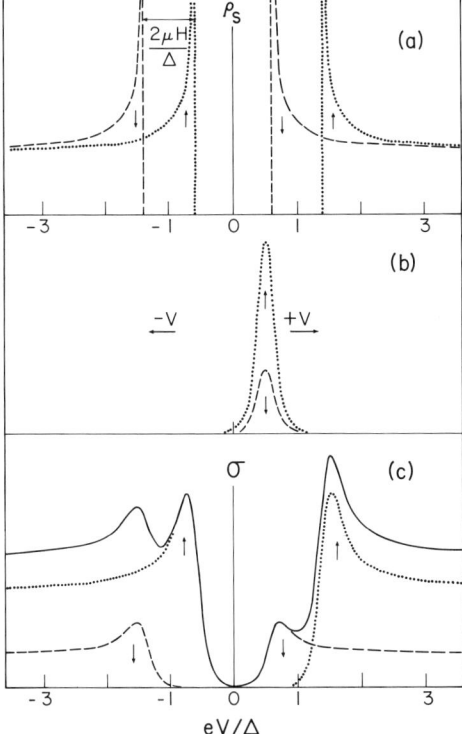

Fig. 2. (a) Superconductor density of states split into spin-up and spin-down components. (b) The sharply peaked function which is the second term in the integral for σ. The unequal heights of the spin-up and spin-down parts represent the polarization of the tunneling electrons. (c) The spin-up (dotted) and spin-down (dashed) contributions to σ (solid line) as a function of voltage.

of applied magnetic field. As shown in Table I, the tunneling results are quite similar to those of the photoemission studies. The absolute magnitude of P differs somewhat, but the direction and relative magnitudes are in good agreement, although presumably electrons with quite different energies are involved in the two measurements.

A third experiment[5] using a field emission technique on single crystal ferromagnets has been done by Gleich et al. Their results are quite different from the thin-film results, yielding negative P in several crystal directions in Ni, with only the [111] direction showing positive P. Also they found $P < 6\%$ for Fe. The only obvious way of reconciling the single-crystal and thin-film results is to assume the thin films formed preferentially in an appropriate crystal orientation ([111] for Ni) to produce a positive P. This explanation seems unlikely, especially since H was parallel to the film plane for the tunneling experiments and normal to the film plane for the photoemission studies.

A number of theoretical investigations[6,7] of electron spin polarization have

Table I. Percent Polarization P Measured in Thin Ferromagnetic Films

	Tunneling P, %	Photoemission P,* %
Fe	+ 44	+ 54
Co	+ 34	+ 21
Ni	+ 11	+ 15
Gd	+ 4.3	+ 5.7

* These values are the largest obtained with films deposited at room temperature; disordered films deposited at low temperatures had lower values of polarization.

been made. However, these theories[6] generally were applied only to the energy range involved in photoemission, but could not be used to explain the tunneling results. A possible explanation[7] for both the tunneling measurements and the photoemission measurements has been proposed by Kim. In his model the energy needed to remove a spin-up electron from the ferromagnet is less than that needed to remove a spin-down electron. In the tunneling case this model would lead to a spin-dependent tunneling matrix element. In the photoemission case it would lead to a spin-dependent work function. Both the direction and magnitude of P could be accounted for with this model.

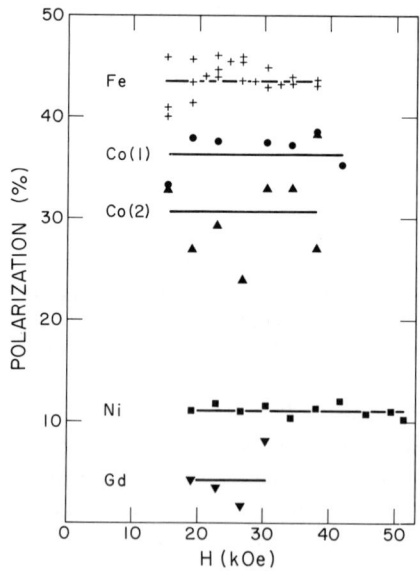

Fig. 3. Measured values of P for films of Fe, Co, Ni, and Gd as a function of H. The Fe data are for four different samples, while two different Co samples differed significantly from each other.

Acknowledgments

We gratefully acknowledge helpful discussions with L. Gruenberg, D. J. Kim, and B. B. Schwartz.

References

1. R. Meservey and P.M. Tedrow, *Sol. St. Comm.* **11**, 333 (1972); P.M. Tedrow and R. Meservey, *Phys. Rev. Lett.* **26**, 192 (1971).
2. R. Meservey, P.M. Tedrow, and P. Fulde, *Phys. Rev. Lett.* **25**, 1270 (1970).
3. I. Giaever and K. Megerle, *Phys. Rev.* **122**, 1101 (1961).
4. V. Bänninger, G. Busch, M. Campagna, and H.C. Siegmann, *Phys. Rev. Lett.* **25**, 585 (1970); G. Busch, M. Campagna, and H.C. Siegmann, *Phys. Rev.* **134**, 746 (1971).
5. W. Gleich, G. Regenfus, and R. Sizmann, *Phys. Rev. Lett.* **27**, 1066 (1971).
6. P.W. Anderson, *Phil. Mag.* **24**, 203 (1971); E.P. Wohlfarth, *Phys. Lett.* **A36**, 131 (1971); N.V. Smith and M.M. Traum, *Phys. Rev. Lett.* **27**, 1388 (1971); B.A. Politzer and P.H. Cutler, *Phys. Rev. Lett.* **28**, (1972).
7. D.J. Kim (to be published).

Proximity Effect for Weak Itinerant Ferromagnets

M.J. Zuckermann

Department of Physics, McGill University
Montreal, Quebec

Although the proximity effect for superconducting sandwiches has been the subject of much research, the proximity effect for itinerant ferromagnets has received very little attention. The reason is that the coherence length of a strong ferromagnet (e. g., Fe) at $T = 0$ is of the order of 10^{-7} cm, thus making the proximity effect unobservable. However, there is now much interest in the properties of weak ferromagnets (e.g., dilute Pd–Ni, $ZnZr_2$) whose Curie temperatures are orders of magnitude lower. Hence itinerant ferromagnets are available whose coherence length is of the order of 10^{-5} cm. In consequence the proximity effect may be observable for such ferromagnets.

In this communication we present a theory for the proximity effect for ferromagnetic/paramagnetic and ferromagnetic/ferromagnetic sandwiches using the Stoner model[1] for ferromagnetism in a metal with a single conduction band. The sandwiches are metallic and the electrical contact between the metals is assumed to be perfect. The Stoner model is equivalent to the Hubbard model[2] for correlations in a single conduction band in the Hartree–Fock approximation. In order to analyze the proximity effect, Hubbard's Hamiltonian[2] H must be extended to the case of a spatially dependent intraatomic Coulomb interaction $I(r)$, i.e.,

$$H = -\frac{1}{2m}\sum_\sigma \int d^3r \psi_\sigma^+(r) \nabla^2 \psi_\sigma(r) + \int d^3r I(r) n_\uparrow(r) n_\downarrow(r) \tag{1}$$

where $n_\sigma(r) = \psi_\sigma^+(r)\psi_\sigma(r)$ and $\psi_\sigma^+(r)$ and $\psi_\sigma(r)$ are field creation and annihilation operators for conduction electrons of mass m. The intrinsic magnetization $\Delta(r)$ at a point r due to $I(r)$ is

$$\Delta(r) = I(r) <n_\uparrow(r) - n_\downarrow(r)> \tag{2}$$

in units of the Bohr magneton. For a proximity sample in which the films on the left- and right-hand sides have Coulomb interactions I_1 and I_2, respectively, $I(r)$ is given by

$$\begin{aligned} I(r) &= I_1, & -d_1 < x < 0 \\ &= I_2, & 0 < x < d_2 \end{aligned} \tag{3}$$

where d_1 and d_2 are the respective film thicknesses. Use of the Green's function method in the Hartree–Fock approximation in conjunction with the Hamiltonian of Eq. (1) gives an integral equation for $\Delta(r)$. This integral equation becomes homogeneous and linear in the neighborhood of the Curie temperature T_C of the proximity sample and is given by

$$\Delta(r) = \int d^3r' \, I(r) \, \chi_{0T}(r - r') \, \Delta(r') \tag{4}$$

where $\chi_{0T}(r - r')$ is the spatially dependent static magnetic susceptibility in the absence of interaction at temperature T. We will use the expression for $\chi_{0T}(r - r')$ in the free-electron approximation.

Let us now write $\chi_{0T}(r - r')$ as follows:

$$\chi_{0T}(r - r') = \chi_{0T}(0)\,\delta(r - r') + \tilde{\chi}_{0T}(r - r') \tag{5}$$

Then direct application of Werthamer's method[3] for the superconducting proximity effect gives the following differential equation for $\Delta(r)$:

$$\tilde{\chi}_{0T_c}(-\nabla_r^2)\,\Delta(r) = \gamma(r)\left[T_C^2 - T_C^2(r)\right]\Delta(r) \tag{6}$$

Here $\tilde{\chi}_{0T_c}(q^2)$ is the value of the Fourier transform of $\tilde{\chi}_{0T}(r)$ defined in Eq. (5) at the Curie temperature T_C and $\gamma(r) = \pi^2 k^2/12\varepsilon_F^2(r)$. Here use has been made of the free-electron form for $\chi_{0T}(r)$ in the limit $kT_C \ll \varepsilon_F$. T_C is the Curie temperature of the proximity sample and $T_C(r)$ and $\varepsilon_F(r)$ are given by

$$\begin{aligned} T_C(r) &= T_{C1}, & \varepsilon_F(r) &= \varepsilon_{F1}, & -d_1 &< x < 0 \\ &= T_{C2}, & &= \varepsilon_{F2}, & 0 &< x < d_2 \end{aligned} \tag{7}$$

T_{C1} and T_{C2} are the bulk Curie temperatures and ε_{F1} and ε_{F2} are the Fermi energies of the ferromagnets composing the left and right films, respectively.

We examine Eq. (6) in the limit of a thick film on the left and a thin film on the right, i.e., $d_1 \gg d_2$. We then assume the solution[3] of Eq. (5) to have the form

$$\begin{aligned} \Delta(r) &= \exp(\pm i q_1 x), & -d_1 &< x < 0 \\ &= \exp(\pm q_\perp x), & 0 &< x < d_2 \end{aligned} \tag{8}$$

Using the continuity of $(d\Delta/dX)/\Delta$ at $X = 0$ and the assumptions $\varepsilon_{F1} = \varepsilon_{F2} = \varepsilon_F$ and $T_{C1} > T_{C2}$ gives from Eqs. (6)–(8) the following equation for T_C for a ferromagnetic/ferromagnetic sandwich:

$$[T_{C1}^2 - T_C^2]^{1/2} \tan\left[\delta_1(T_{C1}^2 - T_C^2)^{1/2}\right] = [T_C^2 - T_{C2}^2]^{1/2} \tanh\left[\delta_2(T_C^2 - T_{C2}^2)^{1/2}\right] \tag{9}$$

where $\delta_i = (12\gamma)^{1/2} d_i q_F$, $i = 1, 2$, and q_F is the Fermi momentum. Figure 1 shows T_C/T_{C1} as a function of $Z = \delta_2 T_{C2}$ for different values of T_{C2}/T_C assuming that the left film is very thick.

Let $N(0)$ be the density of free-electron states at the Fermi level. Then $T_{C1} \neq 0$ and $T_{C2} \neq 0$ implies that $I_1 N(0) > 1$ and $I_2 N(0) > 1$. However, for a ferromagnetic/paramagnetic sandwich the Stoner criterion[1] gives $I_2 N(0) > 1$ and $I_1 N(0) < 1$, i.e., $T_{C2} \neq 0$ and $T_{C1} = 0$. Then we have a thin ferromagnetic film on a thick paramagnetic film which is an enhanced paramagnetic metal (e.g., Pd, Pt). In this case it can be shown to be that Eq. (9) must be replaced by

$$[T_{C2}^2 - T_C^2]^{1/2} \tan\left[\delta_1(T_{C2}^2 - T_C^2)^{1/2}\right] = \xi \tag{10}$$

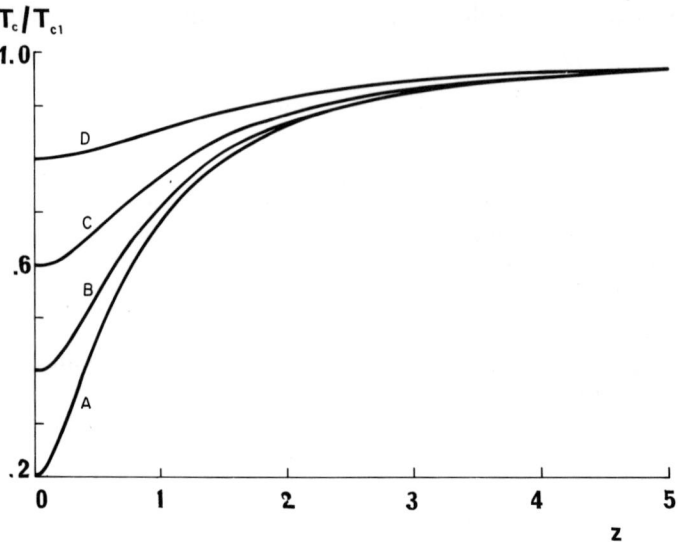

Fig. 1. Curve of T_c/T_{c1} for a ferromagnetic/ferromagnetic sample as a function of Z for the following ratios T_{c2}/T_{c1}: (A) 0.2, (B) 0.4, (C) 0.6, (D) 0.8.

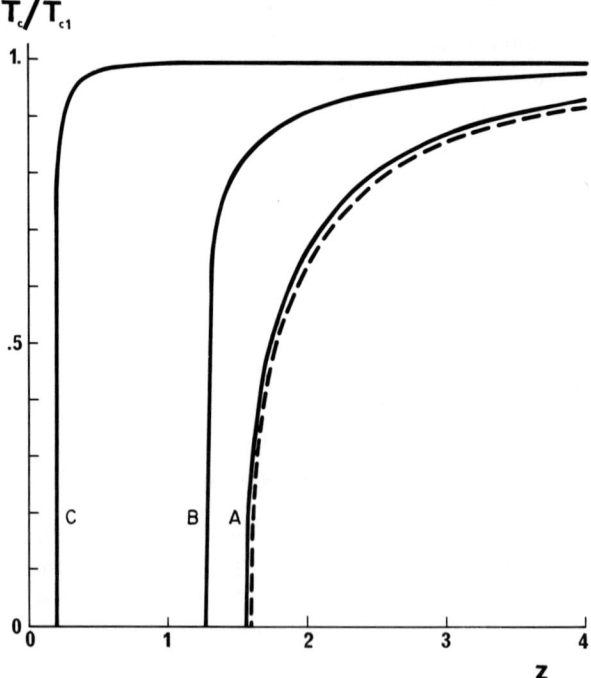

Fig. 2. Curve of T_c/T_{c1} versus Z for a ferromagnetic/paramagnetic sample for the following values of $N(0)I_1$: Dashed curve, zero; (A) 0.9; (B) 0.999; (C) 0.99999.

where $\xi \simeq (1 - N(0)I_1)/N(0)I_1\gamma T_{C2}^2$ as $d_1 \gg d_2$. Figure 2 shows curves of T_C/T_{C1} versus Z for different values of ξ. Note that $\xi^2 > 0$ only if $1 > N(0)I_1$. Note that the sandwich becomes paramagnetic below a critical thickness d_2^c of the ferromagnetic film. It can be shown that $d_2^c \approx \xi_2$, where ξ_2 is the coherence length of the bulk ferromagnet. For weak ferromagnets $\xi_2 \approx 10^{-5}$ cm and d_2^c should be observable, for example in Pd–Ni/Pt films.

Recent experimental work has shown that "dead layers" occur at the surface of strong ferromagnetic films deposited on nonmagnetic substrates.[4] It is unlikely that such dead layers will occur near the junction of the proximity samples under consideration, since there are d electrons (or "magnetic carriers") at the Fermi level in both films.

References

1. E.S. Stoner, *Phil. Mag.* **10**, 27 (1930); **12**, 737 (1931).
2. J. Hubbard, *Proc. Roy. Soc. (London)* **A276**, 238 (1963).
3. N.R. Werthamer, *Phys. Rev.* **132**, 2440 (1963).
4. L. Liebermann, J. Clinton, D.M. Edwards, and J. Mathon *Phys. Rev. Lett.* **25**, 232 (1970).

Onset of Ferromagnetism in Alloys at Low Temperatures

B. R. Coles, A. Tari, and H. C. Jamieson

Department of Physics, Imperial College
London, England

The dilute alloy problem is, in essence, the problem of understanding the appropriate description of an isolated solute atom (which in the free atom possesses a partly filled d or f shell) in substitutional solid solution in a host metal, normally, although not always, a fairly simple metal. One distinguishes three categories of solute: (a) those which exhibit simple local moment behavior over most experimentally accessible ranges of temperature, e.g., Mn in Cu, Gd in La; (b) those which in the same regimes show weak magnetic character (essentially temperature-independent susceptibilities), e. g., Ni in Cu, Fe in Nb, Fe in Al; and (c) those which at some low temperature seem to lose the magnetic character they have at higher temperatures, e. g., Fe in Cu, V in Au. Under the impact of an intensive experimental and theoretical attack this problem (sometimes called the Kondo problem) has yielded a large quantity of interesting physics, some of which even seem to be relevant to the real world.

It has always been hoped by practitioners of these mysteries that this problem will contribute to our understanding of the onset of magnetic order in pure transition metals. We have been acquiring some experimental information of relevance by examining the ways in which magnetic order can appear as the solute concentration is increased. Classifying host metals into (i) simple (nontransition) metals, (ii) transition metals, and (iii) exchange-enhanced transition metals, it is clear that a great deal of information exists about alloys of type (ia). For these systems (**CuMn** and **AuFe** are archetypes) ideas about distributions $p(H)$ of internal fields and "spin glasses" have proved useful. Magnetic order of a random and, on average, antiferromagnetic character sets in at low temperatures even at quite low concentrations because of the long range of the RKKY interaction that couples solute spins. **MoFe**, a (iia) system, seems rather similar. The (iiia) system **PdFe**, with its giant moments, has also been studied in detail. Alloys of type (ib) have been less widely studied, but **CuNi** has yielded evidence[1] of the buildup of spin polarization clouds as the critical concentration for ferromagnetism is approached, and the (iib) alloy **RhNi** has shown related specific heat and susceptibility behavior;[2] it would appear that the onset of long-range ferromagnetism and the stabilization of moments on solute atoms is a single process. (It should be noted that the "local moments" reported[3] within giant polarization clouds from neutron scattering data on these systems are, in fact, long-lived local spin fluctuations; true local moments would order in a magnetic glass at low temperatures.) Few alloys of type (ic) have been examined at larger solute concentrations (there are often metallurgical problems[4]) but it has recently been shown in this laboratory[5] that in the (iic) alloy **RhFe** one can observe first a stabilization of moments on solute atoms, *then* the onset of "spin-glass" ordering as the Fe content increases,

Onset of Ferromagnetism in Alloys at Low Temperatures

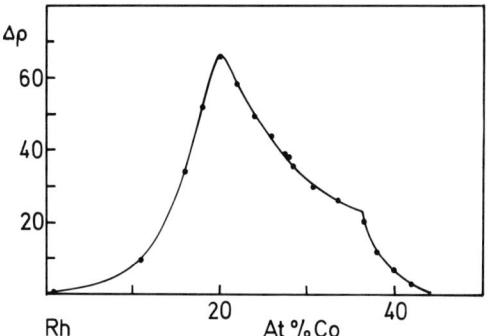

Fig. 1. The resistivity change (in nΩ-cm) between 1.7 and 4.2°K for Rh–Co alloys.

but this does not yield in turn to long-range ferromagnetism even at Fe concentrations of 25%, whereas in AuFe a spin glass/ferromagnetism change is found somewhere between 12%[6] and 20%[7] Fe.

We have now studied the (iiib) system **PdNi** and the (iib) system **RhCo** using electrical resistivity and magnetic susceptibility measurements. In both a striking contribution to the low-temperature resistivity is made by local spin fluctuations in dilute alloys and increases more rapidly than linearly as the concentration increases until at some critical concentration moments are stabilized and this resistivity term decreases again. Details of this behavior have already been reported for **Pd**–Ni,[8] and the critical concentration is found to be at 2.35% Ni. Susceptibility measurements we have made and analyzed in the conventional way (M^2 vs. H/M), as well as susceptibility and neutron scattering data already in the literature, show that the ordered phase is a straightforward ferromagnet.

In Rh–Co the low-temperature contribution (Fig. 1) to the electrical resistivity

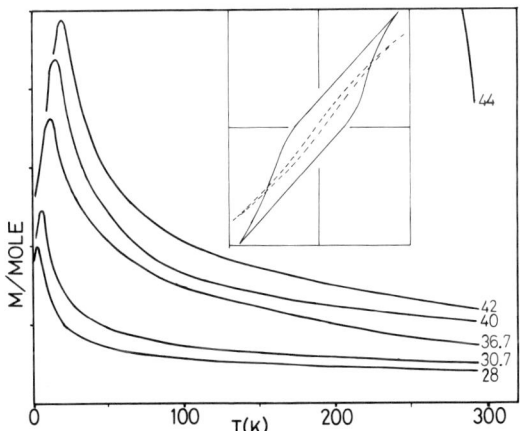

Fig. 2. The magnetization for $H = 500$ Oe of Rh–Co alloys as a function of temperature; the figures against the curves give the Co concentration. Inset: hysteresis curves (up to 8 kOe) for the 36.7% Co alloy at 4.2°K (broken curve) and the 40% alloy at 1.6°K (solid curve).

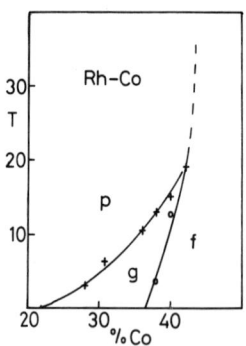

Fig. 3. A tentative magnetic phase diagram for Rh–Co; p, paramagnetic; g, spin glass; f, ferromagnetic. Crosses are susceptibility maxima; circles are resistance anomalies.

from spin fluctuations reveals a critical concentration close to 20% Co, but the rounded maxima in the susceptibility data for these alloys (Fig. 2) seem to indicate that this critical concentration marks the onset of spin-glass behavior very like that of **Rh**–Fe[5] or **Cu**–Mn. The low-temperature magnetization field curves for alloys with 20–30% Co are almost linear and, as shown in the inset to Fig. 2, even for 36.7% Co at 4.2°K the hysteresis curve has the character familiar in **Cu**–Mn. (Field cooling effects can also be produced.)

However, as the concentration is increased above 38% Co, the magnetic scattering term in the resistivity drops abruptly (Fig. 1), and we ascribe this to a rapid rise in a Curie temperature T_C marking the onset of long-range ferromagnetism. The change in character of the hysteresis loop (Fig. 2 inset) is also very striking. By 44%Co T_C is well above 20°K, in agreement with the higher-temperature work of Vogt et al.[9] We show a tentative magnetic phase diagram in Fig. 3, but this must be treated with caution until the role of the martensitic phase change from fcc to cph has been clarified.

Thus there seems to be a significant difference in the way in which long-range ferromagnetic order is established as a function of $3d$ element concentration between nickel alloys, where the transition is a simple para → ferromagnetic one, and cobalt or iron alloys, where stabilization of moments first gives a paramagnetic → spin-glass transition. We believe the difference to be associated with the fact that the number of unpaired d spins cannot exceed one on Ni atoms, while the possibility of two or more such spins exists for Co and Fe. We predict that the conflicting reports of the critical concentration in V–Fe alloys[10] will be found to be associated with the existence of a spin-glass regime.

References

1. T. Hicks, B.D. Rainford, J.S. Kouvel, G.G. Low, and J.B. Comly, *Phys. Rev. Lett.* **22**, 531 (1969).
2. E. Bucher, W.F. Brinkman, J.P. Maita, and H.J. Williams, *Phys. Rev. Lett.* **18**, 1125 (1967).
3. R.W. Houghton, M.P. Sarachik, and J.S. Kouvel, *Sol. St. Comm.* **10**, 369 (1972), and references therein.
4. M. Bancroft, *Phys. Rev.* **2B**, 2597 (1970).
5. A.P. Murani and B.R. Coles, *J. Phys. C. (Met. Phys. Suppl.)* **3**, S159 (1970).
6. J. Crangle and R. Scott, *J. Appl. Phys.* **36**, 921 (1965).
7. M. Ridout, *J. Phys. C* **2**, 1258 (1969).
8. A. Tari and B.R. Coles, *J. Phys. F* **1**, L69 (1971).
9. E. Vogt, F. Bölling, and W. Treutmann, *Ann. der Physik* **25**, 280 (1970).
10. M.V. Nevitt and A.T. Aldred, *J. Appl. Phys.* **34**, 463 (1963); E. Vogt, *Z. Angew. Phys.* **21**, 287 (1966).

Distribution of Atomic Magnetic Moment in Ferromagnetic Ni–Cu Alloys*

A. T. Aldred,† B. D. Rainford,‡ T. J. Hicks,§ and J. S. Kouvel**

Atomic Energy Research Establishment
Harwell, England

Introduction

The results of neutron diffraction experiments have made substantial contributions to our understanding of the complicated problem of atomic magnetic moments in chemically disordered ferromagnetic alloys. In particular, diffuse scattering measurements[1] (which can yield a quantitative measure of the average magnetic moment of each species in a binary alloy) have revealed that the average atomic moment varies widely with the nature and composition of an alloy. This variation suggests that the moment of an individual atom in an alloy should depend on its particular local environment, and such a local effect would produce an additional contribution to the neutron diffuse cross section at small scattering vectors. Low and his associates[2] at Harwell have developed a long-wavelength, low-angle spectrometer which they have used to investigate these local magnetic effects in a wide variety of primarily dilute alloys.

In the present work we have determined the neutron diffuse scattering cross sections of a series of Ni–Cu alloys that extend from the dilute copper region almost completely across the ferromagnetic composition range. A further stimulus for our work arises from subsequent neutron-diffraction experiments[3] at Harwell on Ni–Cu alloys close to the critical concentration for ferromagnetism ($\sim$ 56 at. % Cu), which have shown that the magnetization is distributed in giant polarization clouds with average moment $\sim 10\,\mu_B$. We hoped in the present measurements to determine the magnetic moment variations as they evolve with increasing copper concentration from isolated disturbances in the dilute alloys to a more complex pattern of interacting effects which may be precursory to the polarization clouds that appear in the critical region.

Experimental Procedure

Nickel–copper alloy ingots of nominal composition 2.3, 6.2, 10, 20, 30, and 40 at. % Cu were prepared by induction melting and subsequently were cold-rolled,

---
* Work performed in part under the auspices of the U.S. Atomic Energy Commission.
† On assignment from Argonne National Laboratory, Argonne, Illinois (permanent address).
‡ On attachment from Imperial College, London, England.
§ Present address: Monash University, Clayton, Victoria, Australia.
** On leave from General Electric Research and Development Center, Present address: University of Illinois, Chicago Circle, Chicago, Illinois.

annealed at 1000°C for three days, and water-quenched. This heat treatment was designed to ensure macroscopic chemical homogeneity and to minimize the atomic clustering that cannot be completely suppressed in these alloys.[4,5] Magnetization measurements were made on a Foner-type vibrating sample magnetometer, and the spontaneous moment of each alloy was determined at 4.2°K. The neutron elastic diffuse scattering cross sections were measured at 4.2°K on the diffractometer system previously mentioned[2]; the mean wavelength of the neutron beam was 4.8Å. The nuclear and magnetic parts of the cross section were separated by switching a magnetic field on and off parallel to the scattering vector.

Results and Discussion

To interpret our data, we follow the formal analysis of Marshall[6] and write the magnetic scattering cross section (in mb/sr at) as

$$d\sigma/d\Omega = 48.6c(1-c)S(\kappa)f^2(\kappa)[M(\kappa)]^2 \quad (1)$$

where

$$M(\kappa) = \mu_{Cu} - \mu_{Ni} + (1-c)G(\kappa) + cH(\kappa) + (1-2c)[W(0) + W(\kappa)] \quad (2)$$

In these expressions c is the fractional concentration of copper, κ is the neutron scattering vector ($= 4\pi \sin\theta/\lambda$), $f(\kappa)$ is the atomic $3d$ form factor, and μ_{Cu} and μ_{Ni} are the average $3d$ magnetic moments of the copper and nickel atoms, respectively, in the absence of any short-range order. Furthermore,

$$S(\kappa) = 1 + \sum_{R_i} \alpha(R_i) N(R_i) (\sin \kappa R_i)/\kappa R_i \quad (3)$$

$$G(\kappa) = \sum_{R_i} g(R_i) N(R_i) (\sin \kappa R_i)/\kappa R_i \quad (4)$$

$$H(\kappa) = \sum_{R_i} h(R_i) N(R_i) (\sin \kappa R_i)/\kappa R_i \quad (5)$$

and

$$W(\kappa) = \sum_{R_i} [h(R_i) - g(R_i)] \alpha(R_i) N(R_i) (\sin \kappa R_i)/\kappa R_i \quad (6)$$

where R_i is the radius and $N(R_i)$ is the coordination number of the ith near-neighbor shell, $\alpha(R_i)$ is the corresponding short-range order parameter (as defined by Cowley[7]), $g(R_i)$ represents the disturbance in the $3d$ moment of a nickel atom caused by each additional copper atom at a distance R_i, and $h(R_i)$ represents the corresponding disturbance in the $3d$ moment of a copper atom.

The short-range order function $S(\kappa)$ defined in Eq. (3) differs from that used by Marshall[6] in a systematic manner. This function and the $\alpha(R_i)$ values have been determined for each composition from the nuclear scattering cross sections; a

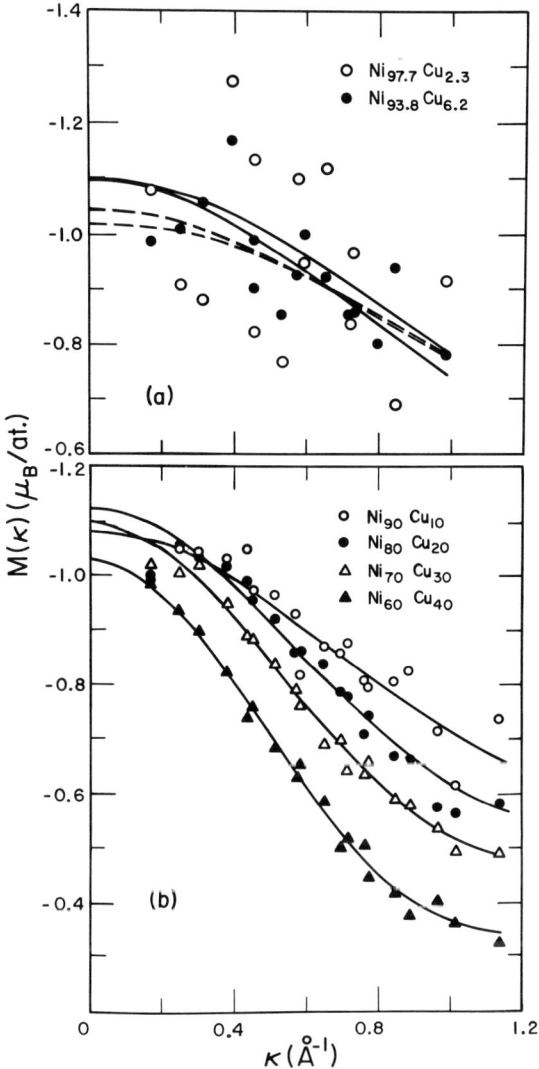

Fig. 1. Magnetic moment density function $M(\kappa)$ as a function of scattering vector κ. The curves represent analytical fits to the data.

similar procedure was used by Cable et al.[5] The weak κ dependence of the atomic $3d$ form factor over our experimental range was approximated by the expression $1 - 0.05\kappa^2$ based on the calculated form factor of Watson and Freeman.[8] Thus, by means of Eq. (1), values of $M(\kappa)$ were determined from the experimental cross sections, their negative sign fixed by the assumption that $\mu_{Ni} > \mu_{Cu}$; these values are presented in Fig. 1. It can be deduced from Marshall's work[6] that $M(0) = d\langle\bar{\mu}\rangle dc$, where $\langle\bar{\mu}\rangle$ is the average atomic moment of an actual alloy with short-range order. Inasmuch as $\langle\bar{\mu}\rangle$ and so $d\langle\bar{\mu}\rangle/dc$ can be determined from the bulk magnetization measurements, this equality fixes the value to which an experimental $M(\kappa)$ vs. κ curve should extrapolate at $\kappa = 0$.

The data in Fig. 1 were analyzed by the least-squares method to determine $\mu_{Cu} - \mu_{Ni}$ and the various κ-dependent terms. To separate the $g(R_i)$, we have assumed, following Cable et al.,[5] that μ_{Cu} is so small that all $h(R_i)$ can be neglected. The results of our analysis are shown by the solid lines in Fig. 1(b) and the dashed lines in Fig. 1(a). Although the lines in Fig. 1(b) extrapolate to values of $M(0)$ in agreement with the results of our bulk magnetization measurements, this is not the case for the dashed lines of Fig. 1(a). Consequently, the fits were constrained to extrapolate to the correct $M(0)$ values for the two dilute alloys; these fits are shown by the solid lines in Fig. 1(a), and they are clearly as consistent with the scattered data as are the original fits.

Our results for $g(R_i)$ are plotted against composition in Fig. 2. Due to the large statistical errors in the higher-order g's, only the $g(R_1)$ values can be taken strictly at face value, and they are seen to be essentially constant over most of the composition range. Nevertheless, the steady increase in magnitude of the higher-order g's with increasing copper concentration reflects the increasingly more rapid decay of $M(\kappa)$ with κ indicated by the data in Fig. 1. This change in shape of $M(\kappa)$ continues into the critical composition range[3] where the Marshall model can no longer provide an adequate representation of the data. The quantity $G(0)$, defined in Eq. (4) as the weighted sum of the g's, is a smooth (essentially linear) function of concentration, which further supports the semiquantitative validity of our higher-order g's. Although our $g(R_1)$ value for the 20 at.% Cu alloy is in good agreement with that reported by Cable et al.,[5] our higher-order g's are larger in magnitude. Inasmuch

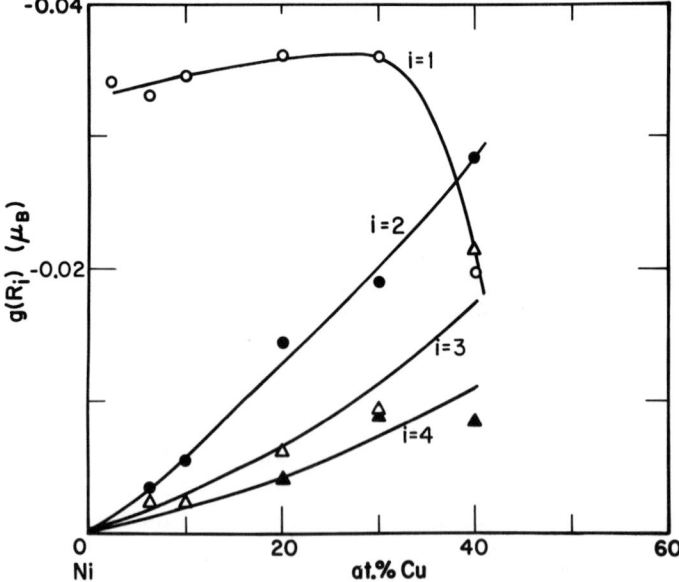

Fig. 2. Composition dependence of the nickel magnetic moment disturbance parameters $g(R_i)$.

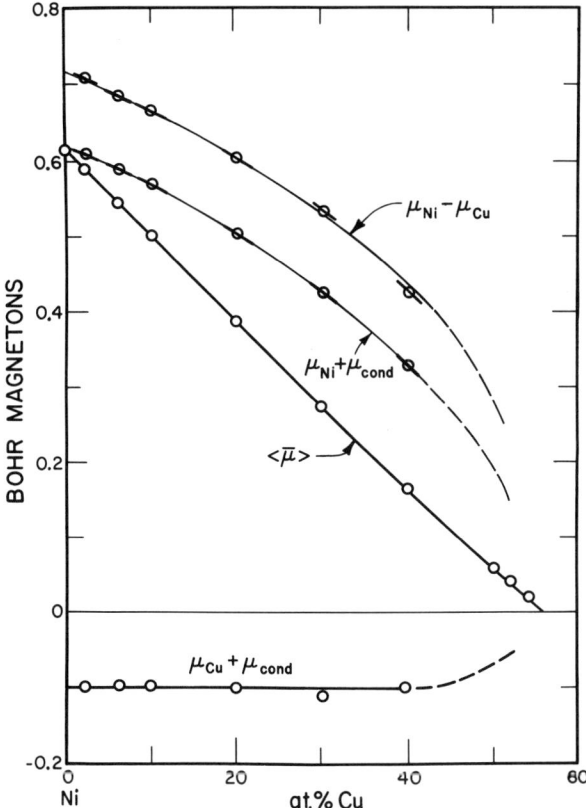

Fig. 3. Composition dependence of various combinations of average magnetic moments.

as our data were obtained at lower κ values, they should yield a more accurate determination of these higher-order components. The constant value of $g(R_1)$ with composition presumably results from a short-range electronic screening around each copper atom, whereas the rapidly changing higher-order g's are probably indicative of cooperative magnetic effects.

The composition dependence of the average moment difference term $\mu_{Ni} - \mu_{Cu}$ determined from our neutron scattering data and the average atomic moment per atom $\langle \bar{\mu} \rangle$ obtained from our bulk magnetization measurements are presented in Fig. 3. According to Marshall,[6]

$$\langle \bar{\mu} \rangle = c\mu_{Cu} + (1 - c)\mu_{Ni} + c(1 - c)W(0) \qquad (7)$$

and thus the values of $\mu_{Ni} - \mu_{Cu}$, when combined with Eq. (7), should yield values of the individual average moments. However, Eq. (7) may also contain a spatially uniform component of any conduction-electron polarization, whose contribution to the neutron scattering cross section would be at κ values below the range of our data. If we add such a moment (μ_{cond}) to the right-hand side of Eq. (7), we find that the moments may be separated into the combinations $\mu_{Ni} + \mu_{cond}$ and $\mu_{Cu} + \mu_{cond}$, which are also plotted in Fig. 3.

The most striking feature of Fig. 3 is that $\mu_{Cu} + \mu_{cond}$ has essentially a constant value over this composition range. It is very unlikely that μ_{Cu} and μ_{cond} have large and compensating composition dependences, and consequently both $d\mu_{cond}/dc$ and $d\mu_{Cu}/dc$ [which, from Marshall,[6] is equal to $H(0)$] are probably close to zero. Therefore our assumption that the individual h's are zero is well justified. Furthermore, $d\mu_{Ni}/dc$ should equal $G(0)$,[6] and values of the latter obtained in the earlier analyses are shown by the short, heavy lines in Fig. 3; the quantitative self-consistency of the data is evident.

Mook[9] has fitted his neutron Bragg scattering data for pure nickel with a $3d$ magnetic form factor calculated for the free nickel atom[8] and deduced a negative uniform polarization of $-0.105\ \mu_B$. A simple extension to our Ni–Cu results (and those of Cable et al.[5]) would strongly imply that the value of about $-0.1\ \mu_B$ for $\mu_{Cu} + \mu_{cond}$ primarily represents a negative uniform polarization (μ_{cond}) and consequently that μ_{Cu} is very small. That this polarization is constant over a composition range where μ_{Ni} decreases by $\sim 40\%$ is intuitively puzzling and clearly calls for further theoretical study. A fuller account of this work will appear elsewhere.

Acknowledgments

We are extremely grateful to Dr. J. W. Garland for many enlightening discussions with regard to the Marshall formalism, and Mr. V. Rainey and Mr. H. F. Burne for their assistance in the neutron and magnetization measurements, respectively.

References

1. C.G. Shull and M.K. Wilkinson, *Phys. Rev.* **97**, 304 (1955).
2. G.G. Low, *J. Appl. Phys.* **39**, 1174 (1968); *Adv. Phys.* **18**, 371 (1969), and references therein.
3. T.J. Hicks, B.D. Rainford, J.S. Kouvel, G.G. Low, and J.B. Comly, *Phys. Rev. Lett.* **22**, 531 (1969).
4. B. Mozer, D.T. Keating, and S.C. Moss, *Phys. Rev.* **175**, 868 (1968).
5. J.W. Cable, E.O. Wollan, and H.R. Child, *Phys. Rev. Lett.* **22**, 1256 (1969).
6. W. Marshall, *J. Phys. C: Proc. Phys. Soc.* **1**, 88 (1968).
7. J.M. Cowley, *Phys. Rev.* **77**, 669 (1950).
8. R.E. Watson and A.J. Freeman, *Acta Cryst.* **14**, 27 (1961).
9. H.A. Mook, *Phys. Rev.* **148**, 495 (1966).

Low-Temperature Resistance Anomalies in Iron-Doped V–Cr Alloys

R. Rusby

*Imperial College, London, England and
National Physical Laboratory, Teddington, Middlesex, England*

and

B. R. Coles

*Department of Physics, Imperial College
London, England*

There is now considerable evidence[1] that interactions between $3d$ solute atoms in dilute alloys can greatly affect the manifestation of local moment (Kondo) properties. Not only does a maximum appear below the resistance minimum in, for example, **Au**–Fe when the iron concentration is increased, but more subtle effects are found in almost every system. (In **Rh**–Fe[2] the transition regime from a T^2 to a T behavior moves from $\sim 2°K$ in the dilute limit to below $0.4°K$ at 0.5% Fe). Striking effects of another kind have for a long time been known to take place in the character of a $3d$ solute when the host is a binary $4d$ alloy (e.g., Nb–Mo) and the relative concentration of the two host components is changed. (The classic observation is that of Matthias concerning the influence of iron on superconductivity in that system, and Sarachik et al.[3] have demonstrated the associated effects in the electrical resistance.)

The analogous system, using $3d$ elements as components of the host, is **V**–**Cr**–Fe, and the much larger solubilities for Fe make it especially convenient for both interaction effect and host variation effect studies. [The concentration range we have used ($\nless 10\%$ V) ensures that no itinerant electron antiferromagnetism complicates the issue.] Our resistivity data for alloys of varying Cr/V ratio and roughly constant Fe content are shown in Fig. 1 and show the onset of resistance minima as the Cr content increases. Figure 2 shows clearly, however, that the Cr content at which such behavior would appear is a function of the amount of Fe used as a probe; Fe–Fe interactions first increase the fraction of anomalous scattering produced by the Fe (lower the Kondo temperature) and later give rise to low-temperature freezing of moments, as in the low-T_K systems like **Cu**–Mn and **Au**–Fe. (In the 70/30 Cr/V alloy, which has no minimum for 0.5% Fe, a well-marked one can be produced by 2.0% Fe.) Since the appearance of a resistance maximum in the ^{4}He temperature range requires more than 5% Fe for the 80/20 Cr/V alloy, as compared with less than 1% for Fe in Mo or in Au, it is clear that the onset of "magnetic glass" type ordering is still inhibited at this Cr/V ratio by an effective loss of moment at low temperatures. In this respect this host material is rather like Rh, where Fe concentrations of about

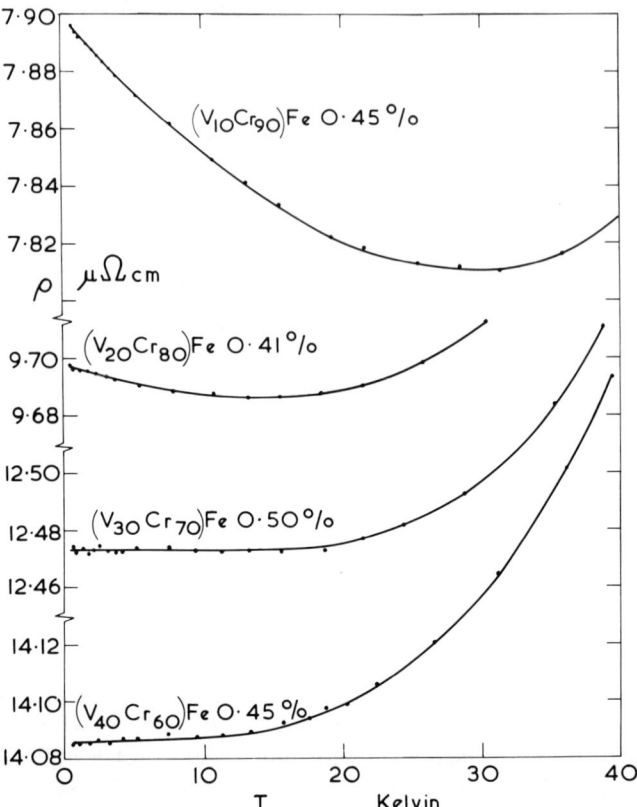

Fig. 1. Resistance–temperature behavior of Cr/V alloys containing about 0.5% Fe. The chromium concentration of the host is indicated.

3% have to be reached[4] before the local spin fluctuations are stabilized enough to yield a susceptibility maximum.

For higher Cr/V ratios resistance anomalies are found even for quite low Fe concentrations, but their character is modified by interactions. In the 90/10 alloy (Fig. 3) *reducing* the Fe content to 0.13% leads to a higher Kondo temperature with greater deviations from log T and a low-temperature flattening out of the $\rho - T$ curve, which is clearly not a magnetic ordering effect but the appearance of a $(1 - \alpha T^2)$ character better described in terms of the local spin fluctuation approach. The relationship of this behavior to the appearance in the same temperature range of the $+ \beta T^2$ term in Rh–Fe at lower Fe contents is in good accord with the indications of theoretical work by Rivier and Zlatic[5] that local spin fluctuation effects have the same *general* character in systems where the absence of marked resonances in the band structure yields resistance contributions that increase with increasing temperature and in those where resonances result in resistance minima.

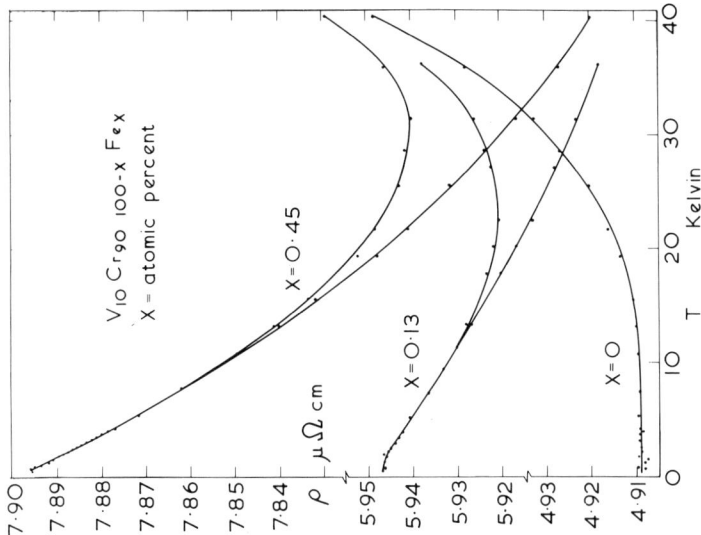

Fig. 3. Resistance–temperature behavior of a 90/10 Cr/V alloy with 0.13% and 0.5% Fe. Lower curves have phonon scattering subtracted.

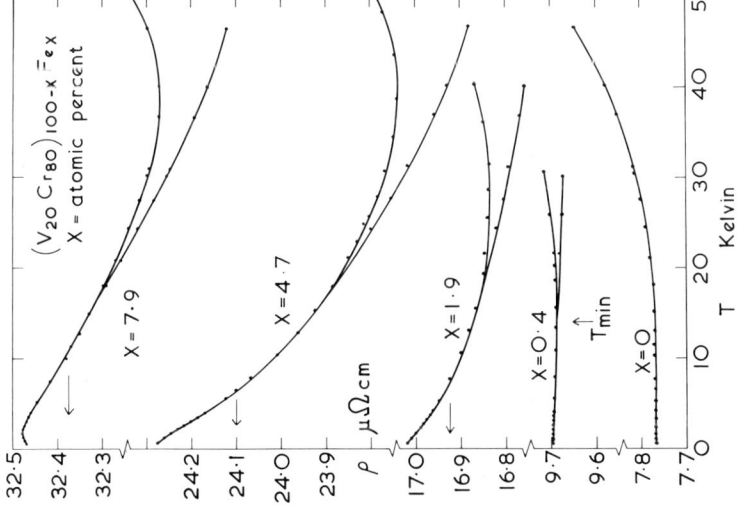

Fig. 2. Resistance–temperature behavior of an 80/20 Cr/V alloy for various (indicated) iron concentrations. Lower curves have phonon scattering subtracted.

References

1. D. Wohlleben and B.R. Coles, in *Magnetism*, Vol. V, G.T. Rado and H. Suhl, eds., Academic Press, New York (1973).
2. R. Rusby, (to be published); and Ph.D. Thesis, Imperial College, London (1973).
3. M.P. Sarachik, E. Corenzwit, and L.D. Longinotti, *Phys. Rev. A* **135**, 1041 (1964).
4. A.P. Murani and B.R. Coles, *J. Phys. C* **3** (Suppl.), 159 (1970).
5. N. Rivier and V. Zlatic, *J. Phys. F* **2**, L87 (1972); *J. Phys. F* 2, L99 (1972).

Magnetic Properties of $(Ge_{1-x}Mn_x)$ Te Alloys[*]

R. W. Cochrane and J. O. Ström-Olsen

Eaton Electronics Research Laboratory
McGill University, Montreal, Quebec

Introduction

In the past several years there has been a continuing effort to examine the effects of the conduction electrons on the magnetic ordering properties of localized magnetic moments in semiconductors and metals. The case of the Eu–Gd monochalcogenides[1,2] provides an excellent example in which the carrier density can be varied from the semiconducting to metallic regimes within the single NaCl crystal structure by alloying across the phase diagram. Moreover, there remains a constant, localized spin ($S = 7/2$) at each site while the carrier concentration is changed. An alternate approach has been to study the magnetic ordering for dilute magnetic impurities doped into a nonmagnetic host whose carrier concentration can be varied in some range. Examples of such studies are Mn doped into SnTe[3] and, to a lesser extent, GeTe.[4,5] The samples with less than 10 at.% Mn order ferromagnetically at low temperatures in both hosts even though MnTe is itself antiferromagnetic with a Néel temperature of 306°K. For Mn-doped GeTe a detailed comparison[5] between the magnetic and magnetotransport properties has been carried out on a 0.9 at.% alloy in our laboratory. In order to extend these results, we have begun a systematic examination of the magnetic and transport properties on a series of $(Ge_{1-x}Mn_y)$ Te alloys.

This paper is a report of magnetization measurements on a set of GeTe alloys with up to 10 at.% Mn, all of which are highly degenerate due to deviations from stoichiometry and have resistivities of order 10^{-4} Ω-cm and carrier concentrations of $\sim 10^{21}$ holes/cm^3,[6] as deduced from Hall effect data. Because the carriers do not freeze out, the system is an excellent host for examining the magnetic interactions down to small impurity concentrations and low temperatures.

Sample Preparation and Measurements

The alloys were prepared by thoroughly mixing the required amounts of finely powdered GeTe and MnTe and sealing in an evacuated ($P \leq 10^{-4}$ Torr.) quartz ampoule. The mixture was maintained above the melting point for 24 hr while it was continuously agitated to ensure a homogeneous ingot; it was then quenched rapidly to room temperature. All the samples were polycrystalline and X-ray powder

---

[*] Research supported in part by the National Research Council of Canada.

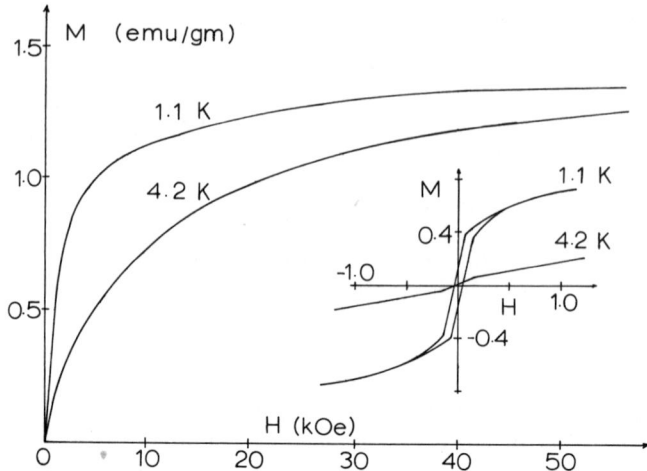

Fig. 1. Magnetization curves for 0.95 at.% Mn in GeTe at 1.1 and 4.2°K up to 55 kOe. The insert is a blowup of the low-field regions for the same two temperatures.

photographs revealed only the rhombohedrally distorted NaCl phase characteristic of pure GeTe.

The measurements were taken on a vibrating sample magnetometer using a Nb–Zr superconducting solenoid for magnetic fields up to 55 kOe. In addition, an insert dewar system permitted temperature variations in the range from 1.1 to 300°K.

Magnetization data have been measured on six samples of GeTe with nominally 0.2, 0.5, 1.0, 2.0, 5.0, and 10 at.% MnTe. Except for the lowest concentration all the samples exhibited ferromagnetic order at low temperatures. Figure 1 shows the magnetization curves for the 1% sample at 4.2 and 1.1°K for fields up to 55 kOe. The distinct curvature at 4.2°K is a result of paramagnetic saturation as expected for high-spin ions. On the other hand, even well below T_c (e.g., 1% sample at 1.1°K) the magnetic moment remains unsaturated at the highest fields. The inserts in Fig. 1 illustrate the low-field behavior of the same alloy at the same temperatures and clearly indicate the onset of the ordered state.

The data for all the samples were analyzed separately in the paramagnetic and ferromagnetic regions to determine the relevant parameters, which are summarized in Table I. The paramagnetic Curie temperature θ was determined by plotting the inverse high-temperature susceptibility χ_e^{-1} against T, where the subscript e (for excess) means that the diamagnetic susceptibility of the GeTe host has been subtracted. Figure 2 presents a number of these plots and shows that the Mn ions follow a Curie–Weiss relation above their ordering temperature. From this it is evident that the Mn moments are localized above T_c. The Curie constants measured from Fig. 2 give $\mu_{eff} \sim 5.4 \pm 0.5$, corresponding to Mn^{2+} with $S = 5/2$. This conclusion has been confirmed by the ESR measurements of Hedgcock et al.,[5] which give $g = 2$, and also from our own unpublished data. Figure 3 is a graph of the θ values plotted against Mn concentration.

In the ferromagnetic regime the ordering temperature T_C has been determined

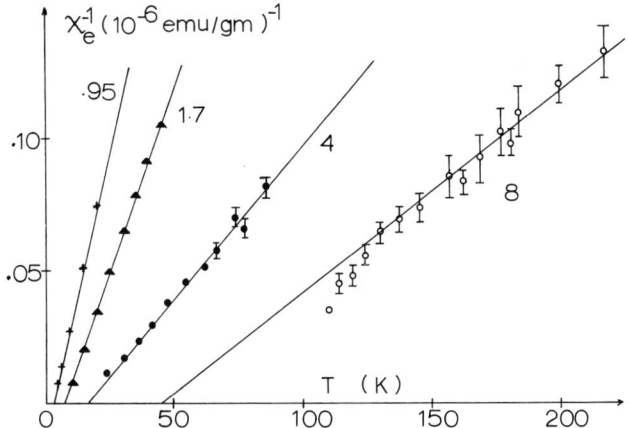

Fig. 2. The inverse of the excess susceptibility plotted as a function of temperature for several GeTe alloys, where the labels show the Mn concentration in atomic percent.

from an analysis of the low-field magnetic moment data near T_C using the relation

$$a(T - T_C)M + bM^3 = H$$

where a and b are constants. In addition, the high-field data well below T_c were extrapolated on an M vs. H^{-1} plot to estimate the saturation moment M_{sat}. However, these values are subject to some uncertainty because the moment is unsaturated even at 55 kOe. Moreover, because the magnetic ions are randomly distributed with corresponding random interactions, complete magnetic saturation even at very low temperatures may be very difficult to achieve. Finally, the coercive field H_c is noted in the final column of Table I.

Discussion

The data presented here are in agreement with previous results[4,5] which indicate that the dilute $(Ge_{1-x}Mn_x)Te$ alloys are ferromagnetic. In contrast with these earlier

Table I. Summary of Data for $(Ge_{1-x}Mn_x)$ Alloys*

x, at.%	n, 10^{21} cm^{-3}	θ, °K	T_C, °K	M_{sat}, emu/g	H_c, Oe
0.25	1.06	0.5 ± 0.05			
0.5	1.20	1.7 ± 0.2			
0.95	1.15	3.3 ± 0.2	2.3 ± 0.2	1.4	30†
1.7	1.33	7.0 ± 0.5	6.5 ± 0.3	2.2	60‡
4	1.23	17 ± 1	11 ± 1	4.5	150‡
8	1.50	45 ± 10	20 ± 5	8.4	500‡

* x, Concentration; n, the number of free carriers (holes) as determined from the Hall voltage; θ and T_C, the paramagnetic and ferromagnetic Curie temperatures, respectively; M_{sat}, the saturated ferromagnetic moment; H_c, coercive field.
† 1.1°K.
‡ 4.2°K.

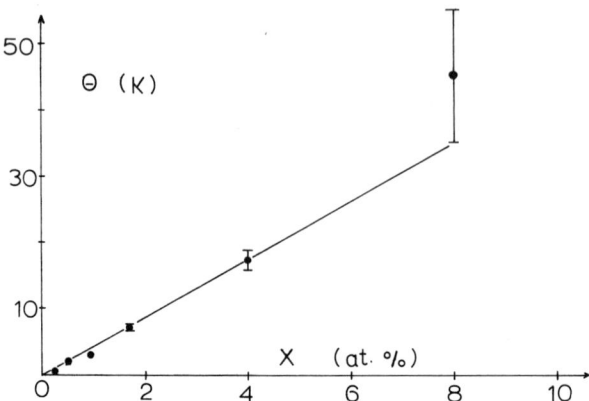

Fig. 3. Paramagnetic Curie temperature plotted as a function of the Mn concentration.

investigations, our alloys exhibit much sharper transitions with considerably lower values of θ, T_C, and H_c. We believe the first results to be characteristic of sizable concentration inhomogeneities in which the effect of the relatively rich manganese regions strongly distorted the bulk parameters. It was for this reason that we took considerable effort to ensure a homogeneous distribution of Mn throughout our samples. Nevertheless, such effects are evident in the 10% Mn sample at 80°K, where small regions of the sample have ordered, as seen from a hysteresis loop at low fields with a coercive field $\sim$ 30 Oe and remanent magnetization $\sim$ 0.02 emu/g. Numerically, this represents considerably less than 1% of all the Mn moments, yet their effect causes χ_e^{-1} to deviate from the high-temperature Curie–Weiss behavior at temperatures well above 2θ, as can be seen in Fig. 2.

The positive θ value of the 0.2% sample implies that even at this low Mn concentration the sample orders magnetically at low enough temperatures. Since the average separation between magnetic ions at this concentration is approximately 30 Å, a long-range interaction via the free carriers is indicated—presumably of the RKKY form. This theory predicts[7] a linear increase in θ with Mn concentration for constant carrier density, which is borne out reasonably well by Fig. 3. This result has also been found for the Mn-doped SnTe alloys studied by Mathur et al.,[3] where the slope of the θ vs. x curve is a factor of four smaller than for the GeTe host, even though the carrier densities are essentially identical ($\sim 10^{21}$ cm^{-3}). SnTe has a cubic NaCl structure and hence is very similar to GeTe (the small distortions from cubic symmetry of GeTe should not effect the θ values to this extent), and furthermore, the lattice parameter a is a mere 5% larger in SnTe.

Calculations by Plischke[8]* indicate that small differences in the carrier concentration of our alloys have only a small effect on θ which would be masked by the experimental uncertainty of the measured values. In detail the calculated θ vs. χ slope

---

* Detailed comparison between the theoretical and experimental results will be presented in a later publication.

depends on the number of free carriers, their mean free path,* and the square of the intraatomic coupling between the carrier and impurity spins. The value of

$$d\theta/dx = 4.4 \pm 0.2°K/at.\%$$

allows us to estimate an effective coupling constant

$$J_{\text{eff}} \sim 0.5 \times 10^{-11} \text{ ergs} \sim 3 \text{ eV}$$

On the basis of the above discussion this would be about twice the effective coupling for Mn in SnTe. Furthermore, the perturbation parameter used in the RKKY approximation,

$$xJ/E_F \sim 10x$$

leads us to expect some deviation from the low-concentration behavior around 10 at.% Mn which is just becoming apparent in our highest-concentration alloy. At present we are pursuing these measurements to higher Mn concentrations in order to examine in detail the deviation from linearity.

Acknowledgments

We are particularly indebted to Dr. M. Plischke for making the results of his calculations available to us and for many profitable discussions of both the theoretical and experimental results. Helpful discussions with Professors F. T. Hedgcock, W. B. Muir, and R. Harris are also gratefully acknowledged.

References

1. F. Holtzberg, T.R. McGuire, S. Methfessel, and J.C. Suits, *Phys. Rev. Lett.* **13**, 18 (1964); *J. Appl. Phys.* **37**, 976 (1966).
2. T.R. McGuire and F. Holtzberg, in *Magnetism and Magnetic Materials 1971*, AIP Conf. Proc. No. 5, p. 855 (1972).
3. M.P. Mathur, D.W. Deis, C.K. Jones, A. Patterson, W.J. Carr, Jr., and R.C. Miller, *J. Appl. Phys.* **41**, 1005 (1970).
4. M. Rodot, J. Lewis, H. Rodot, A. Villers, J. Cohen, and P. Molland, *J. Phys. Soc. Japan* (Suppl.) **21**, 627 (1966).
5. F.T. Hedgcock, J. Lass, and T.W. Raudorf, *J. de Physique* **32** (Cl), 506 (1971).
6. J. Lewis, *Phys. Stat. Sol.* **35**, 737 (1969).
7. D.C. Mattis, *The Theory of Magnetism*, Harper and Row, New York (1965).
8. M. Plischke, McGill University, private communication.

---

* Using the data given above, we estimate the Fermi momentum k_F to be $3 \times 10^7 \text{cm}^{-1}$ and the mean free path to be about 100 Å.

11
Dilute Alloys

Electron Spin Relaxation through Matrix NMR in Dilute Magnetic Alloys

H. Alloul and P. Bernier

Laboratoire de Physique des Solides,* Université de Paris-Sud
Orsay, France

Introduction

Since the discovery of the Kondo effect a large number of experimental works have advanced the understanding of the static magnetic properties of dilute alloys of d elements in noble metal matrices.† The situation regarding dynamic susceptibilities is quite different. Due to the so-called "bottleneck" of relaxation,[3] direct EPR measurements seem to be unable to distinguish the relaxation effects associated with the s–d coupling.[4] Relaxation effects are also involved in Mössbauer spectra,[5] but these are still difficult to understand. The host nuclear spins sense both static and fluctuating local fields associated with the impurity. This has long been known to yield a broadening of the host NMR line, while the nuclear spin–lattice relaxation time T_1 is shortened.[6,7] Though data on the impurity contribution to T_1^{-1} are available in dilute alloys,[9-14] no real understanding has yet been achieved.

In this article we present experimental results on **Cu**–Mn, **Cu**–Cr, and **Cu**–Fe taken by pulse techniques over a large range of fields ($1 < H < 51$ kG) and temperatures ($0.32 < T < 77°$K). It will be shown first that a major difficulty in the interpretation of the data can be removed if nuclear spin diffusion in the matrix is correctly taken into account. The RKY local fields near the impurity strongly decouple the neighboring nuclear spins and then quench spin diffusion processes near the impurity. This will be shown to play a large role in the measured macroscopic relaxation rates. The relaxation mechanisms will be recalled and the applicability of the theoretical calculations to systems with a high Kondo temperature will be discussed. In limiting cases, where simplifying assumptions on the dynamic susceptibilities can be made, experimental results will then be analyzed in terms of a relaxation rate for the impurity electron spin. This electron spin relaxation will then be shown to differ qualitatively from that derived from direct EPR measurements. A number of questions which arise from these experimental results will then be pointed out.

Nuclear Spin Diffusion

The effects related to spin diffusion, which have been studied at length in insulators doped with paramagnetic impurities,[8] have been rather neglected in previous

---

* Associated with CNRS.
† See Ref. 1 for theoretical review, Ref. 2 for experimental review.

studies on dilute alloys.[9-14] The impurity local moment **S** induces a local spin lattice relaxation of nuclear spins **I** which is written

$$[T_1(r)]^{-1} = \mathscr{C} r^{-6} \qquad (1)$$

where $\mathscr{C}$ includes both the magnitude of the impurity fluctuations and the strength of the (**I**, **S**) coupling constant, apart from its distance dependence, which is represented by the r^{-6} factor. If spin diffusion does not take place (diffusionless case), the nuclear magnetization recovery is found to behave as $\exp[-(t/T_1^\circ)^{1/2}]$, where T_1° is related to $\mathscr{C}$.[15] In the opposite case, where spin diffusion equalizes the nuclear spin temperature, a unique exponential relaxation $\exp(-t/T_1)$ holds, where T_1 is an average of expression (1) over the nuclei that contribute to the diffusion processes.

Cu–Mn. In Cu–Mn both types of recoveries, as well as intermediate ones, could be observed, depending on the experimental conditions. This can be thoroughly explained* with a simple model in which spin diffusion is quenched within a critical radius b, called the diffusion barrier. The number of sites n_b within b can be measured directly from the number of nuclear spins that do not contribute to the long-time exponential relaxation. It is found to scale linearly with the magnetization $\langle S_z \rangle$ of the impurities and then with the amplitude of the RKY local fields, as can be seen in Fig. 1. Since an independent measurement of these local fields is given from the linewidth of the ^{63}Cu NMR line, a direct relationship for copper-based alloys can be established:

$$n_b = (14.5 \pm 3) \times 10^{-3} \Delta H_{pp}/c \quad (G) \qquad (2)$$

where c is the concentration of impurities and ΔH_{pp} the peak-to-peak broadening of the absorption line.† Then the impurity-induced long-time exponential recovery of the magnetization is given by[8]

$$\Delta(T_1)^{-1} = (T_1)_{\text{alloy}}^{-1} - (T_1)_{\text{Cu}}^{-1} = \frac{4\pi}{3} Nc \frac{\mathscr{C}}{b^3} = \frac{16\pi^2}{9} N^2 c \frac{\mathscr{C}}{n_b} \qquad (3)$$

where N is the number of lattice sites per unit volume. In the high-temperature limit ($g\mu_B SH \ll kT$), where $n_b \propto \langle S_z \rangle \propto H$ and $\mathscr{C}$ is expected to be field independent, $\Delta(T_1)^{-1}$ has an H^{-1} field dependence which is merely due to the diffusion barrier variation, as can be seen in Fig. 2. The result in Eq. (2) cannot be expected to hold for small values of n_b. It has indeed been found[16] that n_b reaches a lower limit $n_b \sim 50$ at $T = 20°$K, from the saturation of the H^{-1} dependence of $\Delta(T_1)^{-1}$. Such a complete study was possible in **Cu–Mn**, since a rather high impurity concentration could be reached without metallurgical complications.

Cu–Fe. In this system the situation is complicated because the ^{63}Cu linewidth has a peculiar field dependence which has been attributed to an enhancement of the

---

* In Ref. 16 a complete study of the shape of the magnetization recovery and the spin diffusion in **Cu–Mn** is given.
† The linewidth ΔH_{pp} has been measured as indicated in Ref. 7 and yields $\Delta H_{pp}/c = 25 \times 10^4 \langle S_z \rangle/S$, which gives Eq. (2) if combined with the result of Fig. 1.

Electron Spin Relaxation through Matrix NMR in Dilute Magnetic Alloys

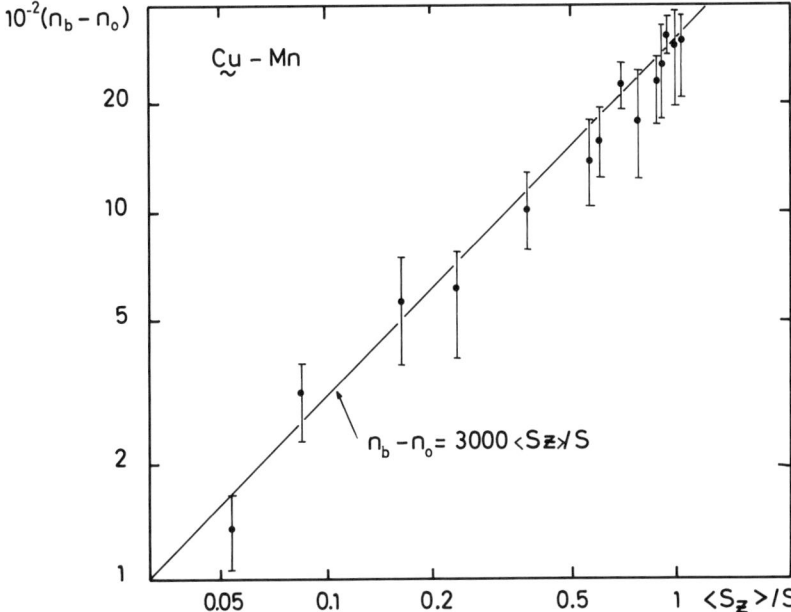

Fig. 1. Plot of $n_b - n_0$ on a log-log scale as a function of the impurity magnetization, which is given by the Brillouin function for $S = 2$. Here n_0 is the number of sites that are not observed in the data and was found[16] to be $n_0 \sim n_b/5$. The data can then be represented by $n_b = (3600 \pm 400)\langle S_z \rangle/S$.

spin polarization[13,17] at low magnetic field and temperature. Moreover, the line broadening does not extrapolate to zero for $H < 2$ kG; this has been related to the presumed existence of ferromagnetic clusters.[17] As suggested by the magnetization measurements of Tholence and Tournier,[8] the low-field anomaly could be associated with the existence of magnetic pairs of impurities with a lower T_K than isolated ones. It must be noted that in the conclusions of Potts and Welsh important contradictions remain because the nonexponential behavior of the relaxation for $c > 300$ ppm is attributed by these authors to "interaction effects." These interaction effects also contribute to the zero-field linewidth while they should not affect it in intermediate fields even for $c = 1260$ ppm (the main argument for this being that the linewidth scales approximately linearly with c). If the line broadening was associated with single impurities, n_b should still increase with ΔH_{pp}. Analysis of our relaxation data as done for **Cu**–Mn yields strong contradictions, since in low fields a significant increase in the nonexponential behavior of the magnetization recovery was observed. This would mean that n_b increases while ΔH_{pp} decreases. The field dependence of $\Delta(T_1)^{-1}$ given in Fig. 3 would be impossible to explain with such a model since, from Eq. (3), this anomalous increase of n_b should show up as a decrease of $\Delta(T_1)^{-1}$ in low fields. Finally, analysis of the data as a function of c shows that an experimental definition of n_b in low fields is no longer possible when the number of nuclear spins that do not contribute to the long-time exponential relaxation does not scale linearly with c. All these results can in fact be explained by assuming that the extra broadening of the ^{63}Cu line and the extra short-time relaxation are associated with nuclei mainly

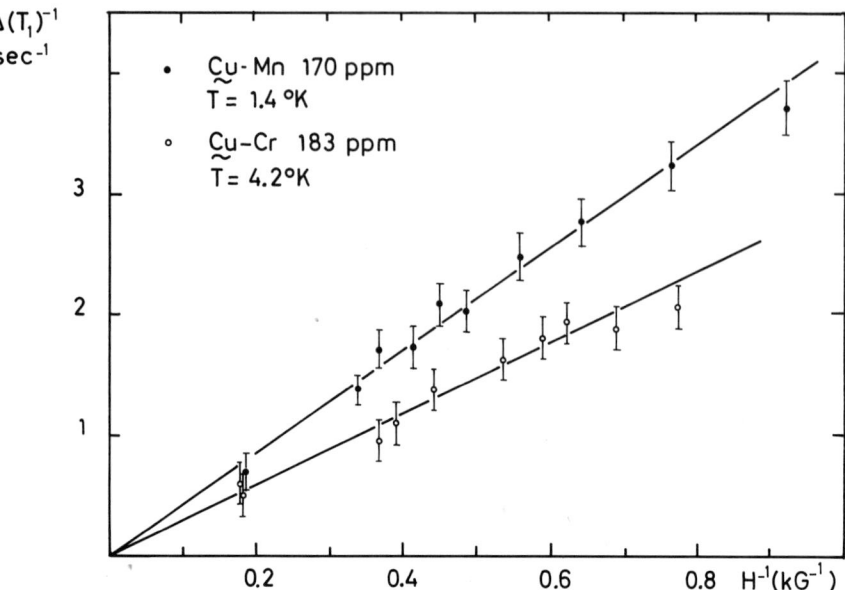

Fig. 2. In the high-temperature range $g\mu_B SH \ll k(T + T_K)$ the impurity contribution to the long-time exponential recovery $\Delta(T_1)^{-1}$ is plotted vs. H^{-1} for one Cu–Mn and one Cu–Cr sample. This H^{-1} dependence is due to the diffusion barrier variation.

affected by clusters of impurities. In high fields the magnetization of such clusters saturates and then gives no further contribution to the nuclear relaxation (as observed for single Mn impurities in Cu[19*]), while they yield a field-independent broadening of the ^{63}Cu resonance. A quantitative analysis can then be done, using the single-impurity broadening as deduced from the high-field, high-temperature results of Potts and Welsh[13]:

$$\frac{\Delta H_{pp}}{H} = \frac{42}{T + 29} c \qquad (4)$$

The expected value for n_b is then, from Eq. (2),

$$n_b = \frac{0.6 \pm 0.15}{T + 29} H \quad (G) \qquad (5)$$

At 4.2°K and 10 kg for $c = 880$ ppm the magnetization of the clusters should be saturated; a measure of n_b from our data yields $n_b = 140 \pm 50$, which compares favorably with the $n_b = 180 \pm 45$ given by Eq. (4). The field dependence of $\Delta(T_1)^{-1}$ shows then from Eq. (3) that Eq. (2) holds,† but that in low fields n_b reaches a lower value, $n_b \sim 40$, which is exactly what has been observed for **Cu**–Mn. All the data can then be explained assuming that n_b and then the local fields around an isolated impurity scale linearly with its magnetization, which implies that the presumed

---

* Reference 20 gives a detailed study of the Cu–Mn data.
† Such an analysis implies that $\mathscr{C}$ is field independent, which will be shown to be a good assumption in such conditions.

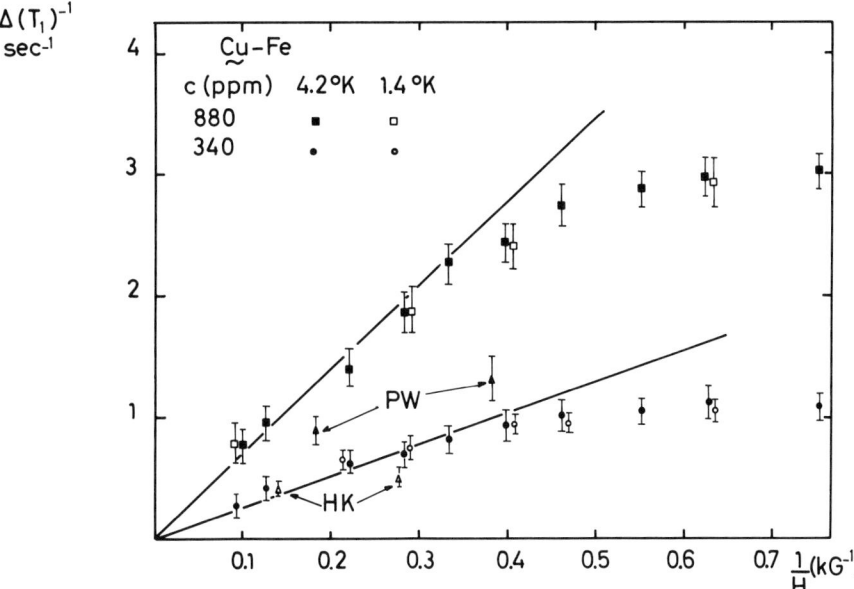

Fig. 3. Plot of $\Delta(T_1)^{-1}$ vs. H^{-1} for two Cu–Fe samples. The saturation of $\Delta(T_1)^{-1}$ in low fields corresponds to the minimum value $n_b \sim 40$ for the diffusion barrier radius. No anomaly for the local fields around a single impurity seem to be present, since such a dependence indicates that they are proportional to H. Between 1.2 and 4.2°K, these results are also temperature independent within experimental error, in agreement with the results of Ref. 14 but in disagreement with those of Ref. 13. Data points from these references, extrapolated for $c = 340$ ppm, $T = 1.4$°K, are plotted as HK and PW.

enhancement of the spin polarization in the Kondo state has no physical reality, as was also the case for Au–V[21] and Al–Mn.[22] The observed linewidth anomalies are mainly associated with interaction effects between impurities.

Cu–Cr. At ^{4}He temperatures the relationship of Eq. (2) between n_b and ΔH_{pp} could be fitted within 20% accuracy, while the long-time relaxation rate has an H^{-1} dependence (Fig. 2), corresponding to the diffusion barrier variation. The situation in the ^{3}He range is even more complicated than for **Cu**–Fe since interaction effects appear, while the range of experimental fields and temperatures just covers the Kondo transition ($T_K \sim 1$°K). The single-impurity magnetization curves are not yet known in this range. Nevertheless, at 0.32°K linewidth measurements and relaxation data show similar effects to those described for **Cu–Fe**, since clustering effects yield an extra short-time relaxation and a field-independent broadening. It can be noted that the low-temperature data of Gladstone[11] also show a field dependence which should be attributed to this diffusion barrier variation rather than to the breakdown of Kondo effect. The main conclusion of this diffusion barrier analysis is that the relationship of Eq. (2) is valid for the three systems provided that only the contribution of the isolated impurities to ΔH_{pp} is taken into account. This, in turn, is merely proportional to the impurity magnetization.

Relaxation Mechanisms

Theory. Now that n_b is well known, measurements of $\Delta(T_1)^{-1}$ allow one to deduce $\mathscr{C}$ from Eq. (3). It is then of interest to recall the nature of the couplings that may induce nuclear spin relaxation.* These can be direct (dipolar) or indirect (pseudo-dipolar, RKY) couplings. Through the scalar $(A_s \mathbf{I} \cdot \mathbf{S})$ or anisotropic $(A_a I^+ S_z)$ parts of these couplings the nuclear spins sense, respectively, the transverse or longitudinal fluctuations of the impurity electron spins. The corresponding results for $\mathscr{C}$ can be written in a simple molecular field model:

$$\mathscr{C}_s = 4\bar{A}_s^2 (g\mu_B)^{-2} k_B T \, \text{Im}[\chi^T(\omega_n)/\omega_n] \tag{6}$$

$$\mathscr{C}_a = 2\bar{A}_a^2 (g\mu_B)^{-2} k_B T \, \text{Im}[\chi^L(\omega_n)/\omega_n] \tag{7}$$

where $\bar{A}_i = A_i r^6$ ($i = s, a$), are mean values arising from the average in A_i of the angular or oscillating dependences; χ^T and χ^L are the transverse and longitudinal dynamic susceptibilities of the impurity; and ω_n is the nuclear frequency. It can be noted that $\bar{A}_s$ is dominated by the RKY coupling constant,[24] while the relative strength of the dipolar and pseudo-dipolar couplings that contribute to $\bar{A}_a$ is unknown. It has been shown that the local fields around the impurity scale with the magnetization, indicating that the coupling constants $\bar{A}_a$ and $\bar{A}_s$ are independent of the state of the impurity, and that all the anomalies associated with the Kondo condensation are included in the localized impurity dynamic susceptibilities. In the free-spin limiting case these quantities can be written

$$\chi^L(\omega) = \frac{\chi_0^L}{1 - i\omega\tau_1} \, ; \qquad \chi_0^L = g\mu_B \frac{\partial \langle S_z \rangle}{\partial H} \tag{8}$$

$$\chi^T(\omega) = \chi_0^T \frac{i - \omega_e \tau_2}{i + (\omega - \omega_e)\tau_2} \, ; \qquad \chi_0^T = g\mu_B \frac{\langle S_z \rangle}{H} \tag{9}$$

where ω_e is the electronic Larmor frequency. The longitudinal and transverse electron spin relaxation times τ_1 and τ_2 should not be confused *a priori* with the EPR relaxation time. Though such formulas should apply for $T \gg T_K$, their validity for $T \ll T_K$ is not warranted in the whole frequency range. Nevertheless, since NMR only senses the low-frequency part of the spectrum, Eqs. (8) and (9) are probably good approximations in this range, though the relaxation times will not have then their usual meaning and may depend on the external field. Equations (6) and (7) become

$$\mathscr{C}_s = 4\bar{A}_s^2 (g\mu_B)^{-2} k_B T \chi_0^T \tau_2 / (1 + \omega_e^2 \tau_2^2) \tag{10}$$

$$\mathscr{C}_a = 2\bar{A}_a^2 (g\mu_B)^{-2} k_B T \chi_0^L \tau_1 / (1 + \omega_n^2 \tau_1^2) \tag{11}$$

Comparison with Experiments. The transverse and longitudinal static susceptibilities of single impurities for the three systems of interest here are not well known.

---

* A review of the relaxation mechanisms can be found in Giovannini et al.[23] In the present article the mechanism first proposed by Giovannini and Heeger is not mentioned since it is found to yield much smaller values for $\Delta(T_1)^{-1}$ than measured experimentally, and also should give a temperature-independent value in the plot of Fig. 4. Quantitative comparisons are detailed in Ref. 20.

Consequently, since those are required for any interpretation of the data, it seems better to investigate only the low-field range

$$g\mu_B SH \ll k(T + T_K) \tag{12}$$

where the static susceptibility, as well as the electron spin relaxation time, should be isotropic and field independent. In such a field range the only field dependence which could be observed (Figs. 2 and 3) is attributed to the diffusion barrier variation. It can then be concluded that no ω^2 dependence such as those included in the denominators of Eqs. (10) and (11) could be detected in our experimental conditions. Since usually $\omega_n \tau_1 \ll 1$ and since the RKY coupling is much greater than the dipolar coupling[24] ($\bar{A}_s \gg \bar{A}_a$), depending on whether $\omega_e \tau_2 \gg 1$ or $\omega_e \tau_2 \ll 1$, the dominant relaxation process should be given by $\mathscr{C}_a$ (dipolar) or $\mathscr{C}_s$ (usually called BGS). This choice cannot be decided solely from our data since the order of magnitude of τ_2 could be required. The quantity $\Delta(T_1)^{-1}$ which is measured is written for $n_b > 50$ from Eqs. (3), (10), and (11) as

$$\Delta(T_1)^{-1} \propto c(T/H)\tau \tag{13}$$

where τ should take the index corresponding to the dominant relaxation process. The quantity τ^{-1} can then be deduced directly from the data for $\Delta(T_1)^{-1}$. Results for the three systems are plotted in Fig. 4. It is clearly shown that τ^{-1} scales linearly with T in the whole temperature range. Only one series of points for each system, taken at constant impurity magnetization (and then constant n_b), has been plotted on Fig. 4. When n_b is changed the data points lie on the same curve, provided that condition (12) is fulfilled. Though the linear dependences have intercepts at $T = 0$, more accurate measurements are needed at low temperatures in order to draw conclusions about their relative order of magnitude in the three systems. In **Cu**–**Mn** a complete quantitative analysis has been developed and allows us to conclude that the BGS relaxation is dominant.[20] Since more experimental data are needed for conclusions about the other systems, we shall limit our interest to the temperature dependence of τ, which is written within the above simplifying assumptions as

$$\tau = \frac{1}{\chi_0} \lim_{\omega \to 0} \frac{\operatorname{Im}\chi(\omega)}{\omega} \tag{14}$$

The Electron Spin Relaxation

At the present stage the physical meaning of the electron spin relaxation time is not clear. The first relationship which can be considered is with direct EPR. This technique is a well-known measure of the relaxation time of the *coupled resonance* of s- and d-electron spins, while the nuclear spins mainly sense the *local* fluctuations of the d-electron spins. This is an important difference and it can be conjectured that the nuclear spins will not be affected by the rate τ_{sl}^{-1} at which the electron spins couple to a lattice. Such a result appears in a theoretical treatment of the bottleneck effect by Göetze and Wölfle.[25] We are also able to demonstrate that point experimentally,

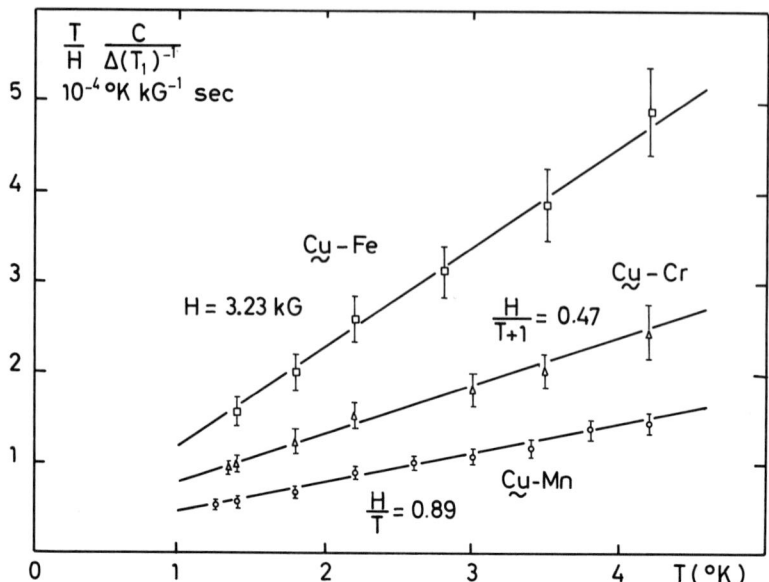

Fig. 4. The quantity plotted vs. T is proportional to the impurity electron spin relaxation rate, as defined in the text. The proportionality constant depends on the system through the corresponding coupling constants A_a or A_s. It is of interest to note that this a fragment of our data, taken at constant $\langle S_z \rangle$ (and thus n_b) for each system. In the three systems τ^{-1} has the same qualitative behavior.

since the introduction of 1000 ppm of Au in **Cu**–Mn does not change the measured $\Delta(T_1)^{-1}$, while it is well known to increase strongly τ_{sl}^{-1} due to the high spin–orbit scattering cross section of Au.[26] Then τ^{-1} should mainly reflect the coupling of the d electrons with the lattice (τ_{dl}^{-1}) and with the s electrons (τ_{ds}^{-1}). MacHenry et al.[27] have shown that in $(\text{La}_{1-x}\text{Gd}_x)\text{Al}_2$ RKY interactions between impurities can also give a contribution to τ^{-1} proportional to c. In our experimental cases this effect appears to be of no importance for temperatures higher than the ordering temperature, since $\Delta(T_1)^{-1}$ is always found to scale linearly with c, which means from Eq. (3) that τ^{-1} is concentration independent. Finally, the T linear relaxation rate should be τ_{ds}^{-1}, while the meaning of the temperature-independent term which could be associated with τ_{dl}^{-1} is still unclear. It is of interest to compare these results with those obtained for the relaxation rate of the impurity nuclear spins in less magnetic systems, such as **Au**–V[21,28] or **Al**–Mn.[22,29] Such measurements yield two contributions to T_1,

$$(T_1)_{(i)}^{-1} = k_B T \mu_B^{-2} \gamma_n^2 (H_{hf}^{(i)})^2 \, \text{Im}\,[\chi_i^T(\omega)/\omega] \tag{15}$$

where i corresponds to d or orbital, depending on whether the susceptibility or hyperfine H_{hf} concerned is of d or orbital character. Since the measurements yield constant values for $T_1 T$ up to 20°K in **Au**–V[28] ($T_K = 290$°K), Eq. (15) shows that $\text{Im}\,[\chi^d(\omega)/\omega] \propto \chi_0 \tau$ is temperature independent in this temperature range. Consequently, τ^{-1} should reach a constant value for $T \ll T_K$. This does not seem to occur down to 1.4°K in **Cu**–Fe ($T_K = 29$°K), while the results of Gladstone[11] would indicate that such a change appears at $T = 0.05$°K for **Cu**–Cr ($T_K = 1$°K). Further experimental data should clarify this point and indicate whether all these systems can be described in a uniform picture.

Conclusion

The difficulties which are encountered in the study of nuclear spin–lattice relaxation in dilute alloys have been partially removed in this work. First, the existence of a spin diffusion barrier in the nuclear spin system has been established. The quantitative knowledge of this phenomenon allows a complete explanation of the low-field dependence of the observed relaxation rates in **Cu–Mn**, **Cu–Cr**, and **Cu–Fe**. The effect of impurity clustering has also been demonstrated, thus leading to an understanding of the nature of the ^{63}Cu linewidth anomaly in **Cu–Fe**. Finally, the s-spin polarization is found to scale linearly with $\langle S_z \rangle$, even for $T \ll T_K$. An important implication of the data is the determination of a local d–s relaxation rate for the impurity electron spin, which is linear in T for the three systems. More quantitative analyses have been done in **Cu–Mn**[20] and are underway for **Cu–Cr** and **Cu–Fe**. Though theoretical and experimental efforts are necessary to achieve a detailed understanding of many points evoked in this paper, further developments appear promising.

References

1. J. Kondo, in *Solid State Physics*, Vol. 23, F. Seitz, D. Turnbull, and H. Ehrenreich, eds., Academic Press, New York (1969).
2. A.J. Heeger, in *Solid State Physics*, Vol. 23, F. Seitz, D. Turnbull, and H. Ehrenreich, eds., Academic Press, New York (1969).
3. H. Hasegawa, *Prog. Theor. Phys.* **21**, 483 (1959)
4. S. Schultz, M.S. Shanabarger, and P.M. Platzman, *Phys. Rev. Lett.* **19**, 749 (1967); P. Monod and S. Schultz, *Phys. Rev.* **173**, 645 (1968).
5. H. Maletta, K.R.P.M. Rao, and I. Nowik, *Z. Physik* **235**, 59 (1970).
6. J. Owen, M. Brown, W.D. Knight, and C. Kittel, *Phys. Rev.* **102**, 1501 (1956).
7. I. Sugawara, *J. Phys. Soc. Japan* **14**, 643 (1959).
8. J.J. Lowe and D. Tse, *Phys. Rev.* **166**, 279 (1968).
9. R.E. Levine, *Phys. Lett.* **28A**, 504 (1969).
10. O.J. Lumpkin, *Phys. Rev.* **164**, 324 (1967).
11. G. Gladstone, *J. Appl. Phys.* **41**, 1150 (1970).
12. P. Bernier, H. Launois, and H. Alloul, *J. Phys. (Paris)* **32** (C1), 513 (1971).
13. J.E. Potts and L.B. Welsh, *Phys. Rev. B* **5**, 3421 (1972).
14. M. Hanabusa and T. Kushida, *Phys. Rev. B* **5**, 3751 (1972).
15. D. Tse and S.R. Hartmann, *Phys. Rev. Lett.* **21**, 511 (1968).
16. P. Bernier and H. Alloul, *J. Phys. F. Metals* **3**, 869 (1973).
17. D.C. Golibersuch and A.J. Heeger, *Phys. Rev.* **182**, 584 (1969); *Sol. St. Comm.* **8**, 17 (1970).
18. J.L. Tholence and R. Tournier, *Phys. Rev. Lett.* **25**, 867 (1970).
19. H. Alloul and P. Bernier, *J. Phys. F. Metals* **2**, L1 (1972).
20. H. Alloul and P. Bernier, (to be published in *J. Phys. F. Metals*).
21. A. Narath and A. C. Gossard, *Phys. Rev.* **183**, 391 (1969).
22. H. Alloul, P. Bernier, H. Launois, and J.P. Pouget, *J. Phys. Soc. Japan* **30**, 101 (1971).
23. B. Giovannini, P. Pincus, G. Gladstone, and A.J. Heeger, *J. Phys. (Paris)* **32**(C1), 163 (1971).
24. R.E. Behringer, *J. Phys. Chem. Sol.* **2**, 209 (1957).
25. W. Göetze and P. Wölfle., *J. Low Temp. Phys.* **6**, 455 (1972).
26. M.R. Shanabarger, private communication.
27. M.R. MacHenry, B.G. Silbernagel, and J.H. Wernick, *Phys. Rev. B* **5**, 2958 (1972).
28. K. Kume, K. Mizuno, S. Kazama, and Y. Nakamura, *J. Phys. Soc. Japan* **27**, 508 (1969).
29. A. Narath and H.T. Weaver, *Phys. Rev. Lett.* **23**, 233 (1969).

NMR Experimental Test for the Existence of the Kondo Resonance

F. Mezei*

Institute Laue–Langevin, Grenoble, France

and

G. Grüner*

Imperial College, London, England

A major part of the information about the behavior of impurities in a metal is contained in the conduction electron scattering on the impurities described, e.g., by a t-matrix. Theories are more or less primarily directly concerned with the t-matrix; experimentally, however, this quantity is not directly accessible. Investigation of the transport properties (mainly resistivity) is the usual way to obtain information about impurity scattering, in which case we are actually concerned with some integrals containing the t-matrix and confined to a narrow energy range around the Fermi energy, characterized by a width scaling with kT. As we will show, the charge density oscillation (Friedel oscillation[1]) around the impurity is also generally related to some integral containing the t-matrix and having its most important contributions from the vicinity of the Fermi energy. This integral is different from those involved in transport phenomena; in particular, it is essentially temperature independent (except for a possible temperature dependence of the t-matrix itself). We will find that the additional information one can obtain from the investigation of the charge density oscillation, combined with the readily available data about the effect of the impurity on the bulk transport properties, yields a powerful tool to study the energy dependence of the t-matrix. It is particularly important to note that this method works at a given temperature; it therefore offers the new possibility of direct separation of the energy and the temperature dependence of the scattering, which is not present in the usual scheme based on transport property studies.

The most prominent common prediction of Kondo-type theories of magnetic impurities[2] is the appearance of a sharp resonance in the scattering around the Fermi energy as $T \to 0$. This involves a drastic variation of the t-matrix around the Fermi energy, and the basic point we will be concerned with is to show how this prediction could be checked by the NMR investigation of the charge density oscillation around the impurities, as a special example of what was stated above.

For the present purpose we can confine our considerations to the $T = 0$ limit.

---

* On leave from the Central Research Institute for Physics, Budapest, Hungary.

This can be easily justified by inspecting the results of the more general derivation, which will be presented elsewhere.[3]

We now deduce an expression for the charge density oscillation around an impurity in a simple metal, in order to generalize Friedel's classical result to the case of an arbitrary, energy-dependent scattering. We apply a simple Green's function method[4] to determine the change in the local density of states of the free-electronlike conduction band for a single spin direction $\Delta\rho(\omega, r)$, which will be a function of the energy ω and of the distance from the impurity r. Assuming that the t-matrix $t_{\mathbf{kk}'}(\omega)$ does not depend explicitly on the absolute value of the moments $\mathbf{k}$ and $\mathbf{k}'$, and that it corresponds to the symmetry of l-wave scattering, we arrive at the following expression for the total charge density oscillation by integrating $\Delta\rho(\omega, r)$ over the occupied states for $T = 0$ and $r \gg k_F^{-1}$:

$$\Delta\rho_l(r) = -(2l+1)\frac{m^{*2}}{4\pi^3 r^2} \operatorname{Im} \int_{-\varepsilon_F}^{0} t_l(\omega) e^{2i[k_\omega r - (l\pi/2)]} d\omega \quad (1)$$

Here $t_l(\omega)$ is the non-spin-flip, purely energy-dependent part of the t-matrix usually considered,[2,4] m^* is the effective mass of the conduction electrons, and $k_\omega = [2m^*(\omega + \varepsilon_F)]^{1/2}$

We recover the well-known Friedel formula[1] by evaluating the integral in Eq. (1) for large enough r values, for which the exponential function in the integral oscillates very rapidly as a function of ω as compared to the changes of $t_l(\omega)$, i.e.,

$$r \gg \xi_c = V_F/2\Delta$$

where $V_F = (dk_\omega/d\omega)^{-1}_{\omega=0}$ is the Fermi velocity and Δ is a characteristic energy giving the width of the structure in $t_l(\omega)$. (Note that quite generally the main contributions to the integral come from the vicinity of the Fermi level, from, say, the upper half of the band.) In this case, by virtue of the Riemann lemma, the integral vanishes, except for the contribution of the sharp discontinuity at the Fermi energy corresponding to the upper limit of integration, which reads

$$\int_{-\varepsilon_F}^{0} t_l(\omega) e^{2i[k_\omega r - (l\pi/2)]} d\omega \simeq \frac{1}{r}\frac{v_F}{2i} t_l(0) e^{2i[k_F r - (l\pi/2)]} \quad (2)$$

Using this approximation and introducing the phase shift $\delta_l(0)$ by the relation $\pi\rho_0 t_l(0) = -\sin\delta_l(0)\exp[i\delta_l(0)]$, Eq. (1) is identical with the Friedel formula:

$$\Delta\rho_l(r) = -(2l+1)\frac{\sin\delta_l(0)}{4\pi^2 r^3}\cos[2k_F r - l\pi + \delta_l(0)]. \quad (3)$$

On the other hand, for distances $r \ll \xi_c$ the function $\exp[2ik_\omega r]$ will change as a function of ω much slower than $t_l(\omega)$ does. Therefore the energy dependence of $t_l(\omega)$ will play the major role in determining the integral, and because of this, it cannot be generally evaluated as simply as above. It is easily seen, however, that for a resonancelike, single-peaked t-matrix the integral will tend to saturate as r goes below ξ_c, instead of continuing to increase according to the r^{-1} function, as given in Eq. (2).

So we conclude that Friedel's expression, Eq. (3), is valid only for distances above a critical length ξ_c, which is determined by the energy variation of the t-

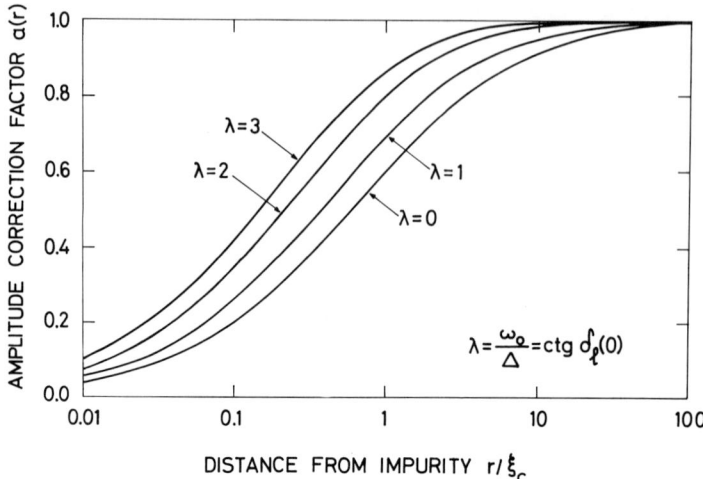

Fig. 1. The amplitude of the charge density oscillation relative to that given by Friedel's expression as a function of the distance r/ξ_C for various resonance energies ω_0/Δ.

matrix; in the special case of a resonant scattering the amplitude of the charge density oscillation at distances below ξ_c tends to stay considerably below the value given by the Friedel formula. This general behavior is well illustrated by the explicit results obtained for the case of a Lorentzian resonance (which, e.g., appears in the Nagaoka theory for $T \to 0$ and is widely used to interpret experimental data[5] on Kondo systems):

$$t_l(\omega) = \frac{1}{\pi \rho_0} \frac{\Delta}{\omega - \omega_0 + i\Delta} \qquad (4)$$

for which the charge density oscillation can be given in the form of a corrected Friedel formula as

$$\Delta \rho_l(r) = - (2l + 1) \frac{\sin \delta_l(0)}{4\pi^2} \frac{a(r)}{r^3} \cos\left[2k_F r - l\pi + \delta_l(0) + \eta(r)\right] \qquad (5)$$

The amplitude and phase correction functions, $a(r)$ and $\eta(r)$ are given explicitly in Ref. 3. For our present purpose it is sufficient to look at the general shape of $a(r)$ given in Fig. 1 for several values of the resonance energy ω_0 measured from the Fermi level.

Now let us turn to the consequences of this analysis. First, we point out that the anomalous behavior predicted by Eq. (5) should be manifested experimentally, e.g., for nonmagnetic transition metal impurities in Cu; in this case the width of the virtual bound-state scattering is believed to be of the order of 0.5 eV, which gives $\xi_c \simeq 10$ Å. Indeed, the charge density oscillation amplitudes around such impurities as measured by the second-order quadrupole wipeout effect in NMR at distances of 6–8 Å from the impurity can be shown to stay significantly below the values that are expected from the first-order quadrupole effect data, which reflect the charge oscillation at distances of 15–20 Å.[3]

Concerning the Kondo effect, let us observe that for the narrow resonances of width $kT_K < 300°K$ the critical length ξ_c will be larger than 200Å; i.e., at the wipe-out distance of about 20 Å for the first-order NMR quadrupole effect the amplitude correction factor shown in Fig. 1 will be about one order of magnitude less than unity. This means that while for $T \to 0$ both the impurity resistivity and the charge density oscillation in the $r \to \infty$ asymptotic limit [Eq. (3)] are directly related to the t-matrix taken at the Fermi energy, in the case of such narrow resonances the charge density oscillation amplitude at a distance accessible to NMR experiments will be anomalously small as compared to the value expected in view of the resistivity data by the use of the Friedel formula. In particular, this implies that the Kondo-type increase of impurity resistivity as $T \to 0$ is not accompanied by a correspondingly pronounced increase of the NMR first-order quadrupole effect. Experimental evidence for this kind of disagreement between the $T \to 0$ limit of the impurity contribution to the resistivity and the $T \to 0$ limit of the charge density oscillation at about 20 Å distance from the impurity would give a direct indication of the occurrence of the expected narrow resonance in the non-spin-flip scattering amplitude. The degree of the disagreement could yield an estimate (probably an upper limit only) for the actual width of the resonance. To be complete, let us make two further remarks. In the $T \to 0$ Kondo nonmagnetic regime the non-spin-flip part of the t-matrix is expected to be spin independent and the spin-flip part to be negligible. Thus only that part of the scattering remains that is relevant to the charge oscillation. The electron–hole symmetry of the s–d model, which makes the net screening charge around the impurity to be necessarily zero,[6] does not imply, however, a vanishing charge density oscillation as well, since it only requires $\text{Re } t_l(0) = 0$, and not $|t_l(0)| = 0$. (In this way the Friedel sum rule is seen to break down for exchange scattering.)

Contrary to the Kondo theories, in the nonmagnetic limit of the Anderson model, which is predicted to be relevant for the $T \to 0$ limit in the local spin-fluctuation theories,[2] one expects the t matrix to show a resonance of a width corresponding to that of the virtual bound state, i.e., of about 0.3–0.5 eV equivalent to $\xi_c \simeq 10-15$ Å. In this model, therefore, no serious depression of the charge density oscillation is expected as sensed by NMR experiments at a distance of 20 Å from the impurity. So we conclude that the predictions of the Kondo and local spin-fluctuation type theories for the anomalous charge density oscillation around the impurities are qualitatively different.

Acknowledgments

The authors are grateful to Drs. V.J. Emery, C. Hargitai, and A. Zawadowski for valuable discussions.

References

1. J. Friedel, *Can. J. Phys.* **34**, 1190 (1956).
2. K. Fischer, *Phys. Stat. Sol.* (b) **46**, 11 (1971).
3. F. Mezei and G. Grüner *Phys. Rev. Lett* **29**, 1465 (1972).
4. F. Mezei and A. Zawadowski, *Phys. Rev. B* **3**, 167 (1971).
5. W.M. Star, Thesis, University of Leiden, 1971.
6. H. Keiter, E. Müller-Hartmann, and J. Zittartz, *Z. Physik* **223**, 48 (1969).

Nuclear Orientation Experiments on Dilute $Au_{1-x}Ag_x$Yb Alloys

J. Flouquet and J. Sanchez

Laboratoire de Physique des Solides
Université de Paris, Orsay, France

The $Au_{1-x}Ag_x$Yb alloys have often been mentioned as the best system in which to study the magnetic–nonmagnetic transition since ytterbium is divalent and nonmagnetic in silver, and trivalent and magnetic in gold.

The observation of a resistivity minimum by Bijvoet et al.[1] for $x \sim 0.7$ is usually interpreted as indicating that the resonant coupling dominates the direct exchange term when x becomes greater than 0.6. Allali et al.[2] have analyzed their high-temperature susceptibility measurements in this view. However, recently the EPR experiments of Tao et al.[3] have shown that the g shift of Yb in Au is negative; this seems to indicate that resonant coupling is the dominant mechanism even for x close to zero.

We have performed nuclear orientation (NO) experiments in order to make magnetization measurements directly on the Yb impurity. The advantage over the high-temperature experiments of Allali et al. lies in the fact that we observe the impurity at a low temperature in a fundamental state defined by the crystal-field splitting. The effects of the crystal field and those of resonant coupling are then well separated. The measured γ anisotropy is related to the population of the nuclear substates of the parent nuclei, which can be computed as a function of the local magnetization in the two extreme cases of a strong or a weak resonant coupling. If we define the strength of the coupling by the Kondo temperature (T_K), these situations correspond to an experimental temperature lower or greater than T_K.[4]

The nuclear parameter of Yb nuclei are now well known, since Spanjaard et al. have recently determined the nuclear moment $\mu = 0.277\mu_n$. The observation of the 283-keV γ rays emitted along the axis of the polarizing field shows all the solid-state information which can be deduced by NO.

The $Au_{1-x}Ag_x$Yb alloys are obtained by ionic implantation of the Yb activity into the $Au_{1-x}Ag_x$ metal target at an energy of 100 keV. This method overcomes solubility problems at the price of a rather high impurity concentration (few hundred ppm within the penetration depth 200 Å) and the introduction of defects into the host lattice. Here most of the implanted impurity stays in a substitutional site, since its solubility is not too low.[5] The greatest difference with the classical preparation is that in the present method the impurity prefers to be near a region rich in gold.[1]

The implanted sample is then soldered at the top of a demagnetization salt and cooled to a temperature near 15 m°K. The anisotropy of the 283-keV γ rays is then measured as a function of the applied field and temperature with fields from 100 to 5000 Oe.

Nuclear Orientation Experiments on Dilute $Au_{1-x}Ag_x$ Yb Alloys

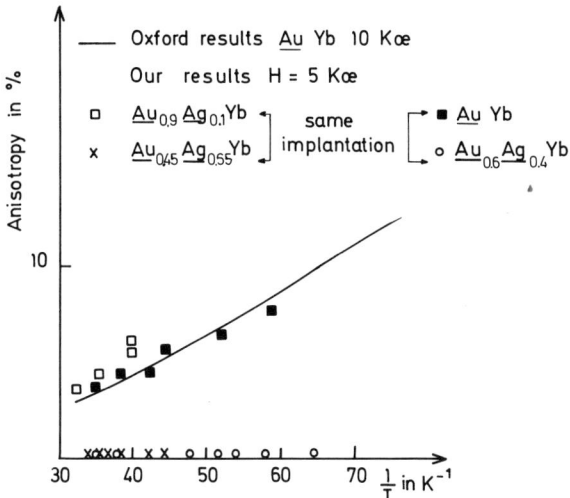

Fig. 1. Gamma-ray anisotropy of oriented Yb as a function of the inverse temperature $1/T$, measured on four different samples ($x = 0, 0.1, 0.4, 0.65$) for an external applied magnetic field of 5 kOe.

The results for the AuYb alloy agree with those of Spanjaard et al.[6] and are consistent with a Γ_7 fundamental state of the Yb impurity. The saturation of the anisotropy in an applied field of 5 kOe shows that an eventual Kondo temperature greater than 100 m°K is ruled out. The magnetic field dependence is close to the weak coupling model. Unfortunately, the small deviation cannot be attributed to a dominant resonant coupling since the sample preparation is not ideal.

However, in the experiments on the binary alloys of $Au_{1-x}Ag_x$ with $x \sim 0.7$ no anisotropy is at first detected. We then simultaneously implant two alloys: one rich in gold ($x = 0, 0.1$), the other with an intermediate binary composition ($x = 0.4, 0.65$). Figure 1 shows the γ anisotropy as a function of the reciprocal temperature $1/T$ for a 5 kOe applied field.

Our results are in complete disagreement with the analysis of Allali et al. who define the dominant resonant coupling for $x > 0.6$. On the other hand, our results agree with the EPR experiments of Tao et al. Thus the main conclusion seems to be that the resonant coupling is the dominant mechanism even for x close to zero. However, it is difficult to understand why no resistivity minimum is observed for the AuYb alloy. This introduces the problem of the description of the coupling of this "abnormal" rare earth with the conduction electrons.

In fact, the derivation of a dominant resonant coupling from the g-shift measurements has been obtained with the classical exchange Hamiltonian:

$$\mathcal{H} = -2\Gamma(g_J - 1)\mathbf{J} \cdot \mathbf{S}$$

where the notations are obvious. Coqblin and Schrieffer[7] have derived another Hamiltonian to take into account the spin and orbit scattering. With this new model it becomes possible to have a transfer momentum $J_z = M$ greater than one.

The relaxation by this interaction is surely more important than the one produced by the exchange Hamiltonian. The Γ constant derived by Tao et al. (-0.4 eV) is quite large and should give a rather high Kondo attenuation in the measured γ anisotropy if we use ordinary models. The NO experiments show clearly the weakness of the correlation of the Yb impurity with the conduction electrons. The d–f mixing is usually omitted. Another interpretation of this disagreement is that this mixing here is more important than is usually supposed.

It must be pointed out that the classical NO method is limited here, since it gives only average information. This difficulty is avoided with a spectroscopic method like, for example, the Mössbauer effect. The γ anisotropy may be reduced by a change of the local symmetry, which can give quadrupolar effects or new fundamental states.

Finally, another attenuation process may occur through the random distribution of the hyperfine field when the resonant coupling increases and the interaction between the impurities becomes important; it is then impossible with an ordinary applied field to polarize the local moment and also the nucleus. In a classical NO experiment we observe only single-impurity effects, since the concentration is often less than one ppm.

These NO experiments show clearly that the description of the $Au_{1-x}Ag_xTb$ alloys is more complicated than the first analyses have claimed. New resistivity measurements on these alloys would be interesting in order to compare the results obtained by different methods.

References

1. J. Bijvoet, Thesis, University of Amsterdam, 1969, unpublished; J. Boes, A.J. Van Dam, and J. Bijvoet, *Phys. Lett.* **28A**, 101 (1968).
2. V. Allali, D. Donze, and A. Treyvaud, *Sol. St. Comm.* **7**, 1241 (1969).
3. L.J. Tao, D. Davidov, R. Orbach, and E.P. Chock, *Phys. Rev. B*, **3**, 45 (1971).
4. J. Flouquet, Thesis, Orsay, France, 1971.
5. H. Bernas, Thesis, Orsay, France, 1972; H. Bernas, M.O. Ruault, and B. Jouffrey, *Phys. Rev. Lett.* **27**, 859 /1971).
6. D. Spanjaard, Thesis, Orsay, France; D. Spanjaard, D. Marsh, and N. Stone, *Hyperfine Interactions and Excited Nuclei*, Gordon and Breach, New York (1971).
7. B. Coqblin and J.R. Schrieffer, *Phys. Rev.* **185**, 847 (1969).

Temperature Dependence of the High-Frequency Resistivity in Dilute Magnetic Alloys

H. Nagasawa

Department of Applied Physics, Tokyo University of Education
Ohtsuka, Bunkyo-ku, Tokyo, Japan

and

T. Sugawara

Institute for Solid State Physics, University of Tokyo
Roppongi, Minato-ku, Tokyo, Japan

Kondo anomalies of the properties of dilute magnetic alloys are theoretically associated with a sharp peak at the Fermi level of the s–d scattering t-matrix. When the t-matrix has a peak at the Fermi level one may expect a deviation of the frequency dependent resistivity from that given by the Drude formula, because the electron relaxation time is no longer constant with respect to frequency. Therefore detailed experimental study of the high-frequency resistivity is expected to give information about the nature of the t-matrix and to contribute to the theoretical treatment of Kondo anomalies. In this paper we report the results of experimental studies on the temperature dependence of the high-frequency resistivity in dilute magnetic alloys, CuCr, CuFe, and CuMn.

The high-frequency resistivity was determined from the microwave power dissipation due to surface impedance of the specimen. All the measurements were carried out at a frequency of 70 GHz and at helium temperatures. The experimental arrangement is illustrated schematically in Fig. 1. The power dissipation was measured as the rise of temperature of the specimen in an evacuated cylindrical cavity of (012) mode. A 0.1-W Allen-Bradley resistor glued to the back of the specimen was used as the temperature sensor. In each run the power dissipation in a pure Cu specimen was measured as the reference. Calibration of the power dissipation was made by a manganin heater also glued to the back of the specimen. Since the energy $\hbar\omega$ of 70-GHz microwave corresponds to 3.4°K, it was comparable with the Kondo energy kT_K for CuCr ($T_K \sim 1°$K), smaller than kT_K for CuFe ($T_K \sim 30°$K), and larger than kT_K for CuMn ($T_K \sim 0.002°$K).

The high-frequency resistivity of the specimen was obtained from the observed power dissipation by the following procedure. According to the theory of surface impedance,[1] the real part $\mathscr{R}$ is given in terms of the mean free path l and the normal skin depth δ as

$$\frac{\mathscr{R}_\infty}{\mathscr{R}} = \frac{2}{3\sqrt{3}} \pi^{-1/3} \alpha^{-1/3} \frac{f'(0)}{f(0)} \qquad (1)$$

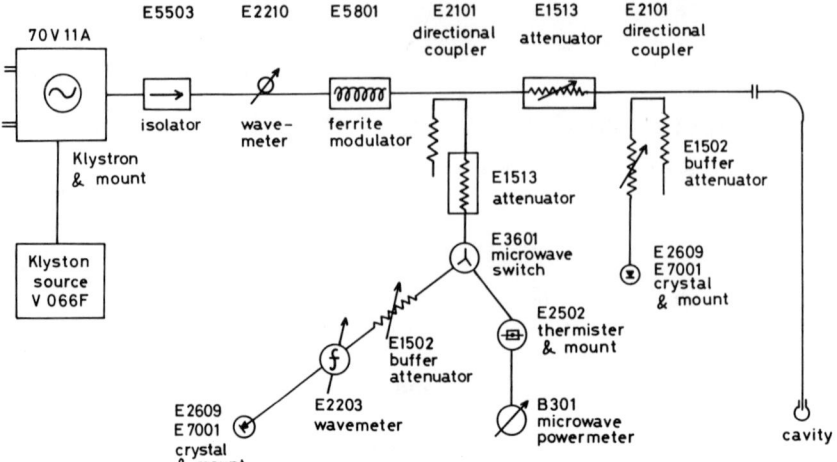

Fig. 1. Experimental arrangement.

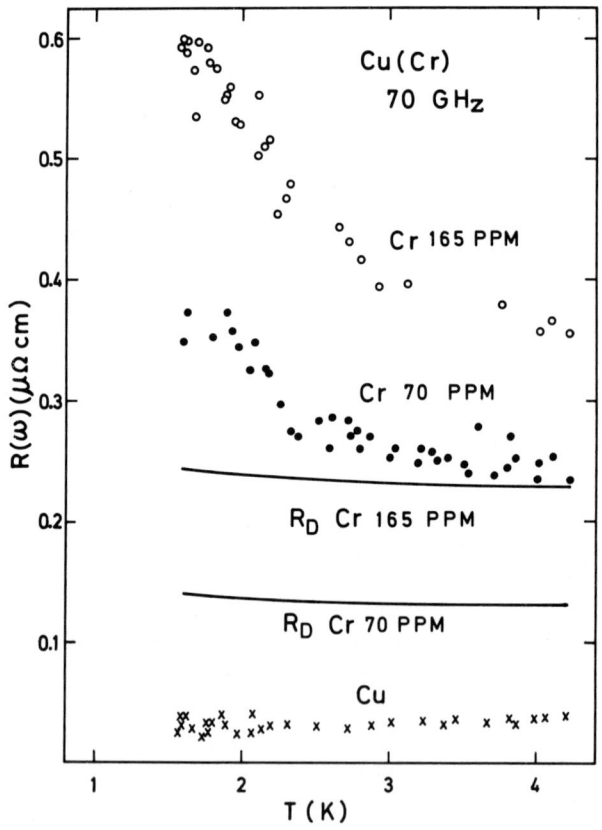

Fig. 2. High-frequency resistivity of CuCr and pure Cu as a function of temperature. R_D is the corresponding Drude resistivity (see text).

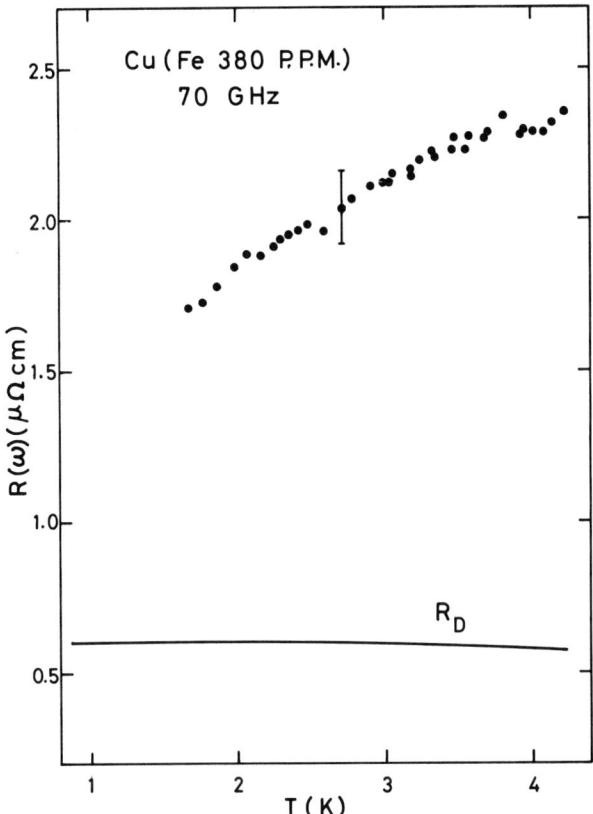

Fig. 3. High-frequency resistivity of CuFe as a function of temperature. R_D is the corresponding Drude resistivity.

with $\alpha = 3l^2/2\delta^2$, where $\mathscr{R}_\infty$ is the $\mathscr{R}$ at the anomalous limit, and $f(0)$ and $f'(0)$ are known functions of α. In Eq. (1) $\mathscr{R}_\infty$ was determined from the observed power dissipation and static resistivity of the pure Cu specimen. Then α was determined from $\mathscr{R}_\infty$ and the observed $\mathscr{R}$ using Eq. (1). Finally, the high-frequency resistivity $R(\omega)$ was calculated from α by

$$R(\omega) = 0.478\,\alpha^{-1/3} \quad \mu\Omega\text{-cm} \qquad (2)$$

The last equation can be obtained from Eq. (1) easily.

Some of the experimental results of the high-frequency resistivity are shown in Figs. 2 and 3, together with the corresponding Drude resistivity. The Drude resistivity was calculated from the observed static resistivity by

$$R_D(\omega) = R(0)\left[1 + (\omega\tau)^2\right] \qquad (3)$$

where $R(0)$ is the static resistivity, ω is the frequency, and τ is the transport relaxation time. The observed high-frequency resistivity in CuCr (Fig. 2) is larger in magnitude than the Drude resistivity and increases with decreasing temperature with a slope much steeper than that of the static resistivity. In CuFe (Fig. 3) the high-frequency resistivity is larger in magnitude than the Drude resistivity, as is the case for CuCr, and decreases with decreasing temperature in contrast to the static resistivity, which increases slightly in the temperature range of the present experiment. In CuMn (19.4 ppm Mn) $R(\omega)$ at first increases with decreasing temperature and below 3°K becomes constant. The static resistivity of this specimen increases like log T between 1.6 and 4.2°K. Since the Mn concentration in this specimen is low, the fact that $R(\omega)$ saturates below 3°K can not be attributed to the effect of interaction between Mn impurities.

According to the recent theory of Moriya and Inoue,[2] the high-frequency resistivity of dilute magnetic alloy is given in terms of the t-matrix as

$$R(\omega) = \frac{3\omega}{2e^2} \frac{1}{\rho(E_F) v_F^2} \left\{ \text{Im} \int_{-\infty}^{\infty} dx \frac{f(x-\omega) - f(x)}{\omega - n_i[t(x+is) - t(x-\omega-is)]} \right\}^{-1} \quad (4)$$

where $\rho(E_F)$ is the density of states, n_i is the impurity concentration, $f(x)$ is the Fermi distribution function, and $t(x)$ is the s–d scattering t-matrix. They have calculated $R(\omega)$ as a function of temperature and frequency for two approximate expressions of the t-matrix, a t-matrix with Lorentzian energy dependence and a solution[3] of the Nagaoka equation. Their result (Fig. 3 of Ref. 2), when applied to the alloys studied here, gives $R(\omega)$ larger than the Drude resistivity, in qualitative agreement with experiment, but gives in general a decrease of $R(\omega)$ with decreasing temperature, in disagreement with the results in CuCr and CuMn. Since the above comparison was made on the basis of theoretical curves for a restricted range of T/T_K and impurity concentration, the results obtained here may not provide any crucial test for the theory of Moriya and Inoue or the formulas of the t-matrix used by them. However, the fact that a considerable discrepancy between the theory and experiment has been found in the temperature dependence of $R(\omega)$ of CuCr seems to suggest a need for reexamination of the theory. In this connection it may be interesting to note that the Kondo energy kT_K of CuCr is comparable with the energy $\hbar\omega$ of the microwave used in this experiment.

A detailed account of this work will be published elsewhere.

References

1. G.H.E. Reuter and E.S. Sondheimer, *Proc. Roy. Soc.* **A195**, 336 (1948).
2. T. Moriya and M. Inoue, *J. Phys. Soc. (Japan)* **27**, 371 (1969).
3. J. Zittartz and E. Müller-Hartmann, *Z. Phys.* **212**, 380 (1968); **217**, 155 (1968).

Spin-Dependent Resistivity in Cu:Mn

G. Toth

Laboratoire de Physique de la Matière Condensée, Ecole Polytechnique
Paris, France

Thus far, spin-dependent transport in dilute magnetic alloys has been measured, to our knowledge, indirectly: It has been extracted from measurements of the total resistivity or magnetoresistance.[1] This extraction is not easy and sometimes necessitates ad hoc assumptions.

We present here a new approach to this kind of problem: It is based on the fact that magnetic resonance provides a means for modifying the spin magnetization of the impurities *without* changing any other relevant physical parameter, such as the dc magnetic field or the temperature. The normal magnetoresistance (non-spin-dependent) is therefore kept constant and no hypothesis is needed to subtract it from the experimental results. Furthermore, this technique is a resonant one and the effect of the spins on the resistivity is quite unambiguous.

Our preliminary experiments measure the variation of the resistivity of a Cu:Mn sample placed in a magnetic field H_0 and irradiated with microwave power at the Larmor frequency of the localized spins.

Two distinct effects can be observed: a spin-dependent collision effect and a bolometer detection of the magnetic resonance. Both can be viewed as a way to detect electron spin resonance (ESR) in various types of alloys.

Spin-Dependent Collision Effect

A well-known property of dilute magnetic alloys is the negative magnetoresistance. This is explained as a quenching of spin-flip collisions between conduction electrons and electrons localized on the magnetic impurities by the application of a magnetic field which polarizes both systems of spins. A reduction of the number of possible collisions leads to a decrease of the resistivity, hence the "negative magnetoresistance."

Note that the spin polarization is the essential factor that will quench or "dequench" the spin-flip collisions. A change of the electronic spin polarization will result in a change of the resistivity. This can be done by application of a resonant rf field at a frequency ω, such that $\omega = \gamma H_0$, where γ is the gyromagnetic ratio of the spins. This resonant field induces transitions between the Zeeman levels and tends to equalize their populations. This so-called "saturation" of the magnetic resonance line (Ref. 2, p. 40) reduces the spin polarization of the system which, in the limit of infinite rf power, vanishes. This reduction of the spin polarization dequenches the spin-flip collision process and leads to a resonant increase of the resistivity.

If the equilibrium magnetization is M_0 in the field H_0, it goes to a value $\langle M_z \rangle$ under irradiation given by (Ref. 2, p. 374)

$$\frac{M_0 - \langle M_z \rangle}{M_0} = \frac{1}{1 + (\Delta H/h_1)^2} \qquad (1)$$

where ΔH is the width of the resonance line and h_1 is the circular component of the rf field (in practice, a microwave field). The maximum effect is obtained for $\langle M_z \rangle = 0$, that is, for $h_1 \gg \Delta H$. However, in our experiments the opposite situation ($h_1 \ll \Delta H$) will generally be the case, due to the very high microwave powers necessary to completely saturate the fairly broad (50 G) resonance lines. The diminishing factor due to partial saturation is seen from Eq. (1) to be $(h_1 \Delta H)^2$.

A straightforward way to measure the variation of the resistivity $\Delta \rho$ is to bias the sample with a constant current I_0 and to measure the variation ΔV of the voltage drop across the sample in a four-terminal setup. The order of magnitude of the experimental signal $\Delta \rho/\rho$ can be calculated using a rough estimate of the ratio $(\rho_{H_0} - \rho_0)/\rho_0$, where ρ_0 is the resistivity at zero field; theoretical[3] and experimental[4] results give $(\rho_{H_0} - \rho_0)/\rho_0 \approx 2 \times 10^{-2}$ for $H_0 = 3$ kG. Apart from the diminishing factor due to partial saturation, one also expects a short-circuiting factor of the order of ten due to the fact that the thickness of the sample $\bar{d}$ is at least ten times greater than the skin depth δ at the frequency used (10 GHz), and that only the spins situated in the skin depth actually "see" the microwave field.* The expected signal is thus

$$\left(\frac{\Delta \rho}{\rho}\right)_{sd} = \frac{\rho_{H_0} - \rho_0}{\rho_0} \left(\frac{h_1}{\Delta H}\right)^2 \frac{\delta}{\bar{d}}$$

In our case $(\Delta \rho/\rho)_{sd} \approx 2 \times 10^{-7}$ for $H_0 = 3$ kG, $h_1 = 0.5$ G, $\Delta H = 50$ G, and $\delta/\bar{d} = 10$. This relative variation of resistivity, although very small, is quite detectable since it can be shown[5] that a reasonable limit with a signal/noise ratio of one is $(\Delta \rho/\rho)_{min} = 10^{-9}$.

Bolometer Detection of the Magnetic Resonance

The experimental conditions described above to detect the spin-dependent collision effect can also lead to the detection of the ESR line through the effect on the resistivity of the power absorbed by the sample at resonance. This method of detection is called bolometer detection.[6] In principle, if a sample is placed in good thermal contact with a bolometer, the power absorbed by the sample at resonance can be measured by the bolometer; it is shown in Ref. 6 that the sensitivity of this type of detection is by no means negligible. In the case of dilute magnetic alloys the sample itself can be its own bolometer, since the resistivity of these alloys is strongly temperature dependent even at very low temperatures. The physical picture in our case is the following: When the resonance field is switched on the spins absorb

---

* This factor of reduction $\delta/\bar{d}$ still holds even if the spin diffusion length is greater than the skin depth, because in this case the same magnetization is averaged over the diffusion length but is still excited only in the skin depth.

a power Π which is given by[6]

$$\Pi = \frac{h_1^2}{h_1^2 + \Delta H^2} \frac{h^2\omega^2}{kT} \frac{1}{T_1} \frac{S(S+1)}{3} N \qquad (2)$$

where T_1 is the longitudinal relaxation time, S is the spin, and N is the number of spins. The variation ΔT of temperature corresponding to the power Π is

$$\Delta T = \Pi \tau / C$$

where τ is the thermal response time and C is the heat capacity of the bolometer. With the same experimental values used to estimate the magnitude of the spin-dependent effect as well as a measured value of $10^{-2}/°K$ for the ratio $(\Delta\rho/\rho)\Delta T$, one obtains

$$(\Delta\rho/\rho)_{bol} \approx 4 \times 10^{-7}$$

The two effects appear to be similar inasmuch as the amplitude of the signal is concerned. They differ, however, in their shape, i.e., the intensity of the signal as a function of the magnetic field (as in conventional ESR, the frequency is kept fixed and the magnetic field is swept through resonance). The power absorbed by a metallic sample containing spins at resonance has been calculated by Dyson[7] to explain the asymmetric lines recorded by classical electromagnetic detection. This calculation is applicable to the "bolometric" signal, since the recorded quantity in this case is precisely the power absorbed by the sample. The fact that the ESR and bolometric line shapes are identical was checked experimentally on a special sample. The lineshape of the spin-dependent collision signal, however, is symmetric: The value of $(\Delta\rho/\rho)_{sd}$ depends only on the absolute value of M_z at each point of the sample, which, in turn, depends only on the separation in field from the resonance, $H - H_0$, through quadratic terms. This property enables one to distinguish between the two effects: A purely symmetric signal would be evidence for a dominant spin-dependent collision effect, while a signal having the same asymmetry as the ESR line would signify that the bolometer detection is much greater.

In our experiments the alloy used is Cu:Mn containing 0.5 at. % Mn prepared in a

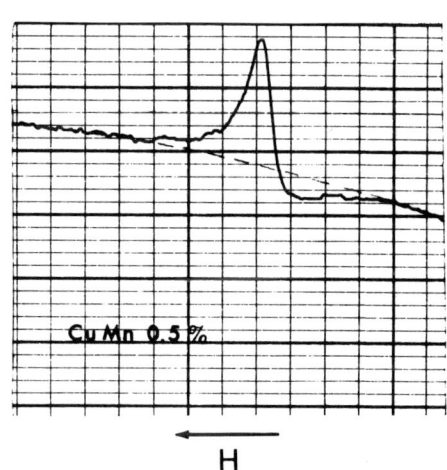

Fig. 1. Recording of the variation of the resistivity of the Cu:Mn sample as the magnetic field is swept through resonance of the localized electronic spins. The curved baseline is due to the bolometer detection of a broad line of the sample holder.

vacuum furnace from 99.999% pure copper (from Asarco Company) and 99.99% pure Mn (from Johnson Matthey). The sample is placed at the bottom of a TE_{102} rectangular microwave cavity in such a way that the solders are outside the microwave field. The microwave power is square-wave-modulated at a frequency of 100 kHz and the signal phase is detected through a lock-in amplifier. The recorded signal is shown in Fig. 1.

The signal is seen to be asymmetrical, but the ratio of the large peak over the small one is larger than that for the "Dysonian" line shape recorded on the same sample by conventional ESR. This shows that in this case, as was estimated, both signals are present and are of the same order of magnitude.

Although this preliminary experiment did not enable us to make quantitative measurements, we believe that the two effects described in this paper are very promising for the study of metallic alloys and even semiconductors, where similar techniques prove to be very successful.[8] By using higher microwave power, it is not impossible to imagine that spin-dependent resistivity could become a new kind of spectroscopy besides the now classical electromagnetic detection.

Acknowledgment

The constant support of Professor I. Solomon is gratefully acknowledged.

References

1. P. Monod, *Phys. Rev. Lett.* **19**, 1113 (1967).
2. A. Abragam, *Principles of Nuclear Magnetism*, Oxford University Press, Oxford, (1961).
3. M.T. Béal-Monod and R.A. Weiner, *Phys. Rev.* **170**, 552 (1968).
4. R.W. Schmitt and I.S. Jacobs, *Can. J. Phys.* **34**, 1285 (1956).
5. I. Solomon, in *Proc. 11th Conf. Physics of Semiconductors, Warsaw, 1972* (to be published).
6. J. Schmidt and I. Solomon, *J. Appl. Phys.* **37**, 3719 (1966).
7. F.J. Dyson, *Phys. Rev.* **98**, 349 (1955).
8. G. Toth, Thesis, Université de Paris VI, 1972, unpublished.

Properties of Dilute Transition-Based Alloys with Actinide Impurities: Evidence of Localized Magnetism for Neptunium and Plutonium

E. Galleani d'Agliano,* A. A. Gomès,† R. Jullien, and B. Coqblin

Physique des Solides, Université de Paris—Sud
Centre d'Orsay, France

A simple model taking into account a $6d$ and a $5f$ virtual bound state with a large d–f hybridization was developed to describe pure actinide metals and has explained the absence of magnetism at the beginning of the actinide series up to americium.[1] We would like to discuss here the behavior of dilute transition-based alloys with actinide impurities of the first half series (up to Am) and especially the possible occurrence of magnetism for neptunium and plutonium impurities. Since the first half series corresponds to the filling up of the $j = \frac{5}{2} 5f$ level, we include here spin–orbit effects in the model.

Model

The starting assumptions of the model are as follows.
(a) We consider a $6d$ and a $5f$ virtual bound state on each independent impurity.
(b) Only the $5f$ level is split by the spin–orbit coupling into two levels $f_1 (j = \frac{5}{2})$ and $f_2 (j = \frac{7}{2})$ separated by the spin–orbit splitting λ_{so}. Therefore we start from three localized levels: $6d$, $5f_1 (j = \frac{5}{2})$, and $5f_2 (j = \frac{7}{2})$ at energies E_{0d}, E_{0f_1}, and E_{0f_2}, with $E_{0f_2} - E_{0f_1} = \lambda_{so}$.
(c) We neglect the orbital degeneracy of each level, and in the case of the f levels we treat them as twofold degenerate levels like those of a 1/2 spin with usual spin components $\sigma = \uparrow$ and $\sigma = \downarrow$.
(d) All the levels are treated inside the resonant scattering mechanism giving three virtual bound states of half-widths Γ_d, Γ_{f1}, and Γ_{f2}.
(e) We introduce the d–f hybridization by a phenomenological one-body Hamiltonian:

$$\sum_\sigma [(V_{df_1} c_{d\sigma}^+ c_{f_1\sigma} + V_{df_2} c_{d\sigma}^+ c_{f_2\sigma}) + \text{h.c.}]$$

where the parameters V_{df_1} and V_{df_2} are assumed to be constant. We develop here only a first approach corresponding to the case of an infinite spin–orbit coupling. The more physical case of a finite spin–orbit splitting, of order 1.5 eV,[2] has been

---
* Present address: University of California, San Diego, La Jolla, California.
† Permanent address: Centro Brasileiro de Pesquisas Fisicas, Rio de Janeiro, Brazil.

previously developed and leads to the same qualitative results for the first half series as those obtained here. Moreover, this more realistic case well describes the magnetic moments of curium and berkelium, which are close to the values of their ionic configurations either in alloys or pure metals.

Within the approximation of an infinite spin–orbit splitting, the study of the first half series is perfectly disconnected from that of the second half series and is described by the simple Hamiltonian which describes only the f_1 levels interacting with the d level:

$$H_0 = \sum_{k,\sigma} \varepsilon_k c^+_{k\sigma} c_{k\sigma} + E_{0d} \sum_\sigma c^+_{d\sigma} c_{d\sigma} + E_{0f_1} \sum_\sigma c^+_{f_1\sigma} c_{f_1\sigma}$$
$$+ U(n_{d\uparrow} + n_{f_1\downarrow})(n_{d\downarrow} + n_{f_1\downarrow})$$
$$+ \sum_{k,\sigma} (V_{kd} c^+_{k\sigma} c_{d\sigma} + \text{h.c.}) + \sum_{k,\sigma} (V_{kf_1} c^+_{k\sigma} c_{f_1\sigma} + \text{h.c.})$$
$$+ \sum_\sigma (V_{df_1} c^+_{d\sigma} c_{f_1\sigma} + \text{h.c.})$$

We have adopted here usual notations and we have taken as equal all the exchange and Coulomb interactions.[1] This Hamiltonian is treated inside the Hartree–Fock approximation, as in the Anderson case.

In order to explain the physical case of actinides, it is natural to choose the different parameters as follows.

(a) E_{0d} is taken equal to Γ_d, in order to fit approximately the total number of d electrons on the impurity site, which is of order two.

(b) Γ_d and Γ_{f_1} are chosen, as in the case of pure actinide metals, by comparison with band calculations,[2] and here we have taken $\Gamma_d = 20\Gamma_{f_1} \sim 2$ eV.

(c) U is chosen in the range $\pi\Gamma_d > U > \pi\Gamma_{f_1}$ in order to have a large f magnetic moment in the case of no d–f hybridization (as in the case of rare earths and a negligible d magnetic moment, so we have taken $U = \Gamma_d \sim 2$ eV.

With this choice of parameters we have calculated the total number of d and f_1 electrons as

$$N = 5(n_{d\uparrow} + n_{d\downarrow}) + 3(n_{f_1\uparrow} + n_{f_1\downarrow})$$

and in Fig. 1 we plot it vs. E_{0f_1}/U for four values of the V_{df_1} parameters. In this figure we have also plotted the magnetic moment

$$M = 5(n_{d\uparrow} - n_{d\downarrow}) + 3(n_{f_1\uparrow} - n_{f_1\downarrow})$$

We see that for a large d–f hybridization, magnetism disappears completely in the entire first half series [case (d)] and if this hybridization is reduced by 30% we obtain [case (c)] a situation where Np and Pu become magnetic (with a larger magnetic moment for Np) and where U is nearly magnetic. If we reduce this hybridization more and more, U can become magnetic (and even Pa, in the case of a zero d–f hybridization).

Comparison with Experiments

Case (d) of a great d–f hybridization can account for the situation of pure actinide metals which are nonmagnetic up to Am as was already described without spin–

Properties of Dilute Transition-Based Alloys with Actinide Impurities

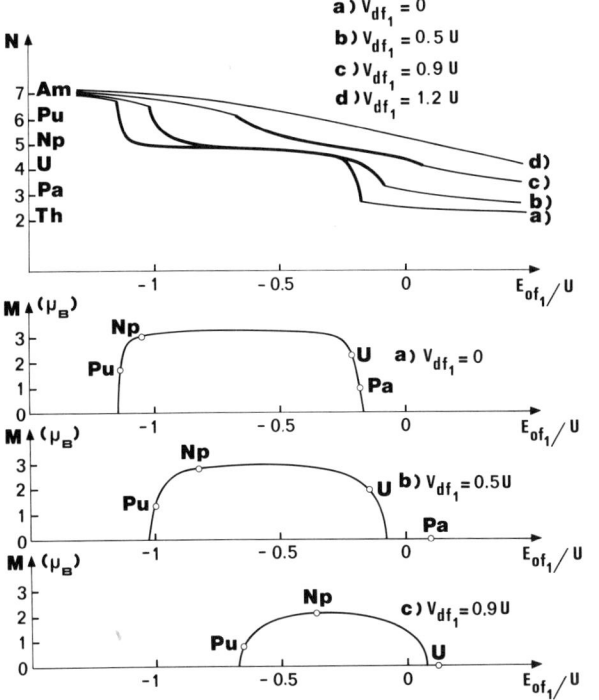

Fig. 1. Plot of N and M vs. E_{0f_1}/U for four values of V_{df_1}.

orbit considerations.[1] But the interesting result is that if we reduce the $d-f$ hybridization Np and Pu become magnetic, as is experimentally observed in transition-based alloys with actinide impurities.[3]

Case of Nonmagnetic Transition Hosts. The superconducting temperature T_c of lanthanum with actinide impurities has been measured by Hill et al.[4] and is plotted in Fig. 2. This experiment is very interesting because it covers the entire first half series up to americium. The depression of T_c is small for thorium, uranium, and americium, indicating that they are not magnetic, while it is very large for neptunium and plutonium, showing a clear magnetic behavior. If we assume that the depression of T_c is proportional to the square of the magnetization, we can deduce that the magnetic moment of LaNp is roughly 50% larger than that of LaPu. These results are checked by the presence of a resistivity minimum in LaPu alloys[5] and by the magnetic susceptibility measurements, which give a 0.75 μ_B value for the magnetic moment of LaPu[6] and a 1.7 μ_B value for that of YNp[6] with a similar host. These results can be accounted for in the theoretical model by case (c) of Fig. 1.

Other experimental data on U, Np, and Pu impurities in various transition hosts can also be accounted for by case (c) of Fig. 1. Uranium impurity can be nearly magnetic, according to the observed spin fluctuations in PdU alloys[7] above 12 at. % U and in dilute ThU alloys.[8] Neptunium impurity has a net magnetic behavior, with a measured magnetic moment of 2.5 μ_B in PdNp,[9] roughly one μ_B in ThNp,[10] and 1.7 μ_B in YNp[6] alloys. Plutonium impurity is on the verge of becoming magnetic, either not magnetic as in Sc, Y, Ti, Th, Hf hosts,[11,12] or weakly magnetic, with a small magnetic moment of 0.75 μ_B in $LaPu^6$ and of one μ_B in $PdPu$,[13] and presents a resistivity minimum at low temperatures in La, Pr, and Zr hosts.[11]

Case of Ferromagnetic Transition Hosts. The hyperfine fields of Ra, Th, U, Pu (in Fe), and Np, Cm (in Ni) are reported in Fig. 3[(14)] and can be analyzed in terms of several contributions.[15]

(a) A first contribution $h_{cep} = A(Z) m_s^{(0)}$ comes from the polarization of the s–p conduction band. By extrapolating the $A(Z)$ values of Campbell[16] and by taking $m_s^{(0)} = -0.2 \mu_B$ as in transition elements, we have $h_{cep} = -3000 - 120(Z - Z_0)$ in kOe [curve (I)], where Z_0 is the atomic number of Ra.

(b) A second contribution $h_{cp}^{(d)} = -\alpha_d m_d^{(0)}$ comes from the d-like electrons. Extending the well-known results of transition elements diluted in Fe or Ni,[17] since the difference of charge between the impurity and iron remains smaller than (or equal to) -2, $m_d^{(0)}$ remains roughly constant and of order $-1\mu_B$. Since α_d remains constant in each d series, h_{cp}^d has to be constant. This is checked by the experimental curve, which is roughly parallel to the curve I up to uranium and verifies the assumption of only the two contributions h_{cep} and $h_{cp}^{(d)}$ to the hyperfine field of Ra, Th, and U. The deduced value of $\alpha_d \sim 3000$ kOe/μ_B is consistent with its values in the other d series. So $h_{cp}^{(d)}$ is of order 3000 kOe for the whole actinide series and is given by the curve II of Fig. 3.

(c) Since the experimental values for Np, Pu, and Cm lie above the curve III giving $h_{cep} + h_{cp}^{(d)}$, it is necessary to invoke for these special impurities an extra contribution coming from an f-localized moment $m_f^{(0)}$:

$$h_{cp}^{(f)} = -\alpha_f m_f^{(0)}$$

The f magnetic moment has both spin and orbital contributions in Np and Pu due to the large spin–orbit coupling, and only a spin contribution in Cm. This is consistent with the positive deduced values of $h_{cp}^{(f)}$, which are roughly 750 kOe for Np, 1500 for Pu, and 1000 kOe for Cm. As explained in Ref. 15, for Np and Pu we can take α_f of order 1500–2000 kOe/μ_B in Fe host and 3–5 times smaller (as for

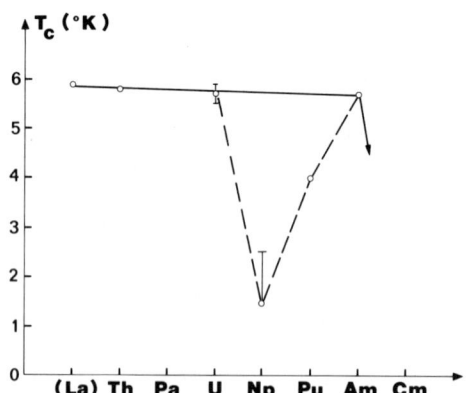

Fig. 2. Superconducting temperature of lanthanum-based alloys with 0.5 at.% actinide impurity.

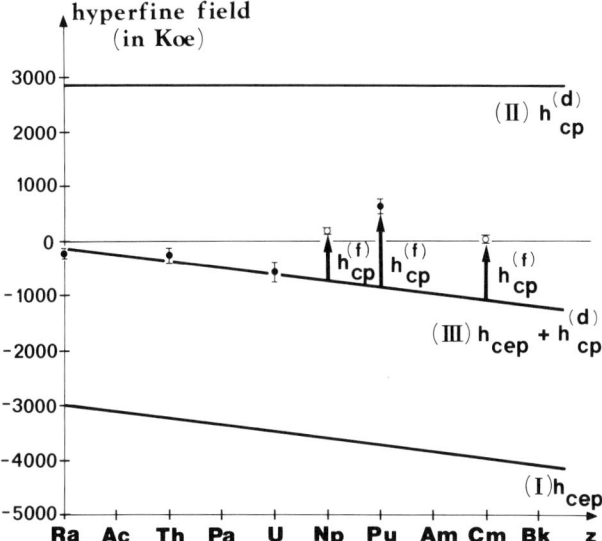

Fig. 3. Hyperfine field of actinides in ferromagnetic hosts and its different contributions I, II, and III, as described in the text. The experimental values (in kOe) of the hyperfine field are given by solid circles for actinides in an iron host and by open circles in a nickel host.

rare earths) in Ni host. On the other hand, α_f has been taken much smaller in the case of Cm, which has only a spin contribution to the f magnetic moment.

Thus the analysis of the hyperfine field data in a ferromagnetic host is consistent with a spin and orbital f moment of order two μ_B for Np and one μ_B for Pu, and with only a spin f moment for Cm.[4]

Conclusion

Both Np and Pu are magnetic in transition-based alloys, while Np and Pu pure metals are not magnetic. This is explained by a decrease of the d–f hybridization. However, it remains a large hybridization, which can be explained by a mechanism involving the d electrons of the conduction band in the case of transition-based alloys. Therefore we suggest a study of the magnetic properties of actinide impurities in hosts without d character in order to see if the magnetic moments of Np and Pu become larger and if uranium can become magnetic as in the cases (b) and (a) of Fig. 1.

References

1. R. Jullien, E. Galleani d'Agliano, and B. Coqblin, *Phys. Rev.* **6**, 2139 (1972).
2. E.A. Kmetko, and H.H. Hill, in *Proc. Rare Earths and Actinide Conf., Durham, 1971*, Institute of Physics, p. 233; A.J. Freeman and D.D. Koelling, *J. Phys. Radium* (to be published); D.D. Koelling, A.J. Freeman, and G.O. Arbman, in *Plutonium 1970*, W.N. Miner, ed. Metallurgical Soc. of AIME, New York, p. 194.
3. B. Coqblin, E. Galleani d'Agliano, and R. Jullien, in *Proc. Superconductivity in d- and f-Band Metals*.

Conf., D.H. Douglass, ed., Rochester, 1971, p. 154; R. Jullien, E. Galleani d'Agliano, and B. Coqblin (to be published).
4. H.H. Hill, J.D.G. Lindsay, R.W. White, L.B. Asprey, V.O. Struebing, and B.T. Matthias, *Physica* **55**, 615 (1971).
5. H.H. Hill, R.O. Elliot, and W.M. Miner, in *Colloque de CNRS no. 180*, Les éléments de terres rares, Paris—Grenoble (1969), p. 541.
6. J.P. Gatesoupe and C.H. de Novion, in *Proc. Rare Earths and Actinide Conf., Durham, 1971*, Institute of Physics, p. 84.
7. W.J. Nellis, M.B. Brodsky, H. Montgomery, and G.P. Pells, *Phys. Rev. B* **2**, 4590 (1970).
8. M.B. Maple, J.G. Huber, B.R. Coles, and A.C. Lawson, *J. Low Temp. Phys.* **3**, 137 (1970).
9. W.J. Nellis and M.B. Brodsky, *J. Appl. Phys.* **41**, 1007 (1970).
10. W.J. Nellis and M.B. Brodsky, *J. Appl. Phys.* **42**, 1463 (1971).
11. R.O. Elliot and H.H. Hill, *J. Less Common Met.* **22**, 123 (1970); R.O. Elliot and H.H. Hill, in *Plutonium 1970*, W.N. Miner, ed., Metallurgical Soc. of AIME, New York, p. 335.
12. J.P. Gatesoupe and C.H. de Novion, *Sol. St. Comm.* **9**, 1193 (1971).
13. W.J. Nellis and M.B. Brodsky, *Phys. Lett.* **32A**, 267 (1970).
14. D.E. Murnick and E.N. Kaufmann, *J. de Phys.* (Suppl.) **32**, 736 (1971); E.J. Ansaldo and L. Grodzins, *Phys. Lett.* **30B**, 538 (1969); **32B**, 479 (1970); **34B**, 43 (1971) and references therein; F. Falk, A. Linnfors, B. Orre, and J.E. Thun, *Physica Scripta* **1**, 13 (1970).
15. R. Jullien, A.A. Gomès, and B. Coqblin (to be published).
16. I.A. Campbell, *J. Phys. C: Sol. St. Phys.* **2**, 1339 (1969).
17. I.A. Campbell and A.A. Gomès, *Proc. Phys. Soc.* **91**, 319 (1967); *Sol. St. Comm.* **6**, 395 (1968).

New Treatment of the Anderson Model for Single Magnetic Impurity*

J. W. Schweitzer

Department of Physics and Astronomy
The University of Iowa, Iowa City, Iowa

This paper describes some results of our recent heuristic treatment[1] of the Anderson model[2] for localized magnetic states in metals. Also, it attempts to demonstrate that these results provide a successful description of many experimental observations for dilute magnetic alloys. It is generally believed that the Anderson model is adequate to describe the total range of magnetic behavior observed in solutions of transition atom impurities in the simple metals. However, a complete solution to the Anderson model showing the transition from weakly to strongly magnetic as the model parameters vary, and the proper relationship between low- and high-temperature regimes, seems to be a prohibitively difficult problem. In contrast to the difficulty of a solution from first principles, we have developed a very simple treatment which is suggested by some general conclusions drawn from various studies of the Anderson model. This treatment is not justified by any detailed analysis of the model, yet its validity might be argued from its simplicity and the reasonableness of the results.

Our treatment is for the special case of the Anderson model having a single extraorbital state with electron–hole symmetry. We write

$$H = \sum_{k\sigma} \varepsilon_k n_{k\sigma} - \tfrac{1}{2} U \sum_{\sigma} n_{d\sigma} + \tfrac{1}{2} U \sum_{\sigma} n_{d\sigma} n_{d-\sigma}$$
$$+ V \sum_{k\sigma} (C^+_{k\sigma} C_{d\sigma} + C^+_{d\sigma} C_{k\sigma}) \qquad (1)$$

The choice of $-\tfrac{1}{2} U$ for the energy of the orbital state relative to the Fermi energy and the use of a flat density of states for the conduction band gives the model electron–hole symmetry in the orbital state. This localizes exactly one electron in the orbital state (i.e., $\langle n_{d\sigma} \rangle = 1/2$), as is expected by screening considerations neglected in the model.

The fundamental assumption of our treatment is that the non-spin-flip scattering of the conduction electrons can be described by an energy-dependent potential of the following simple form:

$$v_{kk'}(\omega) = V^2 \left\{ \frac{A}{\omega} + \frac{1 - \tfrac{1}{2} A}{\omega + \tfrac{1}{2} U} + \frac{1 - \tfrac{1}{2} A}{\omega - \tfrac{1}{2} U} \right\} \qquad (2)$$

---

* Work supported in part by National Science Foundation under Grant No. GP20645.

where A is a positive real parameter which is determined by requiring that the total energy expressed as a functional of v be a minimum. This form for the potential is suggested by the fact that it satisfies certain exact symmetry requirements and has poles at the expected resonances in the model. These symmetry requirements are derived from the fact that the non-spin-flip t-matrix is related to the extraorbital-state Green's function by

$$t_{kk'}(\omega) = V^2 G_d(\omega) \tag{3}$$

and by definition

$$t = v + G^0 v t \tag{4}$$

where G^0 is the Green's function for the host metal. The form we have chosen for v gives the zeroth and first moments of the spectral function of $G_d(\omega)$ correctly, and also $\langle n_{d\sigma} \rangle = 1/2$. There is a pole at $-U/2$ to produce the expected resonance at the orbital state energy and also a pole at $U/2$ as required by the electron–hole symmetry. The remaining pole at the Fermi energy is necessary to produce a Kondo effect. This appears to be the simplest form for the potential which might be capable of producing the essential physics of the Anderson model.

The excess energy due to the impurity can be expressed as a functional of $G_d(\omega)$:

$$E(T) = -\int_{-\infty+i0^+}^{\infty+i0^+} d\omega\, f(\omega) \frac{\operatorname{Im}}{\pi} \left\{ \left(\omega - \frac{U}{2} - i\Delta\right) G_d(\omega) \right\} \tag{5}$$

It is a simple procedure to substitute for $G_d(\omega)$ the approximate expression obtained by the use of Eq. (2) and minimize $E(T)$ in order to evaluate A. In this way, one obtains a parameter $T_c(T) = A(T)\Delta$ which characterizes the width of the resonance in the t-matrix at the Fermi energy. Here Δ is the virtual state width parameter $\Delta = \pi V^2 N(0)$. At zero temperature we find

$$T_c(0) = \begin{cases} \frac{1}{2} U \exp(-\pi U/4\Delta) & \text{for } U/\Delta \gg 1 \\ \Delta & \text{for } U/\Delta \lesssim 1 \end{cases} \tag{6}$$

The value for $T_c(0)$ in the strongly magnetic limit ($U/\Delta \gg 1$) can be identified with the Kondo temperature parameter associated with an antiferromagnetic $J = 4V^2/U$. In this limit there are two broad resonances each of approximate width Δ located at $\pm U/2$, and a narrow resonance at the Fermi energy of width $2T_c(0)$. In the weakly magnetic limit ($U/\Delta \lesssim 1$) there is only a single resonance at the Fermi energy of width 2Δ in agreement with the original Hartree–Fock result of Anderson.[2] The temperature dependence of $T_c(T)$ is that of a function essentially temperature independent up to $T_c(0)$ and decreasing rapidly for $T > T_c(0)$.

The impurity contribution to the specific heat is given by $dE(T)/dT$. Representative curves for the specific heat are shown in Fig. 1. The low-temperature limit is given by

$$C = \frac{\pi}{3}\left(\frac{1}{T_c(0)} + \frac{1}{\Delta}\right) T \tag{7}$$

in units of Boltzmann's constant. The linear temperature dependence at low temperatures and the broad peak near $\frac{1}{2} T_c(0)$ is characteristic of the Kondo effect. The

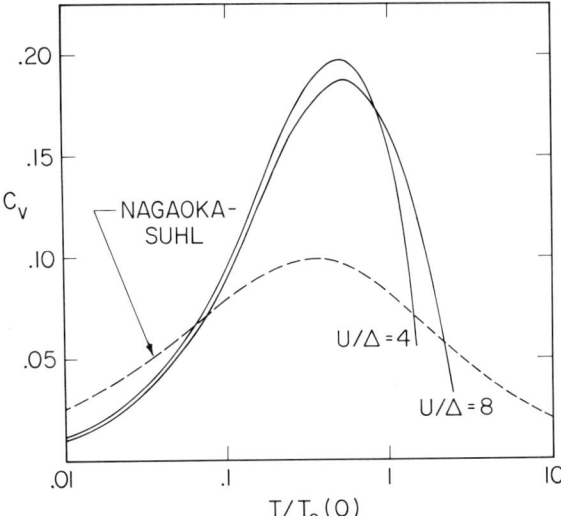

Fig. 1. Specific heat in units of Boltzmann's constant as a function of $T/T_c(0)$ for $U/\Delta = 4$ and 8. Dashed curve is the result of the Nagaoka–Suhl theory of the Kondo model with the Kondo temperature taken to be $T_c(0)$.

entropy change between zero and high temperatures is found to increase as U/Δ increases and approaches ln 2 in the strongly magnetic limit. The ln 2 value indicates a well defined spin degree of freedom at high temperatures which is removed from the system at zero temperature.

The magnetic susceptibility cannot be calculated in our treatment. One needs either the one-particle Green's function for the case of an applied magnetic field or certain two particle Green's functions. But the results for the entropy obtained from the specific heat contribution are not inconsistent with a susceptibility having the correct behavior. The entropy goes to zero as the temperature goes to zero, suggesting that the zero-temperature moment vanishes and the susceptibility is finite. For temperatures greater than $T_c(0)$ the entropy goes to a maximum value which approaches ln 2 as U/Δ increases, suggesting a Curie-law susceptibility with a moment which approaches the spin-1/2 value as U/Δ increases.

The resistivity has been calculated and typical results are shown in Fig. 2. One finds for the low-temperature limit

$$\rho(T) = (1 + \gamma)\rho_0 \left\{ 1 - \frac{\pi^2}{3}\left[\frac{T}{T_c(0)}\right]^2 (1 + \gamma)^{-1} \right\}$$

where ρ_0 is the so-called unitarity limit for the resonant scattering, and γ is the ratio of the relaxation time determined from the Anderson model to the relaxation time associated with scattering processes not included in the model.

These results are certainly reasonable in a qualitative sense and appear quite sound in quantitative detail. The strongly magnetic limit shows a Kondo effect characterized by a scale temperature $T_c(0)$ corresponding to an effective antiferro-

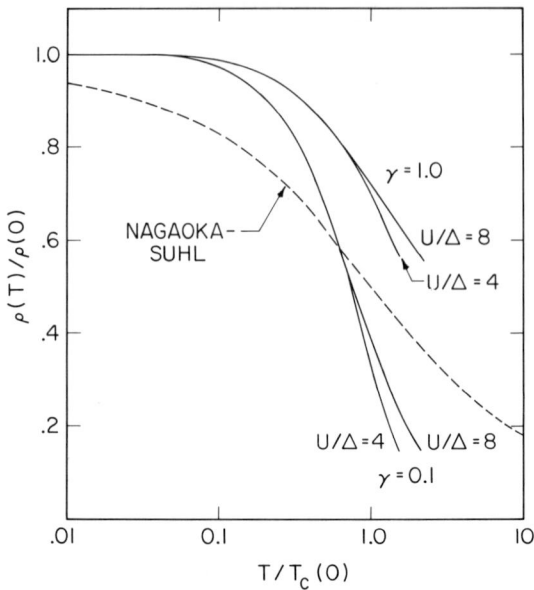

Fig. 2. Resistivity scaled by its value at $T = 0$ as a function of $T/T_c(0)$ for different values of U/Δ and the parameter γ. Dashed curve is the result of the Nagaoka–Suhl theory of the Kondo model with $\gamma = 0$ and the Kondo temperature taken to be $T_c(0)$.

magnetic exchange parameter J having the value predicted by Schotte.[3] Also, there are the broad resonances associated with the orbital state which are seen experimentally in optical and photoemission studies. In the weakly magnetic limit our results reduce to the previous Hartree–Fock results. The specific heat and resistivity have the desired T and T^2 dependences, respectively, as the temperature goes to zero. At temperatures of the order of $T_c(0)$ the behavior is similar to that derived from the Nagaoka–Suhl approximate theory of the Kondo model.[4]

Although the form of the model chosen is certainly an oversimplification for any real system, it is interesting to attempt a comparison with experiment. We have used our results to fit the specific heat of the Cu(Fe) system, which is generally considered to be intermediate between the weakly and strongly magnetic limits. The result is shown in Fig. 3. The data used are that of Triplett[5] for a 195-ppm Cu(Fe) sample. The parameters used in the fit are $U/\Delta = 8$ and $U = 2.5$ eV. Also, the theoretical expression is multiplied by the factor 1.8, which is close to the value $2 = \ln(2S + 1)/\ln 2$ with $S = 3/2$, which suggests itself on the basis of the entropy as a possible means of generalizing to a spin-3/2 impurity, such as Fe in Cu. As can be seen from Fig. 2, the expression for the specific heat as a function of $T/T_c(0)$ is rather insensitive to the value of U/Δ. Therefore quite a range of values for U and Δ could be used to fit the data. Nevertheless, it is encouraging that the data are well fitted using reasonable values for the parameters. This is in marked contrast to the results of the Nagaoka–Suhl approximate theory for the Kondo model, where,

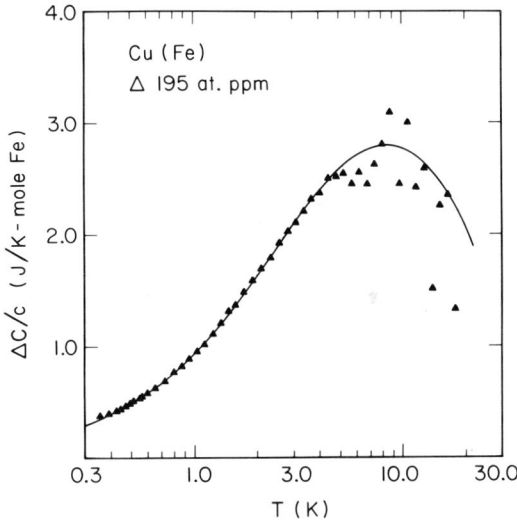

Fig. 3. Excess specific heat per mole of Fe of a 195-at.-ppm Cu(Fe) sample (from Ref. 5). Solid curve is the result of our treatment multiplied by 1.8 and using $U/\Delta = 8$ and $U = 2.5$ eV.

even apart from the incorrect low-temperature behavior, the theory yields much too broad a peak to fit the Cu(Fe) data. It appears that our simple treatment provides an adequate description throughout the entire temperature range of interest.

Acknowledgments

We wish to acknowledge the many contributions made by G. F. Abito and W. G. Delinger.

References

1. J.W. Schweitzer, G.F. Abito, and W.G. Delinger, University of Iowa Research Report 72:10.
2. P.W. Anderson, *Phys. Rev.* **124**, 41 (1961).
3. K.D. Schotte, *Z. Physik* **235**, 155 (1970).
4. K. Fischer, *Phys. Stat. Sol.* (b) **46**, 11 (1971).
5. B.B. Tripplett, Ph. D. Thesis, University of California, Berkeley, 1970, UCRL—19672.

Concentration Dependence of the Kondo Effect in a Random Impurity Potential

Michael W. Klein*

Department of Physics, Bar Ilan University
Ramat Gan, Israel

and

Y. C. Tsay and L. Shen

Department of Physics, Wesleyan University
Middletown, Connecticut

In a recent paper Tsay and Klein[1,2] examined the two-impurity Kondo effect[3] in the presence of an s–d Hamiltonian[4,5] and an impurity–impurity exchange interaction of the form $W_{01}\mathbf{S}_0\cdot\mathbf{S}_1$, where W_{01} is the interaction potential between the impurities located at positions R_0 and R_1 with spins $\mathbf{S}_0$ and $\mathbf{S}_1$. For the two-impurity system the result of our calculation is (in a so-called weak coupling approximation) that all the log T terms in the Kondo effect become renormalized by the second impurity as follows:

$$\log T \xrightarrow{\text{2-imp}} \log(T^2 + W'^2)^{1/2} \qquad (1)$$

where the arrow indicates that the log T term becomes renormalized by the two-impurity interaction. W'^2 in Eq. (1) is

$$W'^2 = \tfrac{2}{3}W_{01}\left\{W_{01}[S(S+1) + \langle\mathbf{S}_0\cdot\mathbf{S}_1\rangle] - \frac{J}{2N}\sum_{k,k'}\langle\sigma_{\alpha\beta}\cdot\mathbf{S}_1 c^+_{k\alpha}c_{k'\beta}\rangle \right.$$
$$\left. \times \exp i(\mathbf{k}-\mathbf{k}')\cdot(\mathbf{R}_0 - \mathbf{R}_1)\right\} \qquad (2)$$

where J is the s–d exchange constant and $c^+_{k\alpha}$ and $c_{k\alpha}$ are the conduction electron creation and annihilation operators with wave vector k and spin α.

Tsay and Klein[1,2] have obtained Eqs. (1) and (2) by decoupling the Green's functions arising from an equation of motion treatment. They used a Nagaoka-like decoupling scheme[6] which included all the most divergent terms of the Kondo effect.

We have extended the two-impurity calculation[2] to the many-impurity system, using a many-impurity s–d Hamiltonian with an interaction term of the form $\tfrac{1}{2}\sum_{i,j}W_{ij}\mathbf{S}_i\mathbf{S}_j$. This calculation was also done in the so-called weak coupling limit,

---

* Part of this work supported by AFOSR Contract No. F44620–71–C–0013 at Belfer Graduate School of Science, Yeshiva University, and part by the Bar Ilan University Research Committee.

Concentration Dependence of the Kondo Effect in a Random Impurity Potential

in which the conduction electron polarizations of the various impurities are assumed to be independent, and will be published elsewhere. We just quote here the essential conclusions arising from the calculation. We find that the renormalization in Eq. (1) is still valid, except that for the many impurity system W' given in Eq. (2) is replaced by

$$W_i'^2 = \tfrac{2}{3}\left\{\sum_j W_{ij}^2[S(S+1) + \langle \mathbf{S}_i \cdot \mathbf{S}_j\rangle] + \sum_j W_{ij}E_{ij}' + \sum_{j,k} W_{ij}W_{ik}\langle \mathbf{S}_j \cdot \mathbf{S}_k\rangle\right\} \quad (3)$$

where

$$E_{ij}' = -\frac{J}{2N}\sum_{k,k'}\langle \sigma_{\alpha\beta}\cdot \mathbf{S}_j c_{k\alpha}^+ c_{k'\beta}\rangle \exp i(\mathbf{k}-\mathbf{k}')\cdot(\mathbf{R}_i - \mathbf{R}_j) \quad (4)$$

The Effective Kondo Temperature

We recall that the Kondo temperature for a single magnetic impurity is defined by the relation $(J\rho/N)\log(T_k/D) = 1$, where ρ is the density of states at the Fermi surface, N is the number of sites in the solid, D is an energy of the order of the Fermi energy, and T_K is the Kondo temperature. In an analogous fashion, we define the effective Kondo temperature in the presence of many impurities by the relation

$$(J\rho/N)\log\{[T_K^E(i)]^2 + W_i'^2\} = 1 \quad (5)$$

where $T_K^E(i)$ is the effective Kondo temperature for an impurity located at site i. We note that since each impurity in the randomly distributed system has a different environment, $T_K^E(i)$ is a random variable and so is $W_i'^2$. Using Eq. (5), we obtain

$$T_K^E(i) = T_K[1 - (W_i'^2/T_K^2)]^{1/2}, \quad |W_i'| < T_K$$
$$= 0, \quad |W_i'| > T_K \quad (6)$$

The meaning of Eq. (6) is that the Kondo temperature is reduced by the impurity–impurity interation, and when $|W'|$ is greater than T_K the effective Kondo temperature becomes imaginary; i.e., the impurity–impurity interaction inhibits the formation of the spin-compensated state; thus a perturbation treatment of the Kondo effect is valid for all temperatures.

Since $T_K^E(i)$ is a random variable, we will be interested in obtaining $\langle T_K^E(i)\rangle$, where $\langle\ \rangle$ indicates an average over all sites in this system. In order to obtain $\langle T_K^E\rangle$, we have to derive the probability distribution of W'^2. This can be done in principle, but becomes very difficult in practice since we have to evaluate the probability distributions of the random spin correlation functions $\langle \mathbf{S}_i\cdot\mathbf{S}_j\rangle$ and $\langle \mathbf{S}_j\cdot\mathbf{S}_k\rangle$ and E_{ij}' in Eq. (3). Later we will find that it is of interest to obtain the effective Kondo temperature at very high temperatures, since this enters into the expression for the high-temperature magnetic susceptibility. We thus evaluate Eqs. (3) and (4) in the limit $T\to\infty$. In this limit $\langle \mathbf{S}_i\cdot\mathbf{S}_j\rangle \to 0$, $\langle \mathbf{S}_j\cdot\mathbf{S}_k\rangle \to 0$, and the approximate value of E_{ij}' becomes

$$E_{ij}' = -\frac{J}{2N}\sum_{k,k'}\frac{f(\varepsilon_k) - f(\varepsilon_{k'})}{\varepsilon_k - \varepsilon_{k'}}[\exp i(\mathbf{k}-\mathbf{k}')\cdot(\mathbf{R}_i - \mathbf{R}_j)S(S+1)] = \tfrac{1}{2}v_{ij}S(S+1) \quad (7)$$

where v_{ij} is the well-known RKKY[4,5,7] interaction, $f(\varepsilon_k)$ is the Fermi function for an electron with energy ε_k, and the temperature under consideration is assumed to be much less than the Fermi energy ε_F. The expression for W'^2 thus becomes

$$W_i'^2 = \tfrac{2}{3} \sum_j (W_{ij}^2 + \tfrac{1}{2} v_{ij}^2) S(S+1) \tag{8}$$

Up to now we have not specified the form of the interaction potential W_{ij}. We now assume that W_{ij} is the RKKY potential. Then $(W'^2) = S(S+1) \sum_j v_{ij}^2$. We write

$$y \equiv W_i'^2 = S(S+1) \sum_j v_{ij}^2 \tag{9}$$

To calculate the probability distribution of y, we fix an impurity at site i and allow all other impurities to be randomly distributed over the sites of the solid. Since it does not matter where in the solid we pick the impurity i, the probability distribution of $W_i'^2$ (but not $W_i'^2$ itself) will be independent of i. We allow the impurities to be randomly, uniformly, and *independently* distributed over the volume of the crystal V, with a probability V^{-1} (this allows for multiple occupancy of the sites, but the error introduced is of order c^2 and is negligible as $c \to 0$, where c is the fractional impurity concentration). To write a formal expression for the probability distribution of $W_i'^2$, we use the statistical model of Margenau[8] with the notation and method used by Klein.[9,10] With the above approximations the expression for the probability distribution of $y, P(y)$, is given in analogy with Eq. (2.12) of Ref. 9 as follows:

$$P(y) = \frac{1}{2\pi} \int_{-\infty}^{\infty} e^{i\rho y} dy \left\{ \frac{1}{V} \int_0^{\infty} 4\pi r^2 \, dr \exp\left[\frac{-i\rho S(S+1) a^2}{r^6}\right] \right\}^N \tag{10}$$

where we are approximating the square of the RKKY interaction $v^2(r)$ by the relation

$$v^2(r) = a^2/r^6 \tag{11}$$

where $|a|$ is the strength of the RKKY interaction at a distance of one lattice constant. (Note that the $\cos^2 2k_F r_{ij}$ term in the square of the RKKY interaction has been replaced by its average value.) Following the steps given in Eqs. (2.12)–(2.15) of Ref. 9, we obtain

$$P(y) = \frac{2}{\pi} \int_0^{\infty} \cos(x^2 y - Bx) e^{-Bx} x \, dx \tag{12}$$

where

$$B = \tfrac{2}{3}\sqrt{2}\pi^{3/2} [S(S+1)]^{1/2} |a| n_0 c \tag{13}$$

where n_0 is the number of sites per unit cell. Equation (13) can be readily integrated[11,12] to give

$$P(y) = (B/\sqrt{2\pi}) y^{-3/2} \exp(-B^2/2y) \tag{14}$$

Using Eq. (14) with Eq. (6) gives

$$\langle T_K^E \rangle = \int_0^{T_K^2} P(y) T_K(y) \, dy$$

$$= T_K [1 - \sqrt{\pi} \alpha + \alpha^2 - 0(\alpha^4)] \tag{15}$$

where $\alpha = (1/\sqrt{2})(B/T_K)$. Equation (15) shows that for any nonzero concentration the effective Kondo temperature differs from the single-impurity Kondo temperature and that for very low concentration $[T_K - \langle T_K^E \rangle]$ is proportional to c.

Comparison With Experiment

An effective Kondo temperature [different from Eq. (15)] has been experimentally obtained from the magnetic susceptibility per impurity χ by Souletie and Tournier.[13] They used the relation $\chi = S(S+1)/(T + T_K^E)$ to determine T_K^E. Now since the impurities are randomly distributed, T_K^E is a random variable and χ is given by

$$\langle \chi \rangle = S(S+1) \int_0^\infty \frac{P(y)\,dy}{T + T_K^E(y)} \tag{16}$$

where $T_K^E(y)$ is given in Eq. (6). Integrating Eq. (16), we obtain for

$$\langle \chi \rangle = \frac{S(S+1)}{T + T_K} \left\{ 1 + \frac{T_K}{T}\operatorname{erf}(\alpha) + \left(\frac{T_K}{T + T_K}\right)\left[1 - \operatorname{erf}(\alpha) - \frac{\langle T_K^E \rangle}{T_K}\right] \right.$$
$$\left. + \left(\frac{T_K}{T_K + T}\right)^2 (2)\left[(1 + \alpha^2)[1 - \operatorname{erf}(\alpha)] - \frac{\langle T_K^E \rangle}{T_K^2} - \frac{\alpha \exp - \alpha^2}{\sqrt{\pi}}\right]\right\} \tag{17}$$

$$\operatorname{erf}(\alpha) = (2/\sqrt{\pi}) \int_0^\infty \exp - t^2\,dt$$

We next compare Eq. (17) with experiment. First we define the quantity U by the relation

$$\frac{U}{T_K} = \frac{S(S+1)}{T_K} \langle \chi \rangle^{-1} - \frac{T}{T_K} \tag{18}$$

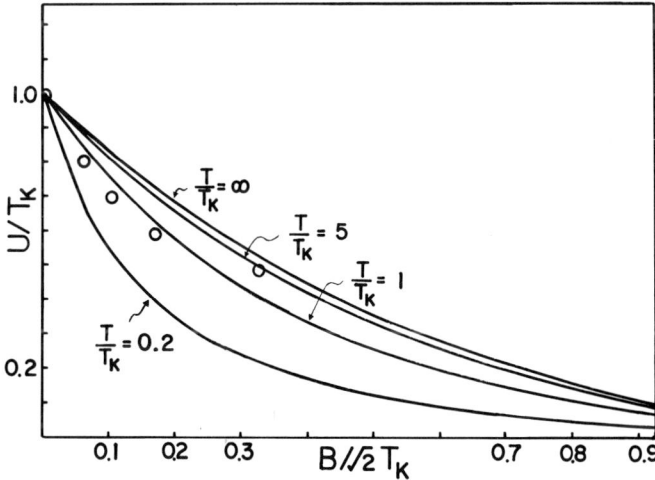

Fig. 1. U/T_K vs. $B/\sqrt{2}T_K$ for different relative temperatures T/T_K. Circles indicate experimental results of Tournier and co-workers[13] for $T \sim T_K$.

Then U/T_K is the effective Kondo temperature (normalized to unity) obtained in Fig. 8 of Souletie and Tournier.[13] We note that U is not to be confused with $\langle T_K^E \rangle$ given in Eq. (15). However, $\lim_{T \to \infty} U \to \langle T_K^E \rangle$, as can be seen from Eqs. (16) and (17). In Fig. 1 we show the quantity U/T_K vs. $(1/\sqrt{2}) B/T_K$ for various values of T/T_K. On the same graph we have also plotted the experimental results of Souletie and Tournier. We thus see from the figure that there is a reasonable qualitative agreement between theory and experiment. The value of the $s-d$ interaction obtained from the value of B is approximately 0.14 eV, which is again not unreasonable.

We note that the theory predicts a change in U with increasing temperature. This prediction could possibly be tested experimentally in the future (At low temperatures ordering sets in, and the spin correlations cannot be neglected.) Our results also show that there is no need to assume a clustering of the impurities, as was done by Souletie and Tournier, to explain the data.

References

1. Y.C. Tsay and M.W. Klein, *1971 Conf. on Magnetism*, H.C. Wolfe, ed. American Institute of Physics, Chicago, Ill., # 5, p. 1145.
2. Y.C. Tsay and M.W. Klein, *Phys. Rev.* (to be published).
3. J. Kondo, *Progr. Theor. Phys. (Kyoto)* **32**, 37 (1964).
4. T. Kasuya, *Progr. Theor. Phys. (Kyoto)* **16**, 45 (1956).
5. K. Yosida, *Phys. Rev.* **106**, 893 (1957).
6. Y. Nagaoka, *Progr. Theor. Phys.* **37**, 13 (1967).
7. M.A. Ruderman and C. Kittel, *Phys. Rev.* **96**, 99 (1954).
8. H. Margenau, *Phys. Rev.* **48**, 755 (1935).
9. M.W. Klein, *Phys. Rev.* **173**, 552 (1968).
10. M.W. Klein, this volume.
11. I.S. Gradsteyn and I.M. Ryzhik, *Table of Integrals Series and Products*, Academic Press, New York, (1966).
12. H. Bateman, *Table of Integral Transforms*, McGraw-Hill, New York (1954), Vol. 1.
13. J. Souletie and R. Tournier, *J. Phys.* **32**, (C2), 172 (1971); J.L. Tholence and R. Tournier, *Phys. Rev. Lett.* **25**, 867 (1970).

Concentration Effects in the Thermopower of Kondo Dilute Alloys

K. Matho

*Centre de Recherches sur les Très Basses Températures
CNRS, Grenoble-Cedex, France*

and

M. T. Béal-Monod

*Department of Physics, University of California—San Diego
La Jolla, California
and
Laboratoire de Physique des Solides
Université de Paris-Sud, Orsay, France*

We have studied conduction electron relaxation in the presence of interacting magnetic impurities by perturbation theory.[1]

The model Hamiltonian[2] $H = H_0 + H'$ has as its unperturbed part

$$H_0 = \sum_{\mathbf{k}\alpha} \varepsilon_\mathbf{k} c^+_{\mathbf{k}\alpha} c_{\mathbf{k}\alpha} - W \mathbf{S}_1 \cdot \mathbf{S}_2 \tag{1}$$

describing (i) a band of Bloch electrons in momentum states $\mathbf{k}$ and spin states α, with one-particle energy $\varepsilon_\mathbf{k}$ and one-particle creation (destruction) operators $c^+_{\mathbf{k}\alpha} (c_{\mathbf{k}\alpha})$; and (ii) an interaction energy W for a pair of localized spins $\mathbf{S}_\nu$, $\nu = 1, 2$. The spatial variation of W with the pair distance $R = |\mathbf{R}_1 - \mathbf{R}_2|$ is supposed to be of the Ruderman–Kittel (RKKY) type. The sign of W may be positive (f, for a ferromagnetic pair) or negative (af, for an antiferromagnetic pair).

The perturbation

$$H' = (1/N) \sum_{\substack{\mathbf{k}'\mathbf{k}\alpha'\alpha \\ \nu=1,2}} (V - J\sigma_{\alpha'\alpha} \cdot \mathbf{S}_\nu) \{\exp[i(\mathbf{k}' - \mathbf{k}) \mathbf{R}_\nu]\} c^+_{\mathbf{k}'\alpha'} c_{\mathbf{k}\alpha} \tag{2}$$

represents local potential (V) and exchange (J) scattering on the two impurity sites. Here N is the number of crystal sites and $\sigma_{\alpha'\alpha}$ the vector of Pauli spin matrices.

Closely following previous resistivity calculations,[2-4] we have studied the diffusion thermoelectric power S_d in this model. An approximae formula

$$S_d \simeq -k_B \langle \beta \varepsilon \tau^{-1}_{\text{th}}(\varepsilon) \rangle_{(\varepsilon)} \Big/ e \langle \tau^{-1}_\rho(\varepsilon) \rangle_{(\varepsilon)}, \quad \beta = (k_B T)^{-1} \tag{3}$$

is employed. Here the total scattering rate of conduction electrons due to the perturbation (2)

$$\tau^{-1}(\varepsilon) = \tau^{-1}_\rho(\varepsilon) + \tau^{-1}_{\text{th}}(\varepsilon) \tag{4}$$

is decomposed into a part varying symmetrically (ρ) and a part varying antisymmetrically (th) in energy across the Fermi surface $\varepsilon = 0$. The low-order perturbational contributions to τ_{th}^{-1} are proportional to VJ^2 and VJ^3. The use of the scattering rates τ^{-1} instead of relaxation time τ in formula (3) is justified when $\tau_\rho^{-1}(\varepsilon)$ contains a large constant contribution due to potential scattering, $V^2 \gg J^2$.[3] Boltzmann's constant k_B over the electron charge $e < 0$ represents a giant negative thermopower. The symbol $\langle \cdots \rangle_{(\varepsilon)} = \int d\varepsilon (-df/d\varepsilon) \cdots$ means an energy integration weighted by the derivative of the Fermi factor.

The antisymmetric contribution

$$\tau_{th}^{-1} = \tau_K^{-1} + \tau_R^{-1} \tag{5}$$

contains two distinct parts causing different effects in the thermopower $S_d = S_K + S_R$. The part S_K is obtained from the single-impurity result[5] in the $W = 0$ limit,

$$S_{0,K} = (k_B \pi^2 \gamma^3 / 2en_0 V) S(S+1) \tag{6}$$

by replacing the spin factor $S(S+1)$ by an effective spin amplitude.[4] In (6), n_0 is the density of states and $\gamma = 2Jn_0 < 0$ is the dimensionless s–d exchange scattering parameter. The effective spin amplitude in the thermopower is given by

$$S_{th}^2(\beta W) = \tfrac{1}{2} \ll \sum_{\Delta = \pm 1} \beta\varepsilon [\tanh \tfrac{1}{2}\beta\varepsilon + \tanh \tfrac{1}{2}\beta(\varepsilon + E_\Delta)] S_{eff}^2(j, \Delta, \varepsilon) >_{(\varepsilon)} > \tag{7}$$

The symbol $\langle \cdots \rangle$ without a subscript means the thermal average over the spin states of the pair, and $\Delta = j' - j$, is the difference in the total spin quantum numbers that may be reached by s–d exchange scattering from a given j. The associated energy difference E_Δ is proportional to W, and $S_{eff}^2(j, \Delta, \varepsilon)$ is the same matrix element derived previously to express the effective spin amplitude in the resistivity,[4]

$$S_{eff}^2(\beta W) = \ll S_{eff}^2(j, \Delta, \varepsilon) >_{(\varepsilon)} > \tag{8}$$

Figure 1 compares $S_{th}^2(\beta W)$ with $S_{eff}^2(\beta W)$ for a spin value $S = 2$ and shows that the reduction of $S(S+1)$ is essentially the same in both cases. In the range $k_B T \simeq |W|$

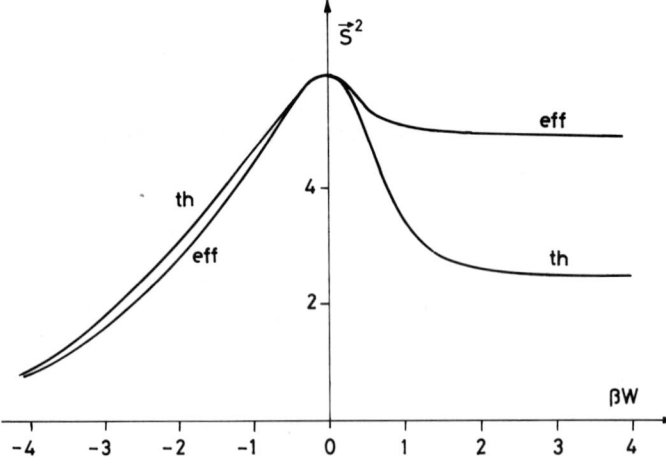

Fig. 1. The effective spin amplitude in the thermopower (th) and in the resistivity (eff), formulas (7) and (8).

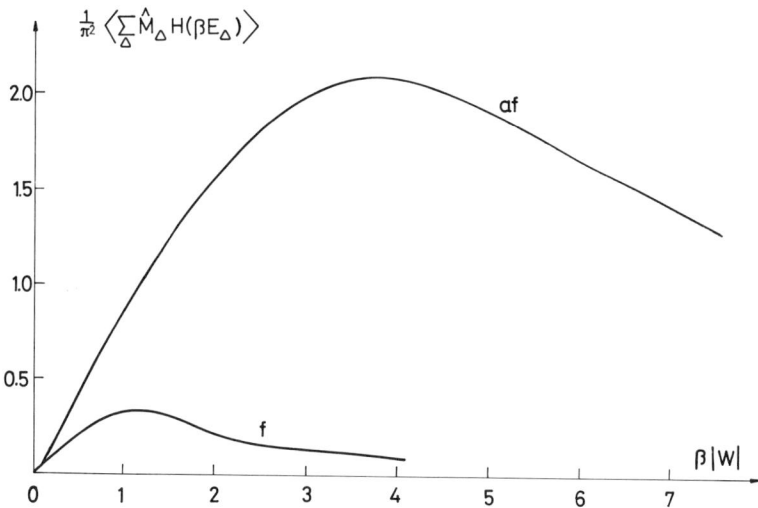

Fig. 2. Thermal average occurring in S, Eq. (10), for ferromagnetic (f) and antiferromagnetic (af) sign of the coupling energy W.

the T dependence thus caused by impurity interaction dissembles the logarithmic T dependence showing up[5] in the isolated $S_{0,K}$ to higher orders in γ.

The second contribution S_R to the diffusion thermoelectric power is due to a resonant effect[1] in the relaxation rate $\tau_R^{-1}(\varepsilon)$ which arises when all three energies $|\varepsilon|$, $|E_\Delta|$, and $k_B T$ are of the same order of magnitude and which vanishes otherwise. Its leading term in the perturbation series occurs already in order VJ^2. A similar effect has been found by Peschel and Fulde[6] for crystal-field-split impurities. Their function

$$H(\beta\delta) \propto \langle \beta\varepsilon [g(\varepsilon + \delta) - g(\varepsilon - \delta)] \rangle_{(\varepsilon)} \qquad (9)$$

with the splitting energy δ and the difference of two anomalous energy functions[7] $g(\varepsilon)$ permits us to write

$$S_R = \frac{k_B \gamma^2}{4en_0 V} \left(1 + \frac{3}{2}\gamma \ln \frac{T}{T_0}\right) \sum_\Delta \langle \hat{M}_\Delta H(\beta E_\Delta) \rangle \qquad (10)$$

T_0 is a temperature of the order of the Fermi temperature. The operators $\hat{M}_\Delta$, $\Delta = \pm 1$, have been defined previously.[4] They have the property

$$\sum_\Delta \hat{M}_\Delta = (\mathbf{S}_1 - \mathbf{S}_2)^2 = 2[S(S+1) - \mathbf{S}_1 \cdot \mathbf{S}_2] \qquad (11)$$

For $\gamma < 0$ S_R is opposite in sign to S_K, irrespective of the sign of V. Equation (10) shows the beginning of a new divergent Kondo series which must be summed to all orders in γ. Again, the $\ln T$ dependence of this factor is dissembled by the T dependence of the thermal average. This average is shown for $S = 2$ in Fig. 2 as a function of βW.

As a further step, the interaction effects have been averaged over the possible couplings W in a dilute random alloy of magnetic impurity concentration C under

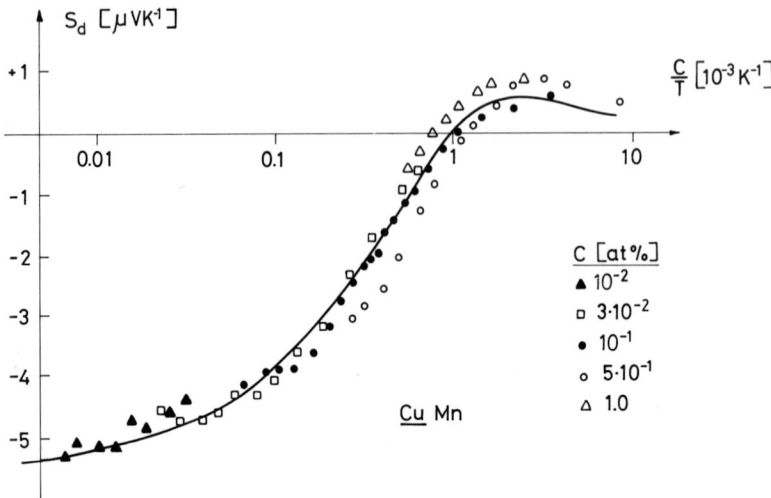

Fig. 3. Thermopower of CuMn as a function of C/T.[8]

the assumptions: (i) The module $|W|$ of the pair coupling energy decays proportional to R^{-3}, and (ii) The probabilities of finding either a ferromagnetic or an antiferromagnetic coupling are $p_f = p_{af} = \frac{1}{2}$, independent of C.

The averaged thermopower $S_d(C, T) = \overline{S_K + S_R}$ is a function of C/T if $\ln T$ variations in the prefactors of (6) and (10) are neglected. Figure 3 shows $S_d(C/T)$ together with experimental points from the measurements of Kjekshus and Pearson[8] on CuMn. The theoretical curve contains three parameters related to $n_0 V$, $n_0 J$, and the amplitude of the RKKY interaction.[4,9]

A very simple model thus explains the principal features of the low-temperature thermoelectric power in dilute magnetic alloys when the electron-phonon interaction is negligible. The details of the theory outlined here will be published elsewhere.[9]

References

1. K. Matho, Thesis, Grenoble, 1972, unpublished.
2. M.T. Béal-Monod, *Phys. Rev.* **178**, 874 (1969).
3. K. Matho and M.T. Béal-Monod, *J. Phys. (Paris)* **32**, 213 (1971).
4. K. Matho and M.T. Béal-Monod, *Phys. Rev. B* **5**, 1899 (1972).
5. K. Fischer, *Z. Physik* **225**, 444 (1969).
6. I. Peschel and P. Fulde, *Z. Physik* **238**, 99 (1970).
7. J. Kondo, in *Solid State Physics*, Vol. 23, Academic Press, New York (1970), p. 183.
8. A. Kjekshus and W.B. Pearson, *Can. J. Phys.* **40**, 98 (1962).
9. K. Matho and M.T. Béal-Monod (to be published).

Resistance and Magnetoresistance in Dilute Magnetic Systems

J. Souletie

Centre de Recheches sur les Très Basses Températures
CNRS, Grenoble-Cedex, France

Within the framework of the Hartree–Fock theory of dilute alloys developed by Friedel and Blandin[1] a different phase shift ($\eta_l^\uparrow \neq \eta_l^\downarrow$) is assumed for electrons parallel (↑) or antiparallel (↓) to **OZ**, the spin axis:

$$\rho^{\uparrow(\downarrow)} = (4\pi c/k_F p) \sum_l (l+1) \sin^2(\eta_l^{\uparrow(\downarrow)} - \eta_{l+1}^{\uparrow(\downarrow)}) \tag{1}$$

We use atomic units $e = \hbar = m = 1$; p is the number of electrons per atom of the matrix; c is the concentration; $4\pi c/k_F p$ is 0.38 nΩ-cm $\times\ c$ (ppm) for a Cu matrix and 0.43 nΩ-cm $\times\ c$ (ppm) for a Au or a Ag matrix.

Following Korringa and Gerritsen[2] we may express the resistivity of up $(+)$ or down $(-)$ electrons when the quantization axis is the field axis **Oz**, and when the impurity is at angle Θ with the field (**Oz, OZ** $-\Theta$):

$$\rho^{\pm} = \tfrac{1}{2}(\rho^\uparrow + \rho^\downarrow) \pm \tfrac{1}{2}(\rho^\uparrow - \rho^\downarrow)\cos\Theta \tag{2}$$

The averaged polarization $\langle\cos\Theta\rangle_{av}$ will be taken as $M/2S$, the ratio of the measured magnetization per impurity atom to the magnetic moment of the impurity. By adding the conductivities of up and down electrons $\sigma = \tfrac{1}{2}\sigma^+ + \tfrac{1}{2}\sigma^-$ the resistivity follows:

$$\rho = \tfrac{1}{2}(\rho^\uparrow + \rho^\downarrow) - [(\rho^\uparrow + \rho^\downarrow)/2]^{-1}[(\rho^\uparrow - \rho^\downarrow)/2]^2 (M/2S)^2 \tag{3}$$

yielding a field-independent value $\rho_{H=0}$ and a negative magnetoresistance $\Delta\rho_H$ proportional to the square of the magnetization.[2] ρ may be expressed in terms of the $\eta_l^{\uparrow(\downarrow)}$. Making the same assumptions as in a previous paper,[3] we will (i) neglect all phase shifts for $l > 2$ when dealing with d impurities; (ii) suppose that magnetic effects ($\eta_l^\uparrow \neq \eta_l^\downarrow$) are restricted to the $l = 2$ partial waves; and (iii) use the sum ($\eta_2^\uparrow + \eta_2^\downarrow = 2\eta_2$) and the difference ($\eta_2^\uparrow - \eta_2^\downarrow = 2\xi\eta_2$) of the phase shifts (rather than the phase shifts themselves) as the variables because they are related to immediately accessible quantities.[3] η_2 is related to the charge difference Z between the matrix and the impurity by the Friedel sum rule:

$$Z = \sum_l [2(2l+1)/\pi]\eta_l = (2/\pi)\eta_0 + (6/\pi)\eta_1 + (10/\pi)\eta_2 + \cdots$$

$$= Z_s + Z_p + Z_d$$

$2\xi\eta_2$ is related through the same relation to the screening charge difference $Z_d^\uparrow - Z_d^\downarrow$ between the parallel and antiparallel directions, i.e., the magnetic moment of the impurity, $2\xi\eta_2 = 2S\pi/5$.

In addition, we will suppose[3] that S or ξ is an effective, temperature- and field-dependent quantity which may eventually be derived from experiment: from the susceptibility in the low H/T limit by letting $\chi_{\text{exp}} = g^2\mu_B^2 S_{\text{eff}}(S_{\text{eff}} + 1)/3kT$, or from the saturation magnetization in the large H/T limit, by taking $M = 2S_{\text{eff}}$. It should be noted that since the experimental susceptibility is observed to increase less rapidly than T^{-1} for temperatures below T_K (the Kondo or the fluctuation temperature) S_{eff} determined in this way, is bound to vary from a maximum value S_{max} for $T \gg T_K$ to zero when $T \ll T_K$. It is hoped in our phenomenological approach that all the difficulties inherent in the description of the magnetic state of the impurity will be contained in the parameter $S(H, T)$, a quantity which will be, when possible, eliminated or, if necessary, derived from the experimental magnetization. This method was successful in a previous attempt,[3] as well as in similar independent approaches[4,5] which, taken together, cover a wide variety of alloys of transition impurities in noble and transition matrices.

Derivation and Comments

From (1) and (3) we get[3] in zero field

$$(4\pi c/k_F p)^{-1} \rho_{H=0} = A - B \cos 2\xi\eta_2 \qquad (4)$$

$$A = \tfrac{5}{2} + \sin^2(\eta_0 - \eta_1), \qquad B = \tfrac{3}{2}\cos 2\eta_2 + \cos 2(\eta_1 - \eta_2) \qquad (5)$$

The resistance of a 17-ppm Au–Fe alloy[6] is shown in Fig. 1 vs. the temperature (50 m°K $< T <$ 5°K). The same data are also represented vs. $\cos(2S_{\text{eff}}\pi/5)$. S_{eff} was deduced from the experimental susceptibility[7] from

$$g^2\mu_B^2 S_{\text{eff}}(S_{\text{eff}} + 1)/3kT = \chi_{\text{exp}} \sim 3.25^2 \mu_B^2/3k(T + 0.4°K)$$

The slope B of the straight line obtained is compared in Table I with the prediction of (5) in the somewhat ideal case where $\eta_1 = \eta_0 = 0$ and $\eta_2 = Z\pi/10$. The agreement is remarkable for this system as well as for other systems for which equivalent results are also listed.[3]

For $H \neq 0$ a magnetoresistance term appears due to the modifications of ξ and M under the action of H. In the $H \ll T$ limit ξ is determined by the temperature and may be considered a constant. The magnetoresistivity $\Delta\rho_H$ is dominated by the freezing of the spin ordering; $\Delta\rho_H$ is negative and proportional to M^2. From (1) and (3)

$$(4\pi c/k_F p)^{-1} (\rho_{H=0}|\Delta\rho_H|)^{1/2} \sim (\pi/5)|D| M (\sin 2\xi\eta_2)/2\xi\eta_2$$

$$\sim |D|(\sin 2\xi\eta_2) M/2S_{\text{eff}} \qquad (6)$$

$$|D| = \tfrac{3}{2}\sin 2\eta_2 - \sin 2(\eta_1 - \eta_2) \qquad (7)$$

The coefficient of M in (6) decreases when T rises from $T \ll T_K$, where $(\sin 2\xi\eta_2)/2\xi\eta_2 = 1$, to $T \gg T_K$, where $2\xi\eta_2 \sim 2S_{\text{eff}}\pi/5$ may be estimated from the susceptibility. Table I shows that the values for D obtained from available data on

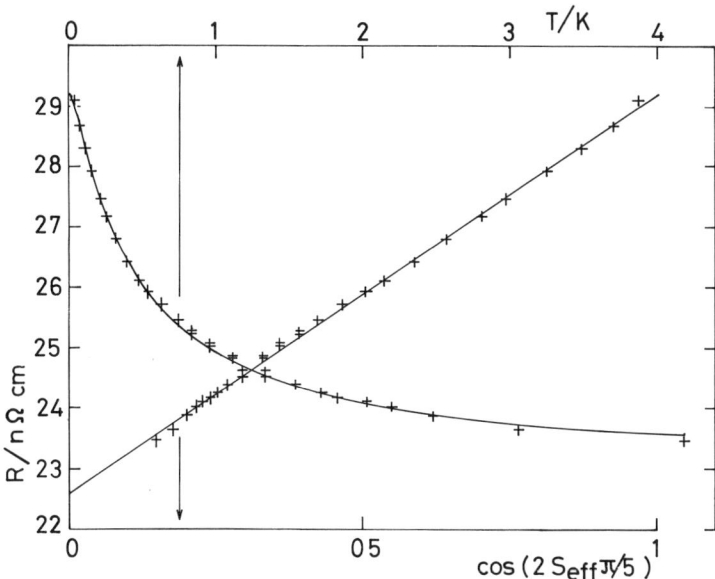

Fig. 1. Resistance of a 17-ppm Au–Fe alloy vs. temperature between 50 m°K and 4.2°K.[6] The same data are also shown vs. $\cos(2S_{\text{eff}}\pi/5)$, S_{eff} being determined from the experimental susceptibility.[7]

Cu–Fe,[8] Au–Fe,[9] and Cu–Mn[8] compare in magnitude with the expectation from (7), assuming $\eta_1 = \eta_0 = 0$, $\eta_2 = Z\pi/10$.

In the $H > T$ limit the freezing of the spin ordering is completed and $M = 2S_{\text{eff}}$. The variations of the resistance (and of the magnetization) show the freezing of the scattering amplitude, i.e., the modifications of ξ due to the field. For sufficient H/T, ξ becomes a function H only and the resistance in a constant field saturates to a temperature-independent value when $T \to 0$, as has been reported, for example, in Cu–Cr[10] and Au–Fe.[11] Both terms on the rhs of Eq. (3) contribute to the magnetoresistance. The total resistance is then $\rho_{H>T}$ such that

$$\left(\frac{4\pi c}{k_F p}\right)^{-1} \rho_{H>T} = \left(A - B\cos\frac{M\pi}{5}\right)\left(1 - \left[\frac{D\sin(M\pi/5)}{A - B\cos(M\pi/5)}\right]^2\right) \quad (8)$$

Table I. Comparison of Experimental Determinations for B (Ref. 3) and D (Refs. 8 and 9) with Expectations from Eqs. (5) and (7) in the Approximation $\eta_1 = \eta_0 = 0$, $\eta_2 = Z\pi/10$

	Au–V	Cu–Cr	Cu–Fe	Au–Fe	Cu–Mn
B	−1.1 to −1.5	−2.5	−0.85	−0.8	—
$\frac{5}{2}\cos(Z\pi/5)$	−2.02	−2.5	−0.772	−0.772	−2.02
D	—	—	1	1.25	0.75
$\frac{5}{2}\sin(Z\pi/5)$	1.47	0	2.38	2.38	1.47

Table II. Phase Shifts, Corresponding Screening Contributions, and Spin Values*

Alloy	η_2	Z_d	η_1	Z_p	η_o	Z_s	S_{max}	S_χ	S_M
Cu–Fe	$3\pi/4$	7.25	$-\pi/3$	-2	0.87π	1.75	1.25	1.27	—
Au–Fe	$3\pi/4$	7.25	$-\pi/3.25$	-1.75	$0.8\ \pi$	1.6	1.25	1.2	1.1

* In the last three columns the maximum value expected for the spin S_{max} is compared with the experimental results[7,12] for the susceptibility at $T \gg T_K$, S_χ, and the saturation magnetization, S_M.

an expression which provides the following relation between $\rho_{H=0}(T=\infty)$ and $\rho_{H=\infty}(T=0)$, the high-temperature, low-field and the low-temperature, high-field limits of the resistivity:

$$(4\pi c/k_F p)^{-2} \rho_{H=0}(T=\infty) [\rho_{H=0}(T=\infty) - \rho_{H=\infty}(T=0)] = +D^2 \sin^2 2\eta_2 \quad (9)$$

It is in principle possible to determine η_2 and η_1 from a couple B, $|D|$; (for example, $3 \cos 2\eta_1 = B^2 + D^2 - 3.25$). Table II shows that for Au–Fe and Cu–Fe this determination did not violate Friedel's sum rule and resulted in values for η_2 in close agreement with the data[7,12] on the susceptibility and the saturation magnetization. [Recall that for $T \gg T_k$, $S_{eff} \sim (5/\pi) \eta_2$ while $M_{sat} \sim (10/\pi) \eta_2$.] The estimates for η_1 and η_0 appear, however, somewhat larger than expected.

The justification of a negative magnetoresistance proportional to M^2, the separation of a contribution due to the freezing of the spin ordering, and of a contribution due to the freezing of the scattering amplitude, has already been derived in terms of the s–d model[13] using perturbation theory. A determination of parameters J and V in agreement with the resistivity results was obtained. It seems worthwhile to point out the fact that the present elementary application of the results of Refs. 1 and 2 is able to qualitatively and quantitatively account for the same results. A deeper comparison of the two approaches could help in giving a more formal justification of the treatment which is presently made in terms of an ad hoc effective spin value. It could give more insight into the physical meaning of the quantities J and V.

The extended validity of this phenomenological model, which does not seem to be restricted to alloys of the first transition row with noble metals,[4,5] argues in favor of more unity in the interpretation of the transition from magnetism to nonmagnetism than is presently the case.

References

1. A. Blandin and J. Friedel, *J. Phys. Radium* **20**, 160 (1959).
2. J. Korringa and A.N. Gerritsen, *Physica* **19**, 457 (1953).
3. J. Souletie, *J. Low Temp. Phys.* **7**, 141 (1972).
4. J.W. Loram, R.J. White, and A.D.C. Grassie, *Phys. Rev. B* **5**, 3659 (1972).
5. H. Nagasawa, *Sol. St. Comm.* **10**, 33 (1972).
6. O. Laborde and P. Radhakrishna, private communication.
7. J.L. Tholence and R. Tournier, 7th Conf. Intern. de Magnétisme, 1970, *J. de Physique Suppl.*, **32**, (2-3), Cl-211 (1971).
8. P. Monod, *Phys. Rev. Lett.* **19**, 113 (1967).
9. F. LaPierre, Thèse de 3ème Cycle, Université de Grenoble, unpublished.

10. M.D. Daybell and W.A. Steyert, *Phys. Rev. Lett.* **20**, 195 (1968).
11. R. Berman, J. Kopp, and C. T. Walker, in *Proc. 11th Intern. Conf. Low Temp. Phys. 1968*, St. Andrews University Press, Scotland (1969), pp. 1238–1241.
12. J.L. Tholence and R. Tournier, *Phys. Rev. Lett.* **25**, 867 (1970).
13. M.T. Béal-Monod and R.A. Wiener, *Phys. Rev.* **170**, 552 (1968).

The Linear Variation of the Impurity Resistivity in AlMn, AlCr, ZnFe, and Other Dilute Alloys

E. Babić

Institute of Physics of the University of Zagreb
Zagreb, Yugoslavia

and

C. Rizzuto

Instituto di Scienze Fisiche e Gruppo Nazionale
Struttura della Materia, Genova, Italy

A number of dilute alloys of transition or rare earth metals in simple hosts have recently been found to show a low-temperature quadratic behavior of the impurity resistivity, among them AlMn,[1] AuV,[2] CuFe,[3] PtCr,[4] YCe,[5] ZnFe,[6] and ZnCr.[7]

In all these cases the low-temperature impurity resistivity can be fitted to a law of the type

$$\rho(T) = \rho(0)\left[1 - (T/\theta_R)^2\right] \quad (1)$$

The values of $\rho(0)$ and θ_R for these alloys have been experimentally evaluated and are given in Table I.

In the AlMn system the quadratic behavior is found to extend up to about 100°K ($\simeq 0.2\theta_R$); above this temperature and up to about 350°K a linear behavior is followed.[8-10]

The accuracy in measuring $\rho(T)$ to higher temperatures in the other alloy systems is in general limited by the inaccurate knowledge of the phonon contribution to the total measured resistivity of the alloy. However, it has recently been suggested[3,10,11] that by comparing alloys with different impurity contents and considering the behavior of the phonon contribution with impurity content, it is possible to account for the deviations from Matthiessen's rule. It has also been found that these deviations are relatively less strong when the impurity resistivity dominates the phonon resistivity.

We have collected from the literature data on systems whose impurity resistivity has been derived from the total measured resistivity in cases which satisfy the above conditions, and we have also extended previous measurements on the resistivity of ZnFe alloys (of concentration 0.1 and 0.3 at. %) and one AlCr alloy ($c = 1$ at. %), which were obtained as described in Ref. 1, to 100°K, and subtracted the phonon contribution as in Ref. 10.

The data are reported in Fig. 1 and are compared by scaling them according to

Linear Variation of Impurity Resistivity in Dilute Alloys

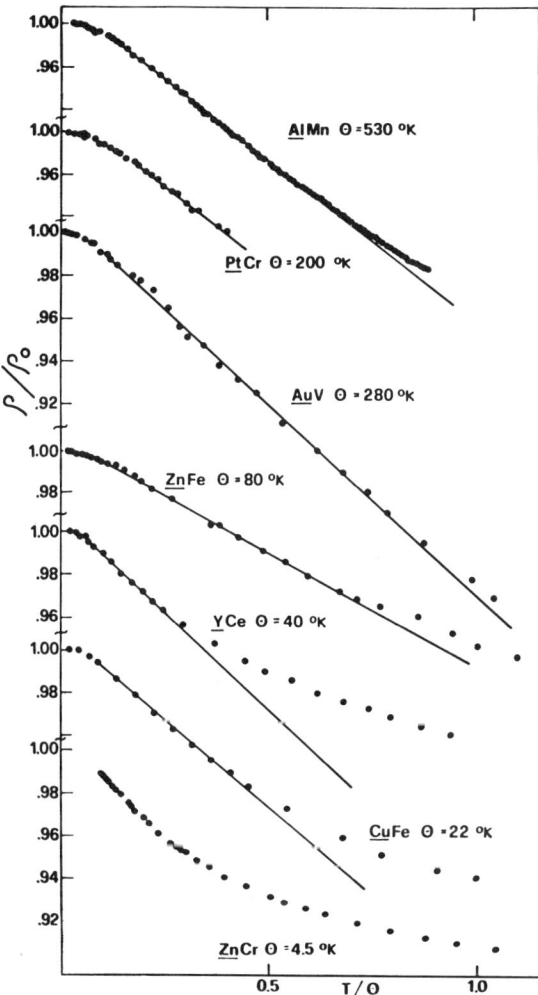

Fig. 1. The resistivities normalized to $\rho(0)$ and to θ_R for the systems described in text. Note the different scale for AlMn and PtCr. The uncertainties in the subtraction of phonons in the temperature range considered are biggest for YCe, CuFe, and ZnFe. Note also that the θ_R value used in this figure for ZnCr is 4.5°K instead of the more recent value of 0.9°K. Also, a value of 23°K for YCe has very recently been reported. Using these two new values in treating the data in the figure would expand the linear part of the curves and increase the similarity of these curves.

Table I. The Zero-Temperature Resistivity and the Characteristic Temperature Fitting the Various Systems to Eq. (1).

Alloy	$\rho(0)$	$\theta_{R2}\,°K$
AlMn	8.05 ± 0.05	530 ± 30
PtCr	6.0 ± 0.5	200
AuV	14 ±	290 ± 10
ZnFe	15.4 ± 0.3	80
YCe	14.1 ± 1	40 ± 1
CuFe	12.1 ± 0.3	21 ± 1
ZnCr	21.7 ± 1	0.9 ± 0.2

their $\rho(0)$ and θ_R values (except for AlCr, where the θ_R value is not yet defined with enough accuracy.[10]

It is noted in Fig. 1 that a linear behavior extending to above $0.5\theta_R$ is observable in all systems with $\theta_R \gtrsim 80°K$, while this behavior exists for systems with $\theta_R \lesssim 40°K$ over a more limited temperature interval as a "flex" region between the low-temperature T^2 behavior and a higher-temperature behavior which is closely logarithmic.[3,6]

The extent of the linear part, especially for systems with $0.1\theta_D \lesssim \theta_R \lesssim \theta_D$ (θ_D is the Debye temperature) is still subject to uncertainty. It is not yet possible to observe the detailed behavior of a sufficient number of systems across and above θ_R to know whether there is a similarity over a more extended temperature interval between the resistivity behavior of all those systems on which a comparison such as that performed in Fig. 1 may be attempted. (Also, the uncertainties in the value of θ_R for the lower-θ_R systems are still relatively large.[7]

Within these limits, however, it seems possible to suggest that the overall shape of the resistivity curve does not change very strongly between systems (called "nonmagnetic") where little or no temperature variation of the impurity magnetic susceptibility is observed (e.g., AlMn) and those (called "magnetic") were a Curie–Weiss-like behavior is observed. Moreover, the behavior shown in Fig. 1 is at least in qualitative agreement with that expected from theoretical calculations based on localized spin fluctuation theories.[12,13]

These results also stress the necessity of improving the present state of resistivity measurements on systems with θ_R or θ_χ (the concentration-independent Curie–Weiss temperature from susceptibility measurements, which is approximately equal to θ_R) between about 0.1 and 100°K to verify whether or not a more direct similarity exists among all these systems.

References

1. A.D. Caplin and C. Rizzuto, *Phys. Rev. Lett.* **21**, 746 (1968).
2. K. Kume, *J. Phys. Soc. Japan* **23**, 1226 (1967).
3. W.M. Star, F.B. Basters, C.M. Nap, E. De Vroede, and C. Van Baarle, *Physica* **58**, 585–622 (1972).
4. H. Nagasawa, *J. Phys. Soc. Japan* **27**, 787.
5. T. Sugawara and S. Yosida, *J. Low Temp. Phys.* **4**, 657 (1971).

6. P.J. Ford, E. Salamoni, and C. Rizzuto, *Phys. Rev. B* **6**, 1851 (1972).
7. A. Pilot, R. Vaccarone, and C. Rizzuto, *Phys. Lett.* **40A**, 405 (1972).
8. E. Babić, R. Krsnik, B. Leontíc, Z. Vučić, I. Zorić, and C. Rizzuto, *Phys. Rev. Lett.* **27**, 805 (1971).
9. E. Kovacs-Csetenyi, F.J. Kedves, L. Gergely, and G. Gruner, *J. Phys. F* **2**, 499 (1972).
10. E. Babić, P.J. Ford, C. Rizzuto, and E. Salamoni, *Sol. St. Comm.* **11**, 519 (1972).
11. A.D. Caplin and C. Rizzuto, *J. Phys. C* **3**, L 117 (1970).
12. M.J. Zuckermann, *J. Phys. F* **2**, L25 (1972).
13. N.Y. Rivier and V. Zlatic, *J. Phys. F* **2**, L87–92 (1972).

Hall Effect Induced by Skew Scattering in *LaCe*

A. Fert and O. Jaoul

Laboratoire de Physique des Solides
Université de Paris-Sud, Orsay, France

Measurements of the Hall effect in gold, silver or copper containing transitional impurities have been made by Hurd and Alderson[1] and Friederich and Monod.[2] The experimental data suggest[2-4] a contribution from skew scattering by the impurities, particularly in the case of the *Au*Fe alloys. However, the field dependence of the Hall resistivity of these alloys is not yet clearly understood.

In this communication we present Hall effect measurements for a *La*Ce alloy where the contribution from skew scattering by the Ce impurities appears more clearly.

The *La*Ce system is known for the mixing of the wave functions of the conduction electrons with the wave function of the f electron localized on the cerium.[9] The f electron of the cerium being in a state of total momentum $j = 5/2$, this leads (for a polarization of the cerium along Oz) to a stronger phase shift for the f partial wave of $j = 5/2$ and $j_z = +5/2$. In a partial wave analysis of the scattering this dependence of the phase shift of the f partial waves on j_z involves the asymmetry of the scattering.[5] For an impurity polarized along Oz and placed in a current flowing along Ox one obtains a deflection of a part of the current along Oy. This deflection can be characterized by a cross section σ_{xy}, and the cross sections $\sigma_{xy}^\uparrow$ and $\sigma_{xy}^\downarrow$ for the conduction electrons of spin parallel or antiparallel to the total momentum of the impurity have been calculated in Ref. 5. They are proportional to $\sin^2\eta_3$, where η_3 is the phase shift of the f partial wave of $j = 5/2$ and should be smaller by about two orders of magnitude compared to the resistivity cross section. The deflection of electrons by a magnetic impurity has been described in a more general way by Kondo[6] and Giovannini.[4]

The Hall resistivity $\Delta\rho_{xy}$ induced by skew scattering is found to be[7]

$$\Delta\rho_{xy} = c\frac{mv_F}{ne}\left[\frac{I_\uparrow^2 + I_\downarrow^2}{(I_\uparrow + I_\downarrow)^2}\frac{\sigma_{xy}^\uparrow(p) + \sigma_{xy}^\downarrow(p)}{2} + \frac{I_\uparrow^2 - I_\downarrow^2}{(I_\uparrow + I_\downarrow)^2}\frac{\sigma_{xy}^\uparrow(p) - \sigma_{xy}^\downarrow(p)}{2}\right]$$

for free conduction electrons of mass m, Fermi velocity v_F, and density n and for impurities of concentration c and magnetic polarization p. Here $I_\uparrow$ and $I_\downarrow$ mean the currents of spin-up and spin-down electrons and $\sigma_{xy}^\uparrow(p)$ and $\sigma_{xy}^\downarrow(p)$ the average deflection cross sections of the impurities for a polarization p. $\Delta\rho_{xy}$ is the sum of two terms linear in p for small values of p ($\mu H \ll kT$). For $p \sim 1$ ($\mu H > kT$), $\Delta\rho_{xy}$ should become saturated. The field dependence of the Hall effect induced by skew scattering is obtained in a different model by Giovannini.[4]

We present the experimental results for the Hall effect of a La–3.4% Ce alloy. The alloy had mainly a cubic structure but some amount of hexagonal structure was

Hall Effect Induced by Skew Scattering in LaCe

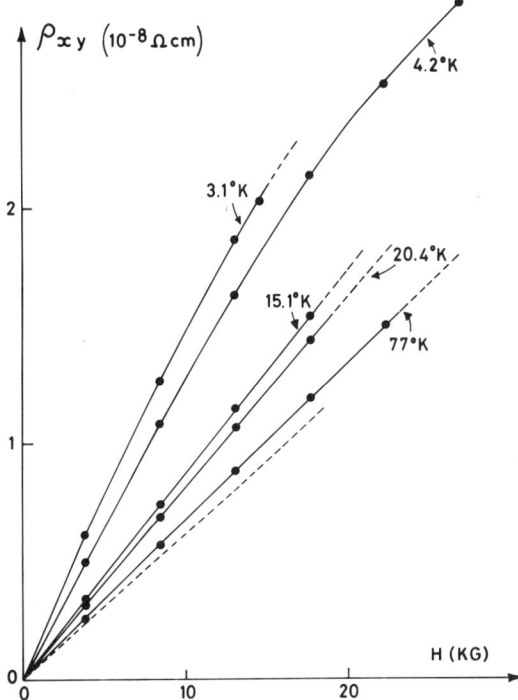

Fig. 1. The experimental Hall resistivity ρ_{xy} of the La–3.4% Ce alloy vs. the magnetic field H for different temperatures.

also present. The Hall resistivity was measured by a classical dc method in a superconductive coil up to 30 kG.

Figure 1 shows that the Hall effect increases strongly from 77°K down to a few °K. The experimental Hall resistivity ρ_{xy} can be fitted with the sum of two terms. First a temperature-independent ordinary Hall resistivity ρ_{xy}^0, which is represented in Fig. 1 by a dashed line. ρ_{xy}^0 was obtained by extrapolation of $\rho_{xy}(H, T)$ to infinite T and is only slightly different from $\rho_{xy}(H, T = 77°K)$. Note that the residual resistivity of our specimen is high enough for the ordinary Hall effect to be almost independent of the temperature and the field ($\omega_c \tau \ll 1$). Second, there is an additional extraordinary term $\Delta\rho_{xy}$—so called by analogy with the extraordinary Hall effect of ferromagnetics —which depends on H and T in the same way as the magnetization of the Ce impurities. The additional slope at low field $R_E = (\Delta\rho_{xy}/H)_{H\to 0}$ varies approximately as T^{-1} (at least for $T > 3°K$), as can be seen in Fig. 2, where R_E^{-1} is plotted as a function of T. Moreover, at low enough temperatures (4.2 and 3.1°K), when $\mu H/kT$ can be larger than unity in the experimental field range, the extraordinary term $\Delta\rho_{xy}$ departs from a linear dependence on H for $\mu H/kT \simeq 0.5$ and bends downward like a magnetization curve (Fig. 1). This dependence on H and T suggests strongly that $\Delta\rho_{xy}$ is due to skew scattering by the Ce impurities. Although the ordinary Hall effect of a dilute magnetic alloy might depend also on the polarization of the impurities and so on $x = \mu H/kT$, it has been shown[8] that the ordinary Hall constant R_0 should possess a term in x^2 at low field. Hence the ordinary Hall resistivity might contain a

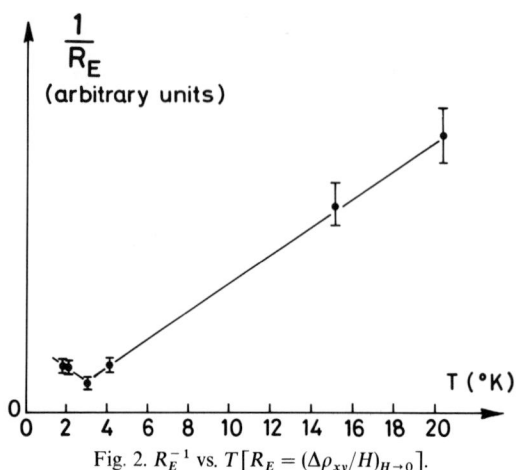

Fig. 2. R_E^{-1} vs. T $[R_E = (\Delta\rho_{xy}/H)_{H \to 0}]$.

term proportional to $x^2 H$ or to H^3/T^2 but $\Delta\rho_{xy}$ does not show this field and temperature dependence and cannot be due to a change of the ordinary Hall effect.

Returning to Fig. 2, one observes below 3°K a deviation from a Curie (or Curie–Weiss) law for R_E. This deviation could be due to a quenching of the orbital moment by the crystal field, but measurements at lower temperature and for alloys of definite hexagonal or cubic structure are needed to clarify this point.

In conclusion, skew scattering should be also observed in other dilute alloys, mainly each time the conduction electrons are scattered by an interaction with localized electrons which bear an orbital magnetic moment. It should be observed not only in the case where the interaction is due mainly to a k–f mixing, as for many compounds with Ce, but also, perhaps to a lesser degree, in the case of other rare earth impurities for which there is simply a classical exchange interaction with the conduction electrons of the host.

References

1. J.E.A. Alderson and C.M. Hurd, *J. Phys. Chem. Solids* **32**, 2075 (1971).
2. A. Friederich, Thèse de 3ème cycle, Orsay, France, 1972; *Proc. 12th Intern. Conf. Low Temp. Phys. 1970*, Academic Press of Japan, Tokyo (1971), p. 755.
3. A. Fert and O. Jaoul, *Phys. Rev. Lett.* **28**, 303 (1972).
4. B. Giovannini, *Phys. Lett.* **42A**, 256 (1972).
5. A. Fert and O. Jaoul, *Sol. St. Comm.* **11**, 759 (1972).
6. J. Kondo, *Prog. Theoret. Phys.* **27**, 772 (1962); **32**, 37 (1962).
7. A. Fert, *J. Phys. F* (to be published).
8. M.T. Béal-Monod and R.A. Weiner, *Phys. Rev. B* **3**, 3056 (1971).
9. B. Coqblin and J.R. Schrieffer, *Phys. Rev.* **185**, 847 (1969).

Kondo Effect in *Y*Ce under Pressure

W. Gey and M. Dietrich

Universität und Kernforschungszentrum Karlsruhe
Karlsruhe, West Germany

and

E. Umlauf

Zentralinstitut für Tieftemperaturforschung
Garching, Germany

Introduction

In an earlier paper we showed that the Kondo temperature T_K of *La*Ce increases upon application of pressure.[1] This result was derived from the pressure dependence of the superconducting transition temperature by applying the theory of Müller-Hartmann and Zittartz.[2] It was independently derived from the temperature dependence of the resistance R at different pressures and the variation of R with pressure at constant temperature, using the calculation of Hamann.[3]

Similar measurements have been reported by other authors, who instead interpreted their results in terms of a magnetic–nonmagnetic transition of the Ce impurities at pressures at which both the slope $dR/d(\ln T)$ and the superconducting pair-breaking effect decrease.[4] Particularly, the disappearance of the resistance minimum of *Y*Ce under pressure has been presented as an argument for the vanishing magnetic moment of the cerium impurities.[5] This explanation has become widely accepted in the literature; review articles have already appeared.[6,7]

We have reinvestigated the pressure dependence of the resistance anomaly of Y 1 at.% Ce. In analogy with the results for *La*Ce and also for *Cu*Fe,[8] we expected a shift of T_K to higher temperatures with pressure. The experimental results clearly show that this is so. The Kondo temperature is raised from 17°K at zero pressure to approximately 110°K at 30 kbar, while the magnitude of the anomaly is unchanged.

Results

Measurements were taken using a piston-cylinder technique in a pair-of-tongs apparatus described earlier.[9] It is sensitive to ± 100 bar at zero pressure and can be cycled repeatedly up to 45 kbar. Steatite has proved to be a sufficiently hydrostatic medium.[10] Samples of pure yttrium and Y 1 at.% Ce were mounted in series in the pressure cell and fed by the same current.

Figure 1 shows the resistance vs. ln T for both Y and *Y*Ce as obtained directly in the low-temperature regime. The Y sample shows normal behavior at all pressures

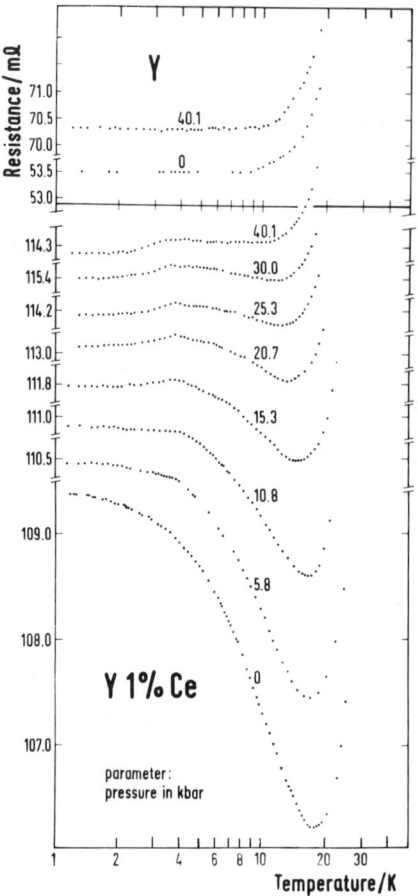

Fig. 1. Low-temperature electrical resistance of Y and Y 1 at.% Ce vs. temperature at various pressures.

up to 40 kbar. At low temperatures the phonon dependence is $T^{4.0}$ for all pressures. Only data for 0 and 40 kbar are given.

The decrease with pressure of the depth of the resistivity minimum in YCe is qualitatively in accord with the results of Maple and Wittig.[5] It led these authors to conclude that the magnetic moment of the Ce impurity vanishes in this pressure regime. In contrast, our results show that this is due to an increase of the Kondo temperature. In order to see this, it is necessary to isolate the magnetic part of the temperature-dependent resistivity from the phonon part at higher temperatures. This can be achieved as follows: Near room temperature the magnetic part of the resistance is negligibly small, at least at low pressure, i.e., low Kondo temperatures. Thus the slope of a R_Y vs. R_{YCe} plot near 300°K determines the geometric factor m between both samples. This factor does not change by more than 0.8% for all pressures up to 25.3 kbar. Assuming the constancy of the residual resistance up to room temperature (i.e., Matthiessen's rule) and that the T dependence of the phonon part of the resistance is identical for YCe and Y, one obtains the magnetic resistance anomaly.

Figure 2 shows the result of this procedure. The main physical result is that an

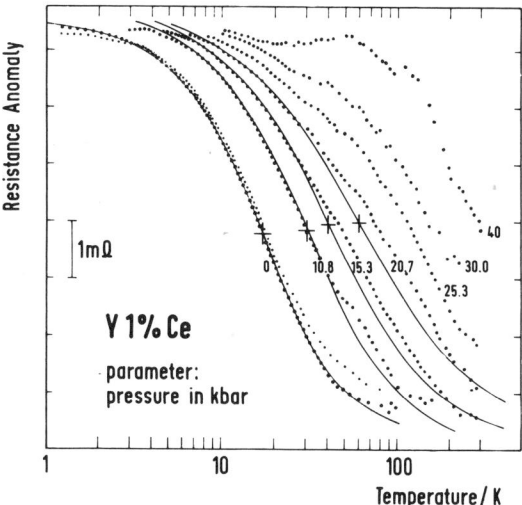

Fig. 2. The magnetic part of the resistance of Y 1 at.% Ce at various pressures and some fitted curves (solid lines) calculated from Hamann's equation.

application of pressure shifts the Kondo anomaly rather drastically to higher temperatures. One notes that for zero pressure the resistance anomaly is rather well described by Hamann's function for $\rho_m(T)$ over two decades of temperature (solid lines). The smaller dots represent zero-pressure data, which were obtained after removal of pressure from 25.3 kbar. With increasing pressure a deviation from the Hamann-type behavior starts to develop above 30°K, which may be due to a deviation from Matthiessen's rule, as suggested by Loram et al. for CuFe, AuFe, and $CuAu$Fe alloys.[11] Hamann fits for higher pressures were thus obtained from the curvature of the data between 8 and 30°K. The fit parameters are given in Table I. While the magnitude of the anomaly $\rho_m(0)$ stays close to 8 mΩ (2.3 μΩcm) up to 30 kbar, the Kondo temperature rises to approximately 110°K.

In conclusion, a monotonic shift of T_K, but no magnetic transition, was observed. The results strongly support our earlier results on LaCe.

Table I

p, kbar	T_K, °K	S	$\rho_m(0)$, μΩ-cm
0	17 ± 0.3	0.11	2.26
5.8	21 ± 1	0.14	2.44
10.8	31 ± 1	0.16	2.46
15.8	40 ± 2	0.20	2.46
20.7	60 ± 3	0.25	2.44
25.3	80 ± 5	0.20	2.55
0	17 ± 0.5	0.11	2.26
30	110 ± 15	0.20	2.32
40	(140 ± 40)	(0.20)	2.32

References

1. W. Gey and E. Umlauf, *Z. Physik* **242**, 241 (1971).
2. E. Müller-Hartmann and J. Zittartz, *Z. Physik* **234**, 58 (1950).
3. D.R. Hamann, *Phys. Rev.* **158**, 570 (1967).
4. M.B. Maple and K.S. Kim, *Phys. Rev. Lett.* **23**, 118 (1969); M.B. Maple, J. Wittig, and K.S. Kim, *Phys. Rev. Lett.* **23**, 1375 (1969); K.S. Kim and M.B. Maple, *Phys. Rev. B* **2**, 4696 (1970).
5. M.B. Maple and J. Wittig, *Sol. St. Comm.* **9**, 1611 (1971).
6. B. Coqblin, M.B. Maple, and G. Toulouse, *Intern. J. Magnetism* **1**, 333 (1971).
7. M.B. Maple, in *AIP Conf. Proc. No. 4, Rochester, 1972.*
8. J.S. Schilling, W.B. Holzapfel, and E. Lüscher, *Phys. Lett.* **38A**, 129 (1972).
9. W. Buckel and W. Gey, *Z. Physik* **176**, 336 (1963).
10. W. Gey, *Phys. Rev.* **153**, 422 (1967); A. Eichler and W. Gey, *Z. Physik* **251**, 321 (1972).
11. J.W. Loram, T.E. Whall, and P.J. Ford, *Phys. Rev. B* **2**, 857 (1970).

The Low-Temperature Magnetic Properties of Zn–Mn Single Crystals*

P. L. Li and W. B. Muir

Eaton Electronics Research Laboratory
McGill University, Montreal, Quebec

Introduction

Recent measurements of electrical resistivity and magnetic susceptibility have shown anisotropic effects in the hexagonal structure ZnMn[1] and Y–Ce[2] Kondo alloy systems. In ZnMn the anisotropy of the temperature dependence of the electrical resistivity of the alloy was reported to be in excess of the anisotropy expected from a crude model of the band structure of Zn. This excess anisotropy was parameterized by using an anisotropic s–d exchange interaction and gave $J_\perp$ about 8% larger than $J_\parallel$.† In the Y–Ce case both the magnetic susceptibility and electrical resistivity were measured. An attempt was made to interpret the observed magnetic anisotropy using crystal-field theory; however, the calculations failed to reproduce the observed temperature dependence of the measurements. No anisotropy in the temperature dependence of the electrical resistivity of the Y–Ce alloys was observed. The present paper presents the results of magnetic anisotropy measurements from 2 to 300°K on several ZnMn alloy crystals having Mn concentrations up to 475 ppm.

Sample Preparation

The single-crystal ingots were grown by slow cooling from the melt in a temperature gradient. They were then annealed in argon for 17 hr at 390°C. The single-crystal ingots were about 5 cm long. The ends of the ingots were cut off, rolled, and annealed, and the residual resistivities compared. All the ingots used had residual resistivities which varied by less than 10% over the length of the ingot. The ingots were oriented using X rays and the 5 × 5 × 10 mm samples were spark-cut from the center of the ingot. The Mn concentration was determined from room-temperature values of the susceptibilities by comparison with the published data.[3] For concentrations below 200 ppm Mn there was good correlation between the Mn concentration obtained from residual resistivity ratio methods[4] and those obtained from susceptibility. For Mn concentrations above 200 ppm the concentration determined by susceptibility was considerably higher (30%) than that obtained by residual resistance methods, suggesting that Mn is precipitated or evaporated from the thin resistivity specimens during annealing.

---

* Research supported in part by the National Research Council of Canada.
† The subscripts $\parallel$ and $\perp$ indicate values determined in directions parallel and perpendicular to the c axis of the hexagonal structure, respectively.

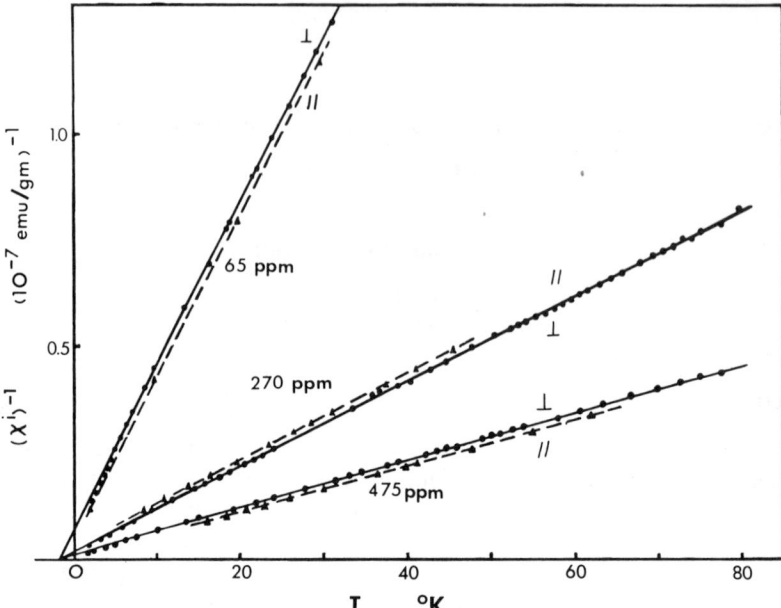

Fig. 1. The excess susceptibility due to Mn, $\chi^i = \chi_{ZnMn} - \chi_{Zn}$ as a function of temperature for field direction perpendicular and parallel to the c axis of the crystal.

Experimental Methods and Results

The magnetic susceptibility was measured using Curie's method on a previously described apparatus.[5] The magnetic anisotropy was measured by the standard torque technique,[6] using a servo torsion balance similar to one previously described.[7] The torsion balance has a sensitivity of 1×10^{-4} dyn cm. The excess susceptibility χ^i due to Mn was obtained by subtracting the susceptibility of pure Zn from the susceptibility of the alloys. Figure 1 shows $\chi^i_\perp$ and $\chi^i_\parallel$ as a function of temperature. $\chi_\perp$ was measured directly using Curie's method and $\chi_\parallel$ was calculated from $\chi_\perp$ by subtracting $\Delta\chi = \chi_\perp - \chi_\parallel$ obtained from torque measurements. Figure 2(a) shows the temperature variation of the total magnetic anisotropy $\Delta\chi$. Figure 2(b) shows the anisotropy per Mn ion $\Delta\chi^i/c$ obtained by subtracting the anisotropy of pure Zn from that of the alloy. Figure 2(c) shows the variation of magnetic anisotropy with Mn concentration at 4.2°K. Figure 3 shows the period of the de Haas–van Alphen oscillations as a function of Mn concentration. The solid lines indicate variation of the period that would be obtained for various valence states of the Mn ion as calculated from a rigid band model, allowing for the known change in axial ratio (c/a) for the ZnMn alloy system.

Discussion

Scalapino's[8] result for the susceptibility of the magnetic ions in a Kondo system can be expressed as

$$\chi^i = (A/T)\{1 - [1/\ln(T/T_K)]\} \tag{1}$$

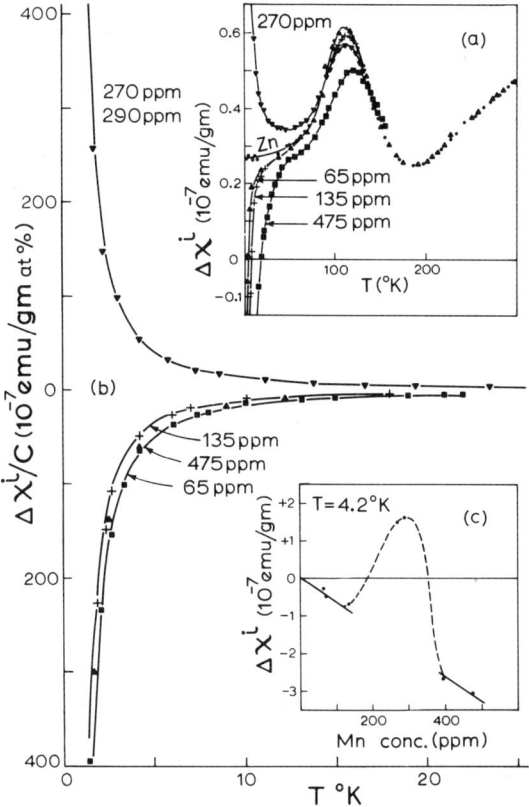

Fig. 2. (a) The magnetic anisotropy $\Delta\chi = \chi_\perp - \chi_\parallel$ of Zn and four ZnMn alloys. The peak at 110°K is a band structure effect (see text). (b) The magnetic anisotropy per Mn ion $(1/c)\Delta\chi^i = (1/c)(\Delta\chi_{ZnMn} - \Delta\chi_{Zn})$ as a function of temperature, where c is the Mn concentration. The solid lines represent the best fit of Eq. (2) to the data. (c) The magnetic anisotropy due to Mn, $\Delta\chi^i$, as a function of Mn concentration at $T = 4.2$°K.

where $A = Ncg^2\mu_B^2 S(S+1)/3k$ for a spin-only ion; N is the number of atoms; c is the atom fraction of magnetic ions; S is the spin quantum number of the magnetic ion; and T_K is the Kondo temperature. This expression was fitted to the experimental values of $\chi_\perp^i$ by adjusting A and T_K (or more exactly $T_{K\perp}$, since we will shortly allow T_K to be anisotropic). The values of A and $T_{K\perp}$ determined by fitting are shown in Table I along with the value of A calculated for various spin values. The experimental value of A corresponds to a spin-only ion, with $S = 2$ indicating a valence of 3 + for the Mn ions.* This valence is consistent with the de Haas–van Alphen effect data shown in Fig. 3.

---

* It is usually found that the moments of transition metal ions dissolved in nonmagnetic metal hosts correspond more closely to the spin-only value for the ion if the alloy is magnetic. For Mn^{3+} the Hund rule ground state would be $^5D_0$, which has a zero magnetic moment.

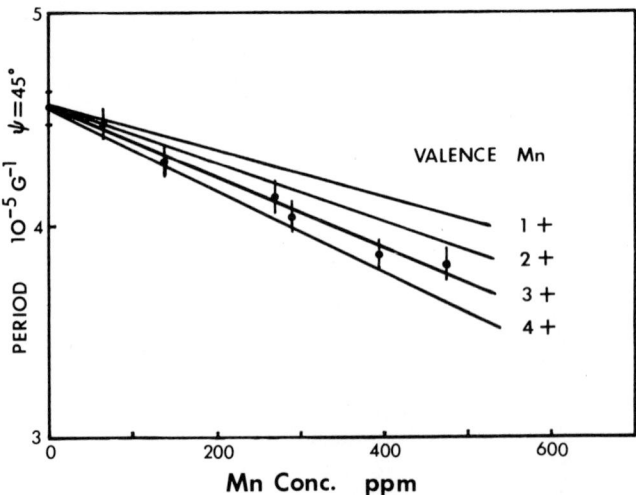

Fig. 3. Period of the de Haas–van Alphen oscillations as a function of Mn concentration. ψ is the inclination of the magnetic field to the c axis during measurement. The solid lines are the expected variation of the period for various valence states of the Mn ion (see text).

The magnetic anisotropy (Fig. 2a) shows a wide variety of effects. The pronounced peak in the neighborhood of 110°K is probably a band structure effect related to the changes which take place in the third zone Fermi surface of Zn with changing axial ratio due to thermal expansion.[9] The magnetic anisotropy per Mn ion (Fig. 2b) can be fitted to Scalapino's expression for the susceptibility if T_K is allowed to be anisotropic. In this case Eq. (1) reduces to

$$\Delta\chi^i = \chi^i_\perp - \chi^i_\parallel = \frac{A}{T} \frac{\ln(T_{K\parallel}/T_{K\perp})}{\ln(T/T_{K\parallel})\ln(T/T_{K\perp})} \qquad (2)$$

Using the values of $T_{K\perp}$ and A given in Table I, Eq. (2) was fitted to the experimental results by varying $T_{K\parallel}$. The results of this procedure are shown as the solid lines in Fig. 2(b) and the values of $T_{K\parallel}$ obtained are tabulated. The average of $T_{K\perp}$ and $T_{K\parallel}$ is independent of Mn concentration and has a value of about 0.25°K, in good agreement with other values obtained on polycrystalline material.[10] In spite of the relatively

Table I. Parameters Obtained by Fitting Eqs. (1) and (2) to the Experimental Data

Alloy conc. ppm	$T_{K\perp}$, °K	$T_{K\parallel}$, °K	$\bar{T}_K$, °K	A_{exp}, 10^{-7} emu °K/g	A_{calc}, 10^{-7} emu °K/g		
					$S = \frac{5}{2}$	$S = 2$	$S = \frac{3}{2}$
65	0.25 ± 0.04	0.20 ± 0.04	0.23 ± 0.04	31.0 ± 0.5	43.5	29.9	18.7
270	0.25 ± 0.04	0.32 ± 0.04	0.27 ± 0.04	116.5 ± 2.2	181	124	77.5
475	0.29 ± 0.04	0.18 ± 0.04	0.25 ± 0.04	215 ± 4	318	218	136
Valence of Mn ion					2	3	4

normal behavior of the average of $T_{K\parallel}$ and $T_{K\perp}$, we are left with the surprising change in the sign of the anisotropy for Mn concentrations of about 270 ppm, which is shown dramatically in Fig. 2(c) and is reflected in the change in relative magnitudes of $T_{K\perp}$ and $T_{K\parallel}$ at the same concentration.

This last fact points out the inadequacy of simply assuming an anisotropic T_K, which is equivalent to assuming an anisotropic J,* and establishes the need for a more complete theoretical understanding of the role of band structure in the Kondo effect.

References

1. M.J. Press and F.T. Hedgcock, *Phys. Rev. Lett.* **23**, 167 (1969).
2. T. Sugawara and S. Yosida, *J. Low Temp. Phys.* **4**, 657 (1971).
3. E.W. Collings, F.T. Hedgcock, and Y. Muto, *Phys. Rev.* **134**, A1521 (1964).
4. G. Boato, M. Bugo, and C. Rizzuto, *Nuovo Cimento* **XLVB**, 226 (1966).
5. F.T. Hedgcock and W.B. Muir, *Rev. Sci. Instr.* **31**, 390 (1960).
6. L.F. Bates, *Modern Magnetism*, Cambridge University Press, Cambridge, Massachusetts (1951), Chapter IV, p. 161.
7. F.T. Hedgcock and W.B. Muir, *Phys. Rev.* **129**, 2045 (1963).
8. D.J. Scalapino, *Phys. Rev. Lett.* **16**, 937 (1966).
9. B.I. Verkin, I.V. Svechkarev, and L.B. Kuz'micheva, *Soviet Phys.—JETP* **23**, 954 (1966).
10. R.S. Newrock, B. Serin, and G. Boato, *J. Low Temp. Phys.* **5**, 701 (1971).

---

* T_K is related to the exchange constant J by $kT_K = E_F e^{-1/|N(0)J|}$, where E_F is the Fermi energy and $N(0)$ is the density of states at the Fermi energy for one spin direction.

Influence of Lattice Defects on the Kondo Resistance Anomaly in Dilute ZnMn Thin-Film Alloys

H. P. Falke, H. P. Jablonski, and E. F. Wassermann

II. Physikalisches Institute RWTH
Aachen, Germany

We have investigated the electrical resistivity of ZnMn (200–3000 ppm Mn) thin films in the temperature range 0.35–350°K. The films are prepared by successive evaporation in uhv ($\sim 10^{-8}$ Torr) of small pellets of premelted alloys onto single-crystalline quartz plates held at 5°K.[1] The thickness of the films is determined from the linear temperature-dependent part of the resistivity and corrected for size effect following the method of v. Bassewitz and v. Minnigerode.[2] A comparison of the resistance between films of different thickness is thus possible.

Figure 1 shows the resistivity vs. temperature curves for a low-concentration (300 ppm Mn) alloy film after stepwise annealing to T_a and recooling to low temperatures. During the annealing procedure the absolute resistivity of the film decreases strongly because of healing out of lattice defects which are present in the layer in high concentration after quench-condensation. The figure shows that the slope of the curves in the range where ρ is proportional to log T ("Kondoslope") increases by 25% while going from $T_a = 15°K$ to $T_a = 150°K$. We believe that the increase in the Kondoslope is caused by the decrease in nonmagnetic defect concentration. This possibility has been considered theoretically by Bohnen and Fischer[3] by taking into account changes in the local density of states at the magnetic impurity site due to a nearby nonmagnetic impurity. This results in additional terms in the total scattering amplitude which are important for all Kondo anomalies. Yet, since the authors have calculated the resistivity only for statistically distributed pairs of magnetic and nonmagnetic impurities with a fixed distance between the pair members, this theory is not adequate for quench-condensed films.

We therefore tried to fit our results to Hamann's[4] expression for the resistivity of Kondo alloys. This means that the Kondo slope is mainly described by the product of the density of states $\rho(0)$ times the coupling constant J. Since this product also defines the Kondo temperature, the changes in the Kondo log T slope would be appropriate to a shift in T_K. Taking $T_K = 1°K$ and $S = 1$[5] for ZnMn, we find that T_K is shifted from the value $T_K = 1°K$ for the 150°K annealed film to $T_K = 0.3°K$ for the quench-condensed film ($T_a = 15°K$).

If the modified Hamann expression as given by Fischer[6] is used, the interference of exchange and potential scattering at the impurity site is respected. The fit of our data to this formula is not satisfying, since a change in potential phase shift δ_v from 37.7° to 39° already covers the total change in Kondo slope (25%) between the

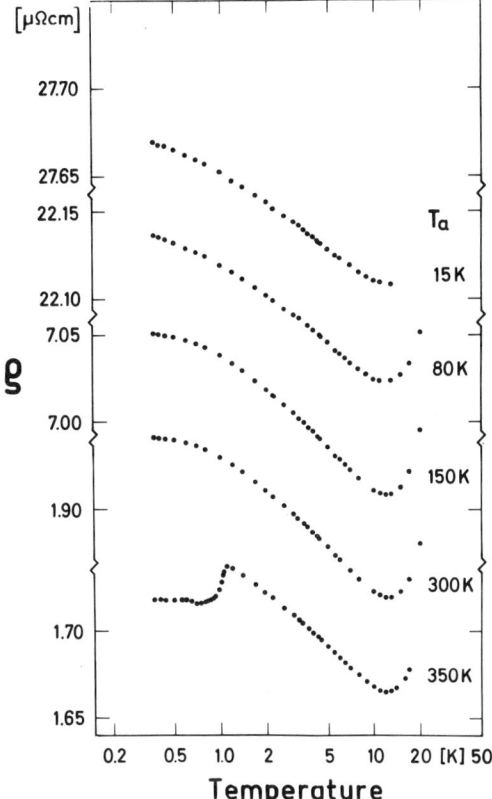

Fig. 1. Resistivity ρ ($\rho_{\text{Zn-matrix}}$ values not subtracted) vs. $\log_{10} T$ for a Zn + 300 ppm Mn film quench-condensed in uhv (10^{-8} Torr) onto a single-crystalline quartz plate held at 5°K. Film thickness 400 Å. The film is stepwise warmed to the listed annealing temperature T_a and then recooled. The hump in the $T_a = 350°$K curve is due to partial superconductivity in Zn grains after precipitation of Mn.

$T_a = 15°$K and $T_a = 150°$K curves. The Kondo temperature, which is also changed through δ_v, is lowered from 1 to 0.7°K.

Figure 1 also shows that the flattening off of the curves (that is, the deviation from Kondo log T behavior) occurs at increasing temperature values during the rise of T_a. Following de Gennes,[7] we attribute this to a strengthening of the Rudermann–Kittel interaction between the magnetic impurities, because of increasing mean free path of the conduction electrons. The effect is more pronounced at higher Mn concentrations, as shown in Fig. 2 for a 3000-ppm ZnMn film. Here we even observe a resistance maximum after 350°K annealing. The latter curve is very similar to a bulk alloy containing an equal amount of Mn.[8] The annealing behavior of the highly concentrated alloys also shows similarities to the systems investigated by Hasegawa and Tsuei,[9] who observed an almost total suppression of the RKKY interactions if the matrix between the magnetic impurities is amorphous.

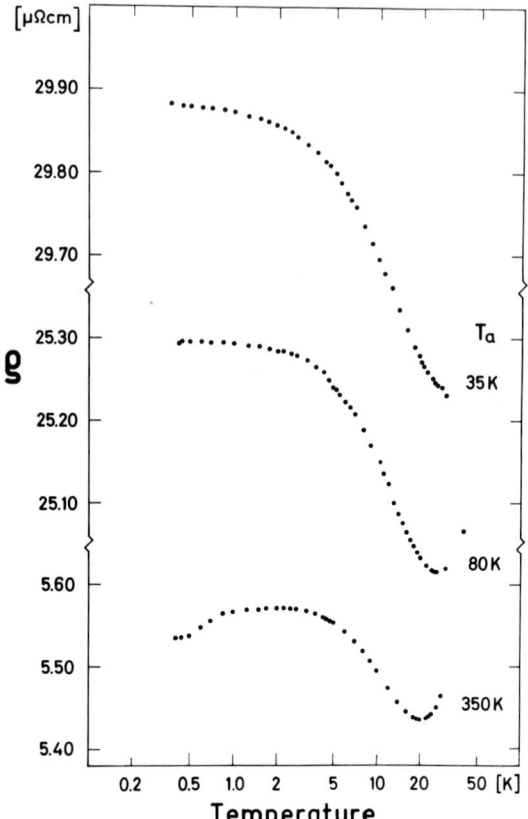

Fig. 2. Resistivity ρ as a function of $\log_{10} T$ for a Zn + 3000 ppm Mn alloy film (400 Å). The film is quench-condensed in uhv at 5°K and stepwise annealed and recooled (for details see caption to Fig. 1). The maximum in the $T_a = 350°K$ curve is due to the onset of RKKY interactions, suppressed after quench condensation.

References

1. F. Fischer, *Z. Physik* **139**, 328 (1954).
2. A. v. Bassewitz and G. v. Minnigerode, *Z. Physik* **181**, 368 (1964).
3. K.P. Bohnen and K. Fischer, *Z. Physik* **248**, 220 (1971).
4. D.R. Hamann, *Phys. Rev.* **158**, 570 (1967).
5. P.J. Ford, C. Rizzuto, and E. Salamoni, *Phys. Rev. B* **6**, 1851 (1972).
6. K. Fischer, Berichte der Kernforschungsanlage Jülich, Nr. 616-FN.
7. P.G. de Gennes, *J. Phys. Radium* **23**, 620 (1962).
8. E.F. Wassermann, *Z. Physik* **234**, 347 (1970).
9. R. Hasegawa and C.C. Tsuei, *Phys. Rev. B* **2**, 1631 (1970).

Evidence for Impurity Interactions from Low-Temperature Susceptibility Measurements on Dilute ZnMn Alloys

W. Schlabitz

Physics Institute, Universität Köln
Köln, Germany

and

E. F. Wassermann and H. P. Falke

Physics Institute, RWTH
Aachen, Germany

The magnetic mass susceptibility of Zn–Mn alloys containing 12, 19, 27, 47, 82, 155, and 275 ppm Mn has been investigated by means of a high-sensitive Faraday balance (sensitivity 1.7×10^{11} emu/g, maximum field 10 kG) in a temperature range from 2.7 to 300°K. The samples are thin strips of cold-rolled alloys, whose low-temperature resistivity anomalies were measured earlier.[1] A pure Zn sample has been submitted to an equal preparation and deformation procedure to allow proper deduction of the influence of the matrix from the alloy data. Within the investigated temperature range the Zn is diamagnetic with a slight tendency to paramagnetism compared to the "as-received" Zn at low temperatures, resulting from the melting process.

Figure 1 shows that in a $1/\Delta\chi$ versus T plot the data for all alloys follow a Curie–Weiss law between 2.7 and $\sim 50°$K, with a concentration-independent paramagnetic Curie temperature $\theta = -1°$K. According to the perturbation theory of the susceptibility of Kondo alloys for $T > T_K$,[2] $1/\Delta\chi$ plotted against T should follow a Curie–Weiss law within the interval from about $7T_K$ to $100T_K$ with a paramagnetic Curie temperature $\theta = -4.5T_K$. This means that for our cold-rolled ZnMn alloys $T_K \sim (0.22 \pm 0.05)°$K. The effective magnetic moment of the Mn is $p_{eff} = 4.28\,\mu_B$ for alloys with concentrations above 27 ppm Mn. This corresponds to a spin of roughly 3/2, a value found in the early investigations of Collings et al.[3]

As can be seen from Fig. 2, the Curie constants of these alloys are almost directly proportional to the impurity concentration ($C = 1.08c$), indicating that isolated impurity atoms are responsible for the Curie behavior. Figure 2 also gives data taken from a paper by Newrock et al.,[4] who found $p_{eff} = 4.66\,\mu_B$ for their ZnMn alloys from Curie behavior between 1.4 and 4.2°K. These authors, however, used samples which were annealed at 410°C.

Our cold-deformed lower concentration alloys (12 and 19 ppm Mn) have smaller Curie constants and smaller magnetic moments (for 19 ppm $p_{eff} = 3.14\,\mu_B$). Surprisingly, annealing of these alloys causes an increase of the effective moment. For

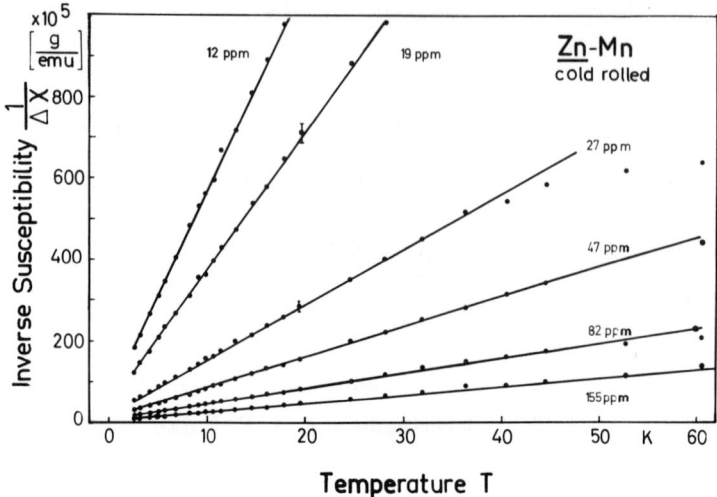

Fig. 1. Inverse susceptibility $1/\Delta\chi = 1/(\chi_{alloy} - \chi_{Zn})$ vs. temperature for different ZnMn alloys. The data follow the Curie–Weiss law down to 2.7°K, the lowest temperature reached with the experimental setup. The extrapolation gives a concentration-independent paramagnetic Curie temperature $\theta = -1°K$.

the 19-ppm sample, annealed 24 hr at 200°C under carefully purified 300-Torr argon, p_{eff} rises to 4.54 μ_B, then to 4.93 μ_B after 24 hr at 300°C, and finally reaches $p_{eff} = 5.32$ μ_B after 24 hr of 400°C annealing. Between 2.7 and 25°K all the samples follow the Curie–Weiss law, yet with increasing θ (cold-rolled alloy, $\theta = -1°K$; 24 hr at 400°C, $\theta = -6°K$).

An increase in the effective magnetic moment could also be observed for the 47-ppm alloy. Here, however, the rise is not so drastic. Disregarding errors in the concentration determination, Fig. 2 shows that after the annealing the Curie constants of our alloys lie close to the values found for annealed alloys by other authors.[4] We think that this agreement has its origin in the sample metallurgy.

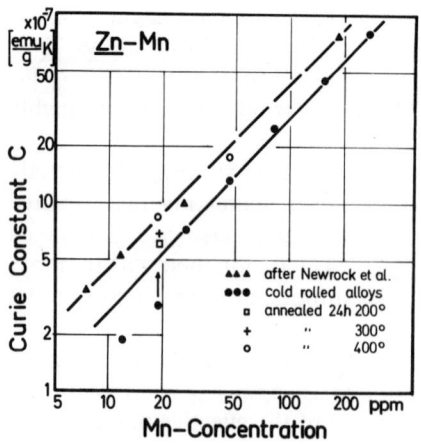

Fig. 2. Curie constant vs. Mn concentration for ZnMn alloys. The data given by the triangles are taken from Ref. 4.

From earlier measurement of the resistivity anomalies of the same ZnMn alloys as investigated in this work[1,5] it is quite evident that during the annealing, formation of pairs, triplets, etc., is possible, and sometimes even clustering and precipitation if the concentration is high enough. An inhomogeneous Mn distribution can also be caused by mechanical deformation.

We conclude that in ZnMn alloys, independent of the metallurgical treatment, impurity–impurity interactions do influence all low-temperature Kondo anomalies down to about 0.5°K if the Mn concentration lies above 20 ppm. Alloys with concentrations below 20 ppm should be given no annealing treatment after casting, quenching, and cold deformation. These samples then represent the "dilute alloys" necessary for comparison with theories. If the anomalies are studied at even lower temperatures, the "critical concentration" for this system goes to much smaller values. Quite recently it was shown by the Rizzuto group[6] that for ZnMn the concentration-independent unitarity limit can be reached below 0.05°K if the Mn concentration is as low as 6 ppm. The samples in Ref. 6 were freshly quenched and were given only a short-time strain relief anneal after mechanical deformation.

From our susceptibility measurements we thus conclude that $p_{eff} = 3.24\,\mu_B$ corresponding to $S = 1.15$ for "dilute" low-concentration ZnMn alloys.

References

1. E.F. Wassermann, *Z. Phys.* **234**, 347 (1970).
2. K. Yoskida and A. Okijii, *Progr. Theor. Phys. (Kyoto)* **34**, 505 (1965); D.J. Scalapino, *Phys. Rev. Lett.* **16**, 937 (1966).
3. E.W. Collings, F.T. Hedgcock, and Y. Muto, *Phys. Rev.* **134**, 1521 (1964).
4. R.S. Newrock, B. Serin, J. Vig, and G. Boato, *J. Low Temp. Phys.* **5**, 701 (1971).
5. E.F. Wassermann, H. Falke, and H.P. Jablonski, in *Proc. 12th Intern. Conf. Low Temp. Phys., 1970*, Academic Press of Japan, Tokyo (1971), p. 243.
6. A. Pilot, R. Vaccarone, and C. Rizzuto, preprint.

Low-Temperature Electrical Resistivity of Palladium–Cerium Alloys

J. A. Mydosh

Institut für Festkörperforschung, Kernforschungsanlage
Jülich, Germany

There has been much recent interest in alloys and compounds containing cerium as a solute.[1,2] In particular, the magnetic state of the cerium is of importance, and if it is magnetic, it allows for a study of the Kondo effect in the presence of crystalline fields.[3] Not long ago an anomalous resistive behavior was reported for the Pd_3Ce intermetallic compound.[4] Here the Ce atoms are considered to be magnetic, exhibiting a valence of about 3+.* For these and other Ce compounds the theory by Maranzana[7] of "Kondo sidebands" is usually applied. In addition, metallurgical investigations of the past few years have shown that a single-phase alloy of Ce in Pd can exist with a solubility limit of ~ 13 at. % Ce at 800°C.[8,9] The question which now arises concerns the magnetic state of the Ce solute randomly distributed in the exchange-enhanced Pd matrix. Early magnetic susceptibility measurements on a 4% Ce in Pd sample have been analyzed in terms of a Curie–Weiss behavior at low temperatures with an effective moment on the Ce impurities of $1.1\mu_B$.[10] Later measurements[5,9,11,12] of the susceptibility have shown no Curie–Weiss-like behavior, and thus consider the Ce to be nonmagnetic, or tetravalent. This conclusion is also supported by measurements of the lattice spacings.[5,9]

In an effort to better understand the nature of Ce in Pd, we have accurately measured the electrical resistivity ρ for three *PdCe* alloys. Using a four-point probe potentiometric technique, changes in ρ were detectable to better than one part in 10,000 in the temperature range 1.5–100°K. The samples were prepared through arc-melting the highest obtainable purity Pd (Johnson Matthey Chemicals, London, England; ~ 1 ppm Fe) and Ce (Rare Earth Products, Lancashire, England; only La, ~ 500 ppm, detected of the rare earths and Ca, ~ 100 ppm, of the common metals). The nominal concentrations by weight for the three alloys were 0.7, 2.2, and 5.7 at. % Ce; however, a more accurate activation analysis revealed concentrations of 0.6, 1.8, and 5.7 at. % Ce, respectively.

The resulting alloys were rolled into foils suitable for resistivity determinations with an average geometry error of about 10%. The samples were measured at low temperatures after first being stored for a few days at room temperature. Figure 1 shows the overall temperature (1–100°K) behavior for a typical cryogenic run. The

---

* A recent X-ray and susceptibility study of $CeRh_{3-x}Pd_x$ alloys by Harris *et al.*[5] indicates a valence of 3.45 for $CePd_3$. The interpretation suggested here is of a nonmagnetic virtual bound state for the 4*f* electrons. Also see Ref. 6.

Low-Temperature Electrical Resistivity of Palladium–Cerium Alloys

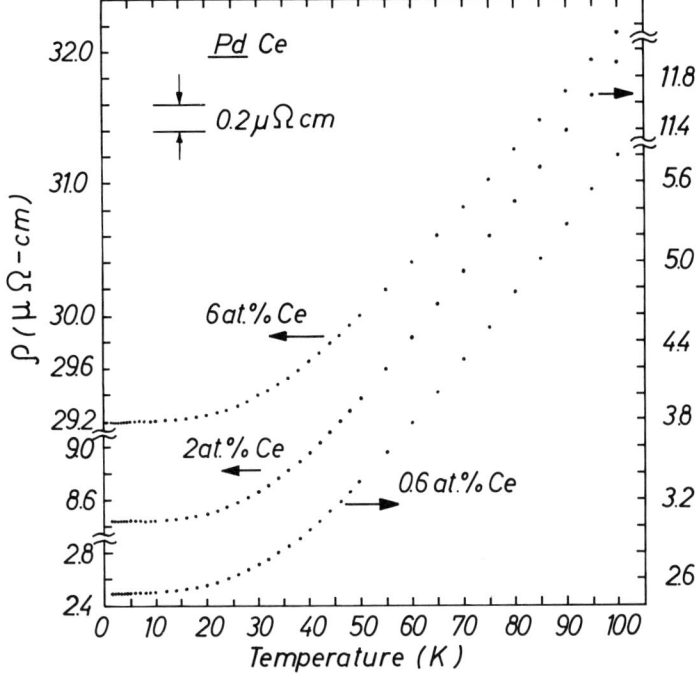

Fig. 1. The total resistivity vs. temperature for three concentrations of PdCe. Each curve has been arbitrarily displaced.

smooth monotonic variation of $\rho(T)$ for the three concentrations presents no indication of magnetic ordering, and in the higher-temperature region ($\sim 80°$K) the resistivity is nearly proportional to T. A subtraction of the Pd host resistivity was attempted over the entire temperature region, but due to the rather large errors in sample geometry, only rough results were obtainable. These did not show anything of great significance, with perhaps some indications of deviations from Matthiessen's rule between 20 and 40°K.

A series of most sensitive resistivity measurements, made in the 1–10K region, did exhibit weak minima in $\rho(T)$ (a few parts in 10,000). The depths of these minima increased somewhat with further room-temperature annealing, while the temperatures of ρ_{min} stabilized at approximately 4.5, 5.5, and 8°K for the 0.6, 2, and 6 at. % Ce samples, respectively. A second set of samples showed essentially the same $\rho(T)$ behavior; these samples are shown by the dots in Fig. 2. A third group was annealed in a partial pressure of helium gas at 800°C for 3 hr, rapidly quenched to room temperature, and immediately measured at low temperatures. For these heat-treated samples, represented by the crosses in Fig. 2, the depths of ρ_{min} were found to decrease and became somewhat harder to discern, due to the intrinsic scatter of the experiment. However, upon allowing these samples to anneal at room temperature for about two months, ρ_{min} were again clearly observed at the above-mentioned temperatures.

The resistivity behavior of these alloys can be interpreted in terms of the Ce impurities being in the nonmagnetic ($\sim 4+$ valence) state. This interpretation is mainly due to the observation of no large anomalies in $\rho(T)$. It should be noted here

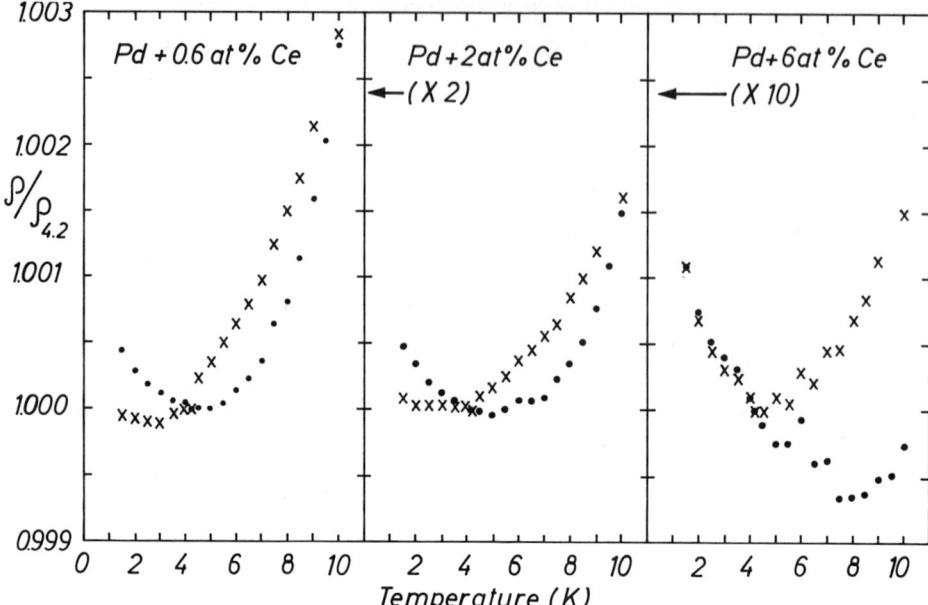

Fig. 2. The normalized (at 4.23°K) variation of the resistivity $\rho/\rho_{4.2}$ at low temperatures. The dots represent a typical set of samples stored for about one month at room temperature, while the crosses represent a set annealed, quenched, and immediately measured at low temperatures. Note the magnified scales for the higher concentrations.

that Ce impurities display dramatic resistive anomalies when in the $\sim 3+$ magnetic configuration.[1-4] Furthermore, it would seem that no great significance can be attributed to the very small ρ_{min} found at low temperatures in these alloys. Perhaps this is caused by band structure effects primarily due to the Pd matrix. Resistivity minima had been previously observed in Pd–Ag and Pd–Au alloys.[13] Another possible explanation for a ρ_{min} is that clusters or precipitations of Ce might exist which, because of a preferred environment, exhibit a magnetic character. These localized regions of short-range order are perhaps related to the Pd_3Ce structure, which shows a ρ_{min} at about 4°K.[4] This explanation would be consistent with the latest susceptibility results of nonmagnetic Ce, particularly since some precipitation might have occurred in the earlier susceptibility studies.[14]

Acknowledgments

I wish to acknowledge stimulating conversations with W. Buckel and E. Bucher. Also, I thank Mrs. S. Britz for assistance in the data analysis, and W. E. Gardner for making a preprint of his work available.

References

1. B. Coqblin, M.B. Maple, and G. Toulouse, *Intern. J. Magnetism* **1**, 333 (1971).
2. H.J. van Daal and K.H.J. Buschow, *Phys. Stat. Sol. (a)* **3**, 853 (1970).
3. B. Cornut and B. Coqblin, *Phys. Rev. B* **5**, 4541 (1972).
4. V.U.S. Rao, R.D. Hutchens, and J.E. Greedan, *J. Phys. Chem. Sol.* **32**, 2755 (1971); R.D. Hutchens,

V.U.S. Rao, J.E. Greedan, W.E. Wallace, and R.S. Craig, *J. Appl. Phys.* **42**, 1293 (1971); R.O. Elliot and H.H. Hill, in *Proc. 9th Rare Earth Research Conf.*, Blacksburg, Virginia (1971), p. 692.
5. I.R. Harris, M. Norman, and W.E. Gardner, *J. Less-Common Metals* **29**, 299 (1972).
6. W.E. Gardner, J. Penfold, T.F. Smith, and I.R. Harris, *J. Phys. F: Metal Phys.* **2**, 133 (1972).
7. F.E. Maranzana, *Phys. Rev. Lett.* **25**, 239 (1970).
8. J.R. Thomson, *J. Less-Common Metals* **13**, 307 (1967).
9. I.R. Harris and M. Norman, *J. Less-Common Metals* **15**, 285 (1968).
10. D. Shaltiel, J.H. Wernick, H.J. Williams, and M. Peter, *Phys. Rev.* **135**, A1346 (1964).
11. J. Crangle and R.B. Layng, in *Proc. 5th Rare Earth Research Conf. Ames, Iowa*, Book Four, Solid State Session P-2, Clearinghouse for Federal Scientific and Technical Information, Springfield, Virginia (1965).
12. R.P. Guertin, et al., *Phys. Rev. B* **7**, 274 (1973).
13. L.R. Edwards, C.W. Chen, and S. Legvold, *Sol. St. Comm.* **8**, 1403 (1970).
14. M. Peter, private communication.

Spin and g Factor of Impurities with Giant Moments in Pd and Pt*

G. J. Nieuwenhuys and B. M. Boerstoel†

*Kamberlingh Onnes Laboratorium der Rijksuniversiteit
Leiden, The Netherlands*

and

W. M. Star‡

*Francis Bitter National Magnet Laboratory§
Massachusetts Institute of Technology, Cambridge, Massachusetts*

The d electrons in Pd and Pt interact strongly with each other, causing the paramagnetic susceptibility to be much larger than might be expected from the electronic density of states. Particularly in Pd, this "exchange enhancement" of the susceptibility is so large that this metal is considered to be nearly ferromagnetic.

When a magnetic impurity is dissolved in a metal the band electron spins are polarized. This polarization is known as the Ruderman–Kittel–Kasuya–Yosida (RKKY) polarization. In hosts with strongly exchange-enhanced susceptibilities the d-spin polarization is large and is a positive function of the distance from the impurity. Thus if the magnetic moment of, say, Fe in Pd is measured, one observes the moment localized at the impurity site plus the polarization of the surrounding d-electrons; the total moment of this impurity–d-electron complex is much larger than the moment expected from Hund's rule for the impurity only. The magnetic moment associated with Fe in Pd is therefore called a giant moment.**

As a consequence of the large d-band polarization, magnetic transition metal impurities in exchange-enhanced hosts interact strongly with each other, causing magnetic ordering of the alloys to persist down to fairly low impurity concentration. (Pd–0.2 at.% Fe is still ferromagnetic at liquid helium temperature.)

In Table I the moments associated with first row transition metal atoms in Pd and

---

* This work is part of the research program of FOM supported by The Netherlands Organization for the Advance of Pure Research and TNO.
† Now at Metaalinstituut TNO, Delft, The Netherlands.
‡ Supported by a NATO Science Fellowship, awarded by The Netherlands Organization for the Advancement of Pure Research. On leave of absence from FOM, Kamerlingh Onnes Laboratorium, Leiden, The Netherlands.
§ Supported by The National Science Foundation.
** Sometimes superparamagnetic clusters, such as in Cu–Ni or Ni–Rh, are also called giant moments. We will reserve this term for an impurity with its surrounding host polarization.

Table I. Magnetic Moments and Spin Values (from Specific Heat) of 3d Atoms Dissolved in Pd and Pt

	In Pd		In Pt	
	Moment*	Spin	Moment*	Spin
Mn	7.5	2.4	?	2.3
Fe	10	1.6	6	?
Co	10	1.4	3.8	1

* In units of the Bohr magneton.

Pt are given.* Spin values are also given. If it is assumed that the g value for the on-site moment is not much different from two, it is obvious that the d-band contribution to the moment is appreciable, particularly in Pd. The spin values of Table I were obtained from specific heat measurements, the magnetic entropy S_m being equal to $cR \log(2S + 1)$, where c is the impurity concentration and S is the impurity spin. A specific heat measurement is the most direct way to obtain the impurity spin. However, other indirect measurements of the impurity spin have been reported in the literature. One possibility is to measure the magnetization σ as a function of applied field H and temperature T and to fit the data to a Brillouin function. The argument of the Brillouin function is known from the saturation moment and the spin is then obtained by choosing the function that best fits the data, including, if desired, a molecular field. Another related method is the Mössbauer effect (which is restricted to ^{57}Fe). Since the hyperfine interaction constant is not known, in this case both the moment and the spin must be obtained by curve fitting (unless the moment is taken from magnetization data and only S is determined from curve fitting). With the Mössbauer effect it is possible to use extremely dilute alloys, thus minimizing impurity interactions.

Some spin values obtained with methods other than specific heat measurements are given in Table II. In all cases the spins are appreciably larger than those given in Table I. According to Craig et al.,[4] for example, the Mössbauer data on ^{57}Fe in Pd can be fitted with a Brillouin function with $S = 13/2$ and $g = 2$. This implies that the giant moment arises from a giant spin. However, the entropy corresponding to this giant spin is not observed in the specific heat. We prefer the S value obtained from the specific heat, since in this case S is obtained directly and no model is involved. It is the purpose of this paper to show that magnetization data are consistent with this S value. This will be done in the following way. We will show that, just as an S value may be obtained indirectly by analyzing magnetization data as a function of temperature and applied field, a value of the magnetic moment μ may be obtained indirectly by analyzing impurity specific heat (ΔC) data as a function of temperature and applied field. A model will be developed to fit specific heat data and the predicted moments will be shown to be close to the measured moments. It will be shown that

---

* Numerous papers on this subject have appeared in the past ten years. Extensive reference to the literature can be found in Refs. 1–3.

Table II. Some Spin Quantum Numbers Determined Using Mössbauer Effect (M), Magnetization (m), and Magnetoresistance (mr) Data

	Pd(Fe)	Pt(Fe)	Pd(Co)
Craig et al.[4] (M)	6.5	—	—
Kitchens et al.[6] (M)	—	2.	—
Maley et al.[7] (M)	3.76	2.98	—
McDougald et al.[8] (m)	10*	—	—
Manuel et al.[9] (m)	6†, 10*	—	—
Grassie et al.[10] (mr)	4.5	—	4.7

* 0.15 at.%.
† 0.05 at.%.

this model also fits magnetization data, with the same spin and a g value equal to $g = \mu/S$. We thus obtain a giant g value instead of a giant spin.*

As has been done by other authors for the magnetization, we will use the Weiss molecular field model (WMFM) to analyze specific heat data. The WMFM is a very simplified model, of course, but it will nevertheless turn out to be a very useful tool. In order to see to what extent the WMFM is applicable, we will now discuss some specific heat data. Figure 1 shows the impurity specific heat of Pd–0.16 at.% Co. The anomaly is strongly field dependent and rather broad, even at zero field. This kind of behavior is typical for the specific heat of dilute ferromagnetic Pd and Pt alloys, with one exception: Pd(Mn). Figure 2 shows the specific heat of two Pd(Mn) alloys. The

Table III. C_{max} As a Function of Temperature for a System of Noninteracting Spins in Magnetic Field (Schottky anomaly) Compared with Experimental Results*

Alloy	Concentration; at.%	C_{max} (Schottky), mJ/mole °K	C_{max}(exp), mJ/mole °K
Pd(Mn)	0.08	5.7	4.7
	0.19	13	14
	0.54	38	38
Pd(Fe)	0.16	10	8.5
Pd(Co)	0.075	4.7	3.0
	0.16	10	6.8
Pt(Mn)	0.36	25.4	20.8
Pt(Co)	0.067	3.55	2.52

* On dilute Pd and Pt alloys in applied fields (27 kOe max).

---

* Takahashi and Shimizu[5] have concluded that the magnetic properties of dilute Pd(Co) and Pd(Fe) alloys are consistent with a "normal" spin value. Their simple molecular field model does not fit the actual data very well, but the equations can be written in a form such that the impurity has a "normal" spin and a "giant" g value. In this sense we arrive at the same conclusion.

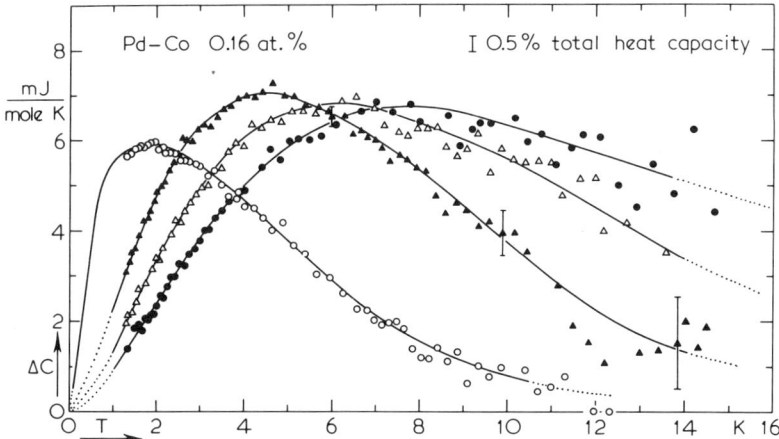

Fig. 1. Impurity contribution to the specific heat (ΔC) of Pd–0.16 at.% Co. Open circles: $H = 0$; black triangles: $H = 9$ kOe; open triangles: $H = 18$ kOe; black circles: $H = 27$ kOe. Points obtained below 1°K at $H = 0$ have been omitted and are replaced by a solid curve.

specific heat is again strongly field dependent, and in this (unique, for dilute alloys) case does show a cusplike anomaly at zero field. The similarity to the WMFM specific heat is only qualitative, however, since Pd(Mn) has a sizable fraction of the entropy above T_c, contrary to the WMFM.

More information about the applicability of the WMFM to dilute Pd and Pt alloys may be obtained from Table III, in which experimental specific heat maxima

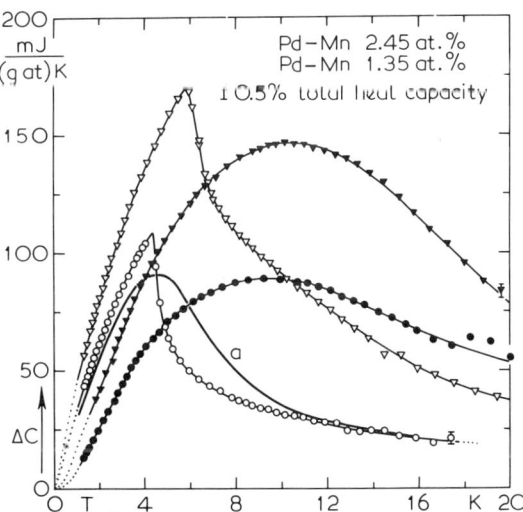

Fig. 2. Impurity contribution to the specific heat of two Pd(Mn) alloys. Pd–1.35 at.% Mn: open circles $H = 0$; black circles $H = 27$ kOe; curve a: $H = 2$ kOe. Pd–2.45 at.% Mn: open triangles $H = 0$; black triangles $H = 27$ kOe.

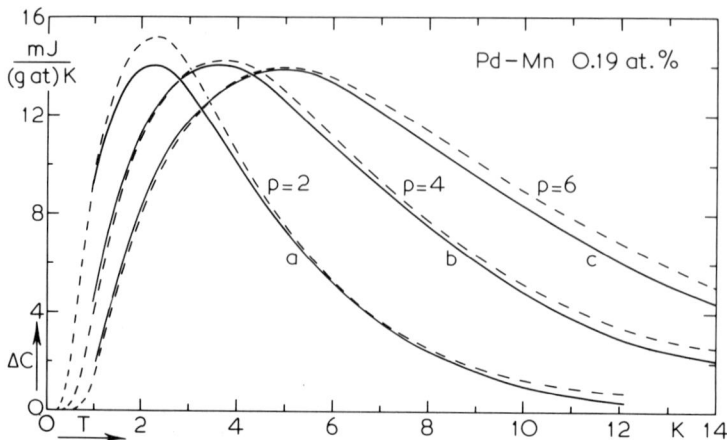

Fig. 3. ΔC vs. T for Pd–0.19 at.% Mn at (a) $H = 9$ kOe, (b) $H = 18$ kOe, (c) $H = 27$ kOe. Full curves: experiments; dashed curve: WMF model calculation; $p = H/H_m(0)$.

are compared with the Schottky anomaly (a system of free spins in an external magnetic field). It can be noted that the WMFM specific heat approaches the Schottky specific heat when $H \gg H_m(0)$, where $H_m(0)$ is the molecular field at $T = 0$. With the exception of Pd(Mn), the experimental specific heat maxima are considerably lower than the Schottky maxima. Since the WMFM specific heat maximum is always larger than the Schottky maximum, it follows that the experimental C vs. T curves are considerably broader than those of the WMFM. Only Pd(Mn) seems to be a promising WMFM candidate.

In Fig. 3 the specific heat of Pd–0.19 at.% Mn in external magnetic fields is presented and compared with the WMFM.[2] The model calculation accounts very well for the experimental data on Pd(Mn) as long as $p = H/H_m(0) \gg 1$. Since the applied field in these experiments was limited to 27 kOe, and $H_m(0)$ increases approximately proportional to c, a good fit of the WMFM to the data was only possible for $c \lesssim 0.5$ at.% Mn. But even above that concentration the WMFM permits an interesting conclusion to be drawn. Boerstoel et al.[2] performed numerical calculations of the WMFM specific heat as a function of temperature in applied magnetic fields. He found that the temperature of the specific heat maximum ($T_{\max,H}$) is approximately a linear function of the applied field. Assuming the spin of Mn in Pd is 5/2 (Table I), one obtains from the model calculations

$$T_{\max,H} = T_C + 0.778\, g\mu_B H/k_B \tag{1}$$

where T_C is the Curie temperature and the other symbols have their usual meaning. Figure 4 shows that $T_{\max,H}$ of Pd(Mn) alloys up to 1.35 at.% Mn satisfies Eq. (1) fairly well. One finds $g \approx 2.9$. Thus, although it was not stated explicitly in Ref. 2, from specific heat data one can predict a giant moment of 7.3 μ_B for Mn in Pd, in very good agreement with magnetization measurements.[11]

In Fig. 5 magnetization data are shown for Pd–0.054 at.% Mn and are compared with the WMFM. $S = 5/2$ was taken from the specific heat, $g = 3.1$ was obtained from the saturation moment, and the molecular field constant was obtained from the

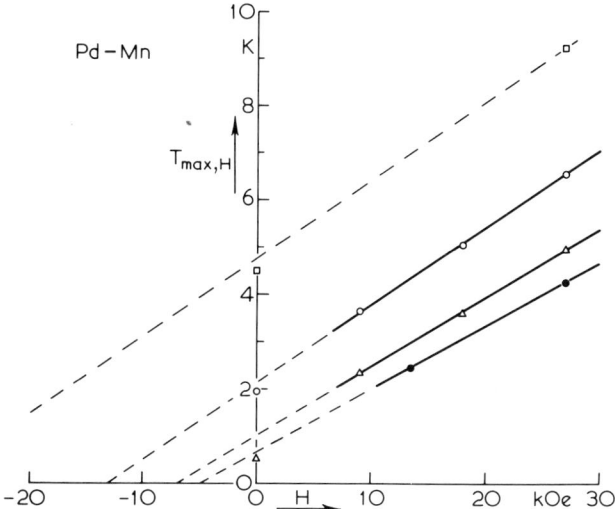

Fig. 4. $T_{\max,H}$ vs. H. Black circles: Pd–0.08 at.% Mn; open triangles: Pd–0.19 at.% Mn; open circles: Pd–0.54 at.% Mn; open squares: Pd–1.35 at.% Mn.

Curie–Weiss temperature θ. [The low-field susceptibility obeys the Curie–Weiss law and the molecular field constant $D = \theta(d\chi^{-1}/dT)$.] Thus the WMFM calculation does not contain any adjustable parameters and fits the experimental data closely.

From the foregoing discussion we conclude that specific heat and magnetization data of dilute Pd(Mn) alloys are consistent with $S = 5/2$ and $g \approx 3$. We will now

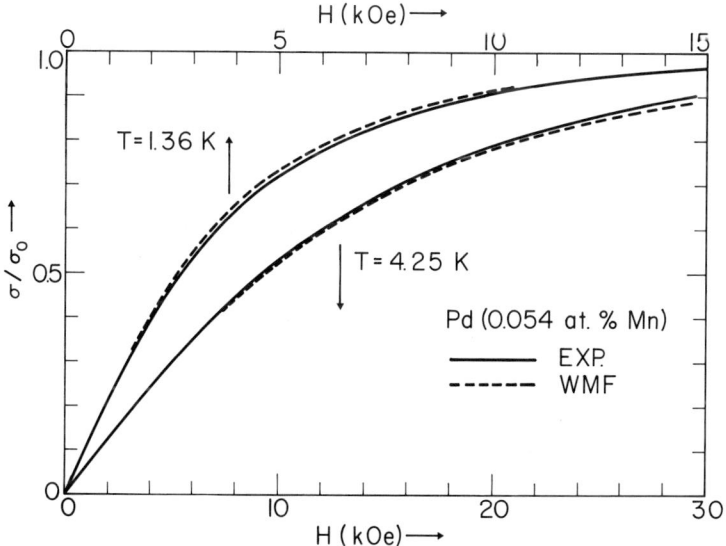

Fig. 5. Impurity magnetization of Pd–0.054 at.% Mn compared with the WMF model, $\sigma_0 = 0.220$ emu/g, $\theta = 0.25°$K.

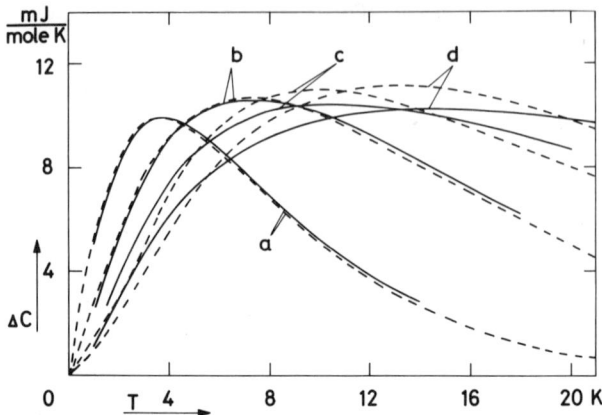

Fig. 6. Comparison between experimental (solid lines) and calculated (dashed lines) specific heat of Pd–0.24 at.% Co. Values of the model parameters: $T_c = 4.16°$K; $F = 0.6$; $H_m(0) = 10$ kOe; $g_0 = 6.7$; $S = 1.5$. (a–d) $H = 0, 9, 18, 27$ kOe, respectively.

demonstrate that specific heat and magnetization data of other dilute Pd and Pt alloys are also consistent with a "normal" spin and a "giant" g value.

The discussion related to Table III indicated that the ΔC vs. T curves of dilute magnetic Pd and Pt alloys other than Pd(Mn) are too broad to be fitted by the WMFM. One way to introduce broadening into the WMFM is to assume a distribution of molecular field constants D rather than one single D. It is quite likely that a broad spectrum of interaction strengths occurs, since the d-band polarization is not uniform and the impurities are randomly distributed in the host. Calculations of the specific heat have been performed using the WMFM and including a Gaussian distribution of D. Comparison with experimental results on Pd–0.16 at. % Fe showed a poor fit, however, even when an unreasonably broad distribution was employed.[3] This poor fit was more or less to be expected, since the WMFM specific heat, even when a distribution of D is included, approaches the Schottky specific heat when $H \gg H_m(0)$ and this is not observed experimentally (Table III).

More effective than a distribution of D in broadening the WMFM specific heat is a distribution of g values. The reason for this is that in the argument of the Brillouin function D only occurs in the molecular field term, whereas g occurs in the external field term (as g) as well as in the molecular field term (as g^2). Furthermore, with a distribution of g values, the model specific heat at high applied fields remains smaller than the Schottky specific heat, which is also observed experimentally (an extensive discussion can be found in Ref. 3).

A Gaussian distribution of g values was used:

$$g \propto \exp[(g - g_0)^2/2(Fg_0)^2] \qquad (2)$$

The WMFM specific heat was computed as a function of reduced temperature and reduced field for various values of F and the parameters were obtained by comparison with experimental data in the manner described in Ref. 3. The results are quite surprising, as shown in Fig. 6 for Pd–0.24 at.% Co. In zero field the model fits the data

Table IV. Parameters Used and g_0 Values Obtained in the Model Calculations.*

Alloy	Concentration, at.%	S	F	g_0	$\frac{Sg_0}{\mu_B}$
Pd(Mn)	0.08	2.5	0	2.6	6.5
	0.19	2.5	0	2.8	7.0
	0.54	2.5	0	3.0	7.5
	1.35	2.5	0	3.2	8.0
Pd(Fe)	0.16	1.5	0.42	6.3	9.5
Pd(Co)	0.16	1.5	0.76	5.3	8.0
	0.24	1.5	0.6	6.7	10.
Pt(Mn)	1.64	2.5	0.57	0.85	2.1
Pt(Co)	0.8	1.	0.7	4.0	4.0

* For the spins the half-integral number closest to the experimental result was chosen.

within experimental accuracy and in applied fields the agreement is quite good. More importantly, for the most probable g value $g_0 = 6.7$ is obtained, which, with $S = 1.5$, leads to a moment of $10 \mu_B$ per Co atom. This is remarkably close to the value obtained from magnetization measurements (Table I). Thus, for Pd(Co) just as for Pd(Mn) it appears to be possible to obtain a value for the magnetic moment from a WMFM analysis of specific heat data. The success of the WMFM including a Gaussian distribution of g values is not confined to Pd(Co). Similar results have been obtained for other Pd and Pt alloys. Some model parameters are given in Table IV. In all cases the predicted moment is close to the measured moment (Table I). A very interesting case is Pt(Mn), where the model predicts $g_0 < 2$, i.e., a "dwarf" moment instead of a giant moment. No magnetization data are available to confirm this prediction.

The present model also accounts for magnetization data. An example is shown in Fig. 7. Maley et al[7] measured $\langle S_z \rangle / S$ for Fe in Pd, using the Mössbauer effect, and obtained $S - 3.76$ and $g = 2.95$ from the best Brillouin fit to the data. As is shown in Fig. 7, a Brillouin function with $S = 3/2$ (from the entropy) gives a much poorer representation of the data ($F = 0$); but if $F = 0.48$, i.e., assuming a distribution of g values, the fit is very good. We therefore generalize our statement regarding Pd(Mn) to all dilute ferromagnetic Pd and Pt alloys: specific heat and magnetization data are consistent with S as obtained from the entropy and $g = \mu/S$, if a distribution of g values is assumed to exist. The d-band electrons apparently do not contribute to the multiplicity and what they contribute to the moment can be accounted for by assuming a g value which differs from 2.

An important question remains to be answered: Is there any physics behind this distribution of g values? We can think of two possibilities. The first is suggested by a well-known formula[5,12] for the giant moment complex: $\mu = 2\mu_B S(1 + \alpha\chi)$, where the g value of the onsite moment is assumed to be 2 and the host polarization is proportional to the host susceptibility χ with a proportionality constant α. The band part of the moment can be included in an effective $g = 2(1 + \alpha\chi)$. It is conceiv-

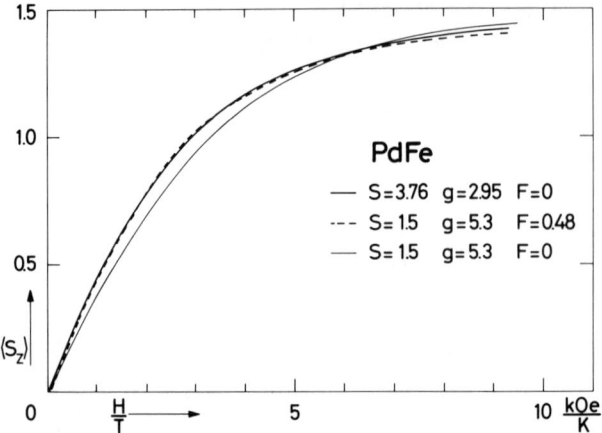

Fig. 7. Magnetization of Pd(Fe). Comparison between the model calculation (dashed line) and a Brillouin function (thick line) with $S = 3.76$ and $g = 2.95$, as obtained by Maley et al.[7] The thin line represents the Brillouin function with $S = 1.5$ and a single g factor $g = 5.3$.

able that α or χ or both depend upon the number and distance of neighboring impurities, leading to a distribution of effective g values.

A second possibility is that spin fluctuation effects play a role, i.e., isolated impurities may be weakly magnetic and pairs or larger groups of atoms may be more strongly magnetic.* Spin fluctuation effects certainly play a role in Pt(Co)[14] and Pt(Fe),[15] but in Pd(Co) the characteristic spin fluctuation temperature was estimated[16] to be much less than 1°K, and the same is probably true for Pd(Fe).

Acknowledgment

The authors are indebted to Drs. S. Foner, G. J. van den Berg, and J. B. Diamond for encouragement and fruitful discussions.

References

1. B.M. Boerstoel, J.E. van Dam, and G.J. Nieuwenhuys, in *Proc. NATO Advanced Study Institute, Magnetism, Current Topics*, S. Foner, ed. Gordon and Breach, New York (1972).
2. B.M. Boerstoel, J.J. Zwart, and J. Hansen, *Physica* **57**, 397 (1972).
3. G.J. Nieuwenhuys, B.M. Boerstoel, J.J. Zwart, H.D. Dokter, and G.J. van den Berg, *Physica* **62**, 278 (1972).
4. P.P. Craig, D.E. Nagle, W.A. Steyert, and R.D. Taylor, *Phys. Rev. Lett.* **9**, 12 (1962).
5. T. Takahashi and M. Shimizu, *J. Phys. Soc. Japan* **20**, 26 (1965).
6. T.A. Kitchens, W.A. Steyert, and R.D. Taylor, *Phys. Rev.* **138**, A467 (1965).
7. M.P. Maley, R.D. Taylor, and J.L. Thompson, *J. Appl. Phys.* **38**, 1249 (1967).
8. M. McDougald and A.J. Manuel, *J. Appl. Phys.* **39**, 961 (1968).
9. A.J. Manuel and M. McDougald, *J. Phys. C* **3**, 147 (1970).
10. A.D.C. Grassie and J.W. Loram, *Phys. Rev. B* **3**, 4154 (1971).

---

* See the paper by Tournier for an extensive discussion of these phenomena.

11. W.M. Star, S. Foner, and E.J. McNiff, Jr. *Phys. Lett.* **39A**, 189 (1972).
12. S. Doniach and E.P. Wohlfarth, *Proc. Roy. Soc. (London)* **A296**, 442 (1967).
13. R. Tournier, this volume.
14. P. Costa Ribeiro, M. Saint Paul, D. Thoulouze, and R. Tournier, this volume.
15. J.W. Loram, R.J. White, and A.D.C. Grassie, *Phys. Rev. B* **5**, 3659 (1972).
16. J.W. Loram, G. Williams, and G.A. Swallow, *Phys. Rev. B* **3**, 3060 (1971).

Specific Heat Anomalies of PtCo Alloys at Very Low Temperatures

P. Costa-Ribeiro,* M. Saint-Paul, D. Thoulouze, and R. Tournier

*Centre de Recherches sur les Très Basses Températures
CNRS, Grenoble, France*

At very low temperatures the magnetic behavior of the PtCo system presents different aspects. For "high" concentrations, $c \geq 2$ at.%, magnetization measurements[1] show the presence of giant moments of 3.6 μ_B per solute atom. For very dilute alloys, $c \simeq 1$ ppm, nuclear orientation measurements[2] give a critical (Kondo or LSF) temperature $T_\kappa \simeq 1.6°$K. It is thus interesting to study the appearance of the magnetic moments on the impurities below T_κ for intermediate concentrations, and also the way in which they appear, smoothly or abruptly, each one carrying its maximum moment.

We have made a systematic study of this system by measuring between 20 m°K and 4°K the specific heat of seven dilute samples of concentration c ranging from 0.1 to 2.6 at.% Co. Magnetization measurements[3] have also been performed on the same samples between 50 m°K and 50°K in fields up to 60 kOe.

The specimens were induction-melted on a watercooled baseplate and cast in a watercooled copper mold.[4] Adsorption spectroscopy analysis gave actual concentrations of 0.08, 0.16, 0.27, 0.31, 0.66, 0.88, and 2.6 at.% Co. A sample of pure platinum was also prepared in the same way and was measured above 0.3°K. The weight of the samples was about 15 g.

The measurements between 4 and 0.3°K were made on a ^{3}He cryostat, using an RC technique.[5] Below 0.3°K an adiabatic method was used with a demagnetization apparatus.[6]

At "high temperatures," i.e., above 0.3°K, for the four more dilute samples the excess specific heat with respect to platinum, ΔC, presents the same type of variation in temperature, particularly a flat maximum at a temperature $T_M \simeq 1.2°$K, nearly independent of the concentration (Fig. 1). The amplitude of this maximum is nearly proportional to the concentration. When the concentration increases, the maximum moves very quickly toward higher temperatures, as can be seen from previous measurements above 4°K.[7]

For the low concentrations the one-impurity effect is associated to the critical (Kondo or fluctuation) temperature of *isolated* impurities which become nonmagnetic. Effectively, the temperature T_M of the maximum is not very far from the value $T_\kappa \simeq 1.6°$K determined by nuclear orientation and susceptibility measurements, but is different from the value $T_\kappa/3$ indicated by the theory[8] and observed in the case of e.g., CuFe.[9] However, near T_M, ΔC fits the theoretical curve very well, with $S = 1$.

---

* CAPES (Brazil) Fellow.

Specific Heat Anomalies of PtCo Alloys at Very Low Temperatures

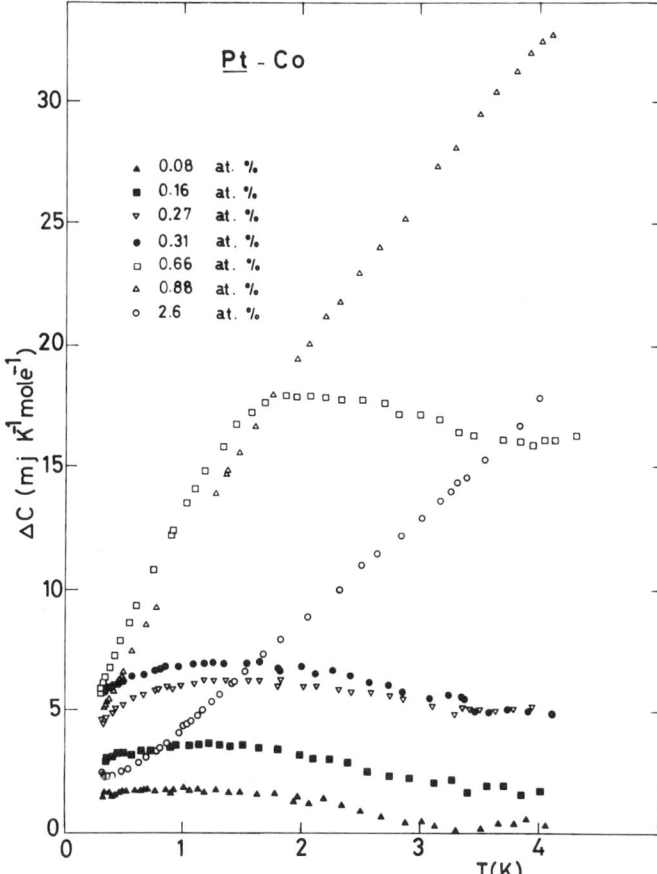

Fig. 1. The excess specific heat ΔC of PtCo alloys vs. temperature for $T > 0.3°K$. The pure platinum contribution has been subtracted from the total specific heat.

The discrepancy between T_M and $T_K/3$ may possibly be related to the fact that when the concentration rises, T_M seems to be slightly shifted toward higher temperatures, from about 1°K for $c = 0.08$ at.% to about 1.4°K for $c = 0.31$ at.%, and the width of the anomaly increases. Effectively, the concentrations considered here are not evanescent and are even relatively close to the magnetic region, so that the local density of states is modified by the presence of the other impurities. The distribution of density effects could lead to an increase of T_M and of the width with the concentration. But these variations may also be due to the high-temperature tail of the order anomaly of the magnetic impurities, as we shall see.

For larger concentrations the magnetic ordering anomaly of the residual magnetic impurities takes place at higher temperatures and overcomes the one-impurity anomaly. These residual magnetic impurities exist well below T_K, as confirmed by the observation of nuclear Schottky anomalies below 0.2°K even for samples which exhibit a one-impurity effect at 1.2°K namely, for $c = 0.27$ at.%

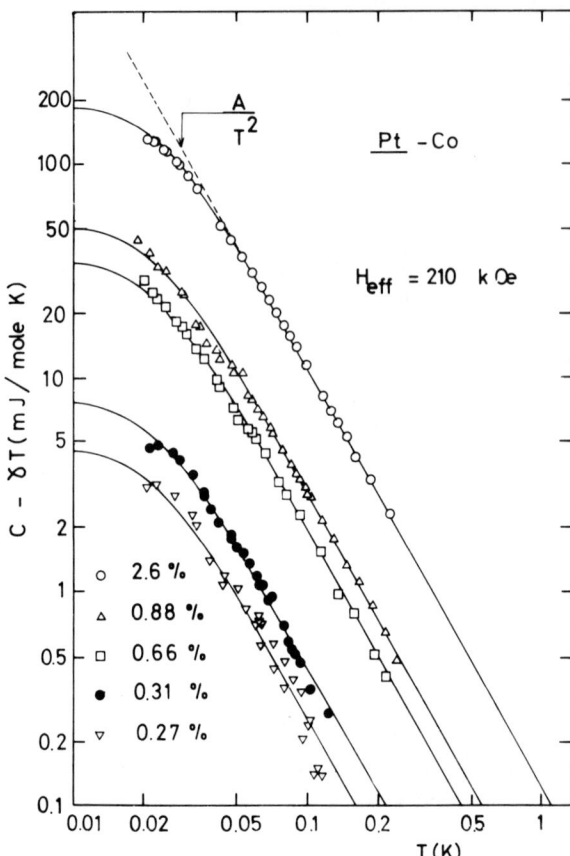

Fig. 2. The nuclear Schottky anomalies vs. temperature for $T < 0.3°K$. The term linear in temperature and the matrix contribution have been subtracted from the total specific heat. The solid lines, representing the calculated Schottky anomalies with $H = 210$ kOe, provide the values of c_m reported in Fig. 3.

and 0.31 at.% Co (Fig. 2).* However, in that range of temperatures, it was not possible to measure the two lowest concentrations since their ordering temperatures would have been too low to observe the hyperfine term in the ordered state.

The high-temperature tail of the nuclear Schottky anomaly is proportional to the number of nuclei affected and to the square of the hyperfine field H_{eff}:

$$AT^2/R = c_m[I(I+1)/3](\gamma^2\hbar^2 H_{\text{eff}}^2/k^2)$$

Assuming a unique value of the moment and of the hyperfine field for all the magnetic

---

* The nuclear specific heat of the matrix, induced by the magnetic impurities, and thus proportional to their concentration, has been estimated from that of the *Pt*Fe alloy containing the same magnetic impurity concentration but in which the contribution of the Fe nuclei is negligible. It amounts to $CT^2 = 35$ ergs °K/mole for $c = 1$ at.% Fe; this corresponds to 7% of the total nuclear contribution of the *Pt*Co alloys.

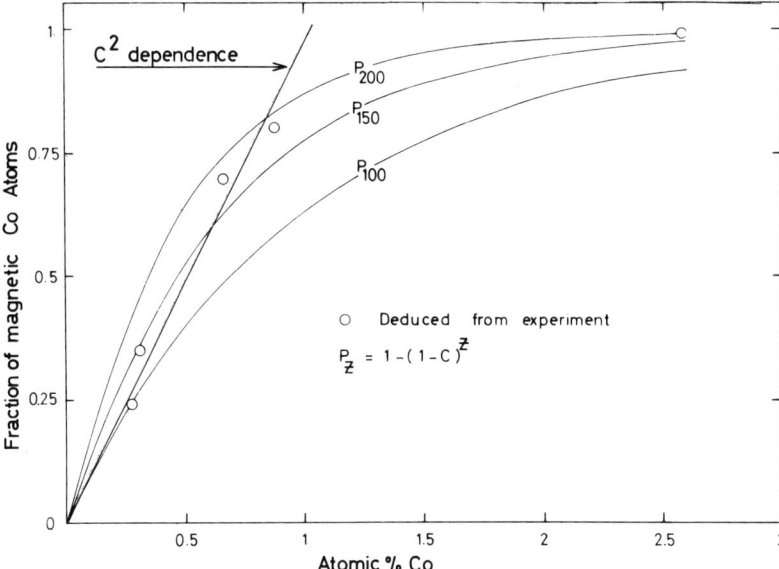

Fig. 3. The ratio c_m/c of magnetic Co atoms to the total concentration as a function of the total concentration c. The full lines represent P_Z for different values of Z. The straight line represents the c^2 dependance of the number of magnetic impurities.

impurities and zero value for the nonmagnetic ones, c_m represents the concentration in magnetic carriers.

The best fit to the T^{-2} high-temperature tail and to the higher-order terms of the Schottky anomalies is obtained for $H_{\text{eff}} = 210$ kOe. For pure fcc cobalt the hyperfine field is 215 kOe,[10] which shows that the magnetic cobalt impurities carry their full magnetic moment (1.6 μ_B). The variable parameter is c_m. For the low concentrations ($c \leq 0.88$ at. %) c_m varies like c^2 (Fig. 3), which suggests that the magnetic atoms are "pairs" of atoms. The same behavior has already been observed in PtCo and in CuFe[10] alloys by magnetization measurements and seems to be a general feature for the appearance of magnetism in dilute alloys having a low critical temperature $T_{\kappa F}$ due to interactions between impurities. Assuming that an impurity is magnetic, if there is another among its Z near neighbors, c_m is equal to $c[1 - (1 - c)^Z]$.

Here we obtain (Fig. 3) $Z \simeq 100$ for the lowest concentration ($c = 0.27$ at. %) and $Z \simeq 200$ for the higher ones, where the major parts of the impurities are magnetic. The magnetization measurements give $Z = 180$ for all the concentrations. This value is obtained for $H \simeq 2$ kOe, while the specific heat measurements are performed in zero field, thus indicating that at low concentrations the difference between $Z = 180$ and $Z \simeq 100$ may be due to some induced moments. The same phenomenon is observed in CuFe alloys for fields of 50 kOe instead of 2kOe, which corresponds roughly to the ratio of the respective critical temperatures, 30 instead of 1.6°K. This allows us to distinguish between two types of magnetic impurities with an equal number $Z \simeq 100$: those that are spontaneously magnetic and those, almost magnetic, on which a magnetic moment may be easily induced by small external fields or by molecular field effects when the concentration rises.

The fact that the nuclear specific heat never provides values of Z higher than those deduced from the magnetization shows that most of the magnetic pairs are ferromagnetic since the magnetization is sensitive to the "ferromagnetic" pairs, while the nuclear specific heat is sensitive to both "ferro-magnetic" and "antiferromagnetic" pairs.

Knowing the proportion of magnetic pairs, it is possible to come back to the results of the more concentrated samples ($c > 0.6$ at.%) above $0.3°K$. The excess specific heat ΔC may be considered as the sum of two terms: ΔC_1 due to the isolated Co atoms, and ΔC_2, due to the magnetic Co pairs: $\Delta C = \Delta C_1 + \Delta C_2$. We take ΔC_1 as proportional to the low-concentration values, for which the magnetic contribution is negligible:

$$\Delta C_1 = \Delta C_{c=0.08} \frac{c(1-c)^Z}{c(1-c)^Z{}_{c=0.08}} \quad \text{with} \quad Z = 180$$

$\Delta C_2 = \Delta C - \Delta C_1$ varies approximately linearly with T, up to a maximum at the ordering temperature T_o; the slope is nearly independent of c_m and equal to 7 mJ $°K^2$/mole. It has the same concentration dependence as c_m. This behavior would be characteristic of RKKY interactions, which is relatively surprising, since for those concentrations the ferromagnetic character seems to be preponderant.

In conclusion, the specific heat of the PtCo system shows the existence of two different types of magnetic behavior: that of the "isolated" impurities, which present a one-impurity effect anomaly near their critical temperature $T_\kappa = 1.6°K$; and the residual magnetism of pairs of cobalt impurities which order much below T_κ, as shown by their nuclear Schottky anomaly, and carry their maximum moment. Those remaining have their local susceptibility enhanced by effects of density of states and carry induced moments.

References

1. J.C. Crangle and W.R. Scott, *J. Appl. Phys.* **36**, 921 (1965).
2. J.C. Gallop and I.A. Campbell, *Sol. St. Comm.* **6**, 831 (1968).
3. B. Tissier and R. Tournier, *Sol. St. Comm.* **11**, 895 (1972).
4. O. Bethoux, M. Ferrari, and B. Cornut, *Rev. de Phys. Appliquée* **5**, 865 (1970).
5. K.H. Gobrecht and M. Saint-Paul, in *Proc. Third ICEC, Berlin*, 1970 p. 235.
6. J.R.G. Keyston, A. Lacaze, and D. Thoulouze, *Cryogenics* **8**, 295 (1968).
7. J.C.G. Wheeler, *J. Phys. C* **2**, 135 (1969); B.M. Boerstoel and C. Van Baarle, *J. Appl. Phys.* **41**, 1079 (1970).
8. P.E. Bloomfield and D.R. Hamann, *Phys. Rev.* **164**, 856 (1970).
9. B.B. Triplett and N.E. Phillips, *Phys. Rev. Lett.* **27**, 1001 (1971).
10. A.C. Gossard, A.M. Portis, M. Rubinstein, and R.M. Lindquist, *Phys. Rev.* **138A**, 1415 (1965).
11. J.L. Tholence and R. Tournier, *Phys. Rev. Lett.* **25**, 867 (1970); E.C. Hirschkoff, M.R. Shanabarger, O.G. Symko, and J.C. Wheatley *J. Low Temp. Phys.* **5**, 545 (1971).

Resistivity of Paramagnetic Pd–Ni Alloys

R. Harris and M. J. Zuckermann

Eaton Electronics Laboratory, McGill University
Montreal, Quebec

In a recent article Harris and Zuckermann[1] showed that a coherent potential approximation (CPA) analogy can be used to describe spin fluctuations in disordered transition metal alloys. In particular, the method was used to explain the nonlinear variation of the inverse spin susceptibility of paramagnetic Pd–Ni alloys with Ni concentration. A critical concentration of 2.3 at.% Ni was predicted for the occurrence of ferromagnetism, which agrees with the analysis of Tari and Coles.[2]

In order to calculate the spin fluctuation contribution to the resistivity of Pd–Ni alloys, it is necessary to calculate the dynamic susceptibility $\chi_{\text{eff}}(q, \omega)$ of the alloy. The CPA equations, because of the self-consistency, prove very intractable in this case and it is necessary to look for a non-self-consistent approximation in which the solutions are close to the CPA solutions. The average t-matrix approximation (ATA) fits this description well, since the frequencies of the localized spin fluctuations in dilute Pd–Ni lie within the band of spin fluctuations in the Pd host.[2]

The model to which the ATA is applied is the following. The alloys are binary AB alloys in which the electronic structure is described by a single band. The conduction electrons interact via intraatomic Coulomb interactions I_A and I_B on the A and B sites, respectively. The application of the ATA to this system gives the following expression for $\chi_{\text{eff}}(q, \omega)$:

$$\chi_{\text{eff}}(q, \omega) = \chi_0(q, \omega)/[1 - I_{\text{eff}}(\omega)\chi_0(q, \omega)] \tag{1}$$

$\chi_0(q, \omega)$ is the dynamic susceptibility of the alloy for $I_A = I_B = 0$ and $I_{\text{eff}}(\omega)$ is the effective intraatomic interaction of the alloy and is given by

$$I_{\text{eff}}(\omega) = I_B + n_A \Delta I \frac{1 + n_A \Delta I \chi^{\text{loc}}(\omega)}{1 - (1 - 2n_A)\Delta I \chi^{\text{loc}}(\omega)} \tag{2}$$

where $\Delta I = I_A - I_B$, n_A is the concentration of A atoms, and

$$\chi^{\text{loc}}(\omega) = \sum_q \chi_0(q, \omega)/[1 - (I_B + n_A \Delta I)\chi_0(q, \omega)] \tag{3}$$

The evaluation of $\chi_{\text{eff}}(q, \omega)$ for Pd–Ni alloys in Eqs. (1) and (2) requires an explicit model expression for $\chi_0(q, \omega)$ for pure Pd. The model used is an extension of the simple model of Lederer and Mills.[2] The magnetic properties of dilute Pd–Ni alloys are assumed to be due to the heavy d holes of Pd, which lie on 12 approximately cylindrical pockets located on the square faces of the fcc Brillouin zone. The static

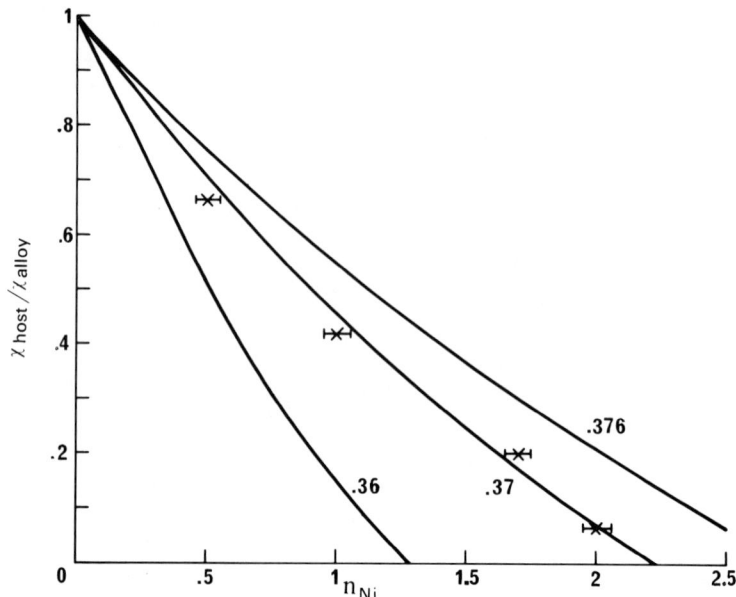

Fig. 1. Static susceptibility of a paramagnetic alloy as a function of concentration: $\alpha = 1.00$; $\gamma = 0.36, 0.37, 0.376$. Crosses give experimental points of Williams for paramagnetic Pd–Ni alloys (see Ref. 4).

q-dependent susceptibility $\chi_0(q, \omega)$ is then approximated by

$$\chi_0(q, \omega) \approx N(0)[1 - \gamma(q/2K_d)^2], \quad q < 2k_d \quad (4)$$
$$= 0 \quad q > 2k_d$$

where $N(0)$ is the density of states at the Fermi level and k_d is the radius of a cylindrical pocket. Now $\chi^{loc}(0)$ can be evaluated from Eqs. (3) and (4) and depends on four parameters, i.e., two band structure parameters—$\alpha = 4\pi R_d^2/(2\pi/a)^2$ and γ—and $N(0)I_{Pd}$ and $N(0)I_{Ni}$. We use $N(0)I_{Pd} = 0.88$ and $N(0)I_{Ni} = 1.16$ (see Ref. 1). An expression for the static spin susceptibility χ^{bulk} is obtained from Eqs. (1) and (2), since $\chi^{bulk} = \chi_{eff}(0,0)$. The best fit to the data of Williams[4] for Pd–Ni alloys gives $\alpha = 1.0$ and $\gamma = 0.37$ (see Fig. 1). Substitution of these values into Eq. (4) yields values for $\chi(q, 0)$ which are close to the computed values of Diamond.[5]

The resistivity due to spin fluctuations[6] ρ_s at a temperature T is given by

$$\rho_s = \frac{\rho_0}{kT} \int_0^\infty \frac{\omega \, d\omega}{(e^{\omega/kT} - 1)(1 - e^{-\omega/kT})} \sum_q q |F(q)|^2 \operatorname{Im} \chi_{eff}(q, \omega) \quad (5)$$

where $F(q)$ is a structure factor. A detailed analysis of Eq. (5) in conjunction with the use of the shape of the Fermi surface for d holes (described above) gives the following expression for the dominant contribution to ρ_s at low temperature in Pd–Ni alloys:

$$\rho_s = AT^2, \quad \frac{A - A_{Pd}}{A_{Pd}} \simeq \frac{n_{Ni}(\Delta I)^2 \sum_q [\chi_{eff}(q, 0)]^2}{[1 - \Delta I \chi_R^{loc}(0)]^2} \quad (6)$$

Resistivity of Paramagnetic Pd–Ni Alloys

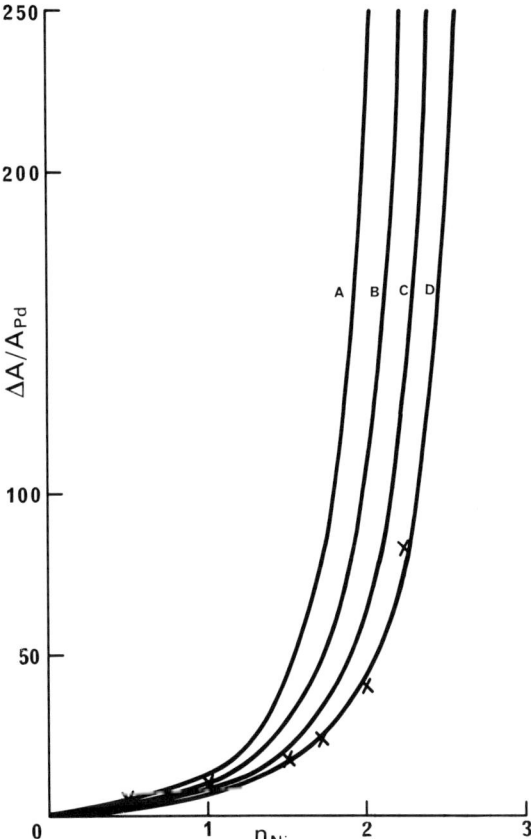

Fig. 2. Fractional change in the coefficient A of the T^2 term in the spin-disorder resistivity with impurity concentration. The parameter α has the value 1.00 for all curves. γ has the values 0.37, 0.372, 0.374, and 0.376 for curves A–D, respectively.

where $\chi_R^{loc}(0)$ is the real part of $\chi^{loc}(0)$. From Eqs. (1)–(4) and Eq. (6) we have $A/A_{Pd} \propto \chi_{Pd-Ni}^{bulk}/\chi_{Pd}^{bulk}$ near the critical point; i.e., A scales with χ^{bulk}. Figure 2 shows A/A_{Pd} as a function of concentration for $\alpha = 1$ and four values of γ close to $\gamma = 0.37$.

Tari and Coles[2] have shown that the resistivity of Pd–Ni near the critical concentration is given by $\rho_s = AT^n$, where $1 < n < 2$. The experimental values of A are given by curves in Fig. 2 and lie below curve A, which corresponds to $\alpha = 1.0$, $\gamma = 0.37$. Kaiser and Doniach[6] have shown that ρ_s has a T^2 behavior for $T < T_s$ and a linear behavior in T for $T \gtrsim T_s$, where T_s is the degeneracy temperature of the spin fluctuations, which tends to zero as the critical concentration n_c is approached. Coles[7] points out that his measurements near n_c may have been made at too high a temperature to observe a T^2 law, so that he sees a power law intermediate between the quadratic and the linear behavior. It can also be shown, using Ref. 5, that the coefficient of T^n ($n < 2$) in ρ_s should be smaller than that for T^2. This is exactly what is predicted in Fig. 2. Experiments at lower temperatures are thus needed to examine the concentration dependence of the T^2 term in ρ_s near n_c.

References

1. R. Harris and M.J. Zuckermann, *Phys. Rev. B* **5**, 101 (1972).
2. A. Tari and B.R. Coles, *J. Phys. F* **1**, L69 (1971).
3. P. Lederer and D.L. Mills, *Phys. Rev.* **165**, 837 (1968).
4. A.I. Schindler and M.C. Mackliet, *Phys. Rev. Lett.* **20**, 15 (1968).
5. J.B. Diamond, *Intern. J. Magnetism* **2**, 241 (1972).
6. A.B. Kaiser and S. Doniach, *Intern. J. Magnetism* **1**, 11 (1970).
7. B.R. Coles, private communication.

Anomalous Low-Temperature Specific Heat of Dilute Ferromagnetic Alloys

M. Héritier and P. Lederer

Laboratoire de Physique des Solides
Université de Paris-Sud, Orsay, France

A few years ago Lederer and Mills [1-3] studied the dilute paramagnetic alloy of nearly ferromagnetic metals containing isoelectronic impurities of ferromagnetic metals. They showed that many properties of these alloys might well be explained if the local properties of the susceptibility in the impurity vicinity were treated in detail. They assumed that the main impurity effect is to enhance the intraatomic Coulomb interaction at the impurity site.[4,5] In this model the free energy has been calculated in the RPA.[6] The paramagnon scattering by the impurity gives a term proportional to the number of paramagnons at temperature T, which corresponds to an additional term in the specific heat linear in temperature. An experimental example is given by the γ measurements on Pd–Ni.[7]

It is interesting to examine the symmetric situation, illustrated by Ni–Pd alloys, of a ferromagnetic host containing dilute isoelectronic impurities of nearly ferromagnetic metal. In the same model (local diminishing of the intraatomic Coulomb potential) one would expect the spin wave scattering by the impurities to add to the free energy of the pure metal a term proportional to the number of magnons, thus to $T^{3/2}$ for a quadratic spin wave dispersion law, leading to a $T^{1/2}$ term in the specific heat.

The d electrons of the pure metal are described by a Hubbard Hamiltonian

$$H_0 = T + U \sum_i n_{i\uparrow} n_{i\downarrow}$$

where T represents the kinetic energy and the second term is the interaction between the d electrons. The impurity at site O adds a new term in the Hamiltonian:

$$H_1 = -(\Delta U) n_{o\uparrow} n_{o\downarrow}, \quad \Delta U > 0$$

We neglect the induced electrostatic potential, assumed to be small in the case of an isoelectronic impurity. We consider alloys dilute enough to neglect the pair effects. Consequently, we only need to treat a one-impurity problem. We are only interested in the magnon scattering, responsible for the $T^{1/2}$ term in the specific heat. Thus we shall assume the existence of an energy gap Δ for the single-particle excitations, and a low temperature compared to Δ.

Ring and Ladder Diagram Contribution

The change in free energy is calculated by summing the diagrams of Fig. 1 to all orders in ΔU. The propagators appearing in these diagrams are the exact propa-

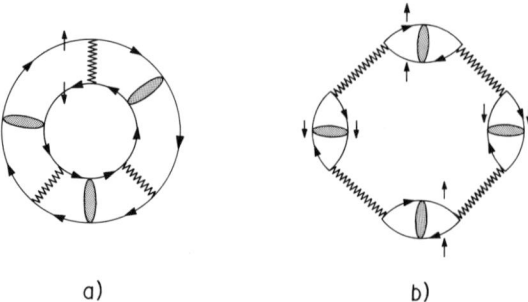

Fig. 1. Free energy expansion when the propagator correction is neglected. Wiggly lines represent the ΔU interaction. The shaded ovals are symbols for the exact particle–hole correlation function of the metal.

gators of the pure metal. The diagrams of Fig. 1(b) correspond to longitudinal excitations, negligible in our model. The diagram summation in Fig. 1(a) gives a correction to the free energy

$$\Delta F_1 = (-2/\pi) \int_0^\infty d\omega f_B(\omega) \operatorname{Im} \ln \left| 1 + \Delta U \sum_{\mathbf{q}} \chi(\mathbf{q}, \omega) \right|$$

$f_B(\omega)$ is the Bose factor, and $\chi(\mathbf{q}, \omega)$ is the exact transverse susceptibility of the pure metal. With a quadratic dispersion law for the magnon

$$\chi(\mathbf{q}, \omega) = \langle S_z \rangle / (Dq^2 - \omega)$$

With this low-q result (obtained, for example, in the RPA), we have, to lowest order in temperature

$$\Delta F_1 = (-2/\pi) \int_0^\infty d\omega f_B(\omega) \alpha n(\omega)$$

$n(\omega) = (a^3/4\pi^2)(\omega^{1/2}/D^{3/2})$ is the magnon density of states, a is the unit cell dimension, and $\alpha = \pi(\Delta U)\langle S_z \rangle / |1 + 4\pi(\Delta U)\langle S_z \rangle / D(2\pi/a)^2|$. As expected, ΔF_1 is proportional to the number of magnons. The change in specific heat due to alloying is, per impurity, to lowest order in T,

$$(\Delta C) v_1 = (3/2) \pi^{1/2} \zeta(3/2) |\alpha k_B / D(2\pi/a)^2| |k_B T/D(2\pi/a)^2|^{1/2}$$

Effect of the Hartree–Fock Correction of Propagators

The preceding calculation is not a self-consistent RPA calculation. Indeed, we should have replaced all the pure metal propagators by the alloy propagators in the Hartree–Fock approximation. This makes the calculation to any order in ΔU complicated because of the wavevector-nonconserving scattering. Thus we have limited the calculation to second order. The nonnegligible diagrams at low temperature are represented in Figs. 2 (first order) and 3 (second order) (assuming all the up-spin states empty). We have computed the new terms in the RPA for U. Thus we should calculate $(\Delta C) v_1$ in the same approximation, which means taking the RPA value of D. The details of calculation will be published elsewhere.

Anomalous Low-Temperature Specific Heat of Dilute Ferromagnetic Alloys

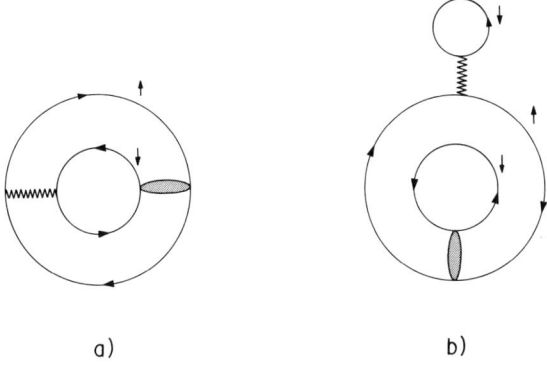

Fig. 2. First-order diagrams for the free energy. (a) Ladder term. (b) Hartree–Fock term.

First Order in ΔU. The new first-order term (Fig. 2b) gives a contribution to the free energy, to lowest order in T,

$$\Delta F_2^{(1)} = 2(\Delta U)\langle S_z\rangle \sum_{\mathbf{q}} f_B(Dq^2)$$

$\Delta F_2^{(1)}$ exactly cancels the first-order term in the expansion of ΔF_1. Because of this fortuitous cancellation, the only effect of the impurity, to first order, is to modify the coefficient of the magnon specific heat.

Second Order in ΔU. The Hartree–Fock correction (Fig. 3b) gives for the free energy, to second order in ΔU and to lowest order in T,

$$\Delta F_2^{(2)} = 2(\Delta U)^2 \sum_{\mathbf{q}} f_B(Dq^2) \Bigg\{ \langle S_z\rangle \sum_{\mathbf{q'}} \chi^0(0,\mathbf{q'}) + U\langle S_z\rangle \sum_{\mathbf{q'}} |\chi^0(0,\mathbf{q'})|^2$$
$$+ \langle S_z\rangle^2 \sum_{\mathbf{q'}} P\cdot \frac{1}{D(q^2-q'^2)} |[U\chi^0(0,q')]^2 - U\chi^0(0,q')| \Bigg\}$$

where $\chi^0(0,q)$ is the Hartree–Fock tranverse susceptibility of the metal at vanishing frequency. By adding to $\Delta F_2^{(2)}$ the second-order term of ΔF_1, we obtain a self-consistent evaluation of the free energy change per impurity, in the RPA for U and to second order in ΔU. The corresponding term in the specific heat for an impurity concentration C is

$$(\Delta C)v = Ck_B\pi^{3/2}(\tfrac{3}{2})\zeta(\tfrac{3}{2})\left[\frac{k_BT}{D(2\pi/a)^2}\right]^{1/2}\frac{(\Delta U)^2}{D(2\pi/a)^2}\Bigg\{\sum_{\mathbf{q'}}\lim_{q\to 0} P\cdot\frac{\langle S_z\rangle^2}{D(q'^2-q^2)}$$
$$\times \{1 - U\chi^0(q') + [U\chi^0(q')]^2\} - U\langle S_z\rangle\sum_{\mathbf{q'}}[\chi^0(q')]^2$$
$$- \langle S_z\rangle\sum_{\mathbf{q'}}\chi^0(\mathbf{q'})\Bigg\}$$

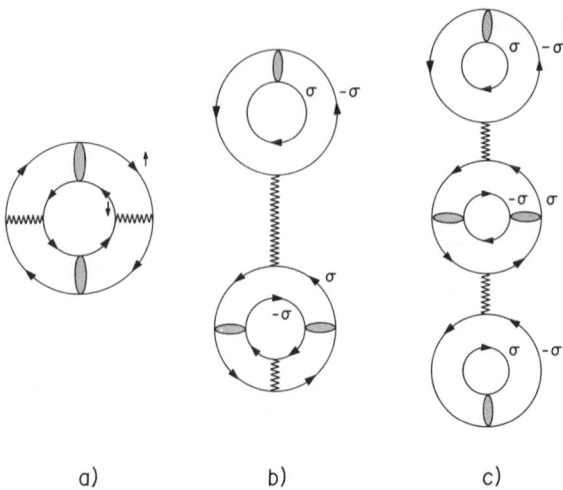

Fig. 3. Second-order diagrams. Ladder term (a) and Hartree–Fock corrections of first-order (b) and zeroth-order (c) ladder terms.

Conditions of Observation

The concept of a spin-wave scattering by one impurity embedded in a pure matrix is meaningless as soon as the magnon wavelength is longer than the mean distance between impurities. The smallest wave vector for which our calculation remains valid is $q_0 \simeq (\pi/a)\,C^{1/3}$. If all the wave vectors of excited magnons are shorter than q_0, the $T^{1/2}$ term is completely suppressed. Thus for this term to be observable, the temperature must satisfy

$$k_B T > k_B T_c = D(\pi/a)^2\,C^{2/3}$$

The second important point is the magnon dispersion law. In fact, because of the magnetocrystalline anisotropy, the true one is, rather than $\omega_q = Dq^2$

$$\omega_q = Dq^2 + E_a$$

Then, ΔF is proportional to

$$\int_0^\infty q^2 f_B(Dq^2 + E_a)\,dq \sim (k_B T)^{3/2} \sum_{n=1}^\infty n^{-3/2} e^{|-E_a/k_B T|}$$

Thus the $T^{1/2}$ term in the specific heat is frozen out below an anisotropy temperature. The second condition is

$$k_B T > k_B T_a = E_a$$

Third, the magnon scattering term must be dominant, compared to the electronic specific heat:

$$T < T_0 = (\beta/\gamma)^2$$

where β is the coefficient of the $T^{1/2}$ law.

T_a is about 0.1°K for Fe and Ni at 0°K, but is two orders of magnitude higher in Co. A low T_c requires very dilute alloy. For example, in nickel, with $D = 0.75 \times 10^{-6}$ erg/cm,[8] a temperature T_c of 1°K demands a Pd concentration not higher than 500 ppm. Even at so low a concentration, still in the nickel case ($\gamma = 7.02$ mJ/mole-deg^2,[9] $\Delta U \sim 1$ eV), $T_0 \sim 15$°K, larger than T_a and T_c.

In conclusion, an experiment on Ni–Pd with 500 ppm Pd should exhibit a $T^{1/2}$ law in the specific heat between 1 and 15°K, although we cannot exclude the possibility that this singular behavior is spurious and would disappear in a higher-order approximation. This effect is analogous to the $T^{3/2}$ term in the resistivity of various dilute ferromagnetic alloys, which has been interpreted as the contribution to the transport relaxation rate of wavevector-nonconserving electron–magnon scattering processes.[10]

References

1. P. Lederer and D.L. Mills, *Sol. St. Commun.* **5**, 131 (1967).
2. P. Lederer and D.L. Mills, *Phys. Rev.* **165**, 837 (1968).
3. P. Lederer and D.L. Mills, *Phys. Rev. Lett.* **20**, 1036 (1968).
4. P. Lederer and A. Blandin, *Phil. Mag.* **14**, 363 (1966).
5. H. Yamada and M. Shimizu, *J. Phys. Soc. (Japan)* **28**, 327 (1970).
6. S. Engelsberg, W.F. Brinkman, and S. Doniach, *Phys. Rev. Lett.* **20**, 1040 (1968).
7. G. Chouteau, R. Fourneau, R. Tournier, and P. Lederer, *Phys. Rev. Lett.* **18**, 1125 (1968).
8. H. Nose, *J. Phys. Soc. (Japan)* **16**, 2475 (1961).
9. W.H. Lien and N.E. Phillips, *Phys. Rev.* **133**, 1370 (1964).
10. D.L. Mills, A. Fert, and I.A. Campbell, *Phys. Rev. B* **4**, 196 (1971).

12
Theory

Charge Transfer and Spin Magnetism of Binary Alloys*

H. Fukuyama

Division of Engineering and Applied Physics
Harvard University, Cambridge, Massachusetts

The spin susceptibility of interacting electrons represented by

$$\mathcal{H} = \sum_{i,j,\sigma} t_{ij} a^+_{i,\sigma} a_{j,\sigma} + U \sum_i n_{i,\sigma} n_{i,-\sigma} \tag{1}$$

is given as follows in a Hartree–Fock approximation[1]:

$$\chi(Q,\omega) = 2\mu_B^2 \frac{\Pi(Q,\omega)}{[1 - U\Pi(Q,\omega)]} \tag{2}$$

$$\Pi(Q,\omega) = -\sum_k \frac{f_k - f_{k+Q}}{\varepsilon(k) - \varepsilon(k+Q) - \omega - i0^+} \tag{3}$$

where $\varepsilon(k) = \sum_j t_{ij} \exp[ik(R_i - R_j)]$, and f_k and μ_B are the Fermi distribution function with energy $\varepsilon(k)$ and the Bohr magneton, respectively.

The aim of the present series of papers[2,3] is to see how Eq. (2) is modified in nondilute alloys represented by

$$\mathcal{H} = \sum_{i,j,\sigma} t_{ij} a^+_{i,\sigma} a_{j,\sigma} + \sum_{i,\sigma} \varepsilon_i^0 n_{i,\sigma} + \sum_i U_i n_{i,\sigma} n_{i,-\sigma} \tag{4}$$

where ε^0 and U_i can take two values ε_A^0 or ε_B^0 and U_A or U_B depending on the sites. t_{ij} is assumed to be independent of these kinds of atoms.

In an earlier paper[2] the spin susceptibility $\chi^0(Q,\omega)$ of the alloys without Coulomb interactions [i.e., $U_i = 0$ but $\varepsilon_A^0 \neq \varepsilon_B^0$ in Eq. (4)] was obtained. By assuming that the total susceptibility $\chi(Q,\omega)$ is given as

$$\chi(Q,\omega) = \chi^0(Q,\omega)[1 - U\chi^0(Q,\omega)/2\mu_B^2]^{-1} \tag{5}$$

for the system where effective Coulomb interactions do not change appreciably between A and B sites, we discussed the possible phase diagram of the kinds of magnetic states in the U–n plane (where n is the carrier number) just as in Penn's argument[4] in pure systems.

However, Eq. (5) is not valid if U_A is much different from U_B. This problem is considered in Ref. 3, where the diagrammatic techniques were for the first time developed to treat two-body interactions in a scheme involving the coherent-potential and Hartree–Fock approximations. By these procedures, we examined the meaning of the two different formulations so far proposed for Eq. (4), one by

---

* Supported in part by Grant No. GP-16504 of the National Science Foundation.

Hasegawa and Kanamori (HK)[5] (Levin et al.[6] adopted the same approximation), and the other by Harris and Zuckermann.[7] These two kinds of approximations are found to be complementary in the sense that the first takes the local energy shift through Coulomb forces, whereas the latter focuses exclusively on the dynamic properties of two-body interactions.

For further investigations we adopted the scheme of HK, where the atomic energy levels $\varepsilon_{i,\sigma}$ in the presence of U_i are taken as

$$\varepsilon_{i,\sigma} = \varepsilon_i^0 + U_i \langle n_{i,-\sigma} \rangle \tag{6}$$

and the $\varepsilon_{i,\sigma}$ are treated in the CPA.[8] In Eq. (6) $\langle n_{i,-\sigma} \rangle$ is the average electron number at ith site and is assumed to take two values $\langle n_{A,-\sigma} \rangle$ and $\langle n_{B,-\sigma} \rangle$ depending on the kind of atom at that site. Since $\langle n_{A,-\sigma} \rangle$ and $\langle n_{B,-\sigma} \rangle$, which are determined self-consistently, are not necessarily equal to the valence number of respective atoms, there exists a finite amount of charge transfer between A and B sites. This charge transfer is absent in a pure system, whereas it is shown to be important in alloys. In order to avoid unnecessary complications, we consider first the Wolff model[9] ($U_B = 0, U_A \neq 0$) although it is straightforward to include U_B. The dynamic spin susceptibility of this system in the paramagnetic state is given by

$$\chi(Q,\omega) = 2\mu_B^2 \left[\Pi_0(Q, i\omega_v) + x^{-1} U_{\text{eff}}(i\omega_v) \frac{\Pi_1(Q, i\omega_v)^2}{1 - x^{-1} U_{\text{eff}}(i\omega_v) \Pi_2(Q, i\omega_v)} \right]_{i\omega_v \to \omega + i0^+}$$

where x is the atomic concentration of magnetic A atoms, and

$$\Pi_0(Q, \omega) = -T \sum_{\varepsilon_n} A\eta \tag{8}$$

$$\Pi_1(Q, \omega) = -T \sum_{\varepsilon_n} A\eta\zeta \tag{9}$$

$$\Pi_2(Q, \omega) = -T \sum_{\varepsilon_n} A\eta\zeta^2 \tag{10}$$

$$U_{\text{eff}}(i\omega_v)^{-1} = U_A^{-1} - \alpha \tag{11}$$

By use of the Green's function G given by

$$G \equiv G(k, i\varepsilon_n) = [i\varepsilon_n - \varepsilon(k) - \Sigma(i\varepsilon_n)]^{-1} \tag{12}$$

and

$$G_+ \equiv G(k+Q, i\varepsilon_n + i\omega_v), \quad \Sigma_+ \equiv \Sigma(i\varepsilon_n + i\omega_v)$$

the quantities A, η, ζ and α appearing in Eqs. (8)–(11) are defined by

$$A = \sum_k GG_+ \tag{13}$$

$$\eta = [1 - \delta^{-1} A(\Sigma - \Sigma_+)(F - F_+)^{-1}]^{-1} \tag{14}$$

$$\zeta = \delta^{-1}(F\Sigma - F_+\Sigma_+)(F - F_+)^{-1} \tag{15}$$

$$\alpha = -(\delta x)^{-1} T \sum_{\varepsilon_n} FF_+(\Sigma - \Sigma_+)(F - F_+)^{-1} \tag{16}$$

where
$$F = \sum_k G(k, i\varepsilon_n) \tag{17}$$
and
$$\delta \equiv \varepsilon_A - \varepsilon_B = \varepsilon_A^0 - \varepsilon_B^0 + U_A \langle n_{A,-\sigma} \rangle$$
$$\equiv \delta_0 + U_A \langle n_{A,-\sigma} \rangle \tag{18}$$

The differences among the three Π_λ, $\lambda = 0, 1, 2$, lie in the factors ζ and ζ^2 in Eqs. (9) and (10), respectively. This comes from the dependence of scattering potential δ, Eq. (18), on degree of disorder through $\langle n_{A,-\sigma} \rangle$, i.e., charge transfer effects. Each of the Π_λ is interpreted as follows by use of two different kinds of spin fluctuation operators:

$$S_z(Q) = \sum_{i \in \text{all sites}} e^{iQR_i} \sigma_z(i), \quad S_z^{(A)}(Q) = \sum_{i \in A \text{ sites}} e^{iQR_i} \sigma_z(i)$$

where $\sigma_z(i)$ is the z component of Pauli spin operator at the ith site. Π_0 is the correlation function of two $S_z(Q)$'s under the assumption that every site is equivalent in the sense of the effective medium. Π_1 is the correlation function between $S_z(Q)$ and $S_z^A(Q)$, or the correlation function between spins at magnetic sites and spins in the effective medium. Π_2 is the correlation function between two $S_z^A(Q)$'s.

Some of the implications of Eq. (7) are as follows.

1. $\chi(Q, \omega)$ reduces to the function for free Bloch electrons at $x = 0$, and to Eq. (2) at $x = 1$, respectively.

2. $\lim_{Q \to 0} \lim_{\omega \to 0} \chi(Q, \omega)$ reduces to the function of Refs. 5 and 6.

3. In the limit of small x the criterion of local moment formation is given by $U_{\text{eff}}(0)^{-1} = 0$. The solution agrees with that of Moriya,[10] but is different from that of Wolff[9] since Wolff does not include the modification of the atomic potential by Coulomb interactions at the magnetic site; i.e., δ_0 appears instead of δ.

4. The results in the case of small x due to Lederer and Mills[11] and Engelsberg et al.[12] do not include the factors ζ and ζ^2 in Π_1 and Π_2, respectively. In order to incorporate the fact that the exchange enhancement exists at magnetic sites only, we need these factors.

Some of the numerical results for alloys at finite concentration including paramagnons are given for simple cubic crystals in the tight-binding approximation.

References

1. T. Izuyama, D.J. Kim, and R. Kubo, *J. Phys. Soc. Japan* **18**, 1025 (1963).
2. H. Fukuyama, *Phys. Rev. B* **5**, 2872 (1972).
3. H. Fukuyama, submitted to *Phys. Rev.*
4. D. Penn, *Phys. Rev.* **142**, 350 (1966).
5. H. Hasegawa and J. Kanamori, *J. Phys. Soc. Japan* **31**, 382 (1971).
6. K. Levin, R. Bass, and K.H. Bennemann, *Phys. Rev. Lett.* **27**, 589 (1971), and preprint.
7. R. Harris and M.J. Zuckermann, *Phys. Rev.* **B5**, 101 (1972).
8. P. Soven, *Phys. Rev.* **156**, 809 (1967); B. Velicky, S. Kirkpatrick, and H. Ehrenreich, *Phys. Rev.* **175**, 747 (1968).
9. P.A. Wolff, *Phys. Rev.* **124**, 1030 (1961).
10. T. Moriya, *Prog. Theoret. Phys.* **34**, 329 (1965).
11. P. Lederer and D.L. Mills, *Phys. Rev. Lett.* **20**, 1036 (1968).
12. S. Engelsberg, W.F. Brinkman, and S. Doniach, *Phys. Rev. Lett.* **20**, 1040 (1968).

Coherent Potential Approximation for the Impure Heisenberg Ferromagnet

R. Harris, M. Plischke, and M.J. Zuckermann

Eaton Electronics Laboratory
McGill University, Montreal, Canada

Recent experiments of Svensson et al.[1] have shown that well-defined spin-wave modes exist in disordered antiferromagnets. The localized modes which exist in dilute alloys persist and become dispersive as the concentration is increased. In this paper we present a theory of the disordered Heisenberg ferromagnet which qualitatively reproduces some of the features found experimentally.

In the linear spin-wave approximation (valid at low temperatures) the nearest-neighbor anisotropic Heisenberg Hamiltonian is

$$H = -\tfrac{1}{2} \sum_{ij} J_{ij}(S_i^x S_j^x + S_i^y S_j^y + \Delta_{ij} S_i^z S_j^z)$$

$$\simeq -\tfrac{1}{2} \sum_{ij} \Delta_{ij} J_{ij} S_i S_j - \sum_{ij} J_{ij}(S_i S_j)^{1/2} (a_i^+ a_j + a_j^+ a_i)$$

$$+ \sum_{ij} \Delta_{ij} J_{ij} S_j a_i^+ a_i \qquad (1)$$

Here the a_i^+ and a_i are the boson spin-wave operators. The coupling constants J_{ij} and anisotropy parameters Δ_{ij} depend on the occupation of sites i and j, i.e., $J_{ij} = J_{AA}, J_{AB}, J_{BB}$ for (i,j) nearest neighbors and the spin size S_i may be S_A or S_B. This Hamiltonian is now formally equivalent to the simple electronic Hamiltonian used in the theory of binary alloys except for the last term. This term depends in detail on the configuration of atoms surrounding the central atom i, and a correct treatment requires a cluster theory. We have chosen, for computational simplicity, to approximate this term by its virtual crystal value

$$\sum_j \Delta_{ij} J_{ij} S_j a_i^+ a_i = \begin{matrix} z\{(1-C)J^{AA}\Delta_{AA}S_A + CJ^{AB}\Delta_{AB}S_B\} a_i^+ a_i \\ z\{(1-C)J^{BA}\Delta_{BA}S_A + CJ^{BB}\Delta_{BB}S_B\} a_i^+ a_i \end{matrix}$$

where z is the coordination number of the lattice and C is the concentration of B-type atoms. This approximation prevents us from recovering the correct dilute limit,[2] but we expect the general features of our results to be correct, especially in the middle concentration range.

We apply the generalized single-site coherent potential approximation of Blackman et al.[3] to the Hamiltonian of Eq. (1) and calculate the density of states, the magnon dispersion relation, and the quasiparticle lifetimes over a wide range of concentrations, first for $J_{ij} = J, \Delta_{ij} = \Delta$ independent of the occupation of sites i, j, and $S_B/S_A = 4$. We find localized states at low concentrations at both levels of

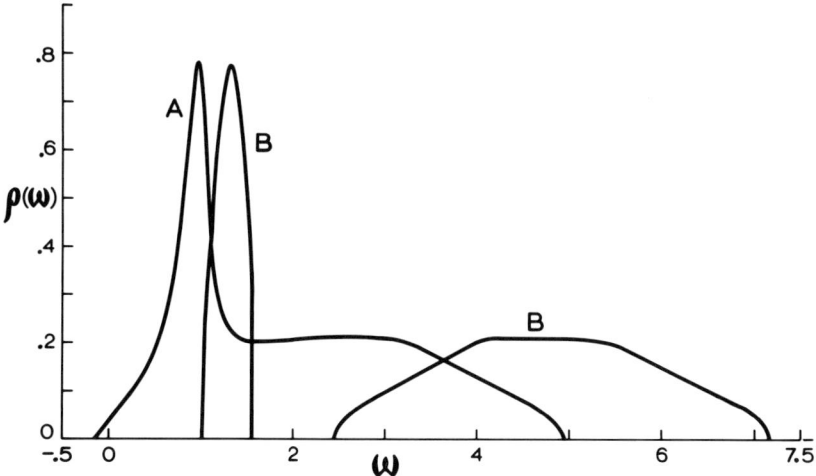

Fig. 1. The density of states $\rho(\omega)$ as a function of ω for $J_{BB}/J_{AA} = 6$, $J_{AB}/J_{AA} = \sqrt{6}$, $S_B = S_A$, $C = 0.7$. (A) $\Delta_{AA} = \Delta_{AB} = \Delta_{BB} = 1$; (B) $\Delta_{AA} = 1$, $\Delta_{AB} = \sqrt{2}$, $\Delta_{BB} = 2$.

the spectrum. These localized states become dispersive at a concentration somewhat greater than the site percolation concentration of the simple cubic lattice. The spin-wave band is shifted and broadened as the concentration of impurities is raised.

We then consider the case of equal spins but different coupling constants. In Fig. 1 the density of states is shown for $J_{BB}/J_{AA} = 6$, $J_{AB} = (J_{AA}J_{AB})^{1/2}$ and for $C = 0.7$, both for $\Delta_{AA} = \Delta_{AB} = \Delta_{BB}$ and $\Delta_{AB} = (\Delta_{AA}\Delta_{BB})^{1/2}$. Both the pure-$A$ spin-wave modes and the pure-B modes are quite recognizable, although they are somewhat shifted. In Fig. 2 we show the spin-wave dispersion relation $\omega(k)$ in the (1, 1, 1) direction along with the two pure spin-wave dispersion curves ($C = 0, C = 1$) for $\Delta_{AB} = (\Delta_{AA}\Delta_{BB})^{1/2}$, $\Delta_{BB} = 2\Delta_{AA}$. Both modes are displaced, but only slightly dis-

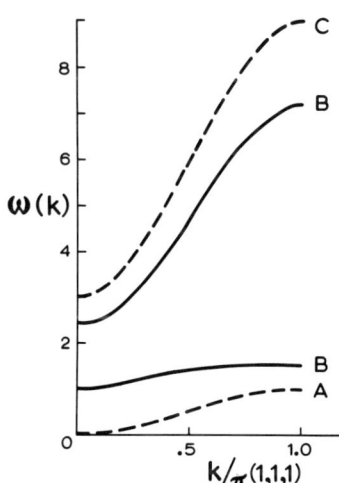

Fig. 2. Plot of the magnon dispersion $\omega(k)$ as function of k in the (1, 1, 1) direction. $J_{BB}/J_{AA} = 6$, $J_{AB}/J_{AA} = \sqrt{6}$, $S_B = S_A$, $\Delta_{AA} = 1, \Delta_{AB} = 2, \Delta_{BB} = 2$. (A) $C = 0$; (B) $C = 0.7$; (C) $C = 1.0$. Curves (A) and (C) are shown dashed.

torted. The displacement is exaggerated because of our virtual crystal approximation of the S^zS^z term. This persistence of the pure modes is one of the experimental results of Svensson et al.[1]

References

1. E.C. Svensson, W.J.L. Buyers, T.M. Holden, R.A. Cowley, and R.W.H. Stevenson, *Can. J. Phys.* **47**, 1983 (1969); W.J.L. Buyers, T.M. Holden, E.C. Svensson, R.A. Cowley, and R.W.H. Stevenson, *Phys. Rev. Lett.* **27**, 1442 (1971).
2. T. Wolfram and J. Calloway, *Phys. Rev.* **130**, 2207 (1963); Yu. A. Izyumov, *Adv. Phys.* **14**, 569 (1965); D. Hone, H. Callen, and L. Walker, *Phys. Rev.* **144**, 283 (1966).
3. J.A. Blackman, N.F. Berk, and D.M. Esterling, *Phys. Lett.* **35A**, 205 (1971); *Phys. Rev. B* **4**, 2412 (1971).

Resistivity of Nearly Antiferromagnetic Metals

P. Lederer, M. Héritier, and A.A. Gomès*

Laboratoire de Physique des Solides
Université de Paris-Sud, Orsay, France

Introduction

The effect of spin fluctuations in nearly ferromagnetic metals (nfm) on such properties as resistivity and electronic specific heat has been discussed recently by several authors.[1,2] In these systems, which are close to satisfying Stoner's criterion for ferromagnetism, large-amplitude, low-frequency spin fluctuations occur for small wave vectors **q**. Although heavily damped, these electron–hole collective excitations renormalize the electron mass, which can become large, inducing important changes in the electronic specific heat. Another effect, in transition metals like Pd, for instance, is conduction electron scattering, induced by the $s-d$ exchange coupling, which connects the spin fluctuations within the nearly ferromagnetic d band and the s-like conduction states. As discussed in detail previously,[3] ferromagnetic spin fluctuations introduce a term proportional to $\chi^{1/2}T^2$ in the resistivity, where χ is the static d-spin susceptibility, which diverges at the ferromagnetic instability. Recently the effect of spin fluctuations in nearly antiferromagnetic metals (nam) on the electronic specific heat was discussed by several authors.[4,5] In these systems the real part $\text{Re}\,\chi(q, \omega)$ of the dynamic susceptibility exhibits a maximum for low frequencies around a finite wave vector $Q \neq 0$. The Stoner criterion $U \text{Re}\,\chi(q, 0) = 1$ then predicts spin fluctuations of wave vector $\mathbf{Q} + \mathbf{q}$, $\mathbf{q}$ being small. The width $\Delta\mathbf{q}$ of the spin fluctuation spectrum depends on the details of the band structure. The antiferromagnetic spin fluctuations occur around a finite wave vector, and are not isotropic in reciprocal space, i.e., $\text{Im}\,\chi(q, \omega)$ depends not only on the magnitude of q, but on its direction. This introduces new features in the scattering properties.

The Model

We consider transition metals where two overlapping s and d bands exist and play a different role: Due to the high effective mass of the d band, s electrons are responsible for electric conduction. Magnetic properties are essentially associated to the narrow d band, which is assumed to be described by a Hubbard-like Hamiltonian.[1] The interaction term between s and d electrons is given by $s-d$ exchange interactions, from which one retains only the spin-flip part[1]:

---

* Permanent address: Centro Brasileiro de Pesquisas Fisicas, Rio de Janeiro, Brazil.

$$H_{s-d} = J \int s_+(\mathbf{r}) S_-(\mathbf{r}) \, d\mathbf{r} + \text{hc} \tag{1}$$

where $s_+(\mathbf{r})$ and $S_-(\mathbf{r})$ are the raising and lowering spin operators for the s and d band, respectively, and J is the coupling parameter. Within this model one calculates the transition amplitude for a spin-flip process using the Fermi golden rule to get

$$W_{sf}(\mathbf{k}, \mathbf{k}') \simeq J^2 f(\varepsilon_\mathbf{k}) [1 - f(\varepsilon_{\mathbf{k}'})] [1 + n_B(\Omega(\mathbf{k}, \mathbf{k}'))]$$
$$\times |F(\mathbf{k} - \mathbf{k}')|^2 \operatorname{Im} \chi(\mathbf{k} - \mathbf{k}', \Omega(\mathbf{k}, \mathbf{k}')) \tag{2}$$

where $f(\varepsilon_\mathbf{k})$ and $f(\varepsilon_{\mathbf{k}'})$ are the Fermi functions, $n(x)$ is the Bose function associated with spin fluctuations, and $F(\mathbf{k} - \mathbf{k}')$ is the form factor for d electrons.

The imaginary part of the susceptibility is approximated for nam and low frequencies ω (such that $\omega/\varepsilon_d F \ll 1$) by

$$\operatorname{Im} \chi(q, \omega) = g(\mathbf{q}, \mathbf{Q}) \omega / [1 - u \operatorname{Re} \chi(\mathbf{Q}, \mathbf{q})]^2 \tag{3}$$

where $g(q, Q)$ is regular for $q = 0$.[5]

Calculation of the Resistivity

In a sample of volume unity, calling $\mathbf{X}$ the wave vector in the direction of the current and $\mathbf{V}(\mathbf{k})$ the velocity of an s electron of wave vector $\mathbf{k}$, the relaxation time τ is given by

$$\frac{1}{\tau} = \frac{m_s}{n_s k_B T} \sum_{\mathbf{k}, \mathbf{k}'} [\mathbf{X} \cdot \mathbf{V}(\mathbf{k})] \times \{\mathbf{X} \cdot [\mathbf{V}(\mathbf{k}) - \mathbf{V}(\mathbf{k}')]\} \times W(\mathbf{k}, \mathbf{k}') \tag{4}$$

We shall only be interested in states $\mathbf{k}$ and $\mathbf{k}'$ that lie close to the s-electron Fermi surface. This expression can be considerably simplified when the Fermi surface is spherical and when $\operatorname{Im} \chi(q, \omega)$ depends only on the magnitude of $\mathbf{q}$ and not its direction, which is a reasonable assumption for a nearly ferromagnetic metal. For a nearly antiferromagnetic metal, however, this simplification does not occur. Let us consider the scattering of a conduction electron of wave vector $\mathbf{k}$ [see Eq. (2)] near the antiferromagnetic instability. The final state for electrons that are critically scattered is necessarily within $\Delta\mathbf{q}$ of a conduction electron state of momentum $\mathbf{k} + \mathbf{Q}$ on the Fermi surface. This implies that only conduction states separated by wave vectors $\mathbf{Q} + \mathbf{q}$, where $-\Delta\mathbf{q} \lesssim \mathbf{q} \lesssim +\Delta\mathbf{q}$, are critically scattered by nearly antiferromagnetic spin fluctuation. This strongly contrasts with the nfm case, where spin fluctuations with small wave vectors $\mathbf{q}$ are present. In this case conduction states $\mathbf{k}$ and $\mathbf{k}'$ close to one another may be scattered into one another wherever they lie on the Fermi surface. The discussion above shows that only parts S_1 of the Fermi surface can be affected by the scattering due to nearly antiferromagnetic spin fluctuations (see Fig. 1).

Discussion

Let us now discuss the implications of this fact. We consider two types of electron scattering: non-spin-fluctuation scattering, for instance, impurity scattering, phonons, etc., and the scattering by spin fluctuations. Let us call $\tau_0(k)$ and $\tau_{sf}(k)$

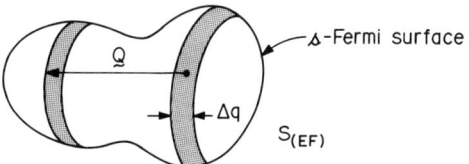

Fig. 1. The shaded areas show the parts of the Fermi surface of the s band that are affected by the scattering off antiferromagnetic spin fluctuations.

the corresponding relaxation times. $\tau_{sf}(k)$ can become very large over the part S_1 of the Fermi surface (Fig. 1) and is of the form αT^2, where α may diverge at the instability. Over S_1

$$1/\tau(\mathbf{k}) = W_{sf}(\mathbf{k}) + W_0(\mathbf{k}) \approx W_{sf}(\mathbf{k}) = 1/\tau_{sf}(\mathbf{k})$$

For spherical Fermi surfaces and assuming that $\tau_i(k)$ is only energy dependent, one gets

$$\rho = \rho_0 S/(S - S_1) \tag{5}$$

S being the total area of the Fermi surface. $S_1 \sim 4\pi k_F \Delta q$ and $S \sim 4\pi k_F^2$. Thus

$$\rho \simeq \rho_0 k_F/(k_F - \Delta q)$$

Thus, contrary to the nfm case, where a large T^2 term appears due to scattering of spin fluctuations, the resistivity in the nam case is enhanced by a temperature-independent factor.

A simple interpretation of this effect is provided by two parallel resistors R_1 and R_2. The resistor R_1 corresponds to impurity and phonon scattering associated with the portion $S - S_1$ of the Fermi surface. The resistor R_2 corresponds to the part S_1 of the Fermi surface and is associated both with spin fluctuation scattering and with the remaining scattering processes.

Then the total resistance for the parallel case is given by

$$R = \frac{R_1 R_2}{R_0 + R_2} = \frac{\alpha \rho_1 (\rho_1 + \rho_2)}{S\rho_1 + S_1 \rho_2} \tag{6}$$

Suppose ρ_2 tends to infinity (as one approaches the antiferromagnetic instability); then one gets from (6)

$$R \simeq \rho_1 \alpha / S_1 = \rho_0 S/(S - S_1)$$

In fact, $\rho_2 = \beta T^2$, with $\beta \to \infty$. At very low temperatures ($\rho_2 < \rho_1$) $R \simeq \rho_0 \{1 + [(S - S_1)/S] \beta T^2/\rho_1\}$.

Our results hold only for the following range of parameters: $\Delta_q \ll k_F, Q < k_F$. Should Q turn out to be larger than $2k_F$, then no part of the s Fermi surface is sensitive to the spin fluctuations in the d band. On the contrary, should Q and Δq both be of the order of k_F, then the whole of the Fermi surface is susceptible to scattering from the spin fluctuations and one should be able to have a sizable T^2 term in a nam. The magnitude of this term is difficult to estimate, though, because of the lack of symmetry in the problem. A large T^2 term in the resistivity has been predicted to exist in nearly antiferromagnetic systems and nearly excitonic systems by other authors.[6] However, the latter work deals with semimetals, where the effective mass of the carriers is equal in the valence and the conduction bands. An average is performed over the Fermi surface for the scattering rate due to spin fluctuations. Such an average is not allowed in the case we consider here.

Experimental Results

Recent measurements were performed on actinides, carbides, nitrides, and admixtures of them.[7] In particular, $U(C_{1-x}N_x)$ shows interesting features. It is known that when alloying from UC to UN an antiferromagnetic ordering sets in at 0°K. This antiferromagnetism is of itinerant character. It seems reasonable that for large concentrations of N in $U(C_{1-x}N_x)$ the system is nearly antiferromagnetic. In this case, however, due to the existence of a delocalized f band, spin fluctuations now occur within this band, in contrast to transition metals, where d states are responsible for them. It is striking that the residual resistivity goes through a maximum in this system for the concentration at which T_N goes to zero, as expected from our model. We would like to emphasize that such an effect is very difficult to understand on the basis of usual ideas about spin fluctuation scattering since the latter, being inelastic, always produces a temperature-dependent resistivity which vanishes at low temperature: spin fluctuations usually are not expected to have an effect on the residual resistivity of impurities.

References

1. D.L. Mills and P. Lederer, *J. Phys. Chem. Solids* **27**, 1805 (1966).
2. S. Doniach and S. Englesberg, *Phys. Rev. Lett.* **17**, 750 (1966); N.F. Berk and J.R. Schrieffer, *Phys. Rev. Lett.* **17**, 433 (1966).
3. P. Lederer, *Lectures on L.S.F.*, Nato Summer School on Magnetism, La Colles/ Loup, 1970 (to be published).
4. J. Castaing, P. Costa, M. Héritier, and P. Lederer, *J. Phys. Chem· Solids* **33**, 533 (1972).
5. T. Moriya, *Phys. Rev. Lett.* **24**, 1433 (1970).
6. D. Jérome, *Sol. St. Comm.* **8**, 1793 (1970); D. Jérome, M. Rieux, and J. Friedel, *Phil. Mag.* **23**, 1061 (1971).
7. C. de Novion, Thèse, CEA, Saclay 1971.

Magnetic Properties of Electrons in a Narrow Correlated Energy Band

C. Mehrotra and K.S. Viswanathan

Department of Physics, Simon Fraser University
Burnaby, British Columbia, Canada

Introduction

The magnetic properties of the nondegenerate-band Hubbard model have been the subject of intensive study in recent years.[1,2] Starting from the atomic limit, Hubbard[3,4] studied the condition for ferromagnetism in this model. Hubbard and Jain[5] derived a general expression for the dynamic spin susceptibility $\chi(q,\omega)$ in a truncation scheme similar to Hubbard's[3] for the single-particle Green's function. Nagaoka[6] studied the nature of the ground state in the $U = \infty$ limit. He showed that the system with one hole will be ferromagnetic at $T = 0°K$. Izuyama[7-9] studied the sum rules for $\chi(q,\omega)$ and pointed out that complete ferromagnetic polarization violates the sum rules. In a forthcoming paper[10] we show that in the $U = \infty$ limit the Hubbard model does not exhibit a spontaneous magnetization at finite temperatures. Functional integral schemes[11-13] have not yielded results beyond the random phase approximation.

The antiferromagnetic properties, in the atomic limit, have not been studied as extensively as the ferromagnetic properties. The HF approximation[14] yields results in agreement with Slater's split band model.[15] Here the insulating nature of the system is tied up with the antiferromagnetic order. Many $3d$ transition metal oxides,[1] such as NiO, exhibit an antiferromagnetic phase transition but remain insulating on either side of the transition temperature. It would thus be interesting to ask if the Hubbard insulator can exhibit an antiferromagnetic ground state as well. Arai[16] calculated the properties of the antiferromagnetic solutions in Hubbard's truncation scheme. However, he failed to calculate the stability of the antiferromagnetic solutions. For an exactly half-filled model, i.e., one electron per atom, it was shown[17-19] that the antiferromagnetic solution in this scheme is unstable. We have been able to extend this result to arbitrary values of n ($0 < n < 2$; n is the number of electrons per atom) and showed[20] that if one considers the hopping term in this model to connect nearest neighbors only, then the Hubbard insulator does not have an antiferromagnetic phase. This result, in our opinion, is interesting, for, as mentioned above, most $3d$ transition metal oxides which are insulating also exhibit an antiferromagnetic phase where the insulating nature has been explained qualitatively on Hubbard's scheme. The question of a reasonable qualitative explanation of this situation, in our opinion, is still open.

Recently the properties of the Hubbard model in $U = \infty$ limit were investigated, especially by Brinkmann and Rice,[21] Sokoloff,[22-24] and Field.[25] Sokoloff studied the antiferromagnetic stability in a truncation scheme which conserves the zeroth and first moments of the spectral density function and concluded that it is unlikely that the system with one band will be antiferromagnetic. Field studied the ferromagnetic stability in the same scheme and showed that ferromagnetism is more likely to occur for a nearly half-filled band, provided the density of states is peaked near the chemical potential. However, as pointed out earlier, we have demonstrated that for $U = \infty$ the Hubbard model does not exhibit a spontaneous magnetization (or sublattice magnetization) for any finite temperature.

In this paper we discuss only the instability of the antiferromagnetic solution in the said truncation scheme.

The Antiferromagnetic Solution

We sketch below the essential steps in showing that within the Hubbard truncation scheme used by Arai[16] the antiferromagnetic ground state is unstable for any value of n ($0 < n < 2$). For details see Ref. 20.

We consider A- and B-type lattice sites such that an A-type lattice is the nearest neighbor to a B-type lattice site and vice versa. We express an antiferromagnetic solution by defining

$$\langle n_A^\uparrow \rangle = \langle n_B^\downarrow \rangle = \tfrac{1}{2}n(1 + \chi), \qquad \langle n_A^\downarrow \rangle = \langle n_B^\uparrow \rangle = \tfrac{1}{2}n(1 - \chi) \tag{1}$$

$n\chi$ is the sublattice magnetization per site, to be determined self-consistently. Assuming the hopping terms (t_{ij}) to connect nearest neighbors only and using a Hubbardlike truncation scheme, we obtain the following for the single-particle Green's functions:

$$\begin{bmatrix} G_{AA}^\sigma(\kappa,\omega) & G_{AB}^\sigma(\kappa,\omega) \\ G_{BA}^\sigma(\kappa,\omega) & G_{BB}^\sigma(\kappa,\omega) \end{bmatrix} = \frac{1}{(\omega - \xi_1)(\omega - \xi_2)(\omega - \xi_3)(\omega - \xi_4)}$$

$$\times \begin{bmatrix} \omega(\omega-U)[\omega-U(1-\langle n_A^{-\sigma}\rangle)] & \varepsilon_\kappa[\omega-U(1-\langle n_A^{-\sigma}\rangle)][\omega-U(1-\langle n_B^{-\sigma}\rangle)] \\ \varepsilon_\kappa[\omega-U(1-\langle n_A^{-\sigma}\rangle)][\omega-U(1-\langle n_B^{-\sigma}\rangle)] & \omega(\omega-U)[\omega-U(1-\langle n_B^{-\sigma}\rangle)] \end{bmatrix} \tag{2}$$

where the $\xi_i(\kappa)$ are the roots of the quartic equation

$$\omega^2(\omega - U)^2 - \varepsilon_\kappa^2[\omega - U(1 - \langle n_A^{-\sigma}\rangle)][\omega - U(1 - \langle n_B^{-\sigma}\rangle)] = 0 \tag{3}$$

and $\omega = \to \omega + \mu$ in the above (μ is the chemical potential).

Case 1: The Limit $U = \infty$ This has been considered by Sokoloff in a different approximation, but the conclusions remain the same. The self-consistency conditions reduce to solving the following pair of equations:

$$\frac{n}{1 - n/2} = \frac{2}{N}{\sum_k}' [f(\omega_1 - \mu) + f(-\omega_1 - \mu)] \tag{4}$$

$$1 = \frac{2}{N}{\sum_k}' [f(\omega_1 - \mu) + f(+\omega_1 - \mu)] \tag{5}$$

where the prime on the sums indicates they are taken over the inner half zone, and

$$\omega_1(k) = \varepsilon_\kappa [(1 - \tfrac{1}{2}n)^2 - \tfrac{1}{4}n^2\chi^2]^{1/2} \tag{6}$$

It is obvious that Eq. (5) can be satisfied at any T only if $\mu = \infty$, i.e., when $n = 1$. Otherwise, Eq. (5) can never be satisfied [the right side of Eq. (5) is always less than unity]. Now, $n = 1$ is an uninteresting case, for the system remains in the ground state at all temperatures. (Note that for $U = \infty$ the maximum number of electrons that can be accommodated is one per atom.)

Case 2: Finite U From Eq. (2) one can easily write down the self-consistency criterion for χ at any temperature T. From this equation, which must be satisfied if a nonzero χ were to exist at $T = 0°K$, the temperature T_N at which χ vanishes can be determined by setting $T = T_N$ and $\chi = 0$. One then gets an equivalent condition for the stability of the antiferromagnetic ground state. It can be further demonstrated that for $T \to \infty$ there is no self-consistent nonzero χ that will satisfy this equation. Hence it is sufficient to establish that there exists no $T_N(0 \leq T_N \leq \infty)$ that will satisfy the self-consistency criterion. The equations to solve are the following:

$$\frac{1}{U} = \frac{1}{N} \sum_k \frac{f_N(\alpha_2 - \mu) - f_N(\alpha_1 - \mu)}{[(U - \varepsilon_\kappa)^2 + 2Un\varepsilon_\kappa]^{1/2}} \tag{7}$$

where the sum is over the full zone, and

$$n = \frac{2}{N} \sum_k \frac{(\alpha_1 - U + \tfrac{1}{2}Un) f_N(\alpha_1 - \mu) - (\alpha_2 + \tfrac{1}{2}Un - U) f_N(\alpha_2 - \mu)}{[(U - \varepsilon_\kappa)^2 + 2Un\varepsilon_\kappa]^{1/2}} \tag{8}$$

where

$$\alpha_{1,2}(\kappa) = \tfrac{1}{2}(\varepsilon_\kappa + U) \pm [(U - \varepsilon_\kappa)^2 + 2Un\varepsilon_\kappa]^{1/2} \tag{9}$$

and

$$f_N(x) = [1 + \exp \beta_N x]^{-1} \tag{10}$$

Equation (7) is the self-consistency condition from which one must solve for a positive, real T_N. Equation (8) determines the chemical potential μ as a function of n. In principle one must solve Eq. (8) for μ and substitute this in Eq. (7). In practice this step can be circumvented, as we show below. We point out here that if one calculates the static paramagnetic spin susceptibility $\chi(Q)$ [$Q = (\pi/a)(111)$] in this approximation,[26] then the condition for the instability of the paramagnetic phase is essentially Eq. (7) with T_N replaced by arbitrary T and the equality sign replaced by $\leq$.

Results

Result 1. For $1 - 1/\sqrt{3} < n < 1 + 1/\sqrt{3}$ there does not exist a real, positive $T_N(0 \leq T_N \leq \infty)$ that will satisfy Eq. (7) for any value of the parameters U and $|t|$. The proof utilizes the fact that $\varepsilon(\mathbf{k}) = \varepsilon(-\mathbf{k})$ and that the integrand in eq. (7) takes its maximum value at $\varepsilon_\kappa = 0$ if n lies between $1 - 1/\sqrt{3}$ and $1 + 1/\sqrt{3}$.

Result 2. For $0 < n < 1 - 1/\sqrt{3}$ and $6|t|/U \leq 2/\sqrt{3}$ there is no solution to Eq. (7) for a real, positive T_N ($0 \leq T_N \leq \infty$).

This can be proved by comparing Eq. (7) with the exact equation (8) and establishing that each term in Eq. (8) [written in a form analogous to Eq. (7)] is larger than a corresponding term in Eq. (7) and hence the right side of Eq. (7) is always less than $1/U$.

Result 3. The third result is an extension for $n > 1 + 1/\sqrt{3}$ and $6|t|/U \leqq 2/\sqrt{3}$. The proof is very similar to the above case.

Result 4. Finally we establish that for $6|t|/U > 2/\sqrt{3}$ the ground state is definitely not antiferromagnetic. As pointed out before, the paramagnetic susceptibility $\chi(Q)$ will have a pole indicating its instability toward antiferromagnetism if the right side of Eq. (7) at $T_N = 0$ is less than $1/U$. But for $n < 1 - 1/\sqrt{3}$ and $n > 1 + 1/\sqrt{3}$ it can easily be demonstrated that the chemical potential μ restricts the sum over **k** to limited regions only and hence it is always less than $1/U$.

References

1. S. Adler, in *Solid State Physics*, Vol. 1, Seitz, Turnbull and Ehrenreich, eds. Academic Press, New York.
2. D.J. Khomsky, *Fiz. Metal. Metalloved.* **29**(1), 31 (1970).
3. J. Hubbard, *Proc. Roy. Soc.* **A276**, 338 (1963).
4. J. Hubbard, *Proc. Roy. Soc.* **A281**, 401 (1964).
5. J. Hubbard and K.P. Jain, *J. Phys. C* **1**, 1650 (1968).
6. Y. Nagaoka, Phys. Rev. **147**(1), 392 (1966).
7. T. Izuyama, *Phys. Rev. B* **5**, 190 (1972).
8. T. Izuyama, *Prog. Theor. Phys. (Kyoto)* **48**, 1106 (1972).
9. T. Izuyama, *Prog. Theor. Phys. (Kyoto)* **47**, 2136 (1972).
10. K.Y. Millard and K.S. Viswanathan (to be published).
11. J.R. Schrieffer, Banff Summer School Lecture Notes, 1969, unpublished.
12. K.S. Viswanathan, *Sol. St. Comm.* **8**, 231 (1970).
13. J.R. Schrieffer and J.C. Kimbal, preprint.
14. W. Langer, M. Pleschke, and D. Mattis, *Phys. Rev. Lett.* **23**, 448 (1969).
15. J.C. Slater, *Phys. Rev.* **82**, 538 (1951).
16. T. Arai, *Phys. Rev. B* **4**, 216 (1971).
17. D.R. Penn, Phys. Lett. **26A**, 509 (1968).
18. A.C. Hewson and V. Lindner, *Phys. Stat. Sol.* **42**, K185 (1970).
19. R.A. Bari and T.A. Kaplan, *Phys. Lett.* **33A**, 400 (1970).
20. C. Mehrotra and K.S. Viswanathan, *Phys. Rev. B* **7**, 559 (1973).
21. W.F. Brinkman and T.M. Rice, *Phys. Rev. B* **2**, 1324 (1970).
22. J.B. Sokoloff, *Phys. Rev. B* **1**, 1144 (1970).
23. J.B. Sokoloff, *Phys. Rev. B* **2**, 3707 (1970).
24. J.B. Sokoloff, *Phys. Rev. B* **3**, 3286 (1971).
25. J.J. Field, *J. Phys. C* **5**, 664 (1972).
26. C. Mehrotra and K.S. Viswanathan, *Sol. St. Comm.* **12**, 129 (1973).

Magnetism and Metal–Nonmetal Transition in Narrow Energy Bands: Applications to Doped V_2O_3

M. Cyrot and P. Lacour-Gayet

Laboratoire de Physique des Solides
Université Paris-Sud, Orsay, France

The purpose of this paper is to present a new way of handling the effects of correlations between electrons in narrow energy bands. We study the Hubbard model,[1] which presents the main physical effects that one can expect. As the ratio U/W of the average Coulomb repulsion between electrons on the same site to the bandwidth increases, one goes from a simple metal to a magnetic insulator. However, Hubbard's mathematical solution has been seriously questioned. Our basic physical approach is a generalization of Slater's idea.[2] The magnetic properties are the first effect of correlations between electrons—metallic or insulating properties are only consequences. For instance, the insulating properties of narrow-energy-band materials are due to a splitting of the band by an antiferromagnetic lattice. We generalize this idea to deal with the paramagnetic state. The mathematical formalism is a functional integral method.[3,4]

Here we give only a simplified version based on a self-consistent approach. Our aim is to construct a phase diagram for the Hubbard model. In order to compare it with the experimental phase diagram of doped V_2O_3, we will be led to discuss the stability of the different phases.

We start from the Hubbard Hamiltonian

$$H = \sum_{ij\sigma} T_{ij} c_{i\sigma}^+ c_{j\sigma} + \sum_i U n_{i\uparrow} n_{i\downarrow} \tag{1}$$

We separate the charge and the magnetization parts of the interaction by writing

$$U n_{i\uparrow} n_{i\downarrow} = -\tfrac{1}{4} U (n_{i\uparrow} - n_{i\downarrow})^2 + \tfrac{1}{4} U (n_{i\uparrow} + n_{i\downarrow})^2 \tag{2}$$

For our purpose charge fluctuations turn out to be unimportant so we replace the last term of the rhs of Eq. (2) by $U/4$ and take into account only the magnetic fluctuations which build up magnetic moments on the sites. We make a Hartree–Fock approximation, which replaces the first term of the rhs by

$$-\tfrac{1}{2} U \mu_i (n_{i\uparrow} - n_{i\downarrow})$$

Thus we have to solve the motion of an electron in a static one-body potential. The magnetic moments μ_i are then determined by the self-consistent equation

$$\mu_i = \langle n_{i\uparrow} \rangle - \langle n_{i\downarrow} \rangle \tag{3}$$

At zero temperature if the self-consistent magnetic moment μ_i is different from zero, we have in the case of one electron per atom

$$\mu_i = \mu \exp(ip \cdot R_i)$$

p is half a reciprocal lattice vector and gives an antiferromagnetic arrangement of the spins. This Hamiltonian can be diagonalized exactly and we obtain two critical ratios for U/W separating three phases: metal, antiferromagnetic metal, and antiferromagnetic insulator.

Above the Néel temperature the μ's are random and we make an analogy with an alloy. If, for instance, μ_i can take only two values $\pm \mu$, we have a binary alloy AB. Component A has a binding energy $-\frac{1}{2}U\mu$ and component B has $+\frac{1}{2}U\mu$. We solve this Hamiltonian with a coherent potential approximation.[5] In this approximation the propagator is given by

$$F(z) = \frac{1}{N} \sum_k \frac{1}{z - \varepsilon_k - \Sigma(z)}$$

and the self-energy obeys the Soven equation:

$$\Sigma = (\tfrac{1}{4}\mu^2 U^2 - \Sigma^2) F$$

knowing the propagator, μ can be calculated through Eq. (3). This leads to the phase diagram of Fig. 1. In this phase diagram T_N represents a real thermodynamic transition. However, the dashed lines, which are the lines of appearance of moments on the sites and of appearance of a gap in the density of states, do not represent a thermodynamic transition. When calculating the free energy taking into account fluctuations no singularity occurs.[4]

At this point it is interesting to compare our approach to that of Slater and Hubbard. The former ascribed the insulating behavior of narrow-energy-band materials to a splitting of the band by an antiferromagnetic lattice. Below the Néel temperature our results are similar. However, in Slater's point of view, the material cannot remain insulating above the Néel temperature. Our formalism recasts the theory in a form appropriate to dealing with the magnetically disordered case because of our alloy analogy: A gap can open up in the density of states even in the

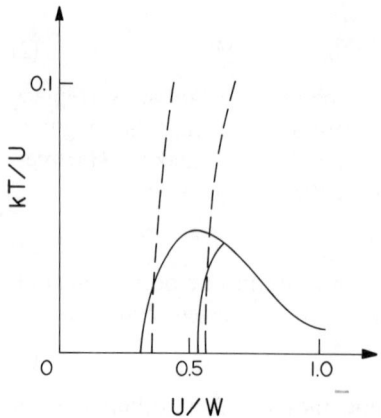

Fig. 1. The theoretically predicted phase diagram for the Hubbard model. The dashed lines correspond to the appearance of a nonzero value for the Hartree–Fock moment and to the opening of a gap in the density of states.

magnetically disordered state. Our analogy is, however, different from Hubbard's, which comes from the fact that if an electron is incident on an atom, the resonant energy is different if an electron is or is not already present: This energy is T_{ii} or $T_{ii} + U$. In our case the analogy comes from magnetism. If the moment is upward (downward), the resonant energy is $-\frac{1}{2}\mu U$ ($+\frac{1}{2}\mu U$). Thus our formulas are identical to Hubbard's only when $\mu = 1$; i.e., when $U/W \to \infty$.

On comparing our theoretical phase diagram to the one obtained for Cr-doped V_2O_3 (Fig. 2)[6] large disagreement seems to exist. The main disagreement is the existence of a first-order transition above the Néel temperature. Also, no metallic antiferromagnetic phase exists. Thus we are led to question the stability of the phases obtained theoretically.

In our phase diagram (Fig. 1), the ratio U/W can be thought as being a function of volume. We calculate the free energy as a function of volume through this ratio. If this free energy has the behavior of Fig. 3, in an experiment under pressure there will be a first-order transition with a discontinuity of volume. We now show that this can happen under certain conditions which are fulfilled for doped V_2O_3.

The change in free energy between the state with μ given by Eq. (3) and the state with $\mu = 0$ is

$$\Delta F = -\tfrac{1}{4} U \int_0^1 d\lambda \mu^2(\lambda) \tag{4}$$

where $\mu(\lambda)$ is the self-consistent solution for the Hamiltonian with U replaced by λU. To calculate the free energy as a function of volume, we assume that U is constant and that only W is a function of volume. The volume dependence of the bandwidth is given by the volume dependence of the overlap integral in a tight-binding approximation

$$W = W_0 \exp\left[-q(V - V_0)/V_0\right] \tag{5}$$

We also make the following assumption for the volume dependence of E when $\mu = 0$:

$$E_0(V) = \tfrac{1}{2} Y (V - V_1)^2$$

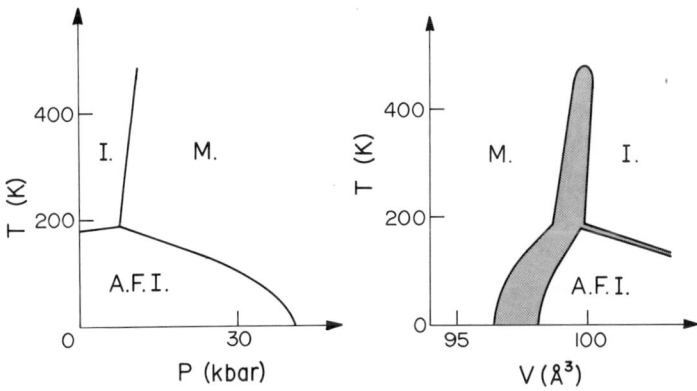

Fig. 2. Phase diagrams of doped V_2O_3.

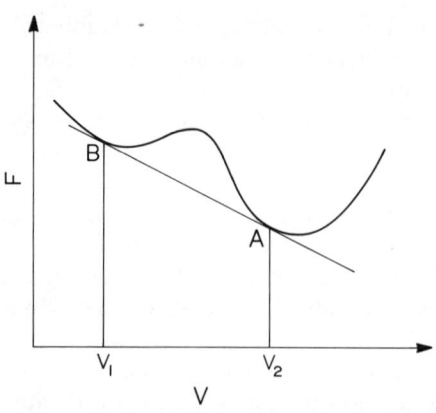

Fig. 3. Schematic curve for the free energy vs. volume, showing a first-order transition.

Y is the bulk modulus per atom in the nonmagnetic phase. The stability requirement is that the compressibility is positive. For instance, at zero temperature, instability occurs for $\partial^2 E/\partial V^2 < 0$. We find for $Y/U < 1.8q^2$ a first-order transition due to an instability of the phase. The change in volume at the transition increases with decreasing value of Y. For Y low enough the whole metallic antiferromagnetic state can be unstable. This appears to be the case in Cr-doped V_2O_3, where we have $Y = 13$ eV,[6] $U = 2$ eV,[7] $W = 3$ eV,[7] and we take $q = 1.9$.

The whole analysis can apply at finite temperature. With increasing temperature the region of instability moves to larger volume and becomes narrower. There exists a critical temperature above which the criterion of instability cannot be fulfilled. Above this temperature no thermodynamic phase transition occurs. The critical temperature is given by

$$Y + \partial^2 \Delta F/\partial V^2 = 0$$

i.e.,

$$\frac{T_c}{U} = \frac{8}{3\sqrt{6\pi^2}}\left(1 - \frac{9\pi^2}{160}\frac{Y}{q^2 U}\right)$$

With the value that we gave for Cr-doped V_2O_3, we obtain $T_c = 480°$K, in good agreement with experiment.

In conclusion, we stress that the simple Hubbard model for correlations between electrons, independent of degeneracy or real band structure, explains the main features of the experimental phase diagram of doped V_2O_3; i.e., the absence of a metallic antiferromagnetic phase, the existence of a first-order transition, and the existence of a critical temperature above which no phase transition occurs.

References

1. J. Hubbard, *Proc. Roy. Soc. A* **276**, 238 (1963); **281**, 401 (1964).
2. J.C. Slater, *Phys. Rev.* **82**, 538 (1951).
3. M. Cyrot, *Phys. Rev. Lett.* **25**, 871 (1970); *J. de Physique* **33**, 125 (1972); *Phil. Mag.* **25**, 1031 (1972).
4. J.C. Kimball and J.R. Schrieffer, in *Intern. Conf. on Magnetism, 1971*.
5. P. Soven, *Phys. Rev.* **156**, 809 (1967).
6. D.B. McWhan and J.P. Remeika, *Phys. Rev. B* **2**, 3734 (1971).
7. J. Wilson (to be published).

Determination of Crossover Temperature and Evidence Supporting Scaling of the Anisotropy Parameter of Weakly Coupled Magnetic Layers*

Luke L. Liu, Fredric Harbus,† Richard Krasnow, and H. Eugene Stanley

Physics Department, Massachussetts Institute of Technology
Cambridge, Massachussetts

Numerous magnetic systems possess directional anisotropy,‡ that is, different coupling strengths in different lattice directions. Quasi-two-dimensional (and quasi-one-dimensional) magnets are in fact magnetic sites on a three-dimensional lattice, with intralayer (or intrachain) coupling strength J much stronger than the interlayer (interchain) coupling RJ. This situation is illustrated schematically in Fig. 1.

We consider first the quasi-two-dimensional case and introduce a spin-1/2 Ising model on a simple cubic (sc) lattice, with Hamiltonian[1,2]

$$\mathcal{H} \equiv -J \sum_{\langle ij \rangle}^{xy} s_i s_j - RJ \sum_{\langle ij \rangle}^{z} s_i s_j \tag{1}$$

The first summation is over pairs of nearest-neighbor (nn) sites whose relative displacement vector $\mathbf{r}_{ij}$ has no z component, while the second summation is over all other pairs of nn sites. When $R = 0$ the system is strictly two-dimensional; when $R = 1$ we have an isotropic three-dimensional system.

If we fix the temperature and allow R to decrease, the correlation between spins of different planes diminishes, and the system becomes more and more two-dimensional. If, on the other hand, we fix R (as is the case with most experiments) at high temperatures the correlation between spins of different planes is destroyed by the fluctuations and the system appears two-dimensional. However, as the temperature is lowered toward $T_c = T_c(R)$ interplanar correlations will become significant and the system will exhibit three-dimensional behavior. Thus as the critical temperature is approached from above the critical exponents will cross over from two-dimensional values to those characteristic of three-dimensional systems (see Fig. 2).

This crossover behavior has long been known to the experimentalist, and is also qualitatively predicted from an extension of the scaling hypothesis.[3] Three questions that concern us here are the following: (1) Given R, what is the temperature

---
* Work supported by NSF, ONR, and AFOSR. Work forms a portion of Ph.D. theses to be submitted by R.K. and F.H. to the MIT Physics Department.
† National Science Foundation Predoctoral Fellow.
‡ Besides those one- and two-dimensional magnets that have $R \leq 10^{-3}$–10^{-4}, we know, e.g., the ideal Heisenberg magnet $CrBr_3$ has $R \sim 1/17$ and the metamagnet $FeCl_2$ has $R \sim -0.05$.

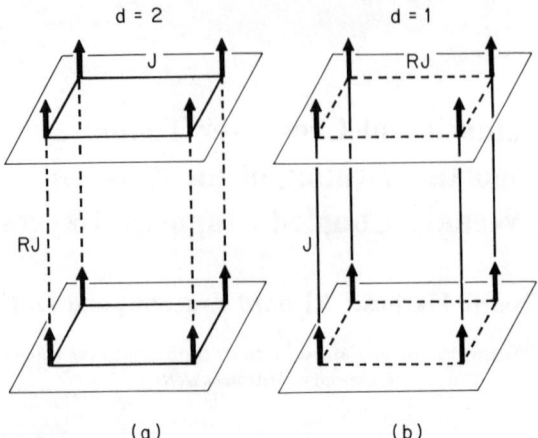

Fig. 1. Schematic representations of (a) quasi-two-dimensional and (b) quasi-one-dimensional systems. Here the solid lines indicate interaction bonds of strength J, while the dashed lines indicate bonds of strength RJ, with $R \ll 1$.

at which the crossover occurs? (2) Is there evidence supporting the validity of the extended scaling hypothesis? (3) If so, then what is the value of the "crossover exponent" φ? [Here φ describes the variation of the critical temperature with R, $T_c(R) - T_c(0) \sim R^{1/\varphi}$.]

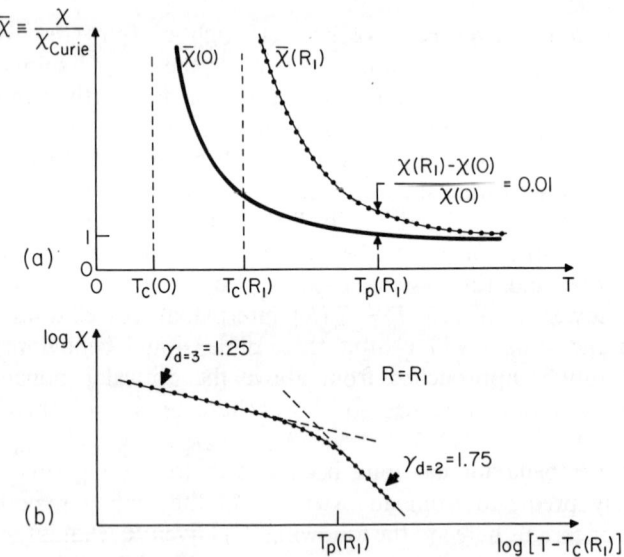

Fig. 2. (a) Dependence of reduced susceptibility $\bar{\chi}$ upon T for $R = 0$ and for $R = R_1$, indicating the definition of $T_p(R)$ (note that this drawing is not to scale); (b) sketch of hypothetical experimental data, plotted in the conventional log-log plot, for a system which is described by the Hamiltonian (1) with $R = R_1$.

Determination of the Crossover Temperature

Consider the reduced susceptibility $\bar{\chi}(R)$ for the Hamiltonian (1). If we consider the coupling in the weak direction as a perturbation, then we may expand $\bar{\chi}(R)$ in a Taylor series about $R = 0$:

$$\bar{\chi}(R) \equiv \bar{\chi}_0 + \bar{\chi}_1(0) R + \bar{\chi}_2(0) \frac{R^2}{2!} + \cdots + \bar{\chi}_n(0) \frac{R^n}{n!} + \cdots \qquad (2)$$

where

$$\bar{\chi}_n(0) \equiv \left. \frac{\partial^n \bar{\chi}(R)}{\partial R^n} \right|_{R=0} \qquad (3)$$

and $\bar{\chi}_0$ is the susceptibility of the two-dimensional system. It can be shown rigorously[4] that for systems described by Eq. (1)

$$\bar{\chi}_1 = 2\beta J \bar{\chi}_0^2 \qquad (4)$$

$$8(\beta J)^2 \bar{\chi}_0^3 \geq \bar{\chi}_2 \geq 4(\beta J)^2 \bar{\chi}_0^3 \qquad (5)$$

Similar relations hold for lattices other than the sc, necessitating only a change in the proportionality factors.

Thus from Eqs. (2)–(5) we have

$$\bar{\chi}(R) = \bar{\chi}_0 [1 + 2R\beta J \bar{\chi}_0 + (2R\beta J \bar{\chi}_0)^2 f(\beta J) + O(R^3)] \qquad (6a)$$

where

$$1 < f(\beta J) \leq 2 \qquad (6b)$$

The effective crossover temperature is now determined by the following argument. If $\bar{\chi}(R)$ vs. T were known exactly for a given R there would be no sudden transition from two- to three-dimensional behavior; rather, as the temperature approached $T_c(R)$ higher-order terms in Eq. (6) would gradually become more significant. Thus we can define a "p-percent crossover temperature" $T_p(R)$ as that temperature at which there is a p-percent deviation of $\bar{\chi}(R)$ from the two-dimensional value $\bar{\chi}_0$. To first order, $T_p(R)$ is given by the solution of

$$2R\beta J \bar{\chi}_0 \simeq 0.01p \qquad (7)$$

where the left-hand side is expressed as a function of T.

Experimentally, this means that if one's confidence limits are p percent, the two-dimensional susceptibility $\bar{\chi}_0$ should be a good enough approximation to $\bar{\chi}(R)$ down to T_p.

Equation (7) is the major result of this section. It is a useful one, since it is generally easier to obtain theoretical information about $\bar{\chi}_0$ than about $\bar{\chi}(R)$. By inverting the above procedure, Eq. (7) may be used to estimate R when R is not known, the crossover temperature $T_p(R)$ being obtained from a plot of $\bar{\chi}(R)$ vs. T. As an illustration, consider the system TMMC,[5] known to be a spin-5/2 quasi-one-dimensional antiferromagnet for $T \geq 1.1°K$, with $J = -6.3°K$. Assuming the experimental error to be 5%, Eq. (6) requires that

$$4R\beta J \bar{\chi}_{1c}(J, T, S) \leq 0.05 \qquad (8)$$

for $T \geq 1.1°K$, where $\bar{\chi}_{lc}$ is the linear-chain susceptibility, known exactly, and evaluated at $J = -6.3°K$ and $S = 5/2$. From this relation we obtain an upper bound on R, $R \leq 10^{-6}$.

Scaling with Respect to R

This crossover behavior is most easily explained in the context of the scaling hypothesis,[3] where we assume that there exist three numbers a_τ, a_H, and a_R such that for all positive λ

$$G(\lambda^{a_\tau}\tau, \lambda^{a_H}H, \lambda^{a_R}R) = \lambda G(\tau, H, R) \tag{9}$$

where G is the Gibbs potential, $\tau \equiv T - T_c(0)$, and H is the magnetic field. It can be shown that Eq. (9) implies that $T_c(R) - T_c(0) \sim CR^{1/\varphi}$, where $\varphi \equiv a_R/a_\tau$ is the "crossover exponent." On differentiating Eq. (9) twice with respect to H and n times with respect to R, we have

$$\bar{\chi}_n(\lambda^{a_\tau}\tau, \lambda^{a_H}H, \lambda^{a_R}R) = \lambda^{1-2a_H-na_R}\bar{\chi}_n(\tau, H, R) \tag{10}$$

where

$$\bar{\chi}_n \equiv \left(\frac{\partial^n \bar{\chi}}{\partial R^n}\right)_{H,\tau}$$

Setting $\lambda \equiv |\tau|^{-1/a_\tau}$, we have

$$\bar{\chi}_n(\tau, H = 0, R = 0) \sim |\tau|^{-\gamma_n} \tag{11}$$

with

$$\gamma_n = \gamma_0 + n\varphi \tag{12}$$

Here $-\gamma_0 = (1 - 2a_H)/a_\tau$ is the susceptibility exponent on a two-dimensional lattice. Similar predictions can be made for other thermodynamic quantities (see Table I).

A *stronger* statement is that the scaling hypothesis for the pair correlation function can be extended to include the parameter R, i.e., there exist *four* numbers b_i such that for all positive λ

$$C_2(\lambda^{b_\tau}\tau, \lambda^{b_H}H, \lambda^{b_r}\mathbf{r}, \lambda^{b_R}R) \equiv \lambda C_2(\tau, H, \mathbf{r}, R) \tag{13}$$

from which the exponent predictions shown in the second column of Table I follow. Here $\mathbf{r}$ is the site separation vector. Since $\int d\mathbf{r} C_2(\mathbf{r}) = \bar{\chi}_T = (\partial^2 G/\partial H^2)_T$, it follows that Eq. (13) implies Eq. (9). The converse is *not* true, and, in particular, Eq. (9) does *not* imply that the second moment $\mu_2 \equiv \int r^2 C_2(\mathbf{r}) d\mathbf{r}$ scales. Scaling also makes the predictions shown in the top line of column 2 of Table I.

We are now able to answer the third question we posed. Clearly, the rigorous results of Eqs. (4) and (5) imply that

$$\gamma_1 = 2\gamma_0 \tag{14}$$

and

$$\gamma_2 = 3\gamma_0 \tag{15}$$

respectively. Coupling these relations together with the scaling prediction, Eq. (12),

Table I

Definition of exponents	Scaling hypothesis plus $\gamma_1 = 2\gamma_0$ proof	Previous numerical results (sc lattice only)	Present work sc	Present work fcc
Test of thermodynamic scaling $T_c(R) - T_c(0) \sim R^{1/\varphi}$ $\bar{\chi}^{(n)} \equiv (\partial^n \bar{\chi}/\partial R^n)_{R=0} \sim \tau^{-\gamma_n}$	$\varphi = \gamma_0$ ($\gamma_0 = 1.75$) $\gamma_n = \gamma_0 + n\varphi$ $= (n+1)\gamma_0$	$\varphi = 1.2^{3,6}$	$\varphi = 1.70 \pm 0.1$	
	$n=1$: 3.50	$3.50^{3,6,7}$	3.500	
	$n=2$: 5.25	$5.0 \pm 0.1^{3,6}$	5.25 ± 0.04	5.25 ± 0.04
	$n=3$: 7.00	$6.5 \pm 0.2^{3,6}$	7.00 ± 0.05	7.00 ± 0.04
	$n=4$: 8.75	$8.0 \pm 0.3^{3,6}$	8.75 ± 0.25	8.75 ± 0.03
	$n=5$: 10.50		$10.5 \pm ^{0.50}_{1.00}$	$10.5 \pm ^{0.1}_{0.2}$
$C_H^{(n)} \equiv (\partial^n C_H/\partial R^n)_{R=0} \sim \tau^{-\alpha_n}$	$\alpha_n = \alpha_0 + n\varphi$ $= \alpha_0 + \gamma_0$ $[\alpha_0 = 0]$			
	$n=2$: 3.50		—	3.5 ± 0.05
	$n=4$: 7.00	5.2 ± 0.1^{7}	—	7.0 ± 0.04
	$n=6$: 10.50	6.9 ± 0.1^{7}	—	10.5 ± 0.05
	$n=8$: 14.00		—	14.0 ± 0.6
Test of correlation function scaling $\mu_2^{(n)} \equiv (\partial^n \mu_2/\partial R^n)_{R=0} \sim \tau^{-(\gamma_0 + 2\nu_n)}$	$\gamma_0 + 2\nu_n = \gamma_0 + 2\nu_0 + n\varphi$ $= 2\nu_0 + (n+1)\gamma_c$ $[\nu_0 = 1]$			
	$n=1$: 5.50		5.500	
	$n=2$: 7.25		$7.25 \pm ^{0.10}_{0.20}$	7.25 ± 0.02
	$n=3$: 9.00		$9.0 \pm ^{0.10}_{0.50}$	9.0 ± 0.1
	$n=4$: 10.75		$10.75 \pm ^{0.10}_{0.25}$	10.75 ± 0.03
	$n=5$: 12.50		$12.85 \pm ^{0.40}_{0.60}$	$12.50 \pm ^{0.10}_{0.05}$

it follows that

$$\varphi = \gamma_0 \tag{16}$$

As for the second question, we cannot prove rigorously that scaling with the parameter R is correct. But, in addition to Eqs. (4) and (5), we are also able to show that

$$48(\beta J)^3 \bar{\chi}_0^4 \geq \bar{\chi}_3(0) \geq 8(\beta J)^3 \bar{\chi}_0^4 \tag{17}$$

Hence

$$\gamma_3 = 4\gamma_0 \tag{18}$$

Furthermore, it can be shown that

$$\mu_2^{(1)} = 2\beta J [\bar{\chi}_0^2 + 2\bar{\chi}_0 \mu_2^{(0)}] \tag{19}$$

which implies

$$\gamma_0 + 2v_1 = 2\gamma_0 + 2v_0 \tag{20}$$

Equations (14), (15), (18), and (20) are in agreement with the scaling predictions (see Table I).

For higher-order derivatives rigorous results are not available. However, for systems described by Eq. (1) we have analyzed general-R series for $\bar{\chi}_T^{(n)}$ and $\mu_2^{(n)}$ on both the sc and fcc lattices, and derivatives of the specific heat $C_H^{(n)}$ on the fcc, using a wide variety of methods (e.g., ratio method, Park's method, Neville tables, and analysis on the series raised to different powers). Our estimates are shown in the last column of Table I. Note that in all cases they do not violate the predictions of the scaling hypothesis.

Summary

We have furnished a way of estimating the crossover temperature for magnetic systems exhibiting directional anisotropy. We have also presented the crossover problem in the language of the scaling hypothesis and obtained numerical and rigorous results supporting the validity of such a hypothesis for the thermodynamic and correlation functions. Further details of this work will be published elsewhere.[4,8]

References

1. G. Paul and H.E. Stanley, *Phys. Lett.* **37A**, 347 (1971); *Phys. Rev. B* **5**, 2578 (1972).
2. J. Oitmaa and I.G. Enting, *Phys. Lett.* **36A**, 91 (1971).
3. E. Riedel and F. Wegner, *Z. Phys.* **225**, 195 (1969); A. Hankey and H.E. Stanley, *Phys. Rev. B* **6**, 3515 (1972).
4. L.L. Liu and H.E. Stanley, *Phys. Rev. Lett.* **29**, 927 (1972); *Phys. Rev. B* **8**, 2279 (1973).
5. R.J. Birgeneau, R. Dingle, M.T. Hutchings, G. Shirane, and S.L. Holt, *Phys. Rev. Lett.* **26**, 718 (1971); R. Dingle, M.E. Lines, and S.L. Holt, *Phys. Rev.* **187**, 643 (1969).
6. I.G. Enting and J. Oitmaa, *Phys. Lett.* **38A**, 107 (1972); *J. Phys. C* **5**, 231 (1972).
7. D.C. Rapaport, *Phys. Lett.* **37A**, 407 (1971).
8. F. Harbus and H.E. Stanley, *Phys. Rev. B* **7**, 365 (1973); R. Krasnow, F. Harbus, L.L. Liu, and H.E. Stanley, *Phys. Rev. B* **7**, 370 (1973).

A Corrected Version of the t-Approximation for Strong Repulsion

G. Horwitz and D. Jacobi

Department of Theoretical Physics
Hebrew University, Jerusalem, Israel

A study has been made of the evolution of localized behavior in the Hubbard model as a function of density for the strongly repulsive case for low to intermediate densities. For few electrons in a band the electrons will be Blochlike no matter how strong the repulsion, while a half-filled band will be localized for very large ratio of repulsion energy to bandwidth. We have studied the situation intermediate between these limits, which can be analyzed clearly in our approach.

Some self-consistent form of the t-matrix (ladder graph) type of approximation is evidently the appropriate approach for relatively low density and strong repulsion. Kanamori's[1] treatment of the low density by calculating the effective interaction, though evidently correct at low densities, cannot readily be extended to higher densities. Horwitz[2] has previously discussed the existence of a pole in the two-particle Green's function calculated in the t-approximation. This pole represents the high-energy state in which two electrons are found on the same site. Although this has no consequences for the effective interaction calculated at low densities, it has drastic effects on the calculation of the renormalized single-particle Green's function. In effect the two-particle pole would appear to carry the projection out of the doubly occupied states in a conventional perturbation theory treatment.

The Hubbard Hamiltonian will be written

$$H = \sum_\sigma \sum_\mathbf{k} \varepsilon(\mathbf{k}) c^\dagger_{\mathbf{k}\sigma} c_{\mathbf{k}\sigma} + U \sum_i n_i^\uparrow n_i^\downarrow \tag{1}$$

the $\mathbf{k}$'s denoting Bloch states and the i's labeling lattice sites. It is essential to summarize the self-consistent formalism involving one- and two-particle Green's functions (using causal and thermal Green's functions). Utilizing the Heisenberg–Bloch representation of operators,

$$c_{\mathbf{k}\sigma} \equiv \exp(uH) c_{\mathbf{k}\sigma} \exp(-uH) \qquad (0 \le u \le \beta = 1/k_\beta T) \tag{2}$$

we define one-particle Green's function

$$G_\sigma(\mathbf{k}, \zeta_r) = \int_0^\beta (du/\beta) \langle T c_{\mathbf{k}\sigma}(u) c^\dagger_{\mathbf{k}\sigma}(0) \rangle e^{u\zeta_r} \tag{3}$$

$$\zeta_r = 2\pi r i / \beta, \qquad r = \pm 1, \pm 3, \ldots$$

and the two-particle Green's functions

$$K(\mathbf{Q}, \eta_v) = \int_0^\beta (du/\beta) e^{u\eta_v}$$

$$\times \sum_{\mathbf{k}} \sum_{\mathbf{k}'} \langle Tc_{(\mathbf{Q}/2)+\mathbf{k},\uparrow}(u) c_{(\mathbf{Q}/2)-\mathbf{k},\downarrow}(u) c_{(\mathbf{Q}/2)-\mathbf{k}',\downarrow}(0) c_{(\mathbf{Q}/2)+\mathbf{k}',\uparrow}(0) \rangle \quad (4)$$

$\eta_v = 2v\pi i/\beta, \quad v = 0, \pm 1, \pm 2, \ldots$

These have, respectively, the Lehman representations

$$G_\sigma(\mathbf{k}, \zeta_r) = \int_0^\beta \frac{d\omega}{2\pi} \frac{A_\sigma(\mathbf{k}, \omega)}{\zeta_r - \omega} \quad (5)$$

and

$$K(\mathbf{Q}, \eta_v) = \int_0^\beta \frac{dE}{2\pi} \frac{B(\mathbf{Q}, E)}{\eta_v - E} \quad (6)$$

The evaluation of K in the t-approximation is a well-known result which we write in the form (on performing analytic continuation to real frequency)

$$K(\mathbf{Q}, E) = \psi(\mathbf{Q}, E)/[1 - U\psi(\mathbf{Q}, E)] \quad (7)$$

where ψ is a convolution of one-particle Green's functions, which, using (5), can be expressed in the form

$$\psi(\mathbf{Q}, E) = N^{-1} \sum_{\mathbf{k}} \int_{-\infty}^\infty \frac{d\omega_1}{2\pi} \int_{-\infty}^\infty \frac{d\omega_2}{2\pi}$$

$$\times \frac{[1 - f(\omega_1) - f(\omega_2)] A_\uparrow(\tfrac{1}{2}\mathbf{Q} + \mathbf{k}, \omega_1) A_\downarrow(\tfrac{1}{2}\mathbf{Q} - \mathbf{k}, \omega_2)}{E + i\eta - \omega_1 - \omega_2} \quad (8)$$

To complete the formal relations, we then obtain the self-consistent self-energy by evaluating the self-energy in the t-approximation,

$$\Sigma_\sigma(\mathbf{k}, \omega) \cong Un_{-\sigma} + N^{-1} \sum_v \sum_{\mathbf{k}'} U^2 K(\mathbf{k} + \mathbf{k}', \eta_v) G_{-\sigma}(\mathbf{k}', \omega - \eta_v) \quad (9)$$

which can be expressed in the form

$$\Sigma_\sigma(\mathbf{k}, \omega) = Un_{-\sigma} + \frac{U^2}{N}$$

$$\times \sum_{\mathbf{k}'} \int_{-\infty}^\infty \frac{d\omega'}{2\pi} \int_{-\infty}^\infty \frac{dE}{2\pi} \frac{B(\mathbf{k} + \mathbf{k}', E) A_{-\sigma}(\mathbf{k}', \omega') [f(\omega') + b(E)]}{\omega + \omega' + i\eta - E} \quad (10)$$

The problem is then solved as follows: The poles of K can be evaluated on the basis of the known structure of single-particle spectral density $A_\sigma(\mathbf{k}, \omega)$ which will differ from zero in two energy ranges, narrow compared to U, whose separation is of order U. Let us call these respectively the lower band (LB) and the upper band (UB). The maximum range of $\varepsilon(\mathbf{k})$ is denoted D, while D_1 and D_2 denote, respectively, the nonzero range of the LB and UB, respectively. Then B will be the sum of pole terms and the conventionally evaluated t-matrix contribution

$$B(\mathbf{Q}, E) = \sum_r 2\pi R_r \delta(E - E_r) + \tilde{B}(\mathbf{Q}, E) \tag{11}$$

with $2\pi R_r$ the residue of the rth pole, and for the limiting case we consider, $U/D \gg 1$,

$$\tilde{B}(\mathbf{Q}, E) \cong \frac{1}{U^2} \frac{-2 \, \text{Im} \, \psi(\mathbf{Q}, E)}{[\text{Re} \, \psi(\mathbf{Q}, E)]^2 + [\text{Im}(\mathbf{Q}, E)]^2} \tag{12}$$

We must insert these results into the expressions for $\Sigma_\sigma(\mathbf{k}, \omega)$ and solve for two distinct domains: the LB case and the UB case. Our principal interest is in the LB case, which are the occupied levels, but we cannot carry out the self-consistent calculation without evaluating the UB case, at least to leading order in U. This being done, we obtain the value of the poles of K, their residues, the location of the bottom of the UB—denoted E^*—and the integrated spectral density for the UB, or, if one prefers, the fractional division between the spectral density in UB and LB. This fraction is expected to go from zero for zero density to a half for a half-filled band. In the low-density limit there is only one pole of K; at higher densities there are to be found two of three possible real poles. The third pole is necessarily at the energy E^* in order to maintain the UB at energy of order U. This low-density pole is found to lie at the energy

$$E_0 = U(1 - \rho) + \hat{E} \tag{13}$$

where

$$\rho \equiv N^{-1} \sum_{\mathbf{k}} \langle n_{\mathbf{k}}^\uparrow + n_{\mathbf{k}}^\downarrow \rangle \tag{14}$$

and

$$\hat{E} = \frac{1}{(1-\rho)N} \sum_{\mathbf{k}} \int_{-\infty}^{\infty} \frac{d\omega_1}{2\pi} \int_{-\infty}^{\infty} \frac{d\omega_2}{2\pi} \\
\times A_\uparrow(\tfrac{1}{2}\mathbf{Q} + \mathbf{k}, \omega_1) A_\downarrow(\tfrac{1}{2}\mathbf{Q} - \mathbf{k}, \omega_2)[1 - f(\omega_1) - f(\omega_2)] \tag{15}$$

with the residue

$$R = (1 - \rho) \tag{16}$$

The evaluation of the self-energy in the lower band then follows:

$$\Sigma_\sigma(\mathbf{k}, \omega) = U n_{-\sigma} + U^2 \int_{-\infty}^{\infty} \frac{d\omega'}{2\pi} N^{-1} \sum_{\mathbf{k}'} \frac{f(\omega') A_{-\sigma}(\mathbf{k}', \omega')(1-\rho)}{\omega + \omega' + i\eta - E} + \tilde{\Sigma}_\sigma(\mathbf{k}, \omega) \tag{17}$$

where $\tilde{\Sigma}_\sigma$ is the conventionally evaluated t-matrix contribution

$$\tilde{\Sigma}_\sigma(\mathbf{k}, \omega) = N^{-1} \sum_{\mathbf{k}} \int_{-\infty}^{\infty} \frac{d\omega'}{2\pi} \int_{-\infty}^{\infty} \frac{dE}{2\pi} \frac{U^2 \tilde{B}(\mathbf{Q}, E) A_{-\sigma}(\mathbf{k}', \omega')[f(\omega') + b(E)]}{\omega + \omega' + i\eta - E} \tag{18}$$

Upon manipulation and discarding terms of order $1/U$, we find that the pole term cancels the Hartree, $\tilde{\Sigma}_\sigma$ is independent of U, and there remains a residual renormalization of $G_\sigma(\mathbf{k}, \omega)$ from the pole, which we can write in the form

$$G_\sigma(\mathbf{k}, \omega) = \gamma \tilde{G}_\sigma(\mathbf{k}, \omega) \tag{19}$$

with the excluded volume factor $\gamma = (1 - \rho)/(1 - \tfrac{1}{2}\rho)$ and

$$\tilde{G}_\sigma(\mathbf{k}, \omega) = \{\omega + \bar{E}_\sigma - [\hat{E}_\rho/(1 - \tfrac{1}{2}\rho)] - \gamma[\varepsilon(k) + \Sigma_\sigma(\mathbf{k}, \omega)]\}^{-1} \quad (20)$$

where

$$\bar{E}_\sigma \equiv [1/(1 - \tfrac{1}{2}\rho)] N^{-1} \sum_{\mathbf{k}'} \int_{-\infty}^{\infty} (d\omega'/2\pi)\, \omega' f(\omega')\, A_{-\sigma}(\mathbf{k}', \omega') \quad (21)$$

The result of this renormalization can be summarized as follows.

(a) The band is contracted by the factor γ, which decreases steadily from one at zero density to zero for the half-filled band. (Actually, the approximation does not remain valid to such high densities.)

(b) The quasiparticle density of states decreases, even though the total density of states at the Fermi energy increases, as the band narrows.

(c) The reference energy $[\hat{E}_\rho/(1 - \tfrac{1}{2}\rho)] - \bar{E}_\sigma$ appears, which may represent a kind of bound-state localization energy.

Thus the result obtained is a kind of superposition of localized and nonlocalized behavior. The implication of a necessary correction, at least at low to moderate densities, for calculations of magnetic states of strongly correlated systems is evident. The approach is almost certainly of interest for the Anderson model of localized moments, and may be of interest in the continuum problems of ^{3}He and nuclear matter, although for the latter cases application is more problematic.

References

1. J. Kanamori, *Prog. Theor. Phys.* **30**, 275 (1963).
2. G. Horwitz, *Rev. Mod. Phys.* **40**, 807 (1968).

13

Magnetism and Superconductivity

Anomalous Behavior of the Kondo Superconductor $(La_{1-x}Ce_x)Al_2$*

G.v. Minnigerode, H. Armbrüster, G. Riblet, F. Steglich, and K. Winzer

Physics Institute
University of Cologne, Germany

At low temperatures two phenomena of collective behavior of the conduction electrons in metals are known: superconductivity and the Kondo effect. With the system (La, Ce) Al_2 we succeeded in demonstrating a strong interaction between these two phenomena as predicted by Müller-Hartmann and Zittartz in 1971.[1] The most unusual behavior of this superconducting alloy is the vanishing of superconductivity below a second transition temperature at very low temperatures. This second transition back into the normal state was first observed by Riblet and Winzer[2] in an adiabatic demagnetization cryostat with a standard ac mutual inductance technique. Figure 1 shows a typical example of these two transitions. The induction signal of the specimen compared to that of a clean $LaAl_2$ probe of the same size is plotted vs. temperature. This specimen with 0.67 at.% Ce substitution of La in $LaAl_2$ becomes superconducting between 1.5 and 1°K and normal again below 0.5°K. As a consequence of this second transition, the temperature dependence of all the other superconducting properties becomes strange. It is therefore proposed to call such an alloy a "Kondo superconductor."

In Fig. 2 a series of such measurements is given to show the influence of the Ce concentration. Here the normalized induction signal is plotted vs. the temperature on a logarithmic scale. The clean $LaAl_2$ sample becomes superconducting with a sharp transition at 3.26°K. With increasing Ce substitution of La the transitions are shifted to lower temperatures and become broadened. Up to Ce concentrations of 0.62 at.% related to the amount of La, no indication of a second transition can be seen. For concentrations larger than 0.66 at.% a partial return to the normal state is observed at very low temperatures. At higher concentrations the two transitions become so broad (and start to overlap) that these specimens show only a maximum in diamagnetic susceptibility at finite temperature. They never become fully superconducting. Samples with a Ce concentration of about 0.9 at.% cause only a small signal at 0.7°K. Above this concentration all superconductivity is destroyed.

The vanishing of superconductivity below a second transition temperature in $(La_{1-x}Ce_x)Al_2$ alloys is restricted to the narrow range $0.0066 < x < 0.009$ of Ce substitution. Such a behavior was predicted to occur under certain conditions due to the strong temperature dependence of the spin-flip scattering amplitude in the Kondo effect.[1] Because of the temperature dependence of this amplitude the

---
* Work sponsored by the Deutsche Forschungsgemeinschaft.

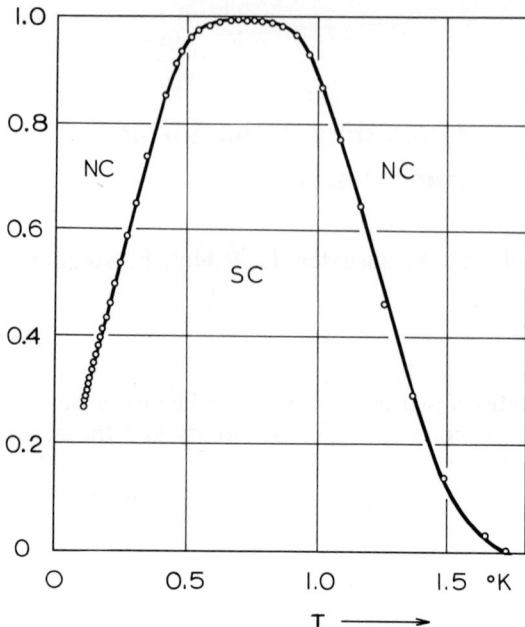

Fig. 1. Induction signal of an $(La_{0.9933}Ce_{0.0067})Al_2$ sample vs. temperature. The signal is given in units of the signal of a clean $LaAl_2$ probe of the same size. The two-phase boundaries are clearly seen.

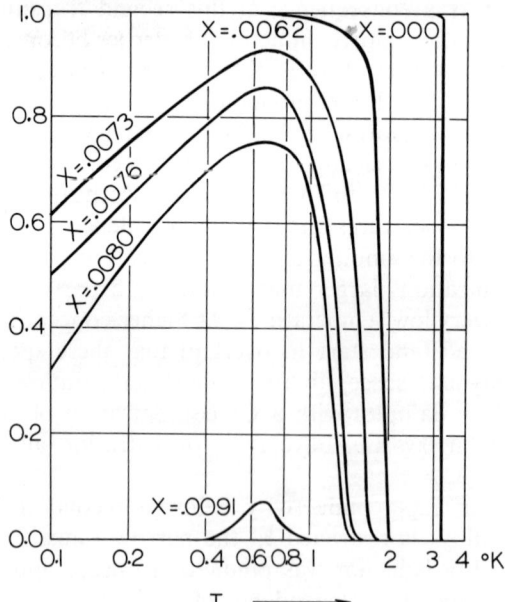

Fig. 2. The transitions of several $(La_{1-x}Ce_x)Al_2$ alloys. The signals, normalized as in Fig. 1, are plotted vs. temperature on a logarithmic scale.

pair-breaking parameter becomes temperature dependent with a maximum at a scaling temperature T_K^*. This scaling temperature is related to the Kondo temperature $T_K = T_F \exp(1/N\mathscr{F})$, where T_F is the Fermi temperature, N is the density of states for one spin direction at the Fermi level, and $\mathscr{F}$ is the conduction electron–impurity spin exchange interaction parameter, which is of a negative sign for the antiferromagnetic interaction in a Kondo system. Below this scaling temperature T_K^* the spin-flip part of the pair-breaking parameter decreases to zero. In a system with $T_K^* \ll T_{c0}$ (where T_{c0} is the transition temperature of the clean superconductor) the transition temperature T_c is lowered by two influences. These are the increasing pair-breaking parameter due to the increasing impurity concentration (as known in the Abrikosov–Gorkov theory), and the increasing pair-breaking force of each impurity, since T_c approaches T_K^*. Therefore superconductivity at lower temperatures can only exist for smaller impurity concentrations c, which means that the T_c vs. c curve turns backward. As soon as T_c becomes smaller than T_K^* the disappearance of the spin-flip part of the pair-breaking force might allow superconductivity at ultra low temperatures for all concentrations. In a certain concentration range, which depends on, among other factors, the ratio T_K^*/T_{c0}, three solutions for T_c are found for a certain impurity concentration c. Up to now no indication for the reappearance of superconductivity at a third transition at ultra low temperatures has been observed.

In order to observe such a Kondo superconductor as described by the Müller–Hartmann–Zittartz (MHZ) theory, at least five conditions have to be fulfilled:

1. The exchange interaction between the impurity spin and the conduction electrons has to be an antiferromagnetic one.
2. The impurity atom is expected to have a spin $S = 1/2$ in the host metal.
3. A statistical distribution of the impurity atoms in the host metal is required.
4. The transition temperature of the clean superconductor T_{c0} has to be sufficiently high so that the three transitions of the impure superconductor can be expected within available temperature.
5. The scaling temperature T_K^* of the system has to be much smaller than T_{c0}.

Our search of the literature suggested that the $(La, Ce)\,Al_2$ system might be favorable. This system is known as a Kondo system; the electronic structure of Ce promises $S = 1/2$, the La atoms should easily be substituted by the Ce impurities, the transition temperature of $LaAl_2$ is about 3.3°K, and, from the initial slope $(dT_c/dc)_{T_{c0}}$, the scaling temperature T_K^* of the system was determined to be $T_K^* = 0.11°K = (1/30)\,T_{c0}$ (see Ref. 2).

As a consequence of the observed two transitions, the critical magnetic field should have two passages through zero. Considering an additive pair-breaking by the impurities and the field, one would expect a maximum of the critical field at finite temperatures even for Ce concentrations too small to show a second transition without field. In order to determine the critical field values and to look for this effect, the magnetic field dependence of the induction signal was measured.[3] In the course of this investigation our samples were found to be type II superconductors with an upper critical field $H_{c2}^0(T = 0) = 2.5\ kG$ for the clean $LaAl_2$ probe. In Fig. 3 the effect of increasing Ce additions on H_{c2} is shown. The measured $H_{c2}(T)$ values of three $(La_{1-x}Ce_x)\,Al_2$ specimens are normalized by $H_{c2}^0(0)$ and plotted versus reduced temperature T/T_{c0}. The dashed curves correspond to a calculation of Maki[4]

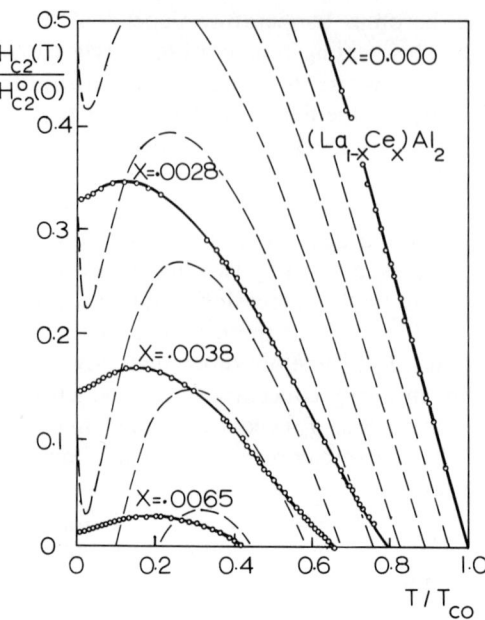

Fig. 3. The upper critical magnetic field $H_{c2}(T)$ of three $(La_{1-x}Ce_x)Al_2$ samples. The field values normalized by $H_{c2}^0(0) = 2.5$ kG of the clean $LaAl_2$ probe are plotted vs. reduced temperature $t = T/T_{c0}$. The dashed curves correspond to a calculation by Maki for $T_{c0} = 30 T_K^*$.

for $T_K^* = (1/30) T_{c0}$ based on the temperature-dependent pair-breaking parameter of the MHZ theory and the additive pair-breaking of the magnetic field. Again, no indication for the reappearance of superconductivity at ultra low temperatures can be seen in the experimental curves. Furthermore, the pair-breaking parameter $\alpha(T)$ does not increase as rapidly beyond the maximum of H_{c2} as is expected for the ratio $T_K^*/T_{c0} = 1/30$. Finally, the slope $(dH_{c2}/dT)_{T_c}$ is less than expected. As long as the residual resistivities of the specimens are not known to be the same, however, it appears premature to draw further conclusions from these preliminary results.

To this point in our studies spherically shaped samples of about 200 mg weight had been prepared in an argon arc furnace from 99.9% pure La and Ce and 99.999% pure Al. Larger specimens of about 10 g weight were prepared for measurements of electrical resistivity and specific heat as described in the following. These samples were shaped like buttons. They were remelted several times in order to homogenize the Ce additions. Pieces of approximately $10 \times 2 \times 2$ mm were cut, by means of a diamond saw, from different parts of the button. They were supplied with four contacts and used for the low-frequency ac resistivity measurements.[5]

The resistivity measurements as plotted in Fig. 4 confirm the two transitions. The Kondo effect in the normal state also is clearly in evidence. Most remarkable is the behavior of the sample with 0.66 at.% Ce substitution of La. This concentration

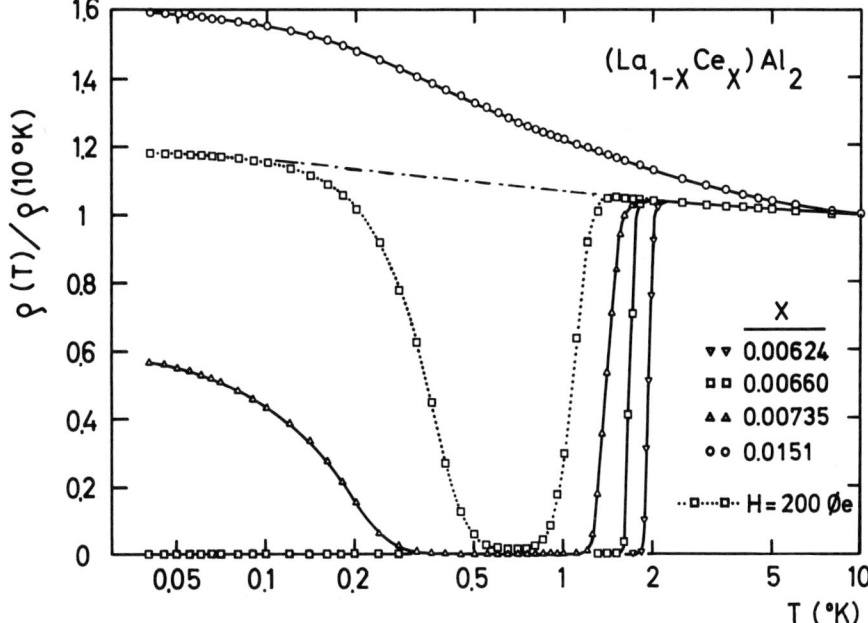

Fig. 4. The low-frequency ac resistivity of several $(La_{1-x}Ce_x)Al_2$ alloys normalized by the value at 10°K vs. temperature in a logarithmic scale. The measurements confirm the two transitions in a narrow range of Ce substitution. The second transition of the sample with a Ce concentration $x = 0.0066$ appears only in an applied magnetic field of 200 G. This demonstrates the increased pair-breaking effect of the Ce impurities at low temperatures and the additive pair-breaking effect of the magnetic field.

is just too small for the second transition to be observed in zero field. With increasing field, though, the transition back into the normal state appears first, followed by increasing suppression of superconductivity, until finally only the Kondo contribution to the resistivity in the normal state remains. In the measurements of resistivity the range of temperature in which superconductivity exists is somewhat wider as compared to the measurements of inductance. One would expect this from superconducting paths acting as short-circuits. Moreover, different probes cut from different parts of the same specimen do not show exactly the same transition temperatures, indicating that the distribution of the Ce additions is not perfect. Annealing of the specimens results in a sharpening of the transitions and may be interpreted by the same assumption. Resistivity measurements on probes cut from the upper and lower parts of the same specimen and inductance measurements on small specimens were used to obtain the T_c vs. c curve in Fig. 5. In this figure the midpoints of the transitions and the transition widths (10:90% and 10:10% of the signal at the turning point) are plotted versus Ce concentration (expressed as at.% of the amount of La).

Measurements of the specific heat[6] were carried out between 0.28 and 5°K in a special ^{3}He cryostat. The sample of about 10 g in weight was connected to the ^{3}He bath by a weak thermal link and heated at a constant rate to a temperature of

0.1°K above the bath temperature. The specific heat was determined from the recorded heating and cooling curves. The results of these measurements are plotted in Fig. 6. In the normal state they reveal a considerable contribution of the Kondo effect to the specific heat. The jump in the specific heat at the superconducting transition is drastically reduced by the Ce additions. The experimental results are in good agreement with recent theoretical calculations of Müller-Hartmann and Zittartz[7] assuming a Kondo temperature T_K of about 1°K in the $(La, Ce)Al_2$ system. Comparing the analytic expression (41) and the asymptotic expansion (42) in Ref. 8, one gets for $S = 1/2$, by neglecting the correction functions,

$$T_K \approx 12 T_{c0} (T_K^* / T_{c0})^{2/\sqrt{3}}$$

Therefore this result for T_K from measurements of the specific heat jumps is in agreement with the previous result for T_K^*.

The experimental results reported here permit a far more exhaustive evaluation than space permits us to present here. Our results have been verified and research is carried on by a research team at La Jolla.[9] They succeeded in obtaining sharp transitions for heat-treated specimens and in extending the measurements down to 6m°K.

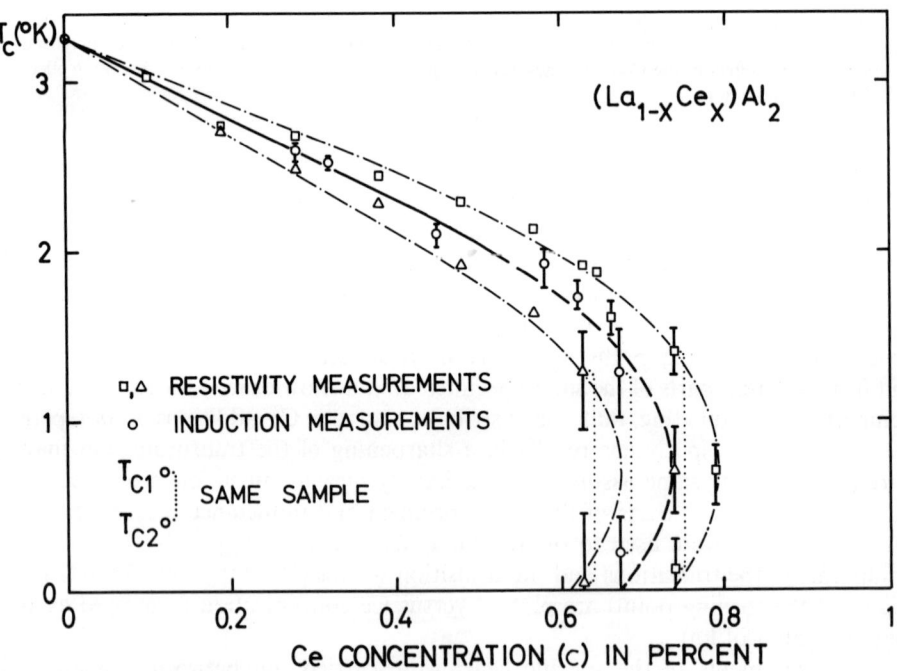

Fig. 5. Transition temperature of $(La, Ce)Al_2$ alloys vs. Ce-concentration expressed as at.% of the amount of La. The midpoints and the widths of the transitions are plotted as determined by resistivity measurements on different parts of the large specimens and by induction measurements on the small specimens.

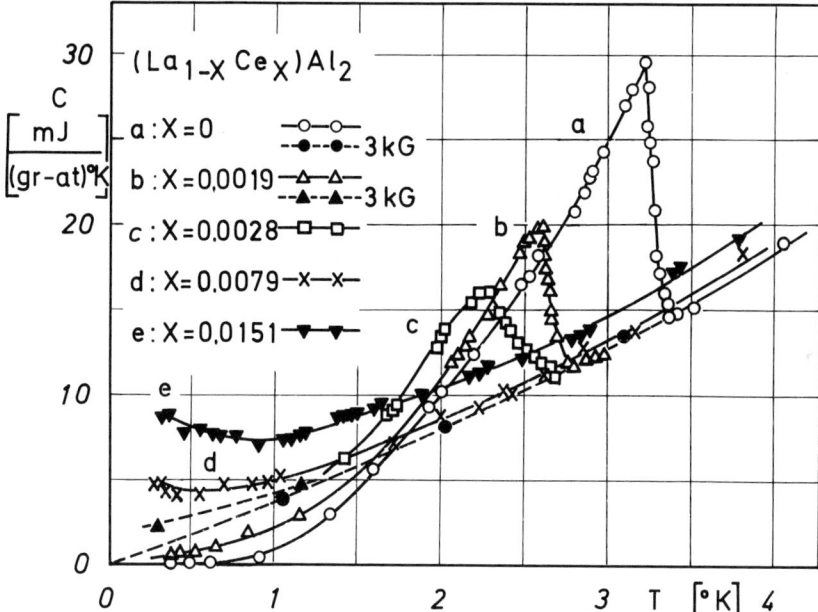

Fig. 6. The specific heat as a function of temperature of a clean LaAl$_2$ probe and of several (La$_{1-x}$Ce$_x$)Al$_2$ alloys. The data in the normal conducting state and in the superconducting state are given by the solid symbols and the open symbols, respectively.

Summary

The vanishing of superconductivity below a second transition temperature has been observed in $(La, Ce)\,Al_2$ alloys due to the Kondo effect. The transitions back into the normal state were measured for the first time, both by a low-frequency mutual inductance technique and by resistivity measurements down to 30 m°K. No indication of the reappearance of superconductivity at a third transition at ultra low temperatures has been found. The strange behavior of the upper critical magnetic field and the specific heat for these alloys has been verified experimentally.

Acknowledgments

We gratefully acknowledge the help received from stimulating and critical discussions with Dr. E. Müller-Hartmann and Professor J. Zittartz, and from the communication of unpublished results by these authors, and the help received from Professor K. Maki.

References

1. E. Müller-Hartmann and J. Zittartz, *Phys. Rev. Lett.* **26**, 428 (1971).
2. G. Riblet and K. Winzer, *Sol. St. Comm.* **9**, 1663 (1971).
3. G. Riblet and K. Winzer, *Sol. St. Comm.* **11**, 175 (1972).
4. K. Maki (to be published).
5. K. Winzer (to be published).
6. H. Armbrüster, G. Riblet, F. Steglich, and K. Winzer (to be published).
7. E. Müller-Hartmann and J. Zittartz (to be published).
8. E. Müller-Hartmann and J. Zittartz, *Z. Physik* **234**, 58 (1970).
9. M.B. Maple, W.A. Fertig, A.C. Mota, L.E. DeLong, D. Wohlleben, and R. Fitzgerald, University of California, San Diego, La Jolla, California, preprint.

Magnetic Impurities in Superconducting La$_3$Al Alloys

Toshio Aoi and Yoshika Masuda

Department of Physics, Nagoya University
Chikusa-ku, Nagoya, Japan

Abrikosov and Gor'kov (AG)[1] showed that magnetic impurities have a number of striking effects on superconductivity. An obvious manifestation of the magnetic impurity effect in the superconducting state is in the depression of superconducting transition temperature. According to the AG theory, the transition temperature T_c of a superconductor with paramagnetic impurities is related to a pair-breaking parameter ρ by the equation

$$\ln(T_c/T_{c0}) = \psi(\tfrac{1}{2}) - \psi(\tfrac{1}{2} + \rho) \tag{1}$$

where T_{c0} is the transition temperature of the pure material ($\rho = 0$) and ψ is the digamma function. To second order in the conduction electron–impurity spin exchange interaction parameter J, ρ is independent of temperature and is given by

$$\rho = (n/k_B T_c) N(E_F) J^2 S(S+1) \tag{2}$$

where n is the impurity concentration, $N(E_F)$ is the density of states at the Fermi level of the conduction band for one spin direction, and S is the impurity spin. In their calculation AG neglected completely any correlation among impurity spins and any dynamic effect associated with impurity spins which may exist in real superconductors.

The superconducting transition temperature T_c for solid solutions of Ce in La$_3$Al alloys was measured as a function of the impurity concentration by means of specific heat measurements. The plots of T_c vs. n are shown in Fig. 1 together with those for Gd in La$_3$Al alloys.[2] The specific heats of nine alloy samples are shown in Fig. 2. For samples containing Ce impurities of less than 1.6 at.%, superconducting transition jumps at T_c are observed. We regarded T_c as the midpoint of the specific heat jump. The change of permeability was also observed to correspond to the appearance of superconductivity.

As can be seen from Fig. 1, the contrasting effect of Ce and Gd impurities on the impurity concentration dependence of T_c of a superconducting La$_3$Al alloy has been observed. From the linear part of these plots we obtain the values $dT_c/dn = -1.97°K/\text{at.}\%$ and $-4.43°K/\text{at.}\%$ for La$_{3-x}$Ce$_x$Al and La$_{3-x}$Gd$_x$Al alloys, respectively. As previously reported, the La$_{3-x}$Gd$_x$Al system is well described in terms of the AG theory.[2] On the other hand, the transition temperatures of the La$_{3-x}$Ce$_x$Al system deviate markedly from the AG theory and show a faster drop than obtained from the AG approach fitted to the same initial slope.

The low-temperature normal-state resistivity measurements of La$_{3-x}$(RE)$_x$Al with RE = Ce and Gd as a function of temperature showed a minimum only for

Magnetic Impurities in Superconducting La$_3$Al Alloys

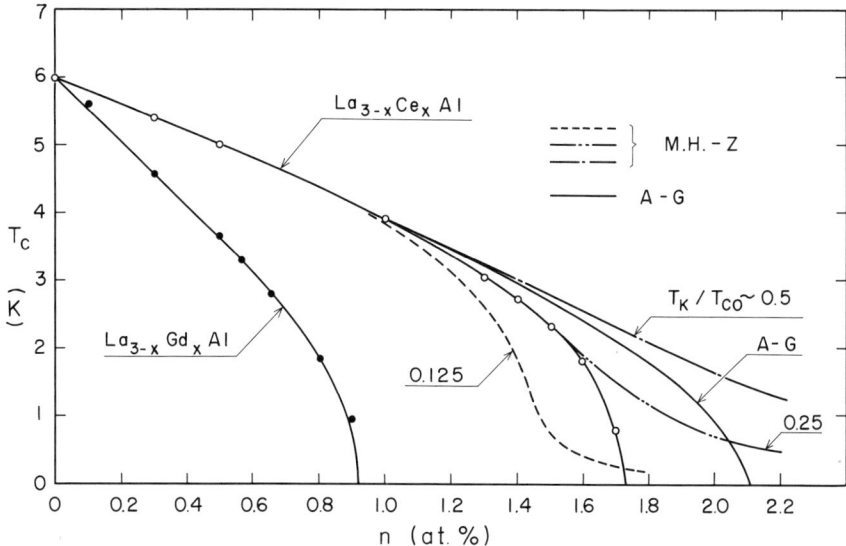

Fig. 1. Transition temperature vs. impurity concentration for the La$_{3-x}$Ce$_x$Al and La$_{3-x}$Gd$_x$Al systems. Theoretical curves for $T_K/T_{c0} = 0.5$, 0.25, and 0.125, calculated by MH-Z are also included. The AG curves are shown for comparison.

Ce. The resistance minimum for the alloys containing 1.6 and 1.8 at.% Ce appeared near 4.7°K, at zero field. To suppress superconductivity of the alloys of low Ce concentrations completely, e.g., a magnetic field was necessary, about 10 kG for the alloy containing 1.0 at.% Ce. Applying higher fields increased the resistance, and the resistivity vs. $(-\ln T)$ curve shifted toward higher temperatures. Therefore it was not possible to precisely determine the resistance anomaly due to resonance scattering at low Ce concentrations. However, it is certain that La$_{3-x}$Ce$_x$Al alloys show an anomaly associated with the Kondo effect.

Recently, Müller-Hartmann and Zittartz (MH-Z)[3] gave a theory explaining the deviation from the AG theory in terms of the Kondo effect. They presented a self-consistent treatment of the Kondo scattering of conduction electrons from magnetic impurities in order to treat finite concentrations of impurities in superconductors. Their theory leads to a concentration dependence for the transition temperature which differs markedly from the AG result. In the MH-Z theory Eq. (1) still describes such a superconducting Kondo system but with ρ dependent on temperature and given by

$$\rho = \frac{n}{(2\pi)^2 \, k_B T_c N(E_F)} \frac{\pi^2 S(S+1)}{\ln^2 T_c/T_K + \pi^2 S(S+1)} \tag{3}$$

This function has maximum at the Kondo temperature $T_K \sim T_F \exp[-1/N(E_F)|J|]$, where T_F is the Fermi temperature. The dependence of T_c on n calculated for $T_K/T_{c0} = 0.50$, 0.25, and 0.125 is depicted in Fig. 1. As can be seen from the figure, the theoretical curve for $T_K/T_{c0} = 0.25$ provides a remarkably good fit to the present experimental results for concentrations below 1.6 at.% Ce. However, it shows a

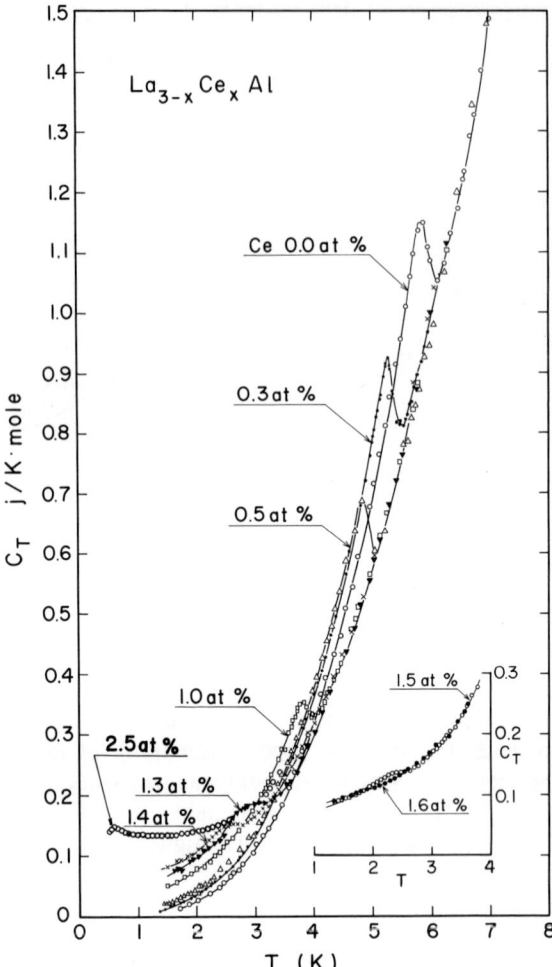

Fig. 2. Total specific heat C_T vs. temperature for eight $La_{3-x}Ce_xAl$ samples containing 0–1.6 at.% Ce.

deviation from the theoretical prediction for concentrations above 1.7 at.% Ce. As the temperature is lowered, a sample containing 1.7 at.% Ce becomes superconducting at 0.77°K and then remains superconducting to 0.02°K. At 1.8 at.% Ce concentration no superconductivity was detected down to 0.02°K, where the theoretical prediction shows the existence of a superconducting transition. Their theory also shows that when $T_K \ll T_{c0}$ alloys within a particular range of impurity concentrations will have three transition temperatures. Experimentally, Riblet and Winzer[4] and Maple et al.[5] observed a maximum in the ac diamagnetic susceptibility of $La_{1-x}Ce_xAl_2$ alloys which is associated with a reentrant superconducting–normal phase boundary. However, in present studies of the $La_{3-x}Ce_xAl$ system by both specific heat and ac mutual inductance techniques we could not observe this peculiar behavior in a range of temperature down to 0.020°K and of concentration below

1.8 at.% Ce, because of the rather high value of T_K/T_{c0}. Although T_K for the $La_{3-x}Ce_xAl$ system has not been derived from normal-state properties precisely, the value $T_K \sim 1.5°K$ does not seem unreasonable compared with that expected from the resistance minimum.

Maki[6] extended the MH-Z approximation to the calculation of the upper critical field $H_{c2}(T)$ for a Kondo superconductor. In this approximation the additive law of different pair-breaking parameters for the transition temperature due to magnetic impurities and due to the magnetic field still holds. Since the pair-breaking parameter arising from magnetic impurities has a maximum at $T \simeq T_K$, it is expected that the upper critical field $H_{c2}(T)$ reflects clearly the Kondo effect in the system. The upper critical field $H_{c2}(T)$ is determined from the relations

$$\rho = \frac{H_{c2}(T_c) T_{c0}}{4\gamma H_{c2}(0) T_c} + \frac{n}{(2\pi)^2 k_B T_c N(E_F)} \cdot \frac{\pi^2 S(S+1)}{\ln^2 T_c/T_K + \pi^2 S(S+1)} \qquad (4)$$

or

$$H_{c2}(T_c) + \frac{4\gamma n H_{c2}(0)}{(2\pi)^2 k_B T_{c0} N(E_F)} \frac{\pi^2 S(S+1)}{\ln^2 T_c/T_K + \pi^2 S(S+1)} \equiv H_{c2}^p(T_c) \qquad (5)$$

where $H_{c2}^p(T_c)$ is the upper critical field in the absence of magnetic impurities, H_{c2}^p is a universal function of the temperature, and $\ln \gamma$ is Euler's constant. In the limit $H_{c2}(T_c) = 0$, Eq. (4) reduces to Eq. (3). The calculated values of $H_{c2}(T)$ for $T_K/T_{c0} = 0.250$ and 0.125 are presented in Fig. 3, which shows that $H_{c2}(T)$ has a plateau

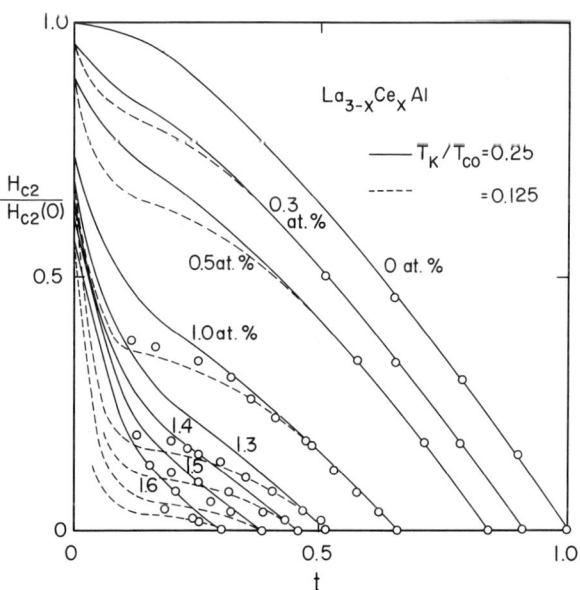

Fig. 3. Reduced upper critical field vs. reduced temperatures at various Ce concentrations. Solid and dashed curves show the theoretical predictions for $T_K/T_{c0} = 0.25$ and 0.125 calculated by Maki using the same approximation used for T_c by MH-Z.

around $T = T_K$. This behavior is easily understood in terms of the temperature-dependent pair-breaking effect. Experimental data are included in Fig. 3 and the agreement with the theory is reasonably good in the vicinity of the $H_{c2}(T)$, where the superconducting order parameter is small.

The evidence of the magnetic character of Ce impurities in the La_3Al alloy is found in the specific heat data. As can be seen from Fig. 2, the specific heat jumps at T_c for five samples containing 0.3–1.4 at.% Ce were observed. The relation between the reduced jump at T_c and the reduced transition temperature derived from critical field measurements shows a faster drop than obtained from the AG approach (as calculated by Skalski et al.[7]) for magnetic impurities with well-defined local moments. This initial depression $d(\Delta C/\Delta C_0)/d(T_c/T_{c0})$ at T_{c0} has the value of 2.3, which may correlate with the Kondo effect.

The specific heat peak at 0.65°K of the sample containing 2.5 at.% Ce in Fig. 2 is not related to superconductivity but to magnetic ordering, which suggests the existence of spin–spin correlations or hopefully a Kondo effect. The measured specific heats of samples with 0.3 and 0.5 at.% Ce in the superconducting state were analyzed to determine whether there was a γT term as previously reported for $La_{1-x}Ce_x$ alloys.[8] Although the data suggest that a γT term may be present, no definite conclusion can be reached until the measurements are extended to lower temperatures.

References

1. A.A. Abrikosov and L.P. Gor'kov, *Soviet Phys.—JETP* **12**, 1243 (1961).
2. T. Mamiya, T. Aoi, K. Iwahashi, and Y. Masuda, *J. Phys. Soc. Japan* **31**, 485 (1971).
3. E. Müller-Hartmann and J. Zittartz, *Phys. Rev. Lett.* **26**, 428 (1971).
4. G. Riblet and K. Winzer, *Sol. St. Comm.* **9**, 1663 (1971).
5. M.B. Maple, W.A. Fertig, A.C. Mota, L.E. Delong, D. Wohlleben, and R. Fitzgerald, *Sol. St. Comm.* (to be published).
6. K. Maki, *J. Low Temp. Phys.* **6**, 505 (1972).
7. S. Skalski, O. Betbeder-Matibet, and P.R. Weiss, *Phys. Rev.* **136**, A1500 (1964).
8. H.V. Culbert and A.S. Edelstein, *Sol. St. Comm.* **8**, 445 (1970).

Ce Impurities in Th-Based Superconducting Hosts*

J. G. Huber and M. B. Maple

Institute for Pure and Applied Physical Sciences
University of California—San Diego, La Jolla, California

Introduction

The experiments reported here were initiated as a study of Ce as an impurity in Th–Y alloy hosts, motivated by an interest in observing in measurements of superconducting transition temperatures T_c an expected transition of the Ce atom from a nonmagnetic configuration in Th[1] to a magnetic configuration in Y.[2] However, proceeding from Th to Y, the zero-pressure T_c of Th–Y alloys falls below 50 m°K (our lower limit of measurement) while the Ce impurities appear to remain nonmagnetic; i.e., T_c vs. Ce concentration n may always be fitted by the modified exponential

$$T_c = T_{c_0} e^{-An/(1-Dn)} \qquad (1)$$

where T_{c_0} is the T_c of the host and A and D are adjustable parameters. This equation was derived by Kaiser[3] for the case of a BCS superconductor into which has been dissolved an impurity element with nonmagnetic Friedel–Anderson resonant d or f electronic states.

The striking fact that Ce remains nonmagnetic in Th–Y hosts to high concentrations of Y prompted us to search for systematics in the depression of T_c by nonmagnetic Ce impurities in a variety of superconducting Th-based solid solution fcc alloys. The elements alloyed with Th were quadrivalent like Th (Zr, Hf) or trivalent (Sc, Y, Lu) and equal to (Y) or less than (Zr, Hf, Sc, Lu) Th in atomic radius. Cerium itself can range between trivalent and quadrivalent. When Ce atoms are dissolved into a metallic host they tend to accept the valence of their host; thus the fraction of an electron in the localized $4f$ level of a Ce atom in solution will depend upon the average number of valence electrons per host atom. Also, the Ce $4f$ level will depopulate when a Ce atom is subjected to pressure, whether externally applied or due to a contraction through alloying of the host lattice. The occupation of the Ce $4f$ level is reflected in the depression of the T_c of the host if it is a superconductor; the more $4f$ electrons, the steeper T_c is depressed.

Experimental Details

Our samples were prepared by arc-melting together the appropriate constituents in an argon atmosphere. The Th–X binary alloys were direct mixtures of the elements

---

* Research sponsored by the Air Force Office of Scientific Research, Air Force Systems Command, USAF, under AFOSR contract No. AFOSR/F-44620-72-C-0017.

involved. Those with Ce impurities originated with a single host which was cut into halves; one-half was doped with Ce and the resulting two alloys were recombined in varying proportions. Weight losses were negligible for all alloys other than those containing Lu.

The samples were measured inductively for superconductivity in their as-cast state. They were cooled in a ^{4}He cryostat, supplemented when necessary by ^{3}He and chromium potassium alum adiabatic demagnetization cooling stages. Temperatures were determined from ^{4}He and ^{3}He vapor pressures or the electrical resistance of calibrated carbon resistors. When measured for T_c under pressure the samples were enclosed in a Be–Cu clamp which retained a pressure applied at room temperature. This pressure was mediated hydrostatically by a 50:50 n-pentane:isoamyl alcohol solution. The absolute value of the pressure was inferred from the T_c of a tin manometer. Our working range extended to over 20 kbar.

For each of our host–Ce impurity alloy systems we have best-fitted Eq. (1) by computer to our T_c vs. n data at zero and other pressures. We transformed Eq. (1) to the linear expression $n/\ln(T_c/T_{co}) = (D/A)n - 1/A$ and wrote a program to find the straight line that passes with least-squared deviation through a set of points described by the coordinates $n/\ln(T_c/T_{co})$ and n. From the slope D/A and the $n = 0$ intercept $-1/A$ we obtain the parameters A and D.

Results and Discussion

Before presenting our study of Ce impurities we note some patterns in the T_c values of the Th–X host alloy systems. As seen in Fig. 1, each element X (X being Sc, Y, Lu, Zr, or Hf, all of which are hcp as pure elements) added to Th in fcc solid solution initially raises T_c. Also, the systems may be grouped. For quadrivalent Zr and Hf the increases in T_c are linear with concentration and extend with similar slopes to the solubility limits. For trivalent Sc, Y, and Lu shallower initial linear increases lead to peaks in T_c at about 20 at.% X. An alloy of 20 at.% trivalent X and 80 at.% quadrivalent Th contains an average of 3.8 valence electrons per atom, e/a, so the aforementioned peaks support the observation of Cooper et al.[4] that for a variety of alloy systems T_c maxima occur at e/a values ranging between 3.7 and 3.9. Beyond 40 at.% Sc, Y, or Lu in Th the measured T_c's broaden and our samples are probably mixtures of fcc and hcp phases. It is only for the Th–Y system, starting at 55 at.% Y, that we feel we have succeeded in making pure hcp phase samples with measurable T_c's. As alluded to in the introduction, even in these Th–Y alloys, whose T_c's fall rapidly to zero with increasing Y concentration, Ce impurities seem to be nonmagnetic. Our reduced T_c, T_c/T_{co}, vs. Ce concentration data for the alloy system $Th_{44}Y_{56}$ Ce may be seen in Fig. 2 fitted by Eq. (1). The steepness of the depression of T_c in comparison with the other representative systems should be noted, but this system will not be discussed further. Our primary observations deal solely with Ce impurities in fcc Th–X alloys with X concentration no greater than 35 at.%.

The purpose of our experiments was to document how the effect of Ce impurities in Th-based superconducting hosts varies when changes are made in (1) the average number of valence electrons per host atom v; (2) the fcc lattice parameter of the host a; and (3) the hydrostatic pressure p to which samples are subjected when their T_c is measured. We therefore fabricated sets of samples with appropriate Ce con-

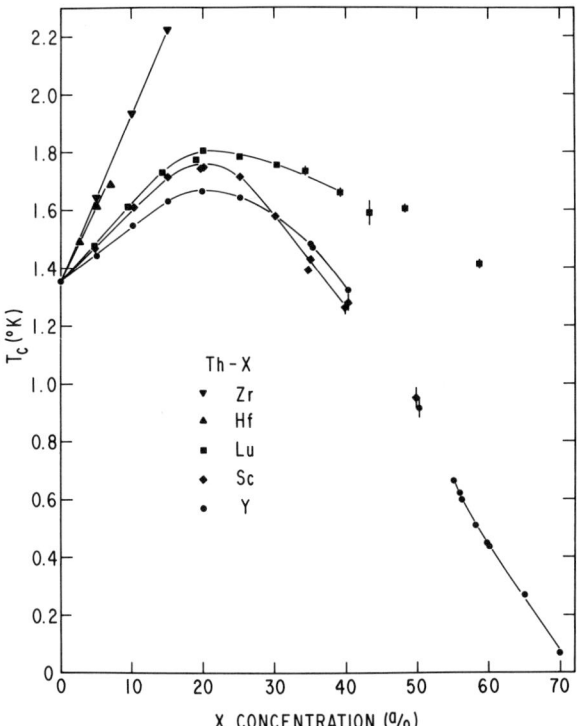

Fig. 1. Superconducting transition temperature T_c vs. X concentration for Th–X alloys, with X being Zr, Hf, Lu, Sc, and Y. The vertical bars show transition widths where they exceed the dimensions of the symbols.

centrations for the following systems: ThCe, $Th_{95}Hf_5$Ce, $Th_{90}Zr_{10}$Ce, $Th_{80}Sc_{20}$Ce, $Th_{80}Sc_{10}Y_{10}$Ce, $Th_{80}Y_{20}$Ce, $Th_{80}Lu_{20}$Ce, $Th_{65}Sc_{35}$Ce, and $Th_{65}Y_{35}$Ce. We then measured their T_c at a variety of pressures. Our results for a selection of these systems at zero pressure are given in Fig. 2, where the reduced T_c are plotted on both linear and logarithmic ordinate axes. We also show the plot of Eq. (1) best-fitted to each set of data. One can appreciate the closeness of the fits as well as the spread in the depressions. We add, though we do not show, that for all of the systems listed above at any pressure, T_c/T_{co} vs. n may be similarly well fitted by Eq. (1).[5]

It is the parameters A and D, then, which characterize the $4f$ configuration of the dissolved Ce atoms. D, being a measure of the deviation from a perfect exponential described by A (compare the dashed and solid lines in the upper half of Fig. 2), is neither as important nor as accurately determinable as A. So we focus our analysis on the parameter A, conscious that beyond its role in Eq. (1) it also represents the initial slope of reduced T_c vs. n; that is, $d(T_c/T_{co})/dn|_{n=0} = -A$, a number for which theoretical derivations abound.

On the far left in Fig. 3 we display the A's evaluated from our zero-pressure data. They are plotted with a logarithmic ordinate against Δa, the difference in host fcc lattice parameter from the pure Th value (about 5.68 Å). The Δa are assumed to

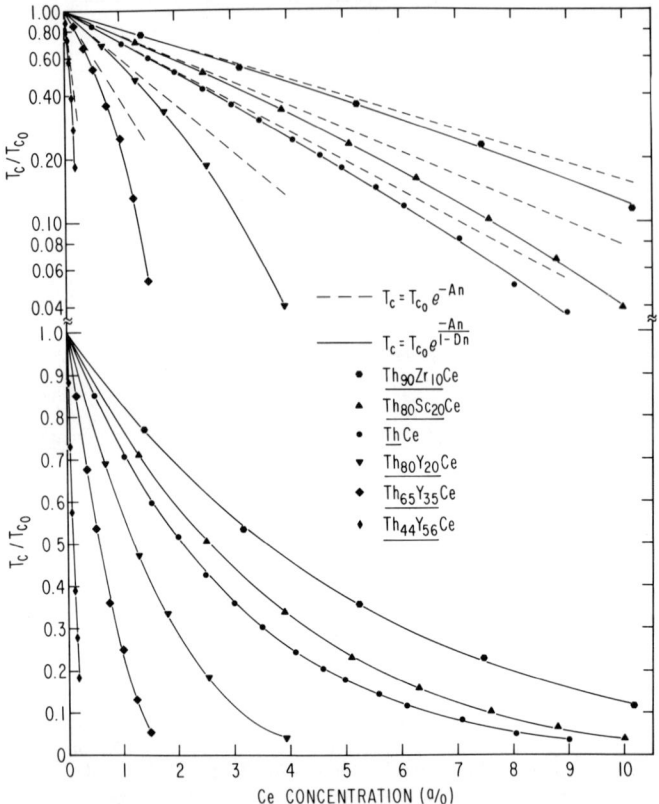

Fig. 2. Reduced superconducting transition temperature T_c/T_{co} at zero pressure (on both logarithmic and linear ordinate axes) vs. Ce concentration for representative systems of Ce impurities in Th-based alloy hosts. The solid lines are best-fitted curves described by $T_c = T_{co}e^{-An/(1-Dn)}$; the dashed lines plot $T_c = T_{cn}e^{-An}$ using the best-fitting A.

be consistent with the findings of other experimenters for fcc $Th_{1-x}X_x$ alloys; da/dx (in Å/at. % at room temperature) for the element X is -0.0030 for Sc (Ref. 6, pp. 5–9), -0.0000 for Y,[7] -0.0013 for Lu (Ref. 6, pp. 11–15), -0.0048 for Zr,[8] and -0.0024 for Hf.[9] It seems significant that the points for the 3.80-valent hosts, $Th_{80}Y_{20}$, $Th_{80}Sc_{10}Y_{10}$, and $Th_{80}Sc_{20}$, fall in a straight line and that lines parallel to this will connect the points for the 3.65-valent hosts, $Th_{65}Y_{35}$ and $Th_{65}Sc_{35}$, and those for the 4.00-valent hosts, Th and $Th_{90}Zr_{10}$. The point for $Th_{80}Lu_{20}$ may be argued closer to the 3.80-valent host line, since we did not account for a nonnegligible Lu boiloff during our sample preparation. Thus, the point for the 4.00-valent host, $Th_{95}Hf_5$, alone is grossly and inexplicably out of place. We deduce that, excepting the case of Hf, there is a linear relationship between log A and Δa, due to a lattice pressure on the Ce if you will, which has a universal slope independent of the value of v. For comparison with the effect of externally applied pressure, we again refer to Fig. 3. In the three segments on the right we have grouped our data according to v, plotting the A's, still logarithmically, against p. Once more, but for the system

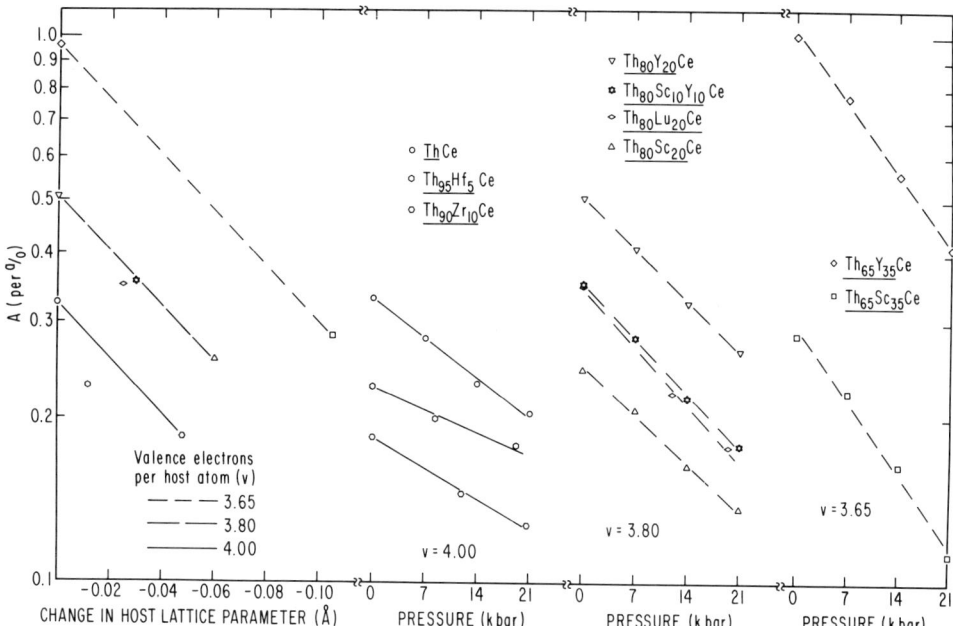

Fig. 3. Plot of A (on a logarithmic ordinate axis) vs. difference in host fcc lattice parameter from the pure Th value Δa and pressure p for systems of Ce impurities in Th-based alloy hosts. The data are grouped according to average number of valence electrons per host atom v.

$Th_{95}Hf_5Ce$, there is a pattern. For a given v, log A vs. p has a general linear slope and the gradient steepens as v decreases. This linearity of log A with p is more convincing than, and lends credence to, that assumed with Δa. In summation, an approximate consolidation of the dependence of A on v, Δa, and p for systems of Ce impurities in superconducting fcc Th–(Sc, Y, Lu, and/or Zr) alloy hosts is given by the equation

$$\log A = -1.11 + f(v)(1 - Bp) + C \Delta a \tag{2}$$

where $C = 5.1$ Å^{-1}, $B = 0.017$ kbar^{-1}, and $f(v) = 0.62, 0.81$, and 1.10 for $v = 4.00$, 3.80, and 3.65, respectively.

Finally, we return to the idea that the depression of T_c, and hence A, is related to the occupation of the Ce 4f level with which one associated the microscopic parameters N_f, the localized 4f electron density of states at the Fermi level, and U_{eff}, the intraatomic Coulomb potential (reduced by correlations) between electrons of opposite spin. In Kaiser's theory[3] there are relatively simple expressions for A and D in terms of these parameters and two characterizing the host; N, the conduction electron density of states at the Fermi level, and V, the BCS conduction electron (phonon mediated) pairing potential. Computations of the microscopic parameters have not been made, however, for several reasons: lack of accurate D's, paucity of host specific heat data necessary for determining N's and V's, and a feeling that Kaiser's derivation may not be as valid as Eq. (1) itself over the obviously wide range of Ce 4f level occupation. (One must appreciate the fact that in our experiments A varies by an order of magnitude.)

References

1. J.G. Huber and M.B. Maple, *J. Low Temp. Phys.* **3**, 537 (1970).
2. N. Nagasawa, S. Yoshida, and T. Sugawara, *Phys. Lett.* **26A**, 561 (1968).
3. A.B. Kaiser, *J. Phys. C* **3**, 409 (1970).
4. A.S. Cooper, E. Corenzwit, L.D. Longinotti, B.T. Matthias, and W.H. Zachariasen, *Proc. Nat. Acad. Sci.* **67**, 313 (1970).
5. J.G. Huber, Ph. D. Thesis, University of California, San Diego, 1971.
6. T.A. Badaeva and R.I. Kuznetsova, *Fiz.-Khim. Splavov Tugoplavkikh Soedin. Toriem Uranom*, O.S. Ivanov, ed., Izd. Nauka, Moscow (1968).
7. D.S. Evans and G.V. Raynor, *J. Nucl. Mater.* **2**, 209 (1960).
8. R.H. Johnson and R.W.K. Honeycombe, *J. Nucl. Mater.* **4**, 59 (1961); D.S. Evans and G.V. Raynor, *J. Nucl. Mater.* **4**, 66 (1961).
9. E.D. Gibson, B.A. Loomis, and O.N. Carlson, *Trans. Am. Soc. Metals* **50**, 348 (1958).

Heat Capacity of ThU Alloys at Low Temperatures

C. A. Luengo,* J. M. Cotignola, and J. Sereni

*Centro Atómico Bariloche and Instituto de Física
"Dr. J. A. Balseiro," Comisión Nacional de Energía Atómica
and Universidad Nacional de Cuyo, Bariloche, Argentina*

A. R. Sweedler †

*Departamento de Física, Facultad de Ciencias
Universidad de Chile, Santiago, Chile*

and

M. B. Maple †‡

*Institute for Pure and Applied Physical Sciences
University of California—San Diego, La Jolla, California*

Measurements of the normal-state electrical resistivity, thermoelectric power, magnetic susceptibility, and concentration dependence of the superconducting transition temperature T_c indicate that U ions are only weakly magnetic when dissolved in Th.[1] This weakly magnetic behavior is associated with U $5f$ local moments which apparently fluctuate in time with a finite frequency $\tau_{sf}^{-1} = k_B T_0 / h$. The characteristic temperature $T_0 \sim 100°$K determined from the normal-state properties implies that the local moments are more long-lived in the ThU system than in the closely related, weakly magnetic systems AlMn ($T_0 = 530°$K)[2] and ThCe (T_0 immeasurably large).[3]

All these systems satisfy the condition $T_0 \gg T_{c_0}$ (T_{c_0} is the T_c of the matrix), and the superconducting properties suggest that the systems are essentially *nonmagnetic* at superconducting temperatures. The depressions of T_c as a function of impurity concentration n[4,5] are described very well by a modified exponential relation recently proposed by Kaiser[6] to account for the effect of nonmagnetic resonant localized impurity states on the superconductivity of the host metal. Moreover, specific heat and/or critical field data satisfy the BCS law of corresponding states in AlMn[7,8] and ThCe.[3]

On the other hand, an attempt has been made to fit the ThU T_c vs. n data by a

---

* John Simon Guggenheim Fellow 1971–1972.
† Supported in part by University of Chile–University of California exchange program financed by the Ford Foundation.
‡ Supported in part by the Air Force Office of Scientific Research, Air Force Systems Command, USAF, under AFOSR grant number AFOSR-631-67-A.

temperature-dependent pair-breaking theory in the limit $T_K \gg T_{co}$ where the Kondo temperature T_K has been identified with the characteristic temperature T_0.[9] Since the thermodynamic properties are not expected to follow the BCS law of corresponding states in the presence of pairbreaking interactions (even for $T_K \gg T_{co}$)[5,10] heat capacity measurements were therefore undertaken in order to obtain additional information concerning the magnetic character of U impurities in a Th matrix. Preliminary results[11] indicated that the specific heat jump ΔC at T_c follows the BCS law of corresponding states, while the normal-state electronic specific heat coefficient γ is greatly enhanced in the presence of the U impurities.

ThU samples of large mass (~ 18 g) were prepared for the present study in the manner previously described.[11] A specimen of pure Th was also remeasured in a newly constructed ^{3}He calorimeter to check on an earlier result[11] which disagreed somewhat with respect to the normal-state parameters previously reported by Gordon et al.[12] For all samples investigated between 60 and 85 experimental points were fitted to the usual relation (Fig. 1)

$$C_v = \gamma T + \beta T^3 \qquad (1)$$

between 1.0 and 5.0°K in the case of the alloys and 1.4 and 4.2°K in the case of the Th matrix.

Values for the normal-state parameters γ and the Debye temperature θ_D are given in Table I. The values for the remeasured Th sample, considered more reliable than our previous values, are in reasonable agreement with those reported by Gordon

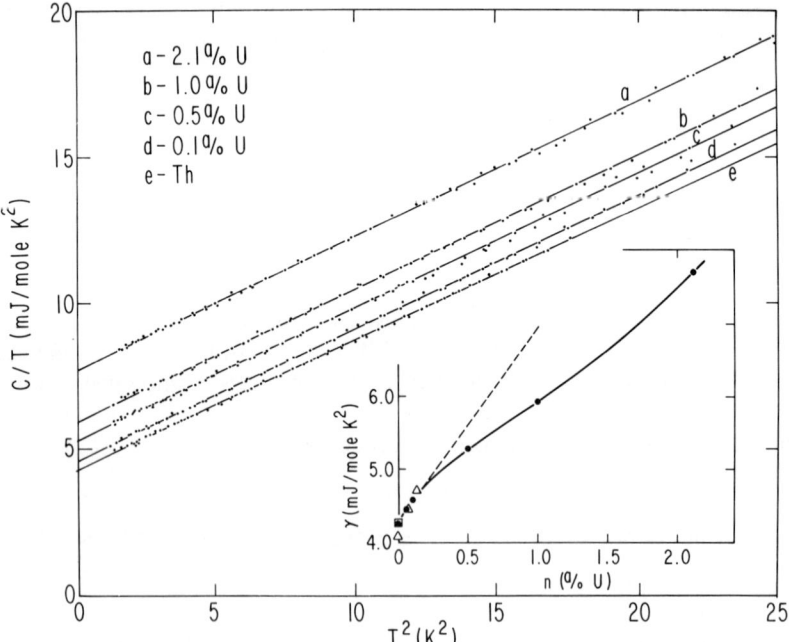

Fig. 1. C/T vs. T^2 in the normal state for several ThU alloys. Inset: Electronic specific heat coefficient as a function of U impurity concentration; solid circles—this work; open triangles—earlier experiments[11]; open square—Gordon et al.[12]

Heat Capacity of ThU Alloys at Low Temperatures

Table I

n, at.% U	γ, mJ/mole °K²	θ_D, °K	$\Delta\gamma/n$, mJ/mole °K² at.% U	T_c, °K	ΔC, mJ/mole °K
0	4.28 ± 0.05	164.1 ± 0.5	—	1.360 ± 0.005	8.50 ± 0.2
0.065	4.45 ± 0.03	161.9 ± 0.3	2.6	0.860 ± 0.005	5.3 ± 0.3
0.075[a]	4.45 ± 0.08	161.3 ± 1.0	2.3	0.785	5.3 ± 0.5
0.110	4.58 ± 0.03	161.7 ± 0.3	2.7	0.550 ± 0.005	3.3 ± 0.3
0.134[a]	4.72 ± 0.04	163.7 ± 0.7	3.3	—	—
0.50	5.28 ± 0.05	162.1 ± 0.5	2.0	—	—
1.00	5.93 ± 0.04	161.9 ± 0.4	1.67	—	—
2.10	7.70 ± 0.06	161.4 ± 0.4	1.62	—	—

[a] From Ref. 11.

et al.[12] The most significant result of the measurements in the normal state is the very large rate at which γ increases with U concentration: At low concentrations $d\gamma/dn = 2.7$ mJ/mole °K² at.% U, while $d\gamma/dn$ shows a significant decrease at higher concentrations (Fig. 1, inset). The θ_D values in Table I indicate that there is no systematic change in θ_D due to alloying.

Measurements were also extended into the superconducting state for pure Th and the three most dilute ThU alloys (Fig. 2). (Measurements become more difficult at the lower temperatures because of the radioactivity of the sample, which gives a heat input of about 150 ergs/min at 0.7°K. This raises the lowest temperature at which specific heat measurements can be made.) The value $\Delta C_0/\gamma T_{c_0} = 1.460 \pm 0.003$ for Th agrees reasonably well with the BCS value of 1.43 and previously reported values.[12,13] As shown in Fig. 3, where the reduced specific heat jump $\Delta C/\Delta C_0$ is plotted vs. reduced transition temperature T_c/T_{c_0}, the ThU specific heat jumps at T_c follow closely the BCS law of corresponding states.

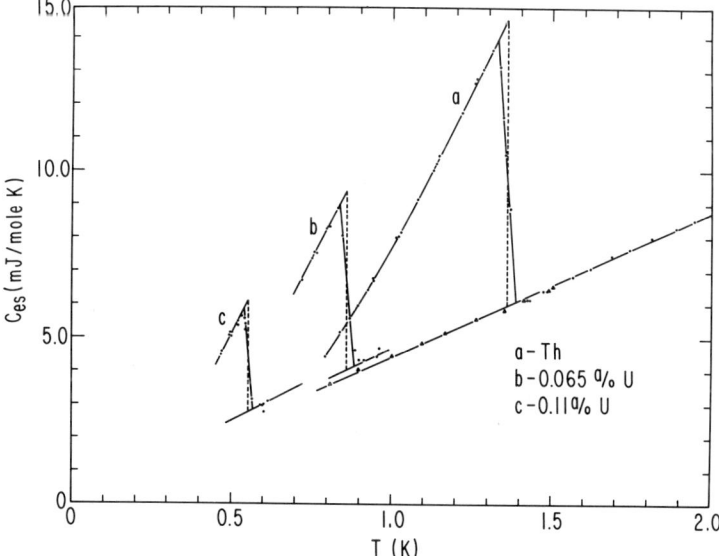

Fig. 2. Electronic specific heat jump ΔC at T_c for Th and ThU alloys.

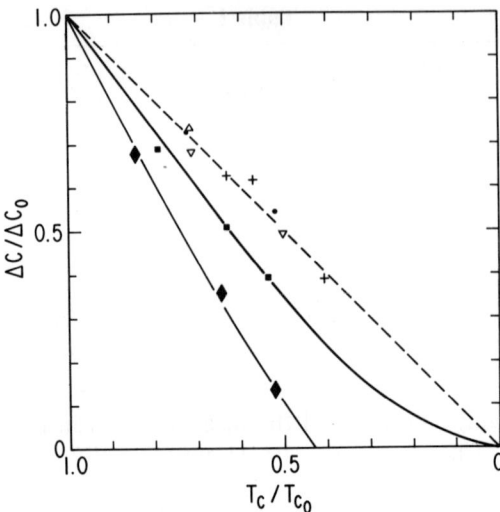

Fig. 3. Reduced specific heat jump $\Delta C/\Delta C_0$ vs. reduced transition temperature T_c/T_{co}. The dashed line represents the BCS law of corresponding states, while the heavy solid line is the AG result. Crosses—ThU (this work); solid circles—ThCe (Dempesy[16]); open triangle—AlMn (Martin[7]); inverted open triangles—AlMn (Smith,[8] from critical field data); solid squares—ThGd (Decker and Finnemore,[13] from critical field data); solid diamonds—$(La, Ce)Al_2$ (Luengo et al.[15])

The close agreement of the ThU specific heat jumps with the BCS law of corresponding states indicates that the rapid modified exponential depression of T_c with U concentration[1,5] is due to *pair-weakening* rather than *pair-breaking* interactions. As shown in Fig. 3, this is also true for the related systems AlMn and ThCe, and, as previously noted,[5] seems to be the rule whenever the characteristic temperature T_0 is much greater than T_{co}. This may be contrasted to systems with well-defined long-lived local moments (e.g., ThGd[13]), where the specific heat jumps at T_c follow instead a relationship given by the pair-breaking theory of Abrikosov and Gor'kov (AG).[14] Even with the pair-breaking theory generalized to include higher-order exchange scattering than J^2 (Kondo effect), the calculated specific heat jumps at T_c in the limit $T_K/T_{co} \gg 1$ tend toward the AG rather than the BCS behavior,[10] and cannot account for our results.

Also shown for comparison in Fig. 3 are specific heat jumps at T_c for the $(La, Ce)Al_2$ system[15] ($T_K \sim 0.1°K^5$); for these alloys $T_K/T_{co} \sim 0.03$, where large departures from the AG theory are expected.[10]

Finally, if it is assumed that the initial increase in γ with U concentration is due to U $5f$ resonant states, the local density of states at the Fermi level (for one spin direction) contributed by each U atom would be 57 states/eV-U atom (rather than the 103 states/eV-U atom previously reported[11]). This value is much larger than obtained by fitting the T_c vs. n curve to the Kaiser theory. However, a proper renormalization of the Kaiser theory to include the effect of spin fluctuations might lead to quantitative agreement.

References

1. M.B. Maple, J.G. Huber, B.R. Coles, and A.C. Lawson, *J. Low Temp. Phys.* **3**, 137 (1970).
2. A.D. Caplin and C. Rizzuto, *Phys. Rev. Lett.* **21**, 746 (1968).
3. J.G. Huber and M.B. Maple, *J. Low Temp. Phys.* **3**, 537 (1970).
4. J.G. Huber and M.B. Maple, *Sol. St. Comm.* **8**, 1987 (1970).
5. M.B. Maple, in *Superconductivity in d- and f- Band Metals*, D.H. Douglass, ed., AIP Conf. Proc. No. 4, (1972), pp. 175–203.
6. A.B. Kaiser, *J. Phys. C* **3**, 409 (1970).
7. D.L. Martin, *Proc. Phys. Soc. (London)* **78**, 1489 (1961).
8. F.W. Smith, *J. Low Temp. Phys.* **6**, 435 (1972).
9. E. Müller-Hartmann and J. Zittartz, *Phys. Rev. Lett.* **26**, 428 (1971).
10. E. Müller-Hartmann and J. Zittartz, *Sol. St. Comm.* **11**, 401 (1972).
11. C.A. Luengo, J.M. Cotignola, J.G. Sereni, A.R. Sweedler, M.B. Maple, and J.G. Huber, *Sol. St. Comm.* **10**, 459 (1972).
12. J.E. Gordon, H. Montgomery, R.J. Noer, G.R. Pickett, and R. Tobón, *Phys. Rev.* **152**, 432 (1966).
13. W.R. Decker and D.K. Finnemore, *Phys. Rev.* **172**, 430 (1968).
14. A.A. Abrikosov and L.P. Gor'kov, *Zh. Eksperim. i Teor. Fiz.* **39**, 1781 (1960); *Soviet Phys.–JETP* **12**, 1243 (1961).
15. C.A. Luengo, M.B. Maple, and W.A. Fertig, submitted for publication.
16. C.W. Dempesy, private communication.

Superconducting Critical Field Curves for Th–U

H. L. Watson, D. T. Peterson, and D. K. Finnemore

Ames Laboratory—USAEC and Department of Physics
Iowa State University, Ames, Iowa

Maple and co-workers[1] have presented strong evidence that the Th–U magnetic impurity system shows localized spin fluctuation (LSF) phenomena.[2] The normal-state low-temperature electrical resistivity data for these alloys obey the $\rho = \rho_0(1 - T^2/T_0^2)$ law, where T_0 is approximately 100°K and the thermoelectric power goes through a maximum at about 80°K. In addition, the superconducting transition temperature vs. concentration curve, T_c vs. n, shows positive curvature, as expected for LSF, and the effective moment determined from susceptibility data decreases rapidly with decreasing temperature below 100°K. All of these measurements point toward a localized spin fluctuation interpretation of the data.

If indeed the magnetic moment on the uranium impurity is temperature dependent in the normal state as predicted by the LSF theory, then one would expect that the depairing parameter for the Cooper pairs[3] of the superconducting host would also be temperature dependent and could be used to probe the excitation spectrum of the uranium impurity. The purpose of this experiment is to measure the superconducting critical field curves of Th–U alloys to look for the deviations from the Abrikosov–Gor'kov (AG) theory[4] which arise because the magnetic impurity has a temperature dependent moment. The magnitude of these deviations can then be used to estimate the temperature dependence of the spin scattering rate τ_s and the changes in the uranium moment with temperature.

Results and Discussion

Details of the experiment are very similar to the magnetization measurements of Th–Gd which were reported earlier.[5] Th–U shows a small irreversibility in the phase transition of about 2% of the critical field at $T = 0$ (H_0). The magnitude of this hysteresis can be reduced by annealing at 1200°C for a week, and we find that both the superheating and the supercooling diminish by about the same amount after annealing. Hence the thermodynamic critical field H_c has been identified as the midpoint of the hysteresis loop as described in detail for the Th–Gd alloys.[5] The accuracy of the H_c measurement is at least 2%, and if our interpretation of H_c is proper, then the accuracy of the critical field curves would be about $\frac{1}{2}$% of H_0.

Critical field curves for Th–0.05 at. % U and Th–0.10 at. % U are shown in Fig. 1. The data lie consistently above the AG predictions shown by the solid lines. For each sample the lifetime broadening constant Γ of the AG theory was selected to match the data at T_c. The magnitude of the deviations from AG are about 10%

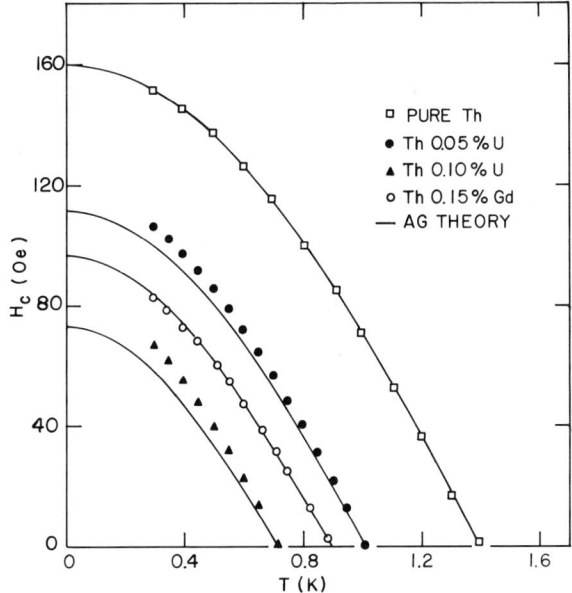

Fig. 1. Superconducting critical field curves for Th–U alloys.

of H_0, so they are easily resolved by the experiment. It should be noted that the Th–U results are quite different from the Th–Gd results shown by the open circles in Fig. 1. Th–Gd follows the AG theory very closely and it appears that the magnetic moment on the Gd ion is independent of temperature. On the other hand, Th–U lies above AG, as would be expected if the spin-scattering rate were decreasing as the temperature decreases. One can, in fact, use the AG theory to determine the effective spin scattering rate $\tau_s^{-1} = \Gamma_e/\hbar$, for each point on the superconducting critical field curves. This is simply a way of parameterizing the data so that deviations from the AG theory appear as a temperature-dependent spin scattering rate. Values of these effective spin scattering rates for both samples are shown in Fig. 2. For the 0.05% U sample $\tau_s^{-1} = \Gamma_e/\hbar = (0.72 + 0.18T) \times 10^{+10}$ sec^{-1} and for the 0.10% U sample $\tau_s^{-1} = \Gamma_e/\hbar = (1.30 + 0.38T) \times 10^{+10}$ sec^{-1}. Both the intercept and the slope are approximately proportional to concentration, but there is some deviation. It will be necessary to measure more samples before anything definite can be said about the concentration dependence. All the data presented in Fig. 2 represent temperatures far below the characteristic temperature for the localized spin fluctuations of about 100°K. Hence one can probably extrapolate the results to $T = 0$ with some confidence.

Conclusions

Superconducting critical field curves of Th–U show marked deviations from the AG theory as would be expected if the U impurity shows localized spin fluctuations. The effective spin scattering rate for the U impurity has a linear rather than quadratic temperature dependence in the range from 0.3 to 1.1°K. This would seem

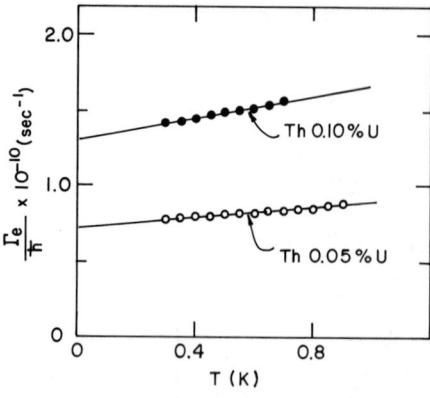

Fig. 2. Temperature dependence of the spin scattering rates for Th–U alloys.

to indicate that the broadening of the impurity uranium state is comparable to the Coulomb repulsion, $U/\pi\Delta \approx 1$.[6] The existence of a residual spin scattering as the temperature approaches zero would seem to imply that U impurities in Th have a small moment even in the limit of $T = 0$.

References

1. M.B. Maple, J.G. Huber, B.R. Coles, and A.C. Lawson, *J. Low Temp. Phys.* **3**, 137 (1970).
2. H. Suhl, *Phys. Rev. Lett.* **19**, 442 (1967); M.J. Levine and H. Suhl, *Phys. Rev.* **171**, 567 (1968); M.J. Levine, T.V. Ramakrishnan, and R.A. Weiner, *Phys. Rev. Lett.* **20**, 1370 (1968); N. Rivier and M.J. Zuckermann, *Phys. Rev. Lett.* **21**, 904 (1968).
3. J. Bardeen, L.N. Cooper, and J.R. Schrieffer, *Phys. Rev.* **108**, 1175 (1957).
4. A.A. Abrikosov and L.P. Gor'kov, *Zh. Eksperim. i Teor. Fiz.* **39**, 178 (1960) [*Soviet Phys.—JETP* **12**, 1243 (1961)]; S. Skalski, O. Betbeder-Matibet, and P.R. Weiss, *Phys. Rev.* **136**,A1500 (1964).
5. W.R. Decker and D.K. Finnemore, *Phys. Rev.* **172**, 430 (1968).
6. K.H. Bennemann, *Phys. Rev.* **183**, 492 (1968).

c^2 Contributions to the Abrikosov—Gor'kov Theory of Superconductors Containing Magnetic Impurities

D. Rainer

Institut für Festkörperforschung der Kernforschungsanlage
Jülich, Germany

In the second-order Born approximation the superconducting transition temperature T_c of conduction electrons coupled to a spin system via a short-range exchange interaction J_0 is determined by

$$\Delta(k, \omega_n) = \frac{\Omega}{(2\pi)^3} \int_{BZ} d^3k' \, T_c \sum_{n'} \left[I_0(\omega_n, \omega_n') - \frac{3}{2} J_0^2 \chi(k - k', \omega_n - \omega_n') \right]$$
$$\times |G(k', \omega_n')|^2 \, \Delta(k', \omega_n') \tag{1}$$

$$G(k, \omega_n) = [i\omega_n - \varepsilon(k) - \Sigma(k, \omega_n)]^{-1}$$

$$\Sigma(k, \omega_n) = \Sigma_0(\omega_n) + \frac{\Omega}{(2\pi)^3} \int_{BZ} d^3k' \, T_c \sum_{n'} \frac{3}{2} J_0^2 \chi(k - k', \omega_n - \omega_n') \, G(k', \omega_n')$$

Here $\varepsilon(k)$ is the band energy of conduction electrons, Σ_0 and I_0 are the ordinary self-energy and interaction vertex, respectively, and $\chi(q, \omega_n)$ is the susceptibility of the spin system.

In order to study the influence of short-range correlations in a dilute, randomly distributed spin system on its pair-breaking effect, we assume that spins sitting on neighboring lattice sites are coupled via an exchange interaction of strength J. Up to second order in the concentration of spins c the susceptibility of such a system is given by (Z = number of nearest neighbors)

$$\chi(q, \omega_m) = c(1 - Zc) \chi_{\text{single}}(q, \omega_m) + Zc^2 \chi_{\text{pair}}(q, \omega_m) \tag{2}$$

$$\chi_{\text{single}} = \frac{1}{4T} \delta_{\omega_m, 0}$$

$$\chi_{\text{pair}} = \frac{1}{2} \frac{1}{3 + \exp(-J/T)} \frac{1}{Z} \sum_{\langle R \rangle} \left\{ \left[\frac{1}{T} \delta_{\omega_m, 0} + \frac{J[1 - \exp(-J/T)]}{\omega_m^2 + J^2} \right] \right.$$
$$\left. + \left[\frac{1}{T} \delta_{\omega_m, 0} - \frac{J[1 - \exp(-J/T]}{\omega_m^2 + J^2} \right] \exp i\mathbf{q} \cdot \mathbf{R} \right\}$$

(for spin = 1/2). We solve Eq. (1) approximately by a variational technique, where we use the order parameter of the Abrikosov–Gor'kov theory,[1] appropriately parameterized, as a trial order parameter.[2] The resulting equation for the transition

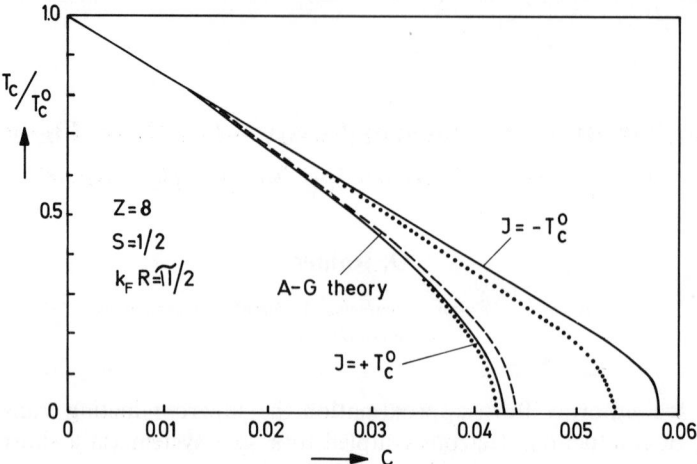

Fig. 1. Concentration dependence of the superconducting transition temperature. Corrections of the Abrikosov–Gor'kov theory (dashed lines) due to spin pairs are shown for antiferromagnetic (upper curve) and ferromagnetic (lower curve) spin–spin interaction.

temperature is

$$1 = N(0)(I_0 - I_{\text{spin}}^{\text{eff}})\pi T_c \sum_n{}' \left(|\omega_n| + |\Sigma(\omega_n)| - \frac{1}{2\tau_0} + \frac{1}{2\tau_s^{\text{eff}}} \right)^{-1} \quad (3)$$

Here the effective pair-weakening parameter $I_{\text{spin}}^{\text{eff}}$ and the effective pair-breaking lifetime τ_s^{eff} are determined by weighted averages of the q- and ω-dependent spin susceptibility (see Ref. 2).

Figure 1 shows characteristic results for the concentration-dependent transition temperature of the above model. To formally separate the role of spatial correlations in the spin system we include in Fig. 1, as dotted lines, the results of a calculation where the energy transfer in inelastic processes is neglected.

References

1. A.A. Abrikosov and L.P. Gor'kov, *Soviet Phys.—JETP* **12**, 1243 (1961).
2. D. Rainer, *Z. Physik* **252**, 174 (1972).

The Effects of Some 3d and 4d Solutes on the Superconductivity of Technetium*

C. C. Koch, W. E. Gardner,† and M. J. Mortimer†

*Metals and Ceramics Division
Oak Ridge National Laboratory, Oak Ridge, Tennessee*

Introduction

Considerable effort[1] has been directed toward obtaining an understanding of the observed variation of the superconducting transition temperature of a metal on the addition of impurities in terms of the details of their magnetic character. Technetium would appear to be an interesting and experimentally convenient superconducting transition metal solvent for comparison with theory. It has both a high transition temperature (7.8°K), thus making unnecessary the very low-temperature techniques needed to study such solvents as ruthenium and iridium, and an apparently high mutual solubility with the "magnetic" 3d elements.[2] The 3d elements are not expected to possess well-defined local moments in solution in technetium due to technetium's position in group VII of the periodic table,[3,4] but we are unable to anticipate how closely they will approach the nonmagnetic limit.

This paper reports the results of an experimental study of the changes in the superconducting transition temperature, the magnetic susceptibility, and the electrical resistivity of technetium with concentrations of up to about 5 at. % Cr, Fe, Co, or Ni (3d "magnetic" solutes) as well as Ru (a 4d "nonmagnetic" solute).

Experiment

The pure technetium used in this work was obtained from the Isotopes Division of Oak Ridge National Laboratory, the preparation and chemical analysis having been described in detail before.[5] Spectrographic analysis of the technetium used in this study indicated < 200 ppm total metallic impurities. The residual resistivity ratio (ρ^{298K}/ρ^{8K}) was 80. The other transition metals used were obtained from Johnson and Matthey Ltd., and described as "spectrographically pure" with < 10 ppm metallic impurities.

The alloys were prepared by arc-melting the constituents together in the conventional manner. The alloys with measurable weight losses, even though these were never excessive, were analyzed by the Analytical Sciences Section, Process

---

* Research sponsored jointly by the United Kingdom Atomic Energy Authority and the U.S. Atomic Energy Commission under contract with the Union Carbide Corporation.
† AERE, Harwell, England.

Technology Division, AERE, Harwell. All alloys were given a homogenization anneal of 6 hr at 900°C in a dynamic vacuum of better than 10^{-7} Torr.

Superconducting T_c measurements were made by a standard inductive technique. The transition temperature was defined at the midpoint of the inductance change between the superconducting and normal states. Typical transition widths for 90% of the total change in inductance were 0.2–0.3°K.

Magnetic susceptibility measurements from 4.2 to 298°K and in applied fields up to 42 kOe were made on samples weighing ~ 100 mg using the Faraday method. Measurements were made below T_c on several samples, including pure technetium, by first quenching the superconductivity with the applied field and then making magnetic measurements by switching on and off a superimposed gradient field.[6]

Electrical resistance measurements were made by the usual four-point method from T_c up to 298°K.

Results

The results of the T_c measurements for all the alloys are listed in Table I together with values of the magnetic susceptibility at 298°K. The initial rate of decrease of T_c with solute addition, i.e., $dT_c/dc|_{c\to 0}$, was 0.18°K/at.% for Ru, ~ 0.47°K/at.% for Cr, Co, and Ni, and 0.85°K/at.% for Fe. However, a clearer picture of the alloying behavior is obtained if, instead of considering solute concentration, T_c is plotted vs.

Table I. Superconducting Transition Temperature, T_c, and $\chi^{298°K}$ for Technetium Alloys

Alloy	AGN	T_c, °K	$\chi^{(298K)}$ × 10^6 emu/g
Tc	7.000	7.85	1.128
Tc–0.19 Cr	6.998	7.72	1.169
Tc–2.3 Cr	6.978	6.97	1.234
Tc–4.4 Cr	6.956	6.60	1.363
Tc–0.84 Fe	7.008	7.15	1.212
Tc–2.01 Fe	7.020	6.15	1.250
Tc–3.73 Fe	7.037	4.90	1.398
Tc–0.22 Co	7.004	7.77	1.150
Tc–1.2 Co	7.024	7.35	1.132
Tc–3.0 Co	7.060	6.65	1.151
Tc–5.9 Co	7.118	5.98	1.180
Tc–0.12 Ni	7.004	7.75	1.124
Tc–0.64 Ni	7.019	7.50	1.116
Tc–1.64 Ni	7.049	7.28	—
Tc–2.33 Ni	7.070	7.25	1.119
Tc–1.9 Ru	7.019	7.50	1.106
Tc–4.3 Ru	7.043	7.07	1.080

Effects of 3d and 4d Solutes on Superconductivity of Technetium

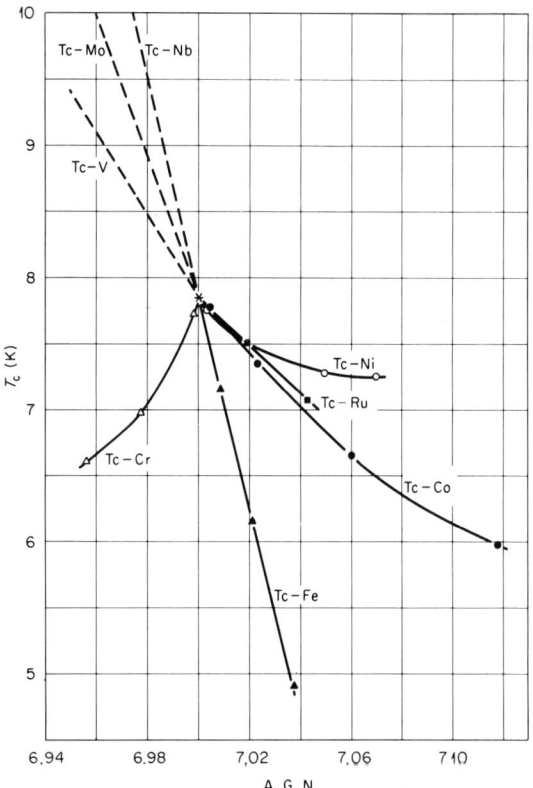

Fig. 1. Transition temperature vs. average group number for technetium with 3d and 4d solutes.

average group number (AGN) in the periodic table. This has been done in Fig. 1, which also includes literature data on Tc–V,[7] Tc–Mo,[8] and Tc–Nb[9] hcp solid solution alloys. It is then evident that for low concentrations Co and Ni are very similar to Ru in lowering the T_c of technetium, whereas iron stands apart, exhibiting a much more drastic influence on T_c. The Tc–Cr alloys with their decreasing T_c values are in notable contrast with the other groups V and VI elements (V, Mo, and Nb), which markedly raise the transition temperature of technetium.

Results of magnetic susceptibility measurements as a function of temperature are presented for pure technetium and representative alloys in Fig. 2. With the exceptions of the low-temperature anomalous behavior in Tc–Ni and Tc–Co, to be considered below, the magnetic susceptibilities are essentially temperature independent. It is concluded that no localized magnetic moments exist on the solute atoms in this series of Tc-base alloys. The value of χ for Tc at 298°K ($= 1.13 \times 10^{-6}$ emu/g) is lower than the value measured previously[10] of 1.22×10^{-6} emu/g. This result may be due to the higher purity of the present sample of Tc. The χ values at 298°K of all the alloys are plotted vs. AGN in Fig. 3 along with data from the literature on Tc–V,[11] Tc–Mo,[12] Tc–Nb,[10] and Tc–Ru[13] normalized to our χ value for pure Tc. For the group VIII solutes Ru lowers χ of Tc; Ni and Co slightly lower and increase

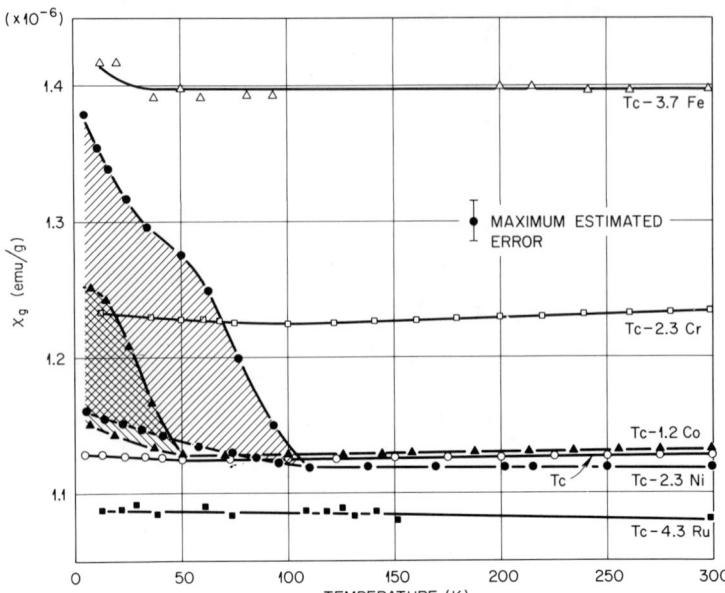

Fig. 2. Magnetic susceptibility vs. temperature.

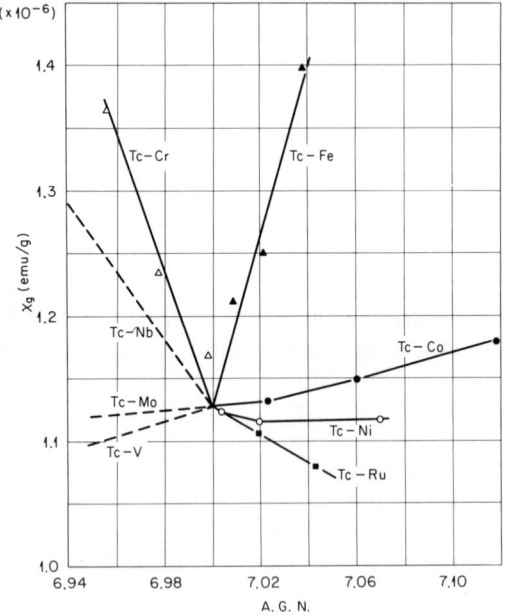

Fig. 3. Magnetic susceptibility at 298°K vs. average group number for technetium with 3d and 4d solutes.

χ, respectively; and Fe dramatically increases χ. The group V and VI solutes in Tc also exhibit varied behavior. Chromium rapidly raises χ; Nb also increases χ; but Mo and V both lower χ of Tc.

The band of χ values observed in the Tc–2.3 at. % Ni and Tc–1.2 at. % Co alloys at low temperatures ($< 100°K$) represents various sequences of measurements, but no clear pattern has yet been established for the apparent irreproducibility. It is possible this behavior may be caused by a martensitic-type phase transformation as, e.g., hcp → fcc. It is consistent that the fcc phase is stable at the Ni- and Co-rich concentrations in these systems but not in the other alloys studied. Low-temperature X-ray diffraction studies are being initiated to study this possibility.

Electrical resistivity was measured as a function of temperature from T_c to 298°K on selected alloys from each system. No resistivity minima with temperature were observed, nor any other obvious anomalies. It is therefore unlikely that any of these Tc-base alloys is a Kondo system.

Discussion

If the concept of a band picture to describe the conduction electrons in a metallic lattice has any meaning in attempts to classify the physical properties of alloys, it should most nearly apply to neighboring elements in the same period, such as Tc–Ru and Tc–Mo. Therefore we shall assume that the decrease of both T_c (Fig. 1) and χ (Fig. 3) of technetium with concentration of "nonmagnetic" Ru approximates the band picture for group VIII solutes. It then follows that the decrease of χ^{Pauli}, which is proportional to the density of electronic states at the Fermi level $N(0)$, would indicate that the decrease of T_c is mainly due to a lower $N(0)$. This is also consistent with measurements of the electronic coefficient of low-temperature specific heat capacity, γ, in this part of the periodic table (e.g., Mo–Ru alloys[14]). If the band model can be extended to the 3d solutes, then both Co and Ni in Tc would also appear to be roughly obeying this concept at low concentrations.

Iron, however, decreases T_c and increases χ of technetium much more rapidly than the other group VIII solutes do. Thus the effect of iron impurities is not fully accounted for by valence changes in the band model, and there must be additional "pair-destroying" or "pair-weakening" interactions present.[15-18] The initial rate of decrease of T_c due to Fe $|\partial T_c/\partial c|_{c\to 0} = 0.85°K/at.\%$ is comparable with the decreases observed in the similar systems Ru–Fe[15] and Ir–Fe[16], with $|\partial T_c/\partial c|_{c\to 0} = 0.21$ and 2.06°K/at. %, respectively. In these systems Fe also depresses T_c more rapidly than other group VIII solutes, and spin fluctuations have been invoked to account for the difference. Unfortunately, it is not sufficient simply to compare these initial rates in order to distinguish the cause of the additional interaction. None of the theories describing the effects of impurities makes quantitative predictions without some assumption about the magnitudes of the interaction. In addition, it only seems to be possible to distinguish between them if the observations are extended close to concentrations at which superconductivity disappears. Since the attractive interaction is much greater in Tc than in Ru or Ir, the concentration of Fe impurities needed will be much larger for an effective test. If the critical concentration is derived from the initial rates of decrease, then it corresponds to iron additions of 9, 2, and

0.06 at.% in Tc, Ru, and Ir, respectively. The present results only extend to an iron addition of 3.73 at.% and further measurements on alloys containing larger concentrations of iron are in progress. However, when comparing the observed behavior over such a large concentration range with spin fluctuation[16,17] or nonmagnetic resonant state[18] theories care will be necessary since the theories are only applicable in the dilute limit.

Of the 3d and 4d solutes from groups V and VI, only Cr decreases the T_c of technetium as shown in Fig. 1. Since the magnetic susceptibility of Tc is also rapidly increased by Cr additions (Fig. 3), as in the Tc–Fe alloys, the decreases in the transition temperature due to Cr additions are also consistent with the need for one of the pair-weakening mechanisms mentioned above.

The literature data for the hcp alloys in the Tc–Nb, Tc–Mo, and Tc–V systems show no obvious or consistent correlation between T_c and χ. Further experimental work is therefore in progress to study these systems in more detail, as well as the Tc–group VIII element systems, in order to clarify the solute effects presented in this paper and to decide which pair-weakening mechanism is responsible for the "anomalous" behavior of Fe and Cr in technetium.

Acknowledgments

One of us (C.C.K.) would like to thank Dr. J. A. Lee and all the members of his section at AERE, Harwell, for encouragement and assistance during the course of this investigation. We would like to thank T. F. Smith and M. B. Maple for useful discussions and J. Penfold and W. Dalzell for help with the magnetic susceptibility measurements. One of us (W.E.G.) would like to thank B. T. Matthias for his kind hospitality during the writing of this paper.

References

1. G. Gladstone, M.A. Jensen, and J.R. Schrieffer, in *Superconductivity*, R.D. Parks, ed., Marcel Dekker, New York (1968); M.B. Maple, *AIP Conf. Proc.* **4**, AIP, New York (1972), p. 175.
2. F.A. Shunk, *Constitution of Binary Alloys*, McGraw-Hill, New York (1969), 2nd Suppl.
3. A. M. Clogston, B.T. Matthias, M. Peter, H.J. Williams, E. Corenzwit, and R.C. Sherwood, *Phys. Rev.* **125**, 541 (1962).
4. B.T. Matthias, *IBM J. Res. Develop.* **6**, 250 (1962).
5. C.C. Koch and G.R. Love, *J. Less Common Metals* **12**, 29 (1967).
6. W.E. Gardner and T.F. Smith, *Progr. Vacuum Microbalance Tech.* **1**, 155 (1972).
7. C.C. Koch, R. Kernohan, and S.T. Sekula, *J. Appl. Phys.* **38**, 4359 (1967).
8. V.B. Compton, E. Corenzwit, J.P. Maita, B.T. Matthias, and F.J. Morin, *Phys. Rev.* **123**, 1567 (1961).
9. A.L. Giorgi and E.G. Szklarz, *J. Less Common Metals* **20**, 173 (1970).
10. D.O. Van Ostenburg, D.J. Lam, M. Shimizu, and A. Katsuki, *J. Phys. Soc. Japan* **18**, 1744 (1963).
11. D.O. Van Ostenburg, D.J. Lam, H.D. Trapp, and D.E. MacLeod, *Phys. Rev.* **128**, 1550 (1962).
12. D.J. Lam, D.O. Van Ostenburg, and D.W. Pracht, *J. Appl. Phys.* **35**, 976 (1964).
13. L.L. Isaacs and D.J. Lam, *J. Phys. Chem. Solids* **31**, 2581 (1970).
14. J.C. Ho and R. Viswanathan, *J. Phys. Chem Solids* **30**, 169 (1969).
15. K. Andres, E. Bucher, J.P. Maita, and R.C. Sherwood, *Phys. Rev.* **178**, 702 (1969).
16. G. Riblet, *Phys. Rev. B* **3**, 91 (1971).
17. G. Riblet and M.A. Jensen, *Superconductivity* (Proc. Intern. Conf. Science of Superconductivity, Stanford, 1969), F. Chilton, ed., North-Holland, Amsterdam (1971), p. 622.
18. A.B. Kaiser, *J. Phys. C* **3**, 409 (1970).

Specific Heat of SnTe between 0.06 and 30°K under Strong Magnetic Field*

M. P. Mathur, M. Ashkin, J. K. Hulm, and C. K. Jones

Westinghouse Research Laboratories
Pittsburgh, Pennsylvania

and

M. M. Conway, N. E. Phillips, H. E. Simon, and B. B. Triplett

Inorganic Materials Research Division
of the Lawrence Berkeley Laboratory and the
Department of Chemistry, University of California
Berkeley, California

Introduction

Tin telluride is an outstanding example of a low-carrier-density superconductor. Extensive work has been carried out on the electrical and superconducting properties of this material,[1] but our knowledge of its electronic band structure is incomplete. In order to remedy this deficiency, we have measured the low-temperature heat capacity of SnTe for various carrier densities and for several doping techniques. Our results for self-doped and Ag-doped samples suggest the presence of two different bands, with two different effective masses. Data for Mn-doped samples indicate an ordering of the Mn spins.

Experimental

In the present work only p-type SnTe was investiaged. Two types of samples were used[2]: self-doped samples, represented by the formula $Sn_{1-x}Te$, and samples in which non-divalent impurities (Mn or Ag) were substituted for Sn.

Self-doped samples were prepared by sintering techniques. Appropriate amounts of Sn and Te were sealed into an evacuated fused quartz tube and melted at 900°C. On removal from the tube the ingot was crushed and ground. The powder was then compacted in a split tungsten carbide die into cylinders typically of 3/4 in. diameter by 1 in. long. Each cylinder was sealed into an evacuated quartz tube and sintered at 500°C for a week. Silver- and Mn-doped samples were prepared by melting the required amount of material in vacuum and electromagnetically stirring during

---

* Work at Westinghouse supported in part by the U.S. Air Force Office of Scientific Research under contract No. F44620-72-C-0035. Work at Berkeley supported by U.S. Atomic Energy Commission.

melting and solidification. This technique was found to prevent oxidation of Mn during preparation.[3] It also yielded samples which were very uniform and homogeneous in phase, as indicated by X-ray lattice parameter studies. A small Hall sample of size 1/8 × 1/2 in. was cut from each ingot to determine the low-field Hall constant R at 77°K. The range of $n_H = 1/Re$ was from 2×10^{20} to 6.5×10^{21} cm^{-3}.

Most heat capacity measurements were performed in ^{3}He and adiabatic demagnetization cryostats[4] covering the temperature span from 0.06 to 20°K. For temperatures from 1.8° to 30°K some measurements were made in a modified pulse-type calorimeter.[5] The accuracy of the measurements was checked by measuring the heat capacity of pure copper. The electronic and lattice terms were both within 0.5% of previously determined values.[6]

Results and Discussion

Nonmagnetic Doping. Our heat capacity data for self-doped and Ag-doped samples can be fitted by the normal expression $C = \gamma T + \alpha T^3$, where γT and αT^3 are the electronic and lattice contributions, respectively. The Debye temperature θ_D has an average value of 147°K and does not vary systematically with dopant or carrier concentration.

In Fig. 1 the electronic heat capacity coefficient is plotted against the inverse Hall coefficient $1/Re = n_H$. A simple, free-carrier model predicts that γ should be proportional to $n_H^{1/3}$. In SnTe this appears to be the case for $n_H > 4 \times 10^{20}$ cm^{-3}, but a definite departure from this behavior occurs at lower carrier densities. The occurrence of γ values substantially below the $n_H^{1/3}$ line for $n_H < 4 \times 10^{20}$ is well outside the limits of experimental error. For $n_H < 4 \times 10^{20}$ cm^{-3} the thermal effective mass is $m_d = 2.1$, and for $n_H > 2.3 \times 10^{20}$ cm^{-3}, $m_d = 1.2$.

Of the valence band models which have been proposed for SnTe,[7] those that include two or more sets of degenerate bands are consistent with these results. The shape of γ vs. $1/Re$ is qualitatively explained by the occupation of a second set of bands above $n_H \sim 4 \times 10^{20}$ cm^{-3}. A more quantitative analysis of the proposed

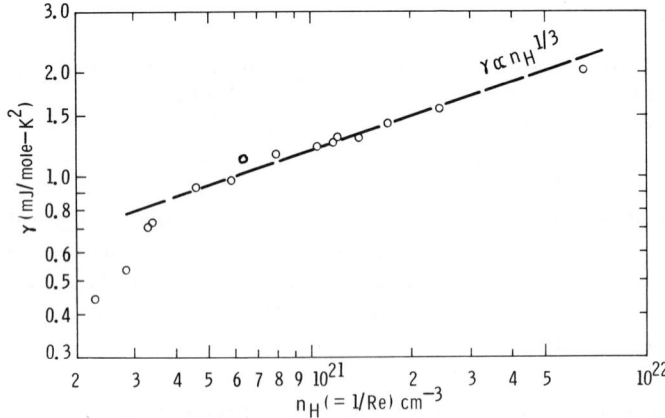

Fig. 1. Dependence of γ on carrier concentration of $Sn_{1-x}Te$ and $Sn_{1-x}Ag_yTe$.

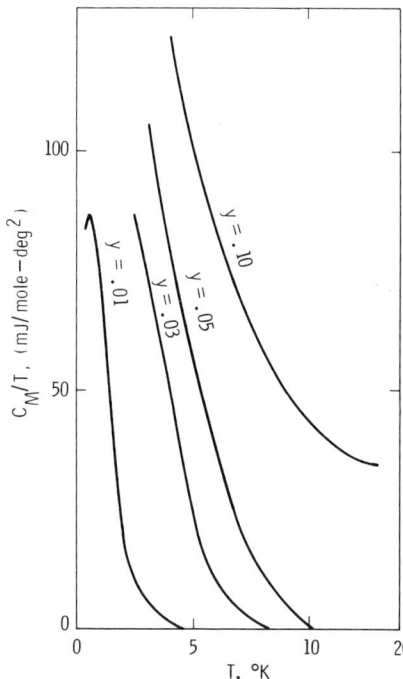

Fig. 2. Magnetic heat capacity C_M of $Sn_{0.97-y}Mn_yTe$ for various y values.

models and the contributions of different bands to γ would require additional electrical measurements on the present samples to obtain the relative populations from $1/Re$. In some of the models of SnTe the upper valence band extrema are placed at the L point of the zone and have a fourfold degeneracy. The resulting spherical band mass for this type of band is 0.25 where the mass enhancement factor $1 + \lambda$ is obtained from previous superconducting transition temperature measurements. This band mass is not grossly inconsistent with the electrical susceptibility mass determined from optical measurements.[8] If the second set of bands starts to fill at $n_H = 4 \times 10^{20}$ cm^{-3}, a band mass of 0.25 gives a 0.3 eV separation of the two bands. This value agrees with those given by many of the models.

Magnetic Doping. It is known[3] that Mn goes substitutionally into the SnTe matrix. Magnetic measurements indicate that the Curie temperature of $Sn_{0.97-y}Mn_yTe$ varies linearly with y. In order to determine the effect of magnetic ordering on the heat capacity, samples with $0 < y < 0.10$ were measured.

It seems reasonable to approximate the total heat capacity of the alloys C_{alloy} by the sum of a magnetic term C_M and a normal heat capacity C of $Sn_{0.97}Te$, as defined earlier. This allows us to separate C_M from C_{alloy}. In Fig. 2 we plot C_M/T vs. T for four representative samples. The heat capacity anomaly associated with the Mn spins is shifted to lower temperatures with decreasing y, as expected.

For $y \lesssim 0.05$, for which C_M/T vanished at high temperature, the extra entropy due to Mn ions was calculated. It was found to be approximately proportional to y, and consistent with $R \ln(2S + 1)$ per mole Mn, where $S = 5/2$ is the Mn ion spin obtained from the magnetic susceptibility measurements.[3] For $y = 0.01$ the peak of the specific heat curve is unusually sharp for a disordered alloy.[10]

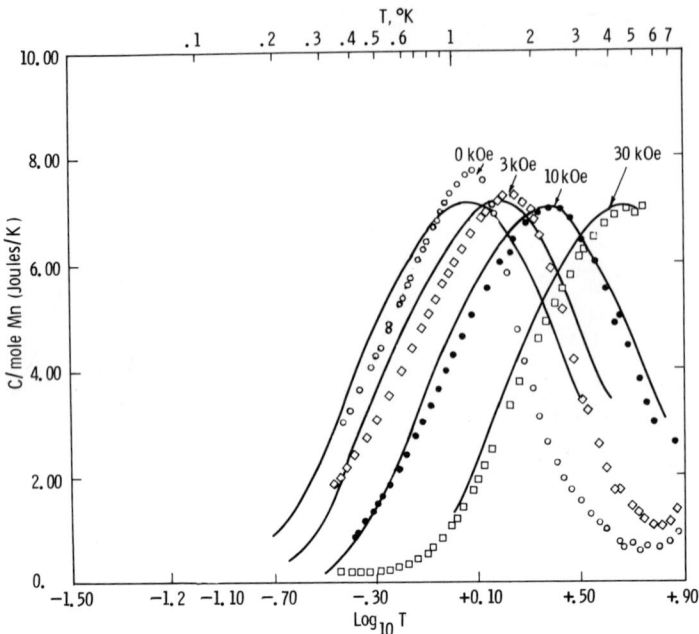

Fig. 3. Magnetic heat capacity C_M of $y = 0.01$ Mn sample in various external fields. Solid lines are theoretical 5/2 Schottky curves.

The nonvanishing C_M/T term for higher manganese contents appears to be essentially constant up to at least 30°K. This effect may be associated with localized enhancement of the density of states, clustering of ferromagnetic impurities, or inadequacy of the approximation that the background heat capacity is independent of y. (No indication of clustering was obtained from susceptibility data.)

The heat capacity of one sample, $Sn_{0.97}Mn_{0.01}Te$, was measured in various applied magnetic fields ranging up to 30 kOe. The excess heat capacity of $Sn_{0.97}Mn_{0.01}Te$ relative to $Sn_{0.97}Te$ is shown in Fig. 3. In general appearance, the observed heat capacity curves closely resemble Schottky curves. The best fit to the data is for spin 5/2, as shown in Fig. 3.

References

1. J.K. Hulm, C.K. Jones, D.W. Deis, H.A. Fairbank, and P.A. Lawless, *Phys. Rev.* **169**(2), 388 (1968).
2. J.K. Hulm, M. Ashkin, D.W. Deis, and C.K. Jones, *Progress in Low Temperature Physics* C.J. Gorter, ed., North-Holland, Amsterdam (1970), Vol. VI.
3. M.P. Mathur, D.W. Deis, C.K. Jones, and W.J. Carr, Jr., *J.Appl. Phys.* **41**(3), 1005 (1970).
4. B.B. Triplett, Ph.D. Thesis, University of California, Berkeley, UCRL-19672, September 1970, unpublished.
5. F.J. Morin and J.P. Maita, *Phys. Rev.* **129**, 1115 (1963).
6. D.W. Osborne, H.E. Flotow, and F. Schreiner, *Rev. Sci. Instr.* **38**, 159 (1967).
7. H. Kochler, *Z. Angew. Phys.* **23**, 270 (1967).
8. R. Tsu, W.E. Howard, and L. Esaki, *Phys. Rev.* **172**, 779 (1968).
9. R.F. Bis and J.R. Dixon, *Phys. Rev. B* **2**, 1004 (1970).
10. W. Marshall, *Phys. Rev.* **118**, 1519 (1960).

Direct Evidence for the Coexistence of Superconductivity and Ferromagnetism*

R. D. Taylor, W. R. Decker,† D. J. Erickson, A. L. Giorgi,
B. T. Matthias,‡ C. E. Olsen, and E. G. Szklarz

*Los Alamos Scientific Laboratory of the University of California
Los Alamos, New Mexico*

Introduction

In 1958 Matthias et al.[1] proposed the simultaneous occurrence of superconductivity and ferromagnetism in certain substituted, cubic Laves-phase, intermetallic compounds. These compounds are generally of the type $Ce_{1-x}R_xRu_2$, where R represents a magnetic rare earth. Their conclusion was based on the slow depression of the superconducting transition temperature $T_c(x)$ of $CeRu_2$ upon addition of RRu_2, the depression of the ferromagnetic ordering temperature $\theta_c(x)$ of RRu_2 upon addition of $CeRu_2$, and the intersection of $T_c(x)$ and $\theta_c(x)$ at some finite temperature and rare earth concentration x^*. The phase diagram for the system $Ce_{1-x}Gd_xRu_2$ is shown in Fig. 1. Note that in the neighborhood of $x = 0.13$ the two critical curves intersect at about 4°K. The solid lines in Fig. 1 were obtained by low-field magnetic susceptibility measurements.[2,3] The temperature at the midpoint of the transition to diamagnetism is taken as T_c; below T_c diamagnetic shielding masks any magnetic behavior of the sample. Values of θ_c for the Gd-rich sample are determined from their paramagnetic behavior at higher T. The dashed lines are extrapolations assuming that the concentration dependences of $T_c(x)$ and $\theta_c(x)$ persist above and below x^*, respectively. One therefore asks whether samples in the concentration region about x^* are either superconducting or ferromagnetic, or both, at temperatures below the dashed lines of Fig. 1. T_c transitions have been observed[1,4] for certain systems with magnetic rare earth concentrations slightly above their x^*. Recently θ_c values below T_c for $x < x^*$ were inferred[3] from susceptibility data taken above T_c. Other techniques have been used to characterize these materials in the regions of the dashed portions of Fig. 1. Measurements of specific heat,[3,5] thermal conductivity,[6] and dilatation[3] sometimes show anomalies associated with the implied transitions, but do not by themselves explicitly demonstrate that magnetic order really appears below T_c.

Although early theoretical predictions[7] suggested that coexistence of ferromagnetism and superconductivity was not possible, a more recent and detailed theory has been given[8] which permits such coexistence. A review of the theory sup-

---

* Work performed under the auspices of the United States Atomic Energy Commission.
† Visiting Staff Member from Western New Mexico University.
‡ Also at the University of California, San Diego and Bell Telephone Laboratories.

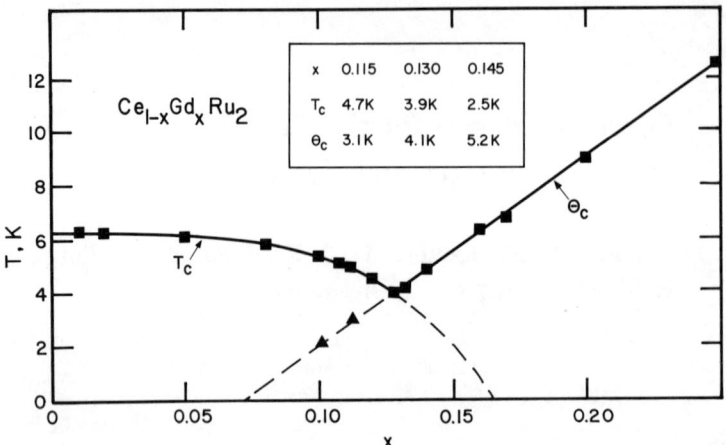

Fig. 1. The superconductivity transition temperature T_c and the ferromagnetic ordering temperature θ_c determined from magnetic susceptibility measurements as a function of mole fraction x of GdRu$_2$ in CeRu$_2$. Data denoted by squares are due to Wilhelm and Hillenbrand.[2] Data denoted by triangles are due to Peter et al.[3] The dashed lines are extrapolations which suggest a region in which coexistence of superconductivity and ferromagnetism is possible.

porting coexistence and its applications has recently been given by Peter et al.[3]

We wish to report here results obtained using a different technique to detect magnetic order in these compounds, a technique which is not sensitive to the presence of superconductivity. The magnetic hyperfine interaction of ^{57}Fe as observed by the Mössbauer effect (ME) has been used as a magnetic probe of host magnetic spin order. From our ME measurements ferromagnetism is shown to exist at low temperatures for R = Gd at concentrations near x^* in a manner consistent with Fig. 1. Magnetic susceptibility measurements on the same samples show transitions to superconductivity in all cases. Similar measurements have been obtained for R = Dy, Tb, and Ho and will be reported elsewhere.

Sample Preparation

High-purity Ce, Gd, and Ru metal powders were initially premelted and purified in an arc furnace. Polycrystalline samples were then prepared by arc-melting the components in an argon atmosphere, first melting the Ce and Gd and then combining with the Ru. Numerous remelts of the button were followed by a 2-hr anneal at about 1400°C in high vacuum.[2] Weight losses in the melting process were determined to be less than 0.3%. Composition of the melts was confirmed by X-ray fluorescence analysis. Inspection of thin disks spark-cut from the buttons revealed no apparent gross inhomogeneities. Metallographic examination of the disks showed very minor amounts of a second phase, determined to be pure Ru by microprobe analysis. As will be discussed below, the presence of this minor second phase does not affect our ME results.

Only the $C15$ cubic phase was detected in the samples from X-ray powder photographs, with the determined lattice constants in agreement with the work of

Evidence for Coexistence of Superconductivity and Ferromagnetism

Wilhelm and Hillenbrand.[2] In this structure the Ce (Gd) atoms occupy positions in a diamond structure, while a tetrahedral network of Ru atoms fills the available space along the body diagonals of the unit cell. All Ru sites are equivalent but may experience an electric field gradient (efg) tensor along the $\langle 111 \rangle$ axes.

Mössbauer sources were prepared by electroplating about 1 mCi of ammoniacal $^{57}CoCl_2$ onto half the area of a 0.34-in.-diameter disk for each composition studied. A second source for each composition was obtained by evaporation during the diffusion-anneal of the plated ^{57}Co sample (15 min at 1400°C, $< 10^{-7}$ Torr). The ME spectra of daughter ^{57}Fe nuclei (from ^{57}Co nuclei which are substitutional[2] at Ru lattice sites) were indistinguishable between plated and evaporated sources.

Results

The Me spectra for ^{57}Fe in $Ce_{1-x}Gd_xRu_2$ at 75, 4.0, and 1.5°K are shown in Fig. 2 for the compositions $x = 0.115, 0.130, 0.145$. Values of T_c and θ_c at these concentrations as obtained from Fig. 1 are listed in the insert in Fig. 1. At 298 and

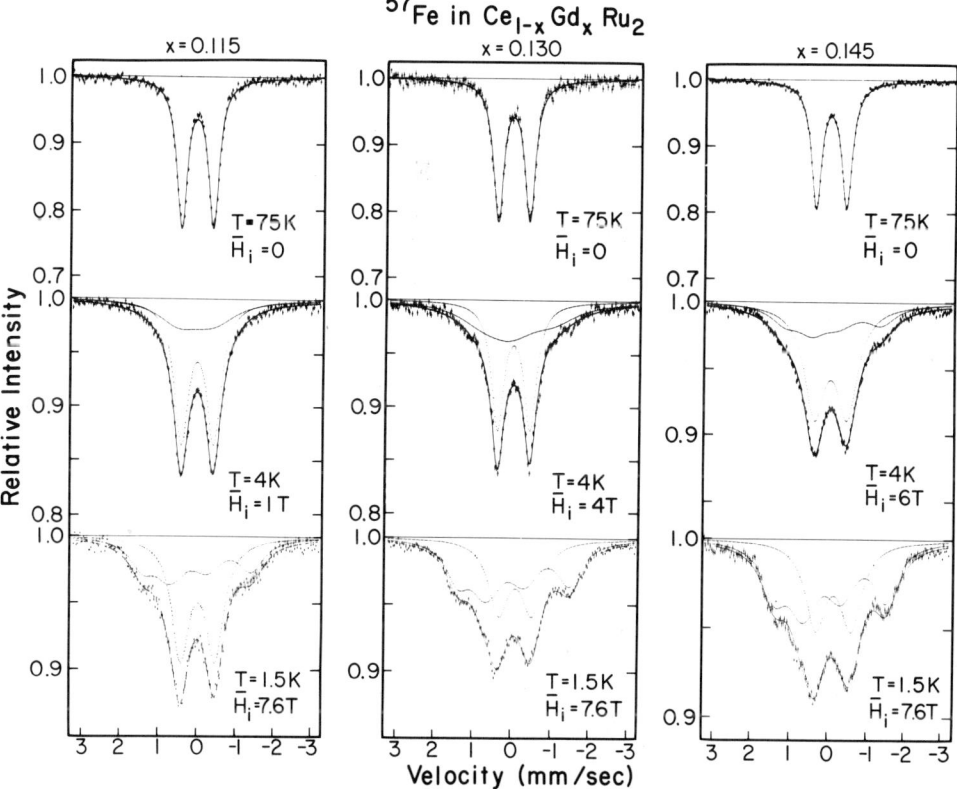

Fig. 2. Mössbauer spectra measured for ^{57}Fe as a source impurity in $Ce_{1-x}Gd_xRu_2$ as a function of x and source temperature T. The absorber used in the measurements was $K_4Fe(CN)_6 \cdot 3H_2O$ at room temperature. As explained in the text, the curves shown for the various spectra are obtained from computer fits assuming a two-site model for the Fe impurities. The average magnetic hyperfine field $\bar{H}_i$ at Fe impurities near ferromagnetically ordered Gd atoms as determined from the computer fits are given for each spectrum.

75°K the ME spectra for all compositions studied are composed of a pure nuclear electric quadrupole split doublet.

For the $x = 0.115$ composition the ME spectrum obtained at 4°K is slightly broadened, due to what we interpret as slight magnetic ordering. At 1.5°K this sample has developed "wings" which we ascribe to the presence of ferromagnetic order at the Gd sites. It is assumed that only those Co (Fe) atoms on Ru sites near Gd atoms will respond fully to the ferromagnetic order. The low-temperature saturation value of the ^{57}Fe magnetic splitting is determined by its electronic state in each particular environment. At higher temperatures a decreased splitting reflects a decreased magnetization at the Gd sites. This magnetic hyperfine splitting is therefore taken as proof that magnetic order can develop below T_c.

For the $x = 0.145$ composition the ME spectrum obtained at 4°K shows substantial magnetic hyperfine structure, as expected from Fig. 1. The 1.5°K spectrum shows that the magnetic order persists as the sample is taken below T_c. For $x = 0.130$, where $\theta_c \sim T_c \sim 4$°K, the magnetic behavior is intermediate and the sample is both ferromagnetic and superconducting at low temperatures.

That the low-temperature magnetic order results from the ordered Gd was verified by lowering the Gd concentration. Sources with $x = 0.08$ show no order at 4°K and a very small amount of order at 1.34°K. Sources of pure CeRu$_2$ show no order to 1.20°K. No change in the ME spectra for any composition is observed when passing through T_c.

A ^{57}Co source of pure Ru, prepared in the same manner as described above, yields a ME spectrum composed of a single, unsplit broadened line having a different isomer (velocity) shift than the spectra shown in Fig. 1. We therefore assert that any contributions to our measured Ce$_{1-x}$Gd$_x$Ru$_2$ spectra resulting from a pure Ru phase are negligible.

The curves drawn in Fig. 2 are obtained from computer fits to the spectra which are based on a two-site model for the ^{57}Fe impurity. Both sites are identical with respect to the efg producing the quadrupole splitting. Iron impurities near the ferromagnetically ordered Gd atoms develop a time-averaged magnetic hyperfine splitting roughly proportional to the average magnetization on a Gd atom. The magnetic axes of the Gd atoms and the Fe impurities are parallel (or antiparallel) and are at some particular orientation with respect to the efg axis. Various orientations corresponding to principal axis directions have been tried in analyzing the spectra. The orientation that gives the best fit to the data and is used in Fig. 2 in calculating the magnetic contribution to the spectra corresponds to a $\langle 111 \rangle$ efg axis and a $\langle 100 \rangle$ magnetization axis. The average magnetic hyperfine fields $\bar{H}_i$ at the Fe nuclei as determined by this analysis are listed in Fig. 2. The contributions of the non-magnetic, pure quadrupole sites are also shown. The ratio of magnetic to nonmagnetic sites increases with x, as expected for this model. The hyperfine field is probably independent of x as $T \to 0$. The line broadening and poor resolution of the spectra at low temperatures imply a somewhat more complicated situation than our rather simple two-site picture.

Discussion

The development of ferromagnetic order in Ce$_{0.885}$Gd$_{0.115}$Ru$_2$ at temperatures well below its T_c is directly demonstrated by the appearance of magnetic hyperfine splitting in the ME spectra. Similarly, it is shown that $Ce_{0.87}Gd_{0.13}Ru_2$ is both ferromagnetic and superconducting below 4°K. A $Ce_{0.92}Gd_{0.08}Ru_2$ sample shows very little ferromagnetism at 1.36°K, in support of the extrapolation of θ_c in Fig. 1. The sharp cutoff at $x = 0.07$ suggests that the Gd–Gd interaction is basically short range.[3] Anomalies in the specific heat for this system[3,9] (as well as[9] $Ce_{0.84}Tb_{0.16}Ru_2$) are also consistent with the system being a homogeneous ferromagnetic superconductor at low temperatures in a certain concentration range.

Interpenetrating magnetic and superconducting sublattices would seem to be a valid picture for a ferromagnetic superconductor, but it is not possible to state categorically that there is coexistence down to the smallest volume element. Short-range magnetic order, however, does not preclude superconductivity over a range of the order of a coherence length. We have initiated a more complete ME study in order to determine the temperature dependence of $\bar{H}_i$ and a better value of $\theta_c(x)$.

Since below θ_c the ^{57}Fe probe technique does not show strong magnetic order at every probe site, then perhaps magnetic order does not exist at every Gd site. We are preparing to measure ME spectra using one of the rare earth ME nuclides to determine the relative degree of magnetic participation of the rare earth component.

References

1. B.T. Matthias, H. Suhl, and E. Corenzwit, *Phys. Rev. Lett.* **1**, 449 (1958).
2. M. Wilhelm and B. Hillenbrand, *Z. Naturforsch.* **26a**, 141 (1971); *J. Phys. Chem. Solids* **31**, 559 (1970).
3. M. Peter, P. Donzé, O. Fischer, A. Junod, J. Ortelli, A. Treyvaud, E. Walker, M. Wilhelm, and B. Hillenbrand, *Helv. Phys. Acta* **44**, 345 (1971).
4. R.M. Bozorth, D.D. Davis, and A.J. Williams, *Phys. Rev.* **119**, 1570 (1961); B. Hillenbrand and M. Wilhelm, *Phys. Lett.* **31A**, 448 (1970).
5. N.E. Phillips and B.T. Matthias, *Phys. Rev.* **121**, 105 (1961).
6. L.J. Williams, W.R. Decker, and D.K. Finnemore, *Phys. Rev. B* **2**, 1287 (1970).
7. V.L. Ginzburg, *Soviet Phys.—JETP* **4**, 153 (1957); G.F. Zharkov, *Soviet Phys.—JETP* **7**, 286 (1958).
8. L.P. Gorkov and A.I. Rusinov, *Soviet Phys.—JETP* **19**, 922 (1964).
9. G. Rupp (private communication, B. Hillenbrand).

14

Small Particles, Heat Capacity, and Paramagnetism

14

Electric and Magnetic Moments of Small Metallic Particles in the Quantum Size Effect Regime

F. Meier and P. Wyder

Fysisch Laboratorium, Katholieke Universiteit
Nijmegen, The Netherlands

For macroscopic metal samples of the size used in most experiments the periodic boundary conditions are the appropriate ones to use in calculating electronic properties. In a naive free-electron model the energy levels are given by $\varepsilon_n = \hbar^2\pi^2 n^2/(2m^* L^2)$, where L is the linear dimension of the system, m^* is the effective mass of the electron, and $n^2 = n_x^2 + n_y^2 + n_z^2$, with the n_i integers. The spacing between two neighboring levels at the Fermi energy ε_F is then given by $\Delta_{\text{deg}} = (\hbar^2\pi^2 n_F/m^* L^2)$, with $n_F = p_F L/\pi\hbar$ (p_F is the Fermi momentum). However, if we consider the electron levels of sufficiently minute metallic particles, imperfections in the shape of such particles will remove the artificial degeneracy of the system due to the periodic boundary conditions. Then the average distance between two levels at the Fermi surface is $\Delta = \Delta_{\text{deg}}/(4\pi n_F^2/8)$, which is, of course, just the inverse of the density of states v at the Fermi surface. In a particle of volume $V = L^3$ containing N electrons we therefore have for spin degenerate levels

$$\Delta = v^{-1} = \pi^2\hbar^3/Vp_F m^* = \tfrac{4}{3}\varepsilon_F/N \qquad (1)$$

For a small gold particle of radius $a = 50$ Å Eq. (1) gives $\Delta = 0.10$ meV. It is obvious and was realized many years ago[1,2] that interesting effects should occur if δ, the average of Δ for an ensemble of small particles, becomes bigger than the thermal energy kT, the Zeeman energy $\mu_B H$ (H is the magnetic field), the electrostatic energy $2eaE$ (E is the electric field), etc. In a remarkable paper Kubo[3] reactivated interest in and triggered recent theoretical[4,5] and experimental[6-9] work on these problems.

The ideas of Kubo,[3] who assumed that the distribution of the electronic energy levels is random, have been generalized by Gor'kov and Eliashberg[4] (GE), who also took into account correlations between levels which manifest themselves mainly in dynamic effects. In order to calculate the static electric dipole moment **d** of a system of small particles in the range of electric field values where perturbation theory can be used, i.e.,

$$2eEa \ll \Delta \qquad (2)$$

GE assume that the particles behave like an atom with a certain polarizability and that they do not manifest their metallic properties at all. Assuming that the field changes little over the dimensions of the particle, the interaction term reduces to the space-independent expression

$$U = -\mathbf{E}\cdot\mathbf{d} \qquad (3)$$

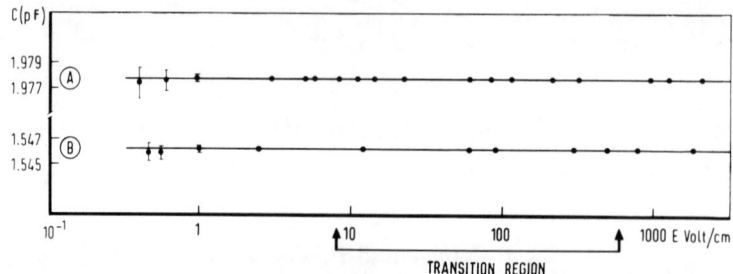

Fig. 1. A: Capacitance of reference sample without gold particles; $T = 1.2°K$. B: Capacitance of sample with gold particles (diameter between 40 and 100 Å); $T = 1.2°K$. Transition region indicates field range where capacitance should decrease by about 6%, i.e., ~ 0.09 pF.

First-order perturbation theory gives for the static electric polarizability at low temperatures $(kT \ll \Delta)$

$$\alpha = \frac{e^2}{\hbar} \frac{2}{3} v \frac{3}{5} a^2 = \frac{4}{15\pi} \frac{e^2}{\hbar^3} a^5 m^* p_F \simeq \left(\frac{p_F a}{\hbar}\right)^2 a^3 \qquad (4)$$

In this way GE arrived at the remarkable theoretical prediction that, for example, a small gold particle with $a = 50$ Å should have a polarizability which is bigger by a factor of 270 than the classical polarizability $\alpha_{cl} = a^3$ (from elementary electrostatics). This result is only valid for weak fields [Eq. (2)] and low temperatures; at higher temperatures or bigger electric fields the different levels start mixing and the usual metallic properties reappear.

In order to verify this GE effect experimentally, we produced small particles of gold in a glass matrix following the method of Maurer.[10] The size distribution of the particles was measured with an electron microscope, and the total number of particles in the sample was determined from the optical Mie absorption, as described by Doremus.[11] From the particle size distribution and their concentration the enhancement of the capacitance of the sample of Fig. 1 due to the GE effect can be calculated to about 6%. To avoid any changes in the dielectric constant of the glass caused by temperature variations, we decided to do the measurements at fixed low temperatures and to observe the capacitance as function of the electric field strength. Many samples were measured between 0.3 and 4.2°K but no change in the capacitance was observed. Figure 1 shows a typical example of these negative findings. We must therefore conclude that for small gold particles no GE effect was detected. Very recently Dupree and Smithard,[8] in a careful study of the temperature variation of the dielectric constant of small metallic particles, also found no GE effect.

These negative experimental results can be analyzed on the basis of classical electrostatics, as has been shown by Strässler and collaborators.[12] It is important that a distinction be made between the applied field E and the local field E_{loc}, even for particles of the size discussed by GE. With the polarization P the local field for a spherical particle is

$$E_{loc} = E - (4\pi/3) P \qquad (5)$$

Equation (4) gives the electric susceptibility χ which relates E_{loc} and P:

$$P = \chi E_{loc} \qquad (6)$$

Therefore the observed polarizability is given by

$$\alpha = \frac{4\pi}{3} a^3 \frac{P}{E} = \frac{4\pi}{3} a^3 \chi \Big/ \left(1 + \frac{4\pi}{3}\chi\right) \qquad (7)$$

If, as GE have shown, χ is very large, Eq. (7) reduces to $\alpha = a^3$, and this is what is observed. It is important to note that χ is essentially given by the Thomas–Fermi screening length,[12] which is of the order of an interatomic distance. This means that even in small particles of the size considered by GE the spatial dependence of E is important, equation (3) is no longer space independent, and a much more elaborate self-consistent field scheme is necessary for calculating the electric polarizability of small metallic particles. On the other hand, it is hopeful to note that the group at the Brown Boveri Research Center in Switzerland has discovered that a GE effect might exist in quasi-one-dimensional metals formed by the mixed-valence planar complex compounds of Pt, where no depolarization gives complications.[13]

The question arises, can the magnetic properties of small metallic systems, for which shielding effects should be less dramatic, show a quantum size effect? This problem was investigated by Kubo[3] and led us to perform some measurements of the static magnetization of small indium particles at low temperatures. Kubo pointed out that the magnetic properties of small metallic particles should be drastically changed in the quantum size limit $kT \ll \delta$. Due to strict charge neutrality at low temperatures (the particles bear no net charge), two categories of particles have to be considered: the ones containing an odd number of electrons ("odd particles") and the ones with an even number of electrons ("even particles"). At temperatures $kT \gg \delta$ the electron spin paramagnetism of small particles is the same as for bulk, whereas for $kT \ll \delta$ the odd particles behave as if they possessed one free spin, leading to a field dependence of the magnetization of the following form:

$$M = \mu_B T gh(\mu_B H/kT) \qquad (8)$$

Apart from this saturating paramagnetism there is also a nonsaturating contribution from the even particles which strongly depends on the particular form of the level correlations of the particle.[4,5]

Small particles of indium were prepared by evaporating indium metal in a helium gas stream which was subsequently led through liquid paraffin, where some of the particles were retained. The size distribution of the particles was determined from electron micrographs; these showed that the large majority of them had a diameter between 20 and 60 Å, which corresponds to an average level separation of about 10–100°K. Most of the particles were of spherical shape. How far the size distribution can be varied as a function of gas pressure, stream velocity, and evaporation temperature still has to be investigated.

At helium temperatures the magnetic moment of the samples was measured with a Foner vibrating sample magnetometer in the field range of 0–50 kG. In order to be sure that the observed paramagnetic behavior is indeed due to the small metallic particles, several checks were made, i.e., measurements at high temperatures,

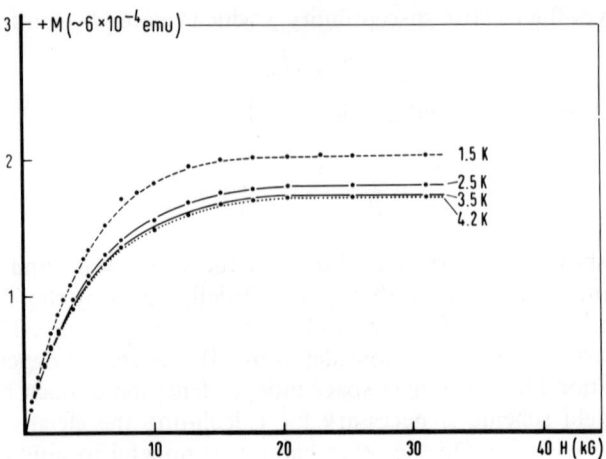

Fig. 2. Nonlinear, saturating part of the magnetization of a paraffin sample containing indium particles of 20–60 Å diameter. The curves are obtained by subtracting from the measured total magnetization the linear diamagnetic part, deduced from the high-field $M(H)$ behavior ($H > 25$ kG).

of pure paraffin, oxidized particles, etc. Figure 2 shows a typical result of our measurements after subtracting the linear part of the magnetization, which is mainly due to paraffin. The following features are clear: The saturation field is within the error limits almost independent of temperature between 1.5 and 4.2°K and has a value of about 20 kG; the saturation moment increases with decreasing T; and the low-field susceptibility seems to be independent of T between 1.5 and 4.2°K. These properties cannot be explained with the Kubo model, where the magnetization curve of the odd particles should look like a hyperbolic tangent of the argument $\mu_B H/kT$, eq. (8). The even particles contribute to the linear term, which has already been subtracted in Fig. 2.

Two other possibilities should be considered. First, Kubo did not take into account possible level correlation effects.[4] These, however, would not furnish the necessary corrections in the field and temperature range of interest. Second, a contribution from paramagnetic impurities such as oxygen can be excluded because their magnetic behavior is also given by a Brillouin function, which is not compatible with the temperature dependence observed.

However, it seems to be possible to explain qualitatively the field and temperature dependences of the magnetization curves by the following simple argument. It is assumed that the particles form a perfectly spherical square well within which the electrons are free to move. By elementary quantum mechanics it is possible to construct the wave functions and eigenvalues of the system. Without a magnetic field the eigenvalues $E_{k^{(i)},l}$ are $(2l + 1)(2s + 1)$-fold degenerate, $s = 1/2$, $k^{(i)}$ denoting a quantum number obtained by the requirement that the wave function vanishes at the wall of the particle. For each l there exists an infinite number of values $k^{(i)}$. Simple considerations show that the level with the highest l-value being occupied by electrons at $T = 0$ has an $l = l_0$ which is not, as Kubo[3] assumed, of the order N (N is the total number of electrons in the particle), but is much less; e.g., for a particle with $N = 10,000$, l_0 is about 30. The magnetic field will split the degenerate levels $E_{k^{(i)},l}$ into $(2l + 1)$ Zeeman terms. The magnetic properties of the particle are determined by the one level $E_{k^{(i)},l}$ that is not completely filled and not completely empty, because the contribution of all the others is obviously zero. Taking all possible occupation numbers of this particular level $E_{k^{(i)},l}$ as equally probable and making the reasonable assumption that Zeeman levels belonging to different l values do not cross each other, one easily calculates the average magnetic moment of the particle; these results are in qualitative agreement with our experimental curves. (The effect of the superconducting transition has not yet been worked out.) We conclude that the spherical symmetry of the particles mainly determines their magnetic properties.

References

1. H. Fröhlich, *Physica* **4**, 406 (1937).
2. F. Hund, *Ann. Physik* **32**, 102 (1938).
3. R. Kubo, *J. Phys. Soc. Japan* **17**, 975 (1962).
4. L.P. Gor'kov and G.M. Eliashberg, *Soviet Phys.—JETP* **21**, 940 (1965).
5. R. Denton, B. Mühlschlegel, and D.J. Scalapino, *Phys. Rev. Lett.* **26**, 707 (1971).
6. C. Taupin, *J. Phys. Chem. Sol.* **28**, 41 (1967).
7. S. Kobayashi, T. Takahashi, and W. Sasaki, *J. Phys. Soc. Japan* **31**, 1442 (1971).
8. R. Dupree and M.A. Smithard, *J. Phys. C: Proc. Phys. Soc., London* **5**, 408 (1972).
9. F. Meier and P. Wyder, *Phys. Lett.* **39A**, 51 (1972).
10. R.D. Maurer, *J. Appl. Phys.* **29**, 1 (1958).
11. R. Doremus, *J. Chem. Phys.* **40**, 2389 (1964).
12. S. Strässler, M.J. Rice, and P. Wyder, *Phys. Rev. B* **6**, 2575 (1972).
13. M.J. Rice and J. Bernasconi, *Phys. Rev. Lett.* **29**, 113 (1972).

High-Frequency Relaxation Measurements of Magnetic Specific Heats*

A. T. Skjeltorp and W. P. Wolf

*Department of Engineering and Applied Science
Becton Center, Yale University, New Haven, Connecticut*

One of the most important quantities characterizing a magnetic system and its microscopic interactions is the magnetic specific heat C_M. In particular, for a paramagnetic system at temperatures much higher than the ordering temperature, the magnetic specific heat can often be expressed as a simple series expansion in $1/T$

$$C_M/R = C_2/T^2 + C_3/T^3 + \cdots \tag{1}$$

whose coefficients $C_2, C_3, \cdots$ can be related directly to traces of different powers of the interaction Hamiltonian describing the system (see, e.g., Ref. 1).†

Unfortunately, C_M is often quite difficult to determine by the usual calorimetric methods, since these measure the *total* specific heat, from which it is necessary to subtract a generally comparable lattice contribution C_L. This results in a correspondingly large uncertainty in C_M, which becomes larger as T increases since C_M falls off (roughly as $1/T^2$) while C_L increases (roughly as T^3). Thus C_M is most difficult to determine in just those regions in which Eq. (1) should be most applicable.

Here we review briefly the use of an alternative method, based on the measurement of the field dependence of the high-frequency susceptibility, in which C_M is determined directly from purely magnetic measurements. The scope and accuracy of this method have recently been improved significantly by several new experimental techniques, primarily the development of stable tunnel diode oscillators suitable for use at low temperatures.

The basic method was first proposed by Casimir and du Pré in 1938,[2] and it was used extensively for a number of years, mainly by Gorter and his co-workers,[3,4] but more recently it seems to have been pursued less actively. The method involves measuring the field dependence of the real part of the ac differential susceptibility $\chi'(H)$ under conditions for which $\chi'(H) = \chi_S(H)$, the thermodynamically defined adiabatic susceptibility. These conditions can be expressed as

$$1/\tau_{SL} \ll 2\pi f \ll 1/\tau_{SS} \tag{2}$$

where f is the measuring frequency and τ_{SL} and τ_{SS} are, respectively, the spin–lattice and spin–spin relaxation times.

---
* Supported in part by the National Science Foundation.
† We follow the usual convention in expressing C_M in units of $R = Nk_B$, where N is the number of magnetic spins.

The field dependence of $\chi'(H)$ can then be related to other magnetic properties using thermodynamics, and in particular

$$\frac{C_M(H, T)}{R} = \left[T \left(\frac{\partial M}{\partial T} \right)_H^2 / R\chi_T(H) \right] \left[\frac{\chi'(0) \chi_T(H)}{\chi'(H) \chi_T(0)} - 1 \right]^{-1} \quad (3)$$

where $\chi_T(H) = (\partial M/\partial H)_T$ is the isothermal differential susceptibility in a field H. Thus $C_M(H, T)/R$ can be found from measurements of a simple ratio of susceptibilities $\chi'(0)/\chi'(H)$, provided only that one knows the static magnetization $M(H, T)$ in the field and temperature regions of interest.

The accuracy with which $M(H, T)$ must be known must be good enough to provide reliable estimates of the derivatives $(\partial M/\partial T)_T$, $\chi_T(H)$, and $\chi_T(0)$, and this together with Eq. (2) establishes the two major conditions for the applicability of the method in any given case.

There is also a third condition determining the strength of the magnetic field used to measure the change in χ'. This depends on the size of the specific heat and the magnetic properties as indicated by Eq. (3), and, of course, it also depends on the accuracy with which a change in χ' can be measured. A rough estimate of the field H' required to reduce χ' to half its zero-field value can be obtained by approximating $M(H, T)$ by Curie's law, $\lambda H/T$, and this gives

$$H' = (C_M T^2/\lambda)^{1/2} \quad (4)$$

where λ is the appropriate Curie constant. In practice maximum fields as small as $H'/5$ have been used, and the limit in any particular case will depend on the accuracy required for the final values of C_M.

In general, we must expect $C_M(H, T)$ to be field dependent, even for small fields, and to extract the quantities most directly related to the intrinsic interactions, it is necessary to analyze the field dependence, and in particular, to extract the specific heat in zero field $C_M(0, T)$. This can be found from the simple thermodynamic relation

$$C_M(0, T) = C_M(H, T) + T \int_0^M (\partial^2 H/\partial T^2)_M \, dM \quad (5)$$

if, as we already suppose, we know $M(H, T)$. Conversely, the observation of the field dependence of $C_M(H, T)$ may in certain cases be used to determine some of the higher-order coefficients in the high-temperature expansion of $\chi_T(0)$ beyond the simple Curie–Weiss law.[5,6]

The central experimental problem is the determination of the adiabatic susceptibility, and to satisfy the relaxation conditions in Eq. (2) over the largest possible temperature range, this calls for a technique for measuring χ' in the region 1–100 MHz. Of the available methods the simplest, and probably also the most sensitive, is to use a tunnel diode oscillator which can be operated inside the cryostat.[7] Such oscillators combine excellent stability over wide ranges of temperature, small power dissipation, small volume, and low cost, and they are relatively independent of the applied field. With our particular system samples as small as 10 mg in weight have been measured at temperatures ranging from 1.5 to 80°K, and it is not difficult to see how the technique could be extended even further.

Examples of systems which have already been studied using this technique include rare earth halides, hydroxides, and garnets, and details of these results have been published elsewhere.[5,6,8-15]

A complete review of the experimental technique, apparatus, and general requirements for using this high-frequency relaxation method is now in the course of publication and preprints may be obtained from the authors on request.*

References

1. J.M. Daniels, *Proc. Phys. Soc. (London) A* **66**, 673 (1953).
2. H.B.G. Casimir and F.K. du Pré, *Physica* **5**, 507 (1938).
3. C.J. Gorter, *Paramagnetic Relaxation*, Elsevier, New York (1947).
4. C.J. Gorter, in *Fluctuation, Relaxation and Resonance in Magnetic Systems*, D. ter Haar, ed., Plenum Press, New York (1962), p. 87.
5. A.T. Skjeltorp and W.P. Wolf, in *AIP Conf. Proc.*, No. 5, p. 695 (1972).
6. A.T. Skjeltorp, C.A. Catanese, H.E. Meissner, and W.P. Wolf, submitted to *Phys.` Rev.* for publication.
7. R.B. Clover and W.P. Wolf, *Rev. Sci. Instr.* **41**, 617 (1970).
8. R.B. Clover, Thesis, Yale University, 1969, unpublished.
9. R.B. Clover and R.J. Birgeneau, *J. Appl. Phys.* **40**, 1151 (1969).
10. R.B. Clover and W.P. Wolf, *Sol. St. Comm.* **6**, 331 (1968).
11. A.T. Skjeltorp and W.P. Wolf, *J. Appl. Phys.* **42**, 1487 (1971).
12. C.A. Catanese, A.T. Skjeltorp, H.E. Meissner, and W.P. Wolf, unpublished.
13. E. Becker and R.B. Clover, *Phys. Rev. Lett.* **21**, 1327 (1968).
14. R.B. Clover, unpublished.
15. R.B. Clover and A.T. Skjeltorp, *Physica* **53**, 132 (1971).

* **Note added in proof:** This work has now been published in *Phys. Rev. B* **8**, 215 (1973).

Some Recent Results on Paramagnetic Relaxations

C. J. Gorter and A. J. van Duyneveldt

Kamerlingh Onnes Laboratorium der Rijksuniversiteit
Leiden, Nederland

In continuation of earlier investigations on paramagnetic relaxation, described and discussed during the last twenty five years in doctor's theses and in the journal *Physica* by Volger, De Vrijer, Van der Marel, Van den Broek, Bölger, Verstelle, Locher, De Vries, Van Duyneveldt, and Roest, recent researches on paramagnetic spin–lattice relaxation have been carried out over a wide range of temperatures and external magnetic fields. By using nonresonant techniques, relaxation times between 10 and 10^{-7} sec have been obtained.

Interesting investigations and theoretical studies in this field have also been carried out in Great Britain, the Soviet Union, and the U. S. In the early sixties the American theoretical physicist Orbach, working in Bleaney's group at Oxford, called attention to three contributions to the spin–lattice relaxation, ascribed to simple absorption and emission of phonons, to quasi-Raman scattering of such quanta, and to a resonant two-phonon scattering.

In the case of an ion with an odd number of spins (Kramers salt) with merely one low-lying doublet, this leads to four terms for the reciprocal relaxation time[1]:

$$\tau^{-1} = ATH^4 + B_1 T^9 + B_2 T^7 H^2 + C\, e^{(\Delta/kT)} \tag{1}$$

The first term represents the direct process, while the second and third terms indicate quasi-Raman processes. The fourth term is due to the "resonant two-phonon process" and is only possible if there exists another level $\Delta < k\theta_D$ (θ_D is the Debye temperature). For Co^{2+} ions in fluosilicate and in Tutton salts (Figs. 1–4) such a level does not occur. The strongly temperature-dependent second term leads to a shortening of the relaxation time τ, e.g., by a factor of 5×10^2 when T increases by a factor two. As can be seen in Fig. 1 ($CoSiF_6 \cdot 6H_2O$) and in Fig. 2 ($CoX_2(SO_4)_2 \cdot 6H_2O$, X = NH_4 or K) this term is observed at low external magnetic fields. The lines drawn in both figures represent the more precise form for the quasi-Raman processes, taking into account the decrease of the slope of the τ vs. T curve if the temperature increases toward θ_D. Contrary to the cobalt Tutton salts, the relaxation times of 18% Co^{2+} in $ZnSiF_6 \cdot 6H_2O$ deviate from those of the concentrated sample, probably in view of a changed crystal structure. Figure 3 gives the relaxation time τ at liquid hydrogen temperatures as a function of the external magnetic field for two cobalt Tutton salts. In low fields the contribution to τ^{-1} from the T^9-Raman process is field independent. In high fields τ^{-1} increases considerably, which may be ascribed to the $T^7 H^2$ term in (1).

At liquid helium temperatures the Raman relaxation times become very long,

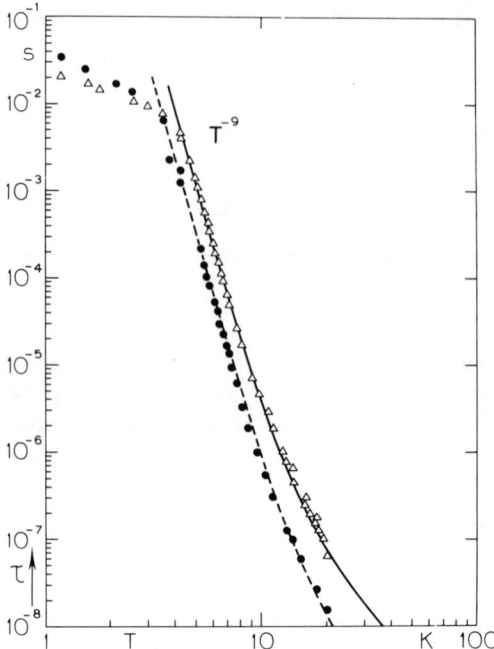

Fig. 1. Relaxation time vs. temperature for powdered cobalt fluosilicate in an external magnetic field of 1 kOe. (●) $CoSiF_6 \cdot 6H_2O$; (Δ) $Co_{0.18}Zn_{0.82}SiF_6 \cdot 6H_2O$; (---) $\tau^{-1} = 3.8 \times 10^{-8} T^9 \int_0^{105/T}[(e^x x^8/e^x - 1)^2] dx \ \text{sec}^{-1}$; (——) $\tau^{-1} = 1.1 \times 10^{-8} T^9 \int_0^{95/T}[e^x x^8/(e^x - 1)^2] dx \ \text{sec}^{-1}$ ($x = \hbar\omega/kT$).

so that the direct process might prevail in the spin–lattice relaxation. However, the measurements on concentrated paramagnetic substances do not show the TH^4 dependence of the direct-process relaxation rate τ_{dir}^{-1}. In low external magnetic fields the direct process is short-circuited by fast relaxation processes involving impurities,[2] lattice imperfections,[3] or exchange interactions.[4] In high fields the system of lattice oscillators cannot maintain the temperature of the helium bath (phonon bottleneck[5]), causing the observed relaxation time to exceed τ_{dir}. At some value of the external magnetic field between these extreme regions the observed relaxation time will be equal to τ_{dir}. Roest et al.[6] developed a phenomenological interpolation method which gives this value of H and therefore leads to an estimate for the coefficient A of the first term in (1). For concentrated, powdered samples of cobalt fluosilicate, cobalt Tutton salt, and cobalt lanthanum double nitrate this procedure yields an identical expression for the direct process: $\tau_{\text{dir}}^{-1} = 0.12 T H^4 \ \text{sec}^{-1}$ (H in kOe). This result coincides very well with the measurements on magnetically diluted samples showing the H^4 dependence of τ_{dir}^{-1} without the influence of a phonon bottleneck [e.g., $Co_{0.017}Zn_{0.983}(NH_4)_2(SO_4)_2 \cdot 6H_2O$, Fig. 4].

The spin–lattice relaxation of ytterbium chloride hexahydrate is also expected to vary with temperature and field as indicated by the first three terms of (1). For a powdered sample the relaxation times observed at liquid helium temperatures

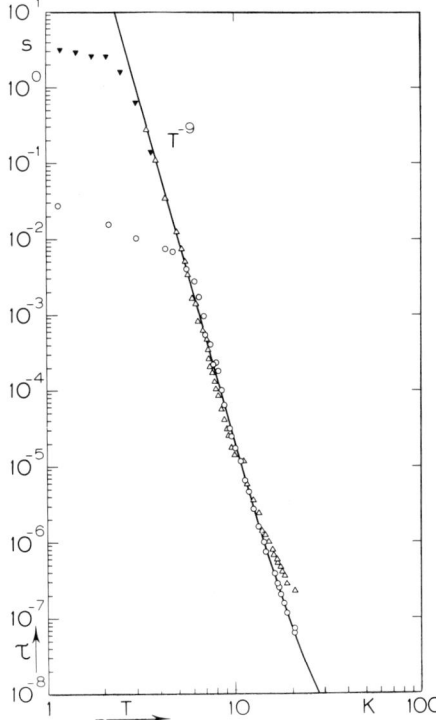

Fig. 2. Relaxation time vs. temperature for powdered cobalt Tutton salts. $Co(NH_4)_2(SO_4)_2 \cdot 6H_2O$: (○) $H = 1$ kOe. $Co_{0.08}Zn_{0.92}K_2(SO_4)_2 \cdot 6H_2O$: (△) $H = 1$ kOe; (▼) $H = 0.3$ kOe. (———) $\tau^{-1} = 1.2 \times 10^{-9} T^9 \int_0^{180/T} [e^x x^8/(e^x - 1)^2] dx$ sec^{-1} ($x = \hbar\omega/kT$).

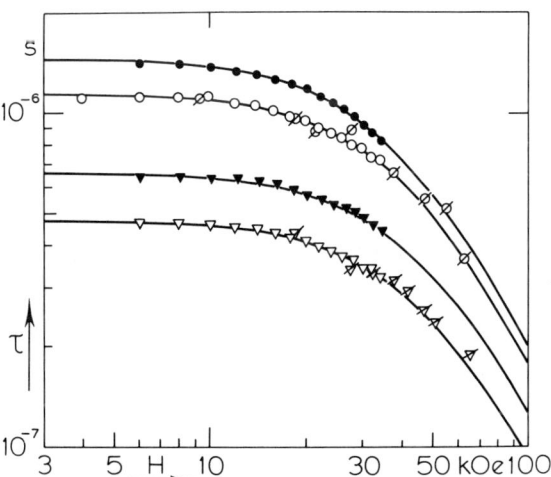

Fig. 3. Relaxation time vs. external magnetic field for powdered $Co(NH_4)_2(SO_4)_2 \cdot 6H_2O$ (open symbols) and $CoK_2(SO_4)_2 \cdot 6H_2O$ (closed symbols) at hydrogen temperatures (circles: $T = 14.1°$K, triangles: $T = 16.0°$K). Measuring points indicated by a bar were obtained in pulsed magnetic fields.

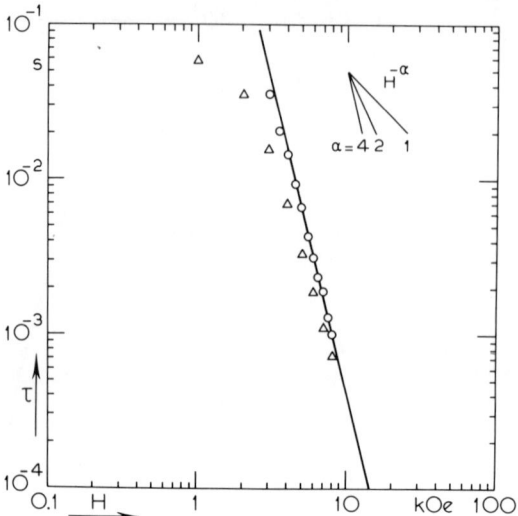

Fig. 4. Relaxation time vs. external magnetic field for powdered $Co_{0.017}Zn_{0.983}(NH_4)_2(SO_4)_2 \cdot 6H_2O$; (○) $T = 2.1°K$; (△) $T = 4.2°K$; (———) $\tau \propto H^{-4}$.

could be analyzed using the above-mentioned interpolation method. In large external magnetic fields we found $\tau^{-1} = 0.9 \times 10^{-2} TH^4$ sec^{-1}(Fig. 5). At low fields the T^9 quasi-Raman process determines the relaxation behavior. At a constant field of 1 kOe, τ^{-1} increases as $T^9: \tau^{-1} = 1.9 \times 10^{-4} T^9$ sec^{-1} (Fig. 6).

Mn^{2+} and Cr^{3+} compounds are Kramers salts with three or two energy doublets, respectively. The splittings between these doublets are small compared to kT. For this case Blume and Orbach[7] predicted

$$\tau^{-1} = A'TH^2 + B'T^5 \tag{2}$$

where the first term is the direct and the second the Raman process. At low magnetic fields the T^5th dependence of τ^{-1} is found in $MnSiF_6 \cdot 6H_2O$, $Mn(NH_4)_2(SO_4)_2 \cdot 6H_2O$ (Fig. 7) and in $CrCs(SO_4)_2 \cdot 12H_2O$ (Fig. 8). At high temperatures we again observe the decrease of the slope of the τ vs. T curve toward two (if $T > \theta_D$), as is to be expected from the more general expression for the quasi-Raman relaxation time.

In the salts mentioned the low-field relaxations at helium temperatures are due to impurities, etc. Above about 10 kOe the direct process is the only effective

Table I. Coefficient A' of the Direct Process*

Sample	A', sec^{-1} kOe^{-4} °K^{-1}
$Mn(NH_4)_2(SO_4)_2 \cdot 6H_2O$	0.13
$MnSiF_6 \cdot 6H_2O$	0.23
$MnSO_4 \cdot 4H_2O$	0.13
$CrCs(SO_4)_2 \cdot 12H_2O$	2.0

*$\tau^{-1} = A'TH^2$.

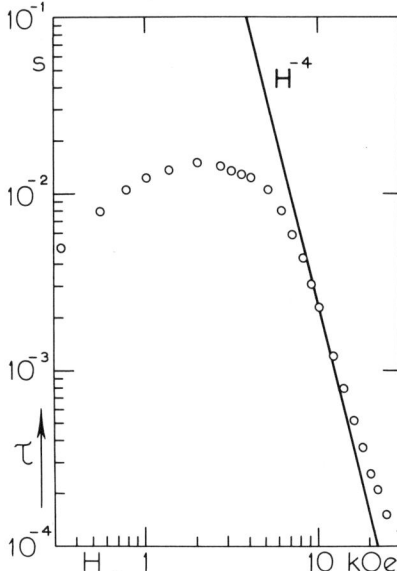

Fig. 5. Relaxation time vs. external magnetic field for powdered $YbCl_3 \cdot 6H_2O$ at $T = 4.2°K$. (———) $\tau^{-1} = 0.9 \times 10^{-2} TH^4$ sec^{-1} (H in kOe).

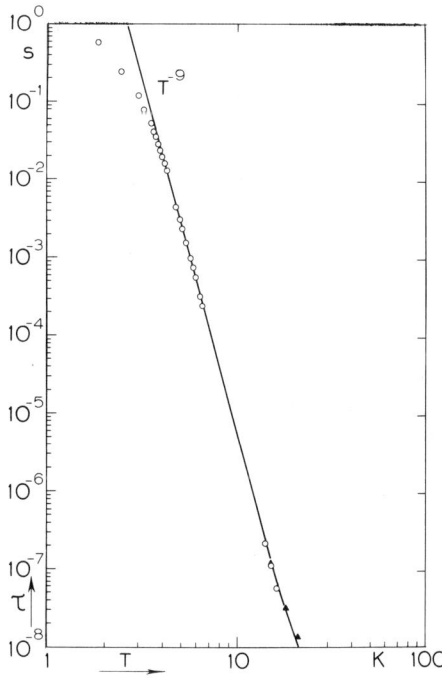

Fig. 6. Relaxation time vs. temperature for powdered $YbCl_3 \cdot 6H_2O$ at an external magnetic field of 1 kOe (different symbols refer to different experimental techniques). The line represents $\tau^{-1} = 1.9 \times 10^{-4} T^9$ sec^{-1}; at high temperatures a deviation from T^9 is taken into account due to $\theta_D = 180°K$.

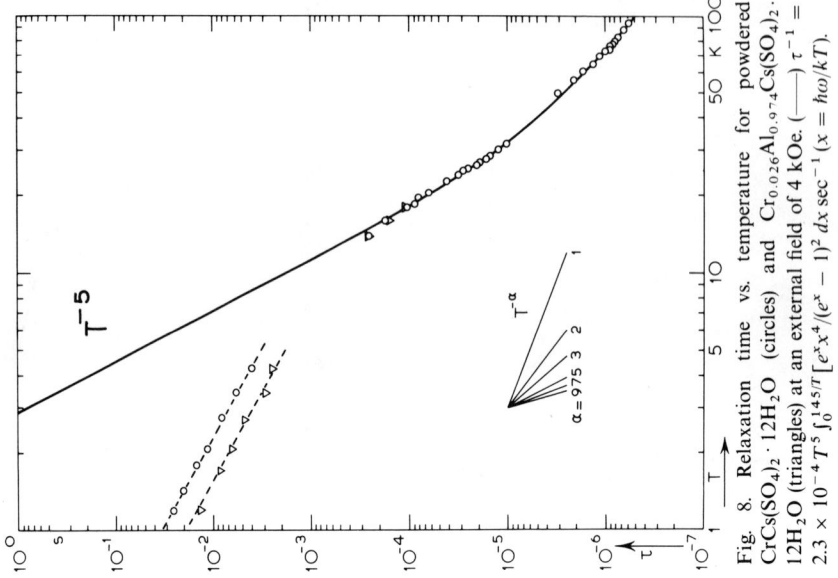

Fig. 8. Relaxation time vs. temperature for powdered CrCs$(SO_4)_2 \cdot 12H_2O$ (circles) and $Cr_{0.026}Al_{0.974}Cs(SO_4)_2 \cdot 12H_2O$ (triangles) at an external field of 4 kOe. (———) $\tau^{-1} = 2.3 \times 10^{-4} T^5 \int_0^{145/T} [e^x x^4/(e^x-1)^2] \, dx \, \sec^{-1} \, (x = \hbar\omega/kT)$.

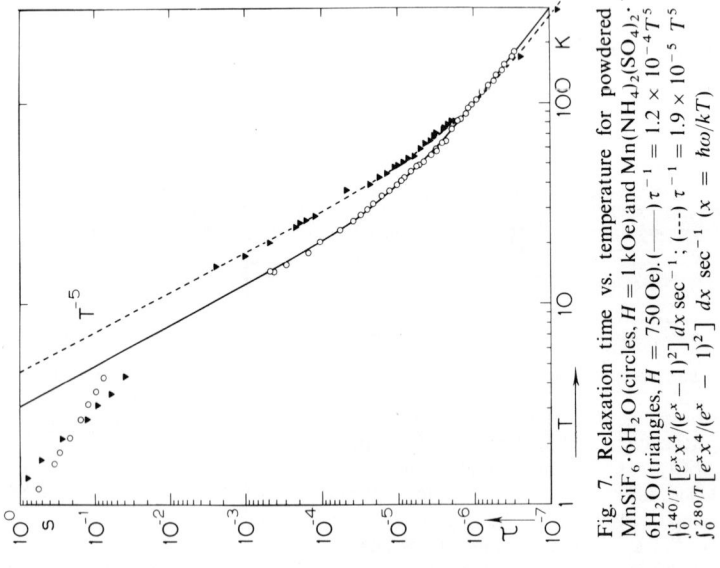

Fig. 7. Relaxation time vs. temperature for powdered MnSiF$_6 \cdot 6H_2O$ (circles, $H = 1$ kOe) and Mn(NH$_4)_2$(SO$_4)_2 \cdot 6H_2O$ (triangles, $H = 750$ Oe). (———) $\tau^{-1} = 1.2 \times 10^{-4} T^5 \int_0^{140/T} [e^x x^4/(e^x-1)^2] \, dx \, \sec^{-1}$; (---) $\tau^{-1} = 1.9 \times 10^{-5} T^5 \int_0^{280/T} [e^x x^4/(e^x-1)^2] \, dx \, \sec^{-1} \, (x = \hbar\omega/kT)$

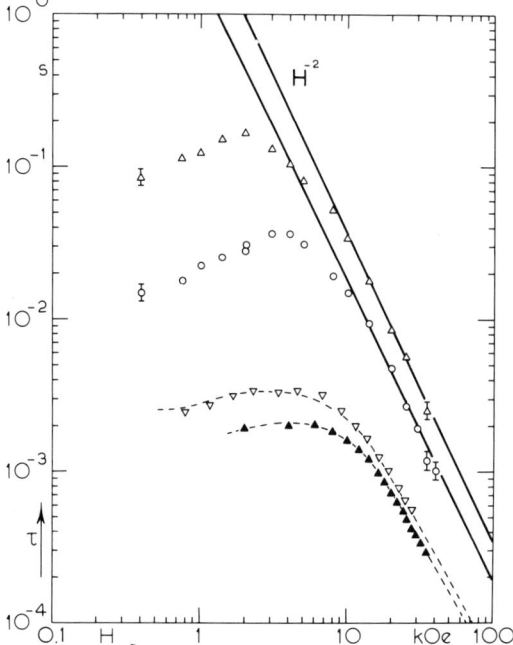

Fig. 9. Relaxation time vs. external magnetic field for powdered $Mn(NH_4)_2(SO_4)_2 \cdot 6H_2O$. ($\triangle$) $T = 2.09°K$, ($\bigcirc$) $T = 4.22°K$, ($\triangledown$) $T = 14.2°K$, ($\blacktriangle$) $T = 16.0°K$. The solid lines represent $\tau \propto H^{-2}$; the dashed lines give a description of τ^{-1} as sum of a term proportional to H^2 (direct process) and one with the Brons–Van Vleck field dependence for the relaxation time of the T^5th Raman process.

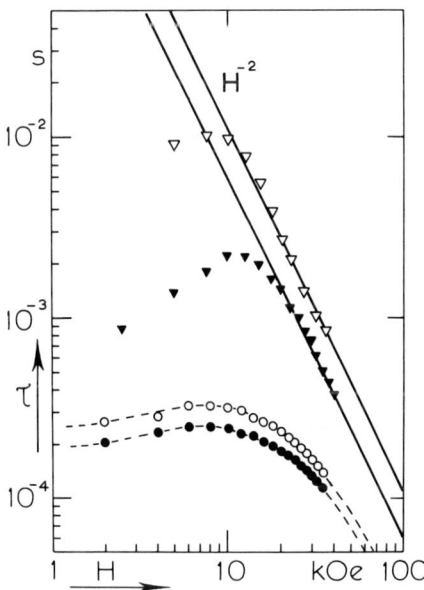

Fig. 10. Relaxation time vs. external magnetic field for powdered $MnSiF_6 \cdot 6H_2O$. ($\triangledown$) $T = 2.06°K$, ($\blacktriangledown$) $T = 4.25°K$, ($\bigcirc$) $T = 14.1°K$, ($\bullet$) $T = 16.0°K$. Lines as in Fig. 9.

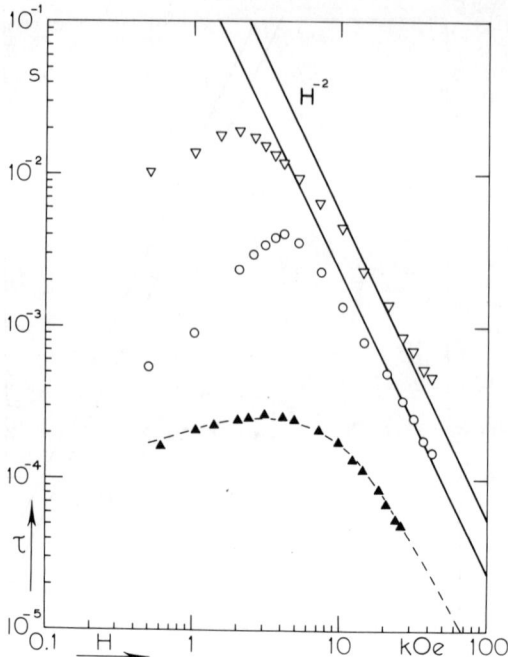

Fig. 11. Relaxation time vs. external magnetic field for powdered $CrCs(SO_4)_2 \cdot 12H_2O$. ($\nabla$) $T = 2.1°K$, ($\bigcirc$) $T = 4.2°K$, ($\blacktriangle$) $T = 14.1°K$. Lines as in Fig. 9.

relaxation mechanism. At liquid helium temperatures we observed the proper H^2 dependence of τ^{-1}, while at liquid hydrogen temperatures the direct and the Raman processes interfere. From a series of measurements similar to the examples shown in Figs. 9–11 we derived the constants A' [Eq. (2)] which are given in Table I. The values for A' of the three manganese samples are remarkably similar. For chromium cesium alum A' is a factor of ten larger. This is not surprising since the manganese ion has no orbital angular momentum in the ground state and it may not be expected to relax effectively by a mechanism involving the influence of the crystalline field on the orbital ionic angular momentum.

References

1. R. Orbach, *Proc. Roy. Soc.* **A264**, 458 (1961).
2. A.J. De Vries, J.W.M. Livius, D.A. Curtis, A.J. Van Duyneveldt, and C.J. Gorter, *Physica* **36**, 65 (1967).
3. A.J. Van Duyneveldt, J. Soeteman, and C.J. Gorter, *Physica* **47**, 1 (1970).
4. S.F.J. Cox, J.C. Gill, and D.O. Wharmby, *J. Phys. C* **4**, 371 (1971).
5. A.M. Stoneham, *Proc. Phys. Soc.* **86**, 1163 (1965).
6. J.A. Roest, A.J. Van Duyneveldt, A. Van der Bilt, and C.J. Gorter, *Physica* **64**, 306 (1973).
7. M. Blume and R. Orbach, *Phys. Rev.* **127**, 1587 (1962); R. Orbach and M. Blume, *Phys. Rev. Lett.* **8**, 478 (1962).

Low-Temperature Heat Capacity of α- and β-Cerium*

M. M. Conway and Norman E. Phillips

*Inorganic Materials Research Division of the Lawrence Berkeley Laboratory
and Department of Chemistry
University of California, Berkeley, California*

Both the β and α phases of Ce are present in samples of the metal at low temperatures. The β phase is essentially trivalent, and has one localized $4f$ electron. The α phase has a valence intermediate between three and four. The fourth electron is no longer in a well-localized $4f$ state,[1] nor is it fully promoted into the conduction band. The valence usually assigned[2] to the α phase is 3.6.

We have measured the heat capacities of mixed-phase samples of different compositions and have extrapolated the data to the pure phases, using the magnetic ordering anomaly in the β phase as a measure of the composition. The standard technique of thermal cycling was used to increase the mole fraction of the β phase χ_β and measurements were obtained on samples whose χ_β ranged from 0.3 to 0.8. The heat capacity observed in experiment i can be represented by $C_i(T) = \chi_{\beta i} C_\beta(T) + (1 - \chi_{\beta i}) C_\alpha(T)$, where C_β and C_α are the heat capacities of the pure β and pure α phases. (This equation neglects the small, χ_β-independent peaks in the heat capacity near 0.15, 0.9, and 6°K. These peaks are sample dependent and probably are impurity related. They do not significantly interfere with the determination of $\chi_{\beta i}$.) C_α can be extrapolated from the mixed-phase data through a series of trials in which different ratios of the $\chi_{\beta i}$ are assigned to the observed C_i. The correct ratio gives a C_α which varies smoothly through the 12.5°K region of the β-phase's antiferromagnetic ordering peak. This C_α agrees with the C_α measured by Panousis and Gschneidner[3] at low temperatures. Above approximately 6°K their sample has an anomalously high heat capacity which cannot be ascribed entirely to the approximately 3% β phase present. Panousis and Gschneidner[4] have also reported the heat capacity of a sample for which χ_β was measured dilatometrically to be 0.91 ± 0.05. From this C_i and from C_α the $\chi_{\beta i}$ of other experiments and C_β can be determined.

The value of γ, 22 mJ/mole°K², obtained for α-Ce is appreciably higher than expected for a metal with a purely $s-d$ conduction band and a valence of 3.6. The excess is attributable to the presence of the $4f$ electron in a state which overlaps the Fermi surface, and is of the order of magnitude predicted by the virtual bound state model.[5] By 10 kbar, γ has decreased[6] to 11.3 mJ/mole-°K². This decrease reflects the increase in energy of the f state relative to the Fermi surface.

The θ_0 of α-Ce is approximately 125°K, lower than the estimated 150°K θ_0 of β-Ce,[7] in spite of the 17% greater density[8] and the higher valence of α-Ce.

---

* Work supported by U.S. Atomic Energy Commission.

Table I. Properties of the Rare Earths

	γ_{300}, mJ/mole°K	Apparent γ_0,[13] mJ/mole°K	T_N,[17]°K	T_C,[17]°K	Low-temperature magnetic structure
La	—	9.4‡	—	—	None
Ce	7.5†	54§	?, 12.5	—	Complex antiferromagnetic
Pr	7.3	26.2	25, —	—	Complex antiferromagnetic
Nd	8.6	58	20, 7.5	—	Complex antiferromagnetic
Sm	11.5	12.4	14.8	—	Complex antiferromagnetic
Eu∗	3.7	12.1	90	—	Complex antiferromagnetic
Gd	9.2	(11)∗∗	292	293	Simple ferromagnetic
Tb	12.2	10.4	229	221	Simple ferromagnetic
Dy	9.1	17.9	178.5	85	Simple ferromagnetic
Ho	7.9	50	132	20	Conical ferromagnetic
Er	11.8	(9.1)#	85	19.6	Conical ferromagnetic
Tm	7.2	22.3	51–60	22	Complex ferrimagnetic
Yb[12]	—	2.9	—	—	None
Lu	—	11.3	—	—	None

* Divalent; the rest are trivalent.
† fcc γ phase.
‡ Ref. 14, dhcp phase.
§ Present work, dhcp β phase.
∗∗Ref. 15, not recalculated.
Ref. 16, not recalculated.

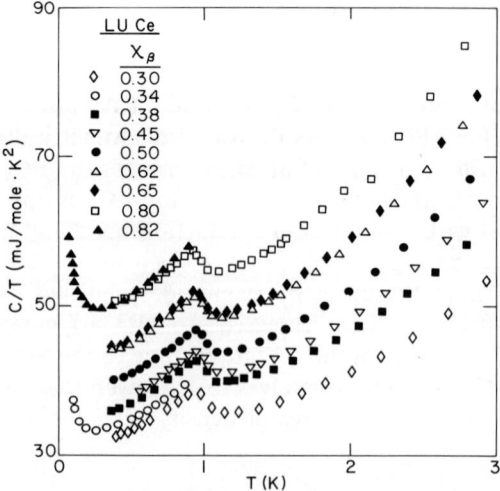

Fig. 1. The low-temperature heat capacity of Ce.

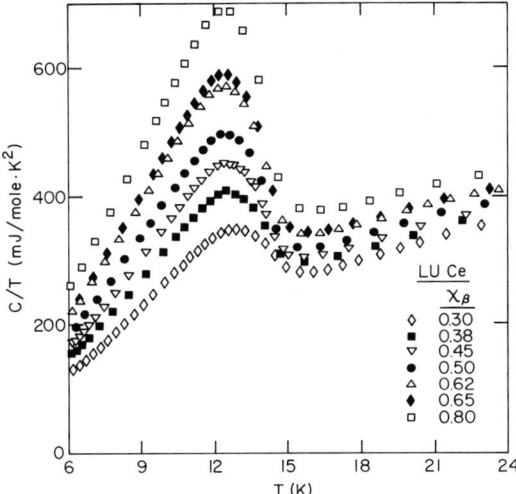

Fig. 2. The high-temperature heat capacity of Ce.

At low temperatures the heat capacity of β-Ce can be represented by $C_\beta = 54T + 5.6T^3$ mJ/mole-°K and is field independent to at least 60 kOe. The T^3 term is an order of magnitude higher than expected for the lattice heat capacity. It is consistent with a spin-wave heat capacity in the absence of an anisotropy field, but in that case it should be substantially reduced by a 60-kOe external field. The linear term is unusually high and is five times higher than would be expected for β-Ce on the basis of its similarity to La. Large linear terms occur for other magnetic rare

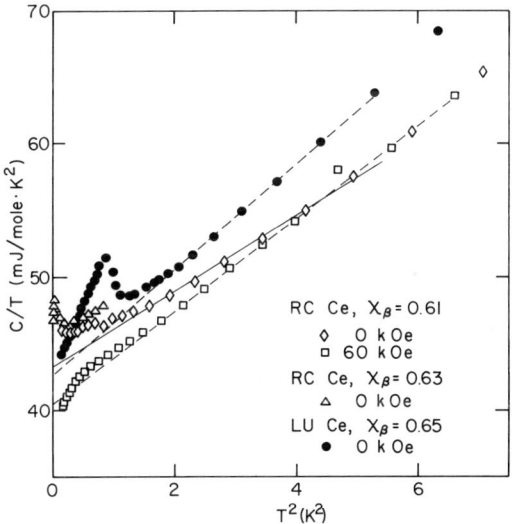

Fig. 3. The effect of a 60-kOe field on the low-temperature heat capacity of Ce. The LU and RC samples are from different suppliers.

earths at low temperatures, but at room temperature the linear terms all have the lower values found in nonmagnetic rare earths, as shown in Table I. These facts suggest the importance of a temperature-dependent magnetic enhancement of the electronic heat capacity.

The entropy associated with the 12.5°K antiferromagnetic ordering peak of β-Ce is approximately $1/2R \ln 2$, indicating that only half of the localized $4f$ electrons order at that temperature. Beta-Ce has a double hexagonal close-packed structure, with half of the atoms in a cubic and half in a hexagonal environment, and it is not surprising that atoms on different sites do not order at the same temperature. The neighboring rare earths Pr and Nd also have this structure. The hexagonal sites of Nd order[10] at 19°K and the cubic ones at 7.5°K. Only the hexagonal sites of Pr order at 25°K.[11] Judging from the trend in transition temperatures, it is the cubic sites that order in β-Ce at 12.5°K.

References

1. M.R. MacPherson, G.E. Everett, D. Wohlleben, and M.B. Maple, *Phys. Rev. Lett.* **26**, 20 (1971).
2. K.A. Gschneidner, Jr., and R. Smoluckowski, *J. Less-Common Metals* **5**, 374 (1963).
3. N.T. Panousis and K.A. Gschneidner, Jr., *Sol. St. Comm.* **8**, 1779 (1970).
4. N.T. Panousis and K.A. Gschneidner, Jr., *Phys. Rev. B* **5**, 4767 (1972).
5. B. Coqblin and A. Blandin, *Advan. Phys.* **17**, 281 (1968).
6. N.E. Phillips, J.C. Ho, and T.F. Smith, *Phys. Lett.* **A27**, 49 (1968).
7. D.H. Parkinson, F.E. Simon, and F.H. Spedding, *Proc. Roy. Soc. (London) A* **207**, 137 (1951).
8. K.A. Gschneider, Jr., R.O. Elliott, and R.R. McDonald, *J. Phys. Chem. Sol.* **23**, 1191 (1962).
9. O.V. Lounasmaa, *Phys. Rev.* **133**, A502 (1964).
10. O.V. Lounasmaa and L.J. Sundstrom, *Phys. Rev.* **158**, 591 (1967).
11. T. Johansson, B. Zebech, M. Nielsen, H.B. Møller, and A.R. Mackintosh, *Phys. Rev. Lett.* **25**, 524 (1970).
12. K. A. Gschneider, Jr., in *Rare Earth Research III*, L. Eyring, ed., Gordon and Breach, New York (1965), p. 153.
13. J.A. Morrison and D.M.T. Newsham, *J. Phys. C* **1**, 370 (1968).
14. D.L. Johnson and D.K. Finnemore, *Phys. Rev.* **158**, 376 (1967).
15. B. Dreyfus, J.C. Michel, and D. Thoulouze, *Phys. Lett.* **24A**, 457 (1967).
16. R.D. Parks, in *Proc. 2nd Conf. Rare Earth Research*, J.F. Nachman and C.E. Lundin, eds., Gordon and Breach, New York (1962), p. 225.
17. W.C. Koehler, *J. Appl. Phys.* **36**, 1078 (1965).

A Measurement of the Magnetic Moment of Oxygen Isolated in a Methane Lattice

J. E. Piott

University of Washington, Seattle, Washington
and
Belfer Graduate School of Science
Yeshiva University, New York, New York

and

W. D. McCormick

University of Texas, Austin, Texas

This study of the proton magnetic resonance line shape of CH_4 examines in detail the broadening by a dilute mixture of unlike spins. Usual treatments of such broadening[1] assume the magnetic moment of an unlike spin to be the same order of magnitude as that of the spin on which the NMR observation takes place. In this case, however, the unlike spin is that of the Σ^3 state of an oxygen molecule. Since its magnetic moment is much larger than that of CH_4, some modifications must be made in the theory.

The secular Hamiltonian of the proton spins in an external magnetic field can be written

$$H = H(\text{Zeeman}) + H(\text{dipole-dipole}) + \sum_{l,k} (\gamma \hbar g \beta / r_{lk}^3) I_z^l \langle S_z^k \rangle (1 - 3\cos^2\theta_{lk}) \quad (1)$$

The first two terms determine the line center and broadening in the absence of paramagnetic impurities. Features of the second moment of pure methane $M_2(d-d)$ have been studied elsewhere.[2] The third term is the interaction of the proton spins I_z^l with the electronic spins S_z^k of the oxygen molecules. During the time T_2 in which the proton spin resonance can be observed the electron spin occupies all of its quantum states. For this reason the third term must use the thermal average of electron spin.

Following the usual procedure, the second moment resulting from the broadening by the paramagnetic impurities can be shown to be

$$M_2(\text{para}) = \gamma^2 g^2 \beta^2 f \, \overline{\langle S_z \rangle^2} \sum_k{}' (1 - 3\cos^2\theta_{lk})^2 / r_{lk}^6 \quad (2)$$

The derivation of this expression makes use of the assumption that the fraction of molecules f that are oxygen occupy random crystalline sites. Note that in this case $\langle S_z \rangle^2$ is the correct factor, rather than $S(S+1)$.

A calculation of the second moment using this expression gives a linewidth

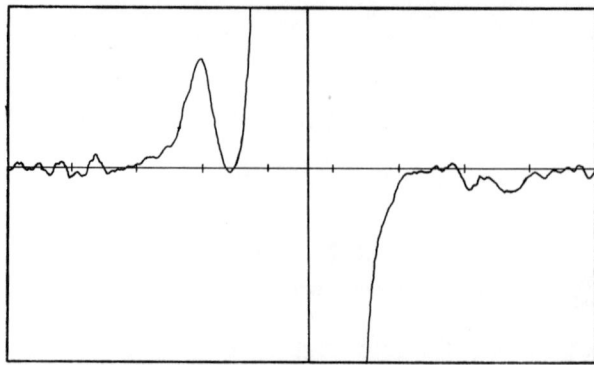

Fig. 1. Anomalies observed at 10.7°K in the wings of the experimental NMR derivative signal of CH_4 containing 0.698% O_2. The size of the center line is approximately ten times the height of the figure.

much larger than the derivative peak-to-peak linewidth observed. A careful examination of the wings of the line, as shown in Fig. 1, reveals that much of the second moment is found in the anomalous peaks of the wings. These peaks result from a large shift of the magnetic field at the site of the methane molecules that are nearest neighbors to an oxygen molecule. The center line of the NMR derivative signal does not involve those oxygen molecules in the first nearest-neighbor shell. Only the second-nearest and further neighbors should be used for the lattice sum in the expression for M_2(para). Observed linewidths then approximately agree with the calculated ones.

Those terms involving nearest neighbors in the interaction between the electron spins and proton spins give the magnetic field shift characterized by the anomalies in the wings. The shift in the magnetic field can be written

$$h = (g\beta/r_1^3)\langle S_z \rangle (1 - 3\cos^2\theta) \qquad (3)$$

In this equation a line connecting the O_2 and CH_4 molecules is at an angle θ with respect to the direction of the magnetic field. In general, for each value of $\langle S_z \rangle$ there are 12 values of h in a single crystallite, one for each of the nearest-neighbor positions.

Calculations of $\langle S_z \rangle$ use the electron spin Hamiltonian

$$H_e = g\beta H_0 S_z + D'[S_{z'}^2 - \tfrac{1}{3}S(S+1)] \qquad (4)$$

The second term results from the axial symmetry of the oxygen molecule.[3] If it were rigidly fixed[4] in the crystalline lattice, we would have $D'/k = D/k = 5.66°K$. When the oxygen molecule is only partially bound[4] $D' = (\tfrac{3}{2}\langle \cos^2\theta \rangle_{av} - \tfrac{1}{2})D$. Here the angle θ is the deviation of the oxygen molecule from its equilibrium position. For a freely rotating molecule D' is zero.

Complications arise as a result of the two different axes of quantized electron spin. The z axis lies parallel to the direction of the external magnetic field, while the z' axis lies along the axis of symmetry of the oxygen molecule in its equilibrium position. Usually, these are not the same. In using the electron spin Hamiltonian to calculate $\langle S_z \rangle$, rotation matrices were used to connect factors involving different axes of quantization.

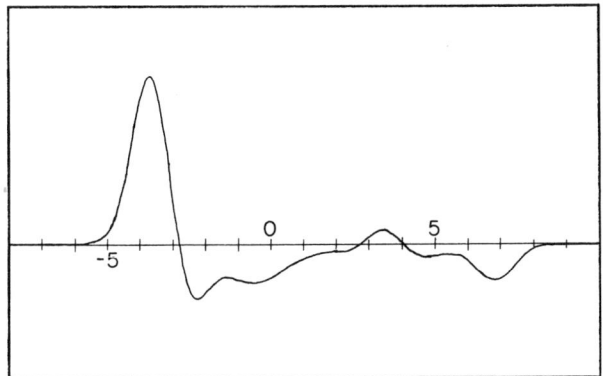

Fig. 2. The calculated polycrystalline NMR derivative line shape of those CH_4 molecules that are nearest neighbors to oxygen molecules. All parameters used correspond to the experimental conditions of Fig. 1.

Steric considerations restrict $\langle S_z \rangle$ to three possible values in a crystallite. In equilibrium the axis of symmetry of the oxygen molecule lies along one of the four-fold axes of symmetry of the crystal. Thirty six resonant frequencies can then be found for the methane molecules that are nearest neighbors to oxygen molecules, by using Eq. (3). The average of all three values of $\langle S_z \rangle^2$, $\overline{\langle S_z \rangle^2}$, is used in Eq. (2).

Each of these 36 lines in a crystallite was broadened by a Gaussian whose second moment is

$$M_2 = M_2(\text{para}) + M_2(d-d) \qquad (5)$$

The line shape obtained was averaged over 100 crystallites selected at random. The typical polycrystalline line shape, shown in Fig. 2, reproduces the anomalies in the experimental line. Both the large peak and dip at one extreme and the smaller peak in the opposite wing always appear. The relative peak heights and relative shifts of the two peaks from the linecenter are reproduced.

The value of D' used is related to the average distribution of electron charge at the site of a molecule during an NMR observation. Two equivalent cases appear to best fit all data. In the first case the oxygen molecule rotates freely, so that $D' = 0$. In the second possibility D' is not zero, but the oxygen molecule jumps among the three equilibrium positions many times during an NMR observation. This second case has a quasispherical symmetry to the charge distribution. Using this average charge distribution, D' would be zero. The two cases become equivalent.

Figure 3 demonstrates how well this model predicts the separation of the large anomaly from the center of the line. The calculated curve lies near but slightly above the data. The discrepancy becomes worse at the lower temperatures. The reason probably lies in the technique used to find the center of the line. By necessity, the large center line was assumed to be symmetric. If this is correct, the position at which the derivative signal crosses the baseline would be the center of the NMR line in the absence of paramagnetic impurities.

By no means should the line be expected to be symmetric. Extrapolating the effect of the paramagnetic impurities on the first-nearest neighbors to other methane

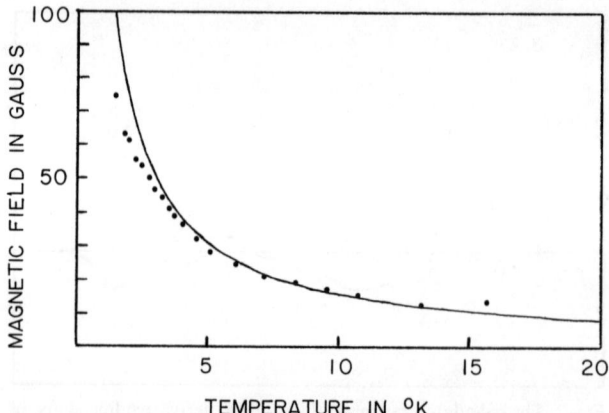

Fig. 3. A comparison of the calculated shift of the large anomalous peak from the center of the line, with the measured shift. Any error in the calculated line is less than 1%, while the statistical error in the experimental points may be as large as 4%.

molecules would predict some asymmetry in the center line. This introduces a systematic error which increases at the lower temperatures as the center line becomes broader. Any further study of this problem must compensate for this effect.

References

1. A. Abragam, *Principles of Nuclear Magnetism*, Oxford University Press, London (1961), p. 122.
2. J.E. Piott, Thesis, University of Washington, 1971, unpublished.
3. A. Carrington and A.D. McLachlin, *Introduction to Magnetic Resonance*, Harper and Row, New York (1967), p. 116.
4. H. Meyer, M.C.M. O'Brien, and J.H. Van Vleck, *Proc. Roy. Soc. London* **243A**, 414 (1957).

Contents
of Volume 1

Fritz London Award Address. *A. A. Abrikosov* 1

Quantum Fluids

1. Plenary Topics

Light Scattering As a Probe of Liquid Helium. *T. J. Greytak* 9
Phase Diagram of Helium Monolayers. *J. G. Dash* 19
Helium-3 in Superfluid Helium-4. *David O. Edwards* 26

2. Quantum Fluids Theory

Some Recent Developments Concerning the Macroscopic Quantum Nature of Superfluid Helium. *S. Putterman* 39
A Neutron Scattering Investigation of Bose–Einstein Condensation in Superfluid ^{4}He. *H.A. Mook, R. Scherm, and M.K. Wilkinson* 46
Temperature and Momentum Dependence of Phonon Energies in Superfluid ^{4}He. *Shlomo Havlin and Marshall Luban* 50
Does the Phonon Spectrum in Superfluid ^{4}He Curve Upward? *S.G. Eckstein, D. Friedlander, and C.G. Kuper* 54
Model Dispersion Curves for Liquid ^{4}He. *James S. Brooks and R. J. Donnelly* 57
Analysis of Dynamic Form Factor $S(k,\omega)$ for the Two-Branch Excitation Spectrum of Liquid ^{4}He. *T. Soda, K. Sawada, and T. Nagata* 61
Excitation Spectrum for Weakly Interacting Bose Gas and the Liquid Structure Function of Helium. *Archana Bhattacharyya and Chia-Wei Woo* 67
Elementary Excitations in Liquid Helium. *Chia-Wei Woo* 72
Long-Wavelength Excitations in a Bose Gas and Liquid He II at $T = 0$. *H. Gould and V. K. Wong* 76
Bose–Einstein Condensation in Two-Dimensional Systems. *Y. Imry, D.J. Bergman, and L. Gunther* 80
Phonons and Lambda Temperature in Liquid ^{4}He as Obtained by the Lattice Model. *P. H. E. Meijer and W. D. Scherer* 84
Equation of State for Hard-Core Quantum Lattice System. *T. Horiguchi and T. Tanaka* 87

Impulse of a Vortex System in a Bounded Fluid. *E. R. Huggins* 92
New Results on the States of the Vortex Lattice. *M. Le Ray, J. P. Deroyon, M. J. Deroyon, M. François, and F. Vidal* 96
Superfluid Density in Pairing Theory of Superfluidity. *W. A. B. Evans and R. I. M.A. Rashid* 101
The ^{3}He–Roton Interaction. *H. T. Tan, M. Tan, and C. -W. Woo* 108
Quantum Lattice Gas Model of ^{3}He–^{4}He Mixtures. *Y. -C. Cheng and M. Schick* 112
The $S = 1$ Ising Model for ^{3}He–^{4}He Mixtures. *W.M. Ng, J.H. Barry, and T. Tanaka* 116
Low Temperature Thermodynamics of Fermi Fluids. *A. Ford, F. Mohling, and J.C. Rainwater* 121
Density and Phase Variables in the Theory of Interacting Bose Systems. *P. Berdahl* 130
Subcore Vortex Rings in a Ginsburg–Landau Fluid. *E. R. Huggins* 135

3. Static Films

Multilayer Helium Films on Graphite. *Michael Bretz* 143
Submonolayer Isotopic Mixtures of Helium Adsorbed on Grafoil. *S. V. Hering, D. C. Hickernell, E. O. McLean, and O. E. Vilches* 147
Spatial Ordering Transitions in Quantum Lattice Gases. *R. L. Siddon and M. Schick* 152
A Model of the ^{4}He Monolayer on Graphite. *S. Nakajima* 156
Thermal Properties of the Second Layer of Adsorbed ^{3}He. *A. L. Thomson, D. F. Brewer, and J. Stanford* 159
Nuclear Magnetic Resonance Investigation of ^{3}He Surface States in Adsorbed ^{3}He–^{4}He Films. *D. F. Brewer, D. J. Creswell, and A. L. Thomson* 163
Nuclear Magnetic Relaxation of Liquid ^{3}He in a Constrained Geometry. *J. F. Kelly and R. C. Richardson* 167
Low-Temperature Specific Heat of ^{4}He Films in Restricted Geometries. *R. H. Tait, R. O. Pohl, and J. D. Reppy* 172
Mean Free Path Effects in ^{3}He Quasiparticles: Measurement of the Spin Diffusion Coefficient in the Collisionless Regime by a Pulsed Gradient NMR Technique. *D.F. Brewer and J.S. Rolt* 177
Adsorption of ^{4}He on Bare and on Argon-Coated Exfoliated Graphite at Low Temperatures. *E. Lerner and J. G. Daunt* 182
Ellipsometric Measurements of the Saturated Helium Film. *C. C. Matheson and J.L. Horn* 185
Momentum and Energy Transfer Between Helium Vapor and the Film. *D. G. Blair and C. C. Matheson* 190
Nuclear Magnetic Resonance Study of the Formation and Structure of an Adsorbed ^{3}He Monolayer. *D. J. Creswell, D. F. Brewer, and A. L. Thomson* 195
Pulsed Nuclear Resonance Investigation of the Susceptibility and Magnetic Interaction in Degenerate ^{3}He Films. *D. F. Brewer and J. S. Rolt* 200

Measurements and Calculations of the Helium Film Thickness on Alkaline Earth Fluoride Crystals. *E. S. Sabisky and C. H. Anderson* 206

The Normal Fluid Fraction in the Adsorbed Helium Film. *L. C. Yang, M. Chester, and J. B. Stephens* 211

4. Flowing Films

Helium II Film Transfer Rates Into and Out of Solid Argon Beakers. *T. O. Milbrodt and G. L. Pollack* 219

Preferred Flow Rates in the Helium II Film. *R. F. Harris-Lowe and R. R. Turkington* 224

Superfluidity of Thin Helium Films. *H. W. Chan, A. W. Yanof, F. D. M. Pobell, and J. D. Reppy* 229

Superfluidity in Thin ^{3}He–^{4}He Films. *B. Ratnam and J. Mochel* 233

On the Absence of Moderate Velocity Persistent Currents in He II Films Adsorbed on Large Cylinders. *T. Wang and I. Rudnick* 239

Thermodynamics of Superflow in the Helium Film. *D. L. Goodstein and P. G. Saffman* 243

Mass Transport of ^{4}He Films Adsorbed on Graphite. *J. A. Herb and J. G. Dash* 247

Helium Film Flow Dissipation with a Restrictive Geometry. *D. H. Liebenberg* 251

Dissipation in the Flowing Saturated Superfluid Film. *J. K. Hoffer, J. C. Fraser, E. F. Hammel, L. J. Campbell, W. E. Keller, and R. H. Sherman* 253

Dissipation in Superfluid Helium Film Flow. *J. F. Allen, J. G. M. Armitage, and B. L. Saunders* 258

Direct Measurement of the Dissipation Function of the Flowing Saturated He II Film. *W. E. Keller and E. F. Hammel* 263

Application of the Fluctuation Model of Dissipation to Beaker Film Flow. *L. J. Campbell* 268

Film Flow Driven by van der Waals Forces. *D. G. Blair and C. C. Matheson* ... 272

5. Superfluid Hydrodynamics

Decay of Saturated and Unsaturated Persistent Currents in Superfluid Helium. *H. Kojima, W. Veith, E. Guyon, and I. Rudnick* 279

Rotating Couette Flow of Superfluid Helium. *H. A. Snyder* 283

New Aspects of the λ-Point Paradox. *Robert F. Lynch and John R. Pellam* 288

Torque on a Rayleigh Disk Due to He II Flow. *W. J. Trela and M. Heller* 293

Superfluid ^{4}He Velocities in Narrow Channels between 1.8 and 0.3°K. *S. J. Harrison and K. Mendelssohn* 298

Critical Velocities in Superfluid Flow Through Orifices. *G. B. Hess* 302

Observations of the Superfluid Circulation around a Wire in a Rotating Vessel Containing He II. *S. F. Karl and W. Zimmermann, Jr.* 307

An Attempt to Photograph the Vortex Lattice in Rotating He II. *Richard E. Packard and Gay A. Williams* 311

Radial Distribution of Superfluid Vortices in a Rotating Annulus. *D. Scott Shenk and James B. Mehl* 314

Effect of a Constriction on the Vortex Density in He II Superflow. *Maurice François, Daniel Lhuillier, Michel Le Ray, and Félix Vidal* 319

AC Measurements of a Coupling between Dissipative Heat Flux and Mutual Friction Force in He II. *Félix Vidal, Michel Le Ray, Maurice François, and Daniel Lhuillier* 324

The He II–He I Transition in a Heat Current. *S.M. Bhagat, R.S. Davis, and R. A. Lasken* 328

Pumping in He II by Low-Frequency Sound. *G. E. Watson* 332

Optical Measurements on Surface Modes in Liquid Helium II. *S. Cunsolo and G. Jacucci* 337

6. Helium Bulk Properties

Measurement of the Temperature Dependence of the Density of Liquid ^{4}He from 0.3 to 0.7°K and Near the λ-Point. *Craig T. Van DeGrift and John R. Pellam* 343

Second-Sound Velocity and Superfluid Density in ^{4}He Under Pressure and Near T_λ. *Dennis S. Greywall and Guenter Ahlers* 348

Superfluid Density Near the Lambda Point in Helium Under Pressure. *Akira Ikushima and Giiuchi Terui* 352

Hypersonic Attenuation in the Vicinity of the Superfluid Transition of Liquid Helium. *D. E. Commins and I. Rudnick* 356

Superheating in He II. *R. K. Childers and J. T. Tough* 359

Evaporation from Superfluid Helium. *Milton W. Cole* 364

Angular Distribution of Energy Flux Radiated from a Pulsed Thermal Source in He II below 0.3°K. *C. D. Pfeifer and K. Luszczynski* 367

Electric Field Amplification of He II Luminescence Below 0.8 K. *Huey A. Roberts and Frank L. Hereford* 372

Correlation Length and Compressibility of ^{4}He Near the Critical Point. *A. Tominaga and Y. Narahara* 377

Coexistence Curve and Parametric Equation of State for ^{4}He Near Its Critical Point. *H. A. Kierstead* 381

Effect of Viscosity on the Kapitza Conductance. *W.M. Saslow* 387

Liquid Disorder Effects on the Solid He II Kapitza Resistance. *C. Linnet, T. H. K. Frederking, and R. C. Amar* 393

The Kapitza Resistance between Cu (Cr) and ^{4}He (^{3}He) Solutions and Applications to Heat Exchangers. *J. D. Siegwarth and R. Radebaugh* 398

Heat Transfer between Fine Copper Powders and Dilute ^{3}He in Superfluid ^{4}He. *R. Radebaugh and J. D. Siegwarth* 401

Thermal Boundary Resistance between Pt and Liquid ^{3}He at Very Low Temperatures. *J. H. Bishop, A. C. Mota, and J. C. Wheatley* 406

The Leggett–Rice Effect in Liquid ^{3}He Systems. *L. R. Corruccini, D. D. Osheroff, D. M. Lee, and R. C. Richardson* 411

Helium Flow Through an Orifice in the Presence of an AC Sound Field. *D. Musinski and D. H. Douglass* 414

7. Ions and Electrons

Vortex Fluctuation Contribution to the Negative Ion Trapping Lifetime. *J. McCauley, Jr.* 421

The Question of Ion Current Flow in Helium Films. *S. G. Kennedy and P. W. F. Gribbon* 426

Impurity Ion Mobility in He II. *Warren W. Johnson and William I. Glaberson* . 430

Measurement of Ionic Mobilities in Liquid ^{3}He by a Space Charge Method. *P.V.E. McClintock* 434

Pressure Dependence of Charge Carrier Mobilities in Superfluid Helium. *R. M. Ostermeier and K. W. Schwarz* 439

Two-Dimensional Electron States Outside Liquid Helium. *T. R. Brown and C. C. Grimes* 443

An Experimental Test of Vinen's Dimensional Theory of Turbulent He II. *D. M. Sitton and F. E. Moss* 447

Measurements on Ionic Mobilities in Liquid ^{4}He. *G. M. Daalmans, M. Naeije, J. M. Goldschvartz, and B. S. Blaisse* 451

Influence of a Grid on Ion Currents in He II. *C.S.M. Doake and P.W.F. Gribbon* 456

Collective Modes in Vortex Ring Beams. *G. Gamota* 459

Tunneling from Electronic Bubble States in Liquid Helium Through the Liquid –Vapor Interface. *G. W. Rayfield and W. Schoepe* 469

Positive Ion Mobility in Liquid ^{3}He. *M. Kuchnir, J. B. Ketterson, and P. R. Roach* 474

A Large Family of Negative Charge Carriers in Liquid Helium. *G. G. Ihas and T. M. Sanders, Jr.* 477

Temperature Dependence of the Electron Bubble Mobility Below 0.3° K *M. Kuchnir, J. B. Ketterson, and P. R. Roach* 482

Transport Properties of Electron Bubbles in Liquid He II. *M. Date, H. Hori, K. Toyokawa, M. Wake, and O. Ichikawa* 485

Do Fluctuations Determine the Ion Mobility in He II near T_λ? *D. M. Sitton and F. E. Moss* 489

8. Sound Propagation and Scattering Phenomena

Absence of a Quadratic Term in the ^{4}He Excitation Spectrum. *P. R. Roach, B. M. Abraham, J. B. Ketterson, and M. Kuchnir* 493

Theoretical Studies of the Propagation of Sound in Channels Filled with Helium II. *H. Wiechert and G. U. Schubert* 497

Developments in the Theory of Third Sound and Fourth Sound. *David J. Bergman* 501

Thermal Excitation of Fourth Sound in Liquid Helium II. *H. Wiechert and R. Schmidt* 510

Inelastic Scattering From Surface Zero-Sound Modes: a Model Calculation. *Allan Griffin and Eugene Zaremba* 515

The Scattering of Low-Energy Helium Atoms at the Surface of Liquid Helium. *J. Eckardt, D. O. Edwards, F. M. Gasparini, and S. Y. Shen* 518

Inelastic Scattering of ^{4}He Atoms by the Free Surface of Liquid ^{4}He. *C. G. Kuper* 522

The Scattering of Light by Liquid ^{4}He Close to the λ-Line. *W. F. Vinen, C. J. Palin, and J. M. Vaughan* 524

Experiments on the Scattering of Light by Liquid Helium. *J. M. Vaughan, W. F. Vinen, and C. J. Palin* 532

Brillouin Light Scattering from Superfluid Helium under Pressure. *G. Winterling, F. S. Holmes, and T. J. Greytak* 537

Brillouin Scattering from Superfluid ^{3}He–^{4}He Mixtures. *R. F. Benjamin, D. A. Rockwell, and T. J. Greytak* 542

Liquid Structure Factor Measurements on the Quantum Liquids. *R. B. Hallock* 547

The Functional Forms of $S(k)$ and $E(k)$ in He II as Determined by Scattering Experiments. *R. B. Hallock* 551

9. ^{3}He–^{4}He Mixtures

Effective Viscosity of Liquid Helium Isotope Mixtures. *D. S. Betts, D. F. Brewer, and R. Lucking* 559

The Viscosity of Dilute Solutions of ^{3}He in ^{4}He at Low Temperatures. *K. A. Kuenhold, D.B. Crum, and R.E. Sarwinski* 563

The Viscosity of ^{3}He–^{4}He Solutions. *D. J. Fisk and H. E. Hall* 568

Thermodynamic Properties of Liquid ^{3}He–^{4}He Mixtures Near the Tricritical Point Derived from Specific Heat Measurements. *S. T. Islander, and W. Zimmermann, Jr.* 571

Dielectric Constant and Viscosity of Pressurized ^{3}He–^{4}He Solutions Near the Tricritical Point. *C. M. Lai and T. A. Kitchens* 576

Critical Opalescent Light Scattering in ^{3}He–^{4}He Mixtures Near the Tricritical Point. *D. Randolf Watts and Watt W. Webb* 581

Second-Sound Velocity, Gravitational Effects, Relaxation Times, and Superfluid Density Near the Tricritical Point in ^{3}He–^{4}He Mixtures. *Guenter Ahlers and Dennis S. Greywall* 586

Excitation Spectrum of ^{3}He–^{4}He Mixture and Its Effect on Raman Scattering. *T. Soda* 591

Contents of Volume I

The Low-Temperature Specific Heat of the Dilute Solutions of ^{3}He in Superfluid ^{4}He. *H. Brucker and Y. Disatnik* 598

Spin Diffusion of Dilute ^{3}He–^{4}He Solutions under Pressure. *D. K. Cheng, P. P. Craig, and T. A. Kitchens* 602

The Spin Diffusion Coefficient of ^{3}He in ^{3}He–^{4}He Solutions. *D.C. Chang and H. E. Rorschach* 608

Nucleation of Phase Separation in ^{3}He–^{4}He Mixtures. *N. R. Brubaker and M. R. Moldover* 612

Renormalization of the ^{4}He λ-Transition in ^{3}He–^{4}He Mixtures. *F. M. Gasparini and M. R. Moldover* 618

The Osmotic Pressure of Very Dilute ^{3}He–^{4}He Mixtures. *J. Landau and R. L. Rosenbaum* 623

Pressure Dependence of Superfluid Transition Temperature in ^{3}He–^{4}He Mixtures. *T. Satoh and A. Kakizaki* 627

Thermal Diffusion Factor of ^{3}He–^{4}He Mixtures: A Test of the Helium Interaction Potential. *W. L. Taylor* 631

Peculiarities of Charged Particle Motion in ^{3}He–^{4}He Solutions in Strong Electric Fields. *B. N. Eselson, Yu. Z. Kovdrya, and O. A. Tolkacheva* 636

First-Sound Absorption and Dissipative Processes in ^{3}He–^{4}He Liquid Solutions and ^{3}He. *N. E. Dyumin, B. N. Eselson, and E. Ya. Rudavsky* 637

Contents of Other Volumes 638
Index to Contributors 661
Subject Index 668

Contents
of Volume 3

Superconductivity

1. Plenary Topics

A Survey of Superconducting Materials. *J. K. Hulm and R. D. Blaugher* 3
Fluctuations in Superconductors. *M. Tinkham* 14
Superconductivity at High Pressures. *J. L. Olsen* 27

2. Type II Superconductors

2.1. Structures

Structure of a Vortex in a Dirty Superconductor. *R. J. Watts-Tobin and G. M. Waterworth* 37
Observation of Landau-Type Branching in the Intermediate State. *J. F. Allen and R. A. Lerski* 42
Attractive Interaction between Vortices in Type II Superconductors at Arbitrary Temperatures. *M. C. Leung and A. E. Jacobs* 46
Calculation of the Vortex Structure at All Temperatures. *L. Kramer and W. Pesch* 49
Magnetic Field Distribution in Type II Superconductors by Neutron Diffraction. *J. Schelten, H. Ullmaier, G. Lippmann, and W. Schmatz* 54
Neutron Diffraction Test of Type II Theories in Relation to Temperature. *R. Kahn and P. Thorel* 64
Nuclear Spin–Spin Relaxation in Superconducting Mixed-State Vanadium. *A. Z. Genack and A. G. Redfield* 69

2.2. Properties

On the Nature of Flux Transport Noise. *C. Heiden* 75
Flux Kinetics Associated with Flux Jumps in Type II Superconductors. *R. B. Harrison, M. R. Wertheimer, and L. S. Wright* 79
Impurity Effect on the Anisotropy of the Upper Critical Field of Nb–Ta Alloys. *N. Ohta, M. Yamamoto, and T. Ohtsuka* 82
Vortex Transport Interference Transitions. *A. T. Fiory* 86
An Alternating Current Investigation of Pinning Sites. *J. Lowell* 90

Effects of Surface Superconductivity in Low κ Al–Ag Alloys. *A. Nemoz and J. C. Solecki* 95

A Model for Flux Pinning in Superconductors. *John R. Clem* 102

Flux Flow of Pure and Dirty Superconductors. *Y. Muto, S. Shinzawa, N. Kobayashi, and K. Noto* 107

The Specific Heat of Very Pure Niobium in the Meissner and Mixed States. *C. E. Gough* 112

Temperature Dependence of Ultrasonic Attenuation in the Mixed State of Pure Niobium. *Frank Carsey and Moises Levy* 116

Mixed-State Ultrasonic Attenuation in Clean Niobium. *M. K. Purvis, R. A. Johnson, and A. R. Hoffman* 120

The Attenuation of Sound, $ql > 1$, in Niobium at Low Flux Densities. *R. E. Jump and C. E. Gough* 125

Thermal Conduction in the Mixed State of Superconducting Niobium at Low Temperatures. *C. M. Muirhead and W. F. Vinen* 130

Entropy of a Type II Superconductor in the Mixed State Close to T_c. *R. Ehrat and L. Rinderer* 134

Flux Pinning in Type II Superconductors. *D. de Klerk, P. H. Kes, and C. A. M. van der Klein* 138

Magnetic Moment in Superconductors Excited by Heat Flow. *N. V. Zavaritsky* 143

Transition to the Mixed State in Lead Films at 4.2°K. *G. J. Dolan and J. Silcox* 149

Observation of Tilted Vortices by Microwave Absorption. *P. Monceau, D. Saint-James, and G. Waysand* 152

Microwave Absorption in Dirty Type II Superconducting Films Around Their Critical Thickness. *Y. Brunet, P. Monceau, E. Guyon, W. Holzer, and G. Waysand* 156

Experimental Evidence for the Thompson Term in the Microwave Conductivity of Type II Superconductors. *Y. Brunet, P. Monceau, and G. Waysand* .. 160

Dynamic Structure of Vortices in Superconductors: Three-Dimensional Features. *Richard S. Thompson and Chia-Ren Hu* 163

Higher-Order Corrections Due to the Order Parameter to the Flux Flow Conductivity of Dirty Type II Superconductors. *Hajime Takayama and Kazumi Maki* 168

The Nascent Vortex State of Type II Superconductors. *Barry L. Walton and Bruce Rosenblum* 172

Hall Effect in Type II Superconductors. *H. Ebisawa* 177

2.3. Dynamics

Current-Induced Intermediate State in Superconducting Strips on Lead and Indium. *R. P. Huebner, R. T. Kampwirth, and D. E. Gallus* 183

Dissipation in a Superconducting Indium Wire. *L. K. Sisemore, K. J. Carroll, and P. T. Sikora* 187

Flux Motion in Lead Indium Wires with Longitudinal Magnetic Fields. *J. E. Nicholson, P. T. Sikora, and K. J. Carroll* 192

Faraday Induction and Flux Flow Voltages in Type II Superconductors: Effect of Magnetic Field and Temperature. *S. M. Khanna and M. A. R. LeBlanc* ... 197

Investigations of Resonances in the RF Absorption of Superconducting Niobium. *P. Kneisel, O. Stoltz, and J. Halbritter* ... 202

Dynamics of the Destruction of Type I Superconductivity by a Current. *H. D. Wiederick, D. C. Baird, and B. K. Mukherjee* ... 207

Influence of Thermal Effects on the Kinetics of the Destruction of Superconductivity by a Current. *E. Posada, D. Robin, and L. Rinderer* ... 212

Fast Neutron Damage in Superconducting Vanadium. *S. T. Sekula and R. H. Kernohan* ... 217

The Effect of an Axial Moment on Normal-Phase Propagation in Type II Superconductors Carrying a Current. *J. F. Bussière and M. A. R. LeBlanc* ... 221

Response of Magnetically Irreversible Type II Cylinders to Currents and Fields Applied in a Parallel Axial Geometry. *D. G. Walmsley and W. E. Timms* ... 227

Limited Flux Jumps in Hard Superconductors. *L. Boyer, G. Fournet, A. Mailfert, and J. L. Noel* ... 232

Boundary Current and Modulated Flux Motion in Superconducting Thin Pb Films. *Y. W. Kim, A. M. de Graaf, J. T. Chen, and E. J. Friedman* ... 238

Effect of Helical Flow on I_c in Cylinders of Type II Superconductors in Axial Magnetic Fields. *R. Gauthier, M. A. R. LeBlanc, and B. C. Belanger* ... 241

Achievement of Nearly Force-Free Flow of Induced Currents in Ribbons of Nb_3Sn. *A. Lachaine and M. A. R. LeBlanc* ... 247

Effect of DC and AC Currents on the Surface Sheath Conductivity of Niobium Single Crystals. *R. C. Callarotti* ... 252

3. Josephson Effect and Tunneling

3.1. Josephson Effect

Thermodynamic Properties of Josephson Junction with a Normal Metal Barrier. *C. Ishii* ... 259

Linewidth of Relaxation Oscillations of a Shunted Superconducting Point Contact. *R. D. Sandell, M. Puma, and B. S. Deaver, Jr.* ... 264

Evidence for the Existence of the Josephson Quasiparticle-Pair Interference Current. *N. F. Pedersen, T. F. Finnegan, and D. N. Langenberg* ... 268

Microwave Emission from Coupled Josephson Junctions. *T. F. Finnegan and S. Wahlsten* ... 272

The Application of the Shunted Junction Model to Point-Contact Josephson Junctions. *Y. Taur, P. L. Richards, and F. Auracher* ... 276

Characteristics of Josephson Point Contacts at the Center of a Spherical Cavity. *A. S. DeReggi and R. S. Stokes* ... 281

Temperature Dependence of the Riedel Singularity. *S. A. Buckner and D. N. Langenberg* ... 285

Fine Structure in the Anomalous DC Current Singularities of a Josephson Tunnel Junction. *J. T. Chen and D. N. Langenberg* ... 289

Temperature Dependence of the Critical Current of a Double Josephson Junction. *M. R. Halse and K. M. Salleh* 293

3.2. Tunneling

Tunneling Measurement of Magnetic Scattering. *N. V. Zavaritskii and V. N. Grigor'ev* 297

Josephson Weak Links: Two Models. *P. K. Hansma and G. I. Rochlin* 301

Anomalous Tunneling Characteristics. *M. H. Frommer, M. L. A. MacVicar, and R. M. Rose* 306

Tunneling and Josephson Experiments in Normal–Superconductor Sandwiches. *A. Gilabert, J. P. Romagnan, and E. Guyon* 312

Bulk Tunneling Measurements of the Superconducting Energy Gaps of Gallium, Indium, and Aluminum. *W. D. Gregory, L. S. Straus, R. F. Averill, J. C. Keister, and C. Chapman* 316

Two-Particle Tunneling in Superconducting PbIn | Oxide | Pb Junctions. *A. M. Toxen, S. Basavaiah, and J. L. Levine* 324

Theory and Measurements on Lead–Tellurium–Lead Supercurrent Junctions. *J. Seto and T. Van Duzer* 328

Tunneling from Pb into Al and Sn in Proximity. *R. Olafsson and S. B. Woods* . 334

Magnetic Field Dependence of the Proximity Effect in the Sn/Pb System. *J. R. Hook* 337

The Superconducting Tunnel Junction as a Voltage-Tunable Source of Monochromatic Phonon Pulses. *H. Kinder* 341

Tunneling Measurements of Electron Spin Effects in Superconductors. *R. Meservey* 345

High-Field Behavior of the Density of States of Superconductors with Magnetic Impurities. *R. Bruno and Brian B. Schwartz* 354

4. Superconductivity Materials

4.1. Elements and Compounds

NMR Anomalies and Superconductivity in Transition Metals: Vanadium. *B. N. Ganguly* 361

Nuclear Magnetic Resonance and Relaxation in $V_3 Ga_{1-x} Sn_x$. *F. Y. Fradin and D. Zamir* 366

Ultrasonic Evidence against Multiple Energy Gaps in Superconducting Niobium. *D. P. Almond, M. J. Lea, and E. R. Dobbs* 367

Superconducting Specific Heat of Nb–Ta Alloys. *T. Satoh, A. Sawada, and M. Yamamoto* 372

Superconductivity of Protactinium. *R. D. Fowler, L. B. Asprey, J. D. G. Lindsay, and R. W. White* 377

The Superconductive Transition in Cadmium. *J. F. Schooley* 382

Contributions at the La and Sn–In Sites to Superconductivity and Magnetism in $LaSn_{3-x}In_x$ Alloys. *L. B. Welsh, A. M. Toxen, and R. J. Gambino* 387

Variation of the Electron–Phonon Interaction Strength in Superconducting $LaIn_{3-x}Sn_x$. *M. H. van Maaren and E. E. Havinga* 392

Superconducting Properties of $LaSn_xIn_{3-x}$. *P. K. Roy, J. L. Levine, and A. M. Toxen* 395

Superconducting Transition Temperatures and Annealing Effect of Single-Crystalline NbN_x Films. *G. Oya, Y. Onodera, and Y. Muto* 399

Calorimetric Studies of Superconductive Proximity Effects in a Two-Phase Ti–Fe (7.5 at. %) Alloy. *J. C. Ho and E. W. Collings* 403

Heat Capacity of Rubidium Tungsten Bronze, *W. E. Kienzle, A. J. Bevolo, G. C. Danielsen, P. W. Li, H. R. Shanks, and P. H. Sidles* 408

Specific Heat, Optical, and Transport Properties of Hexagonal Tungsten Bronzes. *C. N. King, J. A. Benda, R. L. Greene, and T. H. Geballe* 411

High-Transition-Temperature Ternary Superconductors. *B. T. Matthias* 416

Layered Compounds, Intercalation, and Magnetic Susceptibility Measurements. *F. J. Di Salvo* 417

Fluctuation Effects on the Magnetic Properties of Superconducting Layered Compounds. *D. E. Prober, M. R. Beasley, and R. E. Schwall* 428

Studies of the Properties of the System $TaS_{2-x}Se_x$. *J. F. Revelli, Jr., W. A. Phillips, and R. E. Schwall* 433

Nuclear Magnetic Resonance in Layered Diselenides. *B. G. Silbernagel and F. R. Gamble* 438

Microwave Properties of Superconducting Intercalated $2H$-TaS_2. *S. Wolf, C. Y. Huang, F. Rachford, and P. C. W. Chu* 442

The Influence of the Limits of Phase Stability and Atomic Order on Superconductivity of Binary and Ternary A15-Type Compounds. *J. Muller, R. Flukiger, A. Junod, F. Heiniger, and C. Susz* 446

The Importance of the Transition Metal Volume in A15 Superconductors. *F. J. Cadieu and J. S. Weaver* 457

T_c Studies of Nb–Ga Binary and Ternary Compounds. *D. W. Deis and J. K. Hulm* 461

Synthesis of Low-Temperature Stable Superconducting A15 Compounds in the Niobium System. *R. H. Hammond and Subhas Hazra* 465

Studies of Superconducting $Nb_3A_xB_{1-x}$ Alloys. *G. R. Johnson and D. H. Douglass* 468

Low-Temperature Heat Capacity of Nb_3Sn. *R. Viswanathan, H. L. Luo, and L. J. Vieland* 472

Superconductivity in the Alloy System $V_3Ga_xSi_{1-x}$. *B. C. Deaton and D. E. Gordon* 475

Superconducting Properties of A15 Phase V–Ir Alloys. *J. E. Cox, J. Bostock, and R. M. Waterstrat* 480

Superconducting Properties of Some Vanadium-Rich Titanium–Vanadium Alloy Thin Films. *Hermann J. Spitzer* 485

Superconductivity of Lutetium at Very High Pressure: Implications with Respect to the Superconductivity of Lanthanum. *J. Wittig, C. Probst, and W. Wiedemann* 490

Superconductivity of Hafnium and the Dependence of T_c on Pressure. *C. Probst and J. Wittig* 495

Superconducting Properties of Iridium. *R. J. Soulen, Jr. and D. U. Gubser* 498

The Effect of Stress on the Magnetic Superconducting Transitions of Tin Whiskers. *B. D. Rothberg, F. R. N. Nabarro, and D. S. McLachlan* 503

The Strain Dependence of T_c in Sn and In Alloy Whiskers. *J. W. Cook, Jr., W. T. Davis, J. H. Chandler, and M. J. Skove* 507

The Effect of Size and Surface on the Specific Heat of Small Metal Particles. *V. Novotny, P. P. M. Meincke, and J. H. P. Watson* 510

Pressure-Induced Superconductivity in Cesium. *G. M. Stocks, G. D. Gaspari, and B. L. Gyorffy* 515

The Influence of Dissolved Hydrogen on the Superconducting Properties of Molybdenum. *B. D. Bhardwaj and H. E. Rorschach* 517

Low-Temperature Neutron Irradiation Effects in Superconducting Technetium and Niobium. *B. S. Brown, T. H. Blewitt, T. Scott, N. Tepley, and G. Kostorz* 523

The Superconductivity of Elastically Strained Tin Whiskers Near $T_c(\varepsilon)$: A Second-Order Phase Transition with Two Degrees of Freedom. *B. D. Rothberg, F. R. N. Nabarro, M. J. Stephen, and D. S. McLachlan* 528

4.2. Gases, Films, and Granular Materials

Experimental Evidence for an Atomic-like Parameter Characterizing the Systematics of Superconductivity in the Transition Metals. *R. H. Hammond and M. M. Collver* 532

Modification of Surface Mode Frequencies and Superconductivity T_c by Adsorbed Layers. *D. G. Naugle, J. W. Baker, and R. E. Allen* 537

Effect of Surface Charge on the Superconductivity of Vanadium Films. *W. Felsch* 543

Carrier Concentration and the Superconductivity of Beryllium Films. *Kazuo Yoshihiro and Rolfe E. Glover, III* 547

The Upper Critical Field and the Density of States in Amorphous Superconductors. *G. Bergmann* 552

Superconductivity and Metastability in Alloys of the Mo–Re System. *J. R. Gavaler, M. A. Janocko, and C. K. Jones* 558

Superconducting Properties of Crystalline Films of Aluminum on Silicon. *Myron Strongin, O. F. Kammerer, H. H. Farrell, and J. E. Crow* 563

Granular Refractory Superconductors. *J. H. P. Watson* 568

Structural and Superconducting Properties of Granular Aluminum Films. *G. Deutscher, H. Fenichel, M. Gershenson, H. Grunbaum, and Z. Ovadyahu* . 573

Electron Localization in Granular Metals. *B. Abeles and Ping Sheng* 578

5. Phonons

Strong Coupling Superconductivity. *J. P. Carbotte* 587

Phonon Spectrum of La. *L. F. Lou and W. J. Tomasch* 599

A Simple Experiment for the Determination of the BCS Parameter in Normal Metals. *G. Deutscher and C. Valette* 603

Variations of Cutoff Phonon Frequencies in Strong-Coupling Superconductors. *A. Rothwarf, F. Rothwarf, C. T. Rao, and L. W. Dubeck* 607

The Electron–Phonon Enhancement Factor for Some Transition Metals. *G. S. Knapp, R. W. Jones, and B. A. Loomis* 611

Neutron Scattering, Phonon Spectra, and Superconductivity. *H. G. Smith, N. Wakabayashi, R. M. Nicklow, and S. Mihailovich* 615

Superconductivity and Anomalous Phonon Dispersion in TaC. *Philip B. Allen and Marvin L. Cohen* 619

Soft Transverse Phonons in an Amorphous Metal. *B. Golding, B. G. Bagley, and F. S. L. Hsu* 623

Lattice Structure and Instabilities and Electron–Phonon Coupling in Superconductors with High Transition Temperatures. *P. Hertel* 627

6. Fluctuations

One-Dimensional Superconductivity in Bismuth Films. *T. Shigi, Y. Kawate, and T. Yotsuya* 633

Evidence for Magnetic-Field-Induced Reduction of the Fluctuation Dimensionality in Bulk Type II Superconductors Just Above the Upper Critical Field H_{c2}. *R. R. Hake* 638

Fluctuation Conductivity of Superconductors. *Bruce R. Patton* 642

Critical Fluctuation Behavior in the Resistive Transition of Superconducting Bi Films. *M. K. Chien and R. E. Glover III* 649

A Study of Fluctuation Effects on Resistive Transition to Superconductivity in Thin Indium Films. *Anil K. Bhatnagar and Belkis Gallardo* 654

Fluctuation Effects in the AC Impedance of One-Dimensional Superconductors. *John R. Miller and John M. Pierce* 659

Functional Integral Method for Superconducting Critical Phenomena. *Hajime Takayama* 664

Isothermal Superconducting Transitions in Milligauss Fields. *R. Schreiber and H. E. Rorschach* 669

Fluctuation-Induced Diamagnetism above T_c in Al and Al–Ag Alloys. *H. Kaufman, F. de la Cruz, and G. Seidel* 673

Fluctuation-Induced Diamagnetism in Bulk Al and Al Alloys above the Superconducting Transition Temperature. *J. H. Claassen and W. W. Webb* 677

Thermodynamic Fluctuations in "Zero-Dimensional" Superconductors. *R. A. Buhrman, W. P. Halperin, and W. W. Webb* 682

Size Effects in the Fluctuation Diamagnetic Susceptibility of Indium Powders above T_c. *D. S. McLachlan* 687

Fluctuations in a Small Superconducting Grain, *G. Deutscher, Y. Imry, and L. Gunther* 692

Coherent Behavior in Josephson Junction Arrays. *A. Saxena, J. E. Crow, and Myron Strongin* 696

Instabilities in the Voltage–Current Characteristics of Current-Carrying One-Dimensional Superconductors. *J. Meyer and G. v. Minnigerode* 701

Intrinsic Fluctuations in a Superconducting "Flux Detector" Ring Closed by a Josephson Junction: Theory and Experiment, *L. D. Jackel, J. Kurkijärvi, J. E. Lukens, and W. W. Webb* 705

The Pair-Field Susceptibility of Superconductors. *J. T. Anderson, R. V. Carlson, A. M. Goldman, and H.-T. Tan* 709

7. Superconductivity Phenomena

High-Resolution Magnetooptical Experiments on Magnetic Structures in Superconductors. *H. Kirchner* 717

Growth and Current-Induced Motion of the Landau Domain Structure. *R. P. Huebener, R. T. Kampwirth, and David F. Farrell* 728

Investigation of Possibilities for Raising the Critical Temperature of Superconductors. *G. F. Zharkov* 729

Resistance of Superconducting Alloys Near T_c As Caused by Vortex Structure Motion. *L. P. Gor'kov and N. B. Kopnin* 735

The Anisotropy of the Static Magnetic Field Penetration Depth in Superconducting Tin. *P. C. L. Tai, M. R. Beasley, and M. Tinkham* 740

Ultrasonic Attenuation in Superconducting Indium and Indium–Tin Alloys. *F. G. Brickwedde, David E. Binnie, and Robert W. Reed* 745

Low-Temperature Anomalies in Pure Niobium Studied Ultrasonically. *J. R. Leibowitz, E. Alexander, G. Blessing, and T. Francavilla* 750

Ultrasonic Absorption in Superconducting Single Crystals of $Nb_{1-x}Mo_x$. *L. L. Lacy* 756

The Volume Change at the Superconducting Transition of Lead and Aluminum above 0.3°K. *H. R. Ott* 760

Nuclear Spin–Lattice Relaxation in Impure Superconducting Indium. *J. D. Williamson and D. E. MacLaughlin* 763

Electronic Part of the Thermal Conductivity of a Thin, Superconducting Film Composed of Lead and Gadolinium. *D. M. Ginsberg and B. J. Mrstik* 767

Thermoelectrostatic Effects in Superconductors. *A. Th. A. M. de Waele, R. de Bruyn Ouboter, and P. B. Pipes* 772

Structure of Superconductors with Dilute Magnetic Impurities. *Reiner Kümmel* 777

Bose Condensation in Superconductors and Liquid ^{4}He. *M. D. Girardeau and S. Y. Yoon* 781

Enhanced Plasticity in the Superconducting State. *G. Kostorz* 785

Relation between Superconducting Energy Gaps and Critical Magnetic Fields. *D. U. Gubser and R. A. Hein* 790

Theory of Superconductors with Spatially Varying Order Parameter. *Reiner Kümmel* 794

Observation of Pair-Quasiparticle Potential Difference in Non-equilibrium Superconductors. *John Clarke* 798

Electric Potential Near a Superconducting Boundary. *M. L. Yu and J. E. Mercereau* 799

Contents of Other Volumes 805
Index to Contributors 826
Subject Index 833

Contents
of Volume 4

Electronic Properties

1. Plenary Topics

Measurement of Relaxation Times and Mean Free Paths in Metals. *R. G. Chambers* 3

De Haas–van Alphen Effect and Fermi Surface Changes in Dilute Alloys. *P. T. Coleridge* 9

Electronic Structure of Impurities in Transition Metals. *Junjiro Kanamori* 19

2. Electron–Electron Interactions

Surface Paramagnons for Nearly Ferromagnetic Metals. *M. T. Béal-Monod, P. Kumar, D. L. Mills, H. Suhl, and R. A. Weiner* 29

The Dependence of the Electron–Electron Enhancement of the Effective Mass on Magnetic Field. *J. L. Smith and P. J. Stiles* 32

Electron Spin Splitting in the Alkali Metals. *B. Knecht, D. L. Randles, and D. Shoenberg* 37

Correlation Energy of Quantum Plasma in Low Magnetic Field: de Haas–van Alphen Behavior. *M. L. Glasser, R. W. Danz, and N. J. M. Horing* 42

Thermal and Magnetic Properties of Metals at Low Temperatures. *A. Isihara* . 47

Excitonic Instability and Ultrasonic Attenuation under Strong Magnetic Fields in Bi. *H. Fukuyama and T. Nagai* 50

Effects of the Excitonic Fluctuations above the Transition Temperature. *M. T. Béal-Monod, K. Maki, and H. Fukuyama* 54

3. Cyclotron Resonance

Far-Infrared Cyclotron Resonance in Potassium. *S. J. Allen, Jr., L. W. Rupp, Jr., and P. H. Schmidt* 61

Cyclotron Resonances in Graphite by Using Circularly Polarized Radiation. *Hiroyoshi Suematsu and Seiichi Tanuma* 66

Doppler-Shifted Cyclotron Resonance with Helicons and Acoustic Waves in Aluminum. *S. W. Hui and J. A. Rayne* 71

Dopplerons in Tungsten. *J. F. Carolan* 75

Landau Damping in Degenerate Semiconducting Plasmas. *P. Halevi* 79
Oscillatory Magnetooptical Effects in Group 5 Semimetal Alloys. *M. J. Apps* ... 84

4. Fermi Surfaces

The de Haas–van Alphen Spectrum and Fermi Surface of Iron. *David R. Baraff* 91
Some New Results on de Haas–van Alphen Effect in Cobalt. *F. Batallan and I. Rosenman* 97
The de Haas–van Alphen Effect in YZn, an Alloy of the Beta-Brass Type. *J.-P. Jan* 100
De Haas–van Alphen Effect in Technetium. *A. J. Arko, G. W. Crabtree, S. P. Hörnfeldt, J. B. Ketterson, G. Kostorz, and L. R. Windmiller* 104
From Fermi Surface Data to One-Electron Potentials. *J. F. Janak* 109
De Haas–van Alphen Oscillations in Titanium. *G. N. Kamm and J. R. Anderson* 114
The de Haas–van Alphen Effect and the Electronic Structure of the Transition Metals. *L. R. Windmiller, S. P. Hörnfeldt, J. Shaw, G. W. Crabtree, and J. B. Ketterson* 120
Phase Shift Analysis of Iridium. *J. C. Shaw, J. B. Ketterson, and L. R. Windmiller* 131
Hole Octahedron Fermi Surface Curvature in Tungsten. *D. E. Soule and J. C. Abele* 136
Magnetic Breakdown and Quantum Oscillations in the Magnetoresistance of Tin. *J. K. Hulbert and R. C. Young* 142
Quantum Oscillations in the Thermal Transport Properties of Tin. *R. C. Young* .. 146

5. Fermi Surfaces—Stress Effects

The Stress Dependence of the Magnetic Susceptibility in Zinc between 4.2 and 300°K. *H. R. Ott and R. Griessen* 153
Effect of Pressure on the Knight Shift of $AuGa_2$. *H. T. Weaver, J. E. Schirber, and A. Narath* 157
Magnetic-Field-Induced Pseudo Phase Transition in Cadmium. *D. E. Hagen, Wayne E. Tefft, and D. M. Sparlin* 162

6. Lifetimes

Audiofrequency Size Effect and Determination of the Electron Mean Free Path. *H. H. A. Awater and J. S. Lass* 169
De Haas–van Alphen Effect Studies of Electron Scattering in CuH. *B. Lengeler and W. R. Wampler* 173
Quantum Interference Study of Magnetic Breakdown and Scattering Lifetimes in Magnesium. *C. B. Friedberg and R. W. Stark* 177
Electron Scattering in Metals as Seen in the Landau Quantum Oscillations. *R. J. Higgins, H. Alles, Y. K. Chang, and D. W. Terwilliger* 185

Anisotropy of Electronic Lifetime Due to Phonon Scattering in Copper. *David Nowak and Martin J. G. Lee* 195

7. Alloys

Maps of Electronic Lifetime Anisotropy for Dilute Alloys of Ag, Cu, Zn, and Fe in Au from dHvA Dingle Temperature Measurements. *D. H. Lowndes, K. M. Miller, R. G. Poulsen, and M. Springford* 201
The Fermi Surface of Dilute Indium Alloys. *P. M. Holtham and D. Parsons* 222
High-Precision dHvA Studies in Dilute Copper Alloys. *I. M. Templeton* 226
The Electrical Resistivity of Dilute Nickel Alloys. *D. Greig and J. A. Rowlands* 233
The Hall Resistivity in Dilute Crystals of Cd–In and Cd–Ag: An Impurity Problem for Intersheet Scattering. *D. A. Lilly and A. N. Gerritsen* 236
Calculations of de Haas–van Alphen Lifetimes in Copper-Based Alloys. *R. Harris* 241

8. Electron–Phonon Interaction

The Phonon-Exchange Vertex in Transition Metals—A Localized Approach. *H. Gutfreund and A. Birnboim* 247
T_c in Dilute Nb–Ta Alloys Estimated from Resistivity Measurements. *Ö. Rapp and M. Pokorny* 253
Electron–Phonon Interaction in Lead and Indium: Study by High-Frequency Cyclotron Resonance. *P. Goy and B. Castaing* 258
Electron–Phonon Interaction and Electronic Properties of Zinc. *Philip G. Tomlinson and James C. Swihart* 269
The Temperature Dependence of the Electron Mean Free Path in the Radiofrequency Size Effect of Potassium. *P. N. Trofimenkoff* 274
Electronic Mean Free Paths in Cadmium. *P. D. Hambourger* 278
Electromagnetic Generation of Acoustic Waves Using Metal Films at Low Temperatures. *Y. Goldstein and A. Zemel* 282
Magnetoacoustic Effect in Niobium. *J. R. Leibowitz, E. Alexander, G. Blessing, T. Francavilla, and J. R. Peverley* 287

9. Electronic Properties

Electrical and Structural Properties of Amorphous Metal–Metal Oxide Systems. *J. J. Hauser* 295
Lead Phthalocyanine: A One-Dimensional Conductor. *K. Ukei, K. Takamoto, and E. Kanda* 301
Nonequilibrium Effects in Normal Metal–Insulator–Metal Junctions at Low Temperature. *J. G. Adler, H. J. Kreuzer, P. N. Trofimenkoff, and W. J. Wattamaniuk* 306
Low-Temperature Light Scattering Studies of Spin Diffusion and Motional

Narrowing in Wide-Gap Semiconductors. *J. F. Scott, P. A. Fleury, and T. C. Damen* 310

Fine-Particle Size Effects in Metallic Al and Cu. *S. Kobayashi* 315

Ultrasonic Shear Wave Attenuation and the Electronic Structure of Cadmium and Zinc. *D. P. Almond, A. R. Birks, M. J. Lea, and E. R. Dobbs* 320

Electronic Structure of Nickel and Cobalt–Carbon Interstitial Alloys. *M. C. Cadeville, F. Gautier, and C. Robert* 325

10. Transport—Electrical

Induced Torque Anisotropy in Aluminum. *R. J. Douglas, F. W. Holroyd, and W. R. Datars* 335

Magnetoconductivity Formulation for Arbitrary Fermi Surfaces and General Scattering. *Frank E. Richards* 339

The Thermal Resistivity and Thermal Magnetoresistivity of Potassium. *R. S. Newrock and B. W. Maxfield* 343

Magnetoresistance Anisotropy in Potassium. *T. K. Hunt and S. A. Werner* 348

Low-Temperature Magnetoresistance of a Cylinder of Crystalline Lead from Inductive Permeability Measurements. *R. C. Callarotti and P. Schmidt* 353

Quantum Limit Studies in Single-Crystal and Pyrolytic Graphite. *John A. Woollam, David J. Sellmyer, Rodney O. Dillon, and Ian L. Spain* 358

Acoustic Open-Orbit Resonance Line Shape under Diffusive Scattering of the Conduction Electrons. *J. Roger Peverley* 363

Resistivity of n-Type Bi_2Te_3 from 1.3 to 300°K. *W. T. Pawlewicz, J. A. Rayne, and R. W. Ure, Jr.* 368

Effect of Surface Condition on the Lorenz Number and Other Electron Transport Properties of Size-Limited Single-Crystal Tungsten Rods. *D. R. Baer and D. K. Wagner* 373

Observation of a Negative Conduction in Antimony in High Magnetic Fields. *T. Morimoto and K. Yoshida* 379

11. Transport—Thermal

Transport Properties Near Magnetic Critical Points: Thermopower. *S. H. Tang, T. A. Kitchens, F. J. Cadieu, and P. P. Craig* 385

A New Method for the Investigation of Thermoelectric Phenomena in Pure Metals. *J. C. Garland* 399

The Thermoelectric Power of Iron. *F. J. Blatt* 404

Low-Temperature Thermoelectric Power of n-Type GaAs in a Strong Magnetic Field. *J. P. Jay-Gerin* 409

Lattice Thermal Conductivity of Cold-Worked Cu–10 at. % Al between 0.5 and 4°K. *J. A. Rowlands, A. Kapoor, and S. B. Woods* 413

AC Measurement of the Seebeck Coefficient of Nickel Near Its Curie Temperature. *Chester Piotrowski, Craig H. Stephan, and Jack Bass* 417

12. Energy States and Heat

Optical and Heat Capacity Studies of the Solid Solutions $Ti_xTa_{1-x}S_2$. *J. A. Benda, C. N. King, K. R. Pisharody, and W. A. Phillips* 423

Anomalous Specific Heat of Disordered Solids. *Herbert B. Rosenstock* 433

Specific Heat of Strongly Disordered Crystalline and Amorphous Films. *W. Buckel and Ch. Ohlerich* 437

Surface Effects in the Low-Temperature Heat Capacity of Ionic Crystals. *G. P. Alldredge, T. S. Chen, and F. W. de Wette* 441

Heat Capacity Measurements on Isotopes of Molybdenum. *R. Viswanathan, H. L. Luo, and J. J. Engelhardt* 445

Specific Heat and Debye θ of CdS: 0.45% Mn. *M. M. Kreitman, G. J. Svenconis, J. J. Kosiewicz, and J. T. Callahan* 449

Instrumentation and Measurement

13. Plenary Topics

New Techniques in Refrigeration and Thermometry. *John C. Wheatley* 455
Cryogenic Metrology. *B. N. Taylor* 465
Macroscopic Quantum Devices. *J. Clarke* 483

14. Thermometry

Refrigeration and Thermometry in the Microkelvin Range. *R. G. Gylling* 487

The Temperature Scale Defined by $^{60}Co\gamma$-Ray Anisotropy and Noise Thermometry. *H. Marshak and R. J. Soulen, Jr.* 498

Accurate NMR Thermometry below $1°K$. *D. Bloyet, A. Ghozlan, P. Piejus, and E. J.-A. Varoquaux* 503

A Hyperfine Thermometer for Use at Low Temperature. *F. R. Szofran and G. Seidel* 508

Pulsed Platinum NMR Thermometry below $2°K$. *M. I. Aalto, H. K. Collan, R. G. Gylling, and K. O. Nores* 513

Relationship between Noise Temperatures and Powdered CMN Magnetic Temperatures: Application to the Properties of ^3He. *R. A. Webb, R. P. Giffard, and J. C. Wheatley* 517

15. Superconducting Instruments

Magnetochemical Measurements of Biochemical Compounds with a Superconducting Magnetometer. *M. Cerdonio, R. H. Wang, G. R. Rossman, and J. E. Mercereau* 525

Superconducting-Cavity-Stabilized Oscillators of 10^{-14} Stability. *S. R. Stein and J. P. Turneaure* 535

The Relationship of Josephson Junctions to a Unified Standard of Length and Time. *D. G. McDonald, A. S. Risley, and J. D. Cupp* 542

Static Susceptibility of Biological Molecules and SQUID Detection of NMR. *E. P. Day* 550

Gravitational Wave Detection at Low Temperatures. *W. O. Hamilton, P. B. Pipes, and P. S. Nayar* 555

Low-Temperature Gravity Wave Detector. *J. E. Opfer, S. P. Boughn, W. M. Fairbank, M. S. McAshan, H. J. Paik, and R. C. Taber* 559

Large-Diameter Superconducting Quantum Interference Devices for a Measurement of h/m. *C. M. Falco and W. H. Parker* 563

Frequency Pulling in Josephson Junctions. *B. T. Ulrich* 568

The Flux Period of Josephson Junction Interferometers. *D. L. Stuehm and C. W. Wilmsen* 572

Magnetic Field and Q Dependence of Self-Induced Current Peaks in Resonant Josephson Tunnel Junctions. *Klaus Schwidtal and C. F. Smiley* 575

Parametric Oscillations in Superconducting Double Point-Contact Structures. *B. T. Ulrich* 580

Analog Computer Studies of Subharmonic Steps in Josephson Junctions. *C. A. Hamilton* 585

16. Cryogenic Instrumentation and Measurement Techniques

Low-Temperature Susceptibility Measurements of Cerium Magnesium Nitrate. *G. O. Zimmerman, D. J. Abeshouse, E. Maxwell, and D. R. Kelland* 591

Application of ^{4}He Calorimeter to the Specific Heat of Titanium. *K. L. Agarwal and J. O. Betterton, Jr.* 597

A Review of the Effects of High Magnetic Fields on Low-Temperature Thermometers. *H. H. Sample, L. J. Neuringer, and L. G. Rubin* 601

Hydrodynamic Theory Correlation for 72 Helium Evaporators. *Stephan G. Sydoriak* 607

Indium–Barrier–Indium Tunneling Junctions. *M. D. Jack and G. I. Rochlin* 612

Thermal Expansion Coefficient of $La_2Mg_3(NO_3)_{12} \cdot 24H_2O$ and $Ce_2Mg_3(NO_3)_{12} \cdot 24H_2O$ between 10 and 300°K. *J. Carp-Kappen, C. W. de Boom, H. J. M. Lebesque, and B. S. Blaisse* 617

A ^{3}He Thermal Switch for Ultra-Low Temperatures. *L. E. Reinstein and J. A. Gerber* 622

Heat Capacity Measurements of Small Samples–Granular Al Films. *C. N. King, R. B. Zubeck, and R. L. Greene* 626

A 4-MHz Tunnel Diode Oscillator with 0.001-ppm Measurement Sensitivity at Low Temperatures. *Craig T. Van Degrift* 631

Controllable Electron Sources for Cryogenic Temperatures. *David G. Onn, P. Smejtek, and M. Silver* 636

Improved Heat Exchange in Dilution Refrigerators by Use of Continuous Plastic Exchangers. *L. Del Castillo, G. Frossati, A. Lacaze, and D. Thoulouze* 640

Thermal and Electrical Behavior of a Glass-Ceramic Capacitance Thermometer between 0.004 and 0.04°K. *Donovan Bakalyar, Robert Swinehart, Walter Weyhmann, and W. M. Lawless* 646

Electronic Properties of Binary Intermetallic Compounds with Large Nuclear Moments. *E. Bucher, E. Ehrenfreund, A. C. Gossard, K. Andres, J. H. Wernick, J. P. Maita, A. S. Cooper and L. D. Longinotti* 648

Contents of Other Volumes 653
Index to Contributors 676
Subject Index 683

Index to Contributors*
Complete Alphabetical Listing of the Contributors to LT-13, Volumes 1-4

Aalto, M. I. (4) 513
Abele, J. C. (4) 136
Abeles, B. (3) 578
Abeshouse, D. J. (4) 591
Abraham, B. M. (1) 493
Abrikosov, A. A. (1) 1
Adams, E. D. (2) 43, 149
Adler, J. G. (4) 306
Agarwal, K. L. (4) 597
Ahlers, G. (1) 348, 586
Ajiro, Y. (2) 380
Aldred, A. T. (2) 417
Alexander, E. (3) 750, (4) 287
Alldredge, G. P. (4) 441
Allen, J. F. (1) 258, (3) 42
Allen, P. B. (3) 619
Allen, R. E. (2) 245, (3) 537
Allen, S. J. Jr. (4) 61
Alles, H. (4) 185
Alloul, H. (2) 435
Almond, D. P. (3) 367, (4) 320
Amar, R. C. (1) 393
Anderson, C. H. (1) 206
Anderson, J. R. (4) 114
Anderson, J. T. (3) 709
Andres, K. (2) 322, 327, (4) 648
Aoi, T. (2) 574
Apps, M. J. (4) 84
Arko, A. J. (4) 104
Armbrüster, H. (2) 567
Armitage, J. G. M. (1) 258
Ashkin, M. (2) 601
Asprey, L. B. (3) 377
Auracher, F. (3) 276
Averill, F. W. (2) 251
Averill, R. F. (3) 316
Awater, H. H. A. (4) 169

Babic, E. (2) 484
Baer, D. R. (4) 373
Bagley, B. G. (3) 623
Baird, D. C. (3) 207
Bakalyar, D. (4) 646
Baker, J. W. (3) 537
Balzer, R. (2) 115
Baraff, D. R. (4) 91

Barry, J. H. (1) 116
Basavaiah, S. (3) 324
Bass, J. (4) 417
Batallan, F. (4) 97
Béal-Monod, M. T. (2) 475, (4) 29, 54
Beasley, M. R. (3) 428, 740
Belanger, B. C. (3) 241
Benda, J. A. (3) 411, (4) 423
Benjamin, R. F. (1) 542
Berberich, P. (2) 53
Berdahl, P. (1) 130
Bergman, D. J. (1) 80, 501
Bergmann, G. (3) 552
Bernier, M. E. R. (2) 79
Bernier, P. (2) 435
Betterton, J. O., Jr. (4) 597
Betts, D. S. (1) 559
Bevolo, A. J. (3) 408
Bhagat, S. M. (1) 328
Bhardwaj, B. D. (3) 517
Bhatnagar, A. K. (3) 654
Bhattacharyya, A. (1) 67
Bickermann, A. (2) 198
Biem, W. (2) 198
Binnie, D. E. (3) 745
Birgeneau, R. J. (2) 371
Birks, A. R. (4) 320
Birnboim, A. (4) 247
Bishop, J. H. (1) 406
Blair, D. G. (1) 190, 272
Blaisse, B. S. (1) 451, (4) 617
Blatt, F. J. (4) 404
Blaugher, R. D. (3) 3
Blessing, G. (3) 750, (4) 287
Blewitt, T. H. (3) 523
Bloyet, D. (4) 503
Boerstoel, B. M. (2) 510
Bostock, J. (3) 480
Boughn, S. P. (4) 559
Boyer, L. (3) 232
Bozler, H. M. (2) 218
Bretz, M. (1) 143
Brewer, D. F. (1) 159, 163, 177, 195, 200, 559
Brickwedde, F. G. (3) 745
Brooks, J. S. (1) 57

Brown, B. S. (3) 523
Brown, T. R. (1) 443
Brubaker, N. R. (1) 612
Brucker, H. (1) 598
Brunet, Y. (3) 156, 160
Bruno, R. (3) 354
Bucher, E. (2) 322, (4) 648
Buckel, W. (4) 437
Buckner, S. A. (3) 285
Buhrman, R. A. (2) 139, (3) 682
Burgess, A. E. (2) 95
Bussière, J. F. (3) 221

Cadeville, M. C. (4) 325
Cadieu, F. J. (3) 457, (4) 385
Callahan, J. T. (4) 449
Callarotti, R. C. (3) 252, (4) 353
Campbell, L. J. (1) 253, 268
Carbotte, J. P. (3) 587
Carlson, R. V. (3) 709
Carolan, J. F. (4) 75
Carp-Kappen, J. (4) 617
Carroll, K. J. (3) 187, 192
Carsey, F. (3) 116
Castaing, B. (4) 258
Castles, S. H. (2) 43
Cerdonio, M. (4) 525
Chambers, R. G. (4) 3
Chan, H. W. (1) 229
Chandler, J. H. (3) 507
Chang, D. C. (1) 608
Chang, T. S. (2) 343, 349
Chang, Y. K. (4) 185
Chapman, C. (3) 316
Cheeke, J. D. N. (2) 242
Chen, J. T. (3) 238, 289
Chen, T. S. (4) 441
Cheng, D. K. (1) 602
Cheng, Y. C. (1) 112
Chester, M. (1) 211
Chien, M. K. (3) 649
Childers, R. K. (1) 359
Chu, P. C. W. (3) 442
Chung, D. Y. (2) 62
Claassen, J. H. (3) 677
Clarke, J. (3) 798, (4) 483

* Volume number appears in parentheses followed by the page number on which the contribution begins.

Index to Contributors

Clem, J. R. (3) 102
Cochrane, R. W. (2) 427
Cohen, M. L. (3) 619
Cole, M. W. (1) 364
Coleridge, P. T. (4) 9
Coles, B. R. (2) 414, 423
Collan, H. K. (4) 513
Collings, E. W. (3) 403
Collver, M. M. (3) 532
Commins, D. E. (1) 356
Constable, J. H. (2) 223
Conway, M. M. (2) 601, 629
Cook, J. W. Jr. (3) 507
Cooper, A. S. (4) 648
Coqblin, B. (2) 459
Corruccini, L. R. (1) 411
Costa-Ribeiro, P. (2) 520
Cotignola, J. M. (2) 585
Cox, J. E. (3) 480
Crabtree, G. W. (4) 104, 120
Craig, P. P. (1) 602, (4) 385
Creswell, D. J. (1) 163, 195
Crooks, M. J. (2) 95
Crow, J. E. (3) 563, 696
Crum, D. B. (1) 563
Cunsolo, S. (1) 337
Cupp, J. D. (4) 542
Cyrot, M. (2) 551

Daalmans, G. M. (1) 451
Dahm, A. J. (2) 233
Damen, T. C. (4) 310
Danielson, G. C. (3) 408
Danz, R. W. (4) 42
Dash, J. G. (1) 19, 247, (2) 165
Datars, W. R. (4) 335
Date, M. (1) 485
Daunt, J. G. (1) 182
Davis, R. S. (1) 328
Davis, W. T. (3) 507
Day, E. P. (4) 550
Deaton, B. C. (3) 475
Deaver, B. S. Jr. (3) 264
deBoom, C. W. (4) 617
deBruyn Ouboter, R. (3) 772
Decker, W. R. (2) 605
de Graaf, A. M. (3) 238
Deis, D. W. (3) 461
de Klerk, D. (3) 138
de la Cruz, F. (3) 673
Del Castillo, L. (4) 640
DeReggi, A. S. (3) 281
Deroyon, J. P. (1) 96
Deroyon, M. J. (1) 96
Deutscher, G. (3) 573, 603, 692

de Waele, A. Th. A. M. (3) 772
de Wette, F. W. (4) 441
Dietrich, M. (2) 491
Dillon, R. O. (4) 358
DiSalvo, F. J. (3) 417
Disatnik, Y. (1) 598
Doake, C. S. M. (1) 456
Dobbs, E. R. (3) 367, (4) 320
Dolan, G. J. (3) 147
Donnelly, R. J. (1) 57
Douglas, R. J. (4) 335
Douglass, D. H. (1) 414, (3) 468
Dransfeld, K. (2) 53
Dubeck, L. W. (3) 607
Dudko, K. L. (2) 365
Dupas, A. (2) 389
Dupas, C. (2) 355
Dyumin, N. E. (1) 637

Ebisawa, H. (3) 177
Eckardt, J. (1) 518
Eckstein, S. G. (1) 54
Edwards, D. O. (1) 26, 518
Ehrat, R. (3) 134
Ehrenfreund, E. (2) 373, (4) 648
Elgin, R. L. (2) 175
Englehardt, J. J. (4) 445
Eremenko, V. V. (2) 365
Erickson, D. J. (2) 605
Erickson, R. A. (2) 215
Eselson, B. N. (1) 636, 637
Evans, W. A. B. (1) 101

Fairbank, H. A. (2) 85, 90
Fairbank, W. M. (4) 559
Falco, C. M. (4) 563
Falke, H. P. (2) 500, 503
Farrell, D. F. (3) 728
Farrell, H. H. (3) 563
Felsch, W. (3) 543
Fenichel, H. (3) 573
Fert, A. (2) 488
Finnegan, T. F. (3) 268, 272
Finnemore, D. K. (2) 590
Fiory, A. T. (3) 86
Fisk, D. J. (1) 568
Fleury, P. A. (4) 310
Flouquet, J. (2) 448
Flukiger, R. (3) 446
Ford, A. (1) 121
Fournet, G. (3) 232
Fowler, R. D. (3) 377
Fradin, F. Y. (3) 366
Francavilla, T. (3) 750, (4) 287

Franck, J. P. (2) 48
François, M. (1) 96, 319, 324
Fraser, J. C. (1) 253
Frederking, T. H. K. (1) 393
Friedberg, C. B. (4) 177
Friedberg, S. A. (2) 380
Friedlander, D. (1) 54
Friedman, E. J. (3) 238
Frommer, M. H. (3) 306
Frossati, G. (4) 640
Fukuyama H. (2) 537 (4) 50, 54

Gaines, J. R. (2) 223
Gallardo, B. (3) 654
Galleani d'Agliano, E. (2) 459
Gallus, D. E. (3) 182
Gambino, R. J. (3) 387
Gamble, F. R. (3) 438
Gamota, G. (1) 459
Ganguly, B. N. (3) 361
Gardner, W. E. (2) 595
Garito, A. F. (2) 373
Garland, J. C. (4) 399
Gaspari, G. D. (3) 515
Gasparini, F. M. (1) 518, 618
Gauthier, R. (3) 241
Gautier, F. (4) 325
Gavaler, J. R. (3) 558
Geballe, T. H. (3) 411
Genack, A. Z. (3) 69
Gerber, J. A. (4) 622
Gerritsen, A. N. (4) 236
Gershenson, M. (3) 573
Gey, W. (2) 491
Ghozlan, A. (4) 503
Giffard, R. P. (2) 144, (4) 517
Gilabert, A. (3) 312
Gillis, N. S. (2) 227
Ginsberg, D. M. (3) 767
Giorgi, A. L. (2) 605
Girardeau, M. D. (3) 781
Glaberson, W. I. (1) 430
Glasser, M. L. (4) 42
Glover, R. E. III (3) 547, 649
Golding, B. (3) 623
Goldman, A. M. (3) 709
Goldschvartz, J. M. (1) 451
Goldstein, Y. (4) 282
Gomès, A. A. (2) 459, 543
Goodstein, D. L. (1) 243, (2) 175, 180
Gordon, D. E. (3) 475
Gor'kov, L. P. (3) 735
Gorter, C. J. (2) 621
Gossard, A. C. (2) 322, (4) 648

Gough, C. E. (3) 112, 125
Gould, H. (1) 76
Goy, P. (4) 258
Graf, E. H. (2) 218
Green, F. R. Jr. (2) 251
Greene, R. L. (3) 411, (4) 626
Gregory, W. D. (3) 316
Greig, D. (4) 233
Greytak, T. J. (1) 9, 537, 542
Greywall, D. S. (1) 348, 586
Gribbon, P. W. F. (1) 426, 456
Griessen, R. (4) 153
Griffin, A. (1) 515
Grigor'ev, V. M. (3) 297
Grimes, C. C. (1) 443
Grimmer, D. P. (2) 170
Grunbaum, H. (3) 573
Grüner, G. (2) 444
Gubser, D. U. (3) 498, 790
Guggenheim, H. J. (2) 384
Gully, W. J. (2) 134
Gunther, L. (1) 80, (3) 692
Gutfreund, H. (4) 247
Guyer, R. A. (2) 110
Guyon, E. (1) 279 (3) 156, 312
Gylling, R. G. (4) 487, 513
Gyorffy, B. L. (3) 515

Hagen, D. E. (4) 162
Hake, R. R. (3) 638
Halbritter, J. (3) 202
Halevi, P. (4) 79
Hall, H. E. (1) 568
Hallock, R. B. (1) 547, 551
Halperin, W. P. (2) 139, (3) 682
Halse, M. R. (3) 293
Hambourger, P. D. (4) 278
Hamilton, C. A. (4) 585
Hamilton, W. O. (4) 555
Hammann, J. (2) 328
Hammel, E. F. (1) 253, 263
Hammond, R. H. (3) 465, 532
Hankey, A. (2) 349
Hansma, P. K. (3) 301
Harbus, F. (2) 343, 555
Harris, P. M. (2) 215
Harris, R. (2) 525, 540, (4) 241
Harris-Lowe, R. F. (1) 224
Harrison, R. B. (3) 79
Harrison, S. J. (1) 298
Hauser, J. J. (4) 295
Havinga, E. E. (3) 392
Havlin, S. (1) 50
Hazra, S. (3) 465
Hebral, B. (2) 242

Heeger, A. J. (2) 373
Heiden, G. (3) 75
Hein, R. A. (3) 790
Heiniger, F. (3) 446
Heller, M. (1) 293
Herb, J. A. (1) 247
Hereford, F. L. (1) 372
Hering, S. V. (1) 147
Héritier, M. (2) 529, 543
Hertel, P. (3) 627
Hess, G. B. (1) 302
Heubener, R. P. (3) 182, 728
Hewko, R. A. D. (2) 48
Hickernell, D. C. (1) 147
Hicks, T. J. (2) 417
Higgins, R. J. (4) 185
Ho, J. C. (3) 403
Hoffer, J. K. (1) 253
Hoffman, A. R. (3) 120
Holmes, F. S. (1) 537
Holroyd, F. W. (4) 335
Holtham, P. M. (4) 222
Holzer, W. (3) 156
Hook, J. R. (3) 337
Hori, H. (1) 485
Horiguchi, T. (1) 87
Horing, N. J. M. (4) 42
Horn, J. L. (1) 185
Horner, H. (2) 3, 125
Hörnfeldt, S. P. (4) 104, 120
Horwitz, G. (2) 561
Hsu, F. S. L. (3) 623
Hu, Chia-Ren (3) 163
Huang, C. Y. (3) 442
Huber, J. G. (2) 579
Huebener, R. P. (3) 182, 728
Huff, G. B. (2) 165
Huggins, E. R. (1) 92, 135
Hui, S. W. (4) 71
Huiskamp, W. J. (2) 272
Hulbert, J. K. (4) 142
Hulm, J. K. (2) 601, (3) 3, 461
Hunklinger, S. (2) 53
Hunt, T. K. (4) 348

Ichikawa, O. (1) 485
Ihas, G. G. (1) 477
Ikushima, A. (1) 352
Imry, Y. (1) 80, (3) 692
Ishii, C. (3) 259
Isihara, A. (4) 47
Islander, S. T. (1) 571

Jablonski, H. P. (2) 500
Jack, M. D. (4) 612

Jackel, L. D. (3) 705
Jacobi, D. (2) 561
Jacobs, A. E. (3) 46
Jacucci, G. (1) 337
Jamieson, H. C. (2) 414
Jan, J.-P. (3) 100
Janak, J. F. (4) 109
Janocko, M. A. (3) 558
Jaoul, O. (2) 488
Jay-Gerin, J. P. (4) 409
Johnson, G. R. (3) 468
Johnson, R. A. (3) 120
Johnson, R. T. (2) 144
Johnson, W. W. (1) 430
Jones, C. K. (2) 601, (3) 558
Jones, R. W. (3) 611
Jullien, R. (2) 459
Jump, R. E. (3) 125
Junod, A. (3) 446

Kahn, R. (3) 64
Kakizaki, A. (1) 627
Kamm, G. N. (4) 114
Kammerer, O. F. (3) 563
Kampwirth, R. T. (3) 182, 728
Kanamori, J. (4) 19
Kanda, E. (4) 301
Kapoor, A. (4) 413
Karo, D. (2) 295
Kaufman, H. (3) 673
Kawate, Y. (3) 633
Keister, J. C. (3) 316
Kelland, D. R. (4) 591
Keller, W. E. (1) 253, 263
Kelly, J. F. (1) 167
Kennedy, S. G. (1) 426
Kenner, V. E. (2) 245
Kernohan, R. H. (3) 217
Kes, P. H. (3) 138
Ketterson, J. B. (1) 474, 482, 493, (4) 104, 120, 131
Khanna, S. M. (3) 197
Kharchenko, N. F. (2) 365
Kienzle, W. E. (3) 408
Kierstead, H. A. (1) 381
Kim, Y. W. (3) 238
Kinder, H. (3) 341
King, C. N. (3) 411, (4) 423, 626
Kirchner, H. (3) 717
Kirk, W. P. (2) 43, 149
Kitaguchi, H. (2) 377
Kitchens, T. A. (1) 576, 602, (2) 14, 105, 371, (4) 385
Klein, M. W. (2) 397, 470
Knapp, G. S. (3) 611

Index to Contributors

Knecht, B. (4) 37
Kneisel, P. (3) 202
Kobayashi, N. (3) 107
Kobayashi, S. (4) 315
Koch, C. C. (2) 595
Kojima, H. (1) 279
Kopnin, N. B. (3) 735
Kosiewicz, J. J. (4) 449
Kostorz, G. (3) 523, 785, (4) 104
Kouvel, J. S. (2) 417
Kovdrya, Yu. Z. (1) 636
Kral, S. F. (1) 307
Kramer, L. (3) 49
Krasnow, R. (2) 295, 555
Kreitman, M. M. (4) 449
Kreuzer, H. J. (4) 306
Kuchnir, M. (1) 474, 482, 493
Kuenhold, K. A. (1) 563
Kumar, P. (4) 29
Kümmel, R. (3) 777, 794
Kuper, C. G. (1) 54, 522
Kurkijärvi, J. (3) 705

Lacaze, A. (4) 640
Lachaine, A. (3) 247
Lacour-Gayet, P. (2) 551
Lacy, L. L. (3) 756
Lai, C. M. (1) 576
Landan, J. (1) 623
Landesman, A. (2) 73
Langenberg, D. N. (3) 268, 285, 289
Lasken, R. A. (1) 328
Lass, J. S. (4) 169
Lawless, W. M. (4) 646
Lawson, D. T. (2) 85
Lea, M. J. (3) 367, (4) 320
Lebesque, H. J. M. (4) 617
LeBlanc, M. A. R. (3) 197, 221, 241, 247
Lederer, P. (2) 529, 543
Lee, D. M. (1) 411, (2) 25, 134
Lee, M. J. G. (4) 195
Leibowitz, J. R. (3) 750, (4) 287
Leiderer, P. (2) 53
Lengeler, B. (4) 173
Le Ray, M. (1) 96, 319, 324
Lerner, E. (1) 182
Lerski, R. A. (3) 42
Leung, M. C. (3) 46
Levine, J. L. (3) 324, 395
Levy, M. (3) 116
Lhuillier, D. (1) 319, 324
Li, P. L. (2) 495
Li, P. W. (3) 408

Li, Y. (2) 62
Liebenberg, D. H. (1) 251
Lilly, D. A. (4) 236
Lindsay, J. D. G. (3) 377
Linnet, C. (1) 393
Lippmann, G. (3) 54
Litvinenko, Yu. G. (2) 365
Liu, L. L. (2) 555
Longinotti, L. D. (4) 648
Loomis, B. A. (3) 611
Lou, L. F. (3) 599
Lowell, J. (3) 90
Lowndes, D. H. (4) 201
Luban, M. (1) 50
Lucking, R. (1) 559
Luengo, C. A. (2) 585
Lukens, J. E. (3) 705
Luo, H. L. (4) 445, (3) 472
Luszczynski, K. (1) 367, (2) 170
Lynch, R. F. (1) 288

MacLaughlin, D. E. (3) 763
MacVicar, M. L. A. (3) 306
Mailfert, A. (3) 232
Maita, J. P. (2) 322, (4) 648
Maki, K. (3) 168, (4) 54
Manneville, P. (2) 328
Maple, M. B. (2) 579, 585
Marshak, H. (4) 498
Martinon, C. (2) 242
Masuda, Y. (2) 574
Matheson, C. C. (1) 185, 190, 272
Matho, K. (2) 475
Mathur, M. P. (2) 601
Matthias, B. T. (2) 605, (3) 416
Maxfield, B. W. (4) 343
Maxwell, E. (4) 591
McAshan, M. S. (4) 559
McCauley, J., Jr. (1) 421
McClintock, P. V. E. (1) 434
McConville, G. T. (2) 238
McCormick, W. D. (3) 360
McDonald, D. G. (4) 542
McLachlan, D. S. (3) 503, 528, 687
McLean, E. O. (1) 147
McMahan, A. K. (2) 110
Mehl, J. B. (1) 314
Mehrotra, C. (2) 547
Meier, F. (2) 613
Meijer, P. H. E. (1) 84
Meincke, P. P. M. (3) 510
Mendelssohn, K. (1) 298
Mercereau, J. E. (3) 799, (4) 525
Mertens, F. G. (2) 198

Meservey, R. (2) 405, (3) 345
Meyer, H. (2) 189, 194
Meyer, J. (3) 701
Mezei, F. (2) 444
Mihailovich, S. (3) 615
Milbrodt, T. O. (1) 219
Miller, J. R. (3) 659
Miller, K. M. (4) 201
Mills, D. L. (4) 29
Mills, R. L. (2) 203
Milosěvič, S. (2) 295
Minkiewicz, V. J. (2) 14, 105
Minnigerode, G. V. (2) 567, (3) 701
Miyako, Y. (2) 377
Mochel, J. (1) 233
Mohling, F. (1) 121
Moldover, M. R. (1) 612, 618
Monceau, P. (3) 152, 156, 160
Mook, H. A. (1) 46
Morimoto, T. (4) 379
Mortimer, M. J. (2) 595
Moss, F. E. (1) 447, 489
Mota, A. C. (1) 406
Mrstik, B. J. (3) 767
Mueller, K. H., Jr. (2) 90
Muir, W. B. (2) 495
Muirhead, C. M. (3) 130
Mukherjee, B. K. (3) 207
Muller, J. (3) 446
Mullin, W. J. (2) 120
Musinski, D. (1) 414
Muto, Y. (3) 107, 399
Mydosh, J. A. (2) 506

Nabarro, F. R. N. (3) 503, 528
Naeije, M. (1) 451
Nagai, T. (4) 50
Nagasawa, H. (2) 451
Nagata, S. (2) 377
Nagata, T. (1) 61
Nakajima, S. (1) 156
Narahara, Y. (1) 377
Narath, A. (4) 157
Naugle, D. G. (3) 537
Naumenko, V. M. (2) 365
Nayar, P. S. (4) 555
Nemoz, A. (3) 95
Neuringer, L. J. (4) 601
Newrock, R. S. (4) 343
Ng, W. M. (1) 116
Ni, C.-C. (2) 62
Nicholson, J. E. (3) 192
Nicklow, R. M. (3) 615
Nieuwenhuys, G. J. (2) 510

Noel, J. L. (3) 232
Nores, K. O. (4) 513
Noto, K. (3) 107
Novotny, V. (3) 510
Nowak, D. A. (4) 195

Ohlerich, Ch. (4) 437
Ohta, N. (3) 82
Ohtsuka, T. (2) 334, (3) 82
Olafsson, R. (3) 334
Olsen, C. E. (2) 605
Olsen, J. L. (3) 27
Onn, D. G. (4) 636
Onodera, Y. (3) 399
Opfer, J. E. (4) 559
Osgood, E. B. (2) 14, 105
Osheroff, D. D. (1) 411, (2) 134
Ostermeier, R. M. (1) 439
O'Sullivan, W. J. (2) 384
Ott, H. R. (3) 760, (4) 153
Ovadyahu, Z. (3) 573
Oya, G. (3) 399

Packard, R. E. (1) 311
Paik, H. J. (4) 559
Palin, C. J. (1) 524, 532
Parker, W. H. (4) 563
Parsons, D. (4) 222
Patton, B. R. (3) 642
Paulson, D. N. (2) 144
Pawlewicz, W. T. (4) 368
Pedersen, N. F. (3) 268
Pellam, J. R. (1) 288, 343
Pesch, W. (3) 49
Peterson, D. T. (2) 590
Peverley, J. R. (4) 287, 363
Pfeifer, C. D. (1) 367
Phillips, N. E. (2) 601, 629
Phillips, W. A. (3) 433, (4) 423
Piéjus, P. (4) 503
Pierce, J. M. (3) 659
Pincus, P. (2) 373
Piotrowski, C. (4) 417
Piott, J. E. (2) 633
Pipes, P. B. (3) 772, (4) 555
Pisharody, K. R. (4) 423
Plischke, M. (2) 540
Pobell, F. D. M. (1) 229
Pohl, R. O. (1) 172
Pokorny, M. (4) 253
Pollack, G. L. (1) 219
Pope, J. (2) 67
Posada, E. (3) 212
Poulsen, R. G. (4) 201

Prober, D. E. (3) 428
Probst, C. (3) 490, 495
Puma, M. (3) 264
Purvis, M. K. (3) 120
Putterman, S. (1) 39

Rachford, F. (3) 442
Radebaugh, R. (1) 398, 401
Raich, J. C. (2) 227
Rainer, D. (2) 593
Rainford, B. D. (2) 417
Rainwater, J. C. (1) 121
Randles, D. L. (4) 37
Rao, C. T. (3) 607
Rapp, O. (4) 253
Rashid, R. I. M. A. (1) 101
Ratnam, B. (1) 233
Rayfield, G. W. (1) 469
Rayne, J. A. (4) 71, 368
Redfield, A. G. (3) 69
Reed, R. W. (3) 745
Reinstein, L. E. (4) 622
Renard, J. P. (2) 355, 389
Reppy, J. D. (1) 172, 229
Revelli, J. F. Jr., (3) 433
Riblet, G. (2) 567
Richards, F. E. (4) 339
Richards, M. G. (2) 67
Richards, P. L. (3) 276
Richardson, R. C. (1) 167, (2) 134, 139
Rinderer, L. (3) 134, 212
Risley, A. S. (4) 542
Rizzuto, C. (2) 484
Roach, P. R. (1) 474, 482, 493
Robert, C. (4) 325
Roberts, H. A. (1) 372
Robin, D. (3) 212
Rochlin, G. I. (3) 301, (4) 612
Rockwell, D. A. (1) 542
Rollefson, R. J. (2) 161
Rolt, J. S. (1) 177, 200
Romagnan, J. P. (3) 312
Rorschach, H. E. (1) 608, (3) 517, 669
Rose, R. M. (3) 306
Rosenbaum, R. L. (1) 623
Rosenblum, B. (3) 172
Rosenman, I. (4) 97
Rosenstock, H. B. (4) 433
Rossman, G. R. (4) 525
Rothberg, B. D. (3) 503, 528
Rothwarf, A. (3) 607
Rothwarf, F. (3) 607
Rowlands, J. A. (4) 233, 413

Roy, P. K. (3) 395
Rubin, L. G. (4) 601
Rudavsky, E. Ya. (1) 637
Rudnick, I. (1) 239, 279, 356
Rupp, L. W. Jr. (4) 61
Rusby, R. (2) 423
Rybaczewski, E. F. (2) 373

Sabisky, E. S. (1) 206
Saffman, P. G. (1) 243
Sai-Halasz, G. A. (2) 233
Saint-James, D, (3) 152
Saint-Paul, M. (2) 520
Saito, S. (2) 338
Salleh, K. M. (3) 293
Sample, H. H. (4) 601
Sanchez, J. (2) 448
Sandell, R. D. (3) 264
Sanders, T. M., Jr. (1) 477
Sarwinski, R. E. (1) 563
Saslow, W. M. (1) 387
Sato, T. (2) 338
Satoh, T. (1) 627, (3) 372
Saunders, B. L. (1) 258
Sawada, A. (3) 372
Sawada, K. (1) 61
Saxena, A. (3) 696
Schelten, J. (3) 54
Scherer, W. D. (1) 84
Scherm, R. (1) 46
Schick, M. (1) 112, 152
Schirber, J. E. (4) 157
Schlabitz, W. (2) 503
Schmatz, W. (3) 54
Schmidt, P. (4) 353
Schmidt, P. H. (4) 61
Schmidt, R. (1) 510
Schoepe, W. (1) 469
Schooley, J. F. (3) 382
Schreiber, R. (3) 669, (4) 157
Schubert, G. U. (1) 497
Schuch, A. F. (2) 203
Schwall, R. E. (3) 428, 433
Schwartz, B. B. (3) 354
Schwarz, K. W. (1) 439
Schweitzer, J. W. (2) 469
Schwidtal, K. (4) 575
Scott, J. F. (4) 310
Scott, T. (3) 523
Seidel, G. (3) 673, (4) 508
Sekula, S. T. (3) 217
Sellmyer, D. J. (4) 358
Sereni, J. (2) 585
Seto, J. (3) 328
Shanks, H. R. (3) 408

Shaw, J. (4) 120, 131
Shen, L. (2) 470
Shen, S. Y. (1) 518
Sheng, P. (3) 578
Shenk, D. S. (1) 314
Sherman, R. H. (1) 253
Shigi, T. (3) 633
Shinzawa, S. (3) 107
Shirane, G. (2) 14, 105, 371
Shoenberg, D. (4) 37
Siddon, R. L. (1) 152
Sidles, P. H. (3) 408
Siegel, S. (2) 180
Siegwarth, J. D. (1) 398, 401
Sikora, P. T. (3) 187, 192
Silbernagel, B. G. (3) 438
Silcox, J. (3) 147
Silver, M. (4) 636
Simmons, R. O. (2) 115
Simon, H. E. (2) 601
Sisemore, L. K. (3) 187
Sitton, D. M. (1) 447, 489
Skjeltorp, A. T. (2) 618
Skove, M. J. (3) 507
Slusher, R. E. (2) 100
Smejtek, P. (4) 636
Smiley, C. F. (4) 575
Smith, H. G. (3) 615
Smith, J. L. (4) 32
Snyder, H. A. (1) 283
Soda, T. (1) 61, 591
Solecki, J. C. (3) 95
Soule, D. E. (4) 136
Soulen, R. J. Jr., (3) 498, (4) 498
Souletie, J. (2) 479
Spain, I. L. (4) 358
Sparlin, D. M. (4) 162
Spitzer, H. J. (3) 485
Springford, M. (4) 201
Stahl, F. A. (2) 210
Stanford, J. (1) 159
Stanley, H. E. (2) 295, 343, 349, 555
Star, W. M. (2) 510
Stark, R. W. (4) 177
Steglich, F. (2) 567
Stein, S. R. (4) 535
Stephan, C. H. (4) 417
Stephen, M. J. (3) 528
Stephens, J. B. (1) 211
Stewart, G. A. (2) 180
Stiles, P. J. (4) 32
Stocks, G. M. (3) 515
Stokes, R. S. (3) 281
Stoltz, O. (3) 202
Straus, L. S. (3) 316

Ström-Olsen, J. O. (2) 427
Strongin, M. (3) 563, 696
Stuehm, D. L. (4) 572
Suematsu, H. (4) 66
Sugawara, T. (2) 451
Suhl, H. (4) 29
Surko, C. M. (2) 100
Susz, C. (3) 446
Suzuki, H. (2) 334
Svenconis, G. J. (4) 449
Sweedler, A. R. (2) 585
Swihart, J. C. (4) 269
Swinehart, R. (4) 646
Sydoriak, S. G. (4) 607
Szklarz, E. G. (2) 605
Szofran, F. R. (4) 508

Taber, R. C. (4) 559
Tai, P. C. L. (3) 740
Tait, R. H. (1) 172
Takamoto, K. (4) 301
Takayama, H. (3) 168, 664
Tan, H. T. (1) 108, (3) 709
Tan, M. (1) 108
Tanaka, T. (1) 87, 116
Tang, S. H. (4) 385
Tanuma, S. (4) 66
Tari, A. (2) 414
Taur, Y. (3) 276
Taylor, B. N. (4) 465
Taylor, R. D. (2) 605
Taylor, W. L. (1) 631
Tedrow, P. M. (2) 405
Tefft, W. E. (4) 162
Templeton, I. M. (4) 226
Tepley, N. (3) 523
Terui, G. (1) 352
Terwilliger, D. W. (4) 185
Thompson, R. S. (3) 163
Thomson, A. L. (1) 159, 163, 195
Thorel, P. (3) 64
Thoulouze, D. (2) 520, (4) 640
Timms, W. E. (3) 227
Tinkham, M. (3) 14, 740
Tolkacheva, O. A. (1) 636
Tomasch, W. J. (3) 599
Tominaga, A. (1) 377
Tomlinson, P. G. (4) 269
Toms, D. J. (2) 384
Toth, G. (2) 455
Tough, J. T. (1) 359
Tournier, R. (2) 257, 520
Toxen, A. M. (3) 324, 387, 395
Toyokawa, K. (1) 485
Trappe, K. I. (2) 360

Trela, W. J. (1) 293
Trickey, S. B. (2) 251
Triplett, B. B. (2) 601
Trofimenkoff, P. N. (4) 274, 306
Tsay, Y. C. (2) 470
Turkington, R. R. (1) 224
Turneaure, J. P. (4) 535

Ukei, K. (4) 301
Ullmaier, H. (3) 54
Ulrich, B. T. (4) 568, 580
Umlauf, E. E. (2) 491
Ure, R. W. Jr. (4) 368

Valette, C. (3) 603
Van Degrift, C. T. (1) 343, (4) 631
van der Klein, C. A. M. (3) 138
VanderVen, N. S. (2) 380
van Duyneveldt, A. J. (2) 621
Van Duzer, T. (3) 328
van Maaren, M. H. (3) 392
Varoquaux, E. J-A. (4) 503
Vaughan, J. M. (1) 524, 532
Veith, W. (1) 279
Velu, E. (2) 355
Verosub, K. L. (2) 155
Vidal, F. (1) 96, 319, 324
Vieland, L. J. (3) 472
Vilches, O. E. (1) 147
Vinen, W. F. (1) 524, 532, (3) 130
Viswanathan, K. S. (2) 547
Viswanathan, R. (3) 472, (4) 445
Votano, J. (2) 215

Wagner, D. K. (4) 373
Wahlsten, S. (3) 272
Wakabayashi, N. (3) 615
Wake, M. (1) 485
Wallace, B. Jr. (2) 194
Walmsley, D. G. (3) 227
Walton, B. L. (3) 172
Wampler, W. R. (4) 173
Wang, R. H. (4) 525
Wang, T. (1) 239
Wanner, R. (2) 189
Wassermann, E. F. (2) 500, 503
Watanabe, T. (2) 377
Waterstrat, R. M. (3) 480
Waterworth, G. M. (3) 37
Watson, G. E. (1) 332
Watson, H. L. (2) 590
Watson, J. H. P. (3) 510, 568

Wattamaniuk, W. J. (4) 306
Watts, D. R. (1) 581
Watts-Tobin, R. J. (3) 37
Waysand, G. (3) 152, 156, 160
Weaver, H. T. (4) 157
Weaver, J. S. (3) 457
Webb, R. A. (4) 517
Webb, W. W. (1) 581, (2) 139, (3) 677, 682, 705
Weiner, R. A. (4) 29
Welsh, L. B. (3) 387
Werner, S. A. (4) 348
Wernick, J. H. (4) 648
Wertheimer, M. R. (3) 79
Weyhmann, W. (4) 646
Wheatley, J. C. (1) 406, (2) 144, (4) 455, 517
White, R. W. (3) 377
Widom, A. (2) 67
Wiechert, H. (1) 497, 510
Wiedemann, W. (3) 490
Wiederick, H. D. (3) 207

Wilkinson, M. K. (1) 46
Williams, G. A. (1) 311
Williamson, J. D. (3) 763
Wilmsen, C. W. (4) 572
Windmiller, L. R. (4) 104, 120, 131
Winter, J. M. (2) 73
Winterling, G. (1) 537
Winzer, K. (2) 567
Wittig, J. (3) 490, 495
Wolf, R. P. (2) 210
Wolf, S. (3) 442
Wolf, W. P. (2) 618
Wong, V. K. (1) 76
Woo, C.-W. (1) 67, 72, 108
Woods, S. B. (3) 334, (4) 413
Woollam, J. A. (4) 358
Wright, L. S. (3) 79
Wyder, P. (2) 613

Yamadaya, T. (2) 334
Yamamoto, M. (3) 82, 372

Yang, L. C. (1) 211
Yanof, A. W. (1) 229
Yarnell, J. L. (2) 203
Yoon, S. Y. (3) 781
Yoshida, K. (4) 379
Yoshihiro, K. (3) 547
Yotsuya, T. (3) 633
Young, R. C. (4) 142, 146
Yu, M. L. (3) 799

Zamir, D. (3) 366
Zane, L. I. (2) 131
Zaremba, E. (1) 515
Zavaritskii, N. V. (3) 143, 297
Zemel, A. (4) 282
Zharkov, G. F. (3) 729
Zimmerman, G. O. (4) 591
Zimmermann, W. Jr. (1) 307, 571
Zubeck, R. B. (4) 626
Zuckermann, M. J. (2) 410, 525, 540

Subject Index

Anderson model, 465
Anharmonic effects, 62

Correlation length, 272
Coupled magnetic layers, 555
Critical region, 272, 295, 343, 371

D_2, 189, 194, 203, 215
Debye–Waller factor for bcc ^{4}He, 110
Diffusion of ^{3}He in ^{4}He, 73
Dilute magnetic alloys, 257, and Chapter 11
  Anderson model, 465
  de Haas–van Alphen effect, 495
  electron spin relaxation, 435
  Hall effect, 488
  high frequency resistivity, 451
  hyperfine fields, 459
  impurity interactions, 257, 503, 510
  Kondo effect, 435, 444, 470, 475, 491, 500, 567
  nuclear spin diffusion, 435
  pressure effects, 491
  resistance and magnetoresistance, 455, 479, 484, 495, 506, 525
  specific heat anomalies, 520, 529
  superconducting temperature, 459
  susceptibility, 495, 503
  thermopower, 475

Elastic properties of ^{4}He monolayers, 180
Electron paramagnetic resonance, 380, 435
Exchange energy of ^{3}He, 149

Faraday effect, 365
Ferromagnetism in alloys, 414, 417, 423, 427

H_2, 189, 194, 198
Heat of absorption, 175
HD, 218, 223
Heisenberg model, 295, 397
^{4}He, 3, 14, 48, 53, 62, 67, 73, 85, 90, 100, 105, 110, 120, 175, 233
^{3}He, 3, 25, 43, 62, 67, 73, 85, 100, 115, 131, 134, 139, 144, 149, 155, 161, 170
^{3}He, nuclear magnetism (Chapter 3)
  exchange energy of ^{3}He, 144
  magnetic properties of ^{3}He, 134, 139, 149
  melting curve of ^{3}He, 139
  nuclear polarization of ^{3}He, 144
  phonon–nuclear spin interaction in ^{3}He, 155
  spin Hamiltonian for bcc ^{3}He, 131, 149

Helium monolayers (Chapter 4)
  elastic properties, 180
  ^{4}He monolayers, 175, 180
  ^{3}He monolayers, 161, 170
  neon isotopes, 238
  nuclear magnetic resonance in, 161, 170
Hubbard model, 543, 547, 551, 561

Impure Heisenberg ferromagnet, 540
Impurity effects, 223
Ising model, 272, 295, 343

Jahn–Teller effect, 334

Kondo effect, 435, 444, 470, 475, 491, 500, 567

Librons, 198, 227
Light absorption, 365
Lower dimensional systems, 272, 371, 373, 377, 380, 384, 389

Magnetic and thermal properties of ^{3}He, 25
Magnetic moments, 417, 633
Magnetic phase transitions, 272, 295, 343, 349, 360
Magnetic properties of small metallic particles, 613
Magnetic specific heat, 618
Magnetic susceptibility, 334, 343, 380, 495, 503
Magnetism and superconductivity (Chapter 13)
  Abrikosov–Gor'kov theory, 574, 593
  Coexistence of superconductivity and ferromagnetism, 605
  Kondo superconductor, 567
  magnetic impurity effects, 574, 579, 593, 595
  specific heat, 585, 601
  superconducting critical fields, 590
Magnetization, 355, 365
Metamagnets, 343, 365
Molecular field approximation, 397
Motional narrowing, 161

Narrow energy band effects, 547, 551
Neon films, 165
Neutron diffraction, 203, 215
Neutron scattering, 14, 105, 110, 198
Noble gas solids, 245, 251
Nuclear adiabatic cooling, 322
Nuclear magnetic resonance, 67, 144, 161, 170, 218, 338, 389, 435, 444
Nuclear spin–lattice relaxation, 373

Pair potentials, 251
Paramagnetic relaxation, 621
Phonons, 3, 53, 85, 125, 155, 198, 227, 242
Phonon scattering, 85
Poiseuille flow, 85
Polarizability, 194
Pressure studies, 139, 149, 491
Proximity effect for ferromagnets, 410

Quantum solids (Chapters 1 through 6)

Raman scattering, 100, 384
Resistance anomalies, 423
Resistivity of nearly antiferromagnetic metals, 543

Scaling theory, 295, 349, 555
Second sound, 90
Single particle excitations in quantum crystals, 105, 110, 125
Soft modes, 360
Solid helium mixtures, 95

Solid nitrogen, 227
Sound velocity and absorption, 48, 62, 189, 210
Specific heat, 43, 165, 175, 360, 520, 529, 585, 618, 629
Spin fluctuations in one dimensional magnets, 371, 373
Spin polarization, 405
Spin susceptibility of non dilute alloys, 537
SQUID, 334

Thermal boundary resistance, 242
Thermal conductivity, 85, 95, 223
Thermal defects in bcc ^{3}He, 115
Thermal expansion, 245
Tricritical point, 343, 349
Tunneling from ferromagnetic films, 405
Two-roton peak, 100

Vacancy waves in crystalline ^{4}He, 120

X-ray diffraction, 115